HANDBOOK OF SUSTAINABILITY MANAGEMENT

HANDBOOK OF SUSTAINABILITY MANAGEMENT

edited by

Christian N Madu
Chu-Hua Kuei
Pace University, USA

World Scientific

NEW JERSEY • LONDON • SINGAPORE • BEIJING • SHANGHAI • HONG KONG • TAIPEI • CHENNAI

Published by

World Scientific Publishing Co. Pte. Ltd.

5 Toh Tuck Link, Singapore 596224

USA office: 27 Warren Street, Suite 401-402, Hackensack, NJ 07601

UK office: 57 Shelton Street, Covent Garden, London WC2H 9HE

British Library Cataloguing-in-Publication Data
A catalogue record for this book is available from the British Library.

HANDBOOK OF SUSTAINABILITY MANAGEMENT

ISBN-13 978-981-4354-81-3
ISBN-10 981-4354-81-3

Typeset by Stallion Press
Email: enquiries@stallionpress.com

Printed in Singapore.

CONTENTS

PREFACE

Sustainability is development that meets present needs without compromising the ability of future generations to meet their own needs (United Nations 1987). To respond to this challenge, over the last few years, a wide array of visions and executable actions have been developed based on the principles of the triple bottom line (TBL). Basically, TBL looks at the economic, environmental, and social dimensions of businesses. Thus, sustainability issues here recognize the importance of businesses to make profit, proactively manage the environment for sustainability, and perform their social function such as job creation in their immediate and extended communities. Sustainability has taken increased importance in the world today as corporations continue to seek to maximize profits while constrained by the limited earth's resources. There is growing outcry about the depletion of nonrenewable resources and the increased damage that is being done on earth by pollution, deforestation, and incessant demand for earth's resources that may not be renewable or replenishable. A more responsible approach and alternative to the use of these limited resources is required. World communities have taken the lead by building consciousness about the potential dangers of not effectively planning for the future and also the threat that reckless usage of these resources pose for the future generation and for the future of existence on earth. Business organizations are reacting to this call for sustainability and are increasingly finding out that a focus on sustainability is economic wise and could help them to become more competitive. There are several examples where businesses have fared well by streamlining their processes and creating more value while minimizing waste and dependence on natural or nonrenewable resources. Companies have also done well by effectively evaluating and selecting product components that create less environmental hazards.

This book is a handbook of sustainability management. The aim is to develop a comprehensive book that showcases the different strategies and approaches to achieve sustainability. Furthermore, it focuses on the diversity of opinions by including authors from around the globe so we can understand different perspectives and worldviews on this important issue. It will serve as a good reference book on sustainability management. Even though we aim to have a comprehensive handbook, we understand that the topic of sustainability is evolving and quite dynamic so it is difficult to guarantee a book that is exhaustive of all the developments in this area.

Organizations today stand to benefit from knowledge acquisition and knowledge transfer. They need to know best practices and they can benchmark such practices. Some of the best practices and how they were achieved are mentioned in several of the chapters covered in this book. Thus, this book will be a good reference source to both academicians and industry practitioners.

Christian N. Madu
Chu-hua Kuei
Lubin School of Business
Pace University, New York

Chapter 1

INTRODUCTION TO SUSTAINABILITY MANAGEMENT

Christian N. Madu* and Chu-hua Kuei†
Department of Management and Management Science, Lubin School of Business, Pace University, 1 Pace Plaza, New York, NY 10038, USA
** cmadu@pace.edu*
† ckuei@pace.edu

This chapter is an introduction of sustainability management (SM). Two models are used to illustrate the transformation to sustainability, and the interconnectivity of SM. These models offer guidelines to policy and decision makers on how to effectively and successfully implement SM. The basic concepts of SM are also introduced in this chapter.

1. Introduction

Sustainability is development that meets present needs without compromising the ability of future generations to meet their own needs (United Nations, 1987). To respond to this challenge, over the last few years, a wide array of visions and executable actions *have been* developed based on the principles of the triple bottom line (TBL). TBL focuses on economic development, environmental protection, and social equity. This focus directs organizations not only to maximize profit or shareholders' wealth but also to become socially responsible. Hardjono and de Klein (2004) note that this has presented the challenge of innovation to organizations as they strive for survival and also face the reality of economic, ecological, and social balance. Companies such as McDonald's Corp., Hewlett-Packard (HP), and *Wal-Mart have long taken the lead in sustainability management* (SM) and are consequently benefiting from this early lead.

Lo and Sheu (2007, p. 346) define corporate sustainability as "a business approach that creates long-term shareholder value by embracing opportunities and managing risks from three dimensions: economic, environmental, and social dimensions." The aim of such a system is to reach a "sustainable future state" by engaging in voluntary business activities. Nidumolu *et al.* (2009) suggest five distinctive stages of SM: viewing compliance as opportunity, making value chains sustainable, designing sustainable products and services, developing new business models, and creating next-practice platforms. It is apparent from these discussions that sustainability makes economic sense and creates unprecedented marketing opportunities that companies can exploit. Hence, while satisfying the ecological and social needs of the society, businesses can equally maximize shareholders value. Thus, the call for sustainability is not only good for preserving earth's limited and non-renewable resources but is also a good strategy in improving the bottom line of corporations. Innovative companies that exploit these opportunities are likely to be ahead of their competitors. Epstein (2010) also proposes a four-step strategic process for managing sustainability. His approach involves understanding the inputs and the outcomes of a sustainability practice. The inputs look at the business context such as the structure of the organization, the human factor and other resources. The resultant effect is improved sustainability performance and long-term financial performance. Winkler (2010, p. 293) and many others, on the other hand, have suggested focusing on the entire supply chain in managing sustainability. This would have more sustained impact on the organization as the entire value chain is needed to effectively achieve sustainable growth.

1.1. *Sustainability reporting*

Corporate sustainability programs are increasingly being developed. It is the way to the future and may affect the survival of the organization. For example, companies are now creating market niche by differentiating their products on sustainability practices that enable them to produce greener products and services. Companies such as Dell, HSBC, and others have developed "carbon neutral" programs. Industry leaders such as Wal-Mart work with suppliers to control emissions and waste from the source. In addition to these sustainability practices, corporate sustainability indexes are now emerging and measure sustainability performance. One of the popular ones is SAM for the Dow Jones Sustainability Index (DJSL). These indexes provide vital and timely information to both customers and shareholders on the environmental health and sustainability of companies. Such information are crucial in making both purchase and investment decisions. This growing need for information on corporate sustainability practice has also led to more disclosures on environmental risks, which are often published in corporate sustainability or responsibility reports. These reports disclose environmental and social risks. For example, in 2010, the US Securities and Exchange Commission published guidelines for disclosure of climate change risks. There is a growing emphasis on quantifying environmental impacts such as waste production, greenhouse gas emissions, deforestation, and other environmental hazards. Corporate sustainability reporting highlights the role of business community in creating long-term shareholders' value by exploiting opportunities and managing risks associated with economic, environmental, and social development. A UNEP Financial Initiative Report notes the trend toward corporate sustainability report. This report defines sustainability reporting as "a generic term for corporate extra-financial reporting." It further states that this reporting system measures "performance on a number of sustainability dimensions such as economic, environmental, social, and corporate governance performance."

Sustainability could be either internal or external. As outlined in the UNEP report, some benefits of the report include revenue growth, risk management, access to capital, and cost savings and efficiency.[1]

1.2. *Green value stream mapping (GVSM)*

To achieve sustainability, it is important to understand the sequence of activities that lead to the creation of goods and services. Value stream mapping (VSM) is a lean process-mapping method to understand this sequence of activities. The aim of VSM is to identify non-value added activities and to eliminate them and streamline the process. GVSM is an extension that enables us to achieve the same goal by eliminating non-value added time or materials, identifying areas to cut waste, and opportunities to reduce costs and improve performance.

Green Supplier Network has extended the conventional VSM to include both "lean and clean" elements in its mapping technique. Specifically, they focus on environmental wastes such as[2]

- energy, water, or raw material usage beyond the expected needs of the customer;
- the release of pollutants and material wastes such as air emissions, wastewater discharges, hazardous wastes, and solid waste (trash or discarded scrap) into the atmosphere;
- hazardous substances that adversely affect human health and/or environment and their use in the production process or their presence in the product.

It is important to understand where and how these environmental burdens occur in the process. Thus, with VSM, we can

- identify where environmental impacts occur in a process by reviewing the process flow;
- review and quantify material requirements to cut waste and excesses;
- eliminate the activities that create pollution and generate waste;
- identify the root causes of waste and inefficiencies and do something about them.

Through this approach, the major sources of waste are identified and improvement efforts are prioritized. VSM may be complemented by process analyses. The goal in process analyses is to identify process wastes and bottlenecks, and to streamline the process so that it becomes effective and efficient. A sustainability grid may be developed to benchmark world-class industry leaders (see Appendix A).

The goal here is to understand what industry leaders are doing to trim down waste, control emissions and environmental hazards. Next is to position oneself against the leaders in the different activities and identify areas for improvement.

---

[1]Sustainability Management and Reporting — Benefits for financial institutions in developing and emerging economies, UNEP Finance Initiative, December 2006.
[2]https://www.greensuppliers.gov/gsn/htm/gsn/docs/vsm.pdf.

This will present opportunities for copying or benchmarking from the best-in-class. From the sustainability grid, we can create different class of companies namely laggards (L), mediocre (M), and world-class (WW). There are two axis in the grid namely efficiency and effectiveness. Efficiency is about doing things right whereas effectiveness is about doing the right things. The grid has nine cells showing the levels of efficiency and effectiveness a firm has achieved in sustainability practice. Irrespective of the cell a firm is positioned in, it aims to be world-class by attaining the enviable position of WW. Any firm that is not WW has areas that need improvement to achieve the benefits of sustainability.

Laggards: Companies that have laggard in their labels are behind in their VSM. They have not fully exploited the gains from sustainability practices and may have inefficient or ineffective processes or both. They would need a lot of training and learning to modify, rebuild, or re-engineer their processes. There is also a possibility that leadership or organizational culture may not support sustainability practices. The laggards may be in need of re-engineering whereby leadership style, technology, and organizational structure and culture may need to be changed to adapt to sustainability practices.

Mediocre: Companies in this category are average in some aspects. Their processes are not fully optimized and there is still room to grow. They may not be taking full advantage of opportunities available and are still creating wastes since their processes are not yet streamlined or optimized. They need to make more commitment to sustainability and copy the practices of world-class competitors.

World-class: They are the best-in-class. When a company is rated WW, it has both a highly efficient and a highly effective process. These processes are streamlined and the target is zero-waste. They remain innovative and apply continuous improvement strategies. These are companies others are emulating or copying from. These companies have world-class, stable, and predictable processes.

1.3. *Leadership and SM defined*

Sustainability is a key to achieving competitiveness. It is a strategy that proactive companies are adopting knowing that minimization of wastes and prevention of environmental hazards lead to long-term economic and social benefits. Companies with effective sustainability practices have been able to cut down on the cost of raw materials by re-using and recycling without depending on the non-renewable virgin products that are more expensive. Such companies do have stakeholder loyalty and support as their roles in SM are well acknowledged and promoted even by unlikely allies. Companies such as Kodak, Xerox, Dell, HP, Wal-Mart have all achieved sustainability successes through several innovative sustainability programs. Sustainability practices, in fact, maximize both shareholders' and stakeholders' values and provide higher return on investment, and cash flows for companies.

Sustainability cannot be achieved without leadership. It is a strategic initiative that requires the immediate attention and support of top management. Without top management support, involvement, and engagement, sustainable practices cannot be achieved. SM involves reallocation of resources, change of organizational culture, modernization of technology, change of work policies and practices, training, and empowerment of employees through teams. Top management has the authority to effect change as decisions affecting sustainability lay in its purview. Sustainability is a long-term project whose benefits may not be easily captured if the focus is on quick results.

In this chapter, we define SM with two simple equations where each equation represents the letters that make up SM. The definition is as follows:

- sustainability = continuity of economic development, environmental protection, and social equity; and
- management = enabling conditions for sustainability.

This definition of SM stresses that the effective and efficient management of sustainability depends on a systematic and a systemic evaluation of an organization's sustainability competencies. Strategic components and critical success factors must be aligned with sustainability initiatives. Organizations must develop strategic, technical, procedural, and organizational capabilities and capacities to respond to SM. Organizations must, therefore, be able to address the following questions:

- What are the primary dimensions of sustainability?
- What is the dynamic state of modern organizations when moving toward sustainability?
- What are the circumstances, interventions, mechanisms, and outcomes in the implementation of SM?
- What are top management strategies for SM?

2. Dimensions of Sustainability

To drive sustainability through the organization, top management must have a clear idea of corporate changes that must be implemented. These change variables are outlined below as dimensions of sustainability.

2.1. *Economic responsibilities*

Carroll (1991) introduced a "pyramid of corporate social responsibility." According to Carroll's (1991) model, developing and initiating a strategy on corporate social economic responsibility is the foundation on which the other three dimensions are based on. Epstein (2010), in supporting this concept, notes that any plan that calls for a change of strategy on sustainability should both itemize costs of implementing sustainability and identify potential benefits. The ultimate goal of such a process is long-term financial performance. This is possible because a sustainable practice must be economic in the long run and must also serve as the catalyst *to process productivity. It is, therefore, necessary that corporations adopt strategic sustainable programs to prevent potential ecological hazards. The cost of taking corrective actions when ecological problems occur has been shown to be very disastrous and detrimental to the survival of business enterprises. A more proactive response is needed in achieving sustainability.* Sustainability is not only economical but also socially responsible and helps organizations meet the needs of its operating communities. Failure to adhere to sustainability practices may threaten the survival of the organization as it incurs higher costs, lower productivity, and loss of customer and community goodwill. Sustainability has positive effect on economic growth. Wagner (2010) and Wagner (2007) identified the sustainability factors that drive economic performance.

2.2. *Environmental responsibilities*

The biggest challenge today is how to preserve natural resources that provide essential functions to human society (Winkler, 2010). This involves

- efforts to minimize wastes;
- developing new and environmentally friendly technologies;
- creating low carbon/pollution supply chain, effectively using renewable/non-renewable resources;
- using of alternative energy sources;
- fostering harmony between supply chains and nature;
- offering effective ways of cleaning up the environment; and
- enforcing extended producer responsibility through principles such as accountability and process change.

These efforts to achieve sustainability could be further explained using the "4R + 1D" system adopted by HP:

- *Reduce*: *Reducing* the waste generated either through the depletion of limited resources or as by-products of supply networks' operations that end up polluting the earth is important to realize the full potential of SM.
- *Recycle*: Organizations such as HP use open-loop and closed-loop recycling processes to reduce the amount of waste produced for disposal and help protect natural resources.
- *Reuse*: Reusability is a direct function of the frequency of use along the expected life of products. HP normally classifies products at the point of return into three groups: full reusability, partial reusability and recyclability, and zero reusability and recyclability.
- *Report*: Since 2005, HP has published its global citizenship report highlighting its efforts in the following areas: physical, biological, intellectual, social, and economic systems (http://www.hp.com/hpinfo/globalcitizenship/gcreport/).

- *Disposal*: The non-usable and non-recyclable computer components are discarded, according to HP's reuse–recycle–disposal hierarchy, in a safe, secure, environmentally friendly manner.

Being sensitive of the natural system does not imply a neglect of the importance of economic development. The term "sustainability" suggests that the focus on economic development should be cognizant of the interests of future generations and the important needs of the earth such as biodiversity and forests ecosystems. While helping to improve organizations' competitive positions, corporate strategy should also support the preservation of the earth.

2.3. *Social responsibilities*

According to the DJSI, social responsibilities involve corporate citizenship/ philanthropy, labor practice indicators, human capital development, social reporting, talent attraction and retention, and industry specific criteria (http://www.sustainability-indexes.com). Clearly, social responsibility goes deeper than the direct production of goods and services to reflect an organization's concern with the social needs of its employees and extended environment. An organization with a social responsibility function will position itself as one that is ready to contribute resources to the community and improve quality of life (i.e., philanthropic responsibilities), avoid harm and do what is right, be just and fair (i.e., ethical responsibilities), and obey the law (i.e., legal responsibilities) (Carroll, 1991). The fast-food giant McDonald's, for example, is known for its philanthropic activities in the wake of the 2008 earthquake in China. Other well-cited examples include Ronald McDonald House Charities (RMHC) and McDonald's children programs. McDonald's also adopts three strategies to manage and cope with employee competence. They are employee experience, talent management, and commitment enhancement. As a result of those initiatives, employees at McDonald's are motivated to do the "right work" right the first time. To ensure that McDonald's sustainability vision and policies are carried out in an effective and transparent manner, McDonald's includes corporate social responsibility in its organization structure. It maintains, for example, (1) corporate responsibility board of directors, (2) worldwide corporate relations council, and (3) specific governing bodies such as corporate responsibility department, global environmental council, quality assurance board, and sustainable supply steering committee (http://www.crmcdonalds.com/publish/csr/home/report/corporate_ethics.html).

Another aspect of social responsibility that should not be overlooked involves the creation and expansion of the job base. As noted by Madu and Kuei (1995, p. 5), job creation "is perhaps the most important function any firm can render to its community. However, the job base cannot expand if productivity and quality are not simultaneously improved." Therefore, managing quality along a supply chain is critical (Kuei *et al.*, 2008). Supply chain quality management will lead to increases in quality and productivity and subsequently to an expansion of the market share and job base. If indicators with regard to supply chain quality fall, there is need for concern because resources are not being optimized and jobs may be lost. To achieve this, McDonald's uses sustainable supply chain strategies to lead its suppliers. It communicates and works with its suppliers in three areas: ethical responsibility, environmental responsibility, and economic responsibility. Suppliers, as a result of this development, contribute positively to the quality, safety, and integrity of McDonald's final products.

The dimensions of sustainability outlined here expand the focus on traditional business management. These dimensions need to be added and considered simultaneously to deliver a more *sustainable result. They allow organizations of all sizes to understand sustainability from a strategic and a holistic point of view.*

3. The Transformation Process Model

The competitive landscape is changing in the new global economy. It is imperative that organizational-wide changes be initiated to adapt to the changing environment. A discussion of sustainability in the previous section reveals that sustainability strategies and capabilities are increasingly important and complex for enterprises around the globe. To simultaneously achieve excellence in economic, environmental, and social performance is not simple. An organization must undergo a transformation to change from its traditional management approach to SM. In this section, we present a total transformation process that will involve preparing for responsible change. Figure 1 is a graphic depiction of such a model and its components. The aim of this theoretical framework is to offer practical guidelines to sustainability leaders and their value chain partners.

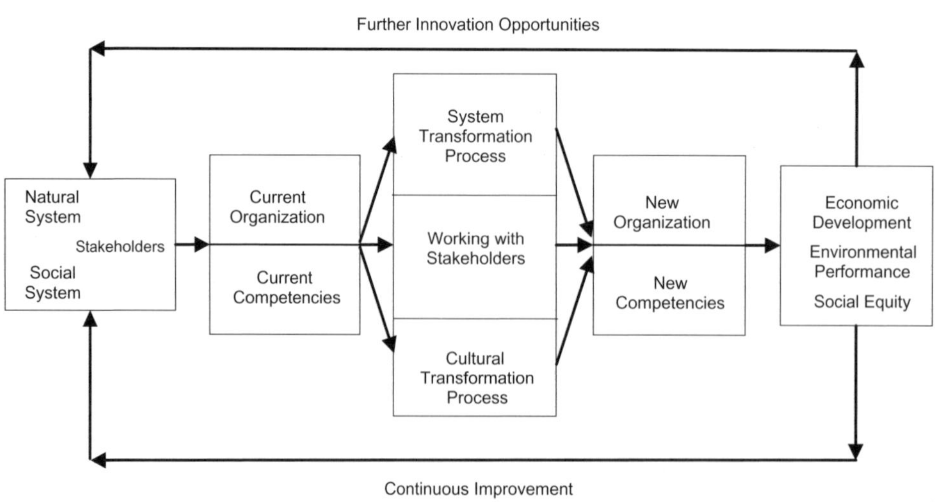

Fig. 1. The transformation process model.

When policy/decision makers are aware of the value of the concept of sustainability, they will encourage process and organizational transformation. Clearly, this transformation is driven by a new vision that understands the interactions of the organization with its natural and social systems and the need to listen to stakeholders. As shown in Fig. 1, the stakeholders interact with both natural and social systems. Stakeholders' needs are holistic as a result of this interaction there must be a balance among economic, environmental, and social impacts in the context of sustainability. To respond to these emerging needs and challenges, core competencies for sustainability need to be recognized. If the core competencies for sustainability are faulty or *insufficient*, course corrections and major changes need to be taken, at one point or another. The aim of such courses of action is to change existing situations into preferred ones. Figure 1 demonstrates that three main areas will be impacted by this process of change: a system transformation process, working with stakeholders, and a cultural transformation process. They are the critical steps required to transform the current organization at the points of reflection. It is, however, important to first distinguish between the notions "system transformation" and "cultural transformation." The former usually results from the process of emergence — the whole is more than the total of its parts. As suggested by Gharajedaghi (2006), it normally involves using system principles (openness, purposefulness, multi-dimensionality, emergent property), system dimensions (membership, decision system, measurement system, organization processes, throughput processes), and system methodology. The latter is also possible due to "the influence the relatedness of the parts of a system has on the behaviors of the parts (Luisi and Houshmand, 2009, p. 101)." In Epstein's (2009, p. 24) words, economic, environmental, and social impacts "are sometimes managed using 'soft' leadership elements such as people and culture along with a variety of informal systems." It thus involves leadership, employee fulfillment, conflict management, and cultural acceptance. Hopefully, all these changes may lead to a new organization with new competencies for sustainability. Along with these transformation processes, organizations must also work with stakeholders on how best to simultaneously achieve excellence in economic development, environmental protection, and social equity. In a supply chain setting, this means product designs, process designs, manufacturing, handling, packaging, and recycling of materials must take into consideration these concerns. Organizations must commit the necessary resources to ensure that the goal of sustainability is achieved.

Clearly, once a transformation is done and a process for sustainability is mature, a new organization is born. At this point, however, the complexity of process change will draw policy and decision makers' attention to the notion of non-linear living systems as they relate to self-organization and emergence. At this point of reflection, they need to ask a number of essential questions.

- Can a system in motion be considered a sustainable operations system?
- Is the new system always willing to do something for its own sake (intrinsic good) or for the sake of something else?

- How can a sustainable operations system stay motivated to achieve its goals?
- Are there leverage points that support a sustainable operations system such that this new organization may continuously improve or should we focus on control mechanisms?
- Is it better to design than to control, or are they equally important?
- To get a system to the top, the transformation process model proposed here encourages goodwill and good behavior. Courses of action, as a result, should always be carried out in a constructive way to change existing situations into preferred ones. As Luisi and Houshmand (2009, p. 100) put it, "it is the whole process that is the foundation."

Figure 1 also shows that the transformation process assumes a continuous cycle that is never ending. The organization operates as an open system that evaluates the process maturity for sustainability at the point of reflection and receives feedback from its internal as well as external environments for innovation and continuous improvement opportunities. This process involves evaluation of value creation relative to risks and costs.

The transformation process model presented here can be used in a cognitive map to demonstrate that the new organization may be more sustainable than the traditional one.

4. The CIMO Model

Building a total sustainability organization requires vision. The vision of the organization should be known to all employees and supply chain partners. We use a strategic cycle to illustrate how SM can be developed. This model adopts a systems approach with a focus on the context–intervention–mechanism–outcome (CIMO) logic (Denyer and Tranfield, 2009). As Fig. 2 demonstrates, there are four major components (Denyer and Tranfield, 2009, p. 683):

- institutional/social/natural setting;
- interventions;
- mechanism;
- relevant outputs and outcomes.

Figure 2, which we call the CIMO model, is built around the concepts and philosophies of SM in order to adopt a holistic perspective of sustainability. We shall briefly discuss each component.

4.1. *Context*

Understanding the complexity of institutional/social/natural systems is critical to successful application of the CIMO model.

First, policy/decision makers must pay attention to the elementary constituents of "forms" in a business context (i.e., aim, function, appearance, and interacting

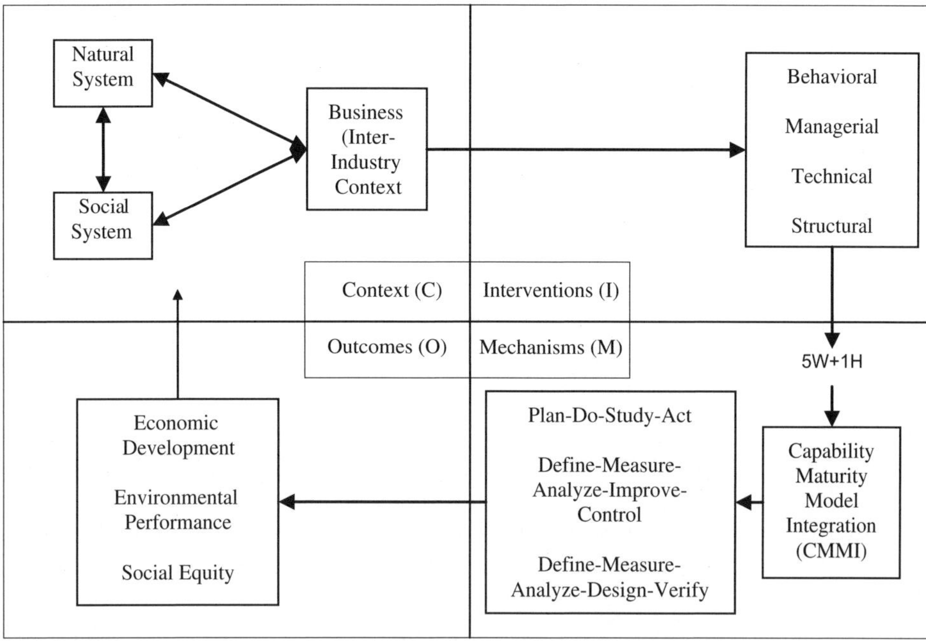

Fig. 2. The CIMO model.

components) at all times. In complex operating environments, this simple exercise separates great policy makers and sustainability managers from the average ones. These four constituents of forms represent a screen through which policy makers and sustainability managers may come to see the complexity of organizations fully and effectively. Systems that are designed and developed on this basis are robust, coherent, and persistent. Boylston (2009, p. 59) noted that "a change in one industry often informs positive change in others." Innovative paperboard CD cases made from 100% recycled content in the CD and DVD industry, for example, may provide inspiration to the packaging industry (Boylston, 2009). The elementary constituents of "forms" in the inter-industry context should be acknowledged when organizations are *making* positive changes to *sustainability.*

Successful policy and decision makers must be active listeners, facilitators, designers, and communicators. They should hear ideas and concepts on philanthropic, ethical responsibilities, and legal responsibilities. They should also employ the 4R + 1D model. They should be willing to define the problem space, focus on sustainability, outline a vision statement, enhance effective communication, encourage everyone to participate, and deliver critical information at the appropriate time to everyone in their internal and extended environments. The four essential *principles* for the *Sustainable* Projects and Activities in *China outlined by* President Hu Jintao of China, at the United Nations Climate Summit in New York on Sep. 22, 2009, is a good example in point. Although the applications and use of sustainability *practices* are not new in China, President Hu still points out that everyone in China needs to pay attention to the following: fulfilling respective responsibilities, achieving mutual benefits and a win–win outcome, promoting common development, and ensuring financing and technology structures. World-wide businesses, communities, and the natural environment may be profoundly affected by President Hu's vision of sustainability in the future.

4.2. *Intervention*

SM is a holistic and a strategic approach to compliance, financial robustness, sustainable operations, social development, environmental management, crisis management, and stakeholder relationship management. All these could be classified into four components: behavioral, managerial, technical, and structural.

The behavioral component calls for intrinsic motivation for sustainability in the workplace. Policy/decision makers need to anticipate social and cultural consequences. To effectively implement sustainability, top management needs to set up sustainability councils or offices, understand the cultural shifts/patterns, encourage entrepreneurial spirit, provide resources, train and educate employees, adopt and fund the change vehicles (e.g., Kaizen workshops), create a favorable social climate, respond positively to innovative ideas and improvement opportunities, work with the critical mass in the organization, treat people just and fair, and manage and measure social and environmental impacts.

The managerial component centers on planning and control methods. There are three hierarchical groupings for this component, namely, strategic, tactical, and operational. The elements of a supply chain system can be strategic. When signaling a commitment to sustainability, it is prudent for the organization to incorporate sustainability policy deployment, sustainability process management, human and financial resources, sustainable/green design, and technology management into any strategic analysis. To accomplish this, top management should use a four-stage process to implement sustainability strategies: (1) strategy formulation, (2) strategy implementation on a small scale, (3) evaluation and control, and (4) strategy implementation on a full scale. The elements of a sustainable operations system can be tactical. Trying to bring all the units with competencies for sustainability within the boundary of systems defined by the policy makers and sustainability manager to make, shape, and transform the organization is not an easy task. Including supplier and deciding on the optimal level of system capacities and capabilities may ensure that sustainable operation processes are operating efficiently and effectively. Members within the boundary of sustainable operation systems should be able to sense new pressures and react proactively to ensure organizational success and survival. A well-designed sustainable operations system should have guidelines for effectively addressing complexity. Complexity could arise from environmentally preferred e-purchasing, cleaner production, sustainable packaging, distribution and transportation, reverse logistics, and waste management.

The technical platform within the boundary of sustainable operations systems offers means of optimizing the use of resources, improving process efficiency and effectiveness, and supporting the preservation of the earth. Winkler (2010) suggests that modern organizations should develop intelligent systems for sustainability. Such a system should be cognizant of the important needs of the earth and its importance and value to life and nature. Two such technical platforms, namely, product life cycle management (PLM) and enterprise resource planning (ERP), are briefly introduced here. PLM can offer "an integrated information management system where enterprise stakeholders can exchange and manage product data on a real-time basis. As PLM is web-enabled, the open standards of the Internet enable seamless collaboration among members of both the internal and extended enterprise (Madu and Kuei, 2004, p. 97)." Once the product passes its maturity stage in its life cycle, an exit strategy can be planned and developed for the product. With the product design completed, an ERP system may be applied to purchasing, manufacturing resource planning, distribution and transportation, cost accounting, human resource planning, and enterprise scorecard reporting. The aim of such is to develop a system whereby all the organizational units are integrated into a single computer system that will serve their different needs. If PLM and ERP are properly implemented, they may help organizations and their supply chain partners to better serve their stakeholders by cutting down on cost, improving productivity and environmental quality, facilitating information acquisition and knowledge management, and enabling stakeholders to collaborate in real time.

Although behavioral changes, managerial components, and technical platforms are important in achieving sustainability, structural change is also critical. The structural component normally includes work flow (or business process) structures, organizational structures, communication infrastructures, product flow structures, and transportation facility structures. Traditionally, they need to be designed and developed to link everyone in the value chain and to deliver results. The focus on sustainability, however, calls for a total redesign of these structures. Reducing the amount of paper waste created in an office setting, for example, can be accomplished by using a computerized communication system. Promoting and implementing executable reverse logistics plans on collection, inspection, selection, sorting, and product recovery are not possible until the structural elements are in place. Policy and decision makers should thus use benchmarking and reengineering techniques to identify the current state of an organizational structure and where to position the organization.

4.3. *Mechanism*

SM relies on applications of proven methods to achieve process improvement or breakthrough. We shall characterize this aspect of SM using "5W + 1H." They are presented as follows:

- Why sustainability? Economic responsibilities, environmental protection, and social equity.
- Who does it? Trained personnel (critical mass).
- What is it? Effective and efficient use of proven techniques and decision sciences models.
- Where is it? Structured horizontal organization with lateral flow of information. How is it done? Standardized frameworks such as capability maturity model integration (CMMI), plan–do–study–act (PDSA), define–measure–analyze–improve–control (DMAIC), and define–measure–analyze–design–verify (DMADV).
- When is it done? No finish line.
- How is it done? Leadership, sustainability reporting, green value chain mapping, and project management.

To improve the decision-making capabilities of the stakeholder team, there are many proven approaches. The approaches are grouped into two and are briefly discussed here.

The first group, called CMMI, is used to guide efforts and help to attain high process maturity. There are five distinctive stages associated with the CMMI model: initial, managed, defined, quantitatively managed, and optimized. The *initial* stage is generally characterized by unpredictable, poorly controlled processes (e.g., forest restoration); in the *managed* stage, programs and projects for sustainability are normally adopted to help improve current processes, while proper management and standard procedures can also be set in motion; in the *defined* stage, guidelines to measure and verify process outputs and outcomes are available and organizational infrastructure to ensure execution and implementation is well-established; in the *quantitatively managed* stage, information systems and models are used to perform tasks in achieving sustainability; in the *optimized* stage, emphasis is on innovative/incremental improvements, carrying everyone along, and world-class delivery processes. In the area of forest restoration, Yuen Foong Yu Paper Manufacturing Co., Ltd. in Taiwan, for example, is now at a solid level four (or five). As a result of such an achievement, policy and decision makers at Yuen Foong Yu Paper Manufacturing Co. have been working on the establishment of consulting businesses in forest management. Organizations can help each other to achieve sustainability. The emphasis of any system capability development is on "potential" (Cheong and Corbitt, 2010).

The second group has two sets of models. One is the PDSA cycle introduced by Dr Walter Shewhart and popularized by Dr Edward Deming, while the other is Six Sigma. The PDSA cycle is used to achieve high process quality. The *plan* stage articulates the mission and purpose of the R&D process. The *do* stage deals with the design of the process and product. Designers need to take into account stakeholder requirements, technical feasibility, labor and cost requirements, and target values or standards. In the *check* stage, a comparative analysis of the green product/process and the design requirements/plans with that of competitors and the best in class is conducted. The *act* stage involves large-scale introduction to the market. The PDSA cycle is a never-ending process with the ultimate goal of continuously improving the quality of the process/product so that stakeholder satisfaction can be achieved. The Six Sigma concept involves two approaches: (1) DMAIC and (2) DMADV. The DMAIC cycle is used to achieve incremental improvement, whereas the DMADV is used to achieve breakthrough results. Both approaches rely heavily on the application of statistical techniques and management theories.

4.4. *Outcome*

Sustainability requires that organizations maintain the integrity of social and environmental systems while reconfiguring human resources, management, technical platforms, and structural components to maximize their financial performances. This change induced by sustainability has instigated the development of new types of performance indicators. Hutchins and Sutherland (2008), for example, identify five themes to evaluate the social performance of organizations: equity (poverty, gender equality), health (mortality, sanitation, drinking water, nutritional status, healthcare delivery), education (literacy, educational level), housing security (crime, living conditions), and population change. Using these five themes, a firm can assess its "impact on society." Brown *et al.* (2009), Moneva and Cuellar (2009), and Skouloudis *et al.* (2009) also encourage organizations to report a wide array of metrics for environmental quality and to focus on financial and non-financial reports such as sustainability reports, environmental statements, and the SA8000 (social accountability standard) report. As performance measures are not all equally important, the relative importance of each performance indicator must be identified first by organizations with different complexities before devoting any resource to the programs for sustainability.

The presented framework in this chapter ensures that sustainability conscious systems are designed and developed. To show the robustness of our proposed framework, we also apply *our* model to the study of Esquer-Peralta *et al.* (2008). Esquer-Peralta *et al.* (2008) carried out interviews with 24 experts in the field of sustainability for the purpose of discovering concepts with respect to SM. We shall organize their findings into a theoretical explanatory scheme. This is presented in Fig. 3.

5. Top Management Strategies for SM

Although the specific strategy used for sustainability policy deployment and process management varies from one organization to another, all generally have a focus on the stakeholders, sustainable operations goals, internal/external enablers, integrated systems, managerial constraints, modern design thinking, business results, innovation opportunities, continuous improvement, learning and growth, enterprise reporting, strategic/tactical choice, and economical/environmental/social impact.

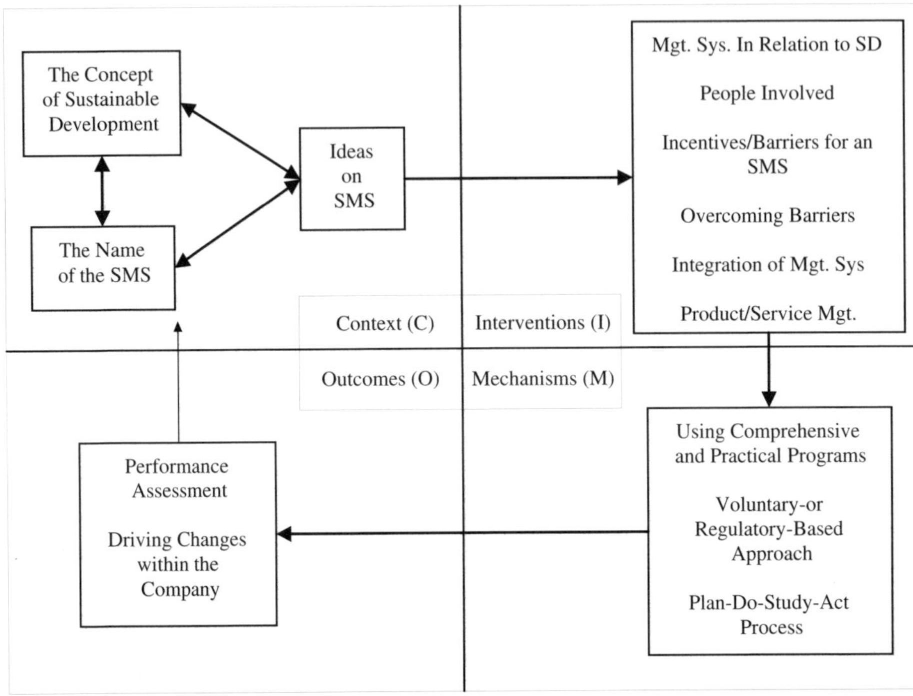

Fig. 3. Linking core elements for sustainability management system (SMS)* to the CIMO model. *Source*: Esquer-Peralta *et al.* (2008).

Top management strategy outlined here is one of the many ways that a firm can use to achieve a competitive advantage.

5.1. *Identifying aspects of managerial and policy challenges and strategic responses*

Globalization has presented new opportunities to modern organizations. These new opportunities are, however, accompanied by a host of new *challenges*. As suggested by Kuei *et al.* (2011), there are five main pressures affecting the functioning and quality of today's global supply chain systems: outsourcing, protecting the environment, reducing waste, developing communities, and adopting advanced technologies. Clearly, organizations with different sizes and complexities are being put under *pressure* to strike a balance between development economics and sustainable economics (Kuei *et al.*, 2011). To achieve this, it requires (1) the understanding of the varying needs of the stakeholders, (2) a strategic change with a set of well-articulated sustainable operations goals and strategies, (3) the commitment of top management and the endorsement of employees and suppliers, (4) restructuring and reengineering the operations system to drive sustainability, (5) adopting and funding change vehicles and programs, (6) building the foundation of SM with suitable technology and science, and (7) maintaining the momentum. Responsible changes are necessary to help transform the traditional firm so that it is supportive of SM initiatives. Such drastic changes, however, could affect other organizational changes and may even be resisted by members of the organization. It will, therefore, require tremendous investment in training and education in the field of sustainability.

5.2. *Developing design system for sustainability leadership*

A sustainable operations system can help to achieve economic development, environmental protection, and social equity. To attain these goals, organizations need to proactively direct their attention and efforts on developing design systems for sustainability. The correct choice of a strategic portfolio and implementation blueprint in the first place can enhance the competitive advantage and position of the organization. In being responsible to the needs of business systems, natural systems, and social systems, designers and sustainability managers have to strive to define the design problem space, find solutions, and develop plans for systemic implementation. Designers today need to be trained and educated in SM. They need to comprehend fundamental issues such as building powerful enabling infrastructures at operational levels. Specifically, there should be a considerable emphasis on the following: designing for sustainability, process changes, outputs, and methods. Design of systems plays an important role when an organization is moving toward sustainability.

5.3. *Investigating and developing competencies to guide the organization into the future*

The new sustainability standards demanded in the global environment today include not only product/process quality and financial performance but also competencies in achieving social and environmental performance. Competence is a portfolio of organizational, managerial, technical, and strategic capabilities and skills developed by organizations over time. As noted by Kuei *et al.* (2008), competence is the collective learning of the organization. Competency development can be classified into four groups: (1) examining the preconditions for competency development, (2) making the right choice and following through, (3) identifying actual competencies, and (4) taking further countermeasures if needed. Organizations that will flourish in the future are those who are able to develop their (as well as their suppliers') competence on economic, legal, ethical, environmental, and philanthropic responsibilities. Although such competence can be initiated through training and planning, it must be refined and extended through practical engagement and on-going learning to drive sustainability through the organization and its supplier bases. As a result of this, Wal-Mart has been working on both the preconditions for competency development and the output dimensions of sustainable operations since 2005. Visionary leaders should develop an innovative plan to build competencies in sustainability and to guide their organizations into the future.

5.4. *Evaluating the effectiveness of specific interventions, mechanisms, and innovative initiatives*

Top management should take the lead in the SM transformation process. The check (study) stage of Deming's plan–do–check (study)–act cycle should also be one of the focal points for evaluating any process of change. Environmental scans and internal scrutiny at both policy development and process management levels are needed to identify an organization's strengths and weaknesses as well as its threats and innovation opportunities. For internal scrutiny at the process level, for example, the five-stage model proposed by Nidumolu *et al.*, (2009) (i.e., viewing compliance as opportunity, making value chains sustainable, designing sustainable products and services, developing new business models, and creating next-practice platforms) will be a very useful tool to apply. By management taking steps to evaluate the effectiveness of interventions, mechanisms, and innovative initiatives along the five stages identified by Nidumolu *et al.* (2009), a linkage can be established between sustainable operations system and a company's vision. If these are not achieved, then the organization is not mature enough to effective sustainability performance.

Top management should identify problems at work, develop corrective measures, and evaluate the effectiveness of specific interventions, programs, and initiatives.

5.5. *Designing a supply chain-wide SM scorecard*

For organizations with global supply networks, a supply chain-wide report card that incorporates the business, social, and environmental contexts and results, must be established. Enterprise reporting using a balanced scorecard approach is strategic. It helps the organization, on a regular basis, to ensure its directional consistency, coordinate its internal functions, verify its operational reality, have tactical flexibility in its organizational routines, and communicate and deliver outputs and outcomes more effectively to its stakeholders. This supply chain-wide report card can also help supply chain partners monitor their systems and structures periodically to ensure that sustainability is an integral component of global supply chain strategy. The quality and integrity of the natural and social systems are the ultimate goal as organizations adopt SM.

5.6. *Validating the theory of sustainability and listening to signals of internal and external communities*

A number of questions need to be effectively answered along the stages of SM and development. Some of the questions are highlighted here.

- Does the cost fit with the budget for sustainability?
- Is it cost effective to integrate sustainability initiatives with existing business processes?

- Will the organization (as well as its supply networks) readily adapt to sustainable operations systems?
- Where does the organization (as well as its supply networks) want to be?
- What are the current states of the organization's processes (e.g., strategic processes, business processes, supporting processes, and continuous improvement processes) and how do they affect the TBL?
- What is the learning curve for the sustainable operations system?
- What are the strengths, weaknesses, opportunities, and threats?
- What are the current/future major trends in the new sustainable economy?
- How do we get to where we want to be?
- What is the level of environmental quality (and social equity) of the organization? And, how does it measure up (or compare with other organizations)?
- What are the metrics for measuring success? And how do organizations assess success?
- How user friendly is the sustainable operations system?

With an understanding of its present position, the organization can further explore other integrated approaches to sustainable development.

5.7. *Understanding the dynamic sustainable future state of organizations and the transformation process for sustainability*

The sustainability transformation process should be supported with strategic thinking, leadership, statistical thinking, system capabilities, employee fulfillment, and cultural adaptation. When the transformation process is present (see Fig. 1), dynamic forms can arise. Rules and guidelines for sustainability enable conditions and produce patterns of behavior. It is of paramount importance to show the members of organizations that new conditions and patterns can emerge by integrating sustainability into all strategic and operational decisions. Policy makers must also understand that while organized structure (pattern) is possible, instability (chaos) is also possible. Course corrections are thus the norm and should be expected. During the transition to dynamic forms (system structures) with many interacting parts, policy/decision makers must be aware that the relationships in a sustainable system are not fixed and non-linear. As explained by Nidumolu *et al.* (2009, p. 62), when moving toward sustainability, "although directional consistency is important, tactical flexibility is critical."

6. Conclusions

"Our planet is a lonely speck in the great enveloping cosmic dark. In our obscurity, in all this vastness, there is no hint that help will come from elsewhere to save us from ourselves." Carl Sagan, Pale Blue Dot (1994)

Why do some societies, organizations, groups, or individuals sometimes make disastrous decisions? This is the question raised by Diamond (2005). Here are some *reasons why* it is still *happening* (Diamond, 2005):

- failure to anticipate,
- failure to perceive that a problem has arisen,
- failure to come up with a solution,
- unworkable solutions,
- poor leadership, and
- undesirable societal values and behaviors.

Kiely (1995) listed three common failures with regards to reengineering initiatives:

- the resistance of employees (especially middle line managers) to change,
- inadequate leadership when implementing the initiatives, and
- unrealistic expectations such as the speed, scope, and benefits of reengineering implementations.

These failure factors are good supplements to this chapter. Any new approach such as reengineering, green supply chain management, or SM that would require significant system-wide changes may likely experience some support and opposition from the critical mass within the boundary of the system. Top management must therefore (1) identify primary dimensions of sustainability and managerial/policy challenges, (2) understand the transformation process, (3) investigate the circumstances, interventions, mechanisms, and outcomes in the implementation of SM, and (4) be active in carrying out strategies for sustainability in order to effectively transform and guide the organization into the future.

References

Boylston, S., 2009, Designing Sustainable Packaging (Laurence King, London).

Brown, H.S., De Jong, M. and Levy, D.L., 2009. Building institutions based on information disclosure: Lessons from GRI's sustainability reporting. *Journal of Cleaner Production* **17**, 571–580.

Carroll, A.B., 1991. The pyramid of corporate social responsibility. *Business Horizons* July–August, 39–48.

Cheong, F. and Corbitt, B.J., 2010. From childhood poverty to catfish: A conceptual participatory modeling framework for strategic decision making. *International Journal of Strategic Decision Sciences* **1**(3), 15–33.

Denyer, D. and Tranfield, D., 2009. Producing a systematic review, Chap. 39. In: Buchanan, D. and Bryman, A. (Eds), *The Sage Handbook of Organizational Research Methods*, pp. 671–689.

Diamond, J., 2005. *Collapse: How Societies Choose to Fail or Succeed.* Penguin Group, NY.

Epstein, M.J., 2009. *Making Sustainability Work: Best Practices in Managing and Measuring Corporate Social, Environmental, and Economic Impacts.* Berrett-Koehler Publishers, San Francisco, CA.

Epstein, M.J., 2010. Thinking straight about sustainability. *Stanford Social Innovation Review* **8**(3), 51–55.

Esquer-Peralta, J., Velazquez, L. and Munguia, N., 2008. Perceptions of core elements for sustainability management systems (SMS), *Management Decision* **46**(7), 1027–1038.

Gharajedaghi, J., 2006. *Systems Thinking: Managing Chaos and Complexity.* Butterworth-Heinemann, Boston, MA.

Hardjono, T. and de Klein, P., 2004. Introduction on the European corporate sustainability framework. *Journal of Business Ethics* **55**, 99–113.

Hutchins, M.J. and Sutherland, J.W., 2008. An exploration of measures of social sustainability and their application to supply chain decisions. *Journal of Cleaner Production* **16**, 1688–1698.

Kiely, T.J., 1995. Managing change: Why reengineering projects fail. *Harvard Business Review* **73**(2), 15.

Kuei, C., Madu, C.N. and Lin, C., 2008. Implementing supply chain quality management. *Total Quality Management* **19**(11–12), 1127–1141.

Kuei, C., Madu, C.N. and Lin, C., 2011. Developing global supply chain quality management (SCQM). *International Journal of Production Research*, forthcoming.

Lo, S. and Sheu, H., 2007. Is corporate sustainability a value-increasing strategy for business? *Journal Compilation* **15**(2), 345–358.

Luisi, P.L. and Houshmand, Z., 2009. *Mind and Life, Discussions with the Dalai Lama on the Nature of Reality.* Columbia University Press, NY.

Madu, C.N. and Kuei, C. 1995. *Strategic Total Quality Management.* Quorum Books, Westport, Connecticut.

Madu, C.N. and Kuei, C. 2004. *ERP and Supply Chain Management.* Chi Publishers, Fairfield, Connecticut.

Moneva, J.M. and Cuellar, B., 2009. The value relevance of financial and non-financial environmental reporting. *Environmental & Resource Economics* **44**(3), 441–456.

Nidumolu, R., Prahalad, C.K. and Rangaswami, M.R., 2009. Why sustainability is now the key driver of innovation. *Harvard Business Review* **87**(9), 57–64.

Shadare, O.A., 2010. Influence of worker's training programs on conflict reduction in industrial organizations in Nigeria. *African Journal of Business Management* **4**(7), 1240–1246.

Skouloudis, A., Evangelinos, K. and Kourmousis, F., 2009. Development of an evaluation methodology for triple bottom line reports using international standards on reporting. *Environmental Management* **44**(2), 298–311.

United Nations, 1987. *Report of the World Commission on Environment and Development.* Retrieved from http://www.un.org/documents/ga/res/42/ares42-187.htm.

Wagner, M., 2007. Integration of environmental management with other managerial functions of the firm: Empirical effects on drivers of economic performance. *Long Range Planning* **40**(5), 611–628.

Wagner, M., 2010. The role of corporate sustainability performance for economic performance: A firm-level analysis of moderation effects. *Ecological Economics* **69**(7), 1553–1560.

Winkler, H., 2010. Sustainability through the implementation of sustainable supply chain networks. *International Journal of Sustainable Economy* **2**(3), 293–309.

Appendix A: Sustainability Grid

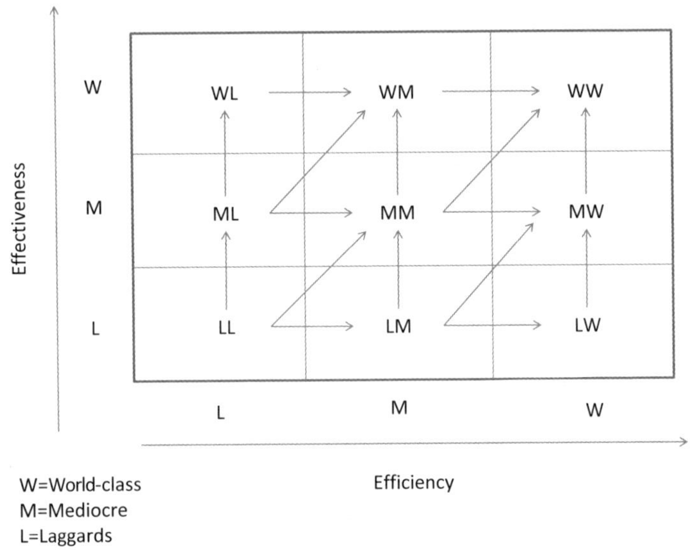

Chapter 2

SUSTAINABLE DEVELOPMENT

Douglas Crawford-Brown
Pell Frischmann, London and EnviroTech,
44 Longworth Avenue, Cambridge CB4 1GU, UK
dcrawfor@email.unc.edu

Sustainable development involves creating an infrastructure of material and energy use in communities that meet human needs while maintaining a wide array of metrics of environmental quality, human health, social equity, and economic vitality. The two key environmental metrics of most interest today are energy use and carbon dioxide emissions, because both act as surrogates for a much larger set of environmental concerns and because rising fuel prices and carbon costs compel their consideration. This chapter explores the primary considerations in developing sustainable communities and activities, using examples addressing questions of whether new construction is needed; where we place it and how we get to it; how we construct it; how efficient it is at energy use; how we power it; how we live within it; and the kinds of economic activity we house there.

1. Framing the Issue

There is little doubt that development is needed globally. A third of the world's population lives in crushing poverty: the "bottom of the pyramid." Their basic human needs — for warmth in the winter and cooling in the summer, for food, for jobs, for housing — are not being met. The answer is to bring in development to meet those needs. Populations are migrating from these poorer nations to the wealthier, and so even these wealthier nations require development, if not to lift themselves out of poverty then at least to provide a built environment for the recent immigrants. And there is a clear link between measures of economic development such as Gross Domestic Product and public health measures such as life span. Hence, there will continue to be a call for economic and infrastructure development in both the developed and developing nations. The key is to manage that development through treating it as part of a system of material and energy use that has increasingly stringent constraints as natural processes are pushed to their limits for assimilating society's waste and supplying materials and energy for economic growth. The collective name for human designs that work within these constraints is sustainability.

It's useful to begin a discussion of sustainable development with an attempt both to make the concept concrete and to narrow its multiple meanings. One of the difficulties with the concept is that it arose, almost simultaneously, from a wide range of fields of study and professional practice, including

- *Resource conservation*, in which sustainability means drawing on a resource (water, oil, etc.) at a rate that equals or is less than the rate of replenishment, so that resources will continue to be available to future generations.
- *Ecological conservation*, in which sustainability means preserving key ecosystem functions such as components of the food chain and natural services such as water purification.
- *Triple bottom line*, in which sustainability means operating businesses in ways that simultaneously promote goals of economic development, environmental quality, and social equity — also going by phrases such as People, Profit and Planet or Environment, Equity and Economy.
- *Public health*, in which sustainability means emission rates for pollution that yields environmental concentrations below those associated with unacceptable risks to human health.
- *Community planning*, in which sustainability means a built environment promoting activities that increase well-being and where decisions reflect the goals and values of all citizens.
- *Energy policy*, in which sustainability means matching the demand for energy to the ability to withdraw that energy without depleting the stock of energy supplies, and developing technologies and behaviors to assist in this balance.

Layered over these multiple traditions is an increasing tendency to try to pack the concept of sustainability with all forms of social ills and their solutions. And so one hears of environmental sustainability, economic sustainability, social sustainability and on and on, all going under the banner of sustainability. With so many traditions and calls for including issues within the umbrella of sustainability, there is danger both that decision makers can be paralyzed by the complexity and that the concept will become so amorphous as to be meaningless — much as value and quality have risen and fallen as inclusive terms of business management and engineering.

This chapter, therefore, examines sustainable development through a particular lens that draws on a number of these traditions and issues, but is rooted ultimately in the twin problems of constrained material and energy supplies and the growing threat of climate change. Sustainable development here will mean constructing communities that meet human needs in ways that greatly increase the efficiency of energy use, reduce the reliance on fossil fuel reserves, and significantly lower emissions of carbon dioxide and other greenhouse gases (by 60–80% from current levels — a daunting technological and behavioral shift). The argument is that meeting these goals will also reduce a wide range of other environmental threats, as these threats share a common cause in patterns of material and energy use in communities. It won't always be the case that strategies optimized to reduce energy use and greenhouse gas emissions will satisfy the demands of other environmental threats such as tropospheric ozone or habitat fragmentation, but this approach is a good place to start so long as these other threats are kept in mind as constraints within which strategies aimed at energy and greenhouse gas reduction can be considered.

2. Seeking Solutions

Energy security and energy prices — which are related to energy conservation and its connections to the built environment — and climate change are dominating both media coverage of sustainability and public policy responses. Patterns of development in the past have been largely disconnected from these resource and environmental issues. With oil, and hence petrol, inexpensive, there is little economic burden in building houses that are dozens of miles from shopping, schools or places of work. However, with the price of oil rapidly approaching $200 per barrel, with calls for cap-and-trade on greenhouse gas emissions and a world-wide carbon pricing scheme, and with the possibility of increasing fragility of communities in the face of climate extremes that can destroy entire cities in days, the world can no longer keep development patterns, energy, and climate change separate.

There is a pressing call for innovative solutions to development that do a better job of balancing demand and the natural limits on supply while meeting reasonable needs of people and communities for a better life. If we're honest with ourselves, we must admit that humans have been pretty sloppy with energy in the past. In the streams of energy in electricity production typical of the developed world (World Business council for sustainable development, 2006), every 100 units of energy extracted by society yields only about 30 units of useful energy that satisfies human needs, whereas 70 units go right back out into the environment as waste — primarily waste heat. Those 70 units provide for no human needs and cannot be counted in any sensible definition of development. This was fine when energy supplies were plentiful and cheap, and when there was no recognition of the environmental consequences of energy use. However, those days are gone and so the task ahead is to meet human needs in ways that greatly reduce the figure of 70 units wasted and replace the sources of the overall flow of energy in society with energy supplies that do not emit greenhouse gases — the decarbonization of the economy at the heart of sustainable development.

This flow of energy is caused by the infrastructure of society. It is caused by meeting human needs through buildings, roads, power stations, stores, etc. There are crucial roles for societal structures (economic, political, social) and superstructures (religious beliefs, scientific knowledge, ethical positions) in policies of sustainable development. These structures and superstructures guide behavior, including behaviors that lead to infrastructure development. However, in the end these structures and superstructures act on the environment through the infrastructure. And so the focus in this paper is on getting the infrastructure right without dictating what the structures and superstructures must be. Successful policies must work in diverse global societal settings with potentially large differences in economies and political systems and modes of living. The focus is on coming to understand infrastructure as a system that takes human needs in at one end, marshals natural resources to meet those needs, and puts out at the other end varying degrees of environmental change, including climate change and resource depletion. Sustainable development is then a form of environmental management system in which the questions asked of the infrastructure are as follows: (i) Is new construction really the answer? (ii) Where do we put it and how do we get to it? (iii) How do we construct it? (iv) What is its energy performance? (v) How do we power it? (vi) How do we live within it? and (vii) What economic activities do we place within it? These 7 questions, and innovations in their answers, form the remainder of this paper.

3. Providing Answers

3.1. *Is new construction the answer?*

The mantra of sustainable or green construction is "the greenest building is one that was never built in the first place because it wasn't needed." While the movement toward improving the energy efficiency of buildings through programs such as LEED (Leadership in Energy and Environmental Design) has been laudable, it has had an unfortunate side effect (although it should be kept in mind that systems such as LEED don't focus solely on energy efficiency, and all of its competitors such as BREEAM in the UK have the same problem described here). Projects have been certified as green based on measures of the energy used per unit floor area. Such measures fail to account for the total floor area produced in a community, and yet the environmental impacts of development are related to the product of floor area and energy consumed per unit floor area, and not to either number separately. The result has been an increase in new floor space to house development activities worldwide, all increasingly energy efficient (and certified as "green") but resulting in ever increasing rates of carbon dioxide emissions and greater demands for energy.

Providing the infrastructure for development, however, need not involve creation of new space. A recent survey by the firm Gensler (2008) suggests that existing office space is often poorly utilized, with up to a third of that space being wasted. This waste both increases the need for constructing new space, resulting in a global glut of unoccupied office space in urban areas, and creates a need to heat and cool the wasted space — energy that fulfills no human need. Through creative office designs, the Gensler team has found that all of the needs of occupants can be met without the wasted space, freeing up space for activities that would otherwise have required new construction. And this creative and flexible use of space results in increased productivity and enjoyment of the workplace. Their analysis goes even further to suggest that companies and municipalities might cooperate to view the available office space throughout a city as a flexible system of facilities and property management, with businesses shifting their operations into spaces that are matched precisely to their needs, placing a premium on construction that provides space that can be reconfigured periodically as those needs change and as companies grow and contract.

3.2. *Where do we put it and how do we get to it?*

Approximately a third of the environmental impacts of energy use in modern societies arises from transport, either to move people to the points of activity (jobs, shopping, schools, etc.) or to bring goods into a community. Focusing here solely on the first issue, the sustainability of the transport system is a function in part of how far people must travel to meet daily human needs. This is in turn a problem of the spatial layout of a community, measured by the average distance between places of residence and the locations of core services.

As just one of many examples that might be given considers a recent study done in Cambridge (UK) of the possible location for a new development of homes (see Fig. 1). The UK has both an expanding population and a social shift that is causing young families to seek their own homes rather than the more traditional pattern of living with parents initially. The result is a call by the UK government for development of millions of new homes to be built over the next decade. Cambridge is to be the site for 18,000 of these homes. Even with the greenest construction practices for the homes, the result will be an increase in carbon dioxide emissions on the order of 100,000 tonnes per year. And so the City Council is asking where these homes might best be located.

A variety of scenarios for home construction were simulated: location of the homes on either the inner or outer rings shown in Fig. 1, and provision of services (shops, etc.) either using the existing services in city center or through development of new services within the community of homes. Travel back and forth to these services was then simulated for home locations around these two rings, assuming the existing mix of transport mode choices (a mix of cars, buses, bikes, and walking). The results suggest that the choice of inner or outer ring, or even direction from the city, was not crucial in determining the carbon footprint of the community (changing that footprint by something on the order of 10%), but that co-location of new services near the homes would reduce transport-related footprint by up to half. The result of studies such as these has been the rise of new urbanism and mixed use communities.

Apart from distance traveled to fulfill a need, spatial locations influence mode choice (car, bus, walking, biking). The decision as to whether walk or ride a bike, on the one hand, or use a personal vehicle, is affected significantly by distance that must be traveled (see Fig. 2), as well as by demographics (e.g., whether one is young or elderly; whether there is a child in tow), availability of alternatives (e.g., easily

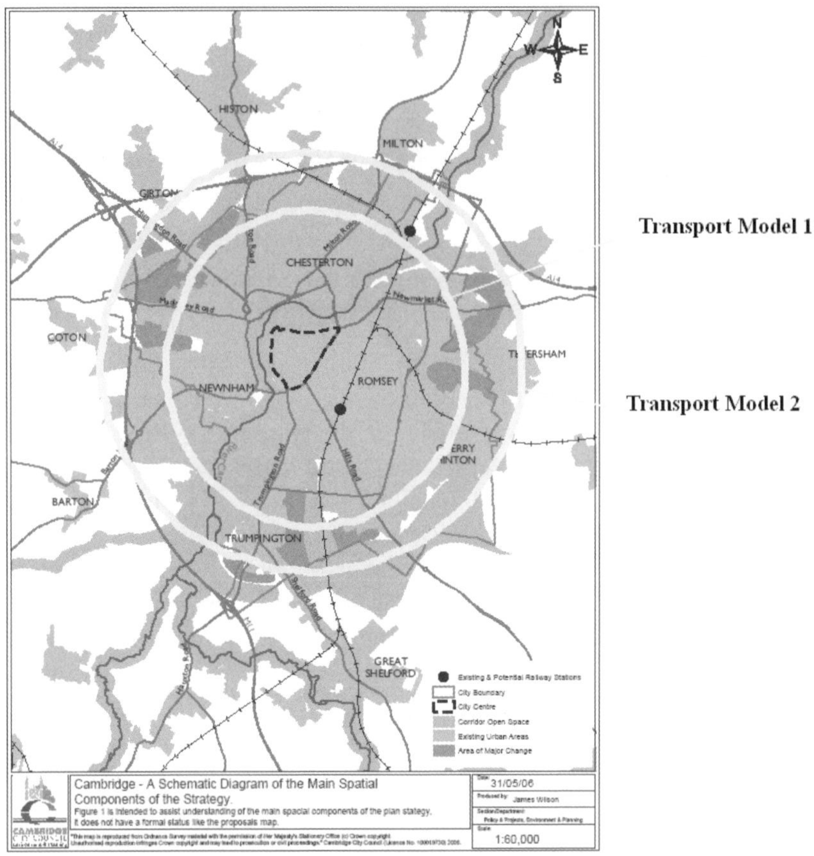

Fig. 1. New homes in Cambridge might be located on either the inner or outer rings above, and serviced either by the existing shops or through construction of new shops near the homes. The different locations have, in turn, different requirements for energy use by transport.

affordable bus routes), and the design of roads and paths (Frumkin *et al.*, 2004). Separation of biking and walking paths from roadways, maintaining the paths, shielding them from the weather with trees, etc. all influence the willingness of someone to walk or use their bike rather than travel by car. And there are follow-on health benefits of such well-designed walking and biking paths, as riding a bike on a roadway shared by cars or walking across busy streets is much more hazardous than driving in a car. In the study on Cambridge, reductions in the transport-related carbon footprint were on the order of 40% when mode choice for shopping switched from personal vehicles to buses, and 70% when biking and walking became a viable option due to proper design of walking and biking paths.

3.3. *How do we construct it?*

This question puts aside for the moment the issue of how a building will perform once it has been constructed, and examines instead the process of construction itself. The

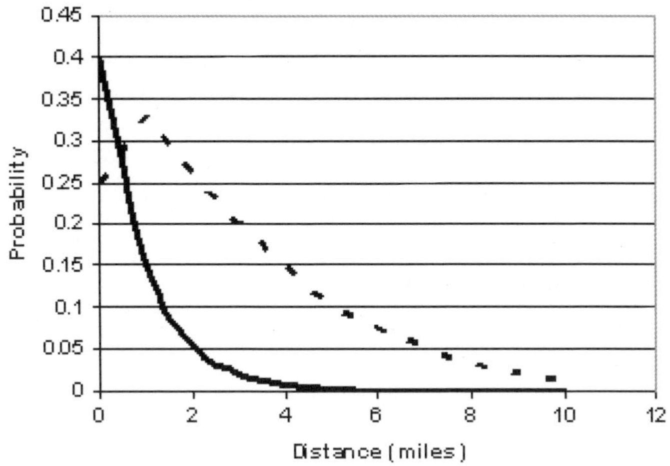

Fig. 2. Change in probability of walking or biking as distance to a need is increased. This figure is for a 34 year-old female in the U.S.

focus here is on the concept of embodied energy and embodied carbon in a building, or the energy used and carbon dioxide released in producing the materials that go into a project. Sustainable construction practices focus on the twin tasks of reducing use of new materials and ensuring that the new materials that do get used have the lowest feasible embodied energy and carbon. Similar principles are now required in almost all municipal planning processes in the UK (the US being a bit behind in this regard), as can be seen in the City of Westminster's Unitary Development Plan and Land Development Framework requirements that projects "minimize the materials, energy and water needed to construct and operate buildings" and "reuse materials and components and use recycled components and materials."

The latter issue is one dealt with by closing the life cycle of materials used in construction, much as is advocated by McDonough and Braungart in Cradle-to-Cradle (McDonough and Braungart, 2002). One of the more important examples here is the recycling of concrete, which has an embodied energy on the order of 1–2 MJ (MJ = millions of joules) per kg of concrete, and an embodied carbon of 0.1–0.2 kg CO_2 per kg of concrete (Hammond and Jones, 2006). Recycling of 1000 cubic meters of concrete from a former building into the foundation of new construction on-site (not an unusual amount of concrete that might be available in a typical project) can therefore reduce the embodied energy of the new building by a million or more MJ, and reduce the carbon footprint by 100,000–200,000 tonnes. These are significant reductions that don't even account for the reduction in transport-related energy and carbon emissions that would otherwise be needed to haul away the demolition materials.

As to new materials, the goal is reduction of material use through innovative construction materials, and replacement of materials with high embodied energy and carbon values with materials having low values. Post-tensioning can reduce concrete use by 10–20% while providing for the same floor area, simultaneously increasing beam spans and making the space more flexible for future uses. And as just one of many examples of material choice, wall insulation might be either fiberglass at an embodied energy of 28 MJ per kg of insulation, or wood wool at an embodied energy of 11 MJ per kg of insulation. The same need for warmth and cooling is met either way, but at significantly reduced embodied energy if wood wool is used.

3.4. *What is its energy performance?*

People don't want to consume energy. They want light for seeing, to be warm in the winter and cool in the summer, to be able to get about to their jobs and shopping and schools, etc. Energy just happens to be the way in which the infrastructure provides for those needs. To the extent a building's or community's energy performance can be Improved through innovative design, the same needs can be met with lower energy and environmental costs.

The solutions here are two-fold. Wall materials with low U values (the U value is a measure of how well the walls or glazing conducts heat) are readily available and reduce energy consumption for heating and cooling by 30% or more compared with traditional construction methods. Occupants are no less comfortable — no colder in winter or hotter in summer — but energy use, energy bills, and carbon dioxide emissions are reduced by similar percentages. Day-lighting of a building through use of modern glass with reflective surfaces to keep heat out in the summer and in during the winter, both reduces the consumption of energy needed to provide artificial light, further reducing the heat load during the summer, and increases the productivity of workers and the ability of students to learn. Such measures, therefore, increase economic development while reducing energy and carbon footprints.

One of the first demonstrations of these design principles was the Elizabeth Fry Building at the University of East Anglia in the UK, where John Miller and Partners have applied sustainable design principles to great effect. In their own words (see the complete description of the building at www.johnmillerandpartners.co.uk/fry.htm): "To achieve low energy comfort within the building, ambient conditions are maximised, by using the building structure as an energy store, combined with low pressure mechanical ventilation, heat recovery and opening windows. To assist low energy consumption, the building envelope is super insulated and is of airtight construction, with no thermal bridging of the fabric, while minimising solar gains externally. The result is a building with a total heat demand from boilers of 24 KW (two domestic wall mounted boilers) and no mechanical cooling, serving a building of 3000 m^2 with a potential occupancy of 850 people."

Second, a properly designed and operated community of buildings makes reuse of waste energy feasible. University campuses usually house highly energy-intensive computing facilities. Some of the larger supercomputing centers can consume the power normally provided to a small town (on the order of many megawatts of power). The heat generated by these activities is usually well beyond the heating needs for the buildings that house them, and so that heat is dumped to the outside atmosphere as waste. The co-location of other buildings with heating needs can create a system in which the waste heat of the one building (the computer lab) becomes the input energy into another. The overall efficiency of energy use in the community of buildings is therefore increased through such a systems approach, mimicking Nature's systems of efficient recycling of nutrients.

3.5. *How do we power it?*

A typical building uses electricity to power the lighting and cooling systems, and natural gas to provide heat. With emissions factors of approximately 0.5 kg of CO_2 per kilowatt-hour of electricity production and 0.2 kg of CO_2 per kilowatt-hour of natural gas use, the carbon footprint of a building can easily exceed several hundred kg of CO_2 per square meter per year if designed and operated poorly.

Improvements in efficiency — discussed previously — can reduce this footprint, but so can careful selection of the means of providing energy. Heating is best supplied through local combustion of natural gas rather than the production of electricity, as the waste energy due to generation, transmission, distribution and use is significantly lower for natural gas than for electricity. However, natural gas is as much a non-renewable source of energy as the coal that generally drives electricity production. And so more modern technologies must be sought, ones that decarbonize energy production and the economic development that depends on it.

It isn't possible here to review the many potential solutions to provision of renewable energy. In any event, there will be multiple solutions at different spatial and temporal scales, a phenomenon Bob Socolow of Princeton has labeled "stabilization wedges" (Pacala and Socolow, 2004). Such an approach, Socolow says in the *Economist*, "decomposes a heroic challenge (eliminating the emissions in the stabilization triangle) into a limited set of merely monumental tasks." He lists 15 potential wedges, each capable of reducing worldwide emissions of carbon dioxide by a billion tonnes per year; some combination of 7 of these 15 will therefore be needed to reduce the threat of global climate change (Fig. 3). In no particular order, the 15 candidates are increased transport efficiency; reducing miles traveled; increased heating efficiency; increased efficiency of electricity generation, transmission, and distribution; fuel switching from coal to gas; carbon capture and storage; coal synthetic fuels; hydrogen fuel; nuclear power; wind-generated electricity; solar electricity; wind-generated hydrogen fuel; biofuels; forest storage of carbon; and soil storage of carbon. The particular combination used will depend on the community in which they are adopted, as environmental conditions needed for each wedge vary and there will be differing demands for the economic development that can be produced through each wedge.

Somewhat surprisingly, the list of wedges provided by Socolow does not include one of the most promising technologies for providing energy to buildings: heat pumps. Advances in the design of both ground source and air source heat pumps,

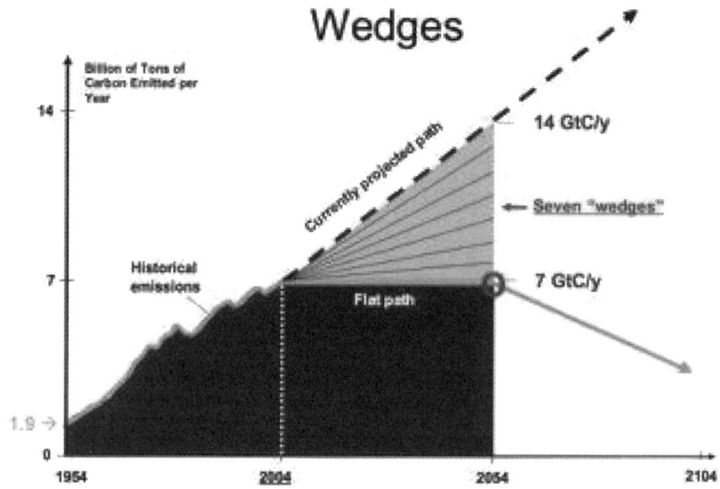

Fig. 3. The wedges approach to energy and climate policy proposed by Pacala and Socolow (2004).

including CO_2-based heat pumps, make these a viable way to extract the energy for heating and cooling a building (which is on the order of 70–80% of the energy use in buildings) from the surrounding environment. Their coefficients of performance are now above 4, meaning there are 4 units of energy extracted from the environment for each 1 unit of energy put into the heat pump system, potentially reducing the energy demand and carbon dioxide emissions to 25% of current values.

An increasingly popular and promising approach to providing power to communities is through municipal or public power systems such as the ones that operate in many European cities. The basis behind public power is to move utilities into the not-for-profit, publicly owned sector operated by municipalities and public bodies. The benefits of offering public power to the community include potentially reduced cost, local ownership, increased reliability, and environmental benefits if public ownership leads more quickly to adoption of low carbon fuels used in generating the power.

More than 2000 public power utilities combine to serve 43 million people in the US, which is approximately 14% of the US electricity consumers. Most public power utilities buy energy on the wholesale market rather than generate energy within the community, but 30% generate at least part of their power through publicly owned facilities (that is, 30% of the 14% of the municipalities with public power). The majority of these have a 5 MW capacity or less, but it is at least a start.

3.6. *How do we live within it?*

While technologies can provide efficient transport systems, and buildings that require less energy to meet a specific human need, the lesson has been that both energy consumption and carbon dioxide emissions continue to rise. Why is this the case? Part of the answer is certainly that the technologies simply are not being deployed at the scale needed to protect the environment and conserve resources. This presumably will happen over time as energy prices continue to rise and the pricing of carbon enters the market. However, the answer also lies in how people choose to use transport systems and occupy buildings. The fuel efficiency of vehicles has increased significantly over the past several decades. However, this has been more than offset by an increase in the number of people owning cars and by a rapid rise in the miles driven per person per year. This problem of transport-related energy use and carbon footprints will resolve itself partially as the cost of fuel rises due to an increasing gap between supply and demand, made all the worse in the developed world by stiff competition from developing nations such as China and India. It also is clear, however, that the willingness to use green transport, especially walking and biking, depends crucially on community designs. Part of the answer here lies in decreasing distances traveled, but it lies equally in designs that encourage these modes of transport through pathways that make such travel safe and convenient: well established and well lit paths, separating these paths from roadways, making the communities esthetically pleasing so people consider the trip itself as important as the destination, etc.

The same processes are at play in the ways we occupy buildings. While the energy efficiency of building envelopes, and the sustainability of energy systems to power buildings, has increased, so has the plug load as televisions, computers and other electrical devices have become more common. This is joined by an increasing desire to control interior temperatures within narrow bands, with many homes in the developed world maintained at temperatures that vary by only a few degrees throughout the seasons, despite large variations in outdoor temperatures. Again, the increase in efficiency has been more than offset by the increased plug load and demand for heating and cooling. The answer to this dilemma, called the "rebound effect," lies in part in building the environmental costs of energy into the price of that energy. It also lies in behavioral change that will be facilitated through energy management systems in buildings that show occupants in real time their energy use and environmental impacts.

A solution lies in intelligent grids and micro-grids for communities. Much of the environmental impacts of energy generation come from the need to provide for peak power rather than baseload. A summer afternoon, with high outside temperatures, can cause large surges in energy demand as air conditioners kick in. This incremental peak energy is often supplied by less efficient and more costly facilities that can be turned on and off as needed but at a sacrifice to efficiency and environmental controls. This can be avoided by systems of demand response, or load leveling, in which surges in demand are in part offset through signals passing from the generator to appliances, signals that cause these appliances to shut themselves down partially during the peak (think of refrigerators that reduce their energy use during such periods, warning occupants of a building to avoid opening the appliance until the peak has passed). This can be coupled to fleets of plug-in electric hybrids that consume electricity during the night when demands are low, store that energy in batteries or fuel cells or capacitors, use the energy during parts of the day when only baseload demands must be met, and then plug back into the grid to discharge energy when peak demands must be met. Through such approaches, a community, including producers and consumers of energy, becomes a system in which energy use is optimized and needs are met while reducing the environmental impacts.

Finally, there is the embodied energy and carbon of the products we bring into a building. Whether office equipment, furniture, cell phones, televisions or food, the choices of consumption made in a building are the driving force behind a market-based system of industrial production that uses up to a third of the energy associated with life in a community. While the economic activities that produce these goods may not take place in a given community (in the West, we tend to have our goods manufactured in China, simply shifting the emissions to that country), the consumption of these goods become part of the energy and carbon footprints of that community as part of the system of communities worldwide. Managing that system requires understanding the energy and carbon implications of market choices that put in motion the wheels of research and development, primary manufacturing, secondary manufacturing, transport, sale and ultimate disposal (or recycling) of goods.

3.7. *What economic activities do we place within it?*

The final piece of the sustainability puzzle is consideration of the kinds of economic development that will take place within sustainable communities. The transition to a decarbonized economy, one much less reliant on cheap fossil fuels and one in which the price of the remaining fuels is increasingly high, will not be without disruption. The economies of the developed world have been built almost entirely on the cheap availability of fossil fuels, whether coal for electricity generation, natural gas for heating or oil for transport fuel and commercial products. While this energy is used with increasing efficiency in generating economic development (see Fig. 4), the rate

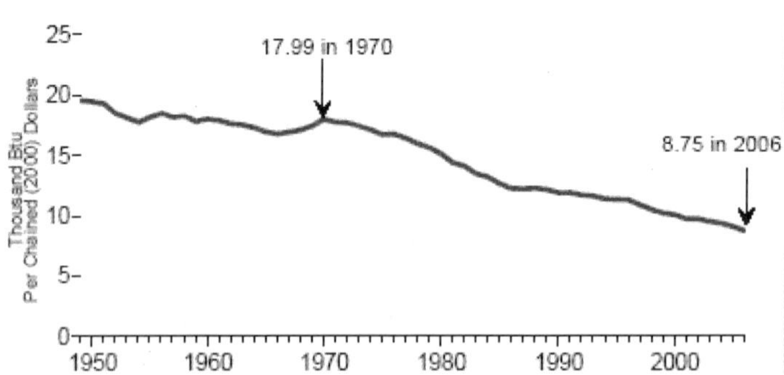

Fig. 4. In the US and throughout the developed world, increases in energy efficiency have driven down the amount of energy consumed per unit economic activity, measured here by gross domestic product. The figure is taken from US Energy Information Administration (2006).

of development has been higher than the rate of improvement in energy intensity. The solution to improved sustainability lies, therefore, not simply in improving the energy intensity of the economy, but in making a transition to a different base of economic activity that produces less carbon dioxide per unit energy consumed.

That transition will involve the evolution of all six of the players or stakeholders that helped develop the current, carbon-based, economy. Their motivations in creating new economic activities will be diverse, ranging from ensuring energy security, to economic development opportunities, to concern for climate change, to worries over a reliance on energy from politically unstable regions, to escalating energy prices. All of these motivations share a common concern for the stability, and hence sustainability, of development. The actions to be coordinated involve

- *Innovators*: Individuals and organizations responsible for the basic research and development of novel ideas for energy production, building materials, energy storage, intelligent grids, etc.
- *Investors*: The source of capital to enable both innovation and the development of products so they can penetrate the market.
- *Providers*: The companies that manufacture and bring these products to the customer, identifying demand and creating the commercial lines to meet that demand through innovations.
- *Adopters*: Individuals, organizations, and communities that make use of the new economic activity, paying for the goods and services and using them to meet their needs.
- *Regulators*: The institutions, primarily government and trade, that set the economic policies, provide public development capital or establish constraints such as carbon cap-and-trade schemes that influence the pace of technological and behavioral change.
- *Framers*: The institutions, largely non-governmental organizations, and think tanks that influence how society views a challenge, and what it takes to be sustainable solutions linked to other issues such as poverty, job creation, and housing.

All six of these actors in a community are needed for sustainable development. It is essential that each is embraced as part of the strategy for a community, or the sustainable solutions of one community may become the unsustainable practices of another. Gone are the days when a community can simply lay the blame for environmental problems on the doorstep of industry, forgetting that industry is part of a complex of material and energy processes responding in large measure to consumer demands. A key example of a potential trap in moving industries away from communities rather than embracing their sustainable forms is the movement toward demonstration communities such as the EcoCities of the UK Such communities may gain their sustainability simply by displacing their industrial activity to other communities, or even other nations. The world doesn't become better in such a chess move, because the atmosphere doesn't care where emissions arise. A truly sustainable community is not only one where the economic activity within that community's boundaries are sustainable, but where that sustainability persists even as one includes the footprint of industrial processes located elsewhere but supplying the goods that flow into that community.

4. Conclusions

The past was dominated by a conceptual and political separation between energy production, economic activity, environmental quality and community design. Rather than being viewed as an interconnected system of issues that must be managed simultaneously, the availability of cheap, fossil fuel-based energy, and a tendency to focus on end-of-pipe solutions to environmental problems created separate systems to manage these aspects of development. Societies can no longer afford this facile separation of what are at heart related issues. Creative solutions will be built on recognition that communities may be different things to different people, but they share a root in material and energy flows that define the relationship between lifestyles and the quality of the environment. These solutions will involve the recreation of the entire material and energy infrastructure of society, building communities that reflect principles of systems management in which human needs are met while closing the cycles of materials and energy, reducing overall consumption of these goods, and keeping waste (including carbon dioxide emissions) at tolerable levels.

Brad Allenby and Tom Graedel formulate this new ethic as Earth Systems Management (Allenby, 2000/2001). Their argument is that a past focus by the environmental community on conservation and preservation of pristine nature is no longer sufficient. Similar to it or not, human society is a very powerful, in fact perhaps the most powerful, force shaping the environment. The new battle ground will not be found in keeping the built environment away from the natural environment, but rather in recognizing that these have merged. There is now one earth system in which the built environment and the development it supports has become part of the metabolism of the earth's systems. The sooner we recognize this, the sooner we can get on with the task of adjusting our material and energy systems within communities so they are suited to their roles in the larger systems of material and energy flows in the atmosphere, lithosphere, hydrosphere, and biosphere.

References

Allenby, 2000/2001. Earth systems engineering and management. *IEEE Technology and Society Magazine* **Winter**, 10–24.

Frumkin, H., Frank, L. and Jackson, R., 2004. *Urban Sprawl and Public Health: Designing, Planning, and Building for Healthy Communities.* Island Press.

Gensler, 2008. *The Gensler Design + Performance Index: The U.S. Workplace Survey*, 17 July 2008, London. Available at www.gensler.com.

Hammond and Jones, 2006. *Inventory of Carbon and Energy (ICE).* University of Bath, Bath, UK.

McDonough and Braungart, 2002. *Cradle-to-Cradle*, North Point Press.

Pacala, S. and Socolow, R., 2004. Stabilization wedges: Solving the climate problem for the next 50 years with current technologies. *Science* **305**(5686), 968–972.

U.S. Energy Information Administration, 2006. *Annual Energy Review 2006: Energy Perspectives.* U.S. Energy Information Administration, Washington D.C. Available at www.eia.doe.gov/emeu/aer/ep/ep_frame.html.

World Business Council for Sustainable Development, 2006. *Making Tomorrow's Buildings More Energy Efficient.* WBCSD News, 26 June 2006. Available at www.wbcsd.org/plugins/DocSearch/details.asp?type=DocDet&ObjectId=MTk1MjU.

Chapter 3

CORPORATE SOCIAL RESPONSIBILITY AND REPORTING

Giannis T. Tsoulfas[a] and Costas P. Pappis[b]

[a] *Department of Business Administration (Tutor in Administration of Businesses and Organizations), School of Social Science, Hellenic Open University, Patras, Greece*
tsoulfas@unipi.gr

[b] *Department of Industrial Management, University of Piraeus, 80 Karaoli & Dimitriou Str., 185 34 Piraeus, Greece*
pappis@unipi.gr

Corporate social responsibility (CSR) has been a subject for debate during the last 60 years and several related matters have been widely discussed. In this chapter, the concept of CSR is briefly presented through its evolution. Particular attention is drawn on the paradigm of stakeholder theory, according to which corporate activities should be regarded in terms of the corporation's ability to satisfy not only its shareholders, but also its various stakeholders and not necessarily only from an economic perspective. CSR, in fact, is correlated with the dominant pattern of sustainable development. In this chapter, CSR reporting is extensively dealt from a Triple Bottom Line perspective, i.e., as a means to articulate the concept of sustainable development in the business world. The most widely accepted guidelines, reporting standards and initiatives that have been developed to facilitate CSR management, are presented as well as various relevant rating and evaluation schemes. A brief discussion also follows about the critique, which has been discoursed concerning CSR and CSR reporting.

1. Introduction

Since its formalization by Bowen (1953), the concept of corporate social responsibility (CSR) has been widely discussed in several scientific disciplines, from ethics and sociology to marketing and management. It has also been commonplace even in politics and of course business practice. The starting point of CSR introduction and development may be regarded the question of whether corporations have any responsibilities other than generating profits for their shareholders by providing products and services to societies. In case of an affirmative reply, another question, which may be posed, is to whom the corporation is responsible other than its shareholders. Naturally, various research issues have arisen over the years and several theories have been proposed and either evolved or abandoned.

In this chapter, the concept of CSR is presented through its configuration in the academia and the business world. Furthermore, it is correlated with the prevailing pattern of sustainable development. Next, CSR reporting is extensively dealt, especially with respect to business guidance, reporting standards and initiatives that have been developed. Moreover, a brief discussion follows about the critique, which has been discoursed concerning CSR and CSR reporting. The chapter ends with some concluding remarks.

A final introductory note should be added regarding the term CSR. CSR is commonly used throughout this chapter. However, in several cases the term Corporate Responsibility (CR) is preferred. These are the cases where this term would be more accurate to use, at least from an etymological perspective and taking into account the bias posed by the word "social." The general rule for using each term is how the term appears in the reference cited.

2. The concept of CSR

Carroll (1979, 1991) addressing the issue of CSR defined the four responsibilities of corporations as economic, legal, ethical, and discretionary (philanthropic). These responsibilities are composed of four discrete layers of the so-called CSR pyramid.

- Economic responsibilities are at the basis (foundation) of the pyramid, since all other business responsibilities are predicated upon the economic responsibility of the firm.
- Legal responsibilities are composed of next layer of the pyramid expressing the historical development of "codified ethics", while they are dealt as coexisting with economic responsibilities.
- Ethical responsibilities, which are the next layer of the CSR pyramid, actually refer to the attitude that society expects from a corporation to have, beyond legal requirements.
- Discretionary responsibilities, which are corporate actions undertaken in response to society's expectations and based upon the goodwill of corporations, are composed of the final layer of the CSR pyramid.

Schwartz and Carroll (2003) proposed an alternative approach to CSR, namely, the intersecting circles model of CSR. According to this framework, the three core domains of economic, legal, and ethical responsibilities are represented in a Venn diagram (a diagram that shows all hypothetically possible logical relations between a finite collection of sets) in order to highlight the overlapping nature of the domains. Thus, seven categories are created, in which CSR may be conceptualized, analyzed, and illustrated. In comparison with the CSR pyramid, the intersecting circles model differs in two main aspects: it recognizes the possibility of interrelationships among CSR domains and rejects the hierarchical order of importance (Geva, 2008).

Wood (1991) distinguishes three levels of CSR analysis: institutional, organizational, and individual. Then, one principle is corresponded to each level.

- The principle of legitimacy that refers to the institutional level states: "Society grants legitimacy and power to business. In the long run, those who do not use power in a manner which society considers responsible will tend to lose it." This principle that applies equally to all corporations clarifies the relationship between society and the corporation.
- The principle of public responsibility that refers to the organizational level states: "Businesses are responsible for outcomes related to their primary and secondary areas of involvement with society." This principle clarifies to whom the corporation is responsible.
- The principle of managerial discretion that refers to the individual level states: "Managers are moral actors. Within every domain of CSR, they are obliged to exercise such discretion as is available to them, toward socially responsible outcomes." This principle refers to the voluntary social involvement of individual managers in the issues of the company's stakeholders.

Garriga and Melé (2004) assumed as a hypothesis that the most important CSR theories and approaches are focused on one of the following aspects: profits, political performance, social demands, and ethical values. Subsequently, they proposed the classification of CSR theories into four groups:

- Instrumental theories, in which the corporation is solely seen as an instrument for wealth creation and its social activities, are only means to achieve economic objectives.
- Political theories that deal with the power of corporations in society and its responsible use in the political arena.
- Integrative theories, in which the corporation is focused on responding to social demands.
- Ethical theories, based on ethical responsibilities of corporations to society.

Apparently, the numerous definitions that have been proposed for CSR are blended with the background of the entities that have suggested them. Through a content analysis of alternative CSR definitions, Dahlsrud (2008) identified five dimensions of CSR: the environmental, the social, the economic, the stakeholder, and the voluntariness. All these dimensions are integrated in the definition of the European Commission according to which CSR is "a concept whereby companies integrate social and environmental concerns in their business operations and in their interaction with their stakeholders on a voluntary basis" (European Commission, 2001). CSR may be regarded as a form of self-regulation, which aims to contribute to social welfare. This could be in terms of compensation for the business footprint or additively to social well-being. Further insights of the concept of CSR and its evolution are given in Carroll (1999), Falck and Heblich (2007), and Cochran (2007).

In practice, CSR is increasingly accepted by modern corporations arguing that they prefer to act as "citizens of the world" that have to share the planet's problems and contribute actively to their solution along with governmental organizations, non-governmental organizations (NGOs), and individual citizens than ignore these problems, which are also their problems (Pappis, 2010).

3. The Stakeholder Theory Paradigm

Freeman (1984) propounded the stakeholder theory, according to which corporations interact in various ways with many entities that constitute its stakeholders. Stakeholders are individuals or groups that affect, or are affected, directly, or indirectly, by a corporation's policies, operations, and decisions. Stakeholder theory has emerged as a dominant paradigm in CSR, when Clarkson (1995) suggested that CSR can be best understood by analyzing the ways in which corporations manage their relationships with their stakeholders. Through this lens, corporate activities should be regarded in terms of the corporation's ability to satisfy not only its shareholders but also its various stakeholders and not necessarily from an economic perspective. As bidirectional causalities between stakeholders and the corporation are developed, the purpose of the corporation would be to create value for its stakeholders and to balance their conflicting requirements by making choices and setting priorities at the strategic level within the organization.

The concept of stakeholder aims at personalizing social responsibilities by foreshowing the specific groups business should consider in its CSR orientation (Carroll, 1991). Frooman (1999) mentioned that in developing response strategies toward stakeholders three primary questions should be answered:

(1) Who are they? (This question concerns their attributes.)
(2) What do they want? (This question concerns their ends.)
(3) How are they going to try to get it? (This question concerns their means.)

According to Vos (2003), the identification of stakeholders is composed of, at least, a modeling and a normative issue. The former refers to questions such as "Who are our stakeholders?" or "To what extent is it possible to draw the line between stakeholders and non-stakeholders?". The latter refers to questions such as "What stakeholders do we take into account?" or "To what stakeholders are we willing to listen?" In any case, stakeholders, acting either formally or informally, individually or collectively, are a key element in the corporation's internal and external environment that can affect it positively or negatively (Murray and Vogel, 1997). Of course, it is upon management to set priorities with respect to which stakeholders should be satisfied and to what extent.

Stakeholders can be categorized in different ways. A rough distinction can be made between business stakeholders and non-business stakeholders. Business stakeholders include the persons, whose connection with the corporation in its responsibility to provide products and services to societies is primarily of economic nature. Owners, shareholders, employees, suppliers, retailers, business partners, and customers are business stakeholders. Non-business stakeholders do not have economic transactions with the corporation on the basis of its responsibility to provide products and services to societies. However, they get affected by and can affect corporate activities. Public authorities, local communities, NGOs, media, and the society at large belong to non-business stakeholders. In order more or less to describe the above two stakeholders' categories, Clarkson (1995) used an alternative terminology: primary stakeholders for business stakeholders and secondary stakeholders for non-business stakeholders. Although the natural environment has a bidirectional causality with corporations, it is generally not considered a stakeholder. Instead, it is considered that it is "represented" by various non-business stakeholders. Henriques and Sadorsky (1999) identified four critical stakeholder groups: regulatory stakeholders (e.g., public authorities, trade associations), organizational stakeholders (e.g., customers, business partners and employees), community stakeholders (e.g., NGOs), and the media. Maessen *et al.* (2007) discussed alternative stakeholder categorizations and suggested that the dichotomy between primary and secondary stakeholders impedes the description of the level of involvement or the relative distance of a stakeholder group to a particular corporation. Thus, they proposed that these relationships can be mapped into a number of concentric circles around the corporation, where each circle would represent different categories of stakeholders. The circle representation is not the same for every corporation and may change over time. Furthermore, extra circles may be used to give a suitable representation of a specific situation.

The relationship of the corporation with each group of stakeholders with respect to its responsibilities may be multi-dimensional. In the sequel a few aspects of such relationships are briefly outlined for several groups of stakeholders, especially with respect to Carroll's definition of responsibilities (1979, 1991).

- *Shareholders*: Their relationship with the corporation is merely of economic nature and, eventually, their primary interest is to gain profits from their investment.
- *Business partners*: Their relationship with the corporation is based on business transactions of economic nature. Furthermore, this relationship is governed by legislation and various regulations but also from the specific codes, which are developed during their co-operation.
- *Employees*: Apart from their obvious economic relationship with the corporation, there is certain legislation that covers the area of labor as well as other relevant issues such as insurance, health and safety, *etc.* Furthermore, the corporation may share responsibilities for other issues beyond legislation as well, such as career and personal development.
- *Consumers*: This is not just a "give and take" economic relationship. Consumers demand that the products they buy were produced according to certain regulations and they usually prefer corporations, which behave ethically. Moreover,

there are cases where dedicated consumers may expect special treatment from the corporation.

- *Local communities*. They may benefit from corporate activities from an economic point of view, since local economy is enhanced, jobs are created, and social life is supported. However, they are the recipients of the undesired effects of corporate activities such as environmental impacts. Thus, they expect minimization of the negative consequences not only by adhering to the related legislation, but also through voluntary actions.
- *Public authorities*: Being responsible for the political, economic, legal, and social structures, public authorities configure the landscape where the corporation operates. Thus, all kinds of relationships are observed.
- *NGOs*: Ideally, a part of their role is to constructively intervene in corporate activities by suggesting or demanding actions to the benefit of societies. These actions may be either implied by legislation or be voluntary.

Maessen *et al.* (2007) conclude that stakeholder relationships are constantly getting more important for any corporation. Therefore, the perspective is changing from stakeholder management to stakeholder dialogue in order to explore potential arrangements and build trust between individuals and organizations as a precondition for the next step of engagement and collaborative actions. In addition, the quality of stakeholder relationships and the ability to create value over a longer period of time are two sides of the same coin. For further insights in the stakeholder theory, we refer to Mitchell *et al.* (1997), Vos (2003), and Agle *et al.* (2008).

4. CSR and Sustainable Development

According to the most-quoted Brundtland Commission's definition, sustainable development is development that "meets the needs of the present without compromising the ability of future generations to meet their own needs" (United Nations, 1987). As then sustainable development has become a salient issue in political, business, and scientific debates, three core dimensions are distinguished: environmental, social, and economic. These have been drawn in a variety of ways (Fig. 1; International Union for Conservation of Nature, 2006):

- *Pillars*: In this view, all three dimensions are interdependent and mutually reinforcing and need to be addressed at the same time in order to make progress.
- *Concentric circles*: In this view, it is emphasized that the three dimensions need to be integrated.
- *Overlapping circles*: This view endorses the envisagement that economic activity should conform with social progress and that this should be achieved within limitations posed by natural environment.

The concept of sustainable development concerns the macro-scale and calls for a set of instruments and policies specialized in all categories of human activities.

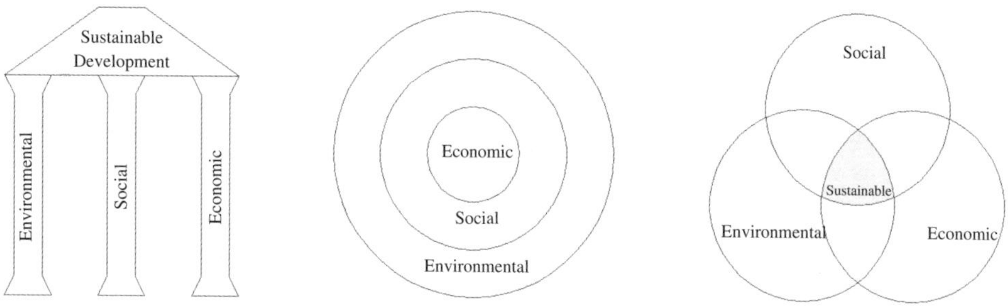

(a) (b) (c)

Fig. 1. Alternative graphical representations for sustainable development: (a) pillars, (b) concentric circles and (c) overlapping circles.

Organization for Economic Co-operation and Development (2001) aptly described CSR as "business's contribution to sustainable development." From this point of view, CSR is most relevant when corporations are considered to share a social contract with the societies where they operate, and thus hold themselves obligated to act in a way that enhances both their own as well as society's best interests (Andreasen and Drumwright, 2001). Moreover, CSR strategies may be regarded as a means for corporations to earn a "license to operate" in diverse settings as a result of globalized activities.

Moon (2007) points out that the criticism is grounded on the fact that sustainable development and CSR are "impossible" concepts, since more or less the economic dimension cannot be integrated with the others. Furthermore, both terms, which are often used vaguely and interchangeably, are both "appraisive" in that they are considered as valued and are not simply empirical concepts. Obviously, being "socially irresponsible" or "not sustainable" would not be something to be proud about. It is considered a good thing for a corporation to be esteemed as socially responsible and promoting sustainable development. As a result, there is a danger — and a real case in many examples — that corporations manipulate these terms simply for image-making purposes.

5. Business Guidance, Reporting Standards and Initiatives

The notion of triple bottom line (TBL) as coined by Elkington (1994) and spread out in his 1997's book "Cannibals with Forks: the Triple Bottom Line of 21st Century Business" has been regarded as a major step forward with respect to the converging scopes of sustainability and CSR (Elkington, 1997). The basic point of TBL is that a corporation should provide publicly insights not only regarding its financial performance, but also with respect to its environmental and social performance. From this point, the expansion of three major issues is implied: (i) the traditional reporting framework, (ii) transparency, and (iii) measurement. Thus, the TBL may be considered as a means to articulate the concept of sustainable development in the business world. Moreover, CSR and stakeholders' concerns seem to be dealt and communicated within a triple dimension perspective. Indeed, corporations that have recognized and accepted their social responsibility are expected to incorporate relevant practices into business operations. At the same time, it is expected that corporations will disclose these practices using credible and globally acknowledged reporting standards (Sutantoputra, 2009).

Nowadays, financial reporting has become a mainstream activity for corporations, regardless their size, since it is usually posed by legislation. On the other hand, the case with respect to reporting on environmental and social issues is neither ubiquitous nor homogeneous among corporations (Martin and Hadley, 2008). Gonzalez-Benito and Gonzalez-Benito (2006) identified several factors that may play a role in this respect, such as the size of the corporation, its position in the value chain, the sector where it belongs, *etc.* For example, financial constraints, time constraints, and lack of knowledge, among others, have been identified as barriers for small- and medium-sized enterprises (SMEs) in their engagement with sustainable development (Rotheroe *et al.*, 2003).

Herzig and Schaltegger (2006) identified the most important goals of corporations with regard to TBL reporting, which are mentioned as follows:

- Social legitimation of corporate activities, products, and services, which have environmental and social impacts.
- Increase in corporate reputation, brand value, and customer loyalty and avoidance of the corresponding risks.
- Gaining or maintaining a competitive advantage.
- Signaling superior competitiveness, with sustainability reporting activities as a proxy indicator for overall performance.
- Comparison and benchmarking against competitors.
- Increasing transparency and accountability within the company.
- Establishing and supporting employee motivation as well as internal information and control processes.

Additional goals include

- Recruitment and retention of more talented employees (Idowu and Towler, 2004);
- Responding to industry codes of conduct (Heslin and Ochoa, 2008);
- Meeting requirements posed by partners (Spar and La Mure, 2003);
- Responding to NGO requests (Spar and La Mure, 2003);
- Aligning with the current trends in management.

From the stakeholders' point of view, a crucial goal of reporting is to contribute to an ongoing dialogue. Indeed, reports alone have little value if they fail to inform adequately the stakeholders for corporate activities. CSR reporting may be done with several ways such as a printed dedicated report, a section in the printed annual report, a mini web site, and an online report. In particular, CSR reports have begun to be published on the web sites of many corporations, duplicating or more often supplanting printed copies, to the point that even when a printed copy exists it is also available electronically (Kolk and Perego, 2010). Scott and Jackson (2002) extensively discuss the pros and cons of reporting via the web with respect to aspects such as visibility and accessibility, timeliness of data, ease of use and quantity of reported data. Furthermore, it is important to point out that web reporting offers fertile ground for interactive communication processes with stakeholders. However, Campbell (2007) stressed the need to distinguish between the rhetoric of socially responsible corporate behavior and substantive action, especially where the corporation's proclamations may be contradicted by its behavior. In such cases, it is unquestionable that reports, advertising, and web pages dedicated to CSR are used by corporations as a means to manipulate their public image.

Over the years, the debate on CSR and sustainability has led to the introduction of management concepts and tools, which aim at assisting corporations in their engagement toward related issues as well as in their reporting activities. In particular, various organizations have developed guidelines, standards, metrics, *etc.* in order to diffuse the concept of CSR among corporations and to facilitate them from a managerial point of view toward identifying their stakeholders and measuring as well as disclosing their corporate environmental, social, and economic performance and impacts. Furthermore, certain initiatives have taken place, whose scope is usually to evaluate independently the overall CSR performance of corporations or a particular dimension of it (e.g., environmental). In these occasions, either the corporations may volunteer to participate in the evaluation or they are selected by the evaluating organization according to criteria such as their revenues, their number of employees, *etc.* In the sequel, the most widely recognized CSR-oriented guidelines, standards, metrics, and initiatives are discussed.

5.1. *The OECD guidelines for multinational enterprises (MNEs)*

The OECD Guidelines for MNEs are recommendations by 42 OECD and non-OECD governments for MNEs operating in or from countries adhered to the OECD Declaration on International Investment and Multinational Enterprises. They provide voluntary principles and standards for responsible business conduct in areas such as employment and industrial relations, human rights, environment, information disclosure, combating bribery, consumer interests, science and technology, competition, and taxation (Organization for Economic Co-operation and Development, 2008). The declaration and the guidelines were adopted by the OECD in 1976 and subsequently revised several times. Strategically, the diffusion of the guidelines takes place on a national level by the so-called National Contact Points. The guidelines aim to ensure that the operations of MNEs are in harmony with government policies, to strengthen the basis of mutual confidence between enterprises and the societies in which they operate, to help improve the foreign investment climate, and to enhance the contribution to sustainable development made by MNEs

(Organization for Economic Co-operation and Development, 2008). Naturally, the guidelines require close communication and co-operation between governmental bodies and MNEs, since both sides undertake certain responsibilities: governmental bodies should ensure the necessary conditions for the operation of MNEs, and MNEs should ensure that timely, reliable, and relevant information is disclosed concerning their activities, structure, performance, and impacts. OECD actually correlates its support to CSR through these guidelines, as they appear to cover a wide spectrum of issues related to business ethics.

5.2. *The united nations global compact (UNGC) and the intergovernmental working group of experts on international standards of accounting and reporting*

The UNGC is a strategic policy initiative for businesses that are committed to aligning their operations and strategies with ten universally accepted principles in the areas of human rights, labor, environment, and anti-corruption (United Nations Global Compact, 2010a). The Compact was first announced by the then UN Secretary-General Kofi Annan in an address to The World Economic Forum in 1999 and was officially launched in 2000 with the support and participation from MNEs, global trade unions and civil society organizations (Kell, 2005). The Compact is today the largest corporate sustainability initiative in the world with more than 7700 participants in 130 countries around the world. The basic scope of the business world through this commitment is to ensure and preserve the conditions under which global economy operates as well as to strengthen their bonds with societies at large. A total of 10 principles are presented in Table 1.

Table 1. A total of 10 UNGC principles.

Area	No.	Principles
Human rights	1	Businesses should support and respect the protection of internationally proclaimed human rights
	2	Make sure that they are not complicit in human rights abuses
Labor standards	3	Businesses should uphold the freedom of association and the effective recognition of the right to collective bargaining
	4	The elimination of all forms of forced and compulsory labor
	5	The effective abolition of child labor
	6	The elimination of discrimination in respect of employment and occupation
Environment	7	Businesses should support a precautionary approach to environmental challenges
	8	Undertake initiatives to promote greater environmental responsibility
	9	Encourage the development and diffusion of environmentally friendly technologies.
Anti-corruption	10	Businesses should work against corruption in all its forms, including extortion and bribery

Source: United Nations Global Compact (2010b).

Although the principles may be considered self-evident, they actually provide a seedbed for multi-stakeholder engagement and collective action within different contexts (e.g., national, cultural, *etc.*). In particular, for the diffusion of UNGC, local networks have been established, which are clusters of participants who come together to advance the Compact and its principles within a particular geographic context (United Nations Global Compact, 2010c). Being an integral part of the overall governance of the Compact, local networks deepen the learning experience of all participants through their own activities and events and promote action in support of broader UN goals. For example, corporations may benefit from each other by interpreting significant achievements as results of the application of the UNGC principles. The objectives of local networks usually include issues such as

- promoting the Compact and its principles in a specific geographic area;
- disseminating 10 principles to corporations and attracting new members;
- creating opportunities for multi-stakeholder engagement and collective activities.

It should be noted that one of the explicit commitments that a corporation makes when it joins the UNGC is to communicate annually to their stakeholders with respect to the progress made in implementing 10 principles of Compact (Communication on Progress, COP). This requirement serves several important purposes such as instilling accountability, driving continuous improvement, safeguarding the integrity of the UNGC as a whole, and contributing to the development of a repository of corporate practices. Failure to meet two COP consecutive deadlines results in a corporation being listed as "inactive" on the UNGC database. Three consecutive failures to submit a COP lead to removal (delisting) of the corporation. While delisted corporations are removed from the UNGC database, they are allowed to return to active status by rejoining the Compact.

Created in 1982, the Intergovernmental Working Group of Experts on International Standards of Accounting and Reporting is another establishment under the broad aegis of UN. The scope of the Group is to spotlight transparency and accounting issues at the corporate level by addressing a variety of issues in corporate accounting and reporting with a view to improving the global comparability and reliability of corporate reports. Among others, the Group focuses on issues such as (United Nations Conference on Trade and Development, 2010):

- the review of developments in the field of international accounting and reporting and the improvement of management standards;
- consulting international bodies with respect to the objective of international harmonization of accounting and reporting, the development of related international standards;
- the enhancement of the availability and comparability of MNEs' disclosures, especially by encouraging the use of internationally recognized standards; and
- the promotion of transparency.

These intents are accomplished through an integrated programme of research, intergovernmental dialogue, consensus building, and technical co-operation. As a result, the Group carries out a number of research and voluntary guidance documents, which are publicly available.

5.3. *The world business council for sustainable development (WBCSD)*

The WBCSD is a global association of some 200 companies dealing with business and sustainable development by focusing on four key areas: energy and climate, development, the business role, and ecosystems. The WBCSD was created in 1995 as a result of the merger of the Business Council for Sustainable Development and the World Industry Council for the Environment. Membership of the WBCSD is by invitation of the Executive Committee to corporations committed to sustainable development, which promote the role of eco-efficiency, innovation, and CSR. The WBCSD maintains projects concerning water and energy use in buildings as well as conducts sector-specific studies in such areas as cement, electric utilities, mining and minerals, mobility, tires, and forest products. The Council, which benefits from a global network of some 60 national and regional business councils and regional partners, provides a platform for corporations to explore sustainable development, share knowledge, experiences, and best practices, and to advocate business positions on these issues in a variety of forums, working with governments, non-governmental, and intergovernmental organizations (World Business Council for Sustainable Development, 2010a). The members of the council are distributed in more than 30 countries and 20 major industrial sectors. The major objectives of the Council are to

- be a leading business advocate on sustainable development;
- participate in policy development to create the right framework conditions for business to make an effective contribution to sustainable human progress;
- develop and promote the business case for sustainable development;
- demonstrate the business contribution to sustainable development solutions and share leading edge practices among members; and
- contribute to a sustainable future for developing nations and nations in transition.

The Council has introduced 10 key messages by which to operate aiming to link aptly corporate activities with sustainable development. These messages are presented in Table 2.

5.4. *CSR Europe and the European Alliance on CSR*

CSR Europe is the leading European business network for CSR. Its predecessor, i.e., the European Business Network for Social Cohesion, was initiated in 1996 as a result of the European Business Declaration against Social Exclusion, which was

Table 2. A total of key messages of WBCSD.

No.	Messages
1	Business is good for sustainable development and sustainable development is good for business: *Business is part of the sustainable development solution, while sustainable development is an effective long-term business growth strategy*
2	Business cannot succeed in societies that fail: *There is no future for successful business if the societies that surround it are not working. Governments and business must create partnerships to deliver essential societal services such as energy, water, health care, and infrastructure*
3	Poverty is a key enemy to stable societies: *Poverty creates political and economic instability, a big threat to business and sustainable development. By contrast, businesses can lift living standards and eradicate poverty*
4	Access to markets for all supports sustainable development: *Sustainable development is best achieved through open, transparent, and competitive global markets.*
5	Good governance is needed to make business a part of the solution: *Supportive frameworks and regulations are needed for business to contribute fully to sustainable development*
6	Business has to earn its license to operate, innovate, and grow: *The way business acts and is perceived is crucial to its success. Accountability, ethics, transparency, social and environmental responsibility and trust are basic prerequisites for successful business and sustainable development*
7	Innovation and technology development are crucial to sustainable development: *They provide key solutions to many of the problems that threaten sustainable development. Business has always been, and will continue to be, the main contributor to technological development*
8	Eco-efficiency — doing more with less — is at the core of the business case for sustainable development: *Combining environmental and economic operational excellence to deliver goods and services with lower external impacts and higher quality-of-life benefits is a key sustainable development strategy for business*
9	Ecosystems in balance — a prerequisite for business: *Business cannot function if ecosystems and the services they deliver, such as water, biodiversity, food, fiber and climate, are degraded.*
10	Co-operation beats confrontation: *Sustainable development challenges are huge and require contributions from all parties - governments, business, civil societies and international bodies. Confrontation puts the solutions at risk. Co-operation and creative partnerships foster sustainable development*

Source: World Business Council for Sustainable Development (2010b).

announced by the then European Commission President Jacques Delors in 1995. This network was renamed CSR Europe in 2000. It is composed of around 75 MNEs as well as 27 individual and independent national partner organizations, which come from 24 European countries (CSR Europe, 2010a). These organizations are membership-based led by business, aiming to promote and facilitate CSR. Together the network numbers more than 2000 corporations, which are engaged in mainstreaming responsible business practice into their activities, its mission is to support member companies in integrating CSR into the way they do business, every day but also to disseminate the concept of CSR in the business world. CSR Europe is a platform for

- connecting companies to share best practice on CSR;
- innovating new projects between business and stakeholders; and

- shaping the modern day business and political agenda on sustainability and competitiveness.

In all its activities and in its interaction with stakeholders, CSR Europe works on the themes and issues that it set out in its European Roadmap for CSR (CSR Europe, 2005). These are

- innovation and entrepreneurship;
- skills and competence building;
- equal opportunities and diversity;
- health and safety;
- environmental protection;
- mainstreaming CSR;
- stakeholder engagement;
- leadership and governance;
- communications and reporting;
- business partnerships.

CSR Europe is one of the three business organizations, which participate in the communication and co-ordination of the European Alliance on CSR (the other two organizations are, namely, Business Europe and the European Association of Craft, Small- and Medium-Sized Enterprises). The Alliance, which was launched in 2006 with strong backing from the European Commission, is a business-lead initiative to promote CSR by mobilizing the resources and capacities of European enterprises and their stakeholders in the interests of sustainable development, economic growth, and job creation. The initiation of the Alliance coincided with the issue of the European Commission entitled "Implementing the partnership for growth and jobs: making Europe a pole of excellence on corporate social responsibility," which proposed a new face of partnership with enterprises. In line with the scope of CSR Europe, the Alliance aims to contribute to the promotion of CSR throughout Europe by demonstrating the significance of voluntary business engagement and the difference it can make. Furthermore, it highlights innovative CSR practices, stimulates new ones, and further eases private initiatives in this area. In December 2008, the Alliance presented its first results, which constitute the CSR Europe's Toolbox with respect to the following major areas: integrated workplace, human capital, sustainable production and consumption, communication and transparency, and (r)evolutionary business models (CSR Europe, 2010b).

5.5. *AccountAbility's AA1000 series of standards*

AccountAbility was founded in the United Kingdom in 1995 with the aim to build and demonstrate the possibilities for tomorrow's global markets and governance through thought-leadership and advisory services (AccountAbility, 2010a). At the core of the organization's work, the AA1000 Series of Standards is based on the principles of

- inclusivity, which prescribes that the stakeholders to whom an organization has an impact, should have a say in the decisions that affect them;
- materiality, which denotes that decision makers should identify, understand, and interpret the relevance and significance of various issues to an organization and its stakeholders; and
- responsiveness, which is the organization's response to stakeholder issues that affect its sustainability performance, expressed through decisions, business strategies, performance objectives actions as well as communication with stakeholders.

The scope of the Standards, which have been developed through multi-stakeholder consultation processes, is to help organizations become more accountable, responsible, and sustainable. They are open source and they are applicable to any type of organization from MNEs, to SMEs, governments, and civil society organizations (AccountAbility, 2010b). In particular, the AA1000 Series of Standards aim to present an integrated solution of non-financial reporting, since they cover a big range of related issues: from the identification of stakeholders to the assurance of a report. Furthermore, they explicitly facilitate the application of other standards or guidelines in order for the organization to cover TBL reporting as thoroughly as possible.

The Standards consist of

- The AA1000 AccountAbility Principles Standard (AA1000APS) 2008, which provides a framework for an organization in order to better identify, understand, prioritize, and respond to its sustainability challenges. Moreover, this Standard provides an organization with a set of principles to guide them to understand, manage, implement, evaluate, and communicate their accountability. It also helps in identifying and acting on opportunities as well as managing non-financial risk and compliance. The Standard is primarily intended for organizations, which develop an accountable and strategic approach to sustainability (AccountAbility, 2010c).
- The AA1000 Assurance Standard (AA1000AS) 2008, which is used to provide assurance on publicly available sustainability information, particularly CSR/Sustainability reports. It requires the assurance provider to provide assurance on the nature and extent of adherence to the AA1000 AccountAbility Principles, and, where applicable, the quality of publicly disclosed information on sustainability performance. The Standard focuses on what is material to the organization and its stakeholders and does not stick to the mere verification of data. Furthermore, it enables the usage of other verification and certification schemes that deal with specific dimensions of sustainability (AccountAbility, 2010d).
- The AA1000 Stakeholder Engagement Standard (AA1000SES) 2005, which provides a framework to help organizations ensure the efficiency of stakeholder engagement processes. This Standard has been under review since 2009 and

an exposure draft has been released for public deliberation. It offers a basis for designing, implementing, evaluating, and assuring the quality of stakeholder engagement (AccountAbility, 2010e).

5.6. *The global reporting initiative (GRI)*

The GRI is a network-based organization that has pioneered the development of the widespread free to use homonym sustainability reporting Framework and is committed to its continuous improvement and application worldwide (Global Reporting Initiative, 2010). The idea of a disclosure framework for sustainability information was conceived in 1997, while the first sustainability reporting Guidelines were released in 2000. This Framework includes the principles and indicators that organizations may use in order to evaluate and report their economic, environmental, and social performance. Brown *et al.* (2009) mention that GRI has been very successful from its infancy in terms of uptake, comprehensiveness, visibility, and prestige, mainly due to good timing and the institutional entrepreneurial tactics of its founders. The same authors condense these tactics in three points:

- the engagement of a broad coalition of actors drawn globally from business, civil society, labor, and professional institutions in a discussion around a set of rules and practices embodied by GRI Reporting Guidelines;
- the creation of a mechanism for maintaining and enhancing the discussion into the future and for building a sense of shared ownership of the new rules and practices; and
- the creation of an organization, which would serve as steward of the Guidelines and of the evolutionary process (GRI Secretariat).

The third version of the Guidelines — known as the G3 Guidelines — was published in 2006 after a two-year development process in which about 3000 stakeholders worldwide got involved. Being relevant to all organizations regardless of size, sector, or location, they may serve as the basis of an organization's annual reporting activities, since they outline a disclosure framework that organizations can voluntarily, flexibly, and incrementally adopt. The G3 Guidelines consist of the following aspects (Global Reporting Initiative, 2006):

Part 1: Reporting principles and guidance:

- *Principles to define report content*: Materiality, stakeholder inclusiveness, sustainability context, and completeness.
- *Principles to define report quality*: Balance, comparability, accuracy, timeliness, reliability, and clarity.
- Guidance on how to set the report boundary.

Part 2: Standard disclosures:

- Strategy and profile;
- Management approach;
- Performance indicators.

Apart from the G3 Guidelines, the Framework contains Protocols and Sector Supplements. Protocols provide the background for each indicator in the Guidelines and include definitions for key terms in the indicator, compilation methodologies, intended scope of the indicator, and other technical references. GRI performance indicators are first organized by general sustainability "category" (economic, environment, labor, human rights, society, product responsibility), and then further arranged under "aspects," which more specifically reflect the issue each indicator is designed to measure. Furthermore, all GRI performance indicators are designated as either "core" or "additional." The former are those identified in the Guidelines to be of interest to most stakeholders and are assumed to be material unless deemed otherwise on the basis of applying the GRI reporting principles. The latter are those identified in the Guidelines that represent emerging practice, or address topics that may be material to some organizations but not generally for a majority. Sector Supplements, which contain further unique indicators, have also been introduced in order to correspond to the specific characteristics of different sectors such as mining, automotive, banking, *etc.* Indicators contained in Sector Supplements are considered core. Furthermore, National Annexes will soon be developed for use in conjunction with the Guidelines and will address country or regional sustainability issues.

All G3 reports must self-declare an Application Level, which communicates to the readers of the report to what extent the G3 Guidelines have been utilized in the sustainability report. Furthermore, they are intended to provide GRI reporters with a pathway in which they can continuously improve their reporting. There are three Application Levels: A, B, and C. These levels can be self-declared, third-party-checked and/or GRI-checked. In cases where external assurance is provided, the symbol '+' is added to the Application Level. Thus, the GRI Framework has offered organizations not only an integrated solution toward CSR reporting, but also a way to begin with and gradually improve. That is why, it is considered the most successful Framework of its kind by both corporations and their stakeholders.

5.7. *Other relevant management systems*

Apart from the instruments, which were presented in Secs. 5.1–5.6 of this chapter, there are some management standards that deal with specific areas such as environmental protection and labor. These standards are presented in the sequel.

ISO 14000 Series of Standards: These widely used Standards, which emerged in 1992, provide guidance to corporations for creating and applying an environmental management system (EMS). They extensively cover the environmental dimension by dealing with issues such as EMS requirements and guidelines, performance evaluation, life-cycle assessment, labeling as well as communication and auditing (International Organization for Standardization, 2010a). It should be noted that the International Organization for Standardization (ISO) has launched the development of an international standard for social responsibility. It is expected to be published in 2010 as ISO 26000 and will be voluntary to use as a guidance standard, which will not include requirements and will not be a certification standard (International Organization for Standardization, 2010b).

Eco-management and Audit Scheme (EMAS): The EMAS is a management tool for organizations to evaluate, report, and improve their environmental performance. EMAS was introduced by EU in 1995 and has been revised for the second time in 2009. Its objective is "to promote continuous improvements in the environmental performance of organizations by the establishment and implementation of environmental management systems by organizations, the systematic, objective and periodic evaluation of the performance of such systems, the provision of information on environmental performance, an open dialogue with the public and other interested parties and the active involvement of employees in organizations and appropriate training" (European Union, 2009). Similar to ISO 14000 Series of Standards, EMAS focuses on the environmental dimension.

Social Accountability 8000 Standard: Published in 1997, the Social Accountability 8000 Standard (SA8000) was developed by Social Accountability International, which is an international NGO dedicated to improving workplaces and communities by developing and implementing socially responsible standards. SA8000 is based on the UN Universal Declaration of Human Rights, the UN Convention on the Rights of the Child and various conventions of the International Labor Organization and focuses on labor issues such as child labor, forced labor, health and safety, freedom of association and right to collective bargaining, discrimination, discipline, working hours, compensation, and management systems (Social Accountability International, 2010). SA8000 has won recognition as one of the strongest workplace standards.

5.8. *Rating and evaluation schemes*

Rating and evaluation schemes are usually initiated by independent organizations aiming at providing a thorough quantitative analysis of CSR activities and reports. In particular, the corporations, which are selected for these processes, are usually selected by the carrier of the analysis under specified criteria. The most relevant schemes are presented in the sequel.

Dow Jones Sustainability Indexes (DJSI): Launched in 1999 with the co-operation of Dow Jones Indexes, STOXX Limited and SAM Group indexes, the DJSI intend to track the financial performance of the leading sustainability-driven companies worldwide and to provide asset managers with reliable and objective benchmarks to manage sustainability portfolios. Thus, DJSI allow them to consider not just an organization's financial status, but also further activities such as corporate conduct, labor practices, and environmental policies. Furthermore, comparisons with other corporations in and across industries are also facilitated. The assessment that is performed includes both general sustainability criteria, which universally applied to all industries as well as specific criteria for industrial sectors. The sustainability criteria reflect opportunities (to assess a company's ability to exploit them) and risks (to assess a company's ability to reduce or avoid sustainability-related costs and risks) (Dow Jones Sustainability Indexes, 2010).

FTSE4GOOD Index Series: The FTSE4Good Index Series, which was launched in 2001, has been designed in order to measure the performance of corporations that meet globally recognized CR standards, and to facilitate investment in those corporations (Financial Times Stock Exchange, 2010). Corporations are included in the Series on the basis of their sustainability record. Thus, the Series can be used in four main ways:

- as a basis for responsible investment, financial instruments, and fund products;
- as a research tool to identify environmentally and socially responsible companies;
- as a reference tool to provide companies with a transparent and evolving global CR standard to aspire and surpass; and
- as a benchmark index to track the performance of responsible investment portfolios.

Accountability Rating: The Accountability Rating is a tool for measuring the extent to which companies have built responsible practices into the way they do business and their impact on the economies, societies, and environments in which they operate. It was developed by the CSR consultancy CSRnetwork (now merged in Two Tomorrows) and AccountAbility (see Sec. 5.5.), and first applied in 2004. Every year the Rating is applied to the world's largest companies (according to the Fortune Global 100) and to the largest companies of several countries. In 2008, the Rating was applied in eight countries: Bulgaria, Greece, Hungary, Italy, Portugal, Russia, South Korea, and Turkey. By drawing on information, which companies put into their public reporting, as well as data on their actual social and environmental performance, the Rating evaluates them across four equally weighted domains: strategic intent, governance and management, engagement, and operational performance (Accountability Rating, 2010).

CR Index: The CR Index is an example of nationally-focused CR instruments. It has been developed by Business in the Community, which is a British business-led coalition with over 800 members, in order to replace its then Environment Index (Business in the Community, 2010). Business in the Community is also one of the charities of The Prince of Wales. The CR Index benchmarks corporations in the following main areas: corporate strategy, integration, management (in regard to community, environment, marketplace and workplace issues), performance and impact (environmental and social), and assurance and disclosure. Although CR Index is strongly associated with the UK, it has lately gone international through associated organizations in Australia, New Zealand, and Greece.

The KPMG International Survey on CR Reporting: This survey, which has been designed in order to examine reporting trends among the world's largest companies, is conducted by KPMG and various partners since 1993 and is issued in a triennial basis. In 2008 the sixth Survey was issued (KPMG, 2008). The goal of the study was to examine CR related information issued by

- the top 250 companies listed on the Fortune Global 500(G250) for the year 2007; and
- the 100 largest companies by revenue (N100) from 22 countries (Australia, Brazil, Canada, Czech Republic, Denmark, Finland, France, Hungary, Italy, Japan, Mexico, Norway, Portugal, Romania, South Africa, South Korea, Spain, Sweden, Switzerland, The Netherlands, United Kingdom, and the United States), except in Sweden where the sample was limited to the largest 70.

6. Critique of CSR and CSR Reporting

The concept of CSR, the corporate activities associated with it as well as CSR reporting, has raised much debate. Levitt (1958) was maybe the first to criticize the concept of CSR, as he stated, among others, that "government's job is not business, and business's job is not government" and that "if it does not pay, there is no game." On the same side, Friedman (1970) popularized this position in his article "The social responsibility of business is to increase its profits," published in New York Times Magazine. Friedman also implied that CSR engagement would be used of a misuse of corporate resources, which could have been employed efficiently in order the corporation to be more profitable. However, he recognized the corporations' obligation to operate according to legal and ethical requirements as well as potential strategic advantages for a corporation engaged in CSR activities. The latter has been supported by Hart (1995) and McWilliams *et al.* (2002), who suggested that CSR engagement may constitute a resource or capability that would lead to a sustained competitive advantage. In addition, Friedman also outlined that the role of managers should be to fulfill the purposes of shareholders or owners merely from an economic perspective. As presented in Sec. 3, Freeman (1984) has pointed out the role of managers to identify and satisfy the corporation's stakeholders. It should be noted though that the interpretation of managers' behavior has been a subject for extended debate in the academia. For example, agency theorists propound a distinct separation of interests between managers and owners/shareholders (Jensen and Meckling 1976), whereas, under the stewardship theory, managers essentially want to do a good job and to be good stewards of the corporate assets (Donaldson and Davis, 1991).

Moreover, critics question the motivation behind CSR and CSR reporting, accusing businesses for corporate hypocrisy and insincerity. Through this lens, CSR is actually a mere marketing "gadget." Thus, not only corporations adopt CSR practices for the commercial benefit they may enjoy by raising their reputation, but also CSR is often used in order to distract the public from ethical questions posed by their core operations (Pappis, 2010). From this perspective, what really matters is better governmental and international regulation and enforcement, rather than voluntary measures, so that companies behave in a socially responsible manner.

Doane (2005) stated that four key drivers would opportunistically impel a company engage in CSR: managing risk and reputation, protecting human capital assets, responding to consumer demands, and avoiding regulation. Thus, CSR policies are focused on business-based incentives and business select to innovate and provide solutions in social and environmental issues when there is a reasonable profit to be made or a loss to be avoided. Doane (2005) also suggested that there are some myths surrounding CSR at the stake of both corporations and governments:

- voluntary reporting improves performance;
- voluntary codes and management systems change corporate behavior;
- the consumer will drive change; and
- the investment industry can provide the strongest incentives.

CSR reports usually consist of "feel-good" cases and anecdotes accompanied by nice visuals, which show happy people and nice natural landscapes that are assumed to contribute to a pleasant feeling for the reader (Cerin, 2004). In addition, the information contained in CSR reports is mostly about legal requirements, whereas the discussion concerning what is done beyond legal requirements is rather limited (Banerjee, 2006).

A significant merit of CSR critique has been devoted to the case of corporations, which either are by definition engaged in harmful industrial activities or are unable to advance the interests of society as a whole. Such corporations are getting more and more involved in several oxymora: tobacco industries appear to support the prohibition of smoking; fast food corporations talk about nutrition; betting companies support fair play; MNEs talk about fair trade; and so on. It should be noted, though, that corporations from specific industries (e.g., tobacco, weapon, nuclear power, *etc.*) are usually excluded from participating in CSR initiatives. Further insights regarding the critique for CSR and CSR reporting are provided by Fougère and Solitander (2009).

7. Concluding Remarks

As Moon (2007) posits, the concepts of CSR and sustainable development, like other essentially contested terms such as justice, address very important issues for people and society at large, particularly because corporations are often the agents by which welfare is attained and preserved and development is enacted. To our point of view, the basic issues are who and why set the agenda; and who and why decide. Obviously, the answers to these questions cannot be unambiguous and permanent as they depend upon the answer to the question who has the power to do so. In an ever-changing political, economic and social landscape, no permanent and unimpaired answer to the latter question may be secured. In addition, our inability to interpret timely and accurately environmental processes and impacts adds to the uncertainty.

Despite arguments and critiques, CSR, in practice, is increasingly accepted by modern corporations. In particular, corporations seek to participate in wider networks and to demonstrate CSR activities to the public. Most major corporate web sites emphasize companies' commitment to promoting CSR and publish regularly reports on related activities. Such activities include launching community-based development projects, developing robust mechanisms to safeguard human health and safety and avoid human or environmental accidents, building a reputation for integrity and best practice, supporting financially local organizations or communities in developing countries, giving charities and donations to national foundations, sponsoring cultural activities, offering grants to students, establishing educational facilities, *etc.* By doing so, corporations respond to various motivations, including self-motivation and conforming to employees and other stakeholders' priorities and expectations, public pressure and regulation (Pappis, 2010).

Initiatives, guidelines, and standards as well as auditing and verification processes definitely provide assistance for management. However, managers are usually faced with dilemmas regarding which to follow, due to frequent inception of new instruments. In any case, these instruments would be worthless unless the link between corporate strategy, information management, and the reporting activities is designed in a systematic manner (Herzig and Schaltegger, 2006). KPMG (2008) concluded that the CSR case has matured with fewer companies issuing reports in the absence of corresponding strategies, but there is still much to be developed on this issue. Furthermore, CSR reporting appears to have left the experimental phase and has been established as a business tool that may add valuable insight into current states of corporate performance. The latter would reveal opportunities for growth, innovation, and learning in an ever-changing world.

References

AccountAbility, 2010a. *AccountAbility — Who We Are.* Retrieved March 8, 2010, from http://www.accountability21.net/default2.aspx?id=54.

AccountAbility, 2010b. *AA1000 Series of Standards.* Retrieved March 8, 2010, from http://www.accountability21.net/aa1000series.

AccountAbility, 2010c. *AA1000 AccountAbility Principles Standard (2008).* Retrieved March 8, 2010 from http://www.accountability21.net/aa1000aps.

AccountAbility, 2010d. *AA1000 Assurance Standard (2008).* Retrieved March 8, 2010, from http://www.accountability21.net/aa1000as.

AccountAbility, 2010e. *AA1000 Stakeholder Engagement Standard.* Retrieved March 8, 2010, from http://www.accountability21.net/aa1000ses

Accountability Rating, 2010. *Accountability Rating — About the Rating.* Retrieved March 11, 2010 from http://www.accountabilityrating.com/about_the_rating.asp.

Andreasen, A. and Drumwright, M., 2001. Alliances and ethics in social marketing. In: Andreasen, A. (Ed.), *Ethics in Social Marketing*, Georgetown University Press, Washington, DC, pp. 95–124.

Agle, B.R., Donaldson, T., Freeman, R.E., Jensen, M.C., Mitchell, R.K. and Wood, D.J., 2008. Toward superior stakeholder theory. *Business Ethics Quarterly*, **18**(2), 153–190.

Banerjee, S.B., 2006. The ethics of corporate social responsibility. In: Clegg, S. and Rhodes, C. (Eds), *Management Ethics: Contemporary Contexts. Routledge*, London.

Bowen, H.R., 1953. *Social Responsibilities of the Businessman.* Harper & Row, New York.

Brown, H.S., de Jong, M. and Levy, D.L., 2009. Building institutions based on information disclosure: Lessons from GRI's sustainability reporting. *Journal of Cleaner Production* **17**, 571–580.

Business in the Community, 2010. *Corporate Responsibility (CR) Index.* Retrieved March 11, 2010, from http://www.bitc.org.uk/integration_and_advice/cr_index/index.html.

Campbell, J.L., 2007. Why would corporations behave in socially responsible ways? An institutional theory of corporate social responsibility. *Academy of Management Review* **32**(3), 946–967.

Carroll, A.B., 1979. A three-dimensional conceptual model of corporate social performance, *Academy of Management Review* **4**(4), 497–505.

Carroll, A.B., 1991. The pyramid of corporate social responsibility: Towards the moral management of organizational stakeholders. *Business Horizons* (July/August), 39–48.

Carroll, A.B., 1999. Corporate social responsibility: Evolution of a definitional construct. *Academy of Management Review* **4**, 268–296.

Cerin, P., 2004. Where is corporate social responsibility actually heading? *Progress in Industrial Ecology: An International Journal* **1**(1–2–3): 307–330.

Clarkson, M.B.E, 1995. A stakeholder framework for analyzing and evaluating corporate social performance. *Academy of Management Review* **20**, 92–117.

Cochran, P.L., 2007. The evolution of corporate social responsibility. *Business Horizons* **50**, 449–454.

CSR Europe, 2005. *European roadmap for businesses — Towards a Sustainable and Competitive enterprise.* Retrieved March 4, 2010, from.

CSR Europe, 2010a. *CSR Europe — About Us.* Retrieved March 4, 2010, from http://www.csreurope.org/pages/en/about_us.html, http://www.csreurope.org/data/files/european_roadmap_for_businesses_2005.pdf.

CSR Europe, 2010b. *CSR Europe's Toolbox.* Retrieved March 4, 2010, from http://www.csreurope.org/pages/en/toolbox.html.

Dahlsrud, A., 2008. How corporate social responsibility is defined: An analysis of 37 Definitions. *Corporate Social Responsibility and Environmental Management* **15**, 1–13.

Doane, D., 2005. Beyond corporate social responsibility: Minnows, mammoths and markets. *Futures* **37**, 215–229.

Donaldson, L. and Davis, J.H., 1991. Stewardship theory or agency theory: CEO governance and shareholder returns. *Australian Journal of Management* **16**(1), 49–64.

Dow Jones Sustainability Indexes, 2010. Retrieved March 11, 2010, from http://www.sustainability-index.com/.

Elkington, J., 1994. Towards the sustainable corporation: Win–win–win business strategies for sustainable development. *California Management Review* **36**(2), 90–100.

Elkington, J., 1997. *Cannibals with Forks: The Triple Bottom Line of 21st Century.* Capstone Publishing, Oxford.

European Commission, 2001. *Promoting a European Framework for Corporate Social Responsibility — Green Paper.* Office for Official Publications of the European Communities, Luxembourg.

European Union, 2009. Regulation (EC) No 1221/2009 of the European Parliament and of the Council of 25 November 2009 on the voluntary participation by organisations in a

Community eco-management and audit scheme (EMAS), repealing Regulation (EC) No 761/2001 and Commission Decisions 2001/681/EC and 2006/193/EC. *Official Journal of the European Union* **L 342/1**. Retrieved March 10, 2010, from http://eur-lex.europa.eu/LexUriServ/LexUriServ.do?uri=OJ:L:2009:342:0001:0045:EN:PDF.

Falck, O. and Heblich, S., 2007. Corporate social responsibility: Doing well by doing good. *Business Horizons* **50**, 247–254.

Fougère, M. and Solitander, N., 2009. Against corporate responsibility: Critical reflections on thinking, practice, content and consequences. *Corporate Social Responsibility and Environmental Management* **16**, 217–227.

Freeman, R.E., 1984. *Strategic Management: A Stakeholder Perspective.* Prentice-Hall, Boston Englewood Cliffs, Pitman, N.J.,

Friedman, M., 1970. The social responsibility of business is to increase its profits. *The New York Times Magazine* 13 September, 122–126.

Financial Times Stock Exchange, 2010. *FTSE4Good Index Series.* Retrieved March 11, 2010, from http://www.ftse.com/Indices/FTSE4Good_Index_Series/index.jsp.

Gao, S.S. and Zhang, J.J., 2006. Stakeholder engagement, social auditing and corporate sustainability. *Business Process Management Journal* **12**(6), 722–740.

Garriga, E. and Melé, D., 2004. Corporate social responsibility theories: Mapping the territory. *Journal of Business Ethics* **53**, 51–71.

Geva, A., 2008. Three models of corporate social responsibility: Interrelationships between theory, research, and practice. *Business and Society Review* **113**(1), 1–41.

Global Reporting Initiative, 2006. *Sustainability Reporting Guidelines.* Retrieved March 8, 2010, from http://www.globalreporting.org/NR/rdonlyres/B52921DA-D802-406B-B067-4EA11CFED835/3882/G3_GuidelinesENU.pdf.

Global Reporting Initiative, 2010. *What is GRI?* Retrieved March 8, 2010, from http://www.globalreporting.org/AboutGRI/WhatIsGRI/.

Gonzalez-Benito, J. and Gonzalez-Benito, O., 2006. A review of determinant factors of environmental proactivity. *Business Strategy and the Environment* **15**, 87–102.

Hart, S., 1995. A natural resource-based view of the firm. *Academy of Management Review* **20**, 986–1014.

Heslin, P.A, and Ochoa, J.D., 2008. Understanding and developing strategic corporate social responsibility. *Organizational Dynamics* **37**, 125–144.

Henriques, I. and Sadorsky, P., 1999. The relationship between environmental commitment and managerial perceptions of stakeholder importance. *Academy of Management Journal* **42**(1), 87–99.

Herzig, C. and Schaltegger, S., 2006. Corporate Sustainability reporting — An overview. In: Schaltegger, S., Bennett, M. and Burritt, R., (Eds), *Sustainability Accounting and Reporting.* Springer, 301–324.

Idowu, S.O. and Towler, B.A., 2004. A comparative study of the contents of corporate social responsibility reports of UK companies. *Management of Environmental Quality: An International Journal* **15**(4), 420–437.

International Organization for Standardization, 2010a. *ISO 14000 Essentials.* Retrieved March 10, 2010, from http://www.iso.org/iso/iso_14000_essentials.htm

International Organization for Standardization, 2010b. *Social Responsibility.* Retrieved March 10, 2010, from http://www.iso.org/sr/.

International Union for Conservation of Nature, 2006. The future of sustainability: Re-thinking environment and development in the twenty-first century. *Report of the IUCN Renowned Thinkers Meeting*, 29–31 January 2006. Retrieved February 20, 2010, from http://cmsdata.iucn.org/downloads/iucn_future_of_sustanability.pdf.

Jensen, M.C. and Meckling, W.H., 1976. Theory of the firm: Managerial behavior, agency costs, and ownership structure. *Journal of Financial Economics* **3**, 305–360.

Kell, G., 2005. The global compact selected experiences and reflections. *Journal of Business Ethics* **59**(1–2), 69–79.

Kolk, A. and Perego, P., 2010. Determinants of the adoption of sustainability assurance statements: An international investigation. *Business Strategy and the Environment* **19**(3), 182–198.

KPMG, 2008. *KPMG International Survey on Corporate Responsibility Reporting 2008.* Retrieved March 11, 2010, from http://www.kpmg.eu/docs/Corp_responsibility_Survey_2008.pdf.

Levitt, T., 1958. The dangers of social responsibility. *Harvard Business Review*, September–October, 41–50.

Maessen, R., Van Seters, P. and Van Rijckevorsel, E., 2007. Circles of stakeholders: Towards a relational theory of corporate social responsibility. *International Journal of Business Governance and Ethics* **3**, 77–94.

Martin, A.D. and Hadley, D.J., 2008. Corporate environmental non-reporting — a UK FTSE 350 perspective. *Business Strategy and the Environment* **17**(4), 245–259.

McWilliams, A., Van Fleet, D.D. and Cory, K., 2002. Raising rivals' costs through political strategy: An extension of the resource-based theory. *Journal of Management Studies* **39**, 707–723.

Mitchell, R.K., Agle, B.R. and Wood, D.J., 1997. Toward a theory of stakeholder identification and salience: Defining the principle of who and what really counts. *Academy of Management Review* **22**, 853–886.

Moon, J., 2007. The contribution of corporate social responsibility to sustainable development. *Sustainable Development* **15**, 296–306.

Murray, K.B. and Vogel, C.M., 1997. Using a hierarchy-of effects approach to gauge the effectiveness of corporate social responsibility to generate goodwill toward the firm: Financial versus Non financial Impacts. *Journal of Business Research* **38**, 141–159.

Organization for Economic Co-operation and Development, 2001. *Corporate Social Responsibility: Partners for Progress.* OECD Publishing, Paris.

Organization for Economic Co-operation and Development, 2008. *OECD Guidelines for Multinational Enterprises.* Retrieved March 2, 2010, from http://www.oecd.org/dataoecd/56/36/1922428.pdf.

Pappis, C.P., 2010. *Climate Change, Supply Chain Management and Enterprise Adaptation: Implications of Global Warming on the Economy.* IGI Global. (ISBN 978-1-61692-800-1) in Press.

Rotheroe, N., Keenlyside, K. and Coates, L., 2003. Local agenda 21: Articulating the meaning of sustainable development at the level of the individual enterprise. *Journal of Cleaner Production* **11**, 537–548.

Schwartz, M.S. and Carroll, A.B., 2003. Corporate social responsibility: A three-domain approach. *Business Ethics Quarterly* **13**(4), 503–530.

Scott, R. and Jackson, R., 2002. Environmental, social and sustainability reporting on the web: Best practices. *Corporate Environmental Strategy* **9**, 193–202.

Social Accountability International, 2010. *SA8000:2008 Standard.* Retrieved March 10, 2010, from http://www.sa-intl.org/index.cfm?fuseaction=Page.viewPage&pageId=937&parentID=479&nodeID=1.

Spar, D.L. and La Mure, L.T., 2003. The power of activism: Assessing the impact of NGOs on global business. *California Management Review* **45**, 78–101.

Sutantoputra, A.W., 2009. Social disclosure rating system for assessing firms' CSR reports. *Corporate Communications: An International Journal* **14**(1), 34–48.

United Nations, 1987. *Report of the World Commission on Environment and Development.* General Assembly Resolution 42/187, 11 December 1987. Retrieved February 20, 2010, from http://www.un.org/documents/ga/res/42/ares42-187.htm.

United Nations Conference on Trade and Development, 2010. *About International Standards of Accounting and Reporting (ISAR).* Retrieved March 2, 2010, from http://www.unctad.org/Templates/Page.asp?intItemID=2905&lang=1.

United Nations Global Compact, 2010a. *Overview of the United Nations Global Compact.* Retrieved March 2, 2010, from http://www.unglobalcompact.org/AboutTheGC/index.html.

United Nations Global Compact, 2010b. *The Ten Principles.* Retrieved March 2, 2010, from http://www.unglobalcompact.org/AboutTheGC/TheTenPrinciples/index.html.

United Nations Global Compact, 2010c. *Local Networks.* Retrieved March 2, 2010, from http://www.unglobalcompact.org/NetworksAroundTheWorld/index.html.

Vos, J.F.J., 2003. Corporate social responsibility and the identification of stakeholders. *Corporate Social Responsibility and Environmental Management* **10**, 141–152.

Wood, D.J., 1991. Corporate social performance revisited. *Academy of Management Review* **16**(4), 691–718.

World Business Council for Sustainable Development, 2010a. WBCSD — About us. Retrieved March 4, 2010, from http://www.wbcsd.org/includes/getTarget.asp?type=p&id=NjA.

World Business Council for Sustainable Development, 2010b. *WBCSD's 10 messages by which to operate.* Retrieved March 4, 2010, from http://www.wbcsd.org/includes/getTarget.asp?type=p&id=MTAyMQ&doOpen= 1&ClickMenu=RightMenu.

Chapter 4

SUSTAINABILITY INDICATORS AND INDICES: AN OVERVIEW

Jianguo Wu[a,b] and Tong Wu[c]
[a]*School of Life Sciences and Global Institute of Sustainability, Arizona State University, P.O. Box 874501, Tempe, AZ 85287, USA*
[b]*Sino-US Center for Conservation, Energy, and Sustainability Science (SUCCESS), Inner Mongolia University, Hohhot 010021, China*
Jingle.Wu@asu.edu
[c]*School of Forestry, Northern Arizona University, Flagstaff, AZ 86011, USA*
Tong.Wu@nau.edu

Human activities have profoundly transformed the biosphere, and various signs — including global climate change, biodiversity loss, and urban pollution — indicate that our world is on an unsustainable trajectory. To insure the well-being of the current and future generations, operationalizing the concept of sustainability has become the most salient challenge of our time. To meet this challenge, a cogent set of metrics — sustainability indicators and indices — is indispensable, which allow us to locate our present condition relative to conditions that have prevailed in the past, and to develop relevant perspectives to frame actions for the future. This, however, can be a difficult task given the diversity of perspectives and disciplines concerned with sustainability, and the complexity of issues that need to be addressed. In this chapter, we present an overview of this broad but still emerging body of research. We discuss the basic principles and scientific perspectives underlying the concept of sustainability or sustainable development. Then, based on these lines of inquiry, a survey of inclusive and influential indicator frameworks is presented. These organizing schemes constitute a foundation that is crucial to the construction, interpretation, and application of sustainability indicators. Then, we examine a selection of commonly used sustainability indices. Finally, we conclude the chapter with some reflections on the nature and future of sustainability measures.

1. Introduction

Mounting evidence indicates that our world has been on an unsustainable trajectory, particularly in the past few centuries. As the most dominant ecosystem engineers, human influences on earth are no longer restricted to taming desired species and extracting valued minerals. We have also begun to domesticate whole ecosystems and landscapes of increasing spatial extent, transforming the entire biosphere in fundamental ways (Kareiva *et al.*, 2007; Wu 2008, 2010). The global human population has increased from fewer than 1 billion before the onset of the Industrial

Revolution in the late 18th century to more than 6.8 billion today. The rate of population growth is astonishing: it took about 10,000 years for human population to increase from 5 million to 1 billion, but only an additional 130 years to reach the third billion and only 40 more years to surpass the sixth billion! The accelerated human dominance of the earth system has led to a number of environmental, as well as social and economic, problems, including biodiversity loss, global climate change, overexploitation of natural resources, degradation of environmental quality, and socioeconomic inequity and instabilities.

Achieving sustainable development is, therefore, is arguably the most pressing issue of our time. One of the main reasons is the simple fact that we have no other viable choice. As Bossel (1999) put it, "There is only one alternative to sustainability: unsustainability." Thus, it is not surprising that sustainability has become a buzzword in both academia and public discourse. There is no doubt that we have entered a sustainability movement of global reach. As with any other human endeavor, ensuring ultimate success through continuing progress means being able to effectively gauge our performance. In the context of sustainable development, these performance metrics are primarily sustainability indicators and indices (SIIs), which have played an increasingly important role in the science and practice of sustainability.

While indicators for assessing and tracking environmental and ecological conditions have been used for more than half a century (Niemi and McDonald, 2004), the development of indicators for gauging sustainable development has a relatively short history. Its major development started after the United Nations Conference on Environment and Development (Earth Summit), held in Rio de Janeiro of Brazil in 1992, which proposed the fundamental principles and the program of action for achieving sustainable development. In particular, the Rio Summit called for the development of sustainability indicators with its Agenda 21, an action plan endorsed by more than 170 national governments. The World Summit on Sustainable Development (Earth Summit, 2002) convened by the United Nations in Johannesburg, South Africa, in 2002 strongly reaffirmed the UN's commitment to the Rio principles and the full implementation of Agenda 21. Consequently, a number of international organizations, governmental agencies, NGOs, local communities and corporations, and academic scholars have devoted significant effort to the design and implementation of indicators that gauge the state and trajectory of environmental conditions and socioeconomic development (see Table 1 for a list of international organizations that have contributed significantly to the development of sustainability indicators). Today, hundreds of indicators and indices of sustainable development have been developed and used at the global, national, and local scales. For example, the Compendium of Sustainable Development Indicator Initiatives, created by the International Institute for Sustainable Development (a Canada-based policy research institute), lists 894 indicator initiatives as of August 2010 (http://www.iisd.org/measure/compendium/).

This chapter's main objectives are to review key conceptual and methodological issues of sustainability measures, examine a selected set of commonly used SIIs, and

Table 1. List of international organizations significantly contributing to the development of sustainability indicators.

Name of organization	Web site
Balaton Group	http://www.balatongroup.org/
International Institute for Sustainable Development (IISD)	http://www.iisd.org/
Organization for Economic Co-operation and Development (OECD)	http://www.oecd.org/
Scientific Committee on Problems of the Environment (SCOPE)	http://www.icsu-scope.org/
United Nations Commission on Sustainable Development (CSD)	http://www.un.org/esa/dsd/csd/csd_index.shtml
United Nations Development Program (UNDP)	http://www.undp.org/
United Nations Environment Program (UNEP)	http://www.unep.org/
United Nations Statistical Division (UNSD)	http://unstats.un.org/unsd/default.htm
World Bank	http://www.worldbank.org/

discuss some of the major challenges and future directions. Owing to the complex nature of sustainability, the diversity of SIIs, and the proliferated literature on both topics, it is impossible to cover the entire range of sustainability measures in great detail. Our strategy here is to focus on the major themes and issues, and illustrate the uses and usefulness of SIIs through specific examples.

2. What is Sustainability?

As SIIs are designed to measure sustainability, any meaningful discussion on SIIs requires a basic understanding of what the concept of sustainability encompasses. In the search for solutions to the apparent conflict between conservation (the protection of natural resources) and development (the exploitation natural resources), the terms "sustainability" and "sustainable development" were coined in the early 1970s, and have acquired a great number of definitions since then (Kidd, 1992; Du Pisani, 2006). The most commonly cited definition of sustainable development or sustainability is the one given in the report by the United Nations World Commission on Environment and Development (WCED) chaired by Gro H. Brundtland (the former Prime Minister of Norway), published in 1987 and subsequently known as the Brundtland Report (WCED, 1987). The general and intentionally vague definition states that sustainable development is development that "meets the needs of the present without compromising the ability of future generations to meet their own needs" (WCED, 1987). Thus, sustainability concerns our ability to maintain a coupled human–nature system at a desirable state for multiple generations in the face of anthropogenic and environmental perturbations and uncertainties. Considering the complexity arising from the multiplicity of components and their intricate interactions, it is difficult to define sustainability in specific terms without controversy.

However, if sustainability is to be more than a purely academic exercise or a political catchphrase, it must be operational. As Bell and Morse (2008) argue, "Sustainable development… embodies an ultimate practicality since it is literally meaningless unless we can 'do' it." At the very least, we need to adequately define what is to be sustained, what is to be developed, and how these components should be balanced or integrated (Kates *et al.*, 2005). Although it is by no means an actionable recipe, the concept of the triple bottom line, or the three pillars, of sustainability defines the major constituent domains of sustainable development (Fig. 1A). This perspective views sustainability as composing of three fundamental dimensions: environment, economy, and society (frequently, the three bottom lines have also been called "planet, people, and profit"). In each of the three domains, certain elements need to be sustained and others developed if sustainability is to be achieved.

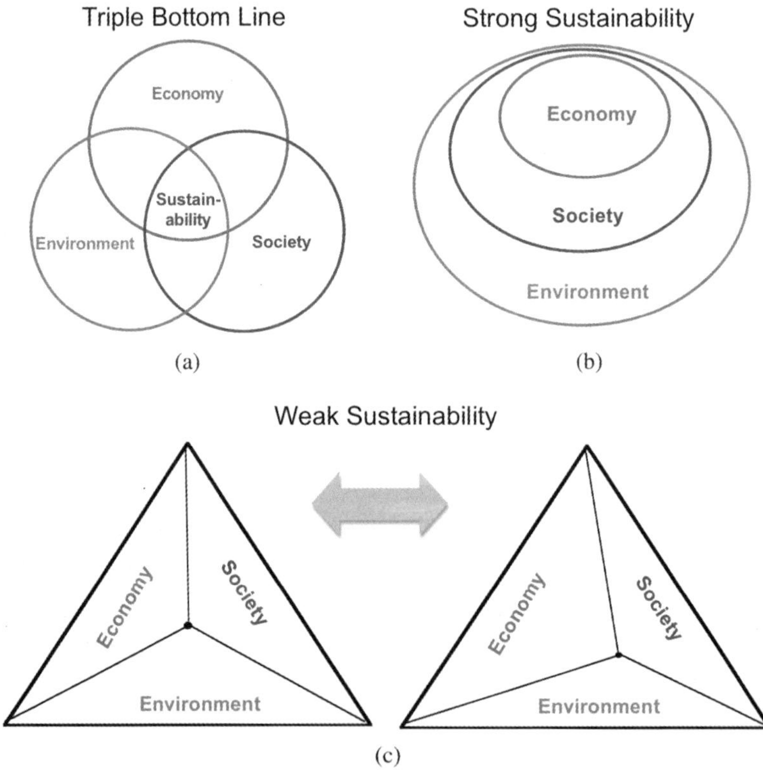

Fig. 1. Illustration of key components of sustainability and their relationships. (A) The triple bottom line concept, most commonly represented as a Venn diagram, implies that the three pillars are all necessary and equally important to sustainability. (B) Strong sustainability is depicted as three nested circles, implying that the environment provides natural resources and ecosystem services necessary for economic and social development — thus mutual substitutability between natural and human-made capital is not sensible. Economic development depends on both social and environmental capital, whereas both economic and social processes influence environmental conditions. (C) Weak sustainability focuses on the attainment of a non-declining level of the overall capital while allowing for mutual substitution between the three pillars of sustainability.

While the three dimensions are widely recognized in the literature, their relationships with each other remain controversial. For instance, can a high level of economic development be substituted for a low level of environmental quality? Whether or not sustainability allows for substitution between natural and human-made (manufactured) capital has been the focus of debate on "weak" *versus* "strong" sustainability (Ayres *et al.*, 1998). Weak sustainability permits mutual substitutability among the three dimensions, whereas strong sustainability does not (Fig. 1B,C). Because of the increased scarcity of natural resources and pressures on the environment imposed by the growing human population, the environmental dimension is increasingly recognized as a fundamental basis for sustainability. A simple yet compelling argument for this is that, without an adequate level of biodiversity and ecosystem functioning and services, no economic or social development is sustainable.

These conceptual issues have important implications for the construction and application of sustainability measures. For example, based on the triple bottom line concept, sustainability-related measures can be classified into environmental, economic, and social indicators, and only those that cover all three dimensions simultaneously can measure sustainability in its entirety. Whether strong or weak sustainability is used as an operational principle will in turn determine how overall sustainability is measured. The development of quantitative measures can help clarify and refine sustainability-related concepts, improve our understanding of the intricate relationships among components of sustainability in practical terms, and thus promote the science and practice of sustainable development. When constructing quantitative measures of sustainability, we have to explicitly specify what aspects of sustainability we want to measure, which ones we want to conserve or develop, and how these different aspects should be related to each other or integrated together. By so doing, the development of indicators can add rigor and accuracy to the notion of sustainability (although being quantitative and precise does not always translate into being rigorous and accurate).

3. What are Sustainability Indicators?

There are numerous definitions of indicator. Broadly speaking, an indicator can be a sign, symptom, signal, tip, clue, grade, rank, object, organism, or warning of some sort — many things in everyday life (Meadows, 1998). In a more restricted sense, as is often used in the scientific literature, an indicator refers to a variable or an aggregate of multiple related variables whose values can provide information about the conditions or trajectories of a system or phenomenon of interest. In other words, an indicator is simply "an operational representation of an attribute (quality, characteristic, property) of a system" (Gallopin, 1997).

For an indicator to be useful, there ought to be an established reference, benchmark, or threshold that represents a normal state, desired behavior, or goal to be achieved. For example, as we know what its normal range is, body temperature or blood pressure has become an indicator of our health conditions; when we know the speed limit of a given street, the speedometer of our car is an indicator that helps us avoid speeding tickets; when we know the atmospheric chemical composition in rural surroundings, the levels of greenhouse gases in a city can be indicators of air pollution due to urbanization. The reference levels or target values of indicators need to be determined based on relevant knowledge of the system of concern or shared understanding of the community that the system involves. Even for indicators whose reference levels or targets are difficult to define, the preferred direction of change should still be specified (e.g., household income or people's happiness).

Sustainability indicators are indicators that provide information on the state, dynamics, and underlying drivers of human–environmental systems. In general, indicators become sustainability (or unsustainability) indicators when time dimension, limits, or targets are associated with them (Meadows, 1998). The values of sustainability indicators are obtained from environmental and socioeconomic data from actual measurements or observations. A group of indicators used together for a particular purpose or project is often referred to as an indicator set. Sustainability includes environmental, economic, and social dimensions (sometimes institutions are listed as the fourth dimension), each of which has a number of components. Accordingly, indicators of sustainability can be as diverse as system components, and vary with regard to worldviews, purposes, and scales of time and space. Many indicators only reflect certain aspects of human–environmental systems, some are more integrative than others, and none is adequate to gauge the multiple dimensions of sustainability by itself.

On the other hand, presenting too many indicators can also be a problem by creating more confusion than understanding, especially when indicators differ in both the magnitude and direction of change. To reduce the number of indicators or to reflect the integrative characteristics of a system, indicators are often combined through mathematical manipulations to produce indices (Fig. 2). In other words, an index is an aggregate of two or more indicators. In reality, the distinction between an indicator and an index can be difficult and unnecessary because both are frequently aggregates of variables themselves, which means that their difference is merely a matter of the degree of aggregation. Thus, in the literature, the phrase "sustainability indicators" often includes both indicators and indices of sustainable development.

Sustainability researchers and specialists, increasingly working with stakeholders of different kinds, produce SIIs from environmental and socioeconomic data to inform policy makers and the public on implementing sustainability actions (Fig. 2). As the leaders in producing SIIs, researchers and specialists are usually responsible for data acquisition and processing as well as the subsequent construction of SIIs. The primary users of SIIs are intended to be policy makers and the public who often do not need, or care about, the details of the data used to generate SIIs. For the public, the most effective sustainability measures are often highly aggregated indices that are easy to understand and directly related to the environment and human well-being (Braat,

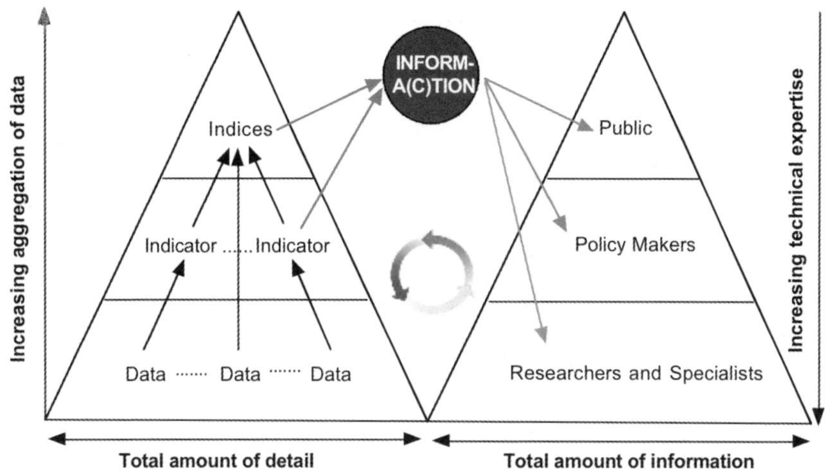

Fig. 2. Illustration of the relationship among data, indicators, indices, and information needed for sustainability actions, as well as the interactions between sustainability measures and their users and producers (based on Braat, 1991).

1991; Meadows, 1998; Bell and Morse, 2003). Such highly aggregated indices usually have the advantage of providing an overall picture of the state or performance of a system of interest in a simple and explicit way.

SIIs arguably represent the most popular approach to gauging sustainable development. Managing human–environmental systems would sustainably be impossible without using some kinds of indicators and indices. It is crucial, however, to understand that SIIs are not always objective, precise, or certain. In fact, many SIIs and the whole process of developing SIIs are to some extent subjective because of the nature of the indicators themselves (e.g., indicators representing social and cultural elements of sustainability) and our choice of specific indicators (e.g., selecting a particular indicator set over another). Also, some indicators are qualitative variables whose values are closely related to human values and perceptions. All SIIs have a degree of uncertainty that arises from the collection and analysis of data, the formulation of sustainability measures, and the unpredictable nature of the human–environmental systems. Thus, SIIs are not a panacea, but an indispensable tool for the science and practice of sustainable development. While SIIs have proven quite useful, utilizing them is still a process with many pitfalls.

4. Criteria and Frameworks for SIIs

4.1. *Criteria for selecting and evaluating indicators*

Determining what aspects of a system to monitor and what variables to use to gauge the state and performance of those aspects is critical to our understanding of sustainability. The principles and standards used in this process are reflective of our worldviews on what should be sustained and developed. Donella Meadows

(1998) said it well, "Not only do we measure what we value, we also come to value what we measure." It also seems true that the aspects that get measured are more likely to get managed and better understood by policy makers and the public. A good example of this measure–value–management phenomenon is Gross Domestic Product (GDP) or Gross National Product (GNP). Although most environmental experts understand that GDP is a poor and potentially misleading indicator of environmental and human well-being, it is still one of the most commonly used (and in many cases misused) indicators of performance by national and local governments (Wu and Wu, 2010). As the choice and use of indicators can profoundly influence our perception of the performance of a system, and as subjectivity cannot be eliminated from measuring sustainability, are there criteria that can guide us to develop more effective SIIs? There are. We discuss some commonly used ones below.

There are general guidelines and specific criteria for indicators. Among the most widely recommended are the "Bellagio Principles," a set of guidelines developed by an international group of sustainability researchers and practitioners from five continents at a conference in Bellagio, Italy, in 1996 (Hardi and Zdan, 1997). These principles can serve as instructions for sustainability assessment, including the selection and design of indicators as well as their interpretation and communication (Table 2). The Bellagio Principles focus on four aspects of sustainability assessment: clear vision and goals (Principle 1), key elements of sustainability assessment (Principles 2–5), issues of the process of assessment (Principles 6–8), and continuing capacity for assessment (Principles 9 and 10).

More specific criteria for selecting and evaluating sustainability indicators have frequently been discussed in the literature (Meadows, 1998; Bell and Morse, 2003; Hak *et al.*, 2007; UN, 2007). Commonly recognized criteria include the following aspects: an indicator set should cover the various dimensions of sustainability and their complex interactions; individual indicators should be indicative of the state and changes of the targeted aspects of sustainability; they should be informative, easy to compile from readily available and lasting data sources, understandable to lay-people, policy-relevant, predictive or leading, and hierarchical in terms of detail and scale; the methods for weighting and aggregating indicators to produce indices should be transparent and unbiased. A representative list of desired characteristics for sustainability indicators is provided in Table 3 (Guy and Kibert, 1998).

4.2. *Indicator frameworks*

An indicator framework is a conceptual structure based on sustainability principles and used to facilitate indicator selection, development, and interpretation. Using indicator frameworks can help us identify gaps in available data, indicator sets, and our overall understanding of the human–environmental system of concern. A number of indicator frameworks have been developed by international organizations (e.g., UNCSD, OECD) and other sources based on diverse core values and sustainable development theories (Meadows, 1998; UN, 2007; Ness *et al.*, 2007).

Table 2. The Bellagio Principles for sustainable development assessment (Hardi and Zdan, 1997).

	Principle	Description
		Assessments of progress toward sustainable development should
1	Guiding vision and goals	be guided by a clear vision of sustainable development and goals that define that vision
2	Holistic perspective	consider the well-being of the whole system and social, ecological, and economic sub-systems, and consider both positive and negative consequences of human activity
3	Essential elements	consider equity and disparity within the current population and between present and future generations, life-support systems, and economic development and non-market activities that contribute to human/social well-being
4	Adequate scope	adopt a time horizon long enough to capture both human and ecosystem time scales; define the space of study large enough to include not only local but also long distance impacts on people and ecosystems; and build on historic and current conditions to anticipate future conditions
5	Practical focus	have an organizing framework that links goals to indicators and assessment criteria, a limited number of key issues for analysis, a limited number of indicators, standardized measurements, and targets and reference values for indicators
6	Openness	make methods and data accessible to all, and all judgments, assumptions, and uncertainties explicit
7	Effective communication	be designed to address the needs of the audience and users; engage decision-makers; and aim for simplicity
8	Broad participation	obtain broad representation of stakeholders, and ensure the participation of decision-makers
9	Ongoing assessment	develop a capacity for repeated measurement to determine trends; be iterative, adaptive, and responsive to change and uncertainty; adjust goals, frameworks, and indicators; and promote development of collective learning and feedback to decision-making
10	Institutional capacity	clearly assign responsibility and provide ongoing support in the decision-making process; provide institutional capacity for data collection, maintenance, and documentation; and support development of local assessment capacity

Major discrepancies among these indicator frameworks arise from the different conceptualizations of, and emphases on, key dimensions of sustainable development and their linkages, as well as the different ways of grouping and aggregating indicators. Here we discuss five types of frameworks that have been widely recognized in the literature.

4.2.1. *Pressure–state–response (PSR) frameworks*

The PSR framework was one of the most popular early indicator frameworks (Fig. 3A) (OECD, 1993). Several versions of PSR frameworks have been developed by different organizations in the recent decades. The most widely recognized has been the Driving force–State–Response (DSR) framework, published in 1996, by the United

Table 3. Desired characteristics of sustainability indicators (Guy and Kibert, 1998).

Criteria	Corresponding questions
Community involvement	Were they developed by and acceptable to the stakeholders of the system of concern?
Linkage	Do they link environmental, economic, and social issues?
Valid	Do they measure something that is related to the state of the system?
Available and timely	Can the data be collected on a regular basis?
Stable and reliable	Are they compiled using a systematic and rigorous method?
Understandable	Are they simple enough to be understood by laypersons?
Responsive	Do they respond quickly and measurably to changes?
Policy relevance	Are they relevant to policy?
Representative	As a group, do they cover the important dimensions of the focus area?
Flexible	Are they important to use even if data are not readily available, considering the fact that data may be available in the future?
Proactive	Do they act as a warning rather than a measure of the existing state?
Long range	Do they focus on the long term?
Act locally, think globally	Do they promote sustainability at the expense of other communities, regions, or countries?

Nations Commission on Sustainable Development (UNCSD) to guide the selection of indicators. In PSR and DSR frameworks, indicators of pressures or driving forces primarily represent anthropogenic processes that have positive or negative impacts on sustainable development. A corresponding question is: Why is the state of our environmental system changing? State indicators focus on the current conditions of the environment and natural resources in terms of both quantity and quality at a given time or during a period of time. A corresponding question then arises: What is happening to the state of our environmental system? Response indicators pertain to societal reactions to changes in the state of the environment and natural resources and the underlying drivers. This prompts another question: What are we doing about the state changes and underlying causes? An expanded version of DSR is the Driving force–Pressure–State–Impact–Response (DPSIR) framework (Fig. 3B), as used in the development of environmental indicators for North America (UNEP, 2006). To increase the relevance to policies, the DSR framework organizes all indicators of drivers, state, and responses according to four pillars of sustainable development (environment, economy, society, and institutions) and in accordance with the categories of sustainable development in Agenda 21.

The World Bank, as well as a number of other international and national organizations, has also adopted PSR-type frameworks (including the PSR, DSR, and DPSIR) in their efforts to develop sustainability indicators. Although they are conceptually appealing, the PSR frameworks have been criticized on several grounds. For example, the classification of indicators into categories such as driving force, state, and response is often ambiguous or impossible in practice because some indicators fall into more than one domain. In addition, the separation of different kinds of indicators makes it rather difficult to adequately consider the multiple

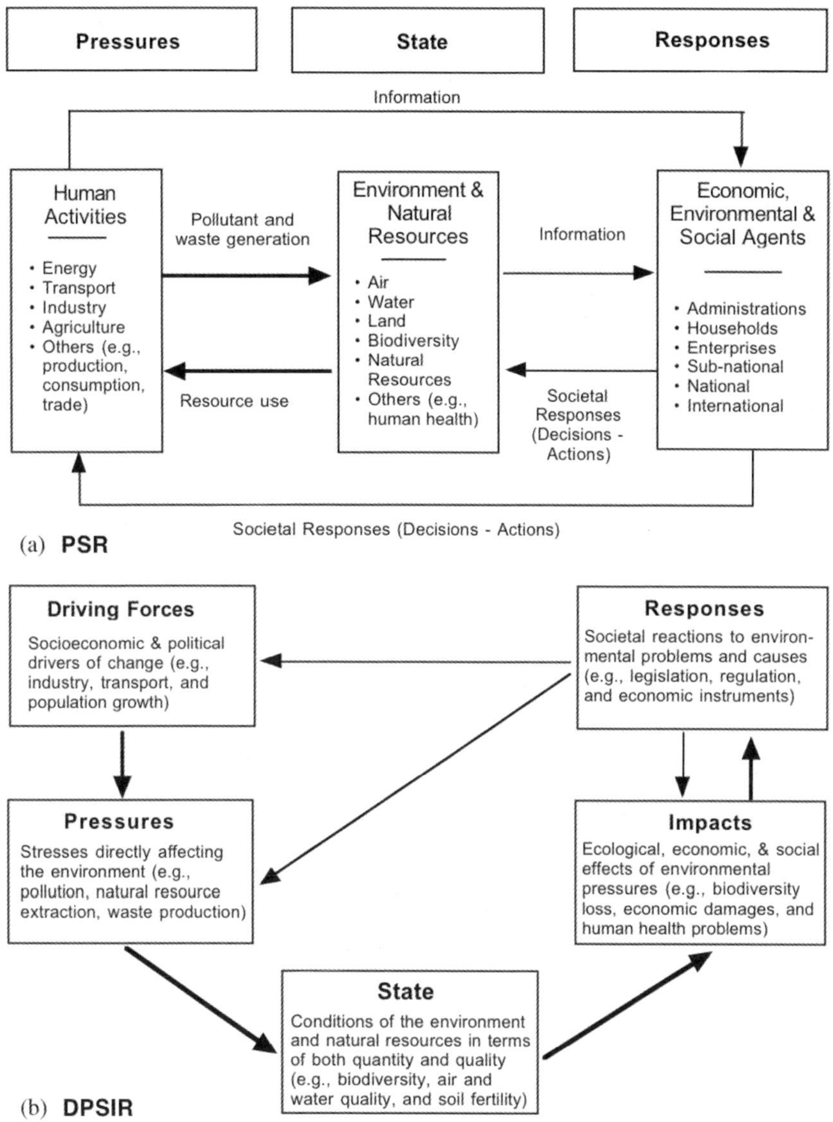

Fig. 3. Illustration of the pressure–state–response type of indicator frameworks: (A) PSR and (B) DPSIR.

causalities among indicators and the complex linkages among issues. Third, the system lacks a clear and close connection between indicators and policy issues. Because of these recognized problems, the UNCSD abandoned the DSR framework in favor of a more flexible theme-based framework that focuses more explicitly on sustainability-related issues. The PSR frameworks seem more appropriate for developing environmentally oriented indicator sets, and because of this, they will likely remain in existence (UNEP, 2006; UN, 2007).

4.2.2. *Theme-based frameworks*

Theme- or issue-based frameworks provide a flexible conceptual structure that organizes indicators around key themes or issues typically determined by policy relevance (UN, 2007). The best-known example has been the 2001 UNCSD theme-based framework that replaced the DSR framework (UNCSD, 2001). The UNCSD theme-based framework has a hierarchical structure: along the four dimensions of sustainable development (social, environmental, economic, and institutional), 15 themes are identified, which in turn are divided into 38 sub-themes, for which a total of 58 indicators are then identified (Fig. 4). The thematic framework has

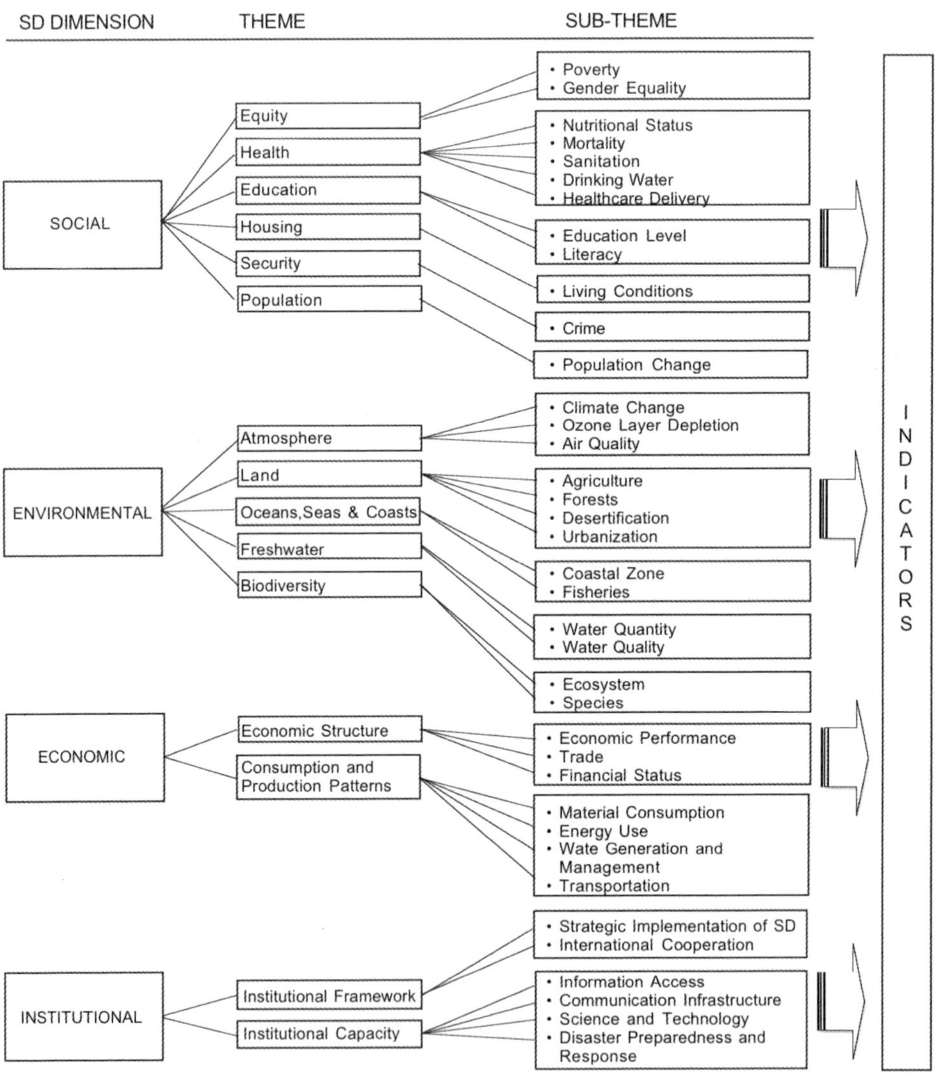

Fig. 4. Illustration of the theme-based indicator framework adopted by UNCSD in 2001.

fewer indicators, but is more policy-oriented than the DSR framework (which has 134 indicators in total).

UNCSD revised the theme-based framework again in 2007 based on evolving perspectives on indicators and the experience in applying the framework at the national and regional levels (UN, 2007). The revised UNCSD theme-based framework retains the thematic and sub-thematic structure, but further emphasizes the multi-dimensional and integrative nature of sustainable development by making the four sustainability dimensions implicit and adding crosscutting themes such as poverty and natural hazards. It also incorporates the eight Millennium Development Goals (MDGs; i.e., eradicate extreme poverty, achieve universal primary education, promote gender equality and empower women, reduce child mortality, improve maternal health, combat HIV/AIDS, malaria and other diseases, ensure environmental sustainability, and develop a global partnership) as well as the Johannesburg Plan of Implementation (UN, 2007). The 2007 UNCSD theme-based framework has 14 themes, 44 sub-themes, 50 core indicators, and a total of 96 indicators. A total of 14 themes are now poverty, governance, health, education, demographics, natural hazards, atmosphere, land, oceans, seas and coasts, freshwater, biodiversity, economic development, global economic partnership, and consumption and production patterns.

4.2.3. *Capital-based frameworks*

Capital frameworks attempt to calculate the wealth of a nation or a region as a function of different kinds of capital (UN, 2007). The capital approach borrows the concept of capital from economics, and expands it to include several types of capital: manufactured or built capital (all produced assets that form the human economy in a traditional sense), natural capital (the natural environment and resources), human capital (capacities of people to work, including knowledge, skills, and health), and social capital (stocks of social networks, trust, and institutional arrangements). An example of a capital-based framework is the Daly's Triangle-based system advocated by the Balaton Group, an international network of researchers and practitioners in the field of sustainable development, in which natural, built, human, and social capitals are identified (Meadows, 1998).

The different forms of capital are usually expressed in the same monetary terms, so that they can be aggregated. Sustainable development, in this context, may be interpreted differently depending on whether a strong or weak sustainability perspective is taken. Can natural capital be replaced by other types of capital at all? What natural resources and ecosystem services are substitutable? What are the limits to such substitutions? These are some of the critical questions that must be addressed when capital frameworks are used. Other challenges to using a capital framework include problems of monetizing different forms of capitals, controversies on substitutability, and issues on intra-generational equity (UN, 2007).

4.2.4. *Integrated accounting frameworks*

Integrated accounting frameworks refer to synthesized economic and environmental accounting systems based on national accounting methodologies. The most prominent example is the System of Integrated Environmental and Economic Accounting (SEEA), developed jointly by United Nations, European Commission, International Monetary Fund, OECD, and World Bank in 2003 (Hecht, 2006; UN, 2007). SEEA integrates environmental accounting with the standard System of National Accounts (SNA). It includes four categories of accounts: (1) physical data on material and energy flows, (2) data on environmental management and environment-related transactions, (3) accounts of environmental assets, and (4) accounts of transactions and adjustments related to the impact of the economy on the environment. Thus, instead of just producing a set of indicators, SEEA provides full accounts of environmental and economic capitals and flows. It has also been used for developing indicators and policy analysis.

However, integrated accounting frameworks such as the SEEA have not taken into consideration the social and institutional dimensions of sustainable development, although efforts to incorporate human and social capital are under way (UN, 2007). Nevertheless, the implementation of SEEA not only serves its own purposes, but also benefits the application of other indicator frameworks, particularly capital-based frameworks. Theme-based frameworks can also profit from SEEA because a consistent database facilitates the development and disaggregation of thematic indicators that can further policy analysis and design. The latest version of the UNCSD indicator framework has, therefore, strengthened its relationship with SEEA (UN, 2007).

4.2.5. *Bossel's orientor framework*

Bossel's orientor framework is a systems-theoretical framework for developing indicators of sustainable development (Bossel, 1977, 1999, 2001). This approach is intended to provide a holistic and comprehensive conceptual structure to guide indicator development. The framework has a nested hierarchical appearance that somewhat resembles Daly's Triangle. The orientor theory assumes that any ecological and socioeconomic system can be characterized by six fundamental environmental properties: a normal environmental state, resource scarcity, variety, variability, change, and other systems. These properties "constrain development possibilities and limit management opportunities on all spatial scales," and render "orientors," which are general categories of our key concerns, values or interests that "orient most of our decisions" (Bossel, 1999, 2001). In the context of sustainable development, Bossel (2001) described seven basic orientors as follows:

(1) *Existence*: The system must be compatible with and able to exist in the normal environmental state. The information, energy, and material inputs needed to sustain the system must be available.

(2) *Effectiveness*: The system should, on balance over the long term, be effective (not necessarily efficient) in its efforts to secure required scarce resources (information, matter, energy) and to exert influence on its environment when necessary.
(3) *Freedom of action*: The system must have the ability to cope in various ways with the challenges posed by environmental variety.
(4) *Security*: The system must be able to protect itself from the detrimental effects of environmental variability, i.e., variable, fluctuating, and unpredictable conditions outside the normal environmental state.
(5) *Adaptability*: The system should be able to learn, adapt, and self-organize to generate more appropriate responses to the challenges posed by environmental change.
(6) *Coexistence*: The system must be able to modify its behavior to respond appropriately to the behavior of the other systems in its environment.
(7) *Psychological needs*: These constitute an additional orientor for sentient beings.

For a system to achieve sustainability, these basic orientors must be satisfied; they usually cannot be measured directly, but their states of fulfillment can be inferred from appropriate indicators. A major advantage of this approach is that it "avoids the problems of incompleteness and double-counting common in ad hoc methods of indicator selection" (Bossel, 2001). Orientor-based indicators are expected to capture the essential aspects of the vitality, performance, and sustainability of human–environmental systems, and their results are typically represented as polygonal graphs known as "orientor stars" (Bossel, 1999, 2001).

5. Composite Indicators — Sustainability Indices

In this section, we discuss a selected set of composite sustainability indices that are usually obtained by aggregating a number of indicators selected from a specific framework. A large number of such indices exist, but we focus on a small set of the most commonly used. As discussed below, some of these indices are more sustainability-relevant than others.

5.1. *Green GDP*

The most popular brand of economics-based measures is the GDP. This procedure of accounting comes in many derivates, including GNP, Net National Product (NNP), and the increasingly common emendation known as "Green GDP." Although the GDP still occupies pride of place in the collection of related measures, conceptual nuances between the metrics reflect attempts to better approximate certain social and financial concerns. For instance, GNP takes into account international income transfers, and NNP is meant to accommodate the economic effects of capital depreciation. Mainly because their scope is delimited completely by the market, these traditional measurements omit many of the important goods and services that humans derive from nature. Thus, these indicators account for economic development but do not reflect actual human or environmental well-bring (Wu and Wu, 2010). Actually, empirical data show that GDP is often negatively correlated with environmental quality, and its positive correlation with social well-being measures disappears after GDP reaches a certain level. Green GDP, a newer addition to this family of indicators in the early 1990s, was developed to factor in the effects of natural resource consumption and pollution on human welfare (Wu and Wu, 2010). Green GDP is meant to correct such biases by expanding the coverage of accounting to include many, but certainly not all, of the values that people derive from nature.

5.2. *Human development index (HDI)*

In an effort to assess the levels of human and social development of different countries around the world, the United Nations Development Program (UNDP) created HDI. A main goal of developing HDI was to "...shift the focus of development economics from national income accounting to people centered policies" (Ul Haq, 1995). Since its inception in the 1990s, HDI has become a standard and widely reported indicator in many official reports and academic publications — most notably the UNDP's Annual Human Development Reports. The index is composed of three primary aspects: life expectancy, education, and standard of living. The education dimension is addressed using a combination of adult literacy rate and enrollment ratio, whereas the standard of living is computed from GDP per capita, adjusted for purchasing power parity. Together with life expectancy calculations, these two measurements are collated into an index taking on a value between 0 and 1 (although no country actually assumes the highest value, and only a handful of countries fall below 0.4). A major criticism of HDI is its abstraction from the environmental dimension of human welfare. As pointed out by many economists and ecologists, human welfare is inevitably a function of natural resources, and the failure to take this fact into account paints a distorted picture of development. In addition, HDI has focused only on the national and global levels of analysis.

5.3. *Inclusive wealth (IW) and genuine savings (GS)*

Unlike GDP and Green GDP, which are "flow" measures, IW and GS are stock-based. IW and GS derive national income and resource use data from the SEEA, designed by the United Nations. It is an emendation of the more standard SNA used in more traditional calculations of economic performance. The economic patterns of production and consumption are necessarily contingent upon the availability and configuration of the base of available resources, or capital. Thus, inter-temporal (e.g., cross-generational) transfers of economic opportunity are best represented by the value of capital stocks. The "inclusive" and "genuine" of the nomenclature derive largely from the inclusion of natural resources into economic accounting.

From this formulation, an intuitive criterion for sustainability follows: a country or region is "sustainable" over a given period if its IW or GS per capita does not decline over that time. A country can grow in both GDP and HDI while declining in wealth per capita (Arrow *et al.*, 2004). When set against this new capital-oriented development standard, and taking into account the role of natural resources, the optimism regarding the expansion of welfare is greatly tempered by much lower (often negative) values. Despite the potential for more accurate approximation of sustainability, it is not standard practice to keep stock-based measurements of economic development, and GDP and HDI still prevail as the most common development indicators at the national and regional levels.

5.4. *Genuine Progress Indicator (GPI) and Index of Sustainable Economic Welfare (ISEW)*

GPI and ISEW are essentially equivalent measures, with the former being more widely recognized and implemented (Lawn, 2003). Both take cues from Green GDP, amending the standard flow-based metric of economic performance to consider the role of environmental well-being. However, unlike Green GDP, which is primarily a project of augmenting the standard national accounting framework, GPI or ISEW divides economic transactions between those that make a positive contribution to human welfare and those that make a negative contribution. A classic example of the rationale underlying this dichotomy is the case of an oil spill. While an environmental disaster of this kind is surely inimical to any coherent sense of welfare, it may actually enhance GDP due to the increased number of transactions it elicits (even if they are only carried out to redress an unfortunate problem that did not have to happen in the first place). Additionally, GPI and ISEW also include imputed values of non-marketed goods and services (both social and ecological), and adjust for income distribution effects. Thus, GPI and ISEW attempt to separate the "goods" from the "bads," and provide a more holistic and sensible assessment of economic activities.

5.5. *Material flows accounting (MFA)*

The flow of materials through an economy can provide an insightful indication of the sustainability of the system *vis-à-vis* its resource base. Unlike Green GDP, which is also a "flow-based" measure, MFA attempts to quantify physical value in weight, not monetary value in currency (Adriaanse *et al.*, 1997; Matthews *et al.*, 2000). It monitors the weight of a number of different material flows, including production inputs and outputs, matter moved in the environment to access resources (such as excavated soil), and residual material from the production process (Hecht, 2006). This focus on the "direct ingredients" of production and consumption cuts out the "middleman" of monetary valuation. MFA produces a single metric of system performance by aggregating different material flows into the total material requirement (TMR) of a given economy. This summary indicator is meant to give a picture of the entire physical metabolism of the economic system. Although monetary accounting is still more widespread, MFA has been expanding and been carried out in a number of countries and regions (Ness *et al.*, 2007). The Statistical Office of European Communities (Eurostat), a research body tasked with *inter alia* the collection and calculation of comparative, performance-related data for European countries, developed an economy-wide MFA that is the most standardized tool for assessment (Ness *et al.*, 2007). This guideline divides material flows into three categories: input, output, and consumption. Further, within each of these categories there are levels indicating whether the flows cover domestic, foreign, and/or hidden — which are materials not included in economic accounting, such as soil erosion (Matthews *et al.*, 2000).

5.6. *Ecological footprint (EF)*

EF is primarily a measure of human appropriation of natural resources, and defined as the land (and water) area that would be required to support a defined human population indefinitely in terms of providing all the energy/material resources consumed and to absorb all the wastes discharged (Wackernagel and Rees, 1996). The basic unit of measurement is the "global hectare," a normalized unit capturing the average biocapacity of all hectares of all biologically productive lands in the world. Consumption patterns of natural resources, from energy to biomass, can be converted to this common metric. This simple but comprehensive measure allows us to compare our demands of the planet's ecosystems to the regenerative capacity of those ecosystems. In this sense, we can create a direct correspondence between our present standard of consumption and the capability of the biosphere to support that standard. According to recent studies using EF analysis, humanity is currently exceeding the biocapacity of the planet by approximately 20% (Wackernagel *et al.*, 2002), or we are consuming our natural capital at 1.2 times the rate at which they are being regenerated. Since its inception in the early 1990s, EF has become a widely influential indicator of human–environment impact, spawning a large body of literature and analysis. The popularity of its application may derive from the fact that it can be employed over a broad range of activities and at varying levels from individuals to the globe. However, EF analysis has not been without criticism, with much of the concerns targeted toward its abstraction from the roles played by technology, land degradation, the demarcation of spatial boundaries (Fiala, 2008; van den Bergh and Grazi, 2010; van den Bergh and Verbruggen, 1999).

5.7. *Happy Planet Index (HPI)*

HPI was developed, by the New Economics Foundation (http://www.happyplanetindex.org/), as an alternative to measures like HDI or GDP. Although still advocating human-oriented measures, HPI directly combines human welfare with human consumption of natural resources. Different from the standard account of welfare, which is largely defined as a simple function of consumption, HPI defines human well-being in "happy life years" — a combination of life expectation and life expectancy. The indicator is then calculated as the ratio between happy life years and environmental impact, which is measured by the EF. HPI is intended to measure "the ecological efficiency with which human well-being is delivered around the world" (New Economics Foundation, 2009).

5.8. *Environmental Sustainability Index (ESI) and Environmental Performance Index (EPI)*

Published between 1999 and 2005, ESI is, like MFA and EF, mainly a biophysical indicator of natural resource use by humanity. ESI was produced by the Center for Environmental Law and Policy at Yale University and Columbia University's Center for International Earth Science Information Network (ICESIN), in collaboration with the World Economic Forum and the European Commission's Joint Research Center. The computational methodology behind ESI involves combining 76 variables into 21 metrics, which are then averaged to yield a single index. ESI was succeeded by EPI, which was developed by the same institutions and has been published in 2006, 2008, and 2010 (http://sedac.ciesin.columbia.edu/es/epi/). Environmental performance, with a focus on assessing current environmental conditions, differs from the original ESI theme of gauging long-term environmental trajectory. While ESI had five assessment themes (i.e., environmental systems states, reducing environmental stress, reducing human vulnerability, institutional capacity, and global stewardship), EPI narrows its aims to two objectives: environmental health and ecosystem vitality. The change in orientation is illustrated by how the two indices differ in analyzing forest management. The 2005 ESI focused on the annual change in forest cover and percentage of total forest area certified for sustainable management. The 2008 EPI, on the other hand, simply used the change in growing stock as a proxy for management performance. EPI is supposed to provide a report of "more immediate value to policymakers" and, as of the 2010 publication, 163 countries are included in the analysis (http://sedac.ciesin.columbia.edu/es/epi/).

6. Conclusion

Is there a comprehensive, all-inclusive measure of sustainability, or must sustainability be measured from multiple aspects? It is not uncommon for the idea of sustainability to be subdivided into more specific, but still broad, categories such as social, ecological, and economic. However, the extents to which such classifications are valid, and the rationale for their construction as opposed to dissections that are even more specific (e.g., from social to cultural and demographic, from ecological to biophysical and geophysical), are inevitably subjective. It may be argued that in any conceptual anatomization of so general an idea as sustainability, there is no principled point at which to draw methodological standards. If cogently formed and supported by rigorous analysis, however, a certain set of metrics may be more appropriate than others for a given set of objectives or a particular region. Our review of the development of SIIs during the past several decades illustrates these points.

Depending on the organizing framework or perspective chosen, one may arrive at vastly different conclusions about the system of interest. For example, interpretations about sustainability that are drawn from Green GDP may not mean much if one rejects the premise of weak sustainability, or, more broadly, any framework of measurement centered on economic valuation. At the same time, while an indicator such as TMR is more "direct" in its treatment of underlying environmental factors, without a way to translate the effects into a language relatable to human welfare, it may be difficult to craft coherent policies. Thus, selecting and evaluating SIIs must begin with a set of guiding sustainability principles, and an understanding of the frameworks that correspond to these criteria. Only then the relevant indicators or indices can be properly contextualized and articulated for policy-making purposes.

It has been increasingly recognized that the most essential value of the term "sustainability" or "sustainable development" lies in its emphasis on unifying the relevant dimensions — of which environmental, economic, and social are common classifications. Accordingly, sustainability measures must focus on the integrative whole of human–environmental systems. As Meadows (1998) pointed out: "Sustainability indicators must be more than environmental indicators; they must be about time and/or thresholds; Development indicators should be more than growth indicators; they should be about efficiency, sufficiency, equity, and quality of life." In our efforts to operationalize sustainable development, indicators and indices are indispensable for creating scientific understanding and shaping policy. As the future unfolds, these metrics must continue to evolve in complexity and nuance to keep pace with the demands of escalating environmental and socioeconomic problems. The search for adequate and effective indicators of sustainability is an evolutionary and learning process (Meadows, 1998). This will require substantial contribution from the social and natural sciences, as well as increasing levels of participation from stakeholders of all kinds.

References

Arrow, K., Dasgupta, P., Goulder, L., Daily, G., Ehrlich, P., Heal, G., Levin, S., Mäler, K.-G., Schneider, S., Starrett, D. and Walker, B., 2004. Are we consuming too much? *Journal of Economic Perspectives* **18**, 147–172.

Ayres, R.U., van den Bergh, J.C.J.M. and Gowdy, J.M., 1998. Viewpoint: Weak versus strong sustainability. *Tinbergen Institute Discussion Papers No 98-103/3*, Tinbergen Institute. URL: http://www.tinbergen.nl/discussionpapers/98103.pdf.

Bell, S. and Morse, S., 2003. *Measuring Sustainability: Learning by Doing.* Earthscan, London.

Bell, S. and Morse, S., 2008. *Sustainability Indicators: Measuring the Immeasurable?* 2nd ed. Earthscan, London.

Bossel, H., 1999. *Indicators for Sustainable Development: Theory, Method, Applications (A Report to the Balaton Group)*. International Institute for Sustainable Development, Winnipeg.

Bossel, H., 2001. Assessing viability and sustainability: A systems-based approach for deriving comprehensive indicator sets. *Ecology and Society* **5**, 12. URL: http://www.consecol.org/vol15/iss12/art12/.

Braat, L., 1991. The predictive meaning of sustainability indicators. In: Kuik, O. and Verbruggen, H. (Eds), *In search of Indicators of Sustainable Development*. Kluwer Academic Publishers, Dordrecht, pp. 57–70.

Du Pisani, J.A., 2006. Sustainable development — historical roots of the concept. *Environmental Sciences* **3**, 83–96.

Gallopin, G.C., 1997. Indicators and their use: Information for decision-making. In: Moldan, B. and Billharz, S. (Eds), *Sustainability Indicators*. Wiley, New York, pp. 13–28.

Guy, G.B. and Kibert, C.J., 1998. Developing indicators of sustainability: US experience. *Building Research and Information* **26**, 39–45.

Hak, T., Moldan, B. and Dahl, A.L. (Eds.), 2007. *Sustainability Indicators: A Scientific Assessment*. Island Press, Washington, D.C.

Hardi, P. and Zdan, T., 1997. *Assessing Sustainable Development: Principles in Practice*. Internation Institute for Sustainablie Development, Winnipeg, Manitoba.

Hecht, J.E., 2006. Can indicators and accounts really measure sustainability? Considerations for the U.S. Environmental Protection Agency. *US EPA Workshop on Sustainability*. URL: http://www.epa.gov/sustainability/other_resources.htm.

Kareiva, P., Watts, S., McDonald, R. and Boucher, T., 2007. Domesticated nature: Shaping landscapes and ecosystems for human welfare. *Science* **316**, 1866–1869.

Kates, R.W., Parris, T.M. and Leiserowitz, A., 2005. What is sustainable development? Goals, indicators, values, and practice. *Environment: Science and Policy for Sustainable Development* **47**, 8–21.

Kidd, C.V., 1992. The evolution of sustainability. *Journal of Agricultural and Environmental Ethics* **5**, 1–26.

Lawn, P.A., 2003. A theoretical foundation to support the Index of Sustainable Economic Welfare (ISEW), Genuine Progress Indicator (GPI), and other related indexes. *Ecological Economics* **44**, 105–118.

Matthews, E., Amann, C., Bringezu, S., Fischer-Kowalski, M., Huüttler, W., Kleijn, R., Moriguchi, Y., Ottke, C., Rodenburg, E., Rogich, D., Schandl, H., Schuütz, H., van der Voet, E. and Weisz, H., 2000. *The Weight of Nations: Material Outflows from Industrial Economies*. World Resources Institute, Washington, D. C.

Meadows, D., (Ed.), 1998. *Indicators and Information Systems for Sustainable Development*. Sustainability Institute, Hartland Four Corners, Vermont.

Ness, B., Urbel-Piirsalu, E., Anderberg, S. and Olsson, L., 2007. Categorising tools for sustainability assessment. *Ecological Economics* **60**, 498–508.

New Economics Foundation, 2009. The (Un)Happy Planet Index 2.0. URL: http://www.happyplanetindex.org/public-data/files/happy-planet-index-2-0.pdf.

Niemi, G.J. and McDonald, M.E., 2004. Application of ecological indicators. *Annual Review of Ecology, Evolution and Systematics* **35**, 89–111.

OECD, 1993. *OECD Core Set of Indicators for Environmental Performance Reviews*. Organization for Economic Co-operation and Development, Paris.

Ul Haq, M., 1995. *Reflections on Human Development*. Oxford University Press, New York.

UN, 2007. *Indicators of Sustainable Development: Guidelines and Methodologies*, 3rd ed. United Nations, New York.

UNCSD, 2001. *Indicators of Sustainable Development: Guidelines and Methodologies.* UN Commission on Sustainable Development.

UNEP, 2006. *Environmental Indicators for North America.* United Nations Environment Programme, Washington, DC.

Wackernagel, M. and Rees, W.E., 1996. *Our Ecological Footprint: Reducing Human Impact on the Earth.* New Society Publishers, British Columbia, Canada.

Wackernagel, M., Schulz, N.B., Deumling, D., Linares, A.C., Jenkins, M., Kapos, V., Monfreda, C., Loh, J., Myers, N., Norgaard, R. and Randers, J., 2002. Tracking the ecological overshoot of the human economy. *Proceeding of the National Academy of Sciences of the United States of America* **99**, 9266–9271.

WCED, 1987. *Our Common Future.* Oxford University Press, New York.

Wu, J., 2008. Making the case for landscape ecology: An effective approach to urban sustainability. *Landscape Journal* **27**, 41–50.

Wu, J., 2010. Urban sustainability: An inevitable goal of landscape research. *Landscape Ecology* **25**, 1–4.

Wu, J. and Wu, T., 2010. Green GDP. In: *Berkshire Encyclopedia of Sustainability, Vol. II — The Business of Sustainability.* Berkshire Publishing, Great Barrington, pp. 248–250.

Chapter 5

BENCHMARKING AND PROCESS CHANGE FOR GREEN SUPPLY CHAIN MANAGEMENT

Joseph Sarkis
Graduate School of Management, 950 Main Street,
Clark University, Worcester, MA 01610, USA
jsarkis@clarku.edu

This chapter introduces issues related to benchmarking, performance measurement, and business process change management within a green supply chain management environment. The chapter provides reflections and insights into the application of these various management tools. Some case study practices of the concepts from actual companies are also summarized. Finally, a set of tools around a general methodology is presented. These tools include the use of rough set theory, DEMATEL, IDEF0, and AHP to aid in the benchmarking, performance measurement, and process change management of green supply chains.

1. Introduction

This chapter introduces various aspects of green supply chain management (GSCM) with a specific focus on performance measurement, benchmarking, and process change management for GSCM.

As supply chains become more integrated and complex in their relationships and more broadly dispersed, the determination of success and effectiveness of supply chains is replete with difficulties. Inter-organizational performance measurement is required for effective benchmarking and improvement of supply chains. When environmental and other sustainability measures are introduced, this difficulty increases due to additional complexities. Especial concerns and issues arise when trying to determine who, how, what, and when to complete performance measures for the supply chain. The reasons for difficulty in managing performance and processes across organizations are multidimensional, including non-standardized data, poor technological integration, geographical and cultural differences, differences in organizational policy, lack of agreed upon metrics, or poor understanding of the need for inter-organizational performance measurement. These barriers are large, but can be overcome with appropriate design, planning, and implementation. Research in this area has been increasing in recent years.

When focus shifts to GSCM, the research on benchmarking, performance measurement, and process change management decreases steeply. That is, there is very little investigation and research on these issues even as the importance of sustainability and greening of organizations increases. Part of the difficulty is that internal organizational environmental practices and operations benchmarking and performance measurement are also lacking. Even when internal environmental management systems such as ISO 14000, which recommend benchmarking and performance measurement, exist, the extension of these systems to the supply chain is difficult. Leap-frogging from benchmarking and performance measurement in internal organizational operations to GSCM is not a feasible endeavor for most organizations. We will make the argument and present some evaluation and discussion that such practices should and, albeit rare, do exist.

This chapter begins with a general discussion of supply chain management and performance measurement principles. This initial discussion is extended by incorporating supply chain or inter-organizational dimensions to performance measurement principles. We then introduce concepts of GSCM, benchmarking, and performance measurement within a process change environment. Integrative and synergistic tools for managing and decision making in this environment are introduced.

2. Supply Chain Management

We provide some fundamental discussion of the supply chain as a general overview in this section. The supply chain can generally be described with upstream, downstream, and internal organizational activities. Figure 1 provides an overview of the major functional activities and relationships in a supply chain with the focal organizational unit at its center.

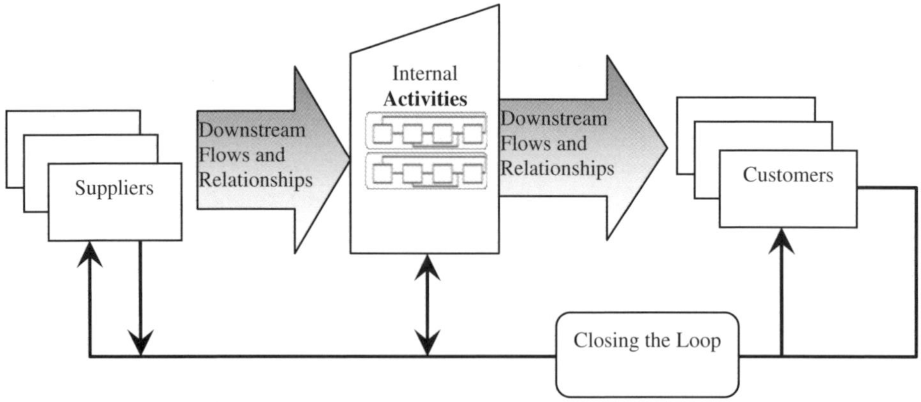

Fig. 1. Generalized diagram of supply chain elements, flows, and relationships.

Upstream activities, flows, and relationships include purchasing and procurement functions. Included amongst these topics are outsourcing, vendor auditing, management and selection, supplier collaboration, and supplier development. The many activities that focus on upstream supply chain management would be encompassed with the traditional purchasing discipline. Each of these activities can be expanded to have greening components. Upstream dimensions of the supply chain, as with activities throughout the supply chain, have related transportation and materials movement concerns. Not only there are operational considerations within the supply chain, but strategic dimensions as well. For example, designing the supplier network with its inbound logistics may include strategic considerations. Managing more than one tier in the supply chain would be part of strategic network design and optimization issues.

Internal organizational supply chain activities are related to traditional production and operations management topics of an organization. Managing the flows, relationships, and resources inside the boundaries of a stand-alone unit or organization, the focal organization, is the scope of this element. Such activities may include research and design, quality, inventory, materials, and technology management within an organization could each influence environmental characteristics of internal organizational processes. These processes could be defined from either a services or manufacturing perspective that would need to be managed. This portion of the supply chain, under the purview of a specifically defined organization or unit, utilizes and transforms the various upstream flows. As we shall see in our later discussion on boundaries, the definition of an internal organization may not always be clear.

The next juncture of a supply chain focuses upon the outbound and downstream relationships and flows. Activities and functions here may include outbound logistics and transportation, marketing, distribution, packaging, and warehousing. The flows are utilized by downstream customers who may be commercial or individual consumers. Some of the greatest pressure for improving environmental performance arrives from external groups and diffuses upstream the supply chain. Thus, organizations need to be available to complete similar supply chain management programs that they would require of their suppliers.

With the "closing of the supply chain loop," supply chain activities have been extended to include reverse supply chains as well. Closing the loop utilizes end-of-life materials that will eventually be consumed back into the system with recycling, remanufacturing, reclamation, and reverse logistics all part of this concept. Even though the environmental implications of closing the loop are obvious, much of the research in this area has been parlaying the economic and business benefits associated with the concept of closing the loop. The closed-loop relationships here may be direct between the organization, its suppliers, and its customers, or internal loops between suppliers, customers, and within the organization.

2.1. *Green supply chain management*

The terminology, and definitions, for the GSCM concept (and its many elements) has had many variations over the years and a comprehensive list includes some of the following terms:

- sustainable supply network management;
- supply and demand sustainability or corporate social responsibility networks;
- supply chain environmental management;
- green purchasing and procurement;
- environmental purchasing;
- green logistics and environmental logistics;
- sustainable supply chains.

From a value chain perspective, GSCM practices can and do occur at the various stages. These stages include inbound logistics (transportation efficiency and emissions reduction, green supplier selection, green purchasing), manufacturing and operations (total quality environmental management, eco-design, cleaner production), outbound logistics and distribution (transportation efficiency, warehousing efficiency, and packaging), and reverse logistics (reclamation, recycling, reuse, remanufacturing) (Sarkis *et al.*, 2006). There are also many other stages and integration across the supply chain (suppliers, internal functions, and customers) that will also need to incorporate environmentally sound practices into their activities. Thus, we can describe GSCM as practices, processes, and activities within the supply chain within and across organizations that minimize the ecological footprint of the supply chain.

3. Benchmarking, Performance Measurement, and Process Change Management

3.1. *Benchmarking*

Benchmarking is a widely used tool in industry to improve performance and manage organizational processes. It is a method of improving performance in a systematic and logical way, by measuring and comparing performance of one unit against other units. Units may be other internal operations and divisions or external organizations and their processes. Benchmarking uses insights from other units' processes and management techniques to make targeted improvements. Benchmarking may be based on "best practices" and comparing an organization's processes to those best practices. It is a core aspect of continuous improvement and organizational learning. It ensures that improvement targets are based on what has been achieved by others, whether it is other individuals, departments, organizations, or supply chains.

3.2. *Performance measurement*

Central to benchmarking is corporate performance measurement. Application of performance measurement continues to grow and encompass both quantitative and qualitative measurements and approaches. The variety and level of performance measures and supporting performance measurement systems (PMS) depend greatly on the goal of the organization or the individual strategic business unit's characteristics (Pinheiro de Lima *et al.*, 2009) and may serve many purposes (Wouters, 2009). For example, when measuring performance, companies must consider existing financial measures such as return on investment, profitability, market share, and revenue growth at a more competitive and strategic level. Other measures such as customer service and inventory performance (supply, turnover) are more operationally focused, but may necessarily be linked to strategic level measures and issues. Performance measures, and benchmarking initiatives, have various taxonomic characterizations including management level to measure — strategic, tactical, or operational; tangible *versus* intangible measures; variations in collection and reporting; an organization's location along the supply chain; or functional differentiation within organizations (e.g., accounting *versus* marketing or operations) (Gunasekaran and Kobu, 2007).

Similar to the performance measurement used, a PMS[a] may be unique to each individual organization, or unit within an organization, reflecting its fundamental purpose and its environment. Several studies have investigated the universal principles of performance measurement (Adams *et al.*, 1995; Gunasekaran and Kobu, 2007; Li and Tang, 2009). These studies have arrived at a number of conclusions related to performance measurement and their systems including

- PMS may have either tangible or intangible measures with a balance of both types used to measure performance; measures should be dynamic and present at multiple levels;
- products and processes need to be included;
- systems and measures are best developed with a team approach with derivation from and links to corporate strategy;
- systems must have effective internal and external communications;
- accountability for results must be clearly assigned and be understood;
- systems must provide intelligence for decision makers and not just compile data;
- various stakeholder interests should be taken into consideration; and
- the system should be capable of linking compensation, rewards, and recognition to performance measurement.

[a]Popular PMS frameworks include the balanced scorecard (Kaplan and Norton, 1996), strategic maps (Kaplan and Norton, 2004), the performance prism (Neely *et al.*, 2001), and the SMART pyramid (Cross and Lynch, 1988).

It has also been argued that performance measurement must evolve to performance management, where the organization develops the appropriate organizational structure and the ability to use performance measurement results to actually bring about change in the organization.

Elements of these efforts are central to total quality and continuous improvement programs and thus process change efforts. Performance measurement has many uses including the determination of the efficiency and effectiveness of an existing system, benchmarking through comparison of competing alternative systems, and data acquisition for strategic planning. Overall, performance measurement is typically used to plan, design, implement, and monitor proposed systems and processes.

3.3. *Benchmarking and performance measurement across the supply chain*

Benchmarking and performance measurement have recently expanded to the supply chain due to the increased complexities of multiple processes and organizations that may be involved in product and service delivery and the role of supply chain relationships and processes in managing organization strategic advantages (Gunasekaran *et al.*, 2004).

The development of supply chain performance measurement and benchmarking systems has not been easy. There has been increasing amount of research on this topic due to the complex nature and need for new models in this area. Even though many PMS models and performance measures exist (Gunasekaran and Kobu, 2007; Kaplan and Norton, 1996; Neely *et al.*, 2001) some have needed adjustment to address supply chain performance measurement. The use of the supply chain operations reference (SCOR) model, developed specifically for supply chain management, has been recommended as an important framework for supply chain PMS from both a research (Gunasekaran and Kobu, 2007; Shepherd and Gunter, 2006) and practical (Chae, 2009) perspective. Thus, SCOR can set the foundation for the environmental and business performance of the supply chain, two important competitive dimensions for modern supply chains (Zhu *et al.*, 2010).

3.4. *Benchmarking and performance measurement for GSCM*

The SCOR model categorizes the processes of five supply chain stages including plan, source, make, deliver, and return. The performance measures within SCOR are categorized on cost, time, quality, flexibility, and innovation dimensions (Shepherd and Gunter, 2006). These performance measures are business oriented, environmental measures are lacking. We can expand SCOR performance measures to incorporate environmental performance examples within the five SCOR measures as shown in Table 1. We focus on the source function in this situation due to this function's direct activities with suppliers, but additional environmental performance measures from all stages of the SCOR model can be introduced.

Table 1. SCOR-based sourcing function business and environmental performance measures categorized on cost, time, quality, flexibility, and innovation dimensions (Sarkis and Talluri, 2002a; Shepherd and Gunter, 2006; Gunasekaran and Kobu, 2007).

Business measures	Environmental measures
Cost	
Supplier cost-saving initiatives, labor efficiency, cost variance from expected costs	Environmental costs savings, energy efficiency of systems, environmental cost performance variance, amount of environmental penalties
Time	
Supplier lead time against industry norm, supplier's booking-in procedures, purchase order cycle time, percentage of late deliveries, information timeliness, efficiency of purchase order cycle time	Length of time to implement environmental programs, meeting environmental program implementation period, speed of acquiring environmental information, communication speed on environmental issues to supplier's suppliers
Quality	
Buyer–supplier partnership level, level of supplier's defect-free deliveries, supplier rejection rate, delivery reliability, percentage of wrong supplier delivery, mutual trust, satisfaction with knowledge transfer, satisfaction with supplier relationship, supplier assistance in solving technical problems, extent of mutual planning cooperation leading to improved quality, extent of mutual assistance leading in problem-solving efforts, distribution of decision competences between supplier and customer, quality and frequency of exchange of logistics information between supplier and customer, quality of perspective taking in supply networks, information accuracy, information availability	Environmental relationship and cooperation level, supplier rejection rate, waste generated from products and materials, percentage recycled material, mutual trust on environmental issues, mutual planning for environmental improvements, mutual assistance for environmental improvements, environmental information accuracy, environmental information availability
Flexibility	
Supplier ability to respond to quality problems, response to product changes, materials variety (number of materials available), product and service variety, product volume variability capabilities, product development time	Amount of environmentally safe alternatives, response to environmental programs for suppliers, response to environmental product requests
Innovation	
Satisfaction with knowledge transfer satisfaction, technological capability levels, involvement in new product design, introduction of new processes	Environmental knowledge transfer satisfaction, environmental technology levels, new environmentally sound processes introduced, new environmentally sound product development

3.5. *Process change management for GSCM*

Business process change management has typically focused on internal processes and activities within organizations. Extending these business process change aspects to the extended enterprise becomes a little more difficult to manage due to a clear lack of authority lines across organizational boundaries. Yet, there have been some, but few, examples where inter-organizational process change management along the supply chain has been modeled (Changchien and Shen, 2002; Presley *et al.*, 2001). There are several steps in the Changchien and Shen (2002) framework including creating vision, identifying core processes to be redesigned, analyzing current core processes, designing for innovation, evaluating the new processes, selecting the best, and transforming and implementing the resulting design. The same approach may be utilized for GSCM, but applications of these models will have to consider the myriad environmental processes and performance measures that are required. Before identifying applicable tools and application of these tools, we describe some actual industrial practices that exemplify benchmarking, performance measurement, and process changes with respect to environmental issues across the supply chain. These exemplary practices set the stage for introducing a toolset of novel techniques that will help in benchmark organizations and help with process change management within GSCM.

4. Some Exemplary GSCM Process Change Requirements in Industry

Practically, organizations are utilizing environmental performance measurement and the various benchmarking and process change management philosophies for their supply chain and environmental purposes. For general performance measurement, Eastman Chemical Company has supplier performance measurement as a key process that procurement groups responsible for sourcing direct raw materials, energy supplies, indirect materials, logistics, and other corporate services. Part of this goal is for Eastman to seek improvement in supplier performance as well as ensure supplier commitment to the environment in their supply chain. Performance measures are used for initial supplier selection criteria that manage service economic, and environmental performance of their global supplier base. For example, they measure transportation incidents for logistics service suppliers and including supplier commitment to Responsible Care (www.eastman.com).

Eastman Chemical Company's sustainable supply chain continual improvement program, part of their corporate process change management program, incorporates a number of process management and changes for its supply chain and partners. A comprehensive program includes

- measure supplier performance;
- maximize payload quantities;
- develop alternate methods of supply;

- develop supplier solutions;
- improve packaging;
- redesign and optimize supply chain network;
- develop customer solutions;
- use "Lean" and "Six Sigma" for process improvement;
- manage investment recovery;
- ensure global compliance;
- use renewable materials.

Advanced Micro-Devices (AMD), a semiconductor manufacturer, found that most of the difficulties encountered in implementation and execution were associated with integrating GSCM into existing operational practices. Operationally, the movement toward GSCM has required a slight cultural change for the company. Integrating both suppliers and environmental health and safety (EHS) into the decision-making and managerial processes was initially challenging. However, with EHS's involvement in various supplier teams, integrating EHS concerns into the supplier selection process has grown easier. Many of AMD's internal groups realize the value of EHS considerations and understand the need for EHS involvement and have been accepting of the change. This situation exemplifies the typical need for organizations to adjust their culture and business processes simultaneously for effective implementation of GSCM.

Boeing, an aircraft manufacturer, has recognized that there is great opportunity to go beyond current information-sharing activities with suppliers. Boeing has developing a more systematic program of *collaboration* and *engagement* with its suppliers in the area of environmental practices. Boeing uses supplier data and supply-base analysis to shape a GSCM program that requires flexibility for its global supply base. Similar to AMD, they explicitly define and use various environmental performance criteria for supplier selection and evaluation purposes. As part of their collaboration with suppliers sharing of better practices for shrinking the environmental footprint of the supply chain is explicitly encouraged (Boeing, 2010).

Pfizer, the world's largest pharmaceutical company, environmentally (and health and safety too) benchmarked its external supply chain that included contract manufacturing and research (CMR) vendors. They took this external benchmarking evaluation of over 100 partners to develop performance measures and standards for their supply chain management operations. The results meant new processes for each operating group. These new processes included (GEMI, 2004)

- Determine whether a new CMR is acceptable before deciding to use it.
- Assure that CMRs receive all relevant EHS information on its proprietary compounds and technology and update it as warranted so that they can manage EHS risks effectively.
- Devote resources to conducting periodic EHS reviews of all CMRs.

- Develop and implement a management system for EHS relationships with CMRs.
- CMR reviews are conducted by the operating group EHS Staff, collaborating with facility and in-country colleagues. Issues covered as part of these reviews included
  - adequacy of permits to operate;
  - site conditions that may reflect environmental sensitivity;
  - emissions and waste management and pollution prevention capabilities.

Nokia (2010), a telecommunications company, has also implemented environmental performance and evaluation requirements for its suppliers, with environmental impacts considered at every stage of the Nokia's product life cycles. These supply chain life cycles have ranged from raw material extraction and processing to component manufacturing and assembly. Part of process changes required in their supply chain included the need to increase visibility of environmental performance and target setting. This effort enabled Nokia to better understand the environmental impacts their supply chain and to help drive improvements. These performance measures and targets for suppliers included

- energy consumption at supplier sites and improvement targets;
- CO_2 emissions from supplier sites and improvement targets;
- water consumption at supplier sites and improvement targets;
- waste generation at suppliers sites and improvement targets.

These examples show the diversity of organizations that are taking the initiative to reengineer their supply chains for environmental purposes using benchmarking and performance measurement as their core elements. Specific tools and techniques for these initiatives were not clearly defined other than general improvement programs such as six-sigma or total quality management efforts to achieve their goals. In the following section a general set of tools to aid in some of these process change initiatives are described.

5. Tools for GSCM Benchmarking and Process Change Management

The above practical case studies exemplify how organizations have been attempting to effectively benchmark and change supply chain processes so that the supply chain can be more environmentally supportive. In theory and practice, numerous quantitative and statistical models exist including standard statistical process control and six-sigma programs that involve tools to aid in green supply chain benchmarking and process change management. In this section, we introduce a number of more recent and complementary developments and tools for managing the GSCM benchmarking and process change management activities. Specifically, we present the following non-standard tools have yet to permeate organizational continuous improvement and business process management tool sets, but could be valuable complements. Included amongst these tools are rough set tools for aiding with data evaluation, a systems process structuring tool called IDEF0, a factors and elements relationship tool called DEMATEL, and a multiattribute decision-making tool, the analytical hierarchy process, which allows for relative evaluation of various processes and factors. We do not provide the mathematical or analytical details of these approaches, but provide general observations on their application to GSCM performance and process change management. In this chapter, we do not delve into the mathematical methodological issues but the many references provided in this section can provide significant mathematical insights. References are provided for each of these tools.

5.1. *Rough set theory*

Rough set theory, proposed by Pawlak (1982), is a mathematical approach to vagueness and ambiguity. The method classifies objects into similarity classes (clusters) containing objects that are indiscernible with respect to previous occurrences and knowledge. These similarity classes are employed to determine hidden data patterns or "rules." Rough set theory has seen significant application in data mining approaches. Approximation vagueness is usually defined by precise values of lower and upper approximations. Lower approximations describe the domain objects, which definitely belong to the subset of interest. Upper approximations describe objects, which may possibly belong to the subset of interest. The difference between the upper and the lower approximations constitutes a boundary region for the vague set. Hence, rough set theory expresses vagueness by employing a boundary region of a set. If the boundary region of a set is empty it means that the set is crisp, otherwise the set is rough (inexact). In many real-world applications, the boundary regions are not always so crisp.

To aid practitioners, and researchers, in offering solutions to the development, selection, and monitoring of supply chain environmental and business performance measures, the rough set approach can prove invaluable. This technique allows for the inclusion of a broad set of measures and considers the significant ambiguity of real-world situations. Rough set methodology is advantageous for a number of reasons. It can incorporate a wide variety of measures while effectively linking the performance measures to outcomes. Unlike regression and other statistical approaches, parametric assumptions are unnecessary, i.e., it is a non-parametric data evaluation methodology (Bai and Sarkis, 2010a, 2010b).

Rough set applications have been increasing as has the methodological developments. For example, the consideration of fuzzy numbers (Wu *et al.*, 2003), incomplete information (Kryszkiewicz, 1998, 1999), and neighborhood distance measures (Hu *et al.*, 2008a, 2008b) allows for significant flexibility for its application.

The basic requirement for a rough set approach is an "information system." Table 2 is composed of three primary components. The first component of the

Table 2. An initial information system table for rough set theory with GSCM elements.

Objects (suppliers)	Conditional attributes (C) (sustainable supply chain performance measures)					Decision attributes (outcomes)	
	C1	C2	C3	...	Cn	D1	D2
Supplier 1	9	40%	VH		5	3	3
Supplier 2	7	39%	L		4	1	2
Supplier 3	6	15%	M		4	3	3
⋮							
Supplier M	8	21%	VL		5	1	2

information system is the set of objects that will be evaluated. As can be seen in Table 2, the objects for a GSCM application would be suppliers. In this case, there would be M suppliers. Alternatively, these objects may be processes that need to be managed.

The next component within the information system is the various "conditional" attributes set that will be associated with each object. These are the n attributes that will help define characteristics of the object. In Table 2, the conditional attributes are represented by the first set of columns ranging from C1 to Cn. In the case of GSCM benchmarking and process change management, these conditional attributes may be characterized by various factors. These factors may be performance measures that we are seeking to evaluate. Alternatively they could represent various characteristics of the suppliers or processes such as length of time in operation, profitability, industry, etc. The values in each of these columns for a conditional attribute may be tangible or intangible. Eventually a valuation needs to be assigned to each conditional attribute either a perceptual or actual value would need to be included.

The decision attributes represent some form of grouping or outcome associated with each object and represent sets to which objects will be assigned. In the case of GSCM, the outcome might be some result such as overall environmental performance or customer satisfaction. The ultimate outcome will be to determine which conditional attributes contribute to an assignment of a supplier to a particular set. Using theories information loss and entropy (Liang *et al.*, 2004), the most pertinent conditional attributes are determined and also rules can be generated that can help predict future occurrences.

Thus, from a business process management perspective specific identified supply chain processes or activities for each supplier and their contribution to an outcome could be determined. These processes can be identified for either further reengineering or improvement, or in the case of poorly performing processes, removed ("disintermediated").

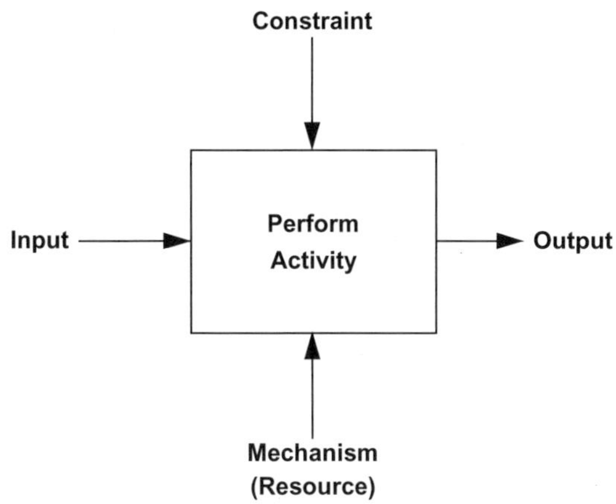

Fig. 2. IDEF0 activity box and elemental general elements.

5.2. *IDEF0 — Business process modeling*

IDEF0 is a systemic functional modeling method developed by the Air Force's Integrated Computer-Aided Manufacturing initiative (Marca and McGowan, 1988). There are five fundamental elements to the IDEF0 functional model (see Fig. 2): activities are represented by boxes; inputs, outputs, controls, and mechanisms are represented by the arrows shown at the left-hand side, right-hand side, top, and bottom of the boxes, respectively. The flow of the arrows shows the flow of information and material from one activity to another.

A characteristic of the IDEF0 modeling technique has activities and arrows that can be decomposed into hierarchical levels of analysis. This characteristic is shown in Fig. 3. Each diagram contains detail about an activity from a more general (or parent) activity. The process or product model that is developed by IDEF0 may be easily and directly mapped to other tools and techniques. In our case, the decompositional characteristics are helpful for linkage to a tool such as AHP. Part of the IDEF0 methodology involves the documentation of each activity and process where descriptive text for each of the activities is developed. Additional information pertaining to environmental aspects should be included in these descriptions and glossaries. IDEF0 has been a popular tool in the business process reengineering of organizations (Gingele *et al.*, 2002; Sarkis, 1995).

For the sake of evaluating processes within a supply chain, we can use IDEF0 to structure various alternative supply chain processes (Figs. 4 and 5). These processes may be identified using the rough set approach. That is, the IDEF0 processes may be evaluated on a number of pertinent performance measures that could have also been identified and selected using the rough set theoretic technique. These alternative processes usually include "as-is" and "to-be" processes that may be evaluated on a variety of performance factors. The development of the processes themselves may require extensive evaluation of comparable processes from other departments or

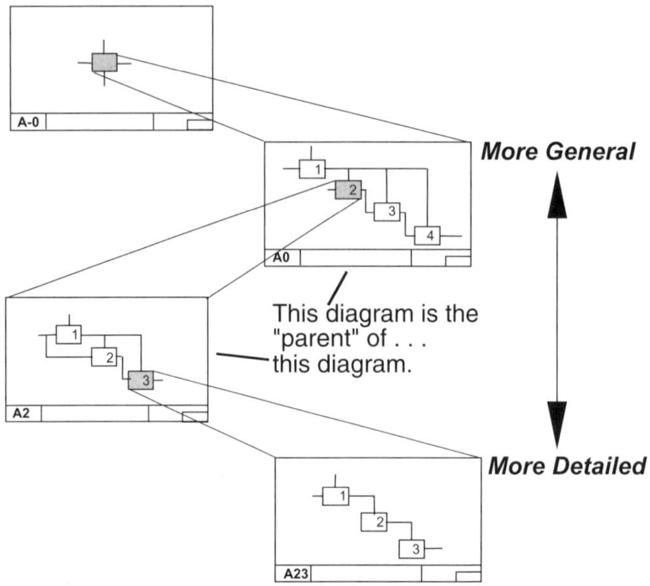

Fig. 3. IDEF0 decompositional characteristic.

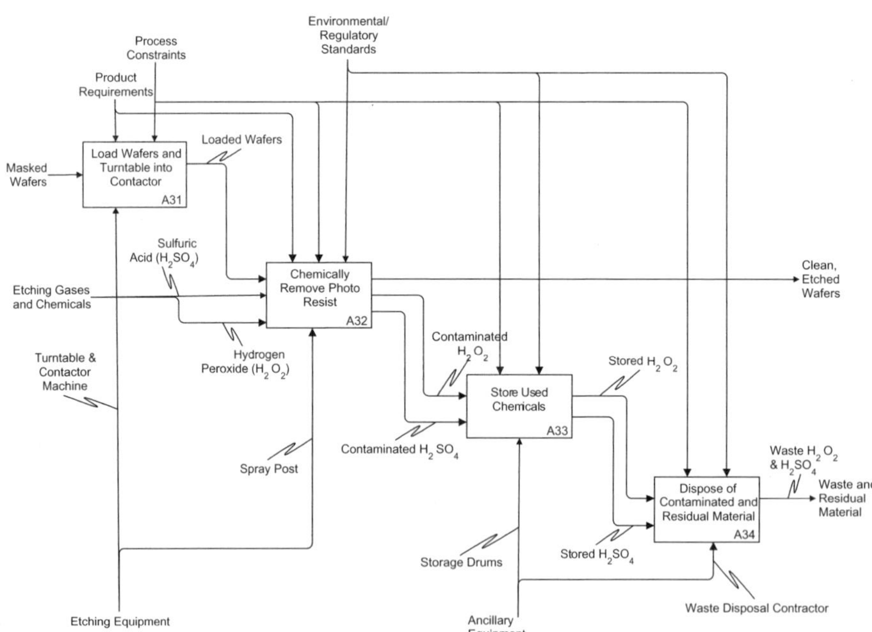

Fig. 4. An example of IDEF0 initial supply chain process for chemical operations. *Source*: Sarkis *et al.* (2006).

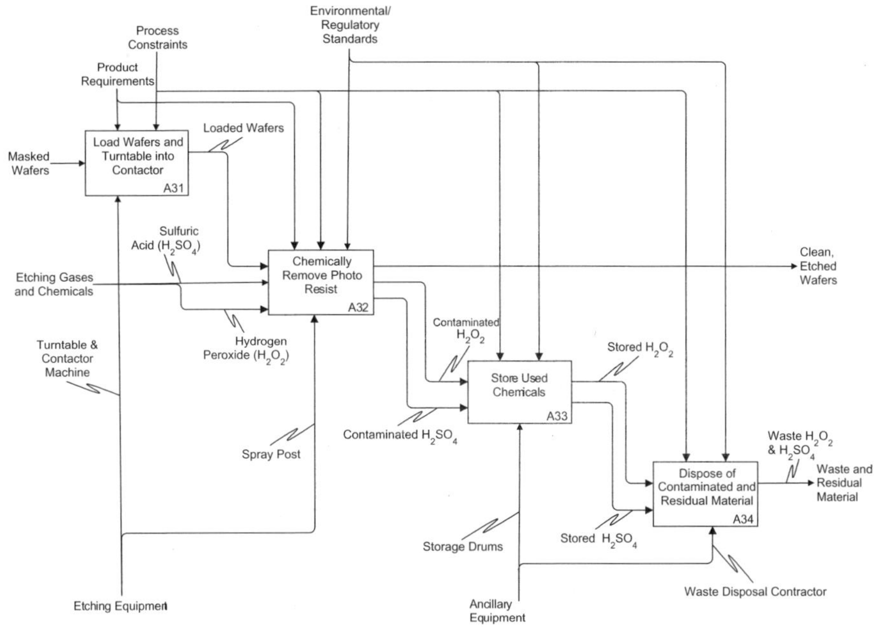

Fig. 5. An example of an alternative IDEF0 supply chain process for chemical operations. *Source*: Sarkis *et al.* (2006).

organizations. This is essentially a knowledge gathering step to be able to generate the "to-be" processes. Within the IDEF0 systems design activities, this knowledge gathering and process design refinement stage is called the "reader–writer" cycle. In this cycle, inputs are given and feedback provided. The readers in this situation are typically experts or process users and the writers may be technical process engineers or systems analysts who complete the actual modeling.

In addition, note that specific processes are considered, the decompositional characteristics of IDEF also allow for broader inter-organizational (extended enterprise) evaluations as well. Thus, environmental implications from strategic engineering of a supply chain, with practices such as network design, intermediation, and disintermediation, may also be modeled using IDEF0 techniques.

IDEF0 tools have been advanced serving as the foundation for development to a broader IDEF family of tools that helps to develop automated simulations to help investigate the various process designs (Jeong *et al.*, 2009). This family of tools represents some of the earliest designed for the purpose of business process reengineering purposes.

As mentioned, IDEF0 requires that you define the relationships amongst the various processes and influences (flows) amongst each. Typically, this is completed using interviews with process designers and users. One way to help formally relate the processes is through structural modeling using a digraph approach called DEMATEL. We now provide an overview of this technique.

5.3. *DEMATEL*

The DEMATEL method originated from the Geneva Research Centre of the Battelle Memorial Institute (Fontela and Gabus, 1976; Gabus and Fontela, 1973). The method is useful to help visualize the structure of complicated causal relationships through the use of matrices or digraphs. The matrices or digraphs portray relationships between system components with strengths of relationships amongst these relationships quantitatively portrayed. The DEMATEL method assumes a system contains a set of components with pairwise relations that can be evaluated.

DEMATEL incorporates four generic stages:

1. Develop a pairwise direct-relation matrix between system components through evaluator or decision maker input.
2. Determine the initial influence matrix, through normalization of the direct-relation matrix.
3. Determine a total relation (influence) matrix.
4. Determine the cause/effect relationships (causal diagram) amongst the components and relative strengths.

DEMATEL aids managers evaluate the relationships amongst various performance measures and processes. Details on DEMATEL and its applications to other tools such as AHP can be found in Liou *et al.* (2007).

5.4. *AHP — Multiattribute decision modeling*

Saaty (1980) developed AHP for decision structuring and decision analysis. AHP allows a set of complex factors that have an impact on an overall objective to be compared with the importance of each factor relative to its impact on the solution of the problem. AHP is a comprehensive framework that is designed to cope with tangible and intangible factors when making multiobjective, multicriterion, and multiactor decisions — exactly the decision-making situation found within environmental management. While AHP is conceptually easy to use, it is decisionally robust so that it can incorporate the complexities of many real-world problems. AHP models a decision-making framework that assumes a unidirectional hierarchical relationship among decision levels. AHP has three basic steps:

1. Development of a decision hierarchy. The top element of the hierarchy is the overall goal for the decision model. The hierarchy decomposes to a more specific attribute until a level of manageable decision criteria is met. The hierarchy is a type of system where one group of factors influences another set of factors.
2. Pair-wise comparisons are conducted to estimate the relative importance weights (or allocations) of the various elements on each level of the hierarchy.

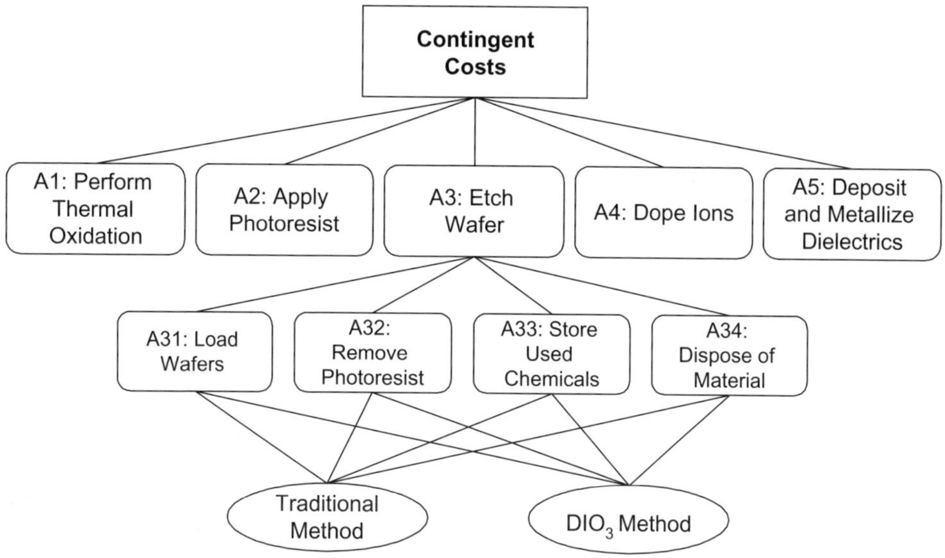

Fig. 6. An AHP cost allocation sub-hierarchy for activity- and process-based resource management approach for contingent costs resource.

3. The weights obtained are integrated to develop an overall ranking of decision alternatives.

In a business process management model, the core factors of the hierarchy will be business processes or activities that are influenced by the greening managerial decision or improvement to the process. The AHP approach will help identify the amount of allocation to be made of whatever resources across activities, facilitating an activity-based management approach.

Figure 6 shows an AHP hierarchy that can be used to structure the processes and relationships between the IDEF0 business processes from Figs. 4 and 5. Note that the goal in this situation is to determine the overall contingent (environmental) costs of the alternative processes for each of the activities. Thus, in this situation two alternative business processes are evaluated for selection, a "traditional method" process and an alternative "DIO_3 method" process. The technique is also useful for a benchmarking or gap analysis between the two processes.

For details on how to complete calculations for AHP as an activity-based management process reengineering tool, refer Sarkis *et al.* (2006). AHP as a benchmarking tool has been posited for a number of years. For example, it has been used to identify the business processes that would need to be benchmarked (Partovi, 1994). AHP has also been integrated with system dynamics (Rabelo *et al.*, 2007), multiattribute utility theory (MAUT) (Sarkis *et al.*, 1997), and data envelopment analysis (Sarkis and Talluri, 2002b) for business process reengineering and improvements along the supply chain. Extensions of these tools to broader GSCM topics and applications have yet to see significant research.

5.5. *Integration of tools within a process change and improvement management framework*

Integration of these various tools can be completed in a process change management framework that is tied to basic continuous improvement cycles from total quality management and six-sigma-like programs. The standard "Plan–Do–Check(Study)–Act" continuous improvement cycle is shown in Fig. 7 and has also been defined as the Deming Wheel. Many tools and approaches are available for this cycle of which performance measurement and benchmarking are core elements at every stage of the process. As an example, in the tools for planning, defining the core problem and describing the current situation can be completed using IDEF0 and AHP approaches at the *Plan* stage. Collection of data and measuring progress within the *Do* stage can be aided through the use of rough set approaches, which help to identify the most pertinent information to gather. The *Study* stage requires a comparative analysis, which can utilize rough set, DEMATEL, and AHP for the evaluation of the specifically identified processes and performance measures. The *Act* stage can utilize IDEF0 to document the best-known methods and AHP to identify new improvement opportunities. Within each of these process change management steps, benchmarking and performance measurement requirements are evident and play critical roles.

Implementing such a continuous improvement model primarily targeted toward internal operations business processes is the extension to the supply chain and focus on environmental implications of these processes. We have primarily focused on the

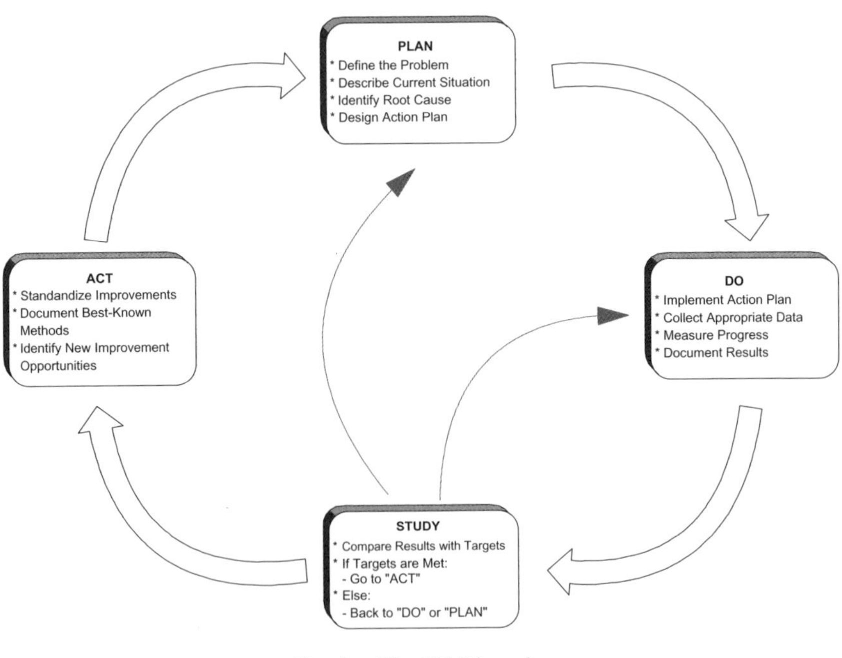

Fig. 7. The PDCA cycle.

development and application of methodologies and tools in this chapter. The roles of employees and management within the GSCM benchmarking and business process change management activities cannot be overstated. Environmental improvements in internal operational settings have shown strong relationships to employee involvement and the "win–win" results of these programs (Hanna *et al.*, 2000). In fact, the role of training of human resources has found to be critical for GSCM activities such as environmental-oriented reverse logistics (Sarkis *et al.*, 2010). Expansion to multi-organizational and multi-stakeholder settings will add to the complexity of these issues.

6. Conclusions and Future Directions

The expansion of environmental issues to the supply chain has caused organizations additional difficulties as inter-organizational, inter-functional, and the incorporation of social issues add complexities that have to be managed. GSCM requires the extension of operations improvement tools to encompass both inter-organizational processes with environmental performance as an integral dimension of analysis and management. The importance of managing GSCM has not gone unnoticed by actual organizations. Examples from actual practice show that organizations are struggling with and implementing various aspects of benchmarking, performance measurement, and business process change management. We did introduce some of the basic concepts of benchmarking, performance measurement, and process change management for the supply chain and green concerns facing these organizations. Supporting mechanisms and tools also exist, with many dispersed amongst disciplines such as information systems management, industrial engineering, management sciences, economics, and operations management. We provided an overview of a number of these tools and their application within a business process improvement cycle that is replete with benchmarking and performance measurement activities. Yet, additional work still remains.

In our evaluation of the issues related to GSCM benchmarking, performance measurement, and business process change, we focused on management decision tools that exist for standard operational and business improvements. We did not intend to shy away from environmental-oriented tools such as mass-balance flow analysis, life-cycle analysis, total cost assessment, and design for the environment (eco-design) tools. In fact, many of these tools have a long history with GSCM and are some of the earliest tools that can be attributed to improving supply chains, albeit from an environmental perspective (Ayres and Kneese, 1969; Stern *et al.*, 1973). The integration of the various business process change management tools presented in this paper, rough set theory, IDEF0, DEMATEL, and AHP with the various environmental-oriented tools, is necessary for acceptance by both internal and external stakeholders.

Expanding these tools to sustainable supply chains would require the inclusion of the third dimension of the triple-bottom-line, social aspects. The inclusion of social aspects would have similar concerns as environmental dimensions. That is, there will be additional complexities, many times with greater qualitative and intangible performance measures and considerations that need to be integrated. In terms of evaluating systems and processes, the number of potential conflicts between environmental, social, and business/economic dimensions of sustainability increases. Thus, the tools discussed in this chapter need to be especially designed to help integrate and manage the tradeoffs associated with the triple-bottom-line dimensions. Dynamic tools to help monitor the designs and measures over time are also required since change is one constant that needs to be considered in this evolving managerial environment.

This chapter provides some insights to integration and application of these various tools to GSCM, but significant opportunities for future research, development, and investigation are still required. Having a toolset in concept is an initial start, some disparate case and practical examples of applications were presented here. Developing a toolset as part of an GSCM performance and decision support system is a necessary step to evaluate the feasibility of the integrated tool set is required. The toolset does not necessarily have to be automated, but requires significant reduction in complexity and ease-of-use improvement to be applicable. However, if organizations seek to spend millions, if not billions, of dollars in response to environmental pressures, having these tools will greatly enhance their probability of competitive and environmental success.

References

Ayres, R.U. and Kneese, A.V., 1969. Production, consumption, and externalities. *The American Economic Review* **59**(3), 282–297.

Adams, S., Sarkis, J. and Liles, D.H., 1995. The development of strategic performance metrics. *Engineering Management Journal* **7**(1), 24–32.

Bai, C. and Sarkis, J., 2010a. Integrating sustainability into supplier selection with grey system and rough set methodologies. *International Journal of Production Economics* **124**(1), 252–264.

Bai, C. and Sarkis, J., 2010b. Green supplier development: Analytical evaluation using rough set theory. *Journal of Cleaner Production*, in Press (Corrected Proof).

Boeing, 2010. http://www.boeingsuppliers.com/environmental/index.html. Last Accessed, March, 2010.

Chae, B., 2009. Developing key performance indicators for supply chain: An industry perspective. *Supply Chain Management: An International Journal* **14**(6), 422–428.

Changchien, S.W. and Shen, H.-Y., 2002. Supply chain reengineering using a core process analysis matrix and object-oriented simulation. *Information & Management* **39**(5), 345–358.

Cross, K.F. and Lynch, R.L., 1988. The "SMART" way to define and sustain success. *National Productivity Review* **8**(1), 23–33.

Fontela, E. and Gabus, A., 1976. *The DEMATEL Observer. DEMATEL 1976 Report.* Battelle Geneva Research Center, Switzerland, Geneva.

Gabus, A. and Fontela, E., 1973. *Perceptions of the World Problematique: Communication Procedure, Communicating with Those Bearing Collective Responsibility (DEMATEL Report No. 1).* Battelle Geneva Research Center, Switzerland, Geneva.

GEMI, 2004. *Forging New Links: Enhancing Supply Chain Value through Environmental Excellence.* http://www.gemi.org/supplychain/ last accessed March, 2010.

Gingele, J., Childek, S.J. and Miles, M.E., 2002. A modelling technique for re-engineering business processes controlled by ISO 9000. *Computers in Industry* **49**(3), 235–251.

Gunasekaran, A. and Kobu, B., 2007. Performance measures and metrics in logistics and supply chain management: A review of recent literature (1995–2004) for research and applications. *International Journal of Production Research* **45**, 2819–2840.

Gunasekaran, A., Patel, C. and McGaughey, R.E., 2004. A framework for supply chain performance measurement. *International Journal of Production Economics* **87**(3), 333–347.

Hanna, M.D., Newman, W.R. and Johnson, P., 2000. Linking operational and environmental improvement through employee involvement. *International Journal of Operations and Production Management* **20**(2), 148–165.

Hu, Q., Yu, D., Liu, J. and Wu, C., 2008a. Neighborhood rough set based heterogeneous feature subset selection. *Information Sciences* **178**(18), 3577–3594.

Hu, Q., Yu, D. and Xie, Z., 2008b. Neighborhood classifiers. *Expert Systems with Applications* **34**(2), 866–876.

Jeong, K.Y., Wu, L. and Hong, J.D., 2009. IDEF method-based simulation model design and development. *Journal of Industrial Engineering and Management* **2**(2), 337–359.

Kaplan, R.S. and Norton, D.P., 1996. *The balanced scorecard: Translating strategy into action.* Harvard Business School Press, Boston, MA.

Kaplan, R.S. and Norton, D.P., 2004. *Strategy Maps: Converting Intangible Assets into Tangible Outcomes.* Harvard Business School Press, Boston, MA.

Kryszkiewicz, M., 1998. Rough set approach to incomplete information systems. *Information Sciences* **112**(1), 39–49.

Kryszkiewicz, M., 1999. Rules in incomplete information systems. *Information Sciences* **113**(3), 271–292.

Li, P. and Tang, G., 2009. Performance measurement design within its organisational context — Evidence from China. *Management Accounting Research* **20**(3), 193–207.

Liang, J.Y., Shi, Z.Z. and Li, D.Y., 2004. The information entropy, rough entropy and knowledge granulation in rough set theory. *International Journal of Uncertainty Fuzziness and Knowledge-Based Systems* **12**(1), 37–46.

Liou, J.H., Tzeng, G.-H. and Chang, H.-C., 2007. Airline safety measurement using a hybrid model. *Journal of Air Transport Management* **13**(4), 243–249.

Marca, D.A. and McGowan, C.L., 1988. *SADT: Structured Analysis and Design Technique.* McGraw-Hill Co., New York, NY.

Neely, A., Adams, C. and Crowe, P., 2001. The performance prism in practice. *Measuring Business Excellence* **5**(2), 6–13.

Nokia, 2010. *2008 Nokia Environmental Report.* http://www.nokia.com/environment/our-responsibility/environmental-report-2008/environmental-management/suppliers-environmental-performance. Last Accessed, March 2010.

Pawlak, Z., 1982. Rough sets. *International Journal of Computer and Information Sciences* **11**(5), 341–356.

Pinheiro de Lima, E., Gouvêa da Costa, S.E. and Reis de Faria, A., 2009. Taking operations strategy into practice: Developing a process for defining priorities and performance measures. *International Journal of Production Economics* **122**(1), 403–418.

Partovi, F.Y., 1994. Determining what to benchmark: An analytic hierarchy process approach. *International Journal of Operations and Production Management* **14**(6), 25–39.

Presley, A., Sarkis, J., Barnett, W. and Liles, D., 2001. Engineering the virtual enterprise: An architecture-driven modeling approach. *International Journal of Flexible Manufacturing Systems* **13**(2), 145–162.

Rabelo, L., Eskandari, H., Shaalan, T. and Helal, M., 2007. Value chain analysis using hybrid simulation and AHP. *International Journal of Production Economics* **105**(2), 536–547.

Saaty, T.L., 1980. *The Analytical Hierarchy Process: Planning, Priority Setting, Resource Allocation.* McGraw Hill, New York, NY.

Sarkis, J., 1995. Manufacturing strategy and environmental consciousness. *Technovation* **15**(2), 79–97.

Sarkis, J., Gonzalez-Torre, P. and Adenso-Diaz, B., 2010. Stakeholder pressure and the adoption of environmental practices: The mediating effect of training. *Journal of Operations Management* **28**, 163–176.

Sarkis, J., Meade, L. and Presley, A., 2006. An activity based management methodology for evaluating business processes for environmental sustainability. *Business Process Management Journal* **12**(6), 751–769.

Sarkis, J., Presley, A. and Liles, D., 1997. The strategic evaluation of candidate business process reengineering projects. *International Journal of Production Economics* **50**(2–3), 261–274.

Sarkis, J. and Talluri, S., 2002. A model for strategic supplier selection. *Journal of Supply Chain Management* **38**(1), 18–28.

Sarkis, J. and Talluri, S., 2002b. A synergistic framework for evaluating business process improvements. *International Journal of Flexible Manufacturing Systems* **14**(1), 53–71.

Shepherd, C. and Gunter, H., 2006. Measuring supply chain performance: Current research and future directions. *International Journal of Productivity and Performance Management* **55**(3/4), 242–258.

Stern, M.O., Ayres, R.U. and Saxton, J.C., 1973. Tax strategies for industrial pollution abatement. *IEEE Transactions on Systems, Man and Cybernetics* **3**(6), 588–603.

Wouters, M., 2009. A developmental approach to performance measures — Results from a longitudinal case study. *European Management Journal* **27**(1), 64–78.

Wu, W.-Z., Mi, J.-S. and Zhang, W.-X., 2003. Generalized fuzzy rough sets. *Information Sciences* **151**, 263–282.

Zhu, Q., Dou, Y. and Sarkis, J., 2010. A portfolio-based analysis for green supplier management using the analytical network process. *Supply Chain Management: An International Journal*, Forthcoming.

Chapter 6

DEVELOPING AN ADAPTIVE FRAMEWORK FOR SUSTAINABLE SUPPLY NETWORKS

Dmitry Ivanov[a] and Boris Sokolov[b]

[a]*Department of Economics and Business Administration, Chemnitz Technical University, D-09107 Chemnitz, Germany*
idm@hrz.tu-chemnitz.de

[b]*Saint Petersburg Institute of Informatics and Automation, Russian Academy of Sciences, 14th Line VO 39, 199178 Saint Petersburg, Russia*
sokol@iias.spb.su

In this chapter, the supply chain management perspective of sustainable management is addressed. Based on the literature analysis, we develop a qualitative adaptive framework for sustainable supply networks that is composed of the integrated elements of environment management, operations economic development and information technologies, product life-cycle management, and social development. Subsequently, we develop a quantitative adaptive framework for sustainable supply networks. As sustainable systems are designed to support decisions on planning, analysis and improving organization's actions and operations to meet competitiveness and sustainability goals, it becomes necessary to interconnect the stages of supply network planning and execution. Indeed, as supply networks are not static but highly dynamic and permanently evolving systems, sustainability of these networks should also be considered as a dynamic category of supply chain (SC) adaptive evolution. Mathematical framework is based on a combined application of operations research and modern control theory. Finally, we address the issue of interrelating environment and economic sustainability and propose to consider the SC property to achieve a balance between the economic output performance and resource consumption as the SC global stability. The proposed framework can be employed as a competitive strategy to increase organizations' wealth and sustainability by producing demand-corresponding products in a cost-effective manner through an increase of SC agility, resilience, stability, and adaptation capabilities.

1. Introduction

Supply networks are one of the main drivers in modern world economy. They include a consistent and interrelated system of production, logistics, and transportation networks. The main method for the handling supply networks is supply chain management (SCM). The term "SCM" was coined in the 1980–1990s. Presently, SCM is considered as the most popular strategy for improving organizational competitiveness along the entire value chain in the twenty-first century. SCM is one of the key components of enterprise management and is responsible for balancing demand and supply along the entire value-adding chain (Christopher, 2005).

A supply chain (SC) is a network of organizations and processes wherein a number of various enterprises (suppliers, manufacturers, distributors, and retailers) collaborate (cooperate and coordinate) along the entire value chain to acquire raw materials, to convert these raw materials into specified final products, and to deliver these final products to customers. According to this definition, we will use the terms "SC" and "supply networks" as synonyms in this paper.

The development of SCM was driven in the 1990s by three *main trends*: customer orientation, markets globalization, and establishing an information society. These trends caused changes in enterprise competitive strategies and required new adequate value chain management concepts.

First, to remain competitive, enterprises concentrated on *product individualization* and maximum meeting of customers' requirements. *Flexibility* and *responsiveness* came to be the key factors in supply management. Second, in the 1990s, new markets in Asia, Eastern Europe, and South America were extensively acquired and the production facilities actively shifted to these regions. Third, remarkable advances in *IT* and establishing the World Wide Web provided the basis for innovative business concepts. On the whole, the focus turned to the consideration of entire value-adding chains, all the elements and links within them and outside the own enterprise to ensure business profitability and competitiveness.

SCM's impact on the enterprise management can be estimated as up to 30% (Ivanov and Sokolov, 2010). From decisions on the SC configuration arise up to 80% of the total SC costs (Harrison, 2005) and up to 75% of the operational costs in SCs (Wannenwetsch, 2005). The increase in sales and reduction in costs in the value-adding chain due to SCM amount to 15–30%. Partial effects such as inventory reductions, an increase in service level (SL), SC reliability and flexibility, a reduction in transaction costs, etc. amount to 10–60%.

These effects occur due to balancing supplies along the entire value-adding chain to ensure mutual iterative balances of production and logistics processes subject to full customer satisfaction. As customer orientation, globalization, and IT advancements are still the ongoing trends in markets, the importance of SCM will become ever greater. The practice of SCM has provided enough evidence that intra-organizational integration, inter-organizational coordination, and collaborative management philosophy along *the entire value-adding chain* have a profounder influence on profitability and competitiveness, rather than the local optimization of intra-organizational functions. SCM will be further developed and there will be ever more investments in SCs (Christopher, 2005).

Krause *et al.* (2009) underline that a company is no more sustainable as its SC. Shepherd and Guenther (2006) treat SCM as key to building a sustainable competitive edge through improved inter- and intra-firm relations. Hence, in developing a sustainable management system, the consideration of SCM is mandatory.

Indeed, the impact of SCM on the enterprise sustainability is becoming more and more profound. With the further development of globalization, worldwide transportation increases drastically. More and more close embedding of suppliers and customers into product development and post-sales stage of the product life-cycle force the integration of SCM and pre- and post-production stages of product life-cycle management (PLM).

Finally, development of supply networks interrelates with many social and political aspects. The danger of production resource deficit challenged modern SCM and requires the development of energy-efficient supply networks (Mueller *et al.*, 2009). According to a recent study by PwC, if oil prices would exceed the level of 200$ pro barrel can launch significant changes in the global supply networks and cause their regionalizing and displacing of production plants. As European legislative has started to control the amount of CO_2-emissions by an emission trading system that will be extended to air transportation in 2012, airlines will be confronted by an additional cost factor. The plant or order volumes displacing from traditional production sites may have negative consequences regarding the unemployment increase in these regions.

The above-mentioned aspects allow concluding that enhancing supply networks through sustainability with the help of establishing interactions between operations, environment, society, policy, and the product life cycle is the critical next step in SCM as driven from recent examinations in relation to operations and sustainability (Carter, 2004; Kleindorfer and Saad, 2005) and operations and the environment (Corbett and Kleindorfer, 2003). In doing so, the focus on environmental management and operations is moved from local optimization of environmental factors to consideration of the entire SC during the production, consumption, customer service, and post-disposal disposition of products. The study by Jaraman *et al.* (2007) emphasize, that sustainable development is a rich area for academic research that is still in its infancy and has the potential to affect future government policy and current production operations, and to identify new business models. It is critical to move forward toward holistic conceptual frameworks and mathematical formulations of the sustainable SCM (Carter and Rogers, 2008; Seurung *et al.*, 2008; Pagell and Wu, 2009).

In this chapter, we develop a framework for sustainable supply networks. We will consider sustainability as a multi-dimensional category regarding the triple bottom line of sustainable management: social development, environmental protection, and economical development (Gladwin *et al.*, 1995). As sustainable systems are designed to support decisions on planning, analysis and improving organization's actions and operations to meet competitiveness and sustainability goals (Russo and Fouts, 1997), it becomes necessary to interconnect the stages of supply network planning and execution. Indeed, as supply networks are not static but highly dynamic and permanently evolving systems, sustainability of these networks should also be considered as a dynamic category of SC evolution. This can be successful realized on the adaptation principles (Ivanov and Sokolov, 2010). Hence, we are going to develop an adaptive framework for sustainable supply networks.

The remainder of this chapter is organized as follows.

2. State of the Art

In analyzing the existing literature on Sustainable SCM, the five major research streams can be distinguished. These are

- SCM and environment;
- SCM and PLM;
- SCM and operations economic development;
- SCM and information technologies; and
- SCM and social development.

2.1. *Supply network sustainability, environment, and PLM*

Environmental aspects have been the most visible and influential topic in research on SC sustainability (Min and Galle, 2001; Matos and Hall, 2007; Seuring *et al.*, 2008). An extensive penetration of SCM into the business practice has a wide overlapping with environmental issues. A rapid increase in transportation has to deal with ecological issues. With the further development of globalization, worldwide increases in transportation can be anticipated. As an example, the increase in defray the biggest container harbors can be brought into discussion (Fig. 1).

To decrease the negative impacts on environment, many companies invest in the development of so-called "green" SCs regarding reduction of CO_2 emissions or products with "Green Label" (Bowen *et al.*, 2001; Kogg, 2003; Rao and Holt, 2005; Zhu *et al.*, 2008). The "greening" is mostly based on re-structuring transportation networks, e.g., by excluding the air transports. In line with the business practices, diverse technological solutions have been developed, e.g., CargoCap systems in Germany for underground transports.

Another important issue in SCM and environment is the danger of production resource deficit challenged modern SCM that requires the development of energy-efficient supply networks (Mueller *et al.*, 2009). First, the existing SCs are frequently designed with significant redundancies in resource consumption that are inadequate the SC output performance. These redundancies should be eliminated and SC resources and output performance should be brought into a mutual correspondence. Second, the technological resource losses, e.g., leakage, should also be eliminated.

A further critical topic in SCM and environment is the problem of taking back products from customers and recovering added value by reusing the entire product, and/or some of its modules, components, and parts. With regard to the above-mentioned issues, the concepts of reverse logistics and closed-loop SCM have been developing over the last few years. Guide and Van Wassenhove (2009) define

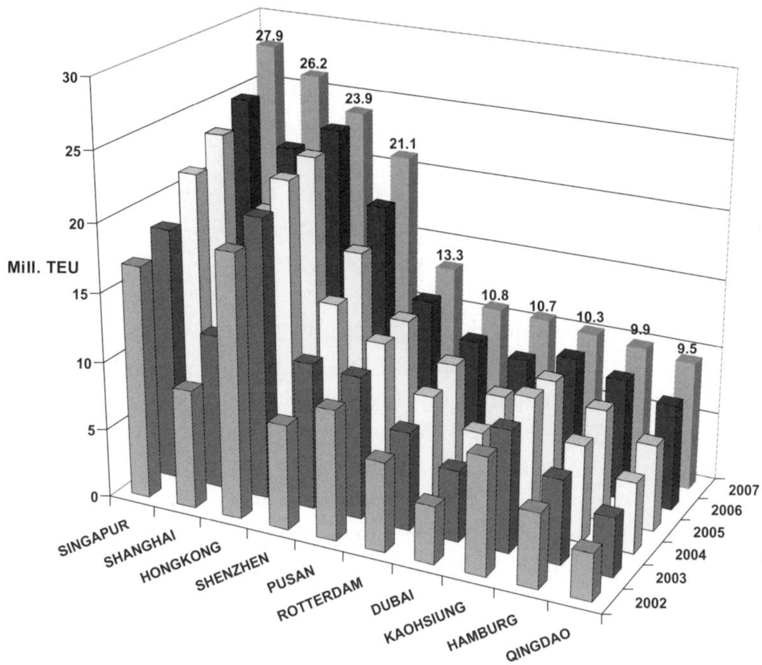

Fig. 1. Dynamics of defray development in the biggest container harbors from 2002–2007 (Boese and Daduna, 2008).

closed-loop SCM as "the design, control, and operation of a system to maximize value creation over the entire life cycle of a product with dynamic recovery of value from different types and volumes of returns over time."

Product recovery activities include used-product acquisition, reverse logistics, product disposition (sort, test, and grade), remanufacturing/repair, utilization, and remarketing. Let us consider some practical examples for the importance of closed-loop SCM. The remanufacturing sector is presently larger than the US domestic steel industry in terms of sales and employment with annual sales in excess of \$53 billion (Lund, 1996). The total value of returns can easily run in the hundreds of millions of dollars for a single retailer. Stock *et al.* (2002) estimate the annual costs of commercial returns in excess of \$100 billion. Guide and Van Wassenhove (2009) showed with a computer network equipment manufacturer that more than \$700 million of perfectly operational recovered products were destroyed. Personal computer manufacturers have short life-cycle products that can lose 1% of their value per week and have high return rates (Guide and Van Wassenhove, 2009). These types of products represent a huge challenge for value recovery. A slow reverse SC that takes 10 weeks to put the returned product back on the market translates to a loss of (approximately) 10% of the total value in that product. Very important part of reverse logistics is the utilization (Pagell *et al.*, 2004). Our own experiences show that a logistically clever utilization of an aircraft can return about \$6 million in to the value chain.

High requirements for sustainability are necessary regarding quality management at production and after-production stages including reverse logistics. For example, because of the problems with accelerators, Toyota recalled more than 4 million cars from US markets, about 2 million cars are anticipated to be recalled from European markets.

In general, three prospectives of interrelation SCM and PLM exist: extension of SCs by means of embedding the PLM, impact of the SCM on the PLM, and impact of the PLM on the SCM. The importance of the SCM and PLM integration is extreme: recent investigations show large efficiency reserves in embedding customers and suppliers in product development (Hoegl and Wagner, 2005). Suppliers' participation in new product development and engineering at early stages may provide an increase in product quality by 30% and the reduction of "time-to-market" by 50% (Chopra and Meindl, 2007). Moreover, for many products, the importance of product utilization and end-of-life stages is not less (or sometimes even more) as the production stage. This includes processes of warranty, update, and recovery services. PLM can also provide valuable information technologies like Product Data Management systems as well as standards of business and technological process documentation at all the product life-cycle stages like the CALS — Continuous Acquisition and Life-Cycle Support — standard.

2.2. *Supply network sustainability, operations management, and information technologies*

Organizations are undergoing significant transformations to answer challenges of customer-driven markets, globalization, and competition pressure. To ensure a long-term competitiveness and survival, companies implement new strategies, based on collaboration with business partners and an advanced application of business information systems and web services with mobile technologies.

Two basic drivers of operational sustainability in SCs are the maximization of economic performance (Chopra and Meindl, 2007) and the reduction of risks (Kleindorfer and Saad, 2005). In order to ensure operational sustainability, SCs are to be agile and resilient. Agility shapes enterprise competitiveness and contributes to increasing SC performance (Christopher and Towill, 2001; Gunasekaran and Ngai, 2009; Swafford *et al.*, 2008). On the other hand, achievement of the planned SC goals can be inhabited by perturbation impacts in a real execution environment (Kleindorfer and Saad, 2005 and Khan and Burnes 2007). SC execution (SCE) is subject to uncertainty at the planning stage and disruptions at the execution stage. Therefore, SCs are to be reliable and flexible enough (6.1) to ensure agility and (6.2) to be able to adapt their behavior in the case of perturbations impacts in order to remain resilient by recovering disruptions once disturbed.

After long-lasting research on SC optimality from the SL's and service costs' points of view, the research community has begun to shift to a paradigm that the performance of SCs is to consider *adaptable, stable and crisis-resistant processes* to compete in a real perturbed execution environment (Kleindorfer and Saad, 2005; Sheffy and Rice, 2005). Modern information technologies such as radio frequency identification (RFID), supply chain event management (SCEM) and mobile business can facilitate these feedbacks in order to incorporate the stages of SC planning (SCP) and SCE.

In line with ongoing economic transformations, SCs are being designed more complex, subject to different redundancies to ensure reliability and flexibility. The reliability is mostly seen as a cost-driven property whereas flexibility is considered from the profitability point of view (Ivanov and Sokolov, 2010; Christopher, 2005). The profit losses through non-purposeful (e.g., demand fluctuations) and purposeful (e.g., terrorism or thefts) perturbation impacts can amount to 30% of the annual turnover. Profit of an SC can decrease by 40% if a disruption in SC becomes public (Hendricks and Singhal, 2003). Mulani and Lee (2002) showed that SC managers spend about 40–60% of their working time handling disruptions. The profit losses through perturbation impacts can amount up to 30% of the annual turnover. For example, in 2000, the material damage to the European retail trade amounted to 13.4 billion euros, and the material damage to the European manufacturers reached 4.6 billion euros. With regard to empirical data of international insurance, companies lose up to 15% of their turnover as a result of threats alone.

The issue of how to design and run SCs with regard to agility and resilience has been a visible and influential topic in recent research. In practice, different policies to mitigate uncertainty are applied. For example, Dell combines the transport and inventory strategies by storing cheap components in Europe and order-driven replenishing of expensive components in Asia. On the contrary, Cisco stores expensive components in the USA and produces cheap components in Asia. Another example is the German automotive supplier MTU Aero Engines Ltd. that is able to run the production three weeks based on the safety stocks.

Recent literature has identified different methods to strengthen SCs to mitigate uncertainty impacts. *First*, different reliability reserves (material inventory, capacities buffers, etc.) can be referred to. The *second* research direction is related to better coordination in SCs. A recent AMR Research study shows the great importance of demand-oriented SC coordination: demand-forecast accuracy creates high responsiveness and cuts costs right through the SC (Friscia *et al.*, 2004). According to the study, the companies that are best at demand forecasting maintain on average 15% less inventory, 17% stronger perfect-order fulfillment, and 35% shorter cash-to-cash life cycles. Different concepts of coordination have been developed over the last two decades, such as efficient consumer response, collaborative planning, forecasting, and replenishment in retail as well as just-in-time and vendor-managed inventory in industries. Enablers of the coordination are information technologies, such as enterprise resource planning, advanced planning systems, electronic data interchange, and RFID. The coordination has become a key factor in mitigating the bullwhip effect and in overcoming information asymmetry (Lee *et al.*, 1997). Moreover, due to Internet technologies, it has become possible to integrate customers into SC considerations, resulting in the development of the build-to-order SCM (Gunasekaran and Ngai, 2009; Sharif *et al.*, 2007).

Third, analysis of SC stability from different perspectives has been applied. In SC management, the issues of stability have attracted increased attention and interest in recent years (Warburton *et al.*, 2004). The study by Ivanov and Sokolov (2010) proposed to consider the SC property to approach the real SC performance with the planned one under the colliding SC processes in the real perturbed execution environment with regard to the variety of execution and goal criteria as the SC global stability.

The literature analysis allows the conclusion that while important contribution has been already made in the direction of supply network sustainability, it is critical to move forward toward holistic conceptual frameworks and mathematical formulations of the sustainable SCM.

3. Qualitative Adaptive Framework of Sustainable Supply Networks

In this section, consideration is given to the conceptual understanding sustainable supply networks as driven from recent advancements in operational management, SC agility, PLM, and environmental management. With increasing transportation, the minimizing of negative impacts on ecology may become one of the primary objectives in SCM and logistics.

The economic environment has changed significantly since the autumn of 2008. Hence, the necessity for new viewpoints on SCM has become even more obvious. The former paradigm of total and unlimited customer satisfaction has naturally failed because of the limited resources for this satisfaction. Therefore, the organizations are moving toward a triangle of goals: economic performance, stability and sustainability with the help of extensive adaptability both in processes and in management organization (see Fig. 2).

Based on the state-of-the-art analysis, let us present the sustainable SCM framework (see Fig. 3).

In the sustainable SCM framework, we do not set off the strategies described in Sec. 2 with each other, but consider them as an integrated framework. The SCM serves as a basis for integration (organizational: suppliers and customer; functional: collaborative business processes; managerial: strategic, tactical, and operative decision-making levels), cooperation, and coordination. The strategies of

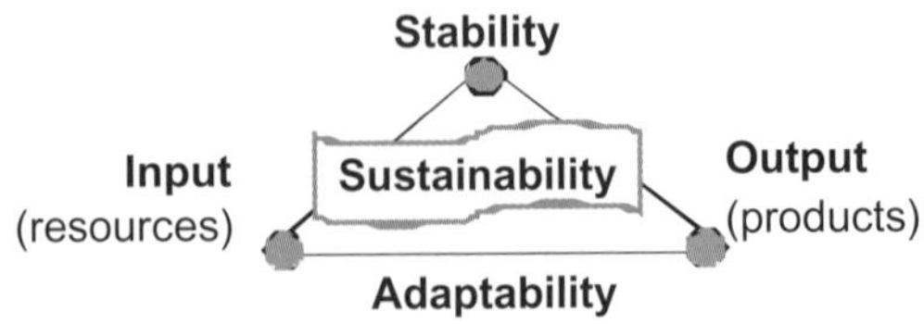

Fig. 2. Drivers of sustainable supply networks.

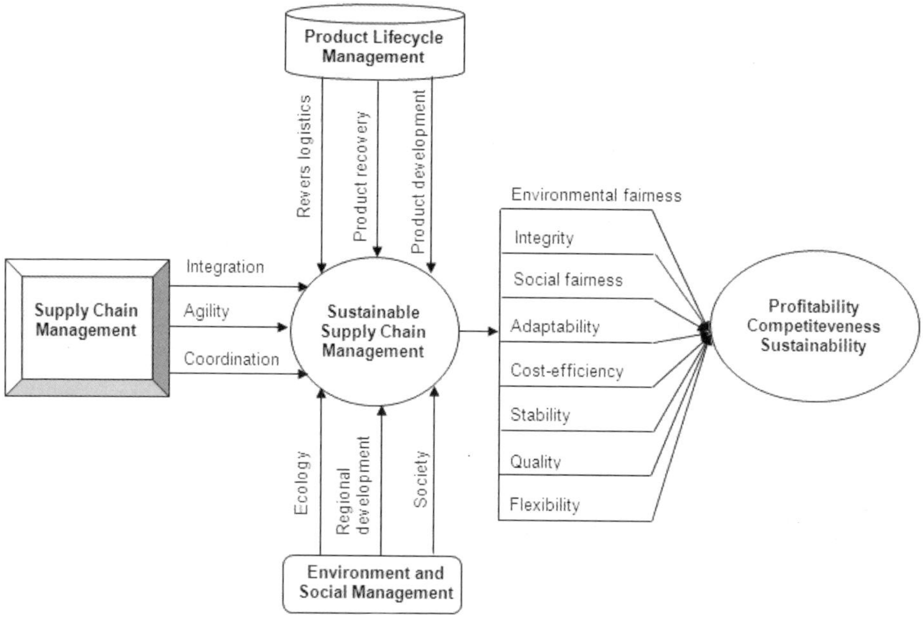

Fig. 3. Qualitative framework of sustainable supply networks.

agility enrich the SCM by means of a general information space with the help of web services and higher flexibility/responsiveness through concentration on core competencies and building virtual alliances/environments. PLM integrates the consideration of the product development, utilization, product end-of-life, and recovery processes. The environment and social management brings into consideration the environments and society issues, which may affect the SCs and which may be affected by SCs. Encapsulation of advantages of the SCM, agility, and sustainability enables the sustainable SCM.

In the sustainable SCM, all three value chain drivers — product and their life cycles, customers and their orders, and suppliers/outsourcers — are enhanced by the combining elements from the SCM, agility, and sustainability. Moreover, these drivers are interlinked within a unified information space. The proposed framework can be employed as an operational competitive strategy to increase organizations' wealth and sustainability by producing demand-corresponding products in a cost-effective manner through an increase of SC responsiveness, reaction speed to market changes, robustness, adaptation capabilities, effectiveness and sustainability of production and pre-/post-production PLM stages.

4. Quantitative Adaptive Framework of Sustainable Supply Networks

While elaborating the mathematical framework, we put particular attention to the comprehensive consideration of SC output performance, resource consumption, and stability. We consider multi-structural SC dynamics along the entire value chain life cycle. As SCM issues are fundamentally interdisciplinary in nature, we apply an interdisciplinary modeling approach that combines elements drawn from operations research (OR), control theory (CT), and distributed artificial intelligence.

4.1. *Particular features of supply network modeling*

With the development of SCM, new problems and integrated problems from operations and logistics management such as distribution–production or production–inventory problems have arisen. In addition, in recent years, the works on SCM have been broadened to cover the SC dynamics. In these setting, SCP and SCE is to consider as an *integrated model on the basis of adaptation principles*. To achieve this integration, static models of SC design (SCD) should be brought in correspondence to the dynamic models of SCP and SCE with establishing feedback loops between the different levels.

Within SCs, there are lots of problems that have been conventionally treated in isolation from each other but which are indeed tightly *interlinked*. In our opinion, these explicit interconnections may potentially provide new quality of decision-making support for SCM. For example, planning and scheduling are an integrated management function. SCs consist of different structures. The planning and execution of SCs are based on the same decision-making procedures that do, however, differ in their decision-making speed. Hence, the planning and execution stages of SCM can be considered within an adaptation framework.

It can be concluded that integration as the reflection of complex interrelations in SCs is the central idea of SCM. Therefore, integration should also become a kernel idea in decision support models. SCM problems are tightly interlinked with each other with regard to different SCM levels, different SC structures, and in SC dynamics and have multi-dimensional characteristics that require the application of different integrated frameworks of decision-making support (Ivanov, 2009; Beamon, 1998; Simchi-Levi, 2004).

Even these links should be considered at the forefront. The further optimization of known referenced problems can hardly provide a significant increase in SC performance and sustainability, whereas the investigation into the links between different problems may lead to a new breakthrough in SCM research. This approach corresponds with the nature of SCM — optimizing links between local problems, structures, and decision makers in SC with regard to total SC performance.

In following these assumptions, we will consider dynamics in SCs at the structural–operational level with concrete processes and parameters rather than at the strategic level, which usually suffers from a lack of links to concrete processes and a high abstraction degree. We will consider SCs as multi-structural systems and propose an SC structure–operations dynamics approach to establish adaptive feedback loops at the strategic, tactical, and operative decision-making levels.

4.2. *Adaptive modeling framework for SCP and SCE*

Let us consider the SC strategy, design, tactics, and operations as a whole system in detail (see Fig. 4).

Based on demand and supply forecasts, SC plans are generated within the designed structures. If concrete orders penetrate an SC, SC operations plans are generated according to the orders' parameters (price, delivery place, batch size, etc.). While running an SC, different disturbances can affect the SC and cause deviations and disruptions.

SC monitoring is meant for the maintenance of output parameter values of an SC in accordance with the required ones of input signals from the SC plans, design, and strategy. The results of the SC monitoring are reflected in the SC performance block.

The SC adaptation serves for implicating monitoring results while the SC is running and for corrective control actions and to SC operative execution, plans, design, and strategy. The adaptation is meant not only for processes but also for SC models, which should cohere with the current execution environment.

In some cases, it is possible to integrate different problems within one of the modeling techniques, e.g., distribution–production planning with mathematical programming or scheduling and execution control with optimal CT. If this is impossible, integrating elements of different techniques should be considered.

In many cases, cross-linked SCP and SC operations control problems require combined application of various modeling techniques (optimization, statistics, heuristics, and simulation). Isolated application of only one solution method leads to a narrowing in problem formulation, overdue constraints, and sometimes unrealistic

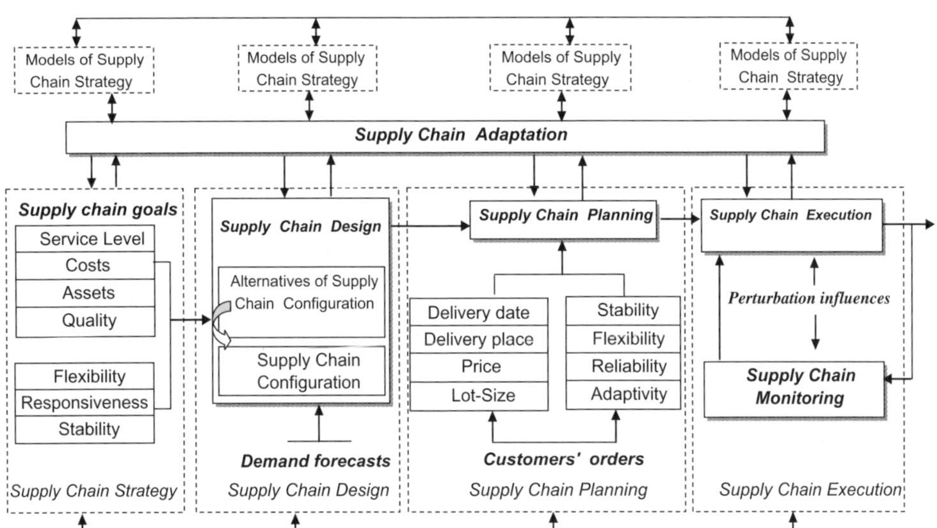

Fig. 4. Conceptual model of linking SC strategy, design, tactics, and operations (Ivanov and Sokolov, 2010).

or impracticable goals. At different stages of the SC life cycle, a particular problem can be solved by means of different modeling techniques due to changeability of data nature, structure and values, as well as requirements for output representation.

Selection of a solution method depends on data fullness, problem scale, one or multiple criteria, requirements on output representation, and inter-connection of a problem with other problems. Different approaches from the OR, CT, systems analysis, and agent-based modeling have a certain application area and a certain solution procedure. In task-oriented integrating their elements, it becomes possible to state and to solve a more complex and near real-world problem compared with their isolated applications.

Even in an integrated model, a leading method is needed. The selection of this method depends on the nature of the problems under integration and on the mathematical and computing implementability. Regarding the problem of integrating planning and execution control decisions on the adaptive principles, it is quite natural to approach the execution control stage from the positions of CT, especially adaptive and optimal control. In this case, to ensure the modeling consistency, the SCD, SCP and SC scheduling should also be based on CT. However, optimal control does not consider sequencing and resource allocation. This would hinder the planning and scheduling models. However, by combining dynamic optimal control model with elements of OR, the scheduling becomes possible (Ivanov and Sokolov, 2010).

In other cases, such a combination could be impossible (e.g., integrating SCD and SCP decisions). Here, using different methods for different parts of the problems and integrating their input–output mathematically are necessary. The modeling consistency in this case can be implemented by functoral integration of static and dynamic models (Okhtilev *et al.*, 2006; Sokolov and Yusupov, 2004).

4.3. *Adaptive framework for SCD*

Conventionally, the investigations into SCP are performed on the structure of physical distribution, manufacturing, and procurement. However, SCs can be considered not only from this organizational point of view, but also from a process point of view (Beamon, 1998). It should be emphasized that SCs consist of different structures: business processes and technological, organizational, technical, topological, informational, and financial structures. All of these structures are interrelated and change in their dynamics (Ivanov *et al.*, 2010). The issue of how to avoid structural incoherency and inconsistency by designing SCs is very important, first for (SCD) itself and second for designing robust SCs and re-designing SCs for new products (Graves and Willems, 2005), new OPP, and a variety of disruptive factors.

Especially in adaptive SCs with high dynamics, the issue of how to achieve structural comprehensiveness, responsiveness, and flexibility as well as to avoid structural incoherency and in-consistency by SCP and SC operations is very important. The adaptation of one structure causes changes in the other related structures. To ensure a high responsiveness level, the SC plans must be formed extremely quickly, but must also be robust. That is why it becomes very important to plan and run SC plans in relation to all the structures. This can be realized if (6.1) different SC structures are considered simultaneously and (6.2) the execution dynamics in all the structures can be reflected to establish bilateral feedback between SC plans and operations.

Let $G = \{G_\chi, \chi \in NS\}$ be the set of structures that are being formed within the SC (Ivanov *et al.*, 2010). To interconnect the structures, let us consider the following DAMG:

$$G^t_\chi = \langle X^t_\chi, F^t_\chi, Z^t_\chi \rangle, \tag{6.1}$$

where the subscript χ characterizes the SCD structure type, $\chi \in NS = \{1, 2, 3, 4, 5, 6\}$, the time point t belongs to a given set T; $X^t_\chi = \{x^t_{\langle\chi,l\rangle}, l \in L_\chi\}$ is a set of elements of the structure G^t_χ (the set of DAMG vertices) at the time point t; $F^t_\chi = \{f^t_{\langle\chi,l,l'\rangle}, l, l' \in L_\chi\}$ is a set of arcs of the DAMG G^t_χ and represents relations between the DAMG elements at time t; $Z^t_\chi = \{z^t_{\langle\chi,l,l'\rangle}, l, l' \in L_\chi\}$ is a set of parameters that characterize relations numerically.

The graphs of different types are interdependent; thus, for each operation the following maps should be constructed:

$$MM^t_{\langle\chi,\chi'\rangle} : F^t_\chi \to F^t_{\chi'}. \tag{6.2}$$

Composition of the maps can also be used at time t:

$$MM^t_{\langle\chi,\chi'\rangle} = MM^t_{\langle\chi,\chi_1\rangle} \circ MM^t_{\langle\chi_1,\chi_2\rangle} \circ \cdots \circ MM^t_{\langle\chi'',\chi'\rangle}. \tag{6.3}$$

A multi-structural state can be defined as the following inclusion:

$$S_\delta \subseteq X^t_1 \times X^t_2 \times X^t_3 \times X^t_4 \times X^t_5 \times X^t_6, \quad \delta = 1, \ldots, K_\sigma. \tag{6.4}$$

Now we obtain the set of the SC multi-structural macro-states (MSMS) in dynamics:

$$S = \{S_\delta\} = \{S_1, \ldots, S_{K_\sigma}\}. \tag{6.5}$$

Allowable transitions from one multi-structural state to another one can be expressed by means of the maps as follows:

$$\Pi^t_{\langle\delta,\delta'\rangle} : S_\delta \to S_{\delta'}. \tag{6.6}$$

Here we assume that each multi-structural state at time $t \in T$ is defined by a composition (Eq. 6.3).

Now, the problem of SC with structure dynamics considerations can be regarded as a selection of MSMS $S^*_\delta \in \{S_1, S_2, \ldots, S_{K_\sigma}\}$ and transition sequence (composition) $\Pi^{t_1}_{\langle\delta_1,\delta_2\rangle} \circ \Pi^{t_2}_{\langle\delta_2,\delta_3\rangle} \circ \cdots \circ \Pi^{T_f}_{\langle\delta',\delta\rangle}$ $(t_1 < t_2 < \cdots < T_f)$, under some criteria of effectiveness, e.g. SL and service costs.

Dynamics of the SCE is presented as a DAMG to relate the above sets and structures. The DAMG is characterized by MSMS. The DAMG and the MSMS have been developed to meet the requirements on multi-structural design and to link planning and execution models, taking into account the structure dynamics.

4.4. *Framework for adaptive SCD and SCP*

SCD is the strategic configuration of supply network structures. SCP is composed of tactical problems of SCP regarding demand forecasts or customers' orders and the most rational use of SC resources to meet the market requirements in the cost-efficient manner (Kreipl and Pinedo, 2004). SCP depends on the SCD. Tactical goals are subject to such SCD goals as SC costs and SL. In these settings, SCD and planning of orders in SCs (SCP) is to consider as an *integrated model on the basis of adaptation principles.*

Of course, it would be mathematically possible to integrate the above-stated models of SC multi-structural design with planning and scheduling models of SCP. However, this would be an enormous mathematical construction with a low practicability. We propose to introduce reality of SCM into the modeling procedure.

Although SCD structures are subject to dynamics, they do not change permanently. There are certain time intervals when these structures remain unchanged. Hence, for these intervals, we can actually apply conventional MP models. To achieve this integration, static models of SCD should be brought in correspondence to the dynamic models of SCP with establishing feedback loops between the SCD and SCP. It becomes to be possible if traditional OR techniques can be extended by dynamics of optimal CT.

In SCD and SCP models, different goals, variables, and constraints can be expressed whether in static or in dynamic form. In applying only one solution method, e.g., mathematical programming or optimal control, significant difficulties in representing both static and dynamic aspects are frequently encountered. Therefore, it can be sensible distribute static and dynamic variables and constraints of SCD and SCP between different models where the corresponding static or dynamic elements can be expressed in the best way.

The basics of the SC multi-disciplinary treatment were developed in the decentralized integrated modeling approach methodology (Ivanov, 2009). The mathematical integration and combined application of various models can be implemented by means of multiple model complexes, which are based on the application of functors. Functors establish relationships between different models. The multiple model complexes allow problem examination and solution in different classes of models, and result representation in the desired class of models. Let us discuss an example of how to apply the multiple model complexes. The problem of the SC analysis and synthesis is mostly formalized using either graph (network) models or models of *linear programming (LP) and integer programming (IP).*

As a rule, the problem of SCE is formalized with the help of *dynamic models*. However, the problems of coordination and consistency of the results remain open. To obtain a constructive solution to these problems, we propose to use a functorial transition from the category of digraphs to the category of dynamic models.

The SCD problem is a multi-criteria LP problem of high dimensionality. Analysis of problem showed that in spite of its high dimensionality it could be solved via special decomposition procedures and the method of successive improvement of plans for LP problems with two-sided constraints. With regard to the SC settings, a special consideration must be given to the fact that real SC operates heterogeneous products of different importance and priority of customer orders. We propose to concretize the obtained plan of SC operation via the differential dynamic model (Ivanov *et al.*, 2010).

Analysis of the dynamic model shows that the coordination of the planning results of the static and the dynamic model can be carried out through the following variables: the amount of products transmitted from one SC node to the next node at a certain time interval and product amounts relating to the concrete origin node and respectively to be stored (according to the plan of SC operation), to be processed, and to be delayed (as caused by limited capacity of SC).

4.5. *Adaptive framework for SC scheduling and SCE*

At the planning stage, SC schedule is generated. In running a scheduling model many times with different input data regarding different time buffers, safety stocks, additional facilities and reservation of capacities, different schedules with different output performance can be generated. A typical performance indicator is the SL. A comprehensive consideration of protection against uncertainty in schedules, regarding both reliability to protect against disruptions and flexibility to handle in the case of changes/disruptions, potentially leads to an increase in SC output SL. However, SC costs increase at the same time. Hence, each of the generated schedules is characterized by different values of SL and costs. This fact makes it necessarily to consider SL together with the SC costs.

In developing a mathematical model for SC scheduling in terms of dynamic analysis of SCE, it is important to ensure mathematical consistency of dynamics to be considered from both a narrow perspective (operations dynamics within a schedule in accordance to a given plan) and a wide perspective (execution dynamics and adaptation of both schedules and plans). Hence, the problems and models of planning, scheduling, and adaptation should be integrated on unified mathematical principles. This can be achieved in applying CT models that consider dynamics at both the planning and execution. In doing so, it is possible to describe both the planning and execution stages in terms of control. However, despite optimal control models make it possible to reflect the optimization of a performance indicator the value of which is accumulated over time, they do not consider sequencing and resource allocation. Hence, scheduling could not be performed. To overcome this shortcoming, the study by Ivanov and Sokolov (2010) developed an SC dynamic scheduling model that is based on a combined application of optimal control and mathematical programming approaches. At each instant of time while calculating solutions in the optimal control dynamic model with the help of the Pontryagin maximum principle, the LP problems to allocate jobs to resources and IP problems for (re)distributing material and time resources are solved with conventional capacitated LP/IP algorithms. The modeling procedure is based on an essential reduction of *problem dimensionality* that is under solution at each instant of time due to connectivity decreases. At each point of time, only variables and constraints that are relevant at this time section are considered for scheduling and transmitted as input data to an LP/IP algorithm. The whole dimensionality and dynamics is kept within the optimal control model.

By applying optimal control models at the planning stage, the coordination and coherency of planning and execution models do not require special efforts and occur in a natural way within the united mathematical axiomatic of CT. Within the scheduling model, we can explicitly change resource consumption for the protection of SC against uncertainty and analyze the impact of these changes at SC output SL and costs.

4.6. *Adaptive framework for SCD, SCP, and SCE*

The main purpose of the adaptation framework is to ensure a dynamic integrated SCD, SCP, and SCE model with a parameter tuning with regard to changes in the execution environment and planned values of performance indicators. In the proposed framework, the process adaptation is connected to the model adaptation. In Fig. 5, the adaptive framework is presented.

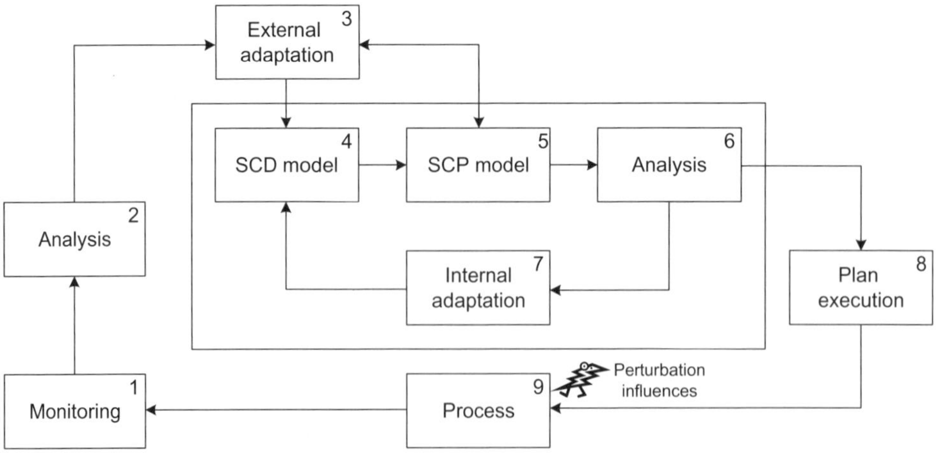

Fig. 5. Integrated adaptive framework of SCD and SCP (Skurihin *et al.*, 1989).

The proposed framework includes the following main phases:

- parametric and structural adaptation of SC processes and models (blocks 1, 2, 3);
- integrated SCD and SCP (blocks 4, 5);
- simulation and analysis of SCE according to different configuration plans and variants of control decisions in real situations (block 6);
- structural and parametric adaptation of the configurations and plans, control inputs, models, algorithms, based on results of simulation experiments (block 7); and
- plan realization in real time, correction of the control action (block 8, 9).

The following parameters can be evaluated through the external adapter:

- the values of SC performance indicators (e.g., minimal admissible SL of 95% can be amended to 93%);
- execution indicators of SC processes (e.g., cycle time); and
- probabilistic characteristics and values of real and observed random processes.

The *second group* of parameters being evaluated through the internal adapter includes such characteristics as

- a redundancy rate for functional, informational, and temporal reserving;
- priority of SC performance indicators; and
- parameters that define the variants of adaptation in the simulation models.

A hierarchy of adjustment actions is brought into correspondence with different deviations in the SCE. A state correction takes place with a time horizon of hours, days, weeks, or months.

5. Adaptive Framework for Interrelating Environment and Economic Sustainability

An important issue in supply network sustainability is the danger of production resource deficit that requires the development of energy-efficient supply networks. First, the existing SCs are frequently designed with significant redundancies in resource consumption that are inadequate the SC output performance. These redundancies should be eliminated and SC resources and output performance should be brought into a mutual correspondence. Second, the technological resource losses, e.g., leakage, should also be eliminated.

In this study, we propose to consider the SC property to achieve a balance between the economic output performance and resource consumption under the colliding SC processes in the real perturbed execution environment with regard to the variety of execution and goal criteria as the *SC global stability.*

Global stability is the ability of an SC to maintain such a resource-efficient execution trajectory, which allows the achievement of the planned output

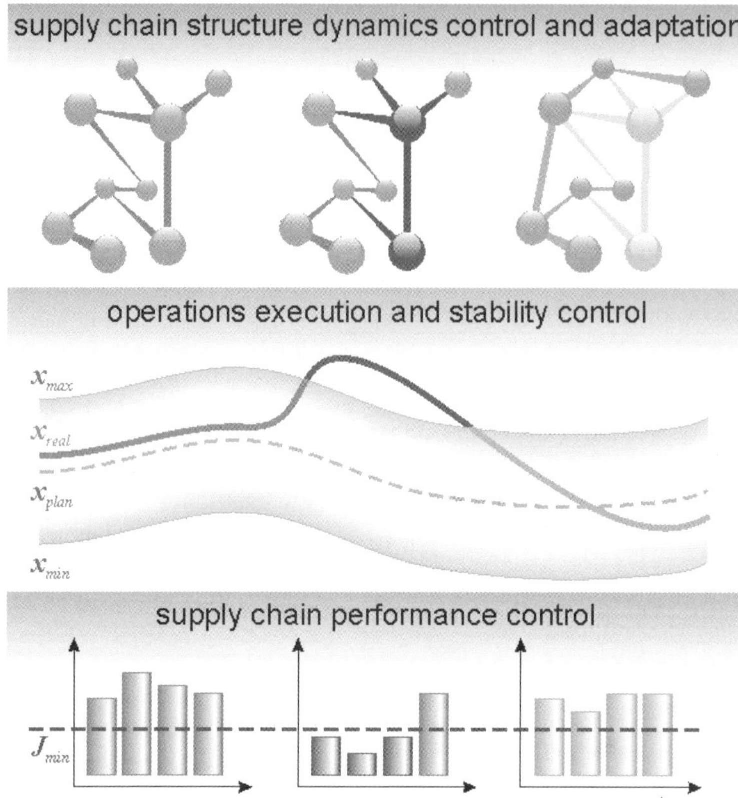

Fig. 6. Perturbation and adaptive recovery of SCE and SC output performance.

performance despite of perturbations in a real execution environment with the help of corresponding control adjustment actions.

This approach commits to principles that are laid down in the *global stability* by Lyapunov, which allows uncertainty in dynamics, the system's parameters, and control actions (Casti, 1979). In this approach, stability is considered as a *dynamic property* that emerges through feedback loops. Hence, stability can be considered as a system behavior property that should be maintained despite perturbation influences by means of corresponding control actions in the feedback loops. As such, stability becomes interconnected with adaptivity within the so-called stabilizing feedback control (see Fig. 6).

In Fig. 6, the influence of perturbation impacts on SCE and output performance is analyzed. SCE trajectory $x(t)$ is constructed at the planning stage and strengthened by the elements of reliability and flexibility. This results in an interval $[x_{\min}; x_{\max}]$ remaining within of which SC achieves its output performance goals J. In the case of global stability with controlled adaptability, perturbation impact causes the decrease in SC operations execution but the output performance is achieved by means of the adaptive recovery as the SC structure is controlled. The quantitative approach and formal tools for global stability analysis are based on the modern CT in combination with OR techniques and use the method of attainable sets (AS).

To consider simultaneously SC performance and stability, a method is required. Here, we propose to use AS. AS is a very useful tool in the study of various problems of dynamic systems optimization. Numerous papers have been devoted to the study of various properties of the AS of the control systems (Guseinov, 2009). In this study, we propose to apply AS to the SCM domain. AS allows the explicit interconnecting of problems of estimation of SC goal abilities and stability. This opens up constructive ways to solve optimal control problems simultaneously with SC stability analysis.

The economic sense of the AS consists of the following: the AS characterizes a set of SC plans and the values of the SC's potential goals corresponding to them. In other words, an AS contains all the possible output performance states in which an SC can be led into regarding all the considered execution scenarios with different resource consumptions. Let us explain the meaning of AS with an example. AS defines the area where permissible solutions exist. For example, we are given the material resources and time to achieve certain goals (e.g., the SC SL) with a forecasted demand. The AS analysis may provide the area in which permissible solutions (SC plans) are included. On the other hand, this may show that, with the given resources and within the given time horizon, it is impossible to achieve the required SL; hence, we should introduce additional resources or expand the SC cycle.

Based on AS, it becomes possible to consider SC stability interconnected with the achievability (or attainability) of SC goals. These goals can be achieved only if there are enough control actions to guide the system to the achievement of its goals in a real execution environment with negative perturbation influences. Indeed, the achievement of the balance between the available control actions subject to a certain scale of perturbation influences makes the SC controllable and leads to its stabilization.

The model is based on the dynamic interpretation of the SC's functioning process (Ivanov and Sokolov, 2010). Based on the quantification of stability with the help of AS, it becomes possible to compare different SC structures with different scope and scale of the resource consumption regarding both the key performance indicators and stability. This analysis can be performed with regard to different execution scenarios and different areas of control impacts in order to achieve the planned output economic performance in a real perturbed execution environment. Hence, the results of the stability analysis can be brought into correspondence with different SC performance and resource consumption scenarios.

6. Conclusions and Outlook

As sustainable systems are designed to support decisions on planning, analysis, and improving organization's actions and operations to meet competitiveness and sustainability goals, it becomes necessary to interconnect the stages of supply network planning and execution. Indeed, as supply networks are not static but highly dynamic and permanently evolving systems, sustainability of these networks should also be considered as a dynamic category of SC adaptive evolution.

Adaptation is composed of the continuous changes in SC functioning and the abilities to function in unpredicted conditions by a goal-oriented adjustment of the process parameters and/or structures to meet the economic and sustainability goals.

From the qualitative point of view, the developed adaptive framework for sustainable supply networks that is composed of the integrated elements of environment management, operations economic development and information technologies, PLM, and social development.

The quantitative adaptive framework for sustainable supply networks aims to interconnect the stages of supply network planning and execution based on a combined application of OR and CT.

Finally, we addressed the issue of interrelating environment and economic sustainability and propose to consider the SC property to achieve a balance between the economic output performance and resource consumption as the SC global stability.

On the whole, the proposed framework can be employed as a competitive strategy to increase organizations' wealth and sustainability by producing demand-corresponding products in a cost-effective manner through an increase of SC agility, resilience, stability, and adaptation capabilities.

In future, the investigations into the triangle "sustainability–economical optimality–complexity" may potentially provide new insights into SCM and engineering. Furthermore, with increasing transportation, the minimizing of negative impacts on ecology may become one of the primary objectives in SCM and logistics. Actually, this is already stated by global automotive companies. Hence, we are moving toward a triangle of goals: profitability, stability, and ecological goals. It may also be beneficiate to incorporate the discussions on sustainable economic development within the framework of system dynamics and CT. Actually, the theory of economic growth has been already using a lot of CT to attain a remarkable scientific status in the domain of economics and social sciences.

In future, the adaptive SC should evolve into self-organizing and self-learning SCs. In self-organizing SCs, both the system and its goals would evolve, unlike in adaptive SCs where the system's design and goals are predetermined. The system's borders would become fuzzy, the system can broaden by "acquiring" a space from the environment, or the system can narrow in the reverse way. However, to approach the self-organizing SCs, time is needed to rethink many crucial management paradigms.

Acknowledgments

The research described in this paper is partially supported by grants from the Russian Foundation for Basic Research (grants 07-07-00169, 08-08-00403,

09-07-00066, 10-07-00311), the Saint Petersburg Scientific Center RAS, and the Department of Nanotechnologies and Information Technologies of the Institute of Systems Analysis RAS (2.3).

References

Boese, J.W. and Daduna, J.R., 2008. Preventing capacity bottlenecks in seaports and the hinterland. *Internationales Verkehrswesen* **60**, 413–414.

Bowen, F.E., Cousins, P.D., Lamming, R.C., and Faruk, A.C., 2001. The role of supply management capabilities in green supply. *Prod Oper Manag* **12**(4), 330–353.

Beamon, B.M., 1998. Supply chain design and analysis: Models and methods. *Int J Prod Econ* **19**(3), 275–292.

Carter, C.R., 2004. Purchasing and social responsibility. *J Supply Chain Manag* **40**(4), 4–16.

Carter, C.R. and Rogers, D.S., 2008. A framework of sustainable supply chain management: Moving toward new theory. *Int J Phys Distrib Mater Manag* **38**(5), 360–387.

Casti, J.L., 1979. *Connectivity, Complexity and Catastrophe in Large-Scale Systems.* Wiley-Interscience, New York London.

Chopra, S. and Meindl, P., 2007. *Supply Chain Management. Strategy, Planning and Operation.* Pearson Prentice Hall, New Jersey.

Christopher, M. and Towill, D.R., 2001. An integrated model for the design of agile supply chains. *Int J Phys Distrib Oper Manag* **31**, 235–244.

Christopher, M., 2005. *Logistics and Supply Chain Management: Creating Value-Adding Networks.* Financial Times, Prentice Hall, Dorchester.

Corbett, C.J. and Kleindorfer, P.R., 2003. Environmental management and operations management: Introduction to the third special issue. *Prod Oper Manag* **Fall**, 1.

Friscia, T., O'Marah, K. and Souza, J., 2004. The AMR research supply chain top 25 and the new trillion-dollar opportunity. *AMR Research Report.*

Gladwin, T.N., Kennelly, J.J. and Krause, T., 1995. Shifting paradigms for sustainable development: Implications for management theory and research. *Acad Manag Rev* **20**(4), 874–907.

Guide, V.D.R. Jr. and Van Wassenhove, L.N., 2009. The evolution of closed-loop supply chain research. *Oper Res* **57**(1), 10–18.

Gunasekaran, A. and Ngai, E.W.T., 2009. Modeling and analysis of build-to-order supply chains. *Eur J Oper Res* **195**(2), 319–334.

Graves, S.C. and Willems, S.P., 2005. Optimizing the supply chain configuration for new products. *Management Science* **51**(8), 1165–1180.

Guseinov, K.G., 2009. Approximation of the attainable sets of the nonlinear control systems with integral constraint on controls. *Nonlinear Anal* **71**, 622–645.

Harrison, T.P., 2005. Principles for the strategic design of supply chains. In: Harrison, T.P., Lee, H.L., Neale, J.J. (Eds), *The Practice of SCM.* Kluwer Academic Publishers, Boston, pp. 3–12.

Hendricks, K.B. and Singhal, V.R., 2003. The effect of supply chain glitches on shareholder wealth. *J Oper Manag* **21**, 501–522.

Hoegl, M. and Wagner, S.M., 2005. Buyer-supplier collaboration in product development projects. *J Oper Manag* **31**(4), 530–548.

Ivanov, D., 2009. DIMA — A research methodology for comprehensive multi-disciplinary modelling of production and logistics networks. *Int J Prod Res* **47**(5), 1133–1155.

Ivanov, D. and Sokolov, B., 2010. *Adaptive Supply Chain Management.* Springer, London.

Ivanov, D., Sokolov, B. and Kaeschel, J., 2010. A multi-structural framework for adaptive supply chain planning and operations with structure dynamics considerations. *Eur J Oper Res* **200**(2), 409–420.

Ivanov, D., Sokolov, B. and Kaeschel, J., 2010. Integrated adaptive design and planning of supply networks. In: Dangelmaier, W., *et al.* (Eds.), *Advanced Manufacturing and Sustainable Logistics (LNBIP 46)*. Springer, Heidelberg, Berlin.

Jaraman, V., Klassen, R. and Linton, J.D., 2007. Supply chain management in sustainable environment. *Prod Oper Manag* **25**(6), 1071–1074.

Khan, O. and Burnes, B., 2007. Risk and supply chain management: Creating a research agenda. *Int J Logist Manag* **18**(2), 197–216.

Kleindorfer, P.R., and Saad, G.H., 2005. Managing disruption risks in supply chains. *Prod Oper Manag* **14**(1), 53–68.

Kogg, B., 2005. Greening a cotton-textile supply chain: A case study of the transition towards organic production without a powerful focal company. *Greener Manager International* **43**, 53–64.

Krause, D.R., Vachon, S. and Klassen, R.D., 2009. Special topic forum on sustainable supply chain management: Introduction and reflections on the role of purchasing management. *J Supply Chain Manag* **45**(4), 18–25.

Kreipl, S. and Pinedo, M., 2004. Planning and scheduling in supply chains: An overview of issues in practice. *Prod Oper Manag* **13**(1), 77–92.

Lee, H.L., Padmanabhan, V. and Whang, S., 1997. Information distortion in a supply chain: The bullwhip effect. *Manag Sci* **43**(4), 546–558.

Lund, R.T., 1996. *The Remanufacturing Industry_Hidden Giant*. Boston University, Boston.

Matos, S. and Hall, J., 2007. Integrating sustainable development in supply chain: The case of life cycle assessment in oil and gas and agricultural biotechnology. *J Oper Manag* **25**(6), 1083–1102.

Min, H. and Galle, W.P., 2001. Green purchasing practices of US firms. *Int J Oper Prod Manag* **21**(9), 1222–1238.

Mueller, E., Engelmann, L., Loeffler, T. and Strauch, J., 2009. *Energie-effizient Fabriken planen und betreiben*. Springer, Dordrecht.

Mulani, N.P. and Lee, H.L., 2002. New business models for supply chain excellence. In: Mulani, N. (Ed.), *Achieving Supply Chain Excellence Through Technology*, Vol. 4. San Francisco, Montgomery Research, Inc.

Okhtilev, M., Sokolov, B. and Yusupov, R., 2006. *Intelligent Technologies of Monitoring and Control*. Nauka, Moscow.

Pagell, M., Yang, C., Krumwiede, D.K. and Sheu, C., 2004. Does the competitive environment influence the efficiency of investments in environmental management? *J Supply Chain Manag* **40**(3), 30–39.

Pagell, M. and Wu, Z., 2009. Building a more complete theory of sustainable supply chain management using case studies of 10 exemplars. *J Supply Chain Manag* **45**(2), 37–56.

Rao, P. and Holt, D., 2005. Do green supply chains lead to competitiveness and economics performance? *Int J Oper Prod Manag* **20**(1), 97–123.

Russo, M.V. and Fouts, P.A., 1997. A resource-based perspective on corporate environmental performance and profitability. *Acad Manag Rev* **40**(3), 534–559.

Seurung, S., Sarkis, J., Muller, M. and Rao, P., 2008. Sustainability and supply chain management — An introduction to the special issue. *Journal of Cleaner Production* **16**(15), 1545–1551.

Sharif, A.M., Irani, Z. and Lloyd, D., 2007. Information technology and performance management for build-to-order supply chains. *Int J Oper Prod Manag* **27**(12), 1235–1253.

Sheffy, J. and Rice, Jr., 2005. A supply chain view of the resilient enterprise. *MIT Sloan Management Review* **47**(1), 45-47.

Shepherd, C. and Guenther, H., 2006. Measuring supply chain performance: Current research and future directions. *Int J Prod Perform Manag* **55**(3–4), 242–258.

Simchi-Levi, D., Wu, S.D. and Zuo-Yun, S., 2004. *Handbook of Quantitative Supply Chain Analysis.* Springer, New York.

Skurihin, V.I., Zabrodsky, V.A. and Kopeychenko, Y.V., 1989. *Adaptive Control Systems in Machinery Industry.* Mashinostroenie, Moscow.

Sokolov, B.V. and Yusupov, R.M., 2004. Conceptual foundations of quality estimation and analysis for models and multiple-model systems. *Int J Comput Syst Sci* **6**, 1–16.

Stock, J.T., Speh, H. and Shear, H., 2002. Many happy (product) returns. *Harvard Bus. Rev* **80**.

Swafford, P.M., Ghosh, S. and Murthy, N., 2008. Achieving supply chain agility through IT integration and flexibility. *Int J Prod Econ* **116**(2), 288–297.

Wannenwetsch, H., 2005. *Vernetztes Supply Chain Management.* Springer, Berlin.

Warburton, R.D.H., Disney, S.M., Towill, D.R. and Hodgson, J.P.E., 2004. Further insights into the stability of supply chains. *Int J Oper Prod Manag* **42**(3), 639–648.

Zhu, Q., Sarkis, J. and Lai, K.-H., 2008. Confirmation of a measurement model for green supply chain management practices implementation. *Int J Prod Econ* **111**, 261–273.

Chapter 7

SUSTAINABLE SUPPLY NETWORKS BY DESIGN

Samir K. Srivastava
Operations Management Group, Indian Institute of Management,
Prabandh Nagar, Lucknow 226013, India
samir@iiml.ac.in

Sustainable supply chain management (SSCM) is the buzzword today after the quality revolution of the 1980s and the supply chain revolution of the 1990s. It is gaining justifiable popularity among society, governments, and industry. In recent years, a rapid increase in corporate initiatives and academic publications on SSCM can be observed. Environmentally conscious actions are no longer seen as challenges and contrary to financial considerations. In fact, they can be the basis for successful companies and supply chains. As supply chains and businesses start focusing on "Planet, People, and Profits" simultaneously, sustainability is likely to emerge as the new metric for the supply chains. This will require monitoring internal operations of firms and those of their suppliers in a different perspective. Nearly everyone agrees that a sustainable supply chain that seizes value-creation opportunities offers significant competitive advantages for early adopters and process innovators. In light of the above, this chapter describes the emergence of SSCM and thereafter tries to suggest models and frameworks for sustainable supply network practices and design.

1. Introduction

The environment is no longer a not-in-my-backyard problem (Srivastava and Srivastava, 2003). Society, governments, and industry are increasingly confronted with the results of our throw-away society (Srivastava and Srivastava, 2003). Natural environment has become a challenging issue to business organizations in recent years as a result of global and local environmental problems. Deepening environmental concerns and perceptions of increased risk to health and safety of community residents from supply chain activities has led to a significant increase in interest in research at the intersection of environmental management and supply chains. Supply chain activities such as product development, sourcing, manufacturing, network design, and logistics are believed to be responsible for most of these problems. Our planet needs a green supply chain revolution to sustain industrial growth that has started affecting the climate. There is a need for fundamental change in the way we produce, consume, and exchange goods. A paradigm change is underway in which the focus is shifting to planet, people, and profits in that order. Most major companies worldwide are now focusing on the environmental burden of their supply chain processes by recognizing some of the contemporary environmental dimensions such as carbon emissions, energy, and natural resources consumption.

Sustainability is the capacity to endure. In ecology, the word describes how biological systems remain diverse and productive over time. For humans, it is the potential for long-term maintenance of well-being, which in turn depends on the well-being of the natural world and the responsible use of natural resources. It is the idea that civilization as a whole can act in a way that supports not only social equality but also protects the earth's ecosystems and natural resources for generations to come.[1] To lower the environmental damage of business, firms and their supply chains need to examine their "ecological footprint."[2] Managers need to integrate sustainability goals, practices, and cognitions into day-to-day supply chain management.

The words "green" and "sustainable" have been used interchangeably in literature and practice. Although, there is a subtle difference,[3] we will also use them interchangeably. Gupta (1996) defines greening as harmonizing corporate environmental performance with stockholders' expectations as well as constituting a significant new source of competitive advantage, such as lower costs and expanded market share. Becoming green is perceived not only good for the environment and a company's public image; it is also perceived as means to improve the company's business operations and reducing its supply chain costs. Companies can adjust multiple levers, such as energy conservation and recycling, to generate positive bottom-line impacts. In the emerging value-creation model, implementing green initiatives along a company's supply chain can raise productivity, enhance customer and supplier relations, support innovation and enable growth.

There has been considerable focus on Green Supply Chain Management (GrSCM) in literature and practice. It is the integration of environmental management principles with the decision-making process for the conversion of resources into usable products. It mainly deals with the global benefit of minimization of environmental impact of production network through co-operation and information sharing among the parties on the supply network. A Green Supply Chain can be thought of as a supply chain that has integrated environmental thinking into core operations from material sourcing through product design, manufacturing, distribution, delivery, and end-of-life recycling. GrSCM involves a paradigm shift, going from end-of-pipe control to meet environmental regulations to the situation of not only minimizing ecological damage, but also increasing overall economic

[1] http://www.worldcentric.org/sustainability/ Accessed on March 20, 2010.

[2] Ecological footprint converts the products and services we use into global hectares and measures the pressure we exert on nature to satisfy our lifestyle demands.

[3] Green has become a common term to portray or convey the environmentally-friendly image of products, processes, systems, and technologies as well as the way business is conducted. Sustainable is mainly related to maintaining ecological balance.

profit. Greening of the supply chain means simply passing on good environmental management system practices to suppliers by the end producer or primary buyer. GrSCM involves a host of issues that deal with "cradle-to-grave" management of resources and goods. Literature (Kopicki *et al.*, 1993) mentions three approaches in GrSCM: the reactive, proactive, and value-seeking. The most far-reaching of them is the value-seeking approach.

A management strategy that follows economic, environmental, and social aims throughout the supply chain may be termed as "sustainable supply chain management (SSCM) (Wittstruck *et al.*, 2009). It focuses toward the broader adoption and development of sustainability throughout the supply chain considering the product from initial processing of raw materials to delivery to the customer as well as closed loops. Further, SSCM must also integrate issues and flows that extend beyond the core of supply chain management: product design, manufacturing by-products, by-products produced during product use, product life extension, product end-of-life, and recovery processes at end-of-life.

Much of the existing research links previously identified best practices to environmental outcomes. Even many of the "new" practices such as design for the environment (DfE) and/or ISO 14000 and other certifications identified in the literature have strong roots in previously identified practices such as design for manufacturing and ISO 9000 certification. Labuschagne and Brent (2004) propose a framework of appropriate criteria to assess the sustainability performances of operational initiatives in industry. Although the current developments in SSCM are encouraging, most of the initiatives and innovations have mainly come in Europe and North America. Other geographical areas and economies also need to become active. This chapter tries to bridge some of this gap.

At the organizational level, responsibility for sustainability cannot be given to a separate entity; it must be part of everyone's job, starting with top management. A tangible commitment to sustainability may be in the form of a written environmental policy (Ramus and Steger, 2000). One of the biggest challenges before the field of SSCM is extending the historical "common wisdom" about managing supply chains. Much research, management education, and practical application have focused on buffering the supply chain operations from external influences, including the natural environment, in order to improve efficiencies, reduce cost and better quality. When the natural environment was considered, it was typically recognized or modeled as an external constraint, requiring operations to work within prescribed limits. Once this basic assumption is relaxed, a fundamental question arises about how best to pursue research on sustainable issues across different stakeholders in a supply chain: should this be considered a separate research stream with its own strategic framework, or should green issues be integrated into existing operations management research frameworks and areas? While the complexity of green issues might favor the former approach, the greatest contributions can be achieved by pursuing opportunities within a more integrative framework.

2. Sustainable Supply Chains

Guide and Van Wassenhove (2009) highlight the relevance of green or sustainable supply chain. A sustainable supply chain is then one that performs well both on traditional measures of profit and loss and on an expanded conceptualization of performance that includes social and natural dimensions. It simultaneously considers and balances economic, environmental, and social goals from a microeconomic standpoint. Such a conceptualization of performance is generally referred to as the triple bottom line (Elkington, 1998). Figure 1 shows a visual representation of these three components.

2.1. *Relationship between different disciplines*

Sustainability is interdisciplinary and has deep roots in both the physical and social sciences as depicted in Fig. 2 (Linton *et al.*, 2007). Understanding the effect and interaction of different activities with the natural environment and its implication,

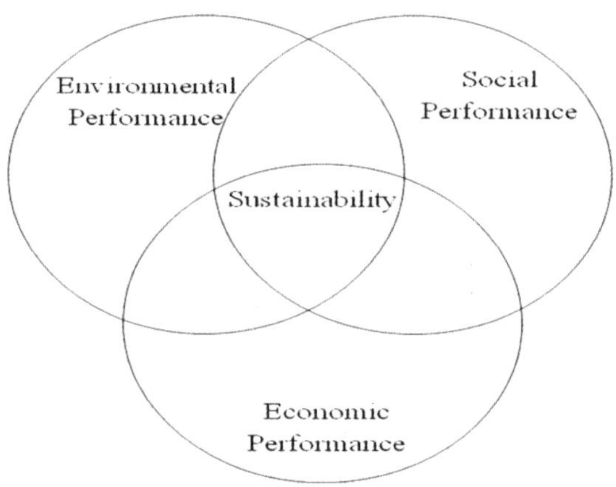

Fig. 1. The triple bottom line framework of sustainable supply chains.
Source: Elkington (1998).

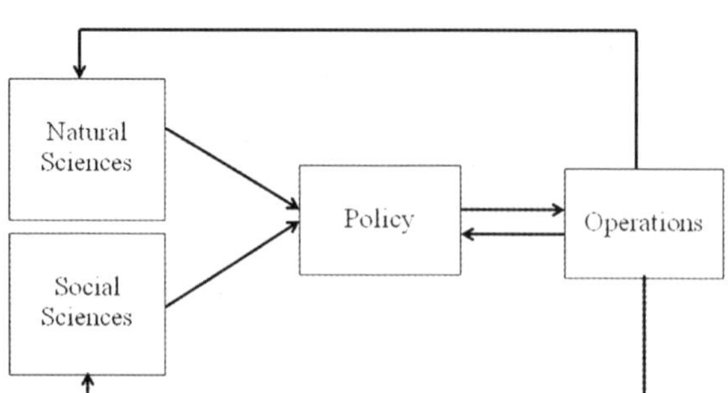

Fig. 2. Relationship between different disciplines in sustainable supply chains.
Source: Linton *et al.* (2007).

current and future quality of life is based on advances in a variety of different areas of the natural sciences. Moreover, the social sciences are critical to our interpretation and attitude toward sustainability, including cultural norms, individual and group behaviors, role of government and community, relationship with science, and relationship with the natural environment. Thus, anthropology, political science, psychology, and sociology interact with the natural sciences and are interpreted and managed through the development of policy.

2.2. *Sustainable supply chain activities and their classification*

There is a need of large improvements in terms of resource productivity — "doing more with less." The challenge is to reduce non-renewable material and energy usage in a fast increasing economic activity. Sustainable supply chains mainly focus their activities toward eco-efficiency, reducing material requirements of goods and services, reducing energy inputs to goods and services, reducing toxic dispersion, enhancing material recyclability, extending product durability, increasing service intensity of goods and services, and maximizing sustainable use of renewable resources. Proactive practices have been implemented through various Rs (Reduce, Re-use, Rework, Refurbish, Reclaim, Recycle, Remanufacture, Reverse logistics, etc.).

Nowadays, most companies follow the environment management principles in the form of standardized Environment Management Systems (EMS) such as British Standard for EMS BS7750, the EU eco-management and audit scheme, and the international standard ISO 14000. Most companies also follow EMAS, Global Sullivan Principle, ICC Charter, and OECD guidelines. Such policies and procedures represent efforts to improve environmental performance only within the organization's operational boundaries rather than being extended throughout the supply chain.

Srivastava (2007) provides a classification of SSCM activities. The same is shown in Fig. 3. Sustainable or green design may be looked into from the viewpoint of DfE taking life-cycle assessment of the product/process into account. The aim is to minimize environmental impacts of products during their entire life cycle. Sustainable or green operations involve all operational aspects related to green manufacturing and remanufacturing (reduce; recycle; production planning and scheduling; inventory management; remanufacturing: re-use, product, and material recovery). Reverse logistics and network design involve collection, inspection/sorting, pre-processing, and detailed network design. Waste management involves source reduction, pollution prevention, environmental remediation, and safe disposal. Although not shown explicitly in Fig. 4, activities related to green logistics, green purchasing as well as activities that aim at improving environmental performance and capabilities of suppliers and customers are also considered within the fold of SSCM activities. Green purchasing includes activities that aim to make sure that purchased items have desirable environmental attributes such as reusability, recyclability, and

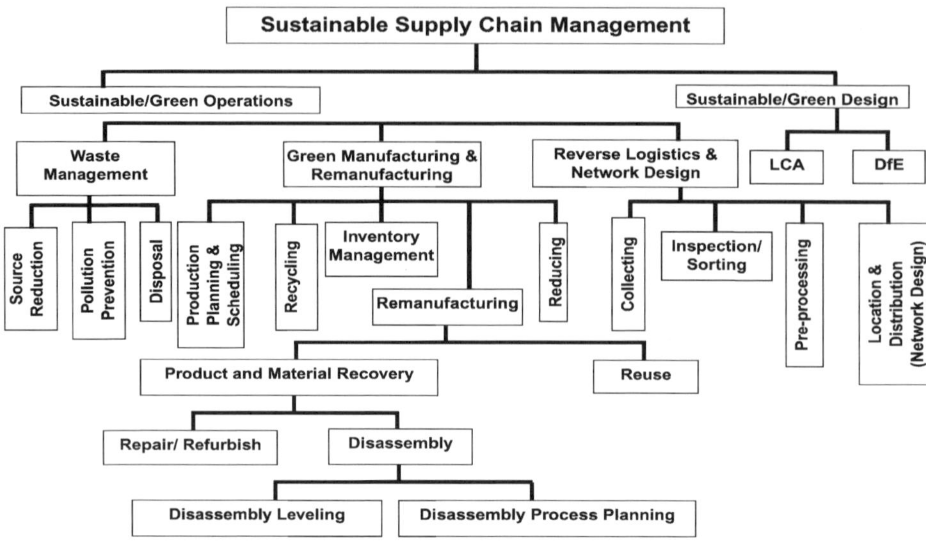

Fig. 3. Classification of SSCM activities (Srivastava., 2007).

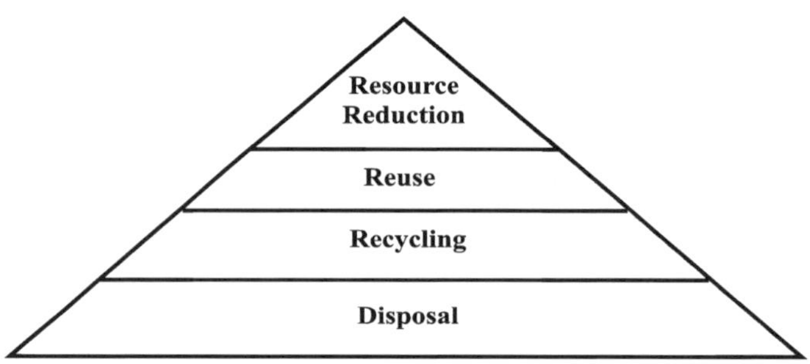

Fig. 4. Greening hierarchy (Srivastava and Srivastava., 2003).

absence of hazardous materials. Similarly, green logistics aims to make sure that transportation and logistics activities have desirable environmental attributes and impacts. Environmental remediation is the removal of pollutants or contaminants for the general protection of the environment. This is accomplished by various chemical, biological, and bulk movement methods, in conjunction with environmental monitoring.

As per greening hierarchy suggested by Srivastava and Srivastava (2003) and shown in Fig. 5, companies can address environmental impact right at the source and throughout the process and migrate from low-end solutions to solutions far higher in the greening hierarchy, with a far more significant impact on greening. The terms are self-explanatory. At the bottom will be disposal without recovery followed by disposal with recovery. Next stage will be recycling of materials. Still upward on the hierarchy is reuse. This includes activities like reusing as-it-is, rework, refurbishing,

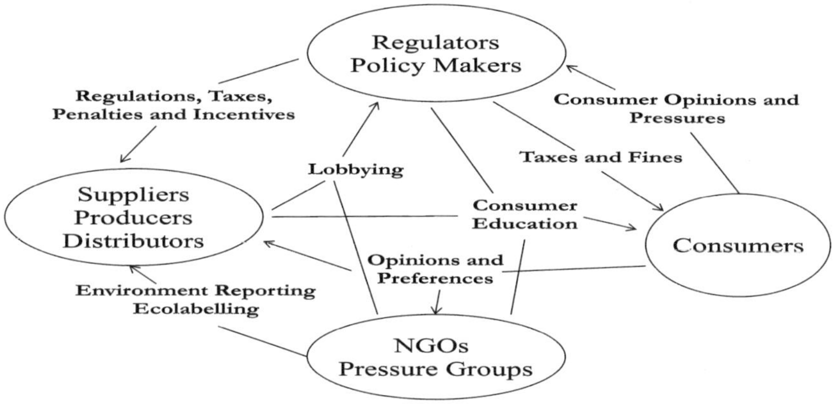

Fig. 5. Interaction among various stakeholders in sustainable supply chains.

remanufacturing and may also involve reverse logistics. Most brownfield ventures fall in this category. At the highest level is Reduce. It may be accomplished by resource reduction through material substitution, better processes, and technologies and other initiatives.

2.3. *Interactions among stakeholders in sustainable supply chains*

Four major inter-organizational forces are driving SSCM activities. These forces are governments, suppliers, customers, and competitors. Globalization, WTO requirements, and depleting natural resources are likely to lead to much stricter regulations and guidelines by the regulatory authorities. The level of consumer awareness of environmental issues is increasing rapidly. This will lead to customers' preference for green products as well as products and services from suppliers with GrSCM processes. More sustainable products by technological innovations are likely to emerge. This, in turn, will force the supply chains to become sustainable. Further, the rules that govern the attractiveness of recovery/reuse of products, materials, and components are undergoing changes at the local, state, national, and global levels. In addition to traditional brokers, many firms are now selling this material through online and traditional auctions. From the most interesting aspect of eco-centricity, a sustainable supply chain would explicitly include non-governmental agencies (NGOs), community members, and even competitors that traditional chains either ignored or treated as adversaries.

From a strategy perspective, four key stakeholder groups are emerging in sustainable supply chains. These are the policy makers and regulators who set the broad macro-rules of the game; the value adders in the overall production and transformation process, namely, the suppliers, producers, and distributors, who cater to consumer demands within the macro-business environment; the all-important consumers who create the actual demand and consume goods and services; and finally, pressure groups and NGOs who of late have been as important. The interaction among these stakeholders is depicted in Fig. 5.

2.4. *SSCM as a competitive advantage strategy*

Environmental issues hold the distinct opportunity for supply chains to give a competitive edge over their rivals. Corporate are quickly learning that SSCM plays an important role in their competitive strategy, which is likely to assume increasing importance in the coming future. Major producers of virgin items, many of whom have not been particularly active in various R's (Reduce, Reuse, Recover, Renew, Rework, Repair, Refurbish, Remanufacture, Recycle), will have to increase their activities in response to public, regulatory, and market forces. In many cases, they will probably want to work in partnership or even joint ventures with entrepreneurial firms. Gold *et al.* (2010) explore the role of SSCM as a catalyst of generating valuable inter-organizational resources and thus possible sustained inter-firm competitive advantage through collaboration on environmental and social issues. A framework for SSCM as a competitive advantage strategy is shown in Fig. 6. The three prime drivers are profits (cost savings through resource savings), changing consumer preferences (sustainability as a unique selling proposition), and innovation and better practices. However, this is not free; substantial investments must be made by the firms. Further, they require substantial stakeholders' support to realize this strategy.

2.5. *Recent initiatives*

The need for sustainability and the globally increasing product demands and manufacturing activities drive a gigantic need for technology and strategies that globally will reduce environmental impact from supply chain activities. For this, both technological choice and innovation must be considered (Debo *et al.*, 2005). Enabling features include manufacturing systems and supply chain structures

Fig. 6. SSCM as a competitive advantage strategy.

for sustainability, e.g., local manufacturing providers, high energy efficiency in manufacturing processes and techniques for replacing fossil-based energy generation by CO_2 neutral generation among others. The aim of a low-carbon economy (LCE) is to integrate all aspects of itself from its manufacturing, agriculture, transportation, power generation, etc. around technologies that produce energy and materials with little Greenhouse Gas (GHG) emission; and thus, around populations, buildings, machines, and devices, which use those energies and materials efficiently, and dispose of or recycle its wastes so as to have a minimal output of GHGs. Furthermore, it has been proposed that to make the transition to an LCE economically viable we would have to attribute a cost (per unit output) to GHGs through means such as emissions trading and/or a carbon tax.

TELCOS Look to a Green Future

Going green' has become a necessity for telecom operators in India. This is because margins are declining due to tariff wars and are denting the profitability of big telecom companies. Energy expenses constitute nearly 25% of the total network operating costs. Therefore, efficient power management holds the key to improving financials. Equipment vendors such as Ericsson, tower companies such as Bharti Infratel, and network services provider GTL are making huge investments in bringing out green products and solutions.

Bharti Infratel Limited is a leading "Passive Telecom Infrastructure" provider company in India. It owns and operates more than 25,000 towers for mobile communications and related telecom infrastructure. The Innovative Infrastructure Company of the Year in Essar Steel Infrastructure Excellence Awards 2010 was bagged by Bharti Infratel Ltd in March 2010 for its Renewable Energy and Energy Efficiency focused P7 Project.

India has adopted a National Plan of Action on Climate Change as a positive step toward mitigating carbon emission levels. Improving energy efficiency and adopting clean coal technology would be a central theme in the low carbon strategy. India has increased its renewable energy target to 14 GW by 2012. Union budget 2010–2011 has a series of "green" initiatives designed to boost environment-friendly energy and emission-free transport. Further, the government wants to use energy-efficient technology and all power plants may have to be based on supercritical technology (Anonymous, 2010).

Johnson & Johnson acknowledged in its 2006 sustainability report that GHG reductions and energy conservation "make good business sense for the company." The firm's energy efficiency program resulted in an estimated $30 million in annualized savings over the 10 years prior to the 2006 report. Johnson & Johnson's GHG reduction projects are achieving an average 16% internal rate of return, according to the report. General electric has received plaudits for reducing its carbon emissions and for calling upon some of its customers, such as utility companies, to reduce their carbon emissions as well. Heineken committed to reduce fuel and electricity use through its "Aware of Energy" program. The company said in its

2006 sustainability report that it aimed to reduce fuel and electricity costs by 15% between 2002 and 2010.

Bayer India has decided to follow a Product Stewardship Code to make health, safety and environment, an integral part of designing, using, recycling, and disposing of products. Amarnath Enviroplast Limited has developed a wood substitute called *Ecowud* fro thermoplastic and jute waste. Maruti Suzuki India Limited has already established India's largest certified used car dealer network. Every *True Value* car is re-conditioned to Maruti Suzuki standards in dealer workshops by trained mechanics using only genuine parts. This is a simplified version of the repair and refurbishment activity (Srivastava, 2008). Recently, Maruti Suzuki India Limited has launched A-star, a truly green car. It reflects a focus on world-class environmental compatibility and comfort. It has an all new Euro 5-compliant 1.0-liter aluminum petrol engine with CO_2 emissions lower than those of European competitors (target: lower than 109 g/km). The ca also meets the European ELV norms, which implies that 85% of the car is recyclable. In addition, it is free from hazardous materials such as lead, cadmium, mercury, and chromium. A-star delivered a record mileage of 19.6 km per liter of petrol, the best in class ARAI certified fuel efficiency, among cars in India. The company has achieved 100 thousand units cumulative export of A-star in less than a year since its first shipment to Europe in January 2009. Tata Group has taken several steps to cut carbon emissions across its companies and value chains. Tata Steel aims to reduce CO_2 emissions at its Jamshedpur plant from 1.8 to 1.7 tonne per tone of liquid steel made by 2012. Global benchmark is 1.5. Tata Motors is introducing Indica EV, an electric car. It is also setting eco-friendly showrooms using natural materials. Indian Hotels is creating eco-rooms, which will have energy-efficient mini-bars, organic bed-linen, and recycled paper napkins.

Initiatives in Indian Railways

Roadways and air travel have a higher carbon footprint than railways. While combustion of diesel oil emits 0.1647 kg of CO_2 per equivalent kW of energy, coal-based thermal power propelled electric trains emit 0.2632 kg CO_2 per kW. Energy consumption can further be reduced through weight reduction of bogies, rolling stock bodies and electrical equipment and aerodynamic designs. Two major initiatives by Indian Railways in the recent past are the successful trials of biofuel blend diesel locomotives and resurrecting production of 3-phase electric locomotives with regenerative braking features. The first successful trials of biofuel blend in diesel locomotives started in 2003 with Southern Railway preparing its own biodiesel using *Pongammia Pinnata.* Used cooking oil was also collected from hotels. The Perambur loco works currently produces 5000 liters of biodiesel/month. All necessary engine test bed trials for use of B10 blend of biodiesel for diesel locomotives are complete. With increased availability of biodiesel fuel, the railways can target a 5% biofuel blend in diesel to reduce CO_2 emissions by 2%.

(*Continued*)

Box (*Continued*)

The WAG 9, WAP 7, and WAP 5 series of locomotives produced by Chittaranjan Locomotive Works and newly proposed 9000/12,000 hp heavy haul locos for DFC are the stars in clean regenerative power. The architecture of these locos enables them to work as generators while the loco brakes. The advanced control algorithm of the front-end Pulse Width Modulation Converters of the loco permits the loco to deliver 6000 hp of power at the wheels even at 100 km/h speed (for WAG 9 freight locos) while pulling/hauling the trailing load. The same equipment permits the locomotive to be used as a mobile generator while braking by converting the traction motors, which normally drive the wheels, to be in turn driven by the wheels and behave as generators. Figures available prove that on level sections, the loco generates around 10% of energy consumed and on graded sections up to 30%. As of now, 300 WAG 9, WAP 7, and WAP 5 locos running on the network annually save about 100 million units of energy, saving 105 tons of CO_2 per annum.

Source: Presentation by Anjali Goyal at the World Transport and Energy Forum 2009, organized by the World Bank in Washington.

Wal-Mart, which in 2005 launched a sweeping business sustainability strategy, recently set the goal of a 5% reduction in packaging by 2013. The retail giant expects the cut in packaging will save 667,000 metric tons of carbon dioxide from entering the atmosphere. Moreover, the company anticipates \$3.4 billion in direct savings and roughly \$11 billion in savings across the supply chain. In July 2009, Wal-Mart announced plans for a worldwide Sustainable Product Index. The company's Sustainable Product Index is an ambitious effort to measure the sustainability of each product the company sells. Implementation of the index involves a three-step approach:

Step 1. Suppliers are asked to complete a self-assessment survey of 15 questions in four key areas: (1) energy and climate, (2) natural resources, (3) material efficiency, and (4) people and community.
Step 2. The company is creating a consortium of universities to work with suppliers, retailers, governments, and non-government organizations to develop a global database of information about the life cycle of products, from raw materials to disposal. Arizona State University and the University of Arkansas are joint administrators of the consortium.
Step 3. The index will be used to create a rating system so customers have what the company anticipates will be accessible and understandable information about the sustainability of the products they purchase.

2.6. *Recent innovations*

There have been many recent innovations that are likely to provide a big boost and impetus to SSCM initiatives. Two of them seem to be the "big ones" and are described briefly below.

2.6.1. *Bloom box — power plant in a box*

Bloom energy has raised nearly \$400 million from investors like Kleiner Perkins for its "a little power plant-in-a-box" machine called the bloom box that presently costs \$700,000 to \$800,000 for 100 kW. The Bloom Energy Server is a solid oxide fuel cell that uses liquid or gaseous hydrocarbons (such as gasoline, biomass gas, landfill gas, ethanol, natural gas, diesel, or propane produced from fossil or biosources) to generate electricity on the site where it will be used. A single cell (one 100 mm × 100 mm metal alloy plate between two ceramic layers) generates 25 watts. Cells can be made out of low-cost materials with high energy efficiencies. The Bloom Energy Server is a distributed power generator. Each "server" consists of thousands of fuel cells.

Current gas-fired power stations convert chemical energy first to thermal energy then to mechanical energy and finally to electrical energy. Bloom Boxes convert chemical energy to electrical energy in one step at 50–55% conversion efficiency and are therefore more fuel efficient. They also reduce transmission/distribution losses by producing power where required. In comparison, solar panels generally produce power at 10–15% efficiency. However, unlike solar panels, the Bloom Energy Server produces CO_2 as a by-product. According to the Energy Collective, "CO_2 emissions when running on natural gas would be just under 0.8 pounds/kW, which compares favorably to electricity from central station coal-fired plants (2 lbs/kW) or natural gas plants (roughly 1.3 lbs/kW) and the national average for on-grid electricity (around 1.3–1.5 lbs/kW). If the box runs on landfill gas or biogas, it produces net zero carbon emissions.

Major companies such as eBay, Google, Staples, Wal-Mart, The Coca Cola Company, Bank of America, and FedEx have already started using bloom boxes. eBay has already saved \$100,000 in electricity costs in nine months since five boxes were installed in mid 2009. As of now, Bloom is not angling for the residential market as the box is quite expensive. Bloom hopes to develop in next few years a scaled-down residential Bloom Box version that would produce 1 kW of power and cost approximately \$3,000 for use in homes.

2.6.2. *Eco-friendly Android applications*

Android applications are the greenest applications for the iPhone. Some of these eco-friendly applications are as follows:

Ecocharging: Lets the user know as soon as the Android phone has finished charging so that one does not uses up too much electricity.
IAmGreen: Available to Android, iPhone, and Blackberry users and helps them reduce the amount of power wasted when using mobile phones.
Print Green: Keeps one up to date with the latest green issues and sustainability news in the printing industry.
GreenYou: Allows one to calculate one's own carbon footprint when doing daily activities, once calculated it works out ways in which one can reduce it.

iGreenCircle: Allows one to connect to a social media network and speak to other green minded people.
Green Mileage: Helps one keep track of how much petrol one is using in one's car and how far one is driving each day.
Seasonal Harvest: Uses one's Android's GPS to work out which foods are local to one's environment during a certain season.
PowerTutor: Works out whether one has too many applications on one's mobile that are just wasting power; helps organize applications to make phone more efficient.

3. Sustainable Supply Networks

The overall objective sustainable supply networks is to reduce impact on the environment, lower supply chain cost, and improve product value. Pagell and Wu (2009) describe case studies of 10 exemplar organizations are doing that is unique in regards to managing their supply chains in a sustainable manner. Exemplars are organizations that are well ahead of their industry on either social and/or environmental performance while still maintaining economic viability. Kogg (2003) describes the greening of a cotton textile supply chain in Peru. By interacting with all actors in the chain, the company has co-ordinated the entire production process from fiber to final product.

Binnenlandse Container Terminal Netherlands
Binnenlandse Container Terminal Netherlands (BCTN) decided to understand more about carbon emissions and to challenge new opportunities in the market. The company is active in distribution and transport business by managing four container terminals in the Netherlands. It transports 350,000 twenty foot equivalent units per year, which constitutes 10% of the Dutch container transport. As a component of transport value system, BCTN believes that intermodal transport is more efficient than other single modes. It also felt the pressure from its customer to limit its environmental impact in terms of greenhouse gases. Hence, BCTN measured its environmental impact to compare it with other modes of transport and to identify opportunities and threats with impending business context to carbon constraining developments.

For this, the value system was analyzed and the part of value system that can be influenced by BCTN was identified. It focused on global warming impact of the determined value system level and made the carbon footprint of the same. The emissions were distributed over the value system and lowest performing links in the value system were identified by using value system specific key performance indicators. Possible effects of different environmental charges on BCTN business were analyzed under different scenarios. BCTN's strategic decisions were re-assessed. In the course of the analysis, operations were identified that could be further optimized to reduce cost and environmental impact. Currently BCTN is developing a program to reduce overall environmental impact of its value system with the co-operation of other parties involved in the system. Moreover, BCTN is considering developing an integrated system to monitor its value system's greenhouse gas impact on a daily basis.

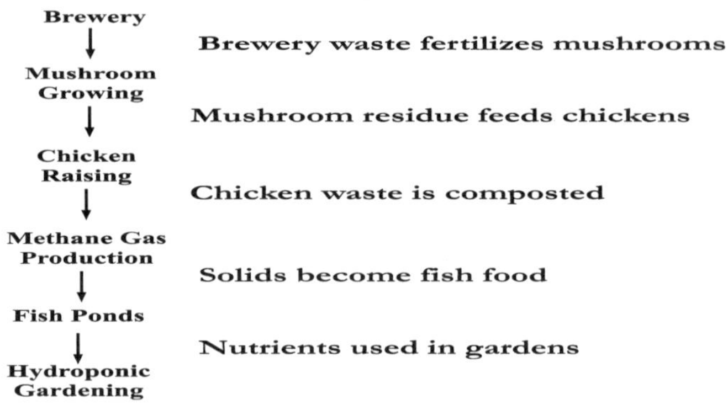

Fig. 7. Concept of industrial eco-systems.

3.1. *Creating sustainable supply networks*

Linton *et al.* (2007) discuss the convergence of supply chains and sustainability. In doing so, the focus on environmental management and operations is moved from local optimization of environmental factors to consideration of the entire supply chain during the production, consumption, customer service, and post-disposal disposition of products.

The origin of sustainable supply networks lies in industrial eco-systems. Figure 7 describes one such simple industrial eco-system. Ideally, one should strive for creating such eco-systems. However, the same may not always be realisable in our present production–consumption-based supply chains. Hence, one needs to explore into certain key requirements of successful sustainable supply chains and possible integration points of sustainable supply networks.

3.1.1. *Key requirements and possible integration points*

While sustainability provides an overarching framework for much of the past and ongoing environmental research in operations, sustainability moves beyond current common practice. Supply chains must be explicitly extended to include by-products of the supply chain, to consider the entire life cycle of the product, and to optimize the product not only from a current cost standpoint but also a total cost standpoint. Total cost must include the effects of resource depletion and the generation of by-products that are neither captured nor used (pollutants and waste). Other two best practices in supply chains that have received significant attention in regard to sustainability are collaboration and supplier certification. Collaborative behaviors with suppliers and customers are a component of creating an environmentally sustainable supply chain (Carter and Carter, 1998; Zhu and Sarkis, 2004). Certification is one of the few areas where social issues such as child labor and unsafe working conditions can be addressed in the SSCM easily. Leadership in Energy and Environmental Design is a third-party certification program that certifies the suppliers' supplier as a sustainable source.

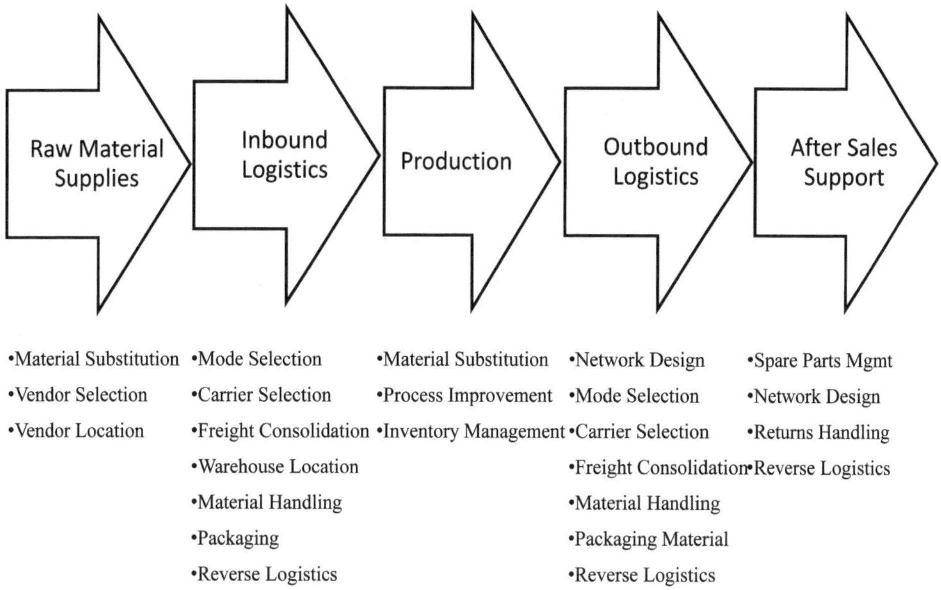

Fig. 8. Possible integration points for sustainable supply chain collaboration.

If we explore from the perspective of better sustainable supply chain collaboration, there could be multiple possible integration points as shown in Fig. 8. It shows various stages in a supply chain as well as the possible potential areas that could be useful for such integrative and collaborative efforts.

3.2. *A model for SSCM practices*

Analysis by Pagell and Wu (2009) suggests that the practices that lead to a more sustainable supply chain are combinations of best practices in traditional supply chain management and certain new behaviors. Based on their findings, they propose a model of SSCM practices as shown in Fig. 9. The ability to be innovative has been linked to sustainability. Hence, the organizational capability to innovate becomes a precursor to successful SSCM. The other organizational attribute that is a precursor to SSCM is managerial orientation. The literature suggests that firms need to be proactive and committed. Pagell and Wu's analysis suggests that being proactive and committed can only be effective if the business model and the environmental and social elements of sustainability are aligned. This business model then guides managerial decision making. When the employees have a way to think about sustainability that is compatible with business goals it is then possible for sustainability to become part of the day-to-day conversation. This can only happen if responsibility for the non-economic components of sustainability is shared across all employees and not housed in a single function or individual.

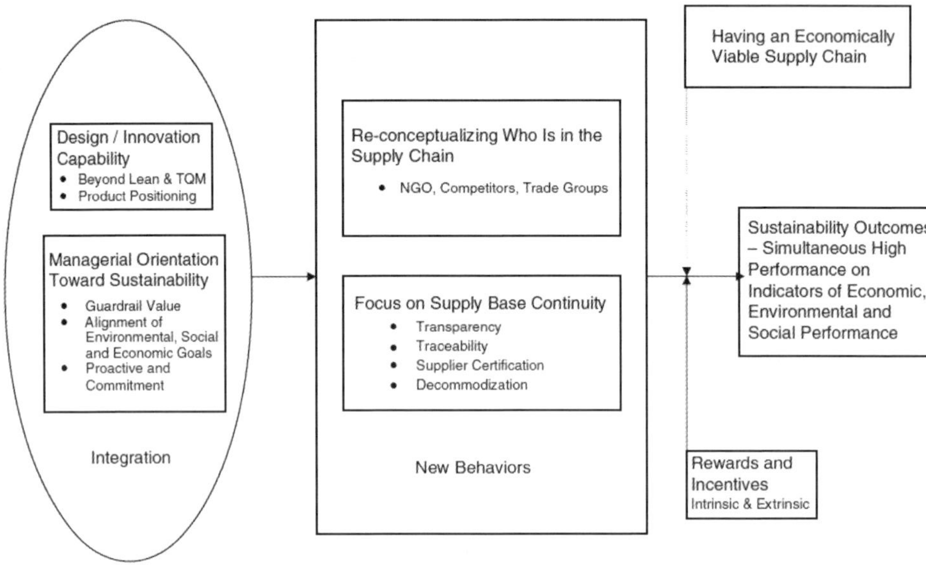

Fig. 9. A model for SSCM practices.
Source: Pagell and Wu (2009).

Organizations that have an innovation capability and a managerial orientation toward sustainability will carry out two unique sets of activities. First, they will reconceptualize who is in the supply chain. Rather than viewing NGOs and the like as adversaries, sustainable supply chains leverage the skills and abilities of these non-traditional chain members. Second, they will try to ensure that this new chain is efficient and effective.

3.3. *Desirable activities in a sustainable supply network design*

Based on all that we have discussed so far, it becomes imperative to develop a model for desirable management activities in a sustainable supply network design. The same have been depicted in Fig. 10. The framework applies to every direct supply chain stakeholder company. The company needs to measure the impact of its supply chain activities in both financial and environmental parameters. The next step would be to analyze the impact of its environmental footprint not only on its own cost structure, but also on the overall supply chain cost structure. This is not easy and new tools and techniques may be required besides the existing ones used in environmental audits. The next stage should decide the most effective supply chain strategy (with concomitant sub-strategies and activities along with their performance metrics), gain the support, and collaborate with other partners to effectively implement that strategy. Finally, as business environments and supply chains are ever dynamic, the partners need to track the changes continuously

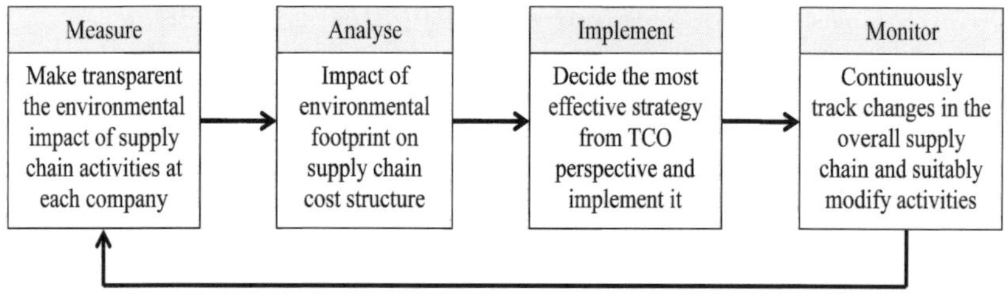

Fig. 10. Desirable management activities in a sustainable supply network design.

and suitably modify certain activities. This desirable management activity is a continuous one.

3.4. *Sustainable supply network design*

Environmental considerations can influence management decisions related to the size, location, design, or specialization of operations of firms. The location of facilities relative to process inputs, customer markets, or waste disposal locations needs to be considered both analytically and empirically. All environmental consequences of a product or a process from initial design to development phases need to be understood and analyzed. Operation and maintenance systems must be improved to increase the efficiency of the materials used. Equipment design must address waste and emission management as well as disposition. Organization culture also needs a paradigm shift and should focus on eliminating waste and looking for reuse/recycling, reverse consumption.

Schultmann *et al.* (2003) show that almost complete recycling of spent batteries can be achieved by transforming the current structure into a modified recovery network. Srivastava (2008) models value recovery network design in the Indian context. It addresses a paradigm shift from end-of-pipe control to not only minimizing ecological damage, but also leading to overall economic profit. It integrates the reverse logistics and value recovery aspects of SSCM. It also indirectly helps in reducing utilization of virgin resources. The proposed model incorporates multiple reuse options for end-of-use products. Various strategic issues in managing returns have also been taken into account. It shows that profit-driven value recovery as a worthwhile proposition even in the context where regulatory and consumer pressures are insignificant.

4. Implications

The emerging trends in SSCM have many policy, managerial and research implications. The policy implications relate to what the government and regulatory authorities, the managerial implications relate to what the practicing mangers at various echelons in the supply chain and research implications relate to what the thinking professionals (academicians and researchers) need to do.

4.1. *Implications on government and regulatory authorities*

National environmental laws and regulations must be globalized. They should be in trend with laws elsewhere because our environment is a global issue rather than an individual nation's problem. Various incentives should be extended to industry consortia and firms for developing/promoting environment conscious products and processes. Even presently, governments are rewarding companies that adopt green strategies and are creating new regulations that penalize those companies that do not. The level of public awareness about sustainability, prevalent and emergent technologies, and the environment needs to be increased. Various fora may be utilized for it. The government and regulatory authorities need to set up and spruce certain regulations to protect the environment (such as for process and product emissions). For example, in Indian context, organizations such as Centre for Science and Environment, Confederation of Industry (CII), and Federation of Indian Chamber of Commerce and Industry (FICCI) may be involved along with NGOs in terms of policy proposal formulation and feedback. Industry specific and national level models may be developed for supplier managed inventories and operations and for waste segregation and recycling. These will result in enhanced co-operation and collaboration and higher efficiencies.

4.2. *Implications on industry*

Process mapping may be done to facilitate environment audit by firms. This will help in eliminating unnecessary sub-processes/processes, on one hand, and improving them in terms of environmental criteria, on the other hand. It will also open an avenue for Process Benchmarking. Similarly, specific energy consumption mapping of various sectors may be done.

The industry and academia should collaborate to come up with technological innovations to address SSCM issues. These resulting will have long-term positive impacts. Effective approaches for data sharing within and across the industries need to be developed. Industry consortia may go for intra-nets. For example, existing networks such as CII, FICCI, etc. may be utilized and Internet may be leveraged to achieve more co-operation and collaboration.

4.3. *Implications on individual firms*

A management process for minimizing ecological impacts of economic production while enhancing competitiveness of firms needs to be promoted. Firms can gain competitive advantage by managing ecological variables. The underlying imperatives driving this should be eco-efficiency, source reduction, and asset recovery. At a more general level, customer concerns about social responsibility must be integrated with other dimensions of value when managing suppliers and contractors.

Individual firms should develop systems for effective collaboration with suppliers and customers/consumers. They should develop integrated SSCM models incorporating the impacts of greening on quality, delivery and price in the decision-making process. Environmental impact assessment, life-cycle costing taking "cradle-to-grave" system into consideration, and waste segregation and recycling may be incorporated in the existing decision-making models of the firms in order to become greener and sustainable for future. The weightage to these considerations in the models will be firm and industry specific.

A component of being sustainable is economic viability. Therefore, a sustainable chain based on the triple bottom line would have to be sustainable from a traditional economic standpoint. In order to successfully manage the supply chain in a more sustainable manner, managers should be willing to work to change their cultures so that good decisions in their organization will be those that enhance natural, social, and economic capital.

4.4. *Implications on research*

Research to date in SSCM compartmentalized it into content areas drawn from operations and supply chain domains. The primary areas of emphasis have been quality, along with operations strategy, supply chain management, and product and process technology, which are collectively beginning to contribute to a more systematic knowledge base. It is reasonable to expect that these research areas will continue to hold the greatest promise for advance in the short term. However, more integrative contributions are needed in the longer term including intra- and inter-firm diffusion of best practices, green technology transfer, and environmental performance measurement indices. Literature on integrated business strategy (composing of product and process design, manufacturing, marketing, reverse logistics, and regulatory compliance) in context of SSCM is to the level of thought papers and perspectives only. Research is needed into how companies should design, procure, process, store, and dispose goods.

For successful implementation of SSCM, it is necessary to develop quantitative decision tools and models. The applicability of traditional tools and models is limited because the objectives, constraints, and other characteristics of the traditional systems are different from those for the SSCM systems. Effort must be made for SSCM systems to be profitable so that the incentive for development and planning of these systems continues. Advanced models, heuristics, tools, and techniques may be developed for impact analysis, improvement analysis, substitution possibilities/material purification, customer satisfaction, product liability, consumer product safety, occupational safety, and hygiene. Mathematical models for sustainable supply network design based on forecasts and changes in sales over the product's life cycle need to be developed. Easy-to-use customized and cost-effective heuristics, simulation programs may be developed to calculate cost savings due to various initiatives/decisions for use at the company level.

5. Conclusions and Outlook

This chapter has focused almost exclusively on the sustainable supply networks. It begins with the introduction to the concept and importance of sustainable supply chain networks and its relationship with other disciplines. The evolution of these networks and related concepts and activities in light of the growing focus on planet, people, and profits is discussed. It then focuses on the interaction between different stakeholders and provides a framework for the same. It highlights the fact that SSCM can be a potential competitive advantage strategy. The chapter also presents a few recent initiatives and innovations to facilitate better understanding and appreciation by the readers. Subsequently, it covers sustainable network design issue along with key requirements for the same. It also suggests possible integration points for better sustainable supply chain collaboration at various stages in a supply chain as well as the possible potential areas that could be useful for integrative and collaborative efforts. It describes briefly a model on SSCM practices and proposes desirable management activities in a sustainable supply network design. Finally, it discusses implications of all this on important stakeholders.

The traditional green initiatives are generally associated with many weaknesses and problems. The end-of-the-pipe approach does not eliminate pollutants, but merely transforms them from one medium to another. Hence, we need to take new initiatives and look for innovative approaches. Research in SSCM to date may be considered compartmentalized. More integrative contributions are needed in the longer term, including intra- and inter-firm diffusion of best practices, green technology transfer, and environmental performance measurement. The focus today is on first finding out "The Right Thing to Do" and subsequently "Doing the Thing Right." Environmentally conscious actions are no longer seen as challenges and contrary to financial considerations. Therefore, stakeholders should look toward converting the imperative greening/sustainability mandate into an opportunity. A good dictum for them could be "Think Global, Act Local" as this may help in addressing avoidable supply chain complexities discussed earlier in the chapter.

Acknowledgments

I would like to thank the corporate executives of few firms for providing inputs about their initiatives, practices, and innovations that have been used here.

References

Anonymous, 2010. Budget's green initiatives. *The Economic Times, New Delhi*, March **05**, 12.

Carter, C.R. and Carter, J.R., 1998. Interorganizational determinants of environmental purchasing: Initial evidence from the consumer products industries. *Decision Sciences* **29**(3), 659–685.

Debo, L.G., Totkay, L.B. and Van Wassenhove, L.N., 2005. Market segmentation and product technology selection for remanufacturable products. *Management Science*, **51**(8), 1193–1205.

Elkington, J., 1998. *Cannibals with Forks: The Triple Bottom Line of the 21st Century.* New Society Publishers, Stoney Creek, CT.

Gold, S., Seuring, S. and Beske, P., 2010. Sustainable supply chain management and inter-organizational resources: A literature review. *Corporate Social Responsibility and Environmental Management* DOI: 10.1002/csr.207.

Guide Jr. V.D.R. and Van Wassenhove, L.N., 2009. The evolution of closed-loop supply chain research. *Operations Research* **57**(1), 10–18.

Gupta, M., 1996. Environmental operations management: An opportunity for improvement. *Production and Inventory Management Journal* **37**(3), 40–46.

Kogg, B., 2003. Greening a cotton-textile supply chain: A case study of the transition towards organic production without a powerful focal company. *Greener Management International* **43**(3), 53–64.

Kopicki, R.J., Legg, L., Berg, M.J., Dasappa, V. and Maggioni, C., 1993. *Reuse and Recycling: Reverse Logistics Opportunities.* Council of Logistics Mgmt., Oak Brook, IL, USA.

Labuschagne, C. and Brent, A.C., 2004. An industry perspective of the completeness and relevance of a social assessment framework for project and technology management in the manufacturing sector. *Journal of Cleaner Production* **16**(3), 253–262.

Linton, J.D., Klassen, R. and Jayaraman, V., 2007. Sustainable supply chains: An introduction. *Journal of Operations Management* **25**(6), 1075–1082.

Pagell, M. and Wu, Z., 2009. Building a more complete theory of sustainable supply chain management using case studies of 10 exemplars. *Journal of Supply Chain Management* **45**(2), 37–56.

Ramus, C.A. and Steger, U., 2000. The role of supervisory support behaviors and environmental policy in employee 'ecoinitiatives' at leading edge European companies. *Academy of Management Journal* **43**(4), 605–628.

Schultmann, F., Engels, B. and Rentz, O., 2003. Closed-loop supply chains for spent batteries. *Interfaces* **33**(6), 57–71.

Srivastava, S.K. and Srivastava, R.K., 2003. How green are Indian firms. *Productivity* **44**(2), 294–302.

Srivastava, S.K., 2007. Green supply chain management: A state-of-the-art literature review. *International Journal of Management Reviews*, **9**(1), 53–80.

Srivastava, S.K., 2008. Network design for reverse logistics, *Omega Special Issue on Logistics: New Perspectives and Challenges* **36**(4), 535–548.

Srivastava, S.K., 2008. Value recovery network design for product returns. *International Journal of Physical Distribution and Logistics Management* **38**(4), 311–331.

Wittstruck, M., Bellgran, M. and Jackson, M., 2009. *Sustainable Manufacturing — Challenges and Possibilities for Research and Industry from a Swedish Perspective, The 41st CIRP Conference on Manufacturing Systems*, p. 119.

Zhu, Q. and Sarkis, J., 2004. Relationships between operational practices and performance among early adopters of green supply chain management practices in Chinese manufacturing enterprises. *Journal of Operations Management* **22**(3), 265–289.

Chapter 8

ENERGY RESOURCE MANAGEMENT

John F. Affisco
Department of Information Technology and Quantitative Methods,
Frank G. Zarb School of Business, Hofstra University,
Hempstead, NY 11549-1340, USA
John.F.Affisco@hofstra.edu

This chapter presents two models for energy resource management that were developed and are in use in the USA. ENERGY STAR is a joint program of the US Environmental Protection Agency and the US Department of Energy whose purpose is to save money and protect the environment through energy-efficient products and practices. ANSI/MSE 2000:2005 is an American standard that applies Deming's PDCA cycle to the problem of energy management. The structure of each approach is presented in detail, and examples of their use in practice are given.

1. Introduction

It has been 35 years since the oil embargo of 1973 resulted in rationing of gasoline in the United States. The proximate result of this embargo was the flight to higher gas mileage automobiles. However, as time marched on and oil became plentiful again at a reasonable price, the lesson of 1973 was quickly forgotten. Fast forward to 2008, it can be argued that the world is in significantly worse condition than in 1973. Demand for oil is rising, prices have passed $145 per barrel, OPEC largely controls price and output on a whim, and the Middle East remains a tinderbox.

As bad as the situation is for the individual consumer, it can be significantly worse for productive organizations. Although in some organizations energy costs may represent a small percentage of the cost of doing business, every organization expends significant funds for energy, and costs continue to rise. Recognizing this, in the late 1990s organizations, some more successfully than others, begin to focus on energy as a resource that needed to be managed. Energy management has progressed to the point that today all organizations recognize its importance, while most are still wrestling with the problem of how best to accomplish it. Thumann and Meta (1995a) define energy management as follows:

> "Energy management is the judicious and effective use of energy to maximize profits (minimize costs) and to enhance competitive positions. A successful energy

management program is more than conservation. It is a total program that involves every area of a business. A comprehensive energy management program is not purely technical. It takes into account planning and communication as well as salesmanship and marketing. It affects the bottom line profits of every business; thus the individual who is assigned the role of "energy manager" has high visibility within the organization."

"Energy management includes energy productivity, which is defined as reducing the amount of energy needed to produce one unit of output. Energy management includes energy awareness, which is essential in motivating employees to save energy. Probably the highest initial rate of return will occur through the establishment of a good maintenance management program."

2. Energy Resource Management

The management of energy resources is a serious problem for most organizations. The basis of the problem is the fact that there are a number of unique aspects associated with the management of energy. Brown (2006a) lists some of these issues as insufficient commitment from executive management, lack of communication across departments, failure to deal with small, recurring energy-related problems, and ineffective response to changing energy markets.

In many organizations, the energy resource has typically been managed by a single mid-level individual. This energy manager is generally situated in a single functional area from which he or she can seldom obtain the attention of senior executives. The impact of the energy resource on the bottom line is simply not easily visible. Hence, energy is often viewed as a minor issue and it is thus deemed to be undeserving of much attention or resources. Further, other organizational issues that compete with energy such as environment, safety, and health are subject to legal mandates that force the allocation of organizational resources to assure compliance.

One common situation in most organizations is the lack of communication between different departments. This is compounded by the cross-organizational presence of energy. Energy is bought by purchasing, used by production, and energy equipment is maintained by maintenance or engineering. Because so many groups with different objectives share the responsibility for energy, communication often breaks down and compromises effective energy management.

A common problem observed in many organizations is the failure to conquer many small, recurring energy-related problems. Because there is no systematic approach to energy management, the same problems occur over and over, with energy engineers never seeming to get ahead enough to achieve real improvement. In the vein of "fighting fires," rising energy costs are generally dealt with as a local crisis rather than a strategic issue. Once this local problem is solved, generally by improved maintenance and operating practices, management is quick to move on to the next pressing problem. In many cases, as there is neither energy monitoring in place nor any procedures to institutionalize any gains made by solving the local problem, similar "local problems" may arise and resources must be allocated to arrive at a solution that may already be in place at other organizational facilities.

The deregulation of energy suppliers has greatly complicated the purchase of energy resources. Furthermore, the increased volatility of energy prices and uncertain availability of adequate resources in the future have added to this complexity. A workable plan for dealing with this volatile environment including the requirement for formal procedures specifying information, data, and processes for effective energy purchasing is an area that is frequently neglected in today's organizations.

Recognizing these problems some organizations have sought to apply structured management systems to the management of the energy resource. Two management systems in particular are the ENERGY STAR program developed by the US Department of Energy (DOE) and Environmental Protection Agency (EPA), and MSE 2000:2005 A management system for energy (MSE) promulgated by the American National Standards Institute (ANSI). These systems are the subject of the next two major sections of this chapter.

3. ENERGY STAR Program for Business Improvement

ENERGY STAR is a joint program of the US, EPA and the US DOE whose purpose is to save money and protect the environment through energy-efficient products and practices. Here, we concentrate on the ENERGY STAR Program for Business Improvement.

In 1992, the EPA introduced ENERGY STAR as a voluntary labeling program designed to identify and promote energy-efficient products to reduce greenhouse gas emissions. Computers and monitors were the first labeled products. Through 1995, EPA expanded the label to additional office equipment products and residential heating and cooling equipment. In 1996, EPA partnered with the DOE for particular product categories. The ENERGY STAR label is now on major appliances, office equipment, lighting, home electronics, and more. EPA has also extended the label to cover new homes and commercial and industrial buildings. Believing that a strategic approach to energy management simultaneously offered savings for the bottom line and improved environmental performance, the EPA offers ENERGY STAR partnerships for business, which provide a proven energy management strategy that assists in measuring current energy performance, setting goals, tracking savings, and rewarding improvements.

Recent benefits of ENERGY STAR as reported in USEPA (2008a) include

- EPA continued to promote the ENERGY STAR Challenge, which calls on US businesses and institutions to reduce energy use by 10% or more. As a result, almost 800 organizations and individuals — including more than 150 local governments — have now joined the Challenge.

- EPA's energy performance rating system experienced tremendous growth and was expanded to include retail space in 2007. Building owners and operators have now used the system to rate the energy efficiency of more than 62,000 buildings — more than double the number since the end of 2006. These buildings represent more than 7.5 billion square feet including 55% of hospital space (acute care), 52% of supermarket space, 31% of office building space, 24% of school space, and 24% of hotel space across the country.
- In 2007, vigorous activity continued for ENERGY STAR's industrial partners, including those in the key focus industries of motor vehicles, cement, wet corn milling, food processing, glass, pharmaceuticals, petrochemicals, petroleum refining, and pulp and paper. EPA released first drafts of plant energy performance rating tools for petrochemical plants, four different kinds of food processing plants, and three types of glass manufacturing plants. Almost 40 plants have now earned the ENERGY STAR label for superior energy performance.
- EPA completed a report to Congress on energy use in US data centers, showing the potential to save up to $4 billion in electricity costs by improving the efficiency of equipment and operations.

3.1. *The ENERGY STAR program (USEPA, 2008b)*

The ENERGY STAR Program consists of seven steps that when implemented will result in continuous improvement of energy resource management. The steps are presented in Table 1.

3.1.1. *Step 1: Commit to continuous improvement*

The critical common element of successful energy management is commitment. Organizations must make a commitment to allocate funds and staff to continuous energy management improvement. This commitment includes the appointment of an energy director, the establishment of an energy team, and the development of an energy policy. The position of energy director can either be full-time or part-time depending on the size of the organization. In either case, the energy director must either report directly to a senior manager or have an "executive ally" in senior management. Some of the responsibilities of the energy director often include coordinating and directing the overall energy program, drafting an energy policy, creating and leading the Energy Team, identifying opportunities for improvement and ensuring implementation, and measuring, tracking, evaluating, and communicating results.

An energy team helps to integrate the process of energy management across the organization. The team is responsible for planning and implementing specific improvements, measuring and tracking energy performance, and communicating with management, employees and other stakeholders. The size of the energy team depends on the size of the organization. In addition to the Energy Director and

Table 1. E ENERGY STAR program for business improvement.

Step	Content
1.0. Commit to continuous improvement	
1.1	Appoint an energy director
1.2	Establish an energy team
1.3	Institute an energy policy
2.0. Assess performance	
2.1	Gather and track data
2.2	Establish baselines
2.3	Benchmark
2.4	Analyze
2.5	Technical assessments and audits
3.0. Set goals	
3.1	Determine scope
3.2	Estimate potential for improvement
3.3	Establish goals
4.0. Create action plan	
4.1	Define technical steps and targets
4.2	Determine roles and resources
5.0. Implement plan	
5.1	Create a communication plan
5.2	Raise awareness
5.3	Build capacity
5.4	Motivate
5.5	Track and monitor
6.0. Evaluate progress	
6.1	Measure results
6.2	Review action plan
7.0. Recognize achievements	
7.1	Providing internal recognition
7.2	Providing external recognition

any other dedicated energy management staff, the energy team should include a representative from each operational area that significantly affects energy use.

An Energy Policy provides the foundation for successful energy management by formalizing senior management's support and articulating the organization's commitment to energy efficiency for employees, shareholders, the community, and other stakeholders. Successful organizations have energy policies that state an objective, establish accountability, ensure continuous improvement, and promote energy goals by linking them to overall financial and environmental goals of the organization. Figure 1 presents 3M Worldwide's Energy Policy, which contains elements discussed above.

Organizational commitment to continuous energy improvement may be found in all types of productive organizations with successful energy programs. Among them are Ford Motor Company, 3M, Virtua Health, and Food Lion.

At more than 100 years old, Ford Motor Company is one of the largest US-based motor vehicle manufacturers. To demonstrate senior management's commitment

3M's Energy Policy

Applicability
This Policy applies to all the 3M operations.

Introduction
The objectives of this policy are to improve energy consumption efficiency, reduce cost, and optimize capital investment for energy efficiency, reduce environmental and greenhouse gas emissions, and conserve natural resources.

Policy Statement
3M will promote the efficient use of energy to produce and deliver products and services to its customers.

Policy Guidelines

- Improve energy efficiency continuously by establishing and implementing effective energy management programs worldwide that support manufacturing capabilities while providing a safe and comfortable work environment.
- Emphasize energy efficiency as a factor in product development and in process and facility design.
- Secure adequate and reliable energy supplies at the most advantageous rates and implement contingency plans to protect operations from energy supply interruptions.
- Encourage continuous energy conservation by employees in their work and personal activities.
- Drive further development of internal and external energy efficient and innovative technologies.
- Cooperate with governmental agencies and utility companies on energy programs.
- Support national energy efficiency policies.

Policy Approval
Corporate EHS Committee, revised Nov. 2004

Fig. 1. 3M's energy policy.

to Ford's energy management program, the CEO gave an address directed to all employees about how Ford's profitability is tied to sustainability. In this briefing, he highlighted energy as a key component of Ford's corporate sustainability strategy and specifically spoke about Ford's progress in using renewable energy sources and improving manufacturing process efficiency (USEPA, 2008C).

Founded over 100 years ago, 3M is a global, diversified, technology company that prides itself on innovation. 3M makes a wide variety of products ranging from post-it notes to pharmaceuticals. 3M has been an ENERGY STAR partner since 1997 and was named an ENERGY STAR Partner of the Year in 2004, 2005, 2006, and 2007.

As a large diversified manufacturer, 3M operates hundreds of different manufacturing facility types. Because of this, the 3M energy program relies on multidisciplinary teaming and coordination to achieve their annual average 7% reduction. The 3M energy team makes use of the engineering, finance, environmental, plant operations, and R&D departments, as well as their suppliers and vendors. This ensures that their energy team covers all stages of the production process, both upstream and downstream. Including the financial division also helps energy programs secure the necessary capital funding for major energy efficiency projects.

3M has created an integrated worldwide energy policy that extends the accountability of the energy program beyond environmental and operation divisions to coordination with their research and development organization, process engineers, and facility design engineers. This policy sets challenging, yet attainable goals that support the broader goals of the company to drive innovation with energy-saving products and optimize capital investment. 3M revised their energy policy in 2004 after soliciting feedback on their policy from manufacturing directors, plant managers, plant operations, and executive management. Energy efficiency improved by 8.3% in 2004 (USEPA, 2008d).

Virtua Health, the largest provider of healthcare in Southern New Jersey, has aggressively pursued energy-efficient upgrades throughout its five hospitals since 1993. The 1100-bed institution went beyond lighting retrofits and adopted an integrated approach to building upgrades with additional work on building automation systems, motors, and air handling units in their newly constructed and existing facilities.

In 1993, Virtua established a new energy program and launched an aggressive energy-efficienty upgrade campaign. As part of this initiative, an Energy Director was appointed to oversee the energy program and to keep senior management involved and supportive. To ensure the continued support of senior management and to drive progress, Virtua Health's Energy Director submits monthly reports directly to the chief executive officer and president. Additionally, the Energy Director keeps Virtua's Board of Directors informed and supportive by providing annual reports on its strategic energy management initiatives.

Maintaining senior management involvement and support has helped to elevate the importance of energy management throughout the organization and has delivered results. Through its investments in energy efficiency and a comprehensive energy management plan, Virtua Health has saved more than $1,064,400 in operating and energy costs (USEPA, 2008e).

Food Lion, one of the largest supermarket chains in the United States, has more than 1220 stores in 11 Southeastern and Mid-Atlantic states. More than 400 Food Lion stores have achieved the ENERGY STAR label designation and the chain has a portfolio-wide energy performance rating above 75.

Energy efficiency is a key focus for Food Lion's upper management. Between 2000 and 2005, Food Lion quadrupled its Energy Department staff to 14 individuals. Most significantly, Food Lion in 2004 added an EPA coordinator, who provides energy analysis and ensures that Food Lion monitors its direct and indirect environmental impacts. Food Lion's upper management encourages the energy team to pursue such projects as load control, peak-shaving generators, and innovative procurement strategies.

Food Lion's aggressive efforts to reduce energy consumption have saved nearly $104.8 million in actual dollars. Food Lion estimates that the company's 2005 energy budget would have been nearly $46 million higher without the energy-efficienty initiatives adopted in the previous four years. Between 2000 and 2005, Food Lion successfully decreased its kBTU/square foot by more than 25%.

Energy costs are a grocer's second-largest expense and can easily consume as much as 10% of each store's costs. Controlling and reducing energy costs is critical as Food Lion works to keep customer prices low. Because of its commitment to energy efficiency and low prices, Food Lion nearly 10 years ago created an energy team to address energy management issues. In 2003, Food Lion reorganized several departments to create a unified department that reported to a single manager. In particular, Food Lion merged the energy management and maintenance departments under a common strategic sourcing manager of energy and maintenance. By bringing energy management decisions under one manager, Food Lion was able to save energy and money. In one year, integrated decision making helped reduce maintenance and energy expenses by several million dollars.

Food Lion's energy policy states

Food Lion has a reputation for providing convenient grocery store locations with products at extra low prices. As part of our commitment to excellence, Food Lion will identify and implement improved financial and operational efficiencies in how we purchase and consume energy, striving to become a world-class leader in energy management within the supermarket industry.

Energy Management will play an increasingly important role in achieving our strategic objectives. Specifically, Food Lion's Energy Management Strategy is to

- support the organization's strategic plan to sharpen our pricing and promotion position, improve convenience of the shopping experience, enhance our fresh product perception, and achieve executional excellence;
- support our commitment to our employees, the environment, and the community in which we conduct business by improving the environment through active efforts to reduce energy consumption and pollution;
- become one of the most efficient grocery stores in the world on a Btu per square foot basis (USEPA, 2008f).

3.1.2. *Step 2: Assess performance*

Assessing performance is the periodic process of evaluating energy use for all major facilities and functions in the organization and establishing a baseline for measuring future results of efficiency efforts. This is accomplished by gathering and tracking data, establishing baselines, benchmarking, analyzing energy use and patterns, and technical assessments and audits.

Collecting and tracking information on how, when, and where energy is being used are necessary for establishing baselines and managing energy use. Important steps when collecting data include determining the appropriate level of detail; accounting for all energy sources whether purchased or generated on-site on both inventory and cost bases; documenting all energy uses by assembling energy bills, meter readings, and other use data; and for normalization and benchmarking purposes, it may be necessary to collect facility and operational data such as building size, operating hours, etc. A system for tracking performance can range from a simple spreadsheet to detailed databases and IT systems. When developing a tracking system one must keep in mind that the design of the system will be shaped by the level and scope of the information that will be tracked and the frequency of data collection; that the system must be easy to use, update, and maintain; and that information must be formatted in such a way that energy performance is expressed in ways that are easily understandable across the organization. Existing tracking systems such as ENERGY STAR's Portfolio Manager and Energy Performance Indicators (EPIs) may be used to organize data and benchmark against the industry.

The energy use of a facility may depend on factors beyond the energy efficiency of the equipment and operations. Such factors may include weather or certain operating characteristics. Hence, to make valid comparisons among facilities, data must be normalized to account for the impact of these factors. To normalize first determine key normalization factors. For commercial and institutional buildings, common normalization factors include climate zone, facility size, fuel choice, price/cost of energy, actual weather history, hours of operation, occupancy levels, and other special features. For industrial facilities, common normalization factors include inputs, product type, outputs, and production processes. When normalizing, we should seek a variable that may be used to correct for differences in facilities. Among suitable metrics that may be used for correction are floor space, amount of product, value of shipments, or number of beds. Finally, after the metrics have been selected they should be weighted by a multiplier that reflects the importance of each variable in relation to its impact on energy use.

A baseline for measuring energy performance should be established. Typically a particular year or average of years serves well as a baseline. Baselines should be established for all appropriate levels of the organization. Once the baseline time period is established, select units of measurement, such as Btu/square foot, Btu/product, or total energy cost/square foot, that effectively and appropriately express energy performance. The performance baseline should be communicated to facilities managers and other key stakeholders in the organization.

Benchmarking means comparing your energy performance against similar facilities. Benchmarking may be done in few ways. These include comparing present performance to past performance of a facility; by comparing performance to the average performance of a peer group; by comparing performance to the best in the industry; and by conducting a qualitative comparison to established practices considered to be best in the industry. Key benchmarking steps include determining at what level to benchmark, for example, equipment, process, facility, or organizational; developing benchmarking metrics; conducting benchmarking comparisons; and tracking performance over time.

For certain types of facilities, benchmarking may be accomplished with the use of ENERGY STAR's Portfolio Manager, which lets you compare the energy performance of your facilities to similar buildings nationwide. Portfolio Manager also normalizes for weather and several other important building and operational characteristics. By inputting energy and building information, this tool provides a benchmark score on a scale of 1–100. Buildings with a score of 75 or over are eligible for the ENERGY STAR label. EPA recognizes buildings with superior performance. Buildings that earn the ENERGY STAR label use about 40% less energy than average buildings, without compromising comfort or services. Benchmarks are currently available for office buildings, K-12 schools, grocery stores, hospitals, and hotels.

The analysis of energy use and patterns may be accomplished by quantitative and qualitative reviews. Quantitative reviews include the development of user profiles, which identify usage patterns and their relationship to operations or key events; the comparison of performance data of similar facilities in a specific industry; assessing the financial impact of energy use to identify areas of high cost; and determining areas where more information is needed. Qualitative reviews are accomplished by interviews, which seek informed opinions from colleagues, specific anecdotes and lessons learned, systems-specific information (e.g., HVAC, lighting, refrigeration), and in-house audits or surveys. Policies and procedures should also be reviewed to determine their impact on energy use.

Periodic technical assessment of the performance of equipment, processes, and systems will help identify opportunities for improvement. Energy audits should be conducted by energy professionals with the purpose of evaluating the actual performance of a facility's systems and equipment against their designed performance level or against best available technology. The difference between these is the potential for energy savings.

Examples of successful performance assessment include the following from 3M, Marriott, and Food Lion.

3M's energy program oversees hundreds of facilities worldwide and all are required to report on energy efficiency. To make this possible, 3M has devised a system to prioritize energy reporting: facility size dictates the level of required reporting. Each of 3M's plants, and any other facility over 30,000 sq. ft, must report energy use, intensity, and cost monthly. Because facilities' products and processes vary greatly, each facility is measured against its own previous performance. While 3M facilities can not benchmark their performance against each other, their performance is closely monitored and benchmarked over time (USEPA, 2008g).

Marriott International, Inc. is a worldwide hospitality company with more than 2600 lodging properties located in the United States and 65 other countries and territories. Marriott Lodging operates and franchises hotels under 16 brands including Marriott, Renaissance, Courtyard, Residence Inn, Fairfield Inn, and the Ritz-Carlton.

To assess performance, Marriott developed an extensive program to ensure that accurate measurement would lead to realistic goal-setting. After partnering with ENERGY STAR in 2001, Marriott embarked on an ambitious, company-wide mission to enhance its energy management systems (EMS) and procedures. To carry out this mission, Marriott needed a uniform auditing and inspection system for each of its properties. Marriott's solution was to collect information at each of the properties through two methods: a bimonthly property audit called an Energy Awareness Measure and daily Routine Inspection Reports, which ensure that all associates are actively participating in the property's energy conservation action plan. At the end of each quarter, the property's General Manager and Energy Conservation Committee meet to review the audit forms and inspection reports that have been filed over previous three months. This review is the foundation for setting goals for the next quarter. This full-circle approach to observing and improving EMS and energy management procedures contributed to Marriott saving $4.5 million annually (USEPA, 2008h).

Food Lion developed an energy program based on internal and external benchmarking. Internally, Food Lion uses normalized quarterly reports, created by Facility IQ, to rank the efficiency of its stores by region. The energy team distributes these rankings to associates across the chain through individualized Performance Scorecards and Maintenance Scorecards, which detail the store's energy use and provide a summary of quarterly energy and year-to-date capital and refrigeration costs. Food Lion posts the scorecards on bulletin boards and its intranet, and sends e-mail notifications to upper management.

Externally, Food Lion compares how its stores are performing against similar stores in the country by using the EPA's energy performance rating system to generate ratings for each store. In 2004, Food Lion expanded its use of the rating system for supermarkets by partnering with Facility IQ and the EPA to generate automated monthly ratings within the Facility IQ environment. With this system, the Energy Team easily views trends on a monthly, quarterly, and yearly basis, both internally and against competitors. Using this automated system has allowed Food Lion to quickly identify energy-wasting problems and see the results of energy-saving initiatives on a much broader level than in the past. Food Lion has used the EPA's energy performance rating system to benchmark 100% of its more than 1200 stores, to identify more than 1000 stores eligible to receive the ENERGY STAR designation, and to complete the ENERGY STAR labeling process for more than 400 stores.

Food Lion collects real-time data to identify opportunities for increased energy efficiency. Collecting and analyzing real-time data allow Food Lion to identify opportunities for savings, including increasing equipment efficiency, real-time testing for system design, and identifying problems in real time. Food Lion uses real-time data to compare the true efficiencies of equipment and sets alarms to automatically notify management when loads exceed normal capacity. This lets Food Lion's management react quickly to changes in energy consumption, such as by sending a maintenance technician to fix equipment within hours or days rather than weeks or months.

Additionally, Food Lion's ability to track data at such levels gives the grocer leverage to convince their refrigeration suppliers to improve the energy efficiency of their cases by 50% resulting in a savings of tens of thousands of dollars per store annually.

To maximize energy performance, Food Lion continually explores opportunities for improvement. Energy audits and technical assessments help Food Lion identify these operational opportunities. At the heart of Food Lion's cutting-edge energy program is a commitment to innovation achieved through hands-on application and testing of existing and new technologies. Food Lion uses benchmarking to identify high-energy-use stores and sends in-house Energy Field Specialists to perform a High Utility Checklist. This checklist is used to verify that all set points in regards to HVAC, lighting, deli equipment, and refrigeration are correct at the store. Stores that underwent energy audits experienced reductions in energy consumption as great as 12% after the checklist was completed (USEPA, 2008i).

3.1.3. *Step 3: Set goals*

Setting goals involves determining scope, estimating the potential for improvement, and establishing the goals. The scope of a goal can include its organizational level and time period. Organization-wide goals provide an encompassing view of how the organization wants to improve. Such goals are useful for communicating the success of energy management to internal and external audiences. Facility level goals may take into account the performance of specific facilities. Goals at this level may be based on the results of benchmarking or energy audits. Process or equipment level goals may be established for specific process lines and equipment when energy use is concentrated in these specific areas. Goals should be expressed for both short-term and long-term time frames. Annual goals provide points for tracking and reporting progress on an on-going basis. Long-term goals may be shaped by financial measures, internal planning horizons and guidelines, organizational strategic plans, and commitments to voluntary environmental initiatives.

Potential for improvement is determined by constraints including the level of performance that is achievable based on the level of resources available to support the improvement effort. Methods that have proved successful in estimating the potential for improvement include reviewing performance data, benchmarking, evaluating past projects and best practices, reviewing technical assessments and audits, and linking to the organization's strategic goals.

Energy performance goals should be formally established and supported by senior management as mission critical for the whole organization. For some organizations, estimating potential for improvement provides a starting point for setting realistic goals; for others, final energy performance goals are based on organizational factors other than what is simply technically feasible. The approach that is adopted will affect how energy performance goals are expressed. Common ways for expressing goals include defined reduction, best-in-class, efficiency improvement, and environmental improvement.

Toyota Motor Manufacturing North America (TMMNA) oversees manufacturing operations for vehicle assembly, parts, and materials in North America. TMMNA operates assembly and other manufacturing plants in nine states as well as Mexico and Canada. "Kaizen," which means continuous improvement, is incorporated into design principles for all new Toyota Motor North America facilities. Kaizen is applied to energy use in all facilities, and so Toyota established the goal that each new production plant must be more efficient than the last plant built. Toyota believes this goal is achievable and a critical part of their corporate energy program. They have also found that implementing the Kaizen objective has lead to new technological improvements in plant design and construction. Setting specific performance targets for new plants has helped Toyota decrease energy intensity by 26% in 8 years (USEPA, 2008j).

Food Lion's organizational goal in relation to energy management is clear to become one of the most efficient supermarket operators in the country on a BTU per square foot basis. Food Lion's goal to be one of the most efficient grocery stores in the nation serves as an umbrella for supporting goals set at various levels within the organization. At the facility level, Food Lion has a goal to ensure all new stores qualify to earn the ENERGY STAR. Food Lion also uses the EPA's energy performance rating system as the basis for the energy portion of the chain's individualized Performance Scorecards and Maintenance Scorecards. At the process and equipment level, energy performance goals drive increased efficiency for refrigeration and HVAC equipment. In addition, Food Lion ties incentive bonuses to goals for the maintenance associates. Food Lion has reduced its energy consumption by more than 2.29 trillion BTUs since 2000, which is the sales equivalent of nearly $1 billion (USEPA, 2008k).

3.1.4. *Step 4: Create an action plan*

With goals in place, a detailed action plan must be developed to guide the organization in achieving them. This is generally an annual plan, which is frequently updated. This plan is developed by analyzing the planning gap, that is the difference between current performance and desired performance as expressed by the goals. Once the planning gap has been determined, the technical steps required to move energy performance from the present state to the desired state should be identified. Performance targets must then be created to allow tracking the progress toward closing the planning gap and achieving goals. These targets must be aligned with specific goals. Hence, they must be stated in terms of facility, process, department, or operation, and on a time line including specific milestones and expected outcomes.

The plan must include specification of which departments and organization members will be responsible for what actions. The degree to which external contractors, if any, will be utilized as part of the plan must be specified. Metrics for gauging the performance of any external contractors must be stated. Further, for each project in the action plan costs, in terms of both financial and human resources, must be estimated and a business case for obtaining these resources from top management must be developed.

Frito-Lay is a division of PepsiCo and manufactures many food products including the Lays, Fritos, and Doritos Brands. PepsiCo divisions include Pepsi-Cola,

Quaker Foods, Gatorade and Tropicana. Frito-Lay operates 33 manufacturing facilities in the US and owns the 6th largest truck fleet in the US.

Facility level targets are critical to meeting energy goals. In Frito-Lay's energy program, each facility must create its own action plan, called a "playbook." Playbooks must be designed to deliver 150% of the facility's energy target, thereby accounting for potential delays or projects that do not deliver sufficient energy savings. Playbooks are further refined with support from regional energy captains and Frito-Lay's energy department. Each item in the playbook states the project's potential for savings, financial value, project leader, completion period, as well as necessary support or funding. An audit of each action item is conducted upon its implementation to verify the delivery of estimated conservation and savings. When projects require capital investments the energy director works with the facility and corporate financing.

While many energy projects guarantee substantial, long-term savings, their payback potential is usually assessed using timelines that are more appropriate for other types of investments. As a result, energy projects often do not get the funding that they need to get off the ground. Frito-Lay's energy manager works closely with senior financial management as part of the company's energy policy and has been able to demonstrate that most energy projects have less risk and higher success rates than other types of projects. As a result, senior management has lowered the hurdle rates for energy projects and increased the acceptable simple payback for energy projects (2008c).

Toyota recognizes that planning for energy efficiency when constructing a new building and installing process equipment is more cost effective than retrofitting after construction. To help ensure energy-efficient designs, the company involves its well-established energy management program team to review projects and provide guidance. Toyota has made energy design feedback an integral part of both new construction and expansion projects. Design feedback sheets are developed by the corporate energy team in response to proposed designs, and provided to plant construction and process installation groups for each project to ensure maximum efficiency within the project budget.

Commitment to continuous improvement of energy performance requires having systems in place to identify opportunities for energy savings. Toyota accomplishes this through their successful plant energy assessment format, which they call a "Treasure Hunt." A Treasure Hunt is a plant-wide assessment conducted to find energy reduction opportunities, and takes place in the span of a weekend. It is Toyota's primary means of identifying opportunities to implement concepts from their energy reduction database at a specific plant.

The Treasure Hunt uses the combined knowledge of the energy team and plant engineering, maintenance, and production staff members. After the hunt, all opportunities identified are prioritized based on payback periods and difficulty of implementation. The three most significant energy-saving opportunities identified are presented in an "out brief" or closing meetings by the Treasure Hunt team. Events like this have allowed Toyota to continually improve their energy performance, decreasing energy intensity 8% while increasing production 4% in 2005 alone (USEPA, 2008m).

3.1.5. *Step 5: Implement plan*

The successful implementation of any plan depends mostly on the people who are charged with implementing it. An energy program, is no exception. To successfully implement an energy program the support and cooperation of key people at different levels within the organization are critical. To accomplish successful implementation, a communication plan must be created, organizational awareness of the energy program must be raised, the capacity of the staff to implement the program must be built, staff must be motivated to achieve energy goals, and progress in implementing the plan must be checked and monitored regularly.

The communication plan must identify key audiences, determine the information that they need, and customize messages for each one. ENERGY STAR provides an online Challenge Toolkit that contains deas, examples, and templates that an organization can customize to help communicate with employees, customers, and other stakeholders.

Raising awareness of energy use, goals, and impacts involves increasing general energy awareness, improving facility energy awareness, and gaining management support. The purpose of increasing general energy awareness is to help people understand how their actions affect energy use. A benefit of this increased awareness is achieving greater support for energy initiatives. Increased awareness may be achieved by a number of tactics including new employee orientation programs, poster campaigns, participation in Earth Day events, publishing information on energy use, environmental impacts, and energy-saving options geared toward a general audience on the organization's web site or intranet site, including energy-savings tips and energy-efficient product information with pay statements, and conducting an energy fair or summit oriented towards employees with information on energy-saving activities and products. Facility energy awareness activities can include providing facility energy summary statistics including overall energy costs, costs to operate equipment, environmental information related to energy use, etc.; providing information on the sources of energy used in the facility along with their associated environmental impacts; providing information on the energy use of equipment and processes that employees regularly use as part of their jobs; developing graphics that illustrate energy performance compared with a national standard. Increasing the awareness of all managers throughout the organization can help build support for energy initiatives. Once again, to be most effective the message must be tailored to key audiences that have been identified. The message must be repeated through regular contact with managers aimed at keeping them informed of progress or changes in performance.

Building staff capacity requires training and the development of knowledge and management information systems. It has been shown that well-trained employees are more likely to contribute ideas, operate equipment properly, and follow procedures, helping to guarantee that capital investments in energy improvements will realize their potential. Training is situational in nature, depending on the requirements of the particular facility or organization. Common training programs include operational and procedural training, which provide instruction on new operating methods or procedures designed to reduce energy use; administrative training including reporting, monitoring, data collection, and other administrative activities that support energy management; and specialized training on using and maintaining equipment or tools to ensure more efficient operation. Computer-based information systems should be developed to manage the energy-related knowledge of the organization and provide management information for decision making. While they are organization-specific, typical information systems contain information on best practices that catalogs successful and effective practices for energy management within an organization; on technologies that contains information on known, used, or recommended technologies, equipment, lighting, HVAC, etc.; and on procedures that houses up-to-date information on specific procedures and operating practices.

Motivating employees to have an interest in energy initiatives is important for the success of the plan. Incentives have been used as a device to foster such motivation. Typical incentives that have been used with some success include fostering internal competition through the use of scorecards to compare the performance of similar facilities; recognizing performance by rewarding the accomplishments of individuals, departments, and facilities; offering financial bonuses and prizes if energy goals are met; using environmental messages to promote a sense of environmental and social responsibility; and tying employee performance standards to energy goals.

The purpose of the tracking system is to provide means for periodic review of the plan to assure that the goals of the energy management program are achieved. For the tracking system to be effective, data need to be updated on a regular schedule, usually weekly or monthly. Monitoring is achieved through a periodic review of the organization's progress in achieving targets and goals. Both the energy team and the management team should be involved in the process of completing these reviews. The reviews should focus on progress made, problems encountered, and potential rewards. Any necessary corrective actions should be identified as a result of this review process.

Food Lion recognizes that employee involvement is critical to the success of energy plan implementation. Food Lion targets energy program efforts to specific employment levels, with the use of Performance Scorecards and financial incentives for maintenance personnel, energy awareness training for managers, and internal recognition for maintenance and retail associates (USEPA, 2008n).

As part of its comprehensive energy management plan, General Motors Corporation (GM) develops "Energy Sufficiency Plans" that focus on individual employees taking responsibility for energy in their own work area. These plans provide detailed information on procedures for turning off equipment, lights, fans, etc. and steps for identifying energy waste in each area of a plant.

The steps and actions outlined in the "sufficiency plans" also identify the person who is responsible for managing a particular piece of equipment and the expected energy performance and savings. GM's energy management program provides training and support as needed to individual operators to help ensure their success. Involving personnel at the operator level has paid off. GM estimates that the sufficiency plans have helped save over $57,000,000 within two years (USEPA, 2008o).

In 2004, Marriott's Energy Conservation Program focused on developing new approaches to preventive maintenance, adopting new technologies where feasible, expanding partnerships, and adopting new procedures. Marriott's preventive maintenance program aimed to maximize the efficiency of the products and procedures already in place. Hotel engineers used handheld electronic devices to track preventive maintenance tasks, and to build accurate engineering databases that yielded invaluable information about equipment performance and maintenance issues. Marriott continues to identify and utilize high-tech approaches to preventive maintenance such as thermoscans to identify leaking roofs or "leaking" electrical panels. Additionally, Marriott regularly participates in the development of new technology such as Ozone Activated Air Laundry systems, which almost eliminates the need for hot water. In addition to this, Marriott is working with their chemical supplier, Ecolab, who has recently produced break through technology reducing cost up to 25% and also newly tested technologies for water reuse.

As a part of Marriott's close partnership with owners and franchisees, Marriott shares long-range energy management plans; provides access to its engineering database; reports regularly through its intranet site, meetings, and presentations; and supports purchasing of new equipment. When Marriott identifies opportunities for increased energy efficiency, it changes the way it does business. For example, Marriott consolidated its laundry service, moving to larger, central facilities that utilize more efficient equipment.

Marriott achieves $4.5 million annually in energy savings through the successful implementation of its Energy Conservation Program (USEPA, 2008p).

At Toyota every plant-level Energy Management Organization (EMO) develops an action plan to meet each specific corporate energy goal and involves all plant team members. These EMOs report to the corporate team on the action plans they develop and identify reduction targets, key improvements to be implemented, and each project's cost and schedule. During the actual implementation stage for energy projects, each action plan is reviewed and updated quarterly with senior management. This multi-level approach to energy initiatives is the backbone of TMMNA's extensive energy program, and the paybacks are not hard to find: TMMNA met its goal of a 15% energy intensity reduction by 2005, two years early.

As one of the activities of their EMO, TMMNA makes educating all employees on energy management a priority. Their goal is to encourage the adoption of the company's resource management philosophy and culture. Toyota has discovered that corporate-wide communication helps to "yokoten" or transfer best practices from one plant to another. For this reason, the EMO takes creative approaches to communicating energy management, including energy fairs where innovative prizes are awarded, monthly awareness campaigns on a chosen subject, quarterly energy newsletters, as well as speeches and presentations given by corporate management.

Toyota goes beyond raising company awareness by sharing best practices and energy messages with suppliers. This helps Toyota reach its goal of reducing energy intensity at all phases of manufacturing. These awareness campaigns lead to strong support of energy programs resulting in one North American plant saving 50,000 MMBtu annually.

TMMNA has developed an internal database where over 14,500 energy-saving ideas are made accessible to all plant energy teams. The database is used by plant energy managers looking for energy-saving projects. Toyota estimates that its North American plants have implemented 40% of these best practices. To demonstrate potential energy savings of the "Kaizen Database," Toyota used their Texas Manufacturing facility as a source for design feedback. If the Texas plant implements all the design feedback ideas the plant's energy would be reduced by 3.18 MMBtu per vehicle compared with the mother plant.

To closely monitor manufacturing energy use, Toyota established a metering standard stating that all plants should record hourly electric data broken down by process equipment. Toyota requires all plants to meter large energy-consuming equipment (such as the paint shop) hourly and analyze energy use trends throughout the day and during downtime. This allows energy teams to identify energy waste, such as during lunch and between shifts. All metering data are available in a central database for employees to review. The metering data are reported to Toyota's North American headquarters, and plant energy teams discuss metering trends at daily and weekly energy briefings to determine where and how energy improvements can be made. These practices helped Toyota to meet their energy goals two years early (USEPA, 2008q).

3.1.6. *Step 6: Evaluate progress*

Evaluating progress involves measuring results by comparing current performance with established goals, and reviewing the action plan based on this comparison. Measurement involves gathering tracking data, reviewing energy use and cost data, organizing reports and data from tracking and monitoring efforts, and analyzing energy-efficiency achievements based on the established performance metrics. In addition, benchmarking should be done including comparing energy performance with baselines, comparing performance against established goals for environmental performance and financial savings, and comparing energy performance with peers and competitors. The action plan review should look at the effectiveness of the action plan. Where activities and projects were successful, best practices should be documented to share throughout the organization. Where goals were not met, the cause must be determined and decisions must be made on what corrective or preventive actions should be taken.

Food Lion needed a way to measure the environmental and financial results of their energy-saving initiatives. Food Lion's Energy Team evaluates progress by integrating data from an outsourced bill handling provider and its internal accounting department to measure the environmental and financial results of their energy-saving initiatives and establish meaningful ways in which to communicate results to associates and stakeholders. For example, Food Lion states that the 312 billion BTU reduction from its energy program in 2005 has saved millions of dollars. In the prior five years, Food Lion reduced energy consumption by more than 2.29 trillion BTUs. In order to earn the same income, Food Lion would have to sell nearly $1 billion worth of merchandise (USEPA, 2008r).

Ford uses ENERGY STAR's EPI to benchmark assembly plants, and Portfolio Manager to benchmark office buildings, which have corporate, research, and engineering facilities. Ford created an EPI results table to compare all facilities, and discovered that the implementation of various energy projects at one of Ford's oldest plants resulted in an efficiency rating typical of new facilities. Ford helps employees to understand benchmarking scores through an "EPI Single Point Lesson," which gives an on-line overview of the tool in one sitting (USEPA, 2008s).

To achieve their goal of "Kaizen" or continuous improvement, Toyota has created an extensive progress evaluation structure that involves goal review, action plan review, and pilot test projects. Goal review happens on weekly, monthly, and annual basis, thereby ensuring that there are no surprises on the road to meeting energy goals. Key Performance Indicators (KPIs) are reviewed monthly at every plant during energy meetings. Next, all plant action plans are reviewed and updated monthly by senior management to evaluate how a plant is meeting its energy reduction goals, target EPI scores, and key project implementation plans. Year-end energy summary reports give plant managers a benchmark of how well they are achieving their energy objectives and the best practices they can adopt from other plants (USEPA, 2008t).

3.1.7. *Step 7: Recognize achievements*

Recognition may come from both internal and external sources. Internal recognition supports the importance of the energy management program. Rewarding particular efforts sets the example for what constitutes success and helps motivate employees through increased job satisfaction. Recognition may be given at the individual, team, or facility level. Criteria for recognition must be established and communicated throughout the organization. The type of recognition must also be decided on. Depending on the purpose of the recognition program and the organizational culture, forms of recognition can range from formal acknowledgements and certificates, to salary increases and cash bonuses, to simple forms of appreciation such as coffee mugs or energy program shirts. External recognition from a third party can provide validation for an organization's energy management program. Some ways to gain external recognition for energy management efforts include participating in established groups, such as government agencies, trade associations, or regional energy conservation groups to demonstrate commitment to achieve results; meeting widely recognized standards of performance, such as those established by ENERGY STAR, that reflect superior performance; qualifying for achievement awards by surpassing a variety of predetermined criteria that identify superior energy management programs; and reporting progress publicly and to targeted stakeholders that monitor and critique energy performance.

Frito-Lay has found that applauding their site energy champion's efforts reinforces the importance of energy management within their large organization including 37 manufacturing plants. Frito-Lay uses its annual energy summit, attended by over 300 employees, as the setting for a National Recognition Dinner. The energy champions of the highest performing plants are invited to this event, dine with senior executives, and are recognized for their contributions towards Frito-Lay's energy goals. Providing this kind of recognition for plant energy managers demonstrates to all employees that their involvement in energy management is highly valued, and crucial in maintaining Frito-Lay's energy performance improvements, which have reduced greenhouse gas emissions every year by over 200,000 metric tons of carbon equivalent compared with emission levels in 1999. These reductions offset emissions equivalent to 140,000 cars (USEPA, 2008u).

Food Lion's energy management strategy is today a core value ingrained in its operations, philosophy, and culture. Receiving recognition from third parties confirmed their position as a leader in their industry and boosts employee morale and company public image. Food Lion has received many awards for its outstanding energy performance, including two ENERGY STAR Partner of the Year awards and three ENERGY STAR awards for Sustained Excellence in Energy Management (USEPA, 2008v).

4. ANSI/MSE 2000–2005 Standard

Believing that traditional management practices are not up to handling the challenges of today's business environment, the Georgia Tech Energy and Environmental Management Center (EEMC) proposed a standard for energy management based on the Deming Plan–Do–Check–Act (PDCA) approach to quality improvement. The main failing of the traditional management approach is in the lack of follow-up of executed plans. Brown (2006) comments on the impact of this failure for energy management

> "From an energy manager's point of view, these inadequacies mean that energy management projects may not be as successful as predicted, project results are not recognized, improvements are never incorporated into daily operations, and savings are gradually lost, or projects may never even be started due

to inadequate resource support. A dynamic environment necessitates flexible management capable of achieving cultural change in the organization. To accomplish this, the simple "plan–do" approach to energy projects must be expanded to encompass a feedback loop that actually requires management involvement and support, and encourages and supports continual improvement."

The PDCA cycle has been used in the ISO 9000 and ISO 14000 standards, which focus on quality improvement and environmental management systems, respectively. The MSE 2000–2005 standard brings this approach to bear on the problem of energy resource management.

4.1. *The PDCA cycle*

In its truest form, the PDCA cycle is a system for managing organizational change. This is accomplished through the cycle's plan and implement elements.

The aim of the Plan phase is to determine an improvement or innovation to a process that will narrow the quality gap between existing process performance and customer and other stakeholder needs and wants. Essentially the plan phase includes problem recognition and the development of actions to solve the problem.

After a plan has been developed, it must be implemented. This begins with the Do phase in which the plan is implemented according to a well-defined process. The implementation continues with the Check phase, the purpose of which is to determine if the plan has been effective in decreasing the difference between process performance and customer and other stakeholder needs and wants. The check phase may include measurement and monitoring activities. Checking is not limited to project review, but also must focus on the review of the management systems that are in place.

Finally, in the Act phase the organization takes action on what has been learned in the Check phase. If the desired result was not achieved, the cycle repeats with a revised plan. If the planned result was achieved, the planned changes must be integrated into the process. This success forms the basis for planning new actions to yield even greater improvements

4.2. *The ANSI/MSE 2000:2005 standard*

The standard originally developed at Georgia Tech was first adopted by the ANSI in 2000. The standard "describes the elements necessary to establish and maintain an effective strategic MSE in a variety of organizations. The management system described by the standard is based on the Deming PDCA management cycle and can be used to establish a management system that actively promotes continual improvement" (2006b). After a required five-year review, a revised standard, MSE 2000:2005 was adopted by ANSI in July 2005.

Table 2 presents the elements of the ANSI/MSE 2000:2005 standard. The standard is organized according to the PDCA cycle as shown by elements 6.0, 7.0,

Table 2. ANSI/MSE 2000:2005 standard elements.

Standard Element	Content
4.0 Management system for energy	
4.1	General requirements
4.2	Documentation requirements
4.3	Record keeping requirements
5.0 Management responsibility	
5.1	Management commitment
5.2	Energy policy
5.3	Strategic planning
5.4	Authority and responsibility
6.0 Energy management planning (PLAN)	
6.1	Energy profile
6.2	External information
6.3	Energy assessment
6.4	Goals and targets
7.0 Implementation and operation (DO)	
7.1	Purchasing
7.2	Facility, equipment, and process control
7.3	Energy management projects
7.4	Control of outsourced energy services
7.5	Communication
7.6	Training, competence, and awareness
8.0 Checking and evaluation (CHECK)	
8.1	Energy monitoring and measuring
8.2	Internal MSE audits
8.3	Corrective and preventive
9.0 Management review (ACT)	

8.0, and 9.0. In addition to these four elements, the standard contains two support elements — 4.0 and 5.0. Element 4.0 contains requirements to support the general management system, whereas element 5.0 details the management attention and responsibility required for the development and operation of a successful EMS.

4.2.1. *Management systems for energy*

Element 4.0 describes the basic requirements of an MSE. This is accomplished through the specification of the documents and record keeping requirements that describe the organization's MSE. General documentation requirements include documented statements of energy policy, energy goals, and targets; an energy manual; documented procedures required by the standard; and documents needed by the organization to ensure the effective planning, operation, and control of its significant energy-related processes and equipment.

The energy manual must document the scope of the MSE and address the requirements of the standard. A procedure for the control of all documents required by the standard must be put in place. Finally, the organization must document procedures for managing energy records.

4.2.2. *Management responsibility*

As everything begins with senior management, let us first more closely look at element 5.0 — Management Responsibility. Management is chiefly responsible for communicating its strong support for the effective management of the energy resource by including energy as a major factor in strategic planning along with the other strategic level functions — operations, marketing, and finance. The management result of this strategic thinking about energy is the development and statement of the organization's energy policy.

The energy policy should describe the overall direction and commitment to energy management by the organization. The energy policy serves to focus management efforts on the effective and efficient use of energy. An important part of this policy is a statement of top management commitment to continuous improvement of energy utilization. In addition, in developing the policy top management should consider efficient and effective use of resources, energy-related environmental impact, available alternate sources of energy, level of quality and appropriateness of sources, and effect on operations. A precursor of energy policy development is the establishment of the organization's energy management team. This team is led by the energy coordinator who is appointed by management and becomes the focal point for all MSE activities. Members of the team are selected from the major functional areas of the organization that have contact, dependence, or responsibility for energy resources. The energy management team, along with executive management, is responsible for developing the energy policy.

4.2.3. *Energy management planning*

Energy management planning is the subject of element 6.0. In this phase, we are initially concerned with learning how energy is used in our facility. We accomplish this by developing energy profiles for our facilities, obtaining information on energy costs from utilities and other providers, completing an energy assessment, and finally developing a set of energy goals and targets.

Energy profiles detail the exact sources of energy consumption in a facility. The energy profile uses energy, production, and financial information to reveal trends, anomalies, price signals, and energy and cost allocations to provide a perspective on the impact of energy on cost of operation (American National Standards Institute, 2005h). The concept of Energy Accountable Centers (EAC) may be used to assist in developing energy profiles. EACs are production areas that are neither too small nor too large, in which energy consumption can be independently measured and reported. For example, it may be revealing to monitor the energy consumption of a separate building, a central boiler house, or a specific production line. For a typical building energy use includes the building envelope, heating and ventilation, air conditioning, and lighting functions. For a manufacturing facility, process use must also be measured. The profile includes energy consumption and cost by fuel type for both primary and secondary energy sources. Additionally, an audit of energy use by function must also be included. To construct valid energy use profiles, a number of audits must be conducted. According to Thumann and Mehta (1995b), they include

1. *Envelope audit*: Surveys the building envelope for losses or gains due to leaks, building construction, doors, glass, lack of insulation, etc.
2. *Functional audit*: Determines the amount of energy required for specific functions including heating, ventilation and air conditioning, building, lighting, domestic hot water, and air distribution.
3. *Process audit*: Determines the amount of energy required for each process function including process machinery and equipment; heating, ventilation and air conditioning process; heat treatment; and furnaces.
4. *Transportation audit*: Determines the amount of energy required for forklift trucks, cars, vehicles, trucks, etc.
5. *Utility audit*: Collects data on and analyzes the monthly, daily or yearly energy usage for each utility.

As part of the construction of energy profiles, KPIs should be developed that relate energy or water cost consumption, or environmental impact to the appropriate organizational output. Typical KPIs for a commercial operation are British thermal units (Btu)/square foot or cost in dollars ($)/square foot. For industrial operations, KPIs may include Btu/unit produced, Btu/$-value added, etc. (American National Standards Institute, 2005a).

External information includes tariffs, alternate suppliers, new technology, and forecasts of energy availability and cost. Energy assessment requires the collection of information on the current state of equipment and processes related to energy purchase, use, storage, and disposal. This information is used in identifying opportunities for continuous improvement (American National Standards Institute, 2005a).

Once policy has been established, a set of energy management goals that are consistent with this policy must be defined. These goals must be specific, cover a defined time frame, and be easily measurable where practical. When establishing these goals the MSE team must consider all aspects of the energy planning that have been previously completed including the information collected from the energy profile, external information that has been collected, and the results of the energy assessment. For each goal, one or more measurable targets must be defined. These targets must be expressed in terms of improvement in a KPI, percentage improvement in cost, or another method of demonstrating measurable improvement toward achieving the associated goa (American National Standards Institute, 2005c).

4.2.4. *Implementation and operation*

Implementation and operation is the subject of element 7.0. Components of this element include purchasing; facility, equipment, and process control; energy management projects; control of outsourced energy services; communication; and training competence and awareness.

The purchasing element requires the organization to consider the effects, energy products and services have on the environment and human health when making a purchasing decision. Purchasing procedures must be put in place to assure that the purchase of energy conforms to the energy policy and is consistent with energy goals and targets. Such procedures related to energy supply include:

- *Energy purchasing specifications*: Specifications must be defined in the following areas as applicable: energy quality, availability, capacity, variation over specified time, and billing parameters. The organization must review these specifications for adequacy before release.
- *Evaluation of energy suppliers*: Suppliers must be evaluated on the basis of their ability to meet the organization's requirements including risk, energy quality, and environmental impact.
- *Energy purchasing bids*: The organization shall evaluate offers, tariffs, and bids as appropriate against the organizations specifications.
- *Energy Purchasing Contracts*: Contracts are required to be reviewed for adequacy prior to approval or acceptance. Contract performance of the suppliers must be evaluated periodically during the contract period, and the results of such evaluations must be recorded.

When dealing with the purchase of energy equipment and systems, the organization must consider energy efficiency, recyclability, durability, reusability, recycled content, and product disassembly potential in the selection process, in addition to operations and maintenance costs (American National Standards Institute, 2005d.)

Facility, equipment, and process control relate to the use and control of energy facilities. Records of costs and benefits must be kept for commissioning, recommissioning, and continuous commissioning of new or existing energy facilities. In addition, the organization must ensure control of equipment and processes related to significant energy uses.

Continuous improvement of energy management is achieved through the judicious selection of energy management projects. Brown (2006a) suggests that energy projects should be selected in a hierarchical order that includes at the lowest level of the hierarchy improved energy and fuel purchasing practices. Projects of this nature generally involve the lowest level of investment, generate significant returns, and payback in a year or less. At the next level of the hierarchy are projects that improve operations and maintenance with a resulting positive impact on the efficient use of energy resources. Operations and maintenance projects typically require a higher level of investment than purchasing projects but also generally result in greater return. Such projects take less than two years to pay back. After a commitment has been made to proper purchasing, operating, and maintenance practices, attention may be turned to long-term capital investment projects, which promise the greatest return and have a payback period of 3–10 years. Successful capital projects are built on the stable foundation created by the continuous improvement in purchasing, operations, and maintenance.

For all types of energy projects, the project selection process must consider alignment with the organization's energy policy, impact on goals, costs, expected results, resource availability, and any other relevant factors. The process must include analysis of projected energy and non-energy benefits and implementation and maintenance costs. Implementation efforts must be monitored and appropriate measurements must be made to ensure that planned results are being achieved. If necessary, changes to relevant documentation and training practices shall be made.

When the organization chooses to outsource any energy service, the organization must ensure control over these services. Such control must be consistent with the energy policy and documented within the MSE (American National Standards Institute, 2005e).

Communication of energy-related data and information is required. Items that must be communicated include the organization's energy policy, energy profile, the importance of complying with energy procedures, and the results of energy management projects (American National Standards Institute, 2005a).

Personnel performing tasks that have a significant impact on energy must demonstrate their competence through proof of suitable education, training, skills, and experience. This means that the organization must develop the capability to deliver training to those personnel that require energy education. First, training needs must be identified. Typically, initial training for all employees will include awareness of the organization's energy policy, energy goals and targets, current energy usage and trends, basic MSE terminology, and energy measures. More specific topics such as process energy management, HVAC energy consumption, etc. should be developed for targeted groups. A major objective of training is to help make employees aware of the importance of their roles and actions to achieving the organization's goals and targets. Finally, appropriate records of all training shall be kept.

4.2.5. *Checking and evaluation*

The subject of element 8 is checking and evaluation. Components of this element include energy monitoring and measuring; internal MSE audits; and corrective and preventive action.

The organization is required to monitor energy performance. Such monitoring includes the collection of energy and related data that are used to track consumption, cost, significant energy uses, and changes to KPIs. Additional data must be collected to verify the effectiveness of energy management projects. Equipment used to collect energy data must be calibrated and well maintained.

Periodic internal audits of the MSE are required to determine whether or not the EMS has been effectively implemented and maintained, conforms to the ANSI standard, and to provide information on the results of the audits to management. Any deficiencies uncovered by internal audits are subject to corrective and preventive action. Procedures for such action must be documented. Changes in documented procedures resulting from the implementation of corrective or preventive action must be recorded.

4.2.6. *Management review*

At specified intervals, top management will review the performance of the MSE. Information required for these reviews will be collected by the energy coordinator and presented to management. Such reviews will be conducted with the principle of continuous improvement in mind. The standard specifies the following inputs to the management review:

- results of MSE audits;
- performance of the MSE;
- changes to KPIs;
- extent to which goals and targets have been met;
- status of corrective and preventive actions;
- follow-up on action items from previous management reviews;
- pertinent changes in the organizational or energy situation;
- recommendations for improvement.

The standard further specifies that outputs from the management review will include any decisions or actions related to

- possible changes to the energy policy;
- changes in KPIs;
- changes to goals or other elements of the MSE consistent with the organization's commitment to continual improvement;
- allocation of resources (American National Standards Institute, 2005f).

4.3. *Implementation of the standard — One example*

Brown (2006a) reports the implementation of ANSI/MSE 2000 at C&A Floorcoverings. C&A Floorcoverings is a major carpet and flooring manufacturer with five plants in the Dalton, Georgia, USA. This industrial sector is highly energy intensive, and the company is concerned with both increasing energy costs and environmental impacts. The company was well known for its commitment to environmental stewardship. The company believed that publicly adopting this energy standard would further illustrate its commitment to continuous improvement of environmental impact and energy usage. Further, management believed that formally registering the ANSI/MSE standard could be used as a marketing tool.

C&A Floorcoverings teamed with the EEMC within the Georgia Tech Enterprise Innovation Institute for the ANSI/MSE 2000 implementation project. Broad support for the implementation was gained from top management and traveled through the organization in a top-down fashion supported by a communication strategy tailored to various organization stakeholders. Owing to resource constraints, the initial decision was made to implement the energy standard only in the two largest plants. The Yarn and Dye plant is 162,000 ft^2 and the Finishing plant is 250,000 ft^2. Both include a variety of highly energy-intensive equipment.

The implementation process proceeded with the writing of the energy manual. The energy coordinator developed policies appropriate to each element of the standard. These policies were codified in the first draft of the energy manual prepared by the energy coordinator. The energy management team reviewed the draft, made recommendations for changes, and participated in the revisions, which resulted in the final version of the energy manual. The energy manual served as the guiding document for the entire EMS for all five plants.

Procedures and work instructions were then developed, which were tailored to the particular facility. Written procedures were initially developed for those operations affecting energy cost, consumption, or disposal. These were then reduced in number by concentrating on significant energy users and those processes necessary for operations. For clarity, each procedure was formatted to include information needed for document control and revision. Many of the system procedures were applicable to all five plants and did not have to be redeveloped.

The first procedure developed was that for completing an energy assessment. This would assure that the energy management organization would hit the ground running. Next the team developed their program for energy management projects. This program ensured that energy projects would address the business objectives and goals of the company and that any changes made as a result of a project would be incorporated into system documentation and training. The corrective and preventive action system provided a method to track and effectively solve any problems with either operations or the MSE itself. The training system pertains only to personnel who significantly affect energy purchasing, storage, usage, or disposal. Thus, the system is not unwieldy; it includes a training needs analysis and documentation that the training was completed. Once a year, C&A Floorcoverings holds a formal management review meeting. This meeting includes upper management and enables them to remain appraised of progress toward goals through energy management projects. It also affords an opportunity for them to examine results from internal audits of the system and review corrective and preventive actions.

The implementation of ANSI/MSE by C & A resulted in a system, which provides consistent sustainable, results from energy management projects. For example, ovens are significant energy users within the carpet industry. Therefore, one of the first projects engaged in was to reduce the burner chamber temperature on the regenerative thermal oxidizer at the recycling facility. The oven had been operating at 15,258 F, but analysis by operators determined that the temperature could be reduced to 12,008 F with no increase in VOC emissions or visible opacity.

Reprogramming costs were just under $10,000, but resulted in annual savings of $9500 and 1600 mcf of natural gas. While the simple payback is one year, the company is confident that the savings will be sustained over the coming years, as appropriate operating changes have been incorporated into operations. Next, the energy team examined the curing ovens in the precoat range at the finishing plant, which was operating at 3258 F. They experimented with lower temperatures and settled on 2708 F, a 558 decrease. This saved $14,100 and 2400 mcf per year, with minimal implementation costs. An added benefit of the lower temperature is longer conveyor belt life.

During the development of work procedures, the energy management team discovered that the thermal oxidizer on the vinyl curing range was running at 14,008 F. Because opacity was the only emissions that concern the MSE team experimented with lower temperatures until 9508 F was determined to be the ideal operating temperature. Annual savings are $55,000 and 9200 mcf of natural gas. An indirect benefit is that internal components will last much longer with this 4508 temperature reduction and maintenance costs should be lower.

Two years after the implementation of the EMS, C&A Floorcoverings experienced an additional 10% reduction in natural gas usage and cost.

5. Conclusion

Two structured approaches to energy resource management are presented in the chapter. ENERGY STAR and ANSI/MSE 2000:2005 have many elements in common. Chief among these are the development of a centralized energy management organizational unit, the critical importance of top management commitment, continuous improvement based on data collection to support accurate energy performance measurement, and the energy project as the locus of energy improvement efforts.

These approaches along with standards developed in other countries including Denmark, Ireland, and Sweden have piqued worldwide interest in the development of an international energy standard. This has led to the 2008 creation of a new project committee (PC) by the International Organization for Standardization (ISO). PC 242 will develop an international standard on energy management. The new standard will provide all types of organizations and companies with a practical and widely recognized approach to increase energy efficiency, reduce costs, and improve their environmental performance. The standard is intended to be pertinent to a number of different industries, including utility, manufacturing, commercial building, general commerce, and transportation. With such broad applicability, estimates indicate that the guidance developed by PC 242 could influence, which has much as 60% of the world's energy demand (American National Standards Institute, 2008).

Acknowledgment

This research was supported by a Summer Research Grant from the Frank G. Zarb School of Business.

References

American National Standards Institute, 2005a. *MSE 2000:2005 A Management System for Energy. Washington, DC.* Retrieved August 8, 2008, from http://webstore.ansi.org/FindStandards.aspx?SearchString=MSE&SearchOption=1&PageNum=0.

American National Standards Institute, 2005b. *MSE 2000:2005 A Management System for Energy. Washington, DC.* Retrieved August 8, 2008, from http://webstore.ansi.org/FindStandards.aspx?SearchString=MSE&SearchOption=1&PageNum=0. p. 4.

American National Standards Institute, 2005c. *MSE 2000:2005 A Management System for Energy. Washington, DC.* Retrieved August 8, 2008, from http://webstore.ansi.org/FindStandards.aspx?SearchString=MSE&SearchOption=1&PageNum=0. p. 5.

American National Standards Institute, 2005d. *MSE 2000:2005 A Management System for Energy. Washington, DC.* Retrieved August 8, 2008, from http://webstore.ansi.org / FindStandards.aspx?SearchString = MSE&SearchOption = 1&PageNum = 0. pp. 5–6.

American National Standards Institute, 2005e. *MSE 2000:2005 A Management System for Energy. Washington, DC.* Retrieved August 8, 2008, from http://webstore.ansi.org/FindStandards.aspx?SearchString=MSE&SearchOption=1&PageNum=0. p. 6.

American National Standards Institute, 2005f. *MSE 2000:2005 A Management System for Energy. Washington, DC.* Retrieved August 8, 2008, from http://webstore.ansi.org/FindStandards.aspx?SearchString=MSE&SearchOption=1&PageNum=0. p. 8.

American National Standards Institute, 2008. *New ISO Project Committee to Development Energy Management Standard.* ANSI News and Publications. Retrieved August 15, 2008, from http://www.ansi.org/news_publications/news_story.aspx?menuid=7&articleid=1781.

Brown, M.L., 2006a. Management systems for energy. In: *Encyclopedia of Energy Engineering*, Marcel Dekker, New York, pp. 1–16.

Brown, M.L., 2006b. Management systems for energy. In: *Encyclopedia of Energy Engineering*, Marcel Dekker, New York, p. 1.

Brown, M.L., 2006c. Management systems for energy. In: *Encyclopedia of Energy Engineering*, Marcel Dekker, New York, p. 6.

Thumann, A. and Mehta, D.P., 1995a. *Handbook of Energy Engineering*, third ed. The Fairmont Press, Liburn, GA, p. 377.

Thumann, A. and Mehta, D.P., 1995b. *Handbook of Energy Engineering*, third ed. The Fairmont Press, Liburn, GA, pp. 43–44.

U.S. Environmental Protection Agency, 2008a. *ENERGY STAR Overview of 2007 Achievements, Washington, DC.* Retrieved August 15, 2008, from http://www.energystar.gov /ia/partners/publications/pubdocs/2007%20CPPD%204pg.pdf.

U.S. Environmental Protection Agency, 2008b. *Guidelines for Energy Management Overview. Washington, DC.* Retrieved August 15, 2008, from http://www.energystar.gov/index.cfm?c=guidelines.guidelines_index.

U.S. Environmental Protection Agency, 2008c. *Ford Motor Company Headquarters Division: In Practice. Washington, DC.* Retrieved August 20, 2008, from http://www.energystar.gov/index.cfm?fuseaction=partners_in_practice.showStory&storyid=1000222&o_id=21883.

U.S. Environmental Protection Agency, 2008d. *3M Company: In Practice. Washington, DC.* Retrieved August 20, 2008, from http://www.energystar.gov/index.cfm?fuseaction=partners_in_practice.showStory&storyid=1000082&o_id=1065235.

U.S. Environmental Protection Agency, 2008e. *Virtua Health System: In Practice. Washington, DC.* Retrieved August 20, 2008, from http://www.energystar.gov/index.cfm?fuseaction=partners_in_practice.showStory&storyid=1000080&o_id=40477.

U.S. Environmental Protection Agency, 2008f. *Food Lion, LLC: In Practice. Washington, DC.* Retrieved August 20, 2008, from http://www.energystar.gov/index.cfm?fuseaction=partners_in_practice.showStory&storyid=1000000&o_id=21884.

U.S. Environmental Protection Agency, 2008g. *3M Company: In Practice. Washington, DC.* Retrieved August 20, 2008, from http://www.energystar.gov/index.cfm?fuseaction=partners_in_practice.showStory&storyid=1000000&o_id=21884.

U.S. Environmental Protection Agency, 2008h. *Marriott Corporation: In Practice. Washington, DC.* Retrieved August 20, 2008, from http://www.energystar.gov/index.cfm?fuseaction=partners_in_practice.showStory&storyid=1000161&o_id=21094.

U.S. Environmental Protection Agency, 2008i. *Food Lion, LLC: In practice. Washington, DC.* Retrieved August 20, 2008, from http://www.energystar.gov/index.cfm?fuseaction=partners_in_practice.showStory&storyid=1000161&o_id=21094.

U.S. Environmental Protection Agency, 2008j. *Toyota Motor Engineering & Manufacturing North America, Inc: In Practice. Washington, DC.* Retrieved August 20, 2008, from http://www.energystar.gov / index.cfm?fuseaction = partners_in_practice.showStory& storyid=1000121&o_id=1004553.

U.S. Environmental Protection Agency, 2008k. *Food Lion, LLC: In Practice. Washington, DC.* Retrieved August 20, 2008, from http://www.energystar.gov/index.cfm?fuseaction=partners_in_practice.showStory&storyid=1000121&o_id=1004553.

U.S. Environmental Protection Agency, 2008l. *Frito Lay, Inc.: In Practice. Washington, DC.* Retrieved August 20, 2008, from http://www.energystar.gov/index.cfm?fuseaction=partners_in_practice.showStory&storyid=1000160&o_id=1045957.

U.S. Environmental Protection Agency, 2008m. *Toyota Motor Engineering & Manufacturing North America, Inc: In Practice. Washington, DC.* Retrieved August 20, 2008, from http://www.energystar.gov/index.cfm?fuseaction=partners_in_practice.showStory&storyid=1000160&o_id=1045957.

U.S. Environmental Protection Agency, 2008n. *Food Lion, LLC: In Practice. Washington, DC.* Retrieved August 20, 2008, from http://www.energystar.gov/index.cfm?fuseaction=partners_in_practice.showStory&storyid=1000160&o_id=1045957.

U.S. Environmental Protection Agency, 2008o. *General Motors Corporation: In Practice. Washington, DC.* Retrieved August 20, 2008, from http://www.energystar.gov/index.cfm?fuseaction=partners_in_practice.showStory&storyid=1000083&o_id=42757.

U.S. Environmental Protection Agency, 2008p. *Marriott Corporation: In Practice. Washington, DC.* Retrieved August 20, 2008, from http://www.energystar.gov/index.cfm?fuseaction=partners_in_practice.showStory&storyid=1000083&o_id=42757.

U.S. Environmental Protection Agency, 2008q. *Toyota Motor Engineering & Manufacturing North America, Inc: In Practice. Washington, DC.* Retrieved August 20, 2008, from http://www.energystar.gov / index.cfm?fuseaction = partners_in_practice.show Story& storyid=1000083&o_id=42757.

U.S. Environmental Protection Agency, 2008r. *Food Lion, LLC: In Practice. Washington, DC.* Retrieved August 20, 2008, from http://www.energystar.gov/index.cfm?fuseaction=partners_in_practice.showStory&storyid=1000083&o_id=42757.

U.S. Environmental Protection Agency, 2008s. *Ford Motor Company Headquarters Division: In Practice. Washington, DC.* Retrieved August 20, 2008, from http://www.energystar.gov/index.cfm?fuseaction=partners_in_practice.showStory&storyid=1000083&o_id=42757.

U.S. Environmental Protection Agency, 2008t. *Toyota Motor Engineering & Manufacturing North America, Inc: In Practice. Washington, DC.* Retrieved August 20, 2008, from http://www.energystar.gov / index . cfm?fuseaction = partners_in_ practice.show Story&storyid=1000083&o_id=42757.

U.S. Environmental Protection Agency, 2008u. *Frito Lay, Inc.: In Practice. Washington, DC.* Retrieved August 20, 2008, from http://www.energystar.gov/index.cfm?fuseaction=partners_in_practice.showStory&storyid=1000083&o_id=42757.

U.S. Environmental Protection Agency, 2008v. *Food Lion, LLC: In Practice. Washington, DC.* Retrieved August 20, 2008, from http://www.energystar.gov/index.cfm?fuseaction=partners_in_practice.showStory&storyid=1000083&o_id=42757.

Chapter 9

GLOBALIZATION, TECHNOLOGY, AND ENVIRONMENT

Jorge Verissimo Pereira
Warwick manufacturing Group, University of Warwick,
CV4 7AL Coventry, UK
jverissimo@jovelogistica.com.br

Globalization is multifaceted and more than a simple movement of goods and investment. It is also about flow of ideas and knowledge. There are different perspectives about the impacts of globalization in the environment and in sustainable development. One point of view says that it deteriorates the environment, especially in developing countries, the other argues that it has promoted the transmission and dissemination of environmental friendly technologies. The influence of globalization and concerns about sustainability have been presented in new management practices, such as the green supply chain. Moreover, new technologies have the potential to reduce CO_2 emission, which is a major responsible for GHG. Indeed, a Los Alamos National Laboratory research has demonstrated that system dynamic simulation is an important tool to implement carbon sequestration technology.

1. Introduction

The balance of the Earth's global system, see Fig. 1, is an interaction among the natural system (environment), built system (economy and infrastructure), and social system (individual, government, and society) (UNESCO, 2002).

This balance has become a prime concern in the world and many international conferences have discussed this issue such as the 1992 Earth Summit in Rio de Janeiro, the Kyoto Conference of 1997, and the United Nations Climate Change Conference, held in Copenhagen in 2009. In all of them, world leaders and civil society have met to find solutions for climate change problem. However, as can be seen in Fig. 1, this is an issue that should be addressed by all the stakeholders, i.e., government, society, and business.

Nowadays, companies must evolve its overall strategy by incorporating environment aspects at the same level of market share and profit considerations (Tsoulfas and Pappis, 2006). This environmental management has the potential for a lasting impact on their performance (Zhu and Sarkis, 2006). In this scenario, cleaner technologies and innovative management practices play a key role.

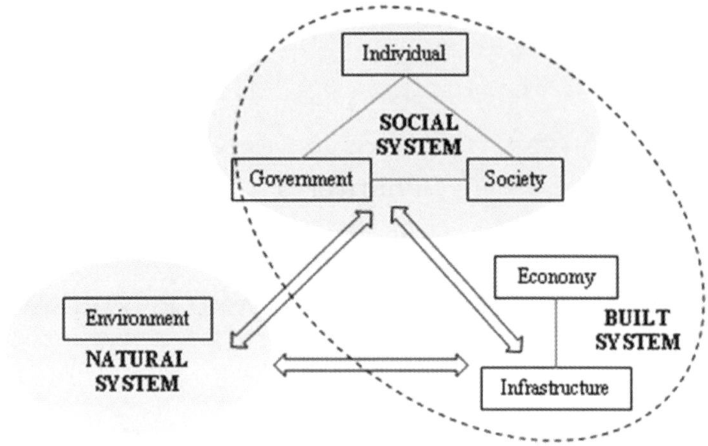

Fig. 1. The balance of the Earth's global system (UNESCO, 2002).

While many advocate that globalization is a power that compels sustainable economic development due to the transmission and dissemination of environmentally friendly technologies, organizational practices, and public policies to developing countries, others argue that it might have negative effects for the environment and sustainable development.

In this chapter, we begin by introducing the concept of globalization and the different viewpoints about its consequences on the environment and on the climate change. Then, climate change and new technologies that have been developed to tackle climate issue are presented. Section 4 portrays how globalization and environment issues have influenced management practices, especially the supply chain (SC) management. Section 5 has a discussion about the potentialities of modeling applied to climate change. Section 6 describes a case study with the application of a methodology developed by the Los Alamos National Laboratory (LANL) that uses the concept of system dynamic (SD) simulation to implement carbon sequestration technology.

2. Globalization: Relevant Issues

Globalization, the integration of national economies and cultures into a world system through movements of goods, services, investment, technology, and (to a lesser extent) labor (Schrecker *et al.*, 2008), was made possible thanks to the development of faster and cheaper communication and transportation and reduction of barriers to trade, especially following WWII (Johnson, 2002).

Advances of transport technologies and infrastructures have supported globalization, reducing the transactional space/time, and enabling the exploitation of new sources of resources, capital, and labor (Rodrigue, 1999). That author suggests the change in production system, from Fordism to Post-Fordism, demands a just-in-time (JIT) strategy with a new kind of terminals, transfer warehouse replacing traditional warehousing with accumulation function, and with small shipments with high frequency instead large shipments with low frequency. Technological improvements and deregulation of financial markets around the word were key factors for the rapid increase in the mobility of investment capital (Schrecker *et al.*, 2008) and in the economic exchanges between nations (Ethier, 2005). Costantini and Monni (2008) define economic globalization as a "process of enlarging opportunities in terms of new technologies, such as information, communication, and competitiveness."

Despite the international movement of goods and investment is the most obvious face of globalization, it is the flow of ideas and knowledge that has had the greater effect on people in the developing world (Johnson, 2002) and creates an interdependence of worldwide networks (McKiea and Galloway, 2007). This vision is shared by Schrecker *et al.* (2008), which says that current globalization, despite being essentially an economic process, is multifaceted and more elaborate than the previous period of globalization (1870–1914), in which trade and colonization were the main aspects.

However, some authors have pointed out negative effects of globalization, especially for developing countries (Perkins and Neumayer, 2009) and criticized the mainstream view of globalization (Dimirovski *et al.*, 2006, Nordas and Gleditsch, 2007). For instance, Milanovic (2003) condemns what he calls a "naïve view" that considers that globalization is a *deus ex machine* for the problems that haunt developing word such as poverty, illiteracy, and inequality. Milanovic (2003) cites and criticizes other authors that consider the period of 1870–1914 and from the 1980s to current days as the periods of universal growth and catch up of poor countries and when globalization mitigated the income gaps between nations without expose its other face. That author agrees that globalization is a multifaceted process that presents different perspectives to different people, depending on their social class, where they live, and where they stand ideologically.

For instance, the rising inequality of income that arisen over the past two or three centuries has been imputed to globalization. What some others ague that inequality has increased owing to globalization has not spread far enough, fast enough (Johnson, 2002). Milanovic (2003) has a comprehensive analysis of globalization in the 19th century and nowadays. Schrecker *et al.* (2008) analyze the correlation between globalization and health after 1980 and give important elements to understand the actual impact of globalization and poverty reduction.

That controversy about the impacts of globalization is also studied by Tisdell (2001). He emphasizes that many believes that globalization is a power that compels sustainable economic development, that has three pillars — environmental, social, and economic (Mobjork and Linner, 2006) — and that have speeded up the transmission and dissemination of environmentally friendly technologies, organizational practices, and public policies to developing countries (Perkins and Neumayer, 2009). Others argue that it might have negative effects for the environment and sustainable development. Tisdell (2001) also presents two points of view when assessing sustainability. One side is analyzed from the individual nation viewpoint; the other is from the global perception. Therefore, even with an improvement in an area due to the globalization process, the environment of other region can worsen and result in deterioration in the global environment.

The competition imposed by globalization is a force to foster the efficiency. Therefore, less material and energy is used per unit of production and there is lower wastage. It is important to highlight that design and process technology usually determine the types of pollutants emitted, the energy used, resources consumed, and solid and hazardous wastes generated (Tsoulfas and Pappis, 2006). For instance, Intel has released chips with an increasing number of transistors, which have resulted in chips with more computational capacity, fewer chips to build, and fewer chips to dispose (Tsoulfas and Pappis, 2006). This results in less pollution emission per unit of product. However, Tisdell (2001) points out that those gains may be offset by the economic scale or economic growth.

Tisdell (2001) lists some benefits that globalization, together with economic growth, brings the following:

1. they may stimulate the development of new resource-saving technologies;
2. the improvement of environment control due to rising incomes; and
3. this economic growth may lead to a new dynamic in the replacement of existing equipments. Generally, these new equipments are more environmental friendly.

Nowadays, for many companies it is strategic and imperative to expand their operations on a world scale (James, 1990). However, at the same time that these companies are striving to maximize their profits, they are stimulated to behave socially responsibly (Dahlsrud, 2008). Globalization has also changed the traditional product-cycle model (Sener and Zhao, 2009). In the traditional model, innovative companies that own the technology of producing the next generation product start the manufacturing process in the North. When products become commodities, those innovative companies look for ways of transferring the production to the South to take advantage of low labor cost at the same time they start a production of a new innovative product in the North. Nowadays, with an efficient global SC this cycle has been broken as innovators have started the production directly in the South. Sener and Zhao (2009) exemplify this new situation with a variety of companies such as Dell, Hewlett-Packard Co., Motorola, Philips, Glaxo-Smith-Kline, Eli Lilly, GAP, and Nike.

Today environment is no longer considered as a limitation to economic growth but a condition to achieve higher living standards, poverty reduction, and increasing human development levels (Costantini and Monni, 2008). Sustainable development was initially applied to natural resource circumstances. Nowadays, it is applied to economic development, environment, energy, social organization among others (Glasser, 2006). Sustainability has many different definitions (Chen *et al.*, 2009). The most cited definition of sustainable development is the development that ensures that the well-being of future generations is at least the same of current generations (Tisdell, 2001; Costantini and Monni, 2008). This definition has been applied in both business and non-business environment (Chen *et al.*, 2009). From his ecological viewpoint, Rees (2006) argues that techno-industrial society is basically unsustainable. He believes that the current consuming rates of materials are made of a continuous extraction of natural resources that cannot be replaced. Moreover, at the end of its life cycle manufactured capital is returned to the ecosystem in a degraded form.

Tisdell (2001) believes that economic growth fostered by globalization might have negative impacts for the environment and for sustainable development. Rees (2006) exemplifies some countries that are running an ecological deficit and need an equivalent surplus somewhere else: UK, Japan, the Netherlands, and United States.

3. Climate Change and Technology

Tisdell (2001) points out that assessing the condition of an environment is not straightforward. However, global environment is a very important issue as global climate change will have vast implications for the quality of life of hundreds of millions of people (Nordas and Gleditsch, 2007). The threats caused by climate change have been assessed by US Department of Defense and regarded as one of US national security challenges. Also, global climate change has many faces and desertification will threat human well-being and the sustainability (Spiess, 2008).

The two major sources of global environment changes are carbon dioxide (CO_2), which is also responsible for greenhouse gas (GHG), and sulfur dioxide (SO_2). The study of CO_2 is important because of its long-term effect in the environment (Tisdell, 2001). In order to assess its impact in climate change, models should be performed with extremes of higher and lower rates of CO_2 in the atmosphere and their effects on temperatures (Nordas and Gleditsch, 2007).

The passenger transport has a significant contribution in terms of GHG and it is a major source of CO_2 emission; therefore, a robust reduction in transport emission will be needed over the next 50 years (Rajan, 2006). This reduction in GHG will call for innovation and large-scale adoption of GHG-reducing technologies all over the global energy system (De Coninck *et al.*, 2008). Current technologies that have the potential to reduce CO_2 emission include nuclear, solar, wind, biomass, geothermal, wave, and tidal.

Moreover, Rajan (2006) points out that relying only in efficient vehicle technology policy may not be enough to meet that goal. This means that other technologies should be developed in order to reduce GHG further.

Two new technologies have the potential to reduce GHG emissions: (1) the use of hydrogen, especially for use in vehicular fuel cells, and (2) the sequestration of carbon dioxide into geological formations. Although sequestration is lesser mature than hydrogen and there are some safety concerns about possible leakage, sequestration is the most promising and relevant technology.

Although geologic sequestration uses technologies from the petroleum industry that has been used since the 1970s, CO_2 is pumped into reservoir for enhanced oil recovery (EOR), it presents some challenges (Stauffer *et al.*, 2009):

Geologic sequestration is more complex than EOR. In the former, CO_2 must be stored for hundreds to thousands years to be effective. In the latter, EOR requires CO_2 storage for only a few decades.
Considering the time period required by geologic sequestration, the processes and interactions among elements must be better analyzed (e.g., the interactions among water–rock–CO_2).
The amount of CO_2 that will need to be sequestrated is greater than that involved in EOR projects (e.g., in 2003, the emissions of CO_2 in the US alone were around 5840 Tg/yr while EOR operations in the US involved only 35 Tg/yr;
In EOR operations, CO_2 is injected into a reservoir whose pressure has been depleted due to oil production, whereas in deep saline reservoirs, which has the greatest potential for geologic sequestration, pore pressures have not been reduced.
EOR project has been done in oil reservoirs that have been studied and with production histories of many years, while in geologic sequestration involves new sites for which comprehensive researches should be perform in order to characterize the reservoir. This task is known as performance assessment (PA).

In another front, technology-oriented agreements (TOAs) has been the focus of greater attention in recent years. Coninck *et al.* (2008) define TOAs as those "international agreements that are aimed at advancing research, development, demonstration, and/or deployment of technologies." In the case of global climate change, these technologies should target at reducing GHG emissions (De Coninck *et al.*, 2008). There are four types of TOAs:

1. knowledge sharing and coordination;
2. research, development, and demonstration (RD&D);
3. technology transfer; and
4. technology deployment mandates, standards, and incentives.

These are end of process technologies that may complement cleaner technologies that promote a more rational use of natural resources and create products with fewer hazardous wastes and the lowest pollutant components (Tsoulfas and Pappis, 2006).

4. SC and Environment

The influence of globalization and environmental concerns in management practices can be seen in the SC. Indeed, global trade has fostered the enforcement of more stringent standards applied in developed countries into lower-regulating countries (Perkins and Neumayer (2009). The authors call this as "SC pressures." In fact, companies are progressively more concerned about their SCs that might have significant impact on their brand image (Ganesan *et al.*, 2009). Zhu *et al.* (2007) give some examples: Bristol-Myers Squibb, IBM, and Xerox have stimulated their Chinese suppliers to evolve environmental management systems in agreement with ISO 14001, and Ford, GM, and Toyota have demanded from their Chinese suppliers the ISO 14001 certification.

Moreover, companies should improve their efficiency in using natural resources in order to contribute for a sustainable development (Vachon and Maoc, 2009). Therefore, their SC must also be redesigned bearing in mind the natural environment and their communities.

Regarding environment management's researches, the study of sustainability is presented in the SC management under terminologies such as green purchasing, reverse logistics and reverse SC product stewardship, and green supply chain (Vachon and Maoc, 2008). Zhu *et al.* (2007) define green SC management (GSCM) as a win–win strategy for companies to balance environment concerns and economic performance by lowering their environment impacts and risks, i.e. becoming more ecological efficient, at the same time they accomplish market share and profit objectives (Zhu and Sarkis, 2006).

Reverse logistics can be defined as "the process of planning, implementing, and controlling the efficient and effective inbound flow and storage of secondary goods and related information opposite to the traditional SC direction for the purpose of recovering value or proper disposal" (Tsoulfas and Pappis, 2006, p. 1). Regarding SC management, it is more than an expression traditionally used to describe the flow of material and service management (Ballou *et al.*, 2000). It involves multiple actors, and flow of materials, information and money (Longo and Mirabelli, 2008). For Lau and Lee (2000), SCM has strategic features as it deals with logistic planning, operation, and performance measures on an integrated basis. Ketikidis *et al.* (2008) and Ballou *et al.* (2000) emphasize that in SC management all organizations are parts of one or more SCs and there must be management and integration of every associated activity of services and product flow, including information flow, either internal or external to the company.

5. Modeling Applied to Climate Change

5.1. *Systems thinking and modeling*

The preceding sections described how challenging it is to manage a company in a time that environment aspects must be incorporated in its strategy, and that assessing the climate change is a complex issue. The study of environment must take into account not only its sustainability but also its resilience (Moser, 2009). A system is resilient when it has the ability to maintain, resume, and restore its original (or desired) state after being disrupted (Pereira, 2009a). Moser (2009) also points out about the lacking of literature on the expected consequences of climate change for small businesses and smaller economic sectors as the winter tourism sector, the wine growing sector, organic farming, or fisheries. However, both large and small companies have the responsibility to contribute to development of equitable and sustainable communities and societies (Hedren, 2009). Moreover, climate change has myriads of consequences, as human well-being can be impacted thorough sea-level rise, human health, and changing weather patterns, and indirectly via factors such as migration. These various causal chains need to be mapped (Nordas and Glenditsch, 2007).

In such a process, a number of different organizations regulate different parts of the process (Wolstenholme, 1990) and nobody is really in charge of the network as a whole (Akkermans and Dellaert, 2005). Wolstenholme (1990) points out that in this situation, where strategies are not integrated, the process will not flow smoothly. This environment is clustered by invisible boundaries with interrelated actions, which take time to fully reveal their effects on each other (Senge, 1992), giving rise to unanticipated side effects (Sterman, 2000). From these countless interactions of a large number of agents, a complex adaptive system emerges (Akkermans and Romme, 2004). For Senge (1992), this increasingly worldwide complexity and our inability to grasp and manage it are behind most of our human problems. It is therefore no surprise that large and complex projects often lead to time and cost overruns (Williams *et al.*, 2003).

When parts interact in a complex way, studying only the behavior of isolated parts can result in a misunderstanding of the whole (Holland, 2000). In order to solve a problem, the focus should take into account the whole system rather than isolated parts of it (Senge, 1992). For instance, a better understanding of the complex dynamics involved in developing new strategies that take into account not only market share and profit but also environment and climate change has become crucial. These complex systems are not easily modeled mathematically, and simulation is a better approach to study these problems (Chen *et al.*, 2007).

This allows comparison of various alternative scenarios, and makes it possible to understand the relative and interactive impact of factors without interrupting the real system (Kull and Closs, 2008; Potter *et al.*, 2007) and it compresses time so that several years of activity can be analyzed in a practicable time (Fishman, 1978). Therefore, it is possible to support decision making at strategic level, for instance, with the redesign of the SC (Kleijnen, 2003). The author points out that this means that simulation does not give a "closed-form" solution but a sensitivity analysis.

When modeling global environment system it is important to bear in mind that it is not static but in constant changes (Dimirovski, 2006). Moreover, it is a complex system with nonlinear feedback structures, multiple loops, and uncertainties. Owing to its dynamics behavior, the study of climate change requires the use of SD approach (Pereira, 2009b).

These features require a new approach of modeling and simulating that takes into consideration quantitative and qualitative methods and also SD and discrete event simulation (DES) theories.

Simulation has been recognized as an increasingly powerful tool for some managerial decisions, such as planning the SC (Murphy and Perera, 2002), and

Table 1. Major steps of SD modeling.

Problem identification, with the definition of key variables, boundary selection, and time horizon
Formulating dynamic hypothesis and model conceptualization. In this qualitative phase, causal loop diagrams are developed
Model formulation using a proper package of SD simulation software. This is the quantitative phase
Testing — model analysis and validation
Policy analysis and design

Source: Adapted from Sterman (2000) and Barlas (1996).

computer-based models have a critical role in the study of dynamics systems (Holland, 2000). Computer simulations are important because they can capture relations between components in complex systems, such as those present in the climate change and sustainable management, providing resources for a better understanding of the impact of changes (Hung *et al.*, 2004). Indeed, simulation does not treat the system as a black box, and therefore it gives insight into cause-and-effect and helps to understand causalities (Kleijnen, 2003).

As in other design laboratory approaches, simulations have some steps to be followed (Forrester, 1962). Barlas (1996) and Sterman (2000) offer some very typical and general steps to be used in models (Table 1). Certainly, in different authors, with different approaches (SD or DES) we can find different processes of modeling. Luna-Reyes and Andersen (2003, p. 275) describe different arrangements of modeling activities, varying from three to seven stages. However, they also point out that despite the ways that grouping activities vary among authors, the different stages remain reasonably constant across them. An important first step is to define the level of system specification. Zeigler *et al.* (2002) define this as the level at which the modeler can describe how the system behaves and the properties responsible for its performance.

Further, it is important to stress that modeling is an interactive process, which adds detail to the small, insight-based model in each interaction (Lyneis, 1999). Indeed, a procedure can be performed more than once, since each turn leads to a better understanding of the system (Wolstenholme, 1990). Moreover, in a cross-disciplinary study, such as climate change problems, the researchers must bear in mind that different disciplines tend to speak different languages (Fowler, 2003).

5.2. *Boundaries*

In addition to the four SD basic components, stock-flow, feedback, delay, and policy, Wolstenholme (2004) cites boundaries as the fifth component. He claims boundaries have the same level of importance as the first four because they

- exist and cannot be neglected, as systems by definition are boundary restricted;
- represent the functions, liability, power and culture of systems, teams, and individuals; and
- are a major key to implementing systemic solutions.

Akkermans and Dellaert (2005) agree and suggest how discussing implicit boundaries can explicitly contribute to an SD approach. However, boundaries could be responsible for the silo mentality, which inhibits the process flow in organizations (Wolstenholme, 2004). He also believes that boundaries can cause boundary management problems as systems boundaries are similar to national boundaries. However, the "principle of the system boundary," proposed by Senge (1992, p. 67), states that "the interactions that must be examined are those most important to the issue at hand, regardless of parochial organizational boundaries."

Boundaries change incessantly (Wolstenholme, 2004). Akkermans and Dellaert (2005) believe that the system boundary must be established according to the problem in question, including all significant elements (Towill, 1996). Then, the laws of change at that level of detail must be captured (Holland, 2000). Holland states that the art of modeling is to find simple laws, yet it is not always possible. For instance, for SC studies, competitors must be considered when this is relevant. Briefly, the boundary map is useful to assess which people and organizations control each variable in the process (Wolstenholme, 1990). Considering that sustainability is an issue that involves varied stakeholders, with different viewpoints, it is fundamental the concept of boundaries.

5.3. *Simulation tool selection*

Simulation is concerned with the handling of real problems, including their uncertainties, through reproduction in an environment controlled by the experimenter (Gamerman and Lopes, 2006). Therefore, to design sustainability management strategy or assessing global climate change, the software should have the capacity to embody uncertainty into the analysis. Simulation can be performed in any computer language. It is certainly better to use packages of software which were built especially for this purpose, for example, Stella, Powesim, Madonna, Arena, and GoldSim. They are able to integrate all the dynamics and data present in a complex system into a computer model. Usually these packages of software have a graphics interface and functionalities, which encompass all the features demanded by these kinds of simulations. In fact, as these software packages are icon-driven, stock and flow diagrams inherit some peculiarities of the tools (Lane, 2000). With the handful of software packages available, choosing one can be a time-consuming and challenging task, as there is little literature concerning this issue (Rizzo *et al.*, 2006). However, on the basis of the evidence indicated in this section, GoldSim has the greater potential for SD–DES modeling, mainly for its flexibility, which Law and Kelton (2000) consider the most important feature for a simulation software product, and because it can incorporate both continuous process and discrete events.

GoldSim is a graphic package, with a drag-and-drop user-friendly interface, which makes it unnecessary for the modeler to have a high level of programming skill. It has an online model example, and a Help Menu with a suitable documentation.

These pop-up menus make it easy for someone to achieve a good level of software understanding. In this software, the programmer/ analyst creates a model of the system using objects called "elements" and "links." Because of this, Miller and Burns (2005) suggest GoldSim must be thought of as a high-level programming language, where the program is the model. That is, all equations and commands are embedded in graphical icons using built-in elements (Rizzo *et al.*, 2006). It is important to highlight that this is the second most important feature for a simulation product according to Law and Kelton (2000).

In fact, GoldSim is more suitable to carry out an SD–DES modeling. For instance, Stella, Powesim, and Madonna were specially designed to be used in SD models. However, they do not satisfy completely some features that Law and Kelton (2000) list a program should have to run a DES model. These requirements include a good random number generation with at least 100 different streams that can be assigned to different sources of randomness, generating random variates from a specific probability distribution, advancing simulated time, collecting output statistics and reporting the results and detecting error conditions. For all of them, GoldSim and Arena satisfy completely, with GoldSim having an unlimited random number of streams, as each element has its own seed and sequence. While with GoldSim is possible to model an SD–DES model, Arena is only suitable for a DES model.

The two sources of instability in complex systems, physical delays and information delays, must be contemplated in a dynamic simulation. GoldSim has "Delay Elements" that deal with this issue. These elements transmit inertia to the system, simulating a dynamic behavior. Treating continuous flows and discrete signals, there are four "Delay Elements" in GoldSim: Material Delay and Information Delay, which delay continuous flows, and Event Delay, and Discrete Change Delay which delay discrete signals.

Sweetser (1999) highlights the differences underlying mathematics that drive software packages. On this point, GoldSim has not only an appropriate method to numerically integrate differential equations that is necessary when modeling continuous systems, but also it has a unique time-stepping algorithm to represent discrete events. Besides Arena, which is not suitable to an SD model, the other software packages do not have an appropriate algorithm to represent discrete events.

The software should also have the capacity to deal with uncertainties. GoldSim has the capability to deal with the uncertainties of sustainability projects, integrating variability and uncertainty into its analysis, and considering disruptive events in the system.

Liu *et al.* (2007) evidence another important characteristic that software simulation should satisfy. It must deal with temporal constraints for a proper study of cause–effect events. GoldSim has the capability to explore this, as exemplified by the authors (event e1 follows event e2 after time T; or N occurrences of event e1 within time T lead to event e2).

In addition, the key tenets of modeling say that a model must be easy to manipulate and that the modeler must be parsimonious, which means starting small and gradually adding more assumptions until the model is more sophisticated (Pidd, 2005). In GoldSim, this is possible because it has a top-down hierarchical model structure, which allows the building of hierarchical multilayer models with great detail at low level. Containers, one of the most important elements in GoldSim and one of its singular features, give to the software the capability to create hierarchical, top-down models. In a multifaceted problem, the fact that GoldSim presentation makes communication easier is very important for the clear understanding of decision makers and the public. Moreover, this brings a high degree of flexibility and scalability that Banks *et al.* (2002) mention as a requirement to address a wide range of strategic, tactical, and operational challenges.

Another important and unique element in GoldSim is the "Triggered Event." Using this element, it is possible to aggregate value to simulation as proposed by Banks *et al.* (2002).

Moreover, dimensional units of measure in equations must be considered carefully when building models. Inconsistencies in dimensions lead to model invalidity. Therefore, a dimensional analysis must be performed and the units of measure of all variables and constants must be examined (Forrester, 1962). GoldSim is dimensionally aware with an extensive internal database of units and conversion factors, which can be customized. GoldSim checks the consistency of each equation and link, issuing a warning when it finds inconsistencies, which minimizes errors.

In climate change projects, the input of data can become a labor-intensive task. GoldSim has the capability to dynamically link to external data repositories as a spreadsheet or database. This data exchange between GoldSim and external data repositories makes the use of large frequently updated data sets direct from the studied company's ERP or governmental database much easier. Indeed, this feature is becoming each day more important in current simulations (Law and Kelton, 2000).

In an important phase in simulations, by means of sensitivity analysis, those parameters to which the model is highly sensitive are determined. A comparison between the real system and the model can then be carried out (Barlas, 1996). Vlachos *et al.* (2007) describe how, in their research, using the dynamic simulation software Powersim, they had to create macros in a commercial spreadsheet in order to carry out the investigation of the system's sensitivity. In GoldSim, there is a function that incorporates sensitivity analysis.

Moreover, Rizzo *et al.* (2006) claim that performance is an important factor in the choice of a software package. In their studies of dynamic systems-based software packages, GoldSim presented one of the highest levels of mathematical power.

In summary, GoldSim has a good relation between programming capabilities and flexible, being more advanced than Stella and Madonna, for instance.

6. Case Study: Geologic Sequestration of CO_2

As highlighted in previous sections, new technologies have the potential to reduce GHG emissions, and sequestration of carbon dioxide is one of the most promising and relevant technology. CO_2 can be stored in carbon "*sinks*" such as oceans, deep geologic reservoirs, forests, and soils.

Scientists at LANL have studied storage carbon sequestration as a solution for reducing GHG emissions and focused on geologic storage. One of their latest case study (Stauffer *et al.*, 2009) is presented in this section.

The process of CO_2 storage is described in Fig. 2 (GoldSim, 2009). This technology contains opportunities and challenges such as the injected CO_2 slips through escape routes. LANL model has shed light in this issue simulating the likely behavior of the site after the introduction of CO_2. This involves questions like the potential impact of the environment in the geologic storage site, how long CO_2 will remain trapped, and if it escapes what will be the consequences. Another matter

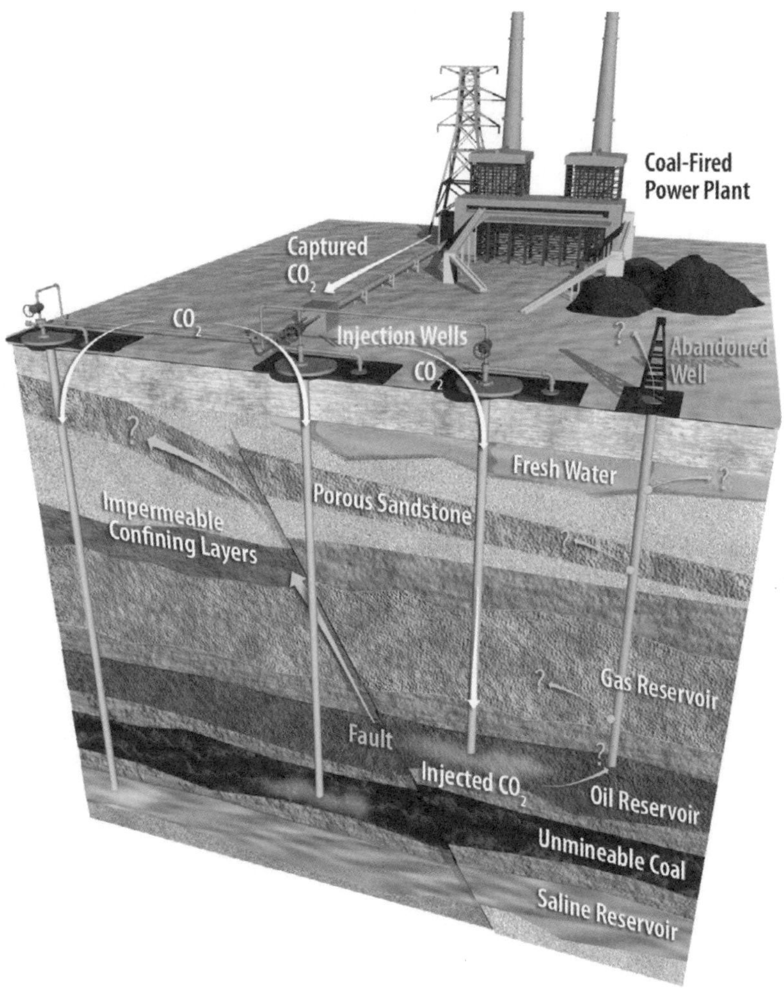

Fig. 2. Process of carbon sequestration in geologic storage (GoldSim, 2009).

addressed by Stauffer *et al.* (2009) is a methodology to analyze storage places that is efficient (economical), effective, safe, and technically-feasible among possible sites that may include depleted gas and oil reservoirs, deep saline aquifers, and deep unmineable coal.

Stauffer *et al.* (2009) have developed a new model CO_2-PENS (Predicting Engineering Natural Systems) that addresses the main issues associated with geologic carbon sequestration. This is a probabilistic model designed to simulate CO_2 capture, transport, and injection in different geologic sites. It also complies with the key tenets of PA while integrating algorithms based on central physics and chemistry concepts.

CO_2-PENS model contemplates uncertainties and the interaction among different processes such as subsurface injection of CO_2, density fingering, and atmospheric mixing that are needed in geologic carbon sequestration. Moreover, an economic module analyzes how different configurations can impact overall costs when selecting the site.

CO_2-PENS is a model that simulates the entire cycle described in Fig. 2. This means that from the encapsulation of CO_2, its injection in the storage reservoir and even possible leaking and faults are simulated. During the preliminary PA process, the model shed light in the following issues:

an estimation of the amount of wells needed to inject a given quantity of CO_2. This number is a function of the temperature and pressure of the target reservoir and the maximum safe injection pressure;
the possible extent of the injected CO_2; and
the cost involved.

The model has a top-down hierarchical structure, which allows the building of hierarchical multilayer models with great detail at low level. For instance, the reduced complexity injection module is designed for use during the initial site selection calculations for the estimations listed above. However, for evaluating performance of specific sites, there is another module that can perform complex calculations employing full multiphase physics.

Moreover, there is an injectivity module that determines whether additional wells are needed to inject a given amount of CO_2. It also tracks the number of active and shut-off wells at any given time step and exports this information back to the system for further analyses.

LANL model was applied in a 1 GW power plant to compare two different target reservoirs: the first case was a cold, shallow reservoir, whereas the second case was a hot, deep reservoir. The former was at 1 km depth and 35°C, whereas the latter was at 3 km depth and 155°C and assuming the same rock properties for both reservoirs. Among others, additional data used in the simulation include rock permeability and porosity, rock conductivity, permeability (kr), viscosity, and density (F) of the fluid. Moreover, the model has overcome the challenge of simulating a system in which the injected CO_2 removes water, whose properties are deduced as a function of pressure and temperature. In each case, 500 realizations were run, what makes the model more robust.

The model gives us the following results for each option (cold-shallow and hot-deep):

the number of boreholes required to inject the CO_2 from a 1000 MW power plant;
the area of CO_2 plume after 50 years of injection;
present value cost per ton associated with wells and local pipelines.

Indeed, geologic sequestration of carbon dioxide is one of the most promising and relevant technology to help combat global warming, and SD simulation is a powerful tool to assist scientists in their climate change researches.

7. Conclusions

In this chapter, a core literature review of globalization, environment, sustainability, and technology was presented. From analyzing the state of globalization, technology, and environment, a detailed picture of the main issues and gaps in knowledge and other problems in these area emerged. There is a greater concern about the impact of globalization and climate change on the human well-being. All the stakeholders, i.e. government, society, and business, must be part of the solutions. One of the major sources of global environment changes is carbon dioxide (CO_2). The most promising and relevant technology to reduce GHG emissions is the sequestration of carbon dioxide into geological formations. The behavioral characteristics of carbon sequestration in geologic storage require a new tool to address these issues, which involve both uncertainties and qualitative measures. System dynamics with a holistic worldview has the capability to identify high leverage points in complex systems. Underpinned in feedback control and nonlinearity, system dynamics can deal with complexity problems characterized by nondeterministic and nonlinear features. Therefore, the use of SD simulation should be a promising and relevant tool for assisting in carbon sequestration technology in the next decade.

Acknowledgments

The author would like to thank those who provided the case study that was used in this paper: Dr. Philip H. Stauffer from the Los Alamos National Laboratory (LANL) and Rick Kossik from GoldSim Technology Group.

References

Akkermans, H.A. and Dellaert, N., 2005. The rediscovery of industrial dynamics: The contribution of system dynamics to supply chain management in a dynamic and fragmented world. *System Dynamics Review* **21**, 173–186.

Akkermans, H.A. and Romme, A.G.L., 2004. *Opportunism and Partnership in a High-tech Supply Network: An Agent-based Model.* Eindhoven University of Technology, The Netherlands.

Ballou, R.H., Gilbert, S.M. and Mukherjee, A., 2000. New managerial challenges from supply chain opportunities. *Industrial Marketing Management* **29**, 7–18.

Banks, J., Buckley, S., Jain, S. and Lendermann, P., 2002. Panel session: Opportunities for simulation in supply chain management. In: Yucesean, E., Chen, C.H., Snowdon, J.L. and Charnes, J.M. (Eds.), *Winter Simulation Conference, San Diego, California.*

Barlas, Y., 1996. Formal aspects of model validity and validation in system dynamics. *System Dynamics Review* **12**, 183–210.

Chen, D., Newburry, W. and Park, S.H., 2009. Improving sustainability: An international evolutionary framework. *Journal of International Management* **15**, 11.

Chen, M.C., Yang, T. and Li, H.C., 2007. Evaluating the supply chain performance of IT-based inter-enterprise collaboration. *Information and Management* **44**, 524–534.

Costantini, V. and Monni, S., 2008. Environment, human development and economic growth. *Ecological Economics* **64**, 14.

Dahlsrud, A., 2008. How corporate social responsibility is defined: An analysis of 37 definitions. *Corporate Social Responsibility and Environmental Management* **15**, 13.

De Coninck , H., Fischer, C., Newella, R.G. and Uenod, T., 2008. International technology-oriented agreements to address climate change. *Energy Policy* **36**, 22.

Dimirovski, G.M., Dinibutun, A.T., Kile, F., Neck, R., Stahre, J. and Vlacic, L.B., 2006. Control system approaches for sustainable development and instability management in the globalization age. *Annual Reviews in Control* **30**, 13.

Ethier, W.J., 2005. Globalization, globalisation: Trade, technology, and wages. *International Review of Economics and Finance* **14**, 22.

Fishman, G.S., 1978. *Principles of Discrete Event Simulation.* John Wiley & Sons, Inc., New York.

Forrester, J.W., 1962. *Industrial Dynamics.* M.I.T. Press, Cambridge, Massachesetts.

Fowler, A., 2003. Systems modelling, simulation, and the dynamics of strategy. *Journal of Business Research* **56**, 135–144.

Gamerman, D. and Lopes, H.F., 2006. *Markov Chain Monte Carlo: Stochastic Simulation for Bayesian Inference.* Taylor & Francis, Boca Raton.

Ganesan, S., George, M., Jap, S., Palmatier, R.W. and Weitz, B., 2009. Supply chain management and retailer performance: Emerging trends, issues, and implications for research and practice. *Journal of Retailing* **85**, 11.

Glasser, J.H., 2006. The evolving implications of health and sustainability in China. *Journal of the Royal Institute of Public Health* **120**, 4.

Goldsim, 2009. Addressing Climate Change from the Ground Up.

Hedren, J., 2009. Shaping sustainability: Is there an unreleased potential in utopian thought? *Futures* **41**, 6.

Holland, J.H., 2000. *Emergence: From Chaos to Order.* Oxford University Press, Oxford.

Hung, W.Y., Kucherenko, S., Samsatli, N.J. and Shah, N., 2004. A flexible and generic approach to dynamic modelling of supply chains. *Journal of the Operational Research Society* **55**, 801–813.

James, B., 1990. Reducing the risks of globalization. *Long Range Planning* **23**, 9.

Johnson, D.G., 2002. Globalization what it is and who benefits. *Journal of Asian Economics* **13**, 13.

Ketikidis, P.H., Koh, S.C.L., Dimitriadis, N., Gunasekaran, A. and Kehajova, M., 2008. The use of information systems for logistics and supply chain management in South East Europe: Current status and future direction. *Omega* **36**, 592–599.

Kleijnen, J.P.C., 2003. Supply Chain Simulation: A Survey.

Kull, T. and Closs, D., 2008. The risk of second-tier supplier failures in serial supply chains: Implications for order policies and distributor autonomy. *European Journal of Operational Research* **186**, 1158–1174.

Lane, D.C., 2000. Diagramming conventions in system dynamics. *Journal of the Operational Research Society* **51**, 241–245.

Lau, H.C.W. and Lee, W.B., 2000. On a responsive supply chain information system. *International Journal of Physical Distribution & Logistics Management* **30**, 598–610.

Law, A.M. and Kelton, W.D., 2000. *Simulation Modeling and Analysis.* McGraw-Hill, Boston.

Liu, R., Kumar, A. and Van Der Aalst, W., 2007. A formal modeling approach for supply chain event management. *Decision Support Systems* **43**, 761–778.

Longo, F. and Mirabelli, G., 2008. An advanced supply chain management tool based on modeling and simulation. *Computers and Industrial Engineering* **54**, 570–588.

Luna-Reyes, L.F. and Andersen, D.L., 2003. Collecting and analyzing qualitative data for system dynamics: Methods and models. *System Dynamics Review* **19**, 271–296.

Lyneis, J.M., 1999. System dynamics for business strategy: A phased approach. *System Dynamics Review* **15**, 37–70.

McKiea, D. and Galloway, C., 2007. Climate change after denial: Global reach, global responsibilities, and public relations. *Public Relations Review* **33**, 9.

Milanovic, B., 2003. The two faces of globalization: Against globalization as we know it. *World development* **31**, 16.

Miller, I. and Burns, A., 2005. Development of a dynamic simulation approach to mission risk and reliability analysis. In: *International Topical Meeting on Probabilistic Safety Analysis, La Grange Park, Illinois, USA.* American Nuclear Society, pp. 809–821.

Mobjork, M. and Linner, B.-O., 2006. Sustainable funding? How funding agencies frame science for sustainable development. *Environmental Science & Policy* **9**, 11.

Moser, S.C., 2009. Now more than ever: The need for more societally relevant research on vulnerability and adaptation to climate change. *Applied Geography* **11**.

Murphy, S.P. and Perera, T., 2002. Successes and failures in UK/US development of simulation. *Simulation Practice and Theory* **9**, 333–348.

Nordas, R. and Gleditsch, N.P., 2007. Climate change and conflict. *Political Geography* **26**, 12.

Pereira, J.V., 2009a. SD–DES model: A new approach for implementing an e-supply chain. *Journal of Modelling in Management* **4**, 15.

Pereira, J.V., 2009b. The new supply chain's frontier: Information management. *International Journal of Information Management* **29**, 8.

Perkins, R. and Neumayer, E., 2009. Transnational linkages and the spillover of environment-efficiency into developing countries. **19**, 9.

Pidd, M., 2005. *Tool for Thinking: Modelling in Management Science.* John Wiley & Sons, West Sussex, England.

Potter, A., Yang, B. and Lalwani, C., 2007. A simulation study of despatch bay performance in the steel processing industry. *European Journal of Operational Research* **179**, 567–578.

Rajan, S.C., 2006. Climate change dilemma: Technology, social change or both? An examination of long-term transport policy choices in the United States. *Energy Policy* **34**, 16.

Rees, W.E., 2006. Globalization, trade and migration: Undermining sustainability. *Ecological Economics* **59**, 6.

Rizzo, D.M., Mouser, P.J., Whitney, D.H., Mark, C.D., Magarey, R.D. and Voinov, A.A., 2006. The comparison of four dynamic systems-based software packages: Translation and sensitivity analysis. *Environmental Modelling and Software* **21**, 1491–1502.

Rodrigue, J.-P., 1999. Globalization and the synchronization of transport terminals. *Journal of Transport Geography* **7**, 7.

Schrecker, T., Labonté, R. and De Vogli, R., 2008. Globalisation and health: The need for a global vision. *Lancet* **372**, 7.

Sener, F. and Zhao, L., 2009. Globalization, R&D and the iPod cycle. *Journal of International Economics* **77**, 8.

Senge, P.M., 1992. *The Fifth Discipline: The Art and Practice of the Learning Organization.* Century Business, London.

Spiess, A., 2008. Developing adaptive capacity for responding to environmental change in the Arab Gulf States: Uncertainties to linking ecosystem conservation, sustainable development and society in authoritarian rentier economies. *Global and Planetary Change* **64**, 9.

Stauffer, P.H., Viswanathan, H.S., Pawar, R.J. and Guthrie, G.D., 2009. A system model for geologic sequestration of carbon dioxide. *Environmental Science & Technology* **43**, 6.

Sterman, J.D., 2000. Business dynamics: Systems thinking and modeling for a complex world. Irwin/McGraw-Hill, Boston, London.

Sweetser, A., 1999. A comparation of system dynamics (SD) and discrete event simulation (DES). In: *The 17th International Conference of the System Dynamics Society and the 5th Australian & New Zealand Systems Conference, Wellington, New Zealand.*

Tisdell, C., 2001. Globalisation and sustainability: Environmental Kuznets curve and the WTO. *Ecological Economics* **39**, 12.

Towill, D.R., 1996. Industrial dynamics modelling of supply chains. *International Journal of Physical Distribution & Logistics Management* **26**, 23–42.

Tsoulfas, G.T. and Pappis, C.P., 2006. Environmental principles applicable to supply chains design and operation. *Journal of Cleaner Production* **14**, 10.

UNESCO, 2002. *Encyclopaedia of Life Support Systems: Knowledge for Peace, Progress and Sustainable Development.* UNESCO and EOLSS, Oxford, UK.

Vachon, S. and Maoc, Z., 2008. Linking supply chain strength to sustainable development: A country-level analysis. *Journal of Cleaner Production* **16**, 9.

Vlachos, D., Georgiadis, P. and Iakovou, E., 2007. A system dynamics model for dynamic capacity planning of remanufacturing in closed-loop supply chains. *Computers and Operations Research* **34**, 367–394.

Williams, T., Ackermann, F. and Eden, C., 2003. Structuring a delay and disruption claim: An application of cause-mapping and system dynamics. *European Journal of Operational Research* **148**, 192–204.

Wolstenholme, E.F., 1990. *System Enquiry: A System Dynamics Approach.* Wiley, Chichester.

Wolstenholme, E.F., 2004. Using generic system archetypes to support thinking and modelling. *System Dynamics Review* **20**, 341–356.

Zeigler, B.P., Praehofer, H. and Kim, T.K., 2002. *Theory of Modeling and Simulation: Integrating Discrete Event and Continuous Complex Dynamics Systems.* Wiley, San Diego.

Zhu, Q. and Sarkis, J., 2006. An inter-sectoral comparison of green supply chain management in China: Drivers and practices. *Journal of Cleaner Production* **14**, 15.

Zhu, Q., Sarkis, J. and Lai, K.-H., 2007. Green supply chain management: Pressures, practices and performance within the Chinese automobile industry. *Journal of Cleaner Production* **15**, 12.

Chapter 10

JAPAN FOR SUSTAINABILITY: OPPORTUNITIES, CHALLENGES, STRATEGIES, AND PRACTICES

Yasuhiko Hotta
Deputy Director, Sustainable Consumption and Production
Institute for Global Environmental Strategies, Japan
hotta@iges.or.jp

This chapter argues that Japan's strategy for sustainability and environmental management can be characterized as an approach to overcome environmental challenges by harmonizing environmental conservation and economic competitiveness. This can be characterized as Ecological Modernization that emphasizes improving efficiency and expansion of voluntary approach by comprehending environmental challenges as opportunities for further technical progress, innovation, and industrial competitiveness. Then, the chapter tries to further clarify such characteristic of the strategy for sustainability through examination of waste management and recycling policies as well as global warming policies of Japan in the 1990s. To follow-up this trend analysis, the chapter overviews and analyzes two recent sustainability strategies developed by Ministry of the Environment of Japan in the 2000s; "Strategy for an Environmental Nation in the 21st Century" in 2007, and "Clean Asia Initiative" in 2008. By doing so, the chapter argues that Japan's sustainable strategy continues to emphasize "increasing efficiency" and "voluntary approaches" but also start to shift its focus on "integration of different policy areas" and "international reach-out." Although, toward true sustainability for Japan, it is necessary to focus on total reduction in environmental impact from production and consumption, there are no clear strategies or visions yet both at governmental level and business level to achieve the next step of sustainability for Japan.

1. Introduction

This chapter tries to discuss characteristics of Japan's strategy for sustainability and environmental management in the 1990–2000s from the perspective of *Ecological Modernization* (EM). Especially, Japanese government has been focused on waste management and recycling as well as global warming as two priority areas in sustainable development since the 1990s. For transforming these challenges into opportunities for innovation and socio-economic reform, Japan's approach was to emphasize "increasing efficiency" and "expanding voluntary approach." Although this approach could improve performance in environmental management at unit level (each industrial facility level, each household level, each product level, etc.) as well as awareness of business and citizens in sustainability, it did not reach into fundamental transformation of the existing socio-economic systems contributing to mass production and consumption. In the 2000s, the trend of Japan's approach for sustainability has not changed from that of 1990s. At the same time, in addition to emphasis in "increasing efficiency" and "voluntary approach," Japan's sustainable strategy, such as Strategy for an Environmental Nation in the 21st Century in 2007, emphasizes "integration of different policy areas" and "international reach-out."

This chapter argues that Japan's strategy for sustainability and environmental management can be characterized as an approach to overcome environmental challenges by harmonizing environmental conservation and economic competitiveness. This can be characterized as EM that emphasizes improving efficiency and expansion of voluntary approach by comprehending environmental challenges as opportunities for further technical progress, innovation, and industrial competitiveness. Then, the chapter tries to further clarify such characteristic of the strategy for sustainability through examination of waste management and recycling policies as well as global warming policies of Japan in the 1990s. To follow-up this trend analysis, the chapter overviews and analyzes two recent sustainability strategies developed by Ministry of the Environment of Japan in the 2000s; "Strategy for an Environmental Nation in the 21st Century" in 2007 and "Clean Asia Initiative (CAI)" in 2008. By doing so, the chapter argues that Japan's sustainable strategy continues to emphasize "increasing efficiency" and "voluntary approaches" but also start to shift its focus on "integration of different policy areas" and "international reach-out." Although, toward true sustainability for Japan, it is necessary to focus on total reduction in environmental impact from production and consumption, there are no clear strategies or visions yet both at governmental level and business level to achieve the next step of sustainability for Japan.

2. EM, Environmental Policy, and Strategies for Sustainability in Japan

Japan's strategy for environmental management and sustainability after the 1990s can be characterized with harmonization of environmental conservation and economic competitiveness through the pursuit of eco-efficiency. This is along the concept of EM, conceptualization of shift in environmental policy, and management from the 1980s to 1990s in Western Europe and Japan. There are several attempts to explain the characteristics of Japan's environmental policy and sustainable strategy after the 1990s through the concept of EM such as Barrett (2005) and Hotta (2004).

EM provides the idea that economic growth and environmental protection are essentially complementary (Dryzek, 1997, p. 15). As noted earlier, in EM discourse, environmental problems are considered as opportunities rather than troubles to "a restructuring of the capitalist political economy along more environmentally sound lines" (Dryzek, 1997, p. 141).

Hotta (2004) summarized that EM characterizes the current trend of environmental policy and sustainable strategy into the following six points:

(1) emphasis on compatibility between economic competitiveness and environmental protection;
(2) emphasis on technological and management innovation in industrial systems;
(3) emphasis on the role of market dynamics and economic agents;
(4) a movement from the "react and cure" principle to the "anticipate and prevent" principle;
(5) new forms of policy making process; and
(6) a shift in the role of science in policy making.

Especially, one of the most important components, which makes EM-type policy concepts appealing to industrial sector, is the idea that *pollution prevention pays* by emphasizing compatibility between economic competitiveness and environmental protection. As Dryzek (1997, p. 142) pointed out, "(f)or the key to ecological modernization is that there is money in it for business. Thus, business has every incentive to embrace rather than resist ecological modernization." The logic of EM-type concepts for business, he continues, is that "(l)ess pollution means more efficient production."

The concept of efficiency is a crucial and most significant notion for businesses and advocates of EM-type policy discourse. Ayers (1997, p. 6), one of the earlier developers of the concept of eco-efficiency, defined eco-efficiency as "the objective of maximizing value added per unit resource input." The approaches on business and environment, such as LCA, environmental management systems, such as ISO 14001, or environmental auditing are considered as tools to achieve this eco-efficiency. As Ayers (1997, p. 6) suggested "(t)his idea is essentially equivalent to maximizing resource productivity at the firm level, rather than simply minimizing wastes or pollution associated with a given product."

The logic of eco-efficiency leads environmental policy and strategies to combine with other logics, including that of "voluntary approach." "Voluntary approach" revolves around the idea that governmental regulations are often inefficient in implementing appropriate measures for environmental problems compared with self-regulation by business and industries. Under the logic of "voluntary approach," governmental regulation is interpreted as the method of encouraging the increase in eco-efficiency.

Japan's environmental policy and sustainable strategy in the 1990s and 2000s can be characterized from pursuit of eco-efficiency and voluntary approach mentioned above. However, this is not a trend arose suddenly but based on a lesson learnt through pollution prevention activities in the 1970s in Japan.

Indeed, through the 1970s and the 1980s, energy-saving and technological innovation in each manufacturer had developed the conditions for EM in Japan. Historically, the response to the energy crisis, for example, seen in the Japanese automobile industry was a successful story of combining energy and resource concerns (thus later found as an environmental benefit) with competitiveness in global market. As Jänicke *et al.* (2000) suggested that, such a shift in the industrial structure with less environmental impact per unit (importantly not overall but "per unit") was mainly "caused by (primarily technological) changes taking place within sectors and enterprises (intrasectoral change....seem for the most part to be autonomous, triggered especially by alternations in price patterns [of energy and resources])" (Jänicke *et al.*, 2000, p. 148). Thus, the story of energy crisis and saving in 1970s and 1980s has contributed to the idea of eco-efficiency and voluntary act as an efficient strategy to tackle global environmental issues. However, because of the serious impact of four major Kougai [industrial pollution] incidents in Japan, pollution prevention was considered as negative activity. That is to say negative in the sense environmental activities in industries are primary connected to the prevention of serious pollution incidents and not contributing to business innovation or opening up a new market opportunity. Thus, such negative image of Kougai and pollution prevention has prevented reimagining the positive business impact of environmental technologies and management for a long time before the arrival of discourse of a potential positive business chance and role in global environmental problems. Some engineers commented that "environmental protection was considered as a way to avoid negative image of business activity. It was difficult for us to imagine that environmental activity has a positive impact on business as a whole." Until recently, business sections dealing with environmental issues are not considered as a main section in business strategy.

In a sense, the emergence of EM discourse in Japan can be observed as the change in this tendency of seeing environment activities from as an obligation and cost into as a new opportunity, challenges, and chances for innovation in technology as well as in management. Indeed, what story line of eco-efficiency and voluntary approach provides to Japanese industry would be an opportunity to repackage both their excellent pollution prevention technologies and resource efficient technologies into driving force of new innovation in technology and management in a "business" challenge of global environmental change (and of course, economic globalization).

3. Characteristics of Environmental Policy and Sustainable Strategies of Japan in the 1990s to Early 2000s

In the 1990s Japanese society faced fundamental changes. First, during the long recession [which is called "Kouzou Fukyou" (structural recession)] of the 1990s, Japanese society started to realize the need for the restructuring of society. Thus, since the end of cold war, with a slowing economy, the atmosphere for the need for the restructuring of Japanese political, economic, and social structure, namely Kouzou-Kaikaku (structural reform), has risen since 1990s. For example, the Prime Minister (2001–2006) Junichiro Koizumi has kept his popularity by advocating the needs for Kouzou-Kaikaku (structural reform). Second, the interest in global environmental problems and waste issues has risen in 1990s after Rio summit in 1992 and especially during the time of COP3 in Kyoto in 1997. There was a growing and common understanding for the need to structure in which Junkan-gata shakai (Sound Material Cycle Society) has now been established. Third, the end of the cold war and IT revolution sped up economic globalization. Japanese social and economic structure, it was argued, should shift from concentrating on the national economy to globalized economy.

At the same time, from 1980s to 1990s, the focus of environmental problems in Japan has shifted from domestic pollution prevention to two further focal points, namely "waste and recycle" and "global environmental problems." This implies, in effect, that the focus of environmental problems has shifted from industrial pollution to broader environmental problems as a result of the restructuring of modern industrial society.[1] Following this important contention, therefore, the following section first overviews two major focal points of environmental problems in Japan since the mid-1980s — waste issues and global environmental issues — and then it examines the implication of these on business responses in the 1990s.

3.1. *Waste management and recycling of Japan after the 1990s*

The Japanese economy of the 1980s has been characterized as a "bubble economy" and as one which resulted in increases in consumption and more and more urbanization of Japanese society. This caused the classic environmental problems of urban lifestyle: waste, lack of waste disposal facilities, water quality, and chemical components in the products of every household, transportation, and destruction of natural environment. The lifestyle of modern society itself, such as prevalence of electronic appliances, automobile, package, instant foods, and using of chemicals, has become the source of problem (IGES 2000: 19). For example, in Japan, household waste has increased from 44 million tons in 1980 to 50 million tons in 1990. Industrial waste emissions increased from 292 million tons in 1980 to 395 million tons in 1990 (Environmental Agency, 1981; Ministry of the Environment, 1992, 2001).

With the lack of final waste disposal facilities, waste problems became a major environmental question in Japanese society. In 1996, the Japanese government calculated that the available capacity of landfill for household waste is about 9.4 years and that of industrial waste is about 3.1 years (Yamaguchi, 2000). Therefore, throughout the 1990s, legislation to promote recycling of waste and products was developed. In 1991, the Law for the Promotion of Utilization of Recyclable Resources (Recycling Law) was enacted. The Containers and Packaging Recycling Law was passed by the Diet in 1995; the Law for Recycling of Specified Kinds of Home Appliances (or Home Appliance Recycling Law) was passed in 1998.

[1] For example, the division of MITI that is in charge of environmental issues has changed its name from "Ricchi *Kougai* Kyoku (Division of Land Plannning and Pollution Prevention)" to "Kankyo Ricchi Kyoku (Division of Environment and Land Planning)" in 1993.

In 2000, six legislative instruments (see Table 1) related to recycling were passed by the Diet: (1) the Basic Law for the Promotion of the Establishment of a Sound Material Cycle Society, which was a law to show the idea and basic framework of Sound Material Cycle Society; (2) the revision of the Law for Waste Management; (3) the revision of the Recycling law through which "reduce," "reuse," and "recycle" became the ideal for waste policy; (4) the Construction Materials Recycling Law; (5) the Food Recycling Law; and (6) the Green Purchasing Law, whereby the national government is advised to purchase recycled goods and environmentally friendly goods.

The aim of these six laws, with that of the Containers and Packaging Recycling Law and Home Appliance Recycling Law, is to promote not only waste processing but also the formation of a sustainable society. The structural recession of 1990s symbolized the limit of the "developmental" regime to keep Japanese economic development successfully through post-world war era. In this context, the "environment" becomes one of the key criteria for restructuring especially among businesses. Therefore, these legislations can be considered as a part of wholesale restructuring of the Japanese economic and social system from a "developmental" one into a "mature" society by taking the examples of Scandinavian and western European countries such as the Netherlands, Germany, Sweden, or Norway. One of the most important principles behind the laws related to a sound material cycle society is to incorporate the Extended Producer Responsibility (EPR) into environmental policy on waste issues. EPR is the policy principle for reducing waste and increasing resource productivity in response to the lack of landfill. EPR is to shift the responsibility of processing general household waste from local governments to

Table 1. Major legislation in relation to waste management and recycling in 1990s and 2000s in Japan.

Law	Year
Recycling Law	1991
Revision of Waste Treatment Law	1991
Basic Environment Law	1993
Container and Package Recycling Law	1995 revised in 2006
Revision of Waste Treatment Law	1997
Environmental Impact Assessment Law	1997
Law for Recycling of Specified Kinds of Home Appliances	1998
PRTR Law	1999
Law for Measurement for Dioxin	1999
Fundamental Law for Establishing a Sound Material-cycle Society	2000
Revision of Waste Treatment Law	2000
Revision of Recycling Law	2000
Construction Material Recycling Law	2000
Food Recycling Law	2000 revised in 2007
Green Purchasing Law	2000
End of Life Vehicle Recycling Law	2002
Fundamental Law for Establishing a Sound Material-cycle Society	2003 and revised in 2008
Revision of Waste Treatment Law	2006

producers of products. For example, the Home Appliances Recycle Law asserts the responsibility of taking back and recycling four home appliances — wasted fridges, washing machines, televisions, and air conditioners — to those of the makers of these products and away from local autonomy. Under this law, the consumer has an obligation to pay for waste processing, shops have an obligation for taking back the waste products and delivering or returning them to the producers. Producers have a responsibility for taking back wasted products from shops and to recycle the products. Therefore, the price of recycling will be reflected in the recycle processing cost of the products of producers. And this will give an incentive for producers to reduce the cost of recycle processing and increase the competitiveness by increasing recyclability of the products (Yamaguchi, 2000, p. 129). By securing this shift in responsibility from local autonomy to the producers, it is expected to achieve minimization of waste generation and recyclability of the products. At first glance, it would seem to be a form of command-and-control type of regulation. However, it is much more a regulation expecting to give incentives to "voluntary action" of technological breakthrough to increase "eco-efficiency."

3.2. *Climate change issues*

Along with economic globalization from mid-1980s, global environmental issues such as global warming and destruction of ozone layer have become major items on the world political agenda. During this period, the role of business to contribute sustainable development was advocated further. For ozone layer protection, the Montreal Protocol was adopted in 1987. It is believed DuPont played a significant role in the direction of this agreement (see Benedick, 1998; Litfin, 1994) resulting in 1992 in the decision to prohibit the production of CFCs. Similarly, global warming was considered as a serious threat to the world and in 1992 the Framework Convention on Climate Change (FCCC) was adopted. In 1992, the Rio summit symbolized the full entry into the era of global environmental issues and it was in Agenda 21 adopted in Rio summit that the positive role of business for sustainable development was emphasized and formalized.

During the 1990s, the notion about the difference between local pollution problems and global environmental issues came to prevail, and this recognition was not limited to Japan. Table 2 by Yamaguchi (2000, p. 5) shows a typical and standard understanding on environmental issues after mid-1980s of bureaucrats and businesses by comparing pollution issues or Kougai.[2]

---

[2]It is rather problematic that Yamaguchi discusses that every actors can be both damaging and damaged party. It is problematic because the damage received from global environmental change varies in each actor. Cause and effect relations are not so linear. Hence, the scale of damage caused by actor A cannot necessarily cause same damage to actor A in global environmental change but more likely to cause severe damage to somewhere and someone else. It is more likely to cause more severe damage to less privileged countries, actors, and residents.

Table 2. Pollution problem and global problem (from Yamaguchi, 2000, p. 5 [my translation]).

	Pollution problems (Kougai)	Global environmental issue
Extent of impact	Specific area	Entire population of the world
Relationship between industry and citizens	Opposing and conflict	Every actor can be both damaging and damaged party
Visibility	Visible	Hard to be visible
Uncertainty	Small	Big
International political aspect	No	North–South relation
Time extent of impact	Present issue	Issue for next generation

The cause of problem in global environmental issues is that of modern industrial society and lifestyle. Therefore, with the shift of environmental issues from pollution problems to global environmental issues and waste issues, the restructuring or restructuring in modern industrial system has become focus of environmental conflict.

To cope with changing times, the Japanese government has started to shift its focus of their environmental policy from pollution prevention to more general relations of "environment, society, and industry" or the realization of a sustainable society. In 1993, in response to Rio summit, the Basic Law for Environment was passed in Japan. This law sought to show that the focus of Japanese environmental policy had shifted from pollution prevention to the "reduction of environmental load" and the "protection of the global environment." In this law, the following three points are considered as ideals for Japanese environmental policy:

(1) maintenance of sound environment over generations;
(2) realization of sustainable society; and
(3) promotion of international cooperation.

To realize these ideals, the Environment Basic Plan was adopted in 1994. Here, four long-term objectives were adopted, namely, recycling, symbiosis with nature, participation, and international cooperation. These ideas are reflected in the environmental laws of the 1990s.

3.3. *Voluntary approach*

Along with the rise of global environmental problems as a central focus of the environment, businesses started to call themselves "global corporate citizens" and emphasize the importance of partnership with "the government, consumers, citizens, and NGOs" (Keidanren Appeal on Environment, July 1996).[3] In international policy circles on the environment, business has shifted from simply being the accused to a being a possible solution provider as well as a cause of the problem.[4]

[3] www.keidanren.or.jp/english/policy/pol046.html.
[4] See Chapter 30 Agenda 21 titled "STRENGTHENING THE ROLE OF BUSINESS AND INDUSTRY" for emphasis in the active role of business in sustainable development.

Behind this, there is an intention of the business sector to promote the idea of voluntary action and partnership as measures to address environmental problems. To observe this process, Keidanren's (the Federation of Economic Organizations) response can be seen as a good example. Keidanren is the largest and most influential business organization in Japan. Most of the large businesses are members of this organization. Keidanren could be a synonym of the Japanese major business sector. Therefore, by observing Keidanren, we can catch the Japanese businesses' major understanding of environmental issues.

In July 1996, Keidanren announced "the Keidanren Appeal on Environment." In this appeal, the following four areas were selected as urgent environmental issues for business sector:

(1) measures to cope with global warming;
(2) structuring of recycle-based society;
(3) restructuring of environmental management system and environmental auditing; and
(4) environmental consideration in evolving overseas projects.

As Yamaguchi argues, this reflects a standard set of environmental issues after 1990s as understood by the business sectors (Yamaguchi, 2000, p. 25). Interestingly, this does not contain specific reference to "Pollution Prevention" or Kougai anymore.

In the 1990s, the Rio Summit in 1992 and the FCCC Conference of the Parties in Kyoto (COP3) in 1997 were the major boost to promote "active response" on global environmental issues from Japanese businesses. Around the 1992 Rio summit, the arguments that businesses should play the significant role in prevention of global environmental issue had risen internationally. In 1991, the International Chamber of Commerce (ICC) proposed Industrial Charter for Sustainable Development. The World Business Council for Sustainable Development (WBCSD) also proposed the realization of Sustainable Development in 1991. In April 1991 in Japan, Keidanren announced the "Global Environment Charter." Yamaguchi analyzes that "[this charter] shows that the companies' recognition has shifted from Pollution Prevention to Environment (Yamaguchi, 2000, p. 24, original in Japanese [my translation])." With the idea of self-responsibility, businesses have started to promote the idea of voluntary environmental action. And it is around this time that companies started to make environmental charters and environmental action plans.

This idea of voluntary approach or government and business partnership is strongly influenced not only among businesses but also by the environmental policy of the Japanese government. The logic to support to this voluntary action from businesses is that global warming and waste issues are modern life-style questions (daily modernity) and that therefore direct regulation can hardly be the appropriate and effective measure for their regulation. A major introduction of economic methods, such as environmental taxation or carbon tax, is still under discussion in

Japan.[5] Therefore, Keidanren discusses that promoting voluntary approach seems "practical." Internationally, the idea of governance[6] and partnership is at the core of the political idea of the realization of sustainable development and environmental management. Domestically in Japan, there has been a tradition of voluntary agreement for the promotion of environmental measures, although there are major differences between voluntary agreement from 1960s and voluntary approaches after the 1990s.

Voluntary agreements were a supplement to fill the defects of national policy by local government where the agreement between local governments and specific company or companies served to address local or area difficulties. By contrast, the intention of voluntary approach is to set voluntary action as the core of national environmental policy.

As noted above, COP 3 held in Kyoto in 1997, following the start of ISO 14001 in 1996 was another major boost for businesses in Japan to respond environmental issues. It attracted a great deal of attention for the potential in responding to environmental problems that the environmental technology which Japanese industry possessed. The potentialities of the environmental technology business sector began to attract attention as the next booming market for Japanese industry, not least as a result of the lead secured with their advanced environmental technology accumulated in course of Kougai (industrial pollution) and Shouene (energy saving) period. In addition to the pressure of high-energy consumption of products and the waste generated in the assembly industry such as automobile and electronics industries, one reason for this attention was that most of the environmental technologies had been accumulated during the decline of heavy industry in Japan. Thus, declining heavy industry had started to seek new business opportunities in environmental technology. As we already noted, in July 1996, one year before COP3, Keidanren announced "the Keidanren Appeal on Environment." To show the responsiveness in activity from businesses, Keidanren made a proposal of Keidanren voluntary action plan in 1997. It was a voluntary action plan with targets for each different industry, and it is around this time that companies started to produce environmental reports and showed their achievement through greater information disclosure.

For reasons of business's preference in applying voluntary act, Yamaguchi discussed that the following three points are important.

By voluntary action, it is easier to pursue the most appropriate activity for different type of industry. Environmental tax is transfer of money from business

[5]This point is important. METI (successor of MITI) and Keidanren are opposing to introduction of environmental tax. Ministry of Environment (successor of Environmental Agency) is postive to introduce. Although Keidanren is opposing to introduction of Environmental tax, the responses of individual companies vary. According to research by NIKKEI, 65% of 1200 companies are positive to introduction of carbon tax (Mitsuhashi, 1999, p. 145).

[6]It is interesting that some of contributors from business sector for the book titled "Private corporations and Environmental Governance" interpret environmental governance as voluntary environmental management (see IGES, 2000).

to government, so industry does not favor economic method such as environmental tax. It is the nature of business not to like regulation by government (Yamaguchi, 2000, p. 115).

This, of course, was not limited to Japan in the 1990s, for in the US, Germany, the Netherlands, it is considered there has also been a major initiative to foster cooperation from business to solve global environmental issues. In the EU, for example, the Environmental Plan from 1993 to 2000 emphasized the shift from "governmental-oriented" to "cooperation with industry." And they wrote the importance of a "conversation with industry" and the need for "voluntary agreement." In 1996, voluntary agreement was set as a most important method.

However, as Usui pointed out, that businesses could not be considered as a solid, unitary group with regard to voluntary action (Usui *et al.*, 1999, p. 155). According to him, "behind the scene of campaign of the Japanese business sector, there are two types of groups of companies: (1) groups of companies without appropriate technology to cope with new regulation and (2) groups of companies with most advanced technological innovation aiming for eco-business [my translation]." He goes on to explain that "The first group has a bigger voice to promote voluntary action. The latter group is waiting for the introduction of new environmental regulation, which has a possibility to open up new business chances for them. But, before the opportunity matures, they keep their mouths shut [my translation]." Therefore, if the timing is right, it is possible that a certain kind of environmental regulation will be welcomed by business.

The above claim that there was increasingly active response to environmental activities from industries in the 1990s was supported by Nikkei's research of 1200 companies in 1997. According to the research, 18% businesses responded that CO2 emission should be reduced by between 10 and 15% compared with 1990 by 2010. For the carbon tax, 65% of companies responded that they would welcome the introduction. It seems that there is more active atmosphere on environmental activity among individual companies than business organization and central government in Japan.

For example, as already discussed extensively in Chap. 3, the ISO 14000 series are typical initiative of voluntary regulation by implementing more efficient business management systems in the area of the environmental management of business. The ISO 14000 series is the voluntary Environmental Management System by a private international organization, the International Organization for Standardization (ISO). The development and implementation of these systems and standards depend on the voluntary action of each organization. In the autumn of 1996, ISO 14001 came into effect. Before that, Keidanren announced the following concerning ISO 14001:

> "We will structure an environmental management system in an effort to address the environmental problem voluntarily, ensure its continuous improvement and perform internal auditing to confirm that the system will steadily work. Keidanren

> has positively participated in the formulation of the ISO environmental management and auditing standards, scheduled to come into effect this fall. It is recommended that Japanese industries, manufacturing or non-manufacturing, should utilize the standards as an effective means of environmental improvement (Keidanren, Keidanren Appeal on Environment, July 16, 1996)."

This gives an impression that ISO 14001 is a legitimizing tool for voluntary action on the part of business sectors. Actually, about half of the chairpersons of working groups preparing for the ISO 14000 series are from major multi-national corporations (Usui *et al.*, 1999, p. 168). However, with ISO 14000 series, businesses have to show accurate information on their environmental activity. The global standardization of environmental management can be result in the availability of the comparison of environmental activities across the world at least among major industries. They cannot use abstract expression such as "environmental-friendly" without accurate data disclosure. Thus, the legitimation of environmental management or activities of industries does not come simply from obedience to the governmental regulation but providing globally accountable information to ensure authenticity and trust in their environmental activities. Since 1996, this environmental management system has explosively spread among Japanese business and other organizations including local governments, municipalities, and even universities. By January 2007, there were 21,799 ISO 14001 certified organization in Japan.

This demand in voluntary environmental management and accountability of their environmental activities are increasingly expanding to the world through global network of supply chains of transnational corporations. In the 1990s to early 2000s in Japan, the "environment" had become more about management issues than any question of technological measurement.

In the 1990s to early 2000s, environmental issues for Japan have shifted from "pollution prevention" to "global environmental issues" and "waste issues." This shift has increased the awareness of environmental issues as those of the restructuring of modern industrial society including its lifestyle. Environmental problems are appreciated not only as an issue the industrial sector, but as a concern for industrial society as a whole.

Eco-efficiency and voluntary action are one of the most prominent interpretations of sustainability and governance to cope with environmental issues "global environmental issues" and "waste issues" in the 1990s to early 2000s in Japan. Environmental issues and sustainable development are interpreted along the logic of increasing efficiency. The effort to solve environmental problems is prompted by the drive for more efficient production and services. Voluntary approach is considered as a better method both from industrial sector and government. Moreover, some businesses think that regulation is welcomed if it opens up new market opportunity and promotes innovation. Introduction of the idea of eco-efficiency and voluntary approach into the central stage of environmental politics is supported by a story that energy saving in 1970s contributed to efficient production as well as less environmental load. To achieve eco-efficiency through voluntary approach, management and evaluation tools such as ISO 14001 and LCA have spread throughout Japanese businesses.

With pressure of globalization and the atmosphere of blockade after end of bubble economy, the "environment" might appeal to Japanese society as an opportunity for restructuring of Japanese society, at least among businesses. However, it cannot be explained only through the experience of energy and resource saving during energy crisis contributing to the idea of eco-efficiency and voluntary act. The interpretation of the environment into the activity of increasing efficiency and productivity is not only limited in Japan but covered at least in the highly industrialized society such as Germany, the Netherlands, Scandinavian countries, and also the United States.

4. Challenges and Opportunities for Sustainability for Japan Since the 2000s

Japan has achieved a rapid economic growth since 1950s. In the 1970s, Japan's engagement in pollution prevention and oil crisis resulted in a source of industrial competitiveness through accumulation of environmental and energy-saving technologies. This experience resulted in the notion behind environmental policy and sustainable strategies in the 1990s and 2000s that improving efficiency and motivating industrial voluntary activities through wise policy intervention could solve challenges for sustainability. On the other hand, it has been widely visible that maintenance of natural environment was in low priority through the excessive pressure of national land development, infrastructure building, and population increase until very recently. For example, the submergence of the countryside for development of large-scale dam or lost of wetlands or natural coastal area from large-scale land reclamation projects has been observed all over in Japan. Land reclamation project in Isahaya bay is a typical of those excessive land development activities. Although they had a certain reason such as flood prevention, this national driving force for land development changed the face of Japanese landscape completely over 50 years. The change in Japanese coastline is one of good examples of such changes. In 1990s, about 44.8% of the mainland Japanese coast line is natural coast and 53.9% of it is covered by artificial or half-artificial coast line. In 1973, about 59.6% of coast line was natural coast. A total of 19.2% was half-natural coast line and 21.2% was artificial coast.[7] Also, in 1990s, 23.5% of river bank is covered by artificial structure such as concrete wall. It was 19.2% in 1973.[8]

---

[7]These figures on coast line in 1990s are based on the 4th National Natural Environment Conservation Basic Research conducted by Environmental Agency in 1990–1992. See Ministry of the Environment of Japan (2003). For the figures in 1973, see Environment Agency Japan (1973).

[8]The figures on river bank are based on researches on river done by Ministry of the Environment in 1979 and 1998. See http://www.biodic.go.jp/kiso/23/23_kasen.html.

Pursuit of efficiency would minimize the environmental impact at unit level of production and consumption activities. However, increasing efficiency cannot be an ultimate approach of the breakthrough for reduction in the total impact of industrial/production/consumption activities. Although it is necessary to design a new national vision for socio-economic development based on sustainability concept, the continuous recession makes difficult for the government to design such comprehensive vision and strategy. Is it possible to direct the Japan's national vision for breakthrough toward sustainability by implementing the vision of low-carbon society, sound material cycle society, and society with harmony with nature? As an initial attempt, let us look at the Strategy for an Environmental Nation in the 21st Century of Japan.

4.1. *A new policy trend for sustainability: A case of Strategy for an Environmental Nation in the 21st Century of Japan*

Strategy for an Environmental Nation in the 21st Century of Japan is a document launched and approved by the Cabinet Meeting in June 2007 to be a guideline of environmental policy for both domestically and internationally in the near future as well as a guidance for Japan's contribution to international efforts to sustainability.

This strategy aims to establish sustainable society, which can overcome global environmental challenges through integration of activities for low-carbon society, sound material cycle society, and society in harmony with nature. Also, it emphasizes "international leadership to overcome climate change challenges," "inherit of natural services by protecting bio-diversity," and "sustainable resource circulation through the 3Rs."

Although there are eight other emphasis in the strategy, the characteristics is embodied in the integrated approach of constructing low-carbon society, sound material cycle society, and society in harmony with nature. The current Japan's strategy for sustainability at least that of Ministry of the Environment of Japan is along this line.

4.1.1. *Low-carbon society*

In the strategy, Japanese government describes low-carbon society as "a society with realization of an affluent lifestyle with reduction in CO_2 emission. With widespread environmental consciousness and technologies in the whole society, through dissemination of conventional as well as new innovative technologies, this concept aims to achieve rich lifestyle and economic growth in harmony with environmental conservation. In concrete, it is a reform in lifestyle and social system such as life in harmony with nature such as forest, efficient transportation system such as public transport, or compact city." This concept represents well the characteristics of EM-type policy discourse. The concept emphasizes social reform through technocentric solution and continues economic growth rather than redefining prosperity. The concept reflects the emphasis in high-energy efficiency of each product such as hybrid car, efficient air conditioning systems, or LED lamps in Japan as well as voluntary approach and technical solution by industrial sector. Therefore, the strategy has a limited strength as a vision due to lack in clear budget allocation, the role setting for stakeholders, and clear target setting.

4.1.2. *Sound material cycle society*

Sound material cycle society is defined under Fundamental Law for Sound Material Cycle Society as "a society where the consumption of natural resources is minimized and the environmental load is reduced as much as possible, by restraining products, etc. from becoming wastes, etc., promoting appropriate recycling of products, etc. when they have become recyclable resources, and securing appropriate disposal of the recyclable resources not recycled, which means the disposal as wastes."

This policy direction of Japan was relatively given more concrete visions such as the role of stakeholders, interaction of legislation and strategies, concrete quantitative targets to be achieved, and constant follow-up process under the Fundamental Plan for Sound Material Cycle Society launched in 2003 and revised in 2008.

In the late 2000s, Japan pushed this strategy forward as an international initiative under the 3R Initiative. The 3R Initiative is a Japan's initiative to promote the "3Rs" (reduce, reuse, and recycle) globally so as to build a sound material cycle society through the effective use of resources and materials under the G8 process. It was agreed upon at the G8 Sea Island Summit in June 2004 as a new G8 initiative and agreed G8's action plan to promote the 3R policy: *Kobe 3R Action Plan* at G8 Environmental Ministers Meeting in May 2008 in Kobe, Japan.

Japan seems to be confident in their achievements in sound material cycle society and shows its ambition of becoming the center of 3R promotion in Asia. The Strategy for an Environmental Nation in the 21st Century of Japan declares to promote the vision of sound material cycle society in accordance with different challenges faced by Asian countries by transferring institutional mechanisms, technological systems, and experience in multi-stakeholders collaborations on the 3Rs and waste management through supporting national strategy development and city development.

4.1.3. *Society in harmony with nature*

The Strategy defines society in harmony with nature as "a society which enjoys the benefits of nature for the current generation and the coming generations with proper conservation of biodiversity, socio-economic activities in harmony with nature such as sustainable agriculture, forestry, and fishery, and enjoying natural environments." This concept is linked with biodiversity conservation and natural landscape conservation including countryside lifestyle and landscape. As this part of the strategy needs a close collaboration among central governmental agencies as well as with local governments, the strategy is not so clear in policy goals to achieve other than at idea/philosophical level.

4.1.4. *Limitation*

Strategy for an Environmental Nation in the 21st Century of Japan can be evaluated as an attempt to show each different component of Japan's sustainable and environmental strategies in integrated and linked manner. Indeed, since the late 2000s, environmental issues and sustainable issues are gotten more and more attention from media and citizens as one of the important strategic policy issues by which Japan may regain their industrial and technical competitiveness in the world market. However, the strategy tended to lack the true strategic vision for socio-economic reform to shift natural landscape into more sustainable manner to lead the country. Therefore, the strategy was not a big leap from pursuits of eco-efficiency and voluntary approach in the 1990s.

Another important character of Japan's sustainable strategy for 2000s is the emphasis in collaboration with developing Asia.

4.2. *Collaboration with Developing Asia: A case of CAI*

CAI was launched in 2008 as a component of international outreach for Strategy for an Environmental Nation in the 21st Century of Japan mentioned above. It is intended to be umbrella strategy for Japan's international cooperation in sustainable and environmental management areas and aims to establish sustainable society in Asia, which promotes economic development in harmony with environmental conservation by transferring Japan's experience, technology, institutions in a packaged manner.

The CAI follows the same emphasis as Strategy for an Environmental Nation in the 21st Century of Japan: low-carbon society, sound material cycle society, and a society in harmony with nature.

Under CAI, Japan's recent emphasis in "internationalization" of sustainable and environmental policy and "integration" of different policy areas are more strongly featured. The followings are the basic strategies and practices mentioned as a part of CAI in its brochure (Clean Asia Initiative Brochure, n.d.).

(1) *Promoting low-carbon/low-pollution society*: Under CAI, the following activities are listed as a part of Japan's action: formulation and share of a vision for a low-carbon society, promotion of co-benefits in development activities to achieve pollution prevention and climate change mitigation, promotion of environmentally sustainable transport in Asian countries, and international cooperation for pollution prevention measures in Asia based on its experience in solving Kougai incidents (pollution problems) in the 1960–1970s.

(2) *Sound material cycle society in Asia*: Under CAI, the following activities are listed as a part of Japan's action to achieve sound material cycle society in Asia: launch of regional 3R forum in Asia, promotion of the 3Rs through technology transfer, prevention of improper transboundary movement of recyclable

resources, promotion of information sharing and data for the 3Rs, and integration of low-carbon and sound material cycle society.

(3) *Adaptation to climate change and achieving a society in harmony with nature*: Under CAI, the following activities are listed as a part of Japan's action to achieve a society in harmony with nature in Asia; conservation and sustainable utilization of natural resources through Satoyama (Japanese terminology for countryside harmonizing agricultural and forestry activities and natural conservation) Initiative, strengthen international network to conserve coral reef, observation and research to understand impacts and vulnerabilities from climate change, and strengthening initiatives at the community level in collaboration with NGOs and NPOs.

(4) *Promoting environmentally conscious economic activities through greening market*: Under CAI, the following activities are listed as a part of Japan's action to expand green market in Asia, promotion of green purchasing and greening the supply chain, utilization of private finance for proliferation of environmental technologies and promote environmental consciousness in project finance, and promotion of research activities for better understanding of mutual relations of environment and economy.

(5) *Establishing of collaborative based for environmental cooperation in Asia*: Under CAI, to develop institutional basis for effective environmental cooperation in the areas mentioned above, the following activities are listed as a part of Japan's effort to strengthen cooperation among Asian countries and related organizations; strengthen partnerships to ensure harmony with nature, promotion of standardization and networking for environmental monitoring and countermeasures, promoting environmental-related human resource development and application, and strengthening mutually supportive relationship between environment and trade.

Under CAI, the measures against climate change are emphasized as one of the most important elements for sustainable development in Asia. Indeed, "promoting low-carbon/low-pollution society" is listed as a top activity among five priority areas of CAI. It emphasizes the transfer of Japan's experience and technology for achieving energy-efficient and low-pollution industrial activities. CAI follows or re-emphasizes the story line documented earlier in this chapter. Japan's EM, sustainable strategy emphasizing eco-efficiency and voluntary approach to harmonize environmental and economic activities, was originated in overcoming *Kougai*, which resulted in energy efficiency and low pollution at unit level of industrial activities and continued to be a major feature in its sustainable strategy.

For the regional cooperation for Sound Material Cycle Society under CAI, this should be understood in the context of the structural change that is taking place in the economic relations among developed and developing countries. The increasing transboundary movement of recyclables can be explained from the following three structural changes in the developed and developing economies: (1) a rapid increase in the recovery of recyclables due to successful implementation of recycling-related legislation in the developed countries; (2) the lowered demands of recyclables in developed countries due to shift in location of manufacturing industries from the developed countries into the Asian developing countries; and (3) the corresponding increase in demands regarding recyclables in developing countries in parallel with improvements in the related economies themselves.

Thus, from the viewpoint of developed countries, the recent interest in the 3R Initiative not only concerns domestic promotion of the 3Rs, but also concerns "a policy response to the possible hollowing out of the domestic recycling industry under globalisation," and "how to establish an environmentally-sound and economically-efficient transnational flow of recyclables." The export of recyclables has started to influence Japan's domestic recycling businesses. For example, because of increasing foreign demand for PET waste as recyclables, PET recycling industries are facing increasing difficulty in securing sufficient PET waste to run their facilities at full capacity. Some are even on the brink of bankruptcy. Therefore, for the developed countries, the establishment of an environmentally sound downstream material flow is difficult to realize domestically without consideration of the international flow of recyclables. This is one of the reasons why developed countries, such as Japan, which already have the capacity for domestic recycling, need to promote resource circulation and waste management policy in collaboration with the developing countries in Asia. In this sense, sustainability issues/environmental challenges are now internationalized even for waste management, which was conventionally considered as domestic and local issues.

In a similar manner, point 4 and 5 above are also related to policy response for economic globalization. Thus, it emphasizes the role of transnational actors such as industrial sector and economic agents as well as international organizations and NGOs.

Among these, the promotion of "society in harmony with nature" is rather different policy in terms of pursuit of sustainability. This is not directly reflecting EM-type policy discourse of "increasing eco-efficiency" and "voluntary approach." For example, the idea of Satoyama Initiative is related to traditional landscape conservation, lifestyle, and preservation of commons. On the other hand, it is not clear how to utilize this concept for international cooperation for bio-diversity conservation or natural landscape conservation in Asia.

Overall, CAI can be evaluated a continuous expression of Japan's EM-type sustainable strategy. Especially, it is now explicit that domestic environmental policy and sustainable strategies are linked to internationalization. This can be explained as follows.

Modern environmental problems, typical in global environmental change, tend to spread, relate, and connect between global warming in global level, national policy, business strategy, local environmental policy, and even to expression of lifestyle of each individual. Thus, each practice for sustainability needs translational linkage to interpret the relations between different levels of actors and issues and decrease the complexity of environmental problematic.

5. Conclusion

The promotion of environmental policy in terms of efficiency as seen in the case of Japan in 1990s and 2000s is intended to include business and industry into effort toward sustainability by using a familiar logic and language.

However, when sustainability debate of businesses, local governments, NGOs, academia, governmental agencies, and international organizations is dominated by the language of efficiency, it may not be the environmental concern that was integrated into the mainstream of business and industry. It may be mainstream industrial and business practices of rationality and efficiency, which tries to transform their practices along the discourse of sustainability and environment.

Environment and sustainability are part of language to set a further target for development or expansion of the logic of efficiency. In this sense, Japan's Sustainable Strategy is essentially techno-corporatist movement to expand rationality into practices to face global issues overspread the domestic governance system. Thus, a spread of industrial and environmental practices based on the idea of eco-efficiency and voluntary regulation could be seen as a part of global industrial and environmental governance. Such process asks participants to raise their technical expertise and technological competence as a source of influence.

The current Japan's Sustainable Strategy is based on its experience attaining technical expertise and industrial competency gained through pollution prevention and energy saving in 1970s as well as the recent success in sound material cycle society in 1990–2000s. However, other than eco-efficiency and voluntary approach, it is not clear yet as a next step of socio-economic reform for sustainability. It is not clear yet what kinds of visions and models of national landscape and infrastructure are expected to be established in the next 10–20 years to come with proper implementation of Japan's sustainable strategy. When a clear and concrete vision of national landscape of sustainable society is established, we will be able to call Japan has the true strategy for sustainability in the 21st century.

References

Ayers, R., 1997. *Toward Zero Emissions: Is There a Feasible Path? — Introduction to ZERI Phase II, Presented as a Working Paper.* UNU/IAS, Tokyo.

Barrett, B., 2005. *Ecological Modernization and Japan.* Routledge, London.

Benedick, R., 1998. *Ozone Diplomacy: New Directions in Safeguarding the Planet.* Harvard University Press, Cambridge, MA, 1991. enlarged edition 1998.

Dryzek, J.S., 1997. *The Politics of the Earth — Environmental Discourses.* Oxford University Press, Oxford.

Environment Agency Japan, 1973. *The 1st Natural Environment Conservation Basic Research*. Available at http://www.biodic.go.jp/reports/1-1/u000.html.

Environment Agency Japan, 1981. *Showa 56 nen-do Kankyo Hakusyo (Environmental White Paper of Japan 1981)*. Environmental Agency of Japan, Tokyo. Available at http://www.env.go.jp/policy/hakusyo/past_index.html#env-kankyo.

Environmental Agency Japan, 1992. *Heisei 4 nendo-ban Kankyou Hakusho (Environmental White Paper, 1992)*. Ministry of the Environment of Japan, Tokyo. Available at http://www.env.go.jp/policy/hakusyo/past_index.html#env-kankyo.

Hotta, Y., 2004. The Transnational Politics of Ecological Modernisation. *An Analysis of the Formation of Transnational Authority in Global Environmental and Industrial Governance, with Special Reference to the Zero Emissions Initiative in Japan*. University of Sussex, Sussex, UK.

IGES (ed.), 2000. *Minkankigyou to Kankyou Gabanansu (Private Corporation and Environmental Governance [translation])*. Chuo-houki, Tokyo.

Jänicke, M., Monch, H. and Binder, M., 2000. Structural change and environmental policy. In: Young, S.C. (Ed.), *The Emergence of Ecological Modernisation — Integrating the Environment and the Economy*? Routledge, London.

Litfin, K.T., 1994. *Ozone Discourse: Science and Politics in Global Environmental Cooperation*. Basic Books, New York.

Ministry of the Environment of Japan, 2001. *Heisei 13 nen-ban Kankyo Hakusho (Quality of the Environment in Japan 2001)*. Ministry of the Environment of Japan, Tokyo. Available at http://www.env.go.jp/policy/hakusyo/past_index.html#env-kankyo.

Ministry of the Environment of Japan, 1992. *Clean Asia Initiative (Brochure)*. Ministry of the Environment of Japan, Tokyo.

Ministry of the Environment of Japan, 2007. *21 Seiki Kankyo Rikkoku Senryaku (Strategy for an Environmental Nation in the 21st Century of Japan)*. Ministry of the Environment of Japan, Tokyo.

Mitsuhasi, T., 1999. *Kankyo-Keizai Nyumon (Introduction to Environmental Economics)*. Nikkei-shimbun, Tokyo.

Yamaguchi, M., 2000. *Chikyu Kankyo Mondai to Kigyou (Global Environmental Issues and Business Actors)*. Iwanami Shoten, Tokyo.

Abbreviation

3Rs: reduce, reuse, recycle
CAI: Clean Asia Initiative
CO2: carbon dioxide
COP:
EM: ecological modernization
EPR: extended producer responsibility
EU: European Union
FCCC: Framework Convention on Climate Change
ICC: International Chamber of Commerce
ISO: International Standardization Organization
LCA: Life Cycle Assessment
NGOs: non-governmental organizations
NPOs: non-profit organizations
WBCSD: World Business Council for Sustainable Development

Chapter 11

SUSTAINABILITY INTEGRATION AND ECONOMIC PERFORMANCE

Marcus Wagner
Lehrstuhl für BWL, insb. Unternehmensgründung und Unternehmenswachstum,
Julius-Maximilians-Universität Würzburg,
Stephanstraße 1/Zi. 301, 97070 Würzburg, Germany
marcus.wagner@uni-wuerzburg.de

The purpose of this chapter is to analyze empirically the link between sustainability integration and economic performance. The chapter takes a comparative perspective by analyzing specific elements of sustainability integration and the usage of management tools, which can explain positive effects on economic performance. The chapter, therefore, provides empirical evidence based on large-scale survey data (rather than individual case studies) that sustainability integration and related management tools help to improve the economic performance of firms. In doing so, it encourages managers to consider more comprehensive application of sustainability integration in their companies. The chapter discusses these findings and the effect of stakeholder constituencies with reference to the broader context of integrating sustainability in business planning.

1. Introduction

Sustainable development and corporate sustainability are increasingly perceived as a long-term trend, rather than a short-term fad. Profit-oriented organizations are, therefore, increasingly expected to provide solutions toward these two objectives (Schaltegger and Burritt, 2005). To enable such solutions, sustainability integration in the widest sense becomes an increasingly important enabling factor, since managers often perceive that it can contribute both to profitability and to corporate sustainability and sustainable development at large.

While sustainability integration can be understood in different ways (Lamberton, 2005), one important element of it that is commonly mentioned is the non-monetary evaluation of performance, i.e., sustainability performance measurement (Yongvanich and Guthrie, 2006). This relates to qualitative and broadly-defined goals, performance measures to assess progress on these goals, target values for quantitative performance indicators, and specific activities that lead to organizational change, which is ultimately reflected in quantitative improvement toward and beyond pre-defined target values for corporate performance.

This chapter, therefore, aims at evaluating the role of integrating corporate strategy and sustainable management. In doing so, it can provide insights that go beyond case studies on individual organizations that to date are predominant in empirical research on sustainability integration. To develop a better understanding of the context of such a study, in the following the literatures on the link between sustainability and economic performance, on the role of integration for sustainability, the relationship with sustainability performance measurement, and the role of management tools (and here particularly the balanced scorecard, BSC), for sustainability integration are reviewed. This will help to identify relevant learnings and will assist the identification of research gaps.

2. Literature Review

2.1. *Linking sustainability and economic performance: The role of contingencies*

As long as one accepts the proposition that more environmental and social management activities increase the level of environmental performance a firm achieves and that this corresponds to increased social welfare, then the issue is reduced to the effect on firms' economic performance, where the latter is understood in a wider sense including more indirect effects on competitiveness such as the corporate or product image, market share effects, and productivity improvements or the firm's ability to recruit excellent staff.

The link between environmental and economic performance has been a topic with a long tradition in the strategic management and sustainable management literatures (Klassen and McLaughlin, 1996; Pava and Krausz, 1996; Waddock and Graves, 1997; Ragothaman and Carr, 2008; Margolis and Walsh, 2001; Konar and Cohen, 2001; Wagner, 2010a, 2010b). Margolis and Walsh (2003), Walsh *et al.* (2003), Orlitzky *et al.* (2003) and Ambec and Lanoie (2008) provide recent reviews and meta-studies summarizing empirical work on the relationship of environmental performance with economic performance. In this, of the 95 studies reviewed by Margolis and Walsh (2001) and discussed in more detail in Margolis and Walsh (2003), the results overall favor a positive relationship between environmental performance and economic performance.

Orlitzky *et al.* (2003) analyze the business case for corporate sustainability but in this keep separate the environmental and social dimensions of the analysis. They find that while there is little evidence of a negative link (and hence against a business case), and some for a weak but statistically significant positive link, there is still significant variation across individual studies, ranging from negative to insignificant to moderately or even strongly positive relationships among environmental, social, and economic performance.

While one commonly agreed feature of all meta-studies is that by and large, firms are not penalized economically for improving their social or environmental performance, the variation in the results indicates that firm-level factors such as the choice of specific tools and their fit to the company matter considerably. Admittedly, other reasons for why this empirical literature has so far not produced a clear consensus exist, such as different measures of environmental and economic performance, different specifications of causal models and control variables, and the assessment of different time periods in different studies. Yet, one additional reason directly related to the choice of management tools that to date is only rarely addressed in empirical work is the effect of integration (i.e., the interlinkage of environmental and social management activities with other strategic considerations).

While the empirical "pays-to-be-green-or-social" literature did not analyze systematically the effect integration has on economic performance, the theoretical literature addressed integration to some degree (Buysse and Verbeke, 2003; Sharma and Vredenburg, 1998), but is limited from a managerial perspective in that it is neither much linked to work on tools that could be used to achieve integration nor to the role of organizational and other contingencies. As a result, important gaps exist in the literature linking sustainability and performance: the limited empirical evidence on the effect of integration from larger surveys and the missing connection between theoretical reasoning on integration, management tools, and contingencies. To shed more light on this gap requires a more detailed inspection of the role of sustainability integration (Wagner, 2007, 2009).

2.2. *Sustainability integration and management*

As highlighted by reviewing the literature on the link between corporate sustainability and performance, an increasingly important question is whether there is actually a direct link between environmental/social and economic performance or if a joint effect of a third variable actually drives the relationship. Judge and Douglas (1998) show in this respect that integration of functional areas within the firm positively affects economic performance, suggesting integration as such a third (and commonly unobserved) variable. Historically, environmental and social management are often conceived as being separate systems with only minimal links (e.g., in terms of personnel or organizational structures and processes) to the general management system of a firm (Hamschmidt and Dyllick, 2002). This also implies a potential for doubling of corporate functions, in turn suggesting inefficiencies. While the general management system is responsible for strategic and operative planning environmental management, for example, is solely responsible for planning the environmental activities of the firm and supporting these by means of guidelines and tools. As a result, economic efficiency is frequently limited, for example, because of the additional coordination efforts needed. Central to the limited efficiency characteristic of this separation is the "parallel," but unconnected management system. As additional resources needed for coordination are usually recorded as indirect or overhead costs or expenditure, they are frequently not traced adequately in the accounts of a firm. This leads to negative effects on economic performance being neglected or, even worse, being overlooked due to limited managerial resources. On the other hand, when firms voluntarily enhance their environmental or social performance beyond the minimum level legally required, they are motivated often by the hope that this produces improvements in their corporate image or similar competitive advantages. Such advantages are, however, difficult to evaluate in terms of their economic value and therefore an assessment of the impact is often not even attempted. Again, limited managerial attention can lead to important benefits resulting from increased integration being overlooked. Overall, these considerations highlight the value of and need for integration, as also identified in earlier work (e.g., Welford and Gouldson, 1993; Burke and Logsdon, 1996; Oktem *et al.*, 2004). In line with this notion, a strong trend has been identified in practice toward integration of internal management systems such as quality and environmental management, health and safety (H&S), and social aspects (Sharma and Ruud, 2003). This is based on the similarity of the underlying management standards such as ISO 9001, ISO 14001, OHSAS 18001, or SA 8000 (Rahimi, 1995). In addition to this, it has been argued that benefits exist from integration, e.g., in terms of scale economies, complementarities, and through supporting a joint approach for ensuring compliance with and beyond environmental, quality, and H&S standards (Matias and Coelho, 2002). Extant work, therefore, supports the notion that integration of social and environmental topics with other strategic aspects (i.e., the design of more closely coupled systems) brings about benefits that exceed cost, and hence should relate positively to firms' economic performance (Wagner, 2007, 2009).

It is stressed that integration is achieved through a process based on tacit capabilities (such as quality management activities or corporate strategy development), which is difficult to imitate (Hart, 1995; Aragon-Correa and Sharma, 2003). This turns the resulting level of integration into a strategic resource. Based on this reasoning, it is also suggested that firm-specific cross-functional coordination integrates environmental management with other strategic aspects in a way that improves economic performance and similarly the argument is made that inclusion of social demands as strategic issues should lead to improved economic performance (Christmann, 2000; Judge and Douglas, 1998; Jansson *et al.*, 2000; Klassen and McLaughlin, 1996).

2.3. *Linking sustainability integration and management tools: The case of the BSC*

Integration can be fostered through the use of specific instruments and this has been discussed in a stream of more applied work developing management tools. For example, the BSC approach has been suggested to assist the integration of environmental, social, and economic objectives (Kaplan and Norton, 1996, 2001; Epstein and Manzoni, 1998). The BSC is for a number of reasons one of the most promising and interesting instruments for better integration of the environmental, social, and economic aspects of sustainable management and hence for fostering sustainability integration in companies.

First of all, this is because the BSC is highly popular amongst managers and has experienced rapid diffusion as a management tool. Because of this and its multi-dimensional conception, it is well placed to efficiently address the major challenges of corporate sustainability, because in business reality environmental, and social performance indicators rarely stand alone but need to be related to economic performance measures in a comprehensive sustainability performance measurement approach that addresses social, environmental, and economic aspects (Schaltegger and Wagner, 2006). Furthermore, sustainability performance measurement requires that environmental and social management activities are integrated with business strategy and management. Hence, integration also has a pivotal role, which underscores the relevance of the BSC.

Finally, since sustainability performance measurement refers to the performance of the organization for a sustainable development of the economy and society as a whole, it is composed of elements of organizational change and thus also relates to specific (environmental or social) management activities at the firm level (Schaltegger and Burritt, 2006). This is particularly relevant since in practice BSCs are not designed *de novo*, but build on existing activities and legacy systems in a firm. For example, specific environmental management activities such as the existence of qualitative and broadly-defined goals, performance metrics to quantify progress toward these goals, target values for quantitative performance indicators and specific activities that lead to organizational change (which is ultimately reflected in quantitative improvement toward and beyond pre-defined target values for corporate performance) can increase the success chances for implementing a BSC. Therefore, existing activities also ultimately impact different dimensions of economic performance and hence a link exists between sustainability and performance.

The starting point of the BSC is the business strategy, which is operationalized through four to five management perspectives based on cause and effect chains (Kaplan and Norton, 2004; Olve *et al.*, 1999). The main focus of implementation is essentially on identifying a suitable BSC of non-monetary (i.e., physical) indicators that in the context of sustainable management also extend beyond purely economic measures to environmental and social aspects (Bennett and James, 1997). This is why the BSC approach seems very suitable for enabling sustainability integration.

3. Research Gaps

Three main areas for research gaps emerge from the review of the extant literature in the domains of sustainability performance, integration, and BSC that was carried out in the previous part of the chapter. The first is whether indeed integration matters for economic performance and competitive advantage as this is a key assumption for promoting sustainability integration as an enabler that simultaneously fosters corporate sustainability and sustainable development at large. As identified in the literature review, this matters because there is some doubt about a direct link between sustainable management and economic performance, with the alternative being that a joint effect of a third variable actually drives the relationship. For example, Wagner (2007) shows that integration of functional areas within the firm positively affects economic performance and is also likely to be related to environmental performance. A positive association of integration (of business strategy with environmental and social aspects) and economic performance in this is a precondition for the existence of a third variable. Therefore, a first set of research questions can be posed:

- Is there a positive association between sustainability integration and different dimensions of economic performance?
- Does the link differ across dimensions?

The second set of research questions that emerges from the literature review is the role of organizational and other contingencies. For both the above research questions, the effect of moderating factors matters and hence several frequently suggested moderator effects need to be analyzed. These organizational contingencies are whether responsibility for corporate sustainability lies at the board level, if firms have implemented a quality management system as a complementarity (Christmann, 2000), whether firms are large or small and if firms are from different regulatory contexts. Therefore, a second set of research questions can be proposed:

- Do contingency factors affect the association of sustainability integration and different dimensions of economic performance?
- Do the contingencies relate more to firm-level organizational or more to other factors or to both equally?

The third set of research questions that emerges from the literature review is whether integration is positively associated with specific activities of sustainable management such as setting goals, deriving performance measures to assess progress on these goals, defining target values for quantitative performance indicators and specific activities that lead to organizational change, which is conducive to achieving or exceeding defined target values. More specifically, in order to evaluate, to what degree the BSC concept discussed in the literature review section has practical viability and to probe deeper into any relevant contingencies identified, the main characteristics of the conventional BSC (objectives, indicators, target values, and measures) are analyzed with regard to the dependence of their usage on the linkage between corporate strategy and sustainability. The expectation based on the literature review is that this drill-down of how firms integrate their environmental and corporate strategy considerations and relate this to the use of core elements of strategic sustainable management as in the BSC method can help explaining differential performance effects and the role of contingent factors for these differences. This expectation is also reflected in the research call by Epstein *et al.* (2000), which further underlines the issue's relevance. Therefore, a third set of research questions can be proposed:

- Are there differences in the association of integration levels and the pursuance of specific activities relating to the main sustainable management elements of the BSC method for any significant contingencies identified?
- Can these differences help to explain any differences in the relationship between integration and different dimensions of economic performance?

Possible answers to the research questions derived from a review of extant literature on sustainability and adjacent fields are illustrated in the following empirically using a data set of firms in the manufacturing sector of two countries.

4. Data and Method

The focus of the illustrative empirical analysis is on environmental aspects based on the assumption that a similar situation can be expected for social aspects and is based on data collected during a mail survey. The questionnaire asked firms for a self-assessment on different aspects of economic performance in terms of benefits from environmental management, about the level of integration between environmental, social, quality, and H&S aspects with the firm's general strategy and for general information about the firm and its structure. Of the 2000 firms contacted in Germany to complete the questionnaire 342 responded, resulting in a response rate of 17.1%. Of the 342 responses, 37 had to be excluded from the analysis due to missing values. In the Netherlands, 2080 firms were contacted, of which 362 responded, equalling a response rate of 17.4%. Of the 362 responses, 10 had to be excluded from the analysis due to missing values. Despite of the exclusions, the overall response rates in both countries are good compared with other firm-level surveys and suggests that it is appropriate to proceed with a more detailed multivariate analysis of the data set.

To assess in more detail the representativeness of the responses data of the German federal agency for employment and the OECD was used (Batenburg, 2006). For Germany, larger firms with more than 500 employees are represented over-proportionally in the responses, whereas firms with 151–500 and less than 150 employees are under-represented in the data set due to fewer responses from smaller firms. For the Netherlands, a decreasing interest of especially large and medium-sized firms in participating in survey research was observed, possibly also impacting representativeness to a small degree.

Concerning response bias, it is possible that the replies received contain over-proportionally many firms that are particularly active in terms of environmental management. Such a bias is a frequent problem of surveys based on written questionnaires (Armstrong and Overton, 1977). However, the broad variability found in the responses indicates that also environmentally inactive firms that participated in the survey.

Next to response bias, self-assessment and the use of only one survey instrument may be a cause for distortions in the data set, in particular concerning common method bias. Common method bias results from variance in the data being attributed more to a measurement method than to the constructs measured. The extent of common method bias varies across fields and is below average in marketing and business studies (Cote and Buckley, 1987).

Self-assessment or soliciting data on independent or dependent variables does not *per se* imply the existence of common method bias since its strength can differ among sub-groups of respondents (e.g. respondents from different countries) and since method-related variance can deflate or inflate the relationships observed (Podsakoff *et al.*, 2003). For the survey data used here, a number of procedural and statistical steps were taken to ensure that common method bias is minimized or at least reduced.

Procedurally, different response formats were used, the anonymity of respondents was ensured, question order was counter-balanced and scale items were improved, especially during the pre-test phase of the survey. All these steps were aimed at reducing socially desirable responses and item ambiguity. For the sake of keeping the anonymity of respondents, it was not generally possible to pursue two other procedural remedies, namely obtaining assessments from different respondents and separating measurements. However, the instructions provided for the survey (in particular to ask for the most knowledgeable person to answer it) and the way the survey was implemented made it possible that even these two remedies could in principle be applied by respondents.

In terms of statistical *ex post* evaluation of the presence of common method bias in the data finally used in the analysis, Harman's single-factor test was applied to establish whether one single factor accounting for most of the variance in the data could be identified from the unrotated solution of a factor analysis. The unrotated factor solution yielded 41 factors of which 11 had eigenvalues larger than unity. The first three factors explain 7.9%, 7.6%, and 7.4%, respectively. All remaining factors with eigenvalues greater than one explain between 6.5% and 2.9% of the variance in the data. This is strong evidence against the existence of one general factor accounting for most of the variance in the data. Therefore, common method variance does not seem to be a critical issue in the data in terms of both *ex ante* procedural precautions and *ex post* statistical evidence. Overall, the analysis of representativeness, response, and common method bias allows concluding that it is save to develop constructs from the data at hand. These are briefly described in the following.

Integration is understood in this chapter in terms of specific patterns of linking environmental, quality, and H&S aspects with the firm's general strategy. This definition includes quality and H&S, because their integration has been proposed by many authors as being a specific stage of integration, i.e., integration proceeds from environment, H&S to total environmental quality management to being integrated with social and strategic issues (Welford and Gouldson, 1993; Benn and Probert, 2006). The specific processes leading to this could not be observed and therefore the process aspect is not elaborated further in this chapter. Instead, the focus rests on the level of integration resulting from the process.

Economic performance is defined in this chapter as that element, which can actually be influenced by environmental management activities and different dimensions are derived as latent variables for the path analysis.[1] This approach was also pursued by Sharma (2001) and seemed to be the most suitable when involving self-assessment by firms. Measurement was thus by means of a set of items asking about the effect of management activities on different economic performance aspects such as effect on market share or cost of insurance to the firm for business liabilities.

Confirmatory factor analyses were carried out for the economic performance and integration constructs prior to involving them in structural equation modeling (SEM). Next to the central variable (integration), this allowed confirming four latent variables for economic performance. Whether or not items were retained for the latent variables was decided on the basis of the change of the Cronbach Alpha for each variable and based on overall model fit. Each latent variable is briefly described in the following.

Integration is measured using three indicators, which are based on respondents rating on a 5-point scale the degree to which environmental management is integrated with quality management, H&S aspects, and corporate strategy, respectively.

The first latent variable identified for economic performance refers to sales, market share, and new market opportunities. Therefore, it was labeled "Markets" since it predominantly addresses the market-related benefits of a company's activities. The relevant items for the second latent variable are corporate image, management satisfaction, and worker satisfaction. It was thus termed "Image" since it mainly refers to internally oriented satisfaction and company image benefits from a company's activities. For the third latent variable identified, the items' short-term profits, cost savings, and productivity are particularly relevant. These predominantly refer to the short-term performance of a company and this factor was, therefore, named "Efficiency." Two items relating to improved insurance conditions and better access to bank loans emerged as the fourth latent variable are related to the financial effects on a company from its chosen level of environmental management activities and are thus labeled "Risk." Table 1 summarizes further validity measures for the above latent variables.

[1] The reason for this is the fact that economic performance in general is determined by many factors, of which environmental management is only a minor one and that hence the chosen definition enables a better focus on this aspect (Lankoski (2000) pursues a similar approach with her concept of environmental profit).

Table 1. Results of confirmatory factor analyses and related indices.

Variable	Indicator no.	Indicator reliability	t-Value (***: $p < 0.01$)	Factor reliability	DEV	Fornell–Larcker ratio
Efficiency	s/t Profits	0.47	—	0.62	0.38	0.46
	Cost savings	0.41	8.02***			
	Productivity	0.39	1.46***			
Image (i.)	Corporate i.	0.43	—	0.75	0.51	0.73
	Mgmt. satis.	0.66	15.11***			
	Empl. satis.	0.44	13.66***			
Risk	Bank loans	0.57	—	0.50	0.36	0.60
	Insur. cond.	0.64	7.55***			
Markets	New. opp.	0.48	—	0.80	0.57	0.28
	Sales	0.56	15.53***			
	Mkt. share	0.65	13.84***			
Integration	Int. safety	0.50	—	0.74	0.49	0.77
	Int. quality	0.45	14.43***			
	Int. strategy	0.53	13.76***			

Note: The first indicator for each latent variable is normalized to unity and therefore t-values are not applicable.

Fornell and Larcker (1981) suggest that the association of a construct with its indicators should be stronger than its association with any other construct. Hence, the Fornell–Larcker ratio is included in Table 1 to address the discriminant validity of each construct. It has to take values below unity to fulfill the outlined condition. Furthermore, indices of local fit suggest that all constructs are reliably measured by their indicators, since for each of the latter the indicator reliability exceeds 0.4 (except for one item on the "Efficiency" construct, which with 0.39 is marginally lower). Also all factor loadings of the manifest items were significant (see t-values in column 4). The factor reliabilities (column 5) and the average share of indicator variance extracted (DEV) by the corresponding latent construct (column 6) are sufficiently high except for a slight deviation on the "Risk" construct. In summary, therefore, all latent constructs are sufficiently separable from one another.

To test the propositions derived above, SEM is employed. Next to an overall path analysis, also moderator analyses were carried out with regard to four important explanatory factors: national policy, firm age, firm size, and the status of sustainable development in the firm. National policy settings and national culture differences (Williams and Seaman, 2001) were proxied by the firm's country location, firm age by the logarithm of the number of years the firm existed, firm size by the logarithm of the size of the number of employees, and the status of sustainable development by a dummy indicating whether the ultimate responsibility for environmental management was with a member of the board or not. I used the median value for each moderator variable to split the whole data set and evaluate the SEM separately for each respective sub-sample.

5. Results

All analyses were carried out with AMOS (Arbuckle, 1999) with raw data as inputs and using full information maximum likelihood estimation of the parameters. The marker variable strategy of fixing the loading of one of the items for each latent variable to 1 was used to achieve model identification (Ullman, 2001). Evaluation of measurement model fit was initially based on the Chi-square value and indicates that the measurement model has a very good fit (Chi-square/df = 1.82; $p < 0.01$). Furthermore, the root mean square error of approximation for the model of less than 0.04, Bentler's (1990) comparative fit index of 0.94 and the Tucker and Lewis (1973) fit index of 0.92 indicate good fit. The overall model fit therefore meets or exceeds the standards of model fit proposed by Hu and Bentler (1999). Table 2 provides an overview of the main effects in the SEM model. As described in Table 1, all loadings of the observable indicators on the latent variables are significant at $p < 0.001$ and are in the right direction, providing confidence in the estimates. Notable are the differing magnitudes of the effects estimated across the four latent variables since this suggests differential performance effects of integration.

As argued when deriving the research questions, it is likely that organizational and other contingencies impact the paths in the model. More particularly, it was argued that the variables firm size, if responsibility for corporate sustainability lies at the board level, whether firms have implemented a quality management system and the regulatory context (i.e., country location) of the firm are the relevant and often-cited moderator variables. To test for any moderating effects of these factors, specific constraints are imposed on the model reported above and the appropriateness of these constraints is tested separately for sub-groups in the data set through hierarchical model comparisons.

More specifically, multi-group AMOS initially constrains all parameters to be equal between the two sub-groups identified based on the median values of the moderator variables. Only if this constrained model is significantly worse in terms of fit than the unconstrained model it makes sense to test individual paths for equality across groups. For the latter, multi-group AMOS then compares two models that are different only with regard to the denominated path. These are nested in that one model restricts the parameter corresponding to the denominated path to be equal across groups whereas the other, more general model allows this parameter to vary across the two groups. For firm size, level of responsibility assignment and quality management system implementation, the differences between constrained

Table 2. Overall results (unstandardized coefficients; CR: critical ratio; *p*: significance level).

Denominated path	Coefficient (coef.)	CR	*p*
From integration to efficiency (integration–efficiency)	0.06	4.80	<0.001
From integration to risk (integration–risk)	0.06	4.80	<0.001
From integration to image (integration–image)	0.29	9.77	<0.001
From integration to markets (integration–markets)	0.05	3.14	0.002

Table 3. Group comparison for country moderation effects (N: Netherlands, G: Germany).

Denominated path	Coef.	CR(N)	P(N)	Coef.	CR(G)	P(G)	χ^2	p
Integration–efficiency	0.01	0.41	0.68	0.12	4.36	<0.001	9.49	<0.01
Integration–image	0.24	7.48	<0.01	0.36	7.75	<0.001	6.36	0.01
Integration–risk	0.17	1.29	0.20	0.11	3.71	<0.001	11.92	<0.01
Integration–markets	0.03	1.16	0.25	0.12	3.98	<0.001	6.57	0.01

Note: A significant χ^2 test result indicates that the coefficients differ significantly between the two countries.

and unconstrained models at the aggregate level are all insignificant (at $p = 0.41$, 0.19, and 0.92, respectively).

The results of the moderator analysis of country location (where the difference between the constrained and unconstrained models is significant at $p = 0.001$) are shown in Table 3.

Given that only country moderation effects could be identified in the SEM multi-group comparison, the attempt is made in the following to probe deeper into what may drive the differing association of integration with the economic performance dimensions in terms of specific management activities that are carried out by firms. Comparing firms in this way between the two countries can provide clues what the crucial activities are to realize a positive effect of integration on performance. The management activities particularly relate to the BSC model (i.e., goals, indicators, targets, actions) since the features of the BSC (Kaplan and Norton, 2001; Olve *et al.*, 1999) such as different perspectives and causal linkages seem to position it very well for a structured firm-level process of integration with regard to environmental, social, and economic aspects that ultimately has a positive impact on performance.[2]

The focus is on this last linkage dimension because in stage models of strategy development these are considered more advanced than total environmental quality management or EHS systems (Hunt and Auster, 1990; Welford and Gouldson, 1993; Oktem *et al.*, 2004). Tables 4 and 5 provide summaries of activities for Germany and the Netherlands.[3]

A first insight from Table 4 is that indeed in German firms higher levels of integration associate significantly positively with management activities conducive for BSC implementation as one specific form of sustainable performance measurement and an important tool of sustainable management. Also for the Dutch firms in the data set, the basic association of integration levels with the probability of pursuing specific management activities conducive for the application of sustainable

[2]It is interesting to note in this respect that the core BSC literature (e.g., Kaplan and Norton, 2001; Olve *et al.*, 1999) does so far not explicitly link integration and the balanced scorecard.

[3]Aggregate figures for the whole data set are not provided here since the focus is on a comparative analysis, but can be easily recalculated using country total response figures provided above.

Table 4. Cross-tabulation of written environmental policy and environmental management activities integrated with corporate strategy for German.

		Activities integrated with corporate strategy						Tests of significant association		
		Not at all (%)	Somewhat (%)	Partly (%)	Very much (%)	Fully integrated (%)	Total (%)	Chi-squared	Fisher exact	Spearman
Written environmental policy	Yes	1.7	4.2	14.6	22.9	12.8	56.3	60.39	62.57	0.45
	No	8.7	9.0	16.7	6.9	2.4	43.8	$p < 0.01$	$p < 0.01$	$p < 0.01$
Use of measureable environmental targets	Yes	1.4	3.2	14.5	22.3	13.4	54.8	72.13	76.12	0.50
	No	8.8	10.2	17.3	7.1	1.8	45.2	$p < 0.01$	$p < 0.01$	$p < 0.01$
Use of quantitative env. performance indicators	Yes	1.8	2.2	12.7	14.9	10.5	42.0	32.20	33.27	0.33
	No	8.7	11.6	18.5	14.1	5.1	58.0	$p < 0.01$	$p < 0.01$	$p < 0.01$
Program to achieve environmental targets	Yes	1.1	3.2	12.8	20.9	12.8	50.7	68.29	72.04	0.49
	No	9.6	10.3	18.8	8.2	2.5	49.3	$p < 0.01$	$p < 0.01$	$p < 0.01$

Note: Small deviations to 100% may exist due to rounding of figures to one decimal digit.

Table 5. Cross-tabulation of written environmental policy and environmental management activities integrated with corporate strategy for Dutch firms.

		Activities integrated with corporate strategy						Tests of significant association		
		Not at all (%)	Somewhat (%)	Partly (%)	Very much (%)	Fully integrated (%)	Total (%)	Chi-squared	Fisher exact	Spearman rank
Written environmental policy	Yes	8.4	14.7	19.8	23.1	8.7	74.8	52.49	61.76	0.38
	No	7.2	10.2	6.9	0.3	0.6	25.2	$p < 0.01$	$p < 0.01$	$p < 0.01$
Use of measureable environmental targets	Yes	5.1	9.6	15.6	19.2	8.4	58.0	57.60	60.48	0.41
	No	10.5	15.0	11.4	4.2	0.9	42.0	$p < 0.01$	$p < 0.01$	$p < 0.01$
Use of quantitative env. performance indicators	Yes	1.2	5.5	8.5	11.5	6.1	32.7	44.55	45.80	0.36
	No	14.5	19.4	18.5	11.8	3.0	67.3	$p < 0.01$	$p < 0.01$	$p < 0.01$
Program to achieve environmental targets	Yes	3.9	9.9	15.7	17.2	7.8	54.5	49.45	50.97	0.39
	No	11.7	15.1	11.1	6.3	1.2	45.5	$p < 0.01$	$p < 0.01$	$p < 0.01$

Note: Small deviations to 100% may exist due to rounding of figures to one decimal digit.

management tools such as the BSC is confirmed. Yet, across countries the strength of association can still differ.

In order to evaluate, to what degree the BSC and the arguments identified in the literature review relate to the findings of the multi-group comparison the main characteristics of the conventional BSC (objectives, indicators, target values, and measures) can be evaluated in terms of the strength of association (measured by means of Spearman rank correlation) to provide insights about the relationship of their usage.

As can be seen, in almost all cases the strength of association is larger for German firms than for Dutch firms in the order of magnitude of 20–25%. The only exception is the use of quantitative performance indicators, but the difference of 0.03 (corresponding to 10%) is much smaller than the differences for the other three binary variables where the German firms have in all cases much stronger associations.[4]

6. Conclusion and Discussion

This research set out to answer a number of research questions on the link of sustainability integration and economic performance. As concerns the first set of research questions, the empirical analysis confirms a positive association between sustainability integration and different dimensions of economic performance. Interestingly, judging from the standardized estimates the effects are strongest for the image- and market-related dimensions. However, also for the efficiency- and risk-related dimensions integration has a positive effect.

This can be understood as an indication that integration helps especially when, e.g., deciding on which new product developments to pursue and which markets to target with which approach. Also it means that integration helps companies to act consistently over a wider range of business activities, in turn significantly improving image-related benefits. It seems that integration especially ensures that there are no gaps or inconsistencies in a firm's sustainability-related activities.

Future empirical research based on large samples of firms should attempt to confirm and corroborate these findings of the analysis and in doing so, amongst other things, help to gain a better understanding whether these are confined largely to Europe or even only to some countries within the European Union. For example, surveys on this topic in the United States, the BRIC countries or other non-European OECD countries would be of considerable value in order to develop a more comprehensive picture of the link between integration and performance, and the role which management tools and sustainable management have in this.

[4]The differences may also be explainable by the fact that variations of the BSC, which incorporate environmental and social aspects in different ways, have been proposed (Epstein and Manzoni, 1998).

As concerns the second set of research questions, country location and national regulation emerge as a strong contingency factor and moderator variable. Opposed to this, for the structural level to which responsibility for corporate sustainability is assigned, firm size and the existence of a certified quality management system do not affect the association of sustainability integration and different dimensions of economic performance. This means, for example, that whether responsibility for sustainability is assigned to the board level does not affect the different paths from integration to the various economic performance dimensions. That is, no moderation effect of board responsibility is identified in the multi-group comparison. One important implication of this is that leadership through board responsibility is insufficient, but that leadership in processes and policies is needed additionally.[5]

Concerning the country moderator effect, the positive effect of integration on different dimensions of economic performance is considerably stronger and more significant in Germany as compared with the Netherlands. This result suggests that firm-level organizational contingencies matter less and that firm-external contingency factors are more important for variations in the relationship between integration and performance.

The role of country-level factors also indicates that regulatory frameworks and national culture most probably impact on the association of integration and performance for firms in a given country. However, there is also some interrelatedness with firm-level organizational contingencies. For example, a survey of the OECD (Johnstone, 2007) that collected data from 4200 facilities in seven countries (including Germany) finds that the level in the organization to which of the individual who has overall responsibility for environmental matters belongs varies across countries. The study suggests that this has strong influence on the type of corporate environmental strategy pursued. For example, in terms of the (normalized) level of integration of environmental management with other functional areas, the level of integration with quality management is 0.72 and with H&S management 0.71. This compares with figures for Germany and the Netherlands of 0.66 (both countries) and 0.73 and 0.67 (for H&S management), respectively, in the research reported in this chapter.

Such differences in the level of integration, while potentially driven by differences in country-level regulatory frameworks, of course also represent firm-level differences and hence indirectly firm-level contingencies may well matter, even though their direct effect was shown to be rather weak given their insignificance in the multi-group comparisons reported above.

As concerns the last set of research questions, some significant differences can be identified between firms in the two countries (which was the only significant

[5]Interestingly, it also seems that such leadership in processes and policies is not necessarily driven by assigning board level responsibility for corporate sustainability since when comparing the latter with levels of integration for general strategy ($p = 0.12$), quality assurance ($p = 0.84$), and improvement and health and safety ($p = 0.24$). This could be indicative of different capabilities being relevant for processes *versus* structures. Detailed test statistics are available on request.

moderator identified) in the association of integration levels and the pursuance of specific activities that relate to the main. Overall, there is more usage of the four activities relating to the core sustainable management aspects of the BSC method in Germany for a given level of integration than in the Netherlands. Relating this back to the multi-group country comparison, this result suggests that the higher occurrence of these activities related to sustainable management and the BSC enables German firms better to reap economic benefits from integration, whereas Dutch firms may pursue a different and somewhat less effective approach. Still, given the SEM results, firms that pursue activities related to the core sustainable management aspects of the BSC method more strongly as part of the integration of corporate strategy and sustainable management perform better. Furthermore, as can be seen from Tables 4 and 5, for quantitative environmental targets (i.e., existence of target values for the indicators used) the usage probability is highest for German firms that strongly link environmental with corporate strategy considerations. While the absolute level of association differs across countries, this linkage is in both countries of similar strength as the one for a written environmental policy, which may also be due to the fact that often such targets are associated closely with the existence of a policy. The likely mechanism behind this is that targets are usually derived from policies and that hence existence of targets is closely associated with the availability of a written policy and that firms are more likely to have an implementation program when they have environmental targets. The somewhat weaker linkage in both countries (relative to the other three binary variables) of quantitative indicators with the integration of environmental, economic, and possibly also social aspects also points to scope for improvement because it suggests that firms having a written policy, targets, and a program to implement the latter in some cases do not use quantitative indicators for monitoring. While this could be due to the fact that targets can simply be achieved by means of appropriate actions without using sustainable management tools to continuously record and track quantitative performance indicators, it could also indicate that a state of fully developed sustainability performance measurement has not yet been reached. Being a core element of sustainability performance measurement based on the BSC method this weaker adoption of quantitative indicators somewhat limits the ability of firms to benefit from implementing a BSC. One important reason for this weakness could be that firms are endowed with capabilities and resources to a different degree. Relating this back to the natural-resource-based view, Marcus and Anderson (2006) in a similar context note that not only one general dynamic capability but also different ones for business and social responsibility objectives are at work in some industries.[6] In the context of the empirical study reported here, this may mean that those firms

[6]While the empirical evidence for one generic dynamic capability affecting both business and social responsibility objectives seems limited such a capability has been proposed frequently in the conceptual literature, e.g., by Hart (1995) who suggested that the capability of establishing a shared vision could be one such generic dynamic capability.

that have strong capabilities with regard to business objectives are well-positioned to make use of performance indicators, but are not the ones that can handle well the more specific elements of sustainable management. This is another issue that should be analyzed in more detail in future research. Even though the core BSC literature does not explicitly suggest the value of this management tool for integration, precisely this is proposed here because the suitability of the BSC to achieve integration is high, linkages between specific sustainable management activities related to this tool and integration levels could be established and also empirically integration supports economic performance in different dimensions. In addition to this also from a managerial point of view, a BSC is well placed to efficiently address the challenge of integration because of its high popularity and rapid diffusion as a management tool and due to its multi-dimensional conception that has already been extended into the sustainability realm (Johnson, 1998; Epstein and Wisner, 2001; Epstein and Roy, 2001).

Furthermore, the BSC approach also fosters involvement of middle managers in strategy making and should thus reduce detrimental effects from micro-politics and the inertia of corporate culture (Collier *et al.*, 2004; Grant and Visconti, 2006; Scullen *et al.*, 2005). This chapter has responded to a gap in extant research about the role of integration of environmental, social, and economic aspects at the firm level in potentially bringing about a positive relationship with economic performance. It has shown that this presumed link can be detected in large-scale survey data and that contingencies exist, which moderate this link.

The chapter also shows that one important explanatory factor for the association of integration and performance can be the degree to which key aspects of sustainable managements are related to the BSC method in practice by firms.

References

Ambec, S. and Lanoie, P., 2008. Does it pay to be green? A systematic overview. *Academy of Management Perspectives* **22**(4), 45–62.

Aragon-Correa, J.A. and Sharma, S., 2003. A contingent resource-based view of proactive environmental strategy. *Academy of Management Review* **28**, 71–88.

Armstrong, J.S. and Overton, T.S., 1977. Estimating nonresponse bias in mail surveys. *Journal of Marketing Research* **14**, 396–402.

Bagozzi, R.P., Yi, Y. and Phillips, K.W., 1991. Assessing construct validity in organizational research. *Administrative Science Quarterly* **36**, 421–458.

Batenburg, R.S., 2006. The mutual relationship between the environmental context and management in the Netherlands. In: Schaltegger, S. and Wagner, M. (Eds.), *Managing the Business Case for Sustainability*. Greenleaf, Sheffield, pp. 312–327.

Benn, S. and Probert, J., 2006. Incremental change towards sustainability. Integrating human and ecological factors for efficiency. In: Schaltegger, S. and Wagner, M. (Eds.), *Managing the Business Case for Sustainability*. Greenleaf, Sheffield, pp. 542–552.

Bennett, M. and James, P., 1997. *Environment-related Performance Measurement: Current Practice and Trends*, Ashridge Management College, Ashridge.

Bentler, P.M., 1990. Comparative fit indexed in structural models. *Psychological Bulletin* **107**, 238–246.

Boons, F. and Wagner, M., 2009. Assessing the relationship between economic and ecological performance: Distinguishing system levels and the role of innovation. *Ecological Economics* **68**(7), 1908–1914.

Burke, L. and Logsdon, J., 1996. How corporate social responsibility pays off. *Long Range Planning* **29**(4), 495–502.

Buysse, K. and Verbeke, A., 2003. Proactive environmental strategies: A stakeholder management perspective. *Strategic Management Journal* **24**, 453–470.

Christmann, P., 2000. Effects of 'Best Practices' on environmental management. *Academy of Management Journal* **3**(4), 663–680.

Collier, N., Fishwick, F. and Floyd, S.W., 2004. Managerial involvement and perceptions of strategy process. *Long Range Planning* **37**(1), 67–83.

Cote, J.A. and Buckley, M.R., 1987. Estimating trait, method and error variance: Generalizing across 70 construct validation studies. *Journal of Marketing Research* **24**, 315–318.

Epstein, M.J. and Manzoni, J.F., 1998. Implementing corporate strategy: From tableau de bord to balanced scorecard. *European Management Journal* **16**(2), 190–203.

Epstein, M.J. and Roy, M.-J., 2001. Sustainability in action: Identifying and measuring key performance drivers. *Long Range Planning* **34**(5), 585–604.

Epstein, M.J. and Wisner, P., 2001. Using a balanced scorecard to implement sustainability. *Environmental Quality Management* **11**(2), 1–10.

Epstein, M.J., Kumar, P. and Westbrook, R.A., 2000. The drivers of customer and corporate profitability: Modeling, measuring, and managing the causal relationships. *Advances in Management Accounting* **9**, 43–72.

Fornell, C. and Larcker, D.F., 1981. Evaluating structural equation models with unobservable variables and measurement error. *Journal of Marketing Research* **18**, 39–50.

Gerbing, D.W. and Anderson, J.C., 1988. An updated paradigm for scale development incorporating unidimensionality and its assessment. *Journal of Marketing Research* **25**, 186–92.

Grant, R.M. and Visconti, M., 2006. The strategic background to corporate accounting scandals. *Long Range Planning* **39**(4), 361–383.

Hamschmidt, J. and Dyllick, T., 2002. ISO 14001: Profitable — Yes! But is it eco-effective? *Greener Management International* **34**, 43–54.

Hart, S.L., 1995. A natural-resource-based view of the firm. *Academy of Management Review* **20**, 874–907.

Hu, L. and Bentler, P.M., 1999. Cutoff criteria for fit indexes in covariance structure analysis: Conventional criterial versus new alternatives. *Structural Equation Modeling* **6**, 1–55.

Hunt, C.B. and Auster, E.R., 1990. Proactive environmental management: Avoiding the toxic trap. *Sloan Management Review* Winter, 7–18.

Jansson, A., Nilsson, F. and Rapp, B., 2000. Environmentally driven mode of business development: A management control perspective. *Scandinavian Journal of Management* **16**, 303–33.

Johnson, H. and Kaplan, R., 1987. *Relevance Lost: The Rise and Fall of Management Accounting*, Harvard Business School Publishing, Boston.

Johnson, S., 1998. Identification and selection of environmental performance indicators: Application of the balanced scorecard approach. *Corporate Environmental Strategy* **5**(4), 34–41.

Johnstone, N., 2007. *Environmental Policy and Corporate Behaviour*, Edward Elgar, Cheltenham.

Judge, W.Q. and Douglas, T.J., 1998. Performance implications of incorporating natural environmental issues into the strategic planning process: An empirical assessment. *Journal of Management Studies* **35**, 241–262.

Kaplan, R. and Norton, D., 1996. *The Balanced Scorecard*. Harvard Business School Press, Boston.

Kaplan, R. and Norton, D., 2001. *The Strategy-Focused Organization*. Harvard Business School Press, Boston.

Kaplan, R. and Norton, D., 2004. *Strategy Maps*. Harvard Business School Press, Boston.

Klassen, R.D. and McLaughlin, C.P., 1996. The impact of environmental management on firm performance. *Management Science* **42**(8), 1199–1214.

Konar, S. and Cohen, M.A., 2001. Does the market value environmental performance? *Review of Economics and Statistics* **83**(2), 281–289.

Lamberton, G., 2005. Sustainability accounting — A brief history and conceptual framework. *Accounting Forum* **29**(1), 7–26.

Lankoski, L., 2000. *Determinants of Environmental Profit*. Helsinki University of Technology, Helsinki.

Marcus, A. and Anderson, M., 2006. A general dynamic capability: Does it propagate business and social competencies in the retail food industry? *Journal of Management Studies* **43**(1), 19–46.

Margolis, J.D. and Walsh, J.P., 2001. *People and Profits? The Search for a Link between a Company's Social and Financial Performance*. Lawrence Erlbaum Associates, Mahwah.

Margolis, J. and Walsh, J., 2003. Misery loves companies: Rethinking social initiatives by business. *Administrative Science Quarterly* **48**(2), 268–305.

Matias, J.C. and Coelho, D.A., 2002. The integration of the standards systems of quality management, environmental management and occupational health and safety management. *International Journal of Production Research* **40**(15), 3857–3866.

Nunally, J.C., 1978. *Psychometric Theory*. McGraw-Hill, New York.

Oktem, U., Lewis, P., Donovan, D., Hagan, J.R. and Pace T., 2004. MS and sustainable development: A model and comparative studies of integration. *Greener Management International* **46**, 11–28.

Olve, N., Roy, J. and Wetter, M., 1999. *Performance Drivers*. Wiley, Chichester.

Orlitzky, M., Schmidt, F. and Rynes, S., 2003. Corporate social and financial performance: A meta-analysis. *Organization Studies* **24**(3), 403–441.

Pava, M.L. and Krausz, J., 1996. The association between corporate social responsibility and financial performance: The paradox of social cost. *Journal of Business Ethics* **15**, 321–357.

Peter, J.P., 1979. Reliability: A review of psychometric basics and recent marketing practices. *Journal of Marketing Research* **16**, 6–17.

Podsakoff, P.M., MacKenzie, S.B., Lee, J.Y. and Podsakoff, N.P., 2003. Common method biases in behavioral research: A critical review of the literature and recommended remedies. *Journal of Applied Psychology* **88**(5), 879–903.

Ragothaman, S. and Carr, D., 2008. The impact of environmental information disclosures on shareholder returns in a company: An empirical study. *International Journal of Management* **25**(4), 613–620.

Rahimi, M., 1995. Merging strategic safety, health and environment into total quality management. *International Journal of Industrial Ergonomics* **16**(2), 83–94.

Schaltegger, S. and Burritt, R.L., 2005. Corporate sustainability. In: Folmer, H. and Tietenberg, T. (Eds.), *The International Yearbook of Environmental and Resource Economics 2005/2006*. Edward Elgar, Cheltenham, pp. 185–222.

Schaltegger, S. and Burritt, R.L., 2006. Corporate sustainability accounting — A catchphrase for compliant corporations or a business decision support for sustainability leaders? In: Schaltegger, S., Bennett, M. and Burritt, R.L. (Eds.), *Sustainability Accounting and Reporting*. Springer, Dordrecht, pp. 37–59.

Schaltegger, S. and Wagner, M., 2006. Integrative management of sustainability performance, measurement and reporting. *International Journal of Accounting, Auditing and Performance Evaluation* **3**(1), 1–19.

Scullen, S.E., Bergey, P.K. and Aiman-Smith, L., 2005. Forced distribution rating systems and the improvement of workforce potential: A baseline simulation. *Personnel Psychology* **58**, 1–32.

Sharma, S. and Ruud, A., 2003. On the path to sustainability: Integrating social dimensions into the research and practice of environmental management. *Business Strategy and the Environment* **12**, 205–214.

Sharma, S. and Vredenburg, H., 1998. Proactive corporate environmental strategy and the development of competitively valuable organizational capabilities. *Strategic Management Journal* **19**, 729–753.

Sharma, S., 2001. Different strokes: Regulatory styles and environmental strategy in the North-American oil and gas industry. *Business Strategy and the Environment* **10**, 344–364.

Tucker, L.R. and Lewis, C., 1973. A reliability coefficient for maximum likelihood factor analysis. *Psychometrika* **38**, 1–10.

Ullman, J.B., 2001. Structural equation modeling. In: Tabachnick, B.G. and Fidell, L.S. (Eds.), *Using Multivariate Statistics*. Allyn & Bacon, Boston, pp. 653–771.

Waddock, S.A. and Graves, S.B., 1997. The corporate social performance — financial performance link. *Strategic Management Journal* **18**, 303–319.

Wagner, M., 2007. Integration of environmental management with other managerial functions of the firm: Empirical effects on drivers of economic performance. *Long Range Planning* **40**(5), 611–628.

Wagner, M., 2009. Innovation and competitive advantages from the integration of strategic aspects with social and environmental management in European firms. *Business Strategy and the Environment* **18**(5), 291–306.

Wagner, M., 2010a. The role of corporate sustainability performance for economic performance: A firm-level analysis of moderation effects. *Ecological Economics*, **69**(7), 1553–1560.

Wagner, M., 2010b. Corporate social performance and innovation with high social benefits: A quantitative analysis. *Journal of Business Ethics*, **94**(4), 581–594.

Walsh, J., Weber, K. and Margolis, J., 2003. Social issues and management: Our lost cause found. *Journal of Management* **29**(6), 859–881.

Welford, R. and Gouldson, A., 1993. *Environmental Management & Business Strategy*. Pitman, London.

Williams, J.J. and Seaman, A.E., 2001. Predicting change in management accounting systems: National culture and industry effects. *Accounting, Organization & Society* **26**, 443–460.

Yongvanich, K. and Guthrie, J., 2006. An extended performance reporting framework for social and environmental accounting. *Business Strategy and the Environment* **15**(5), 309–321.

Chapter 12

TOTAL ENVIRONMENTAL QUALITY MANAGEMENT

Johannes Freiesleben
Credit Suisse AG, Loorenstrasse 37, 8053 Zurich, Switzerland
Johannes.freiesleben@credit-suisse.com

In this chapter, we focus on the interconnectivity between quality and the environment, and the means of jointly improving both, in the context of *total environmental quality management* (TEQM). We take a purely economic perspective, benefitting from the communicability of an economic model and its usefulness for formulating political and normative proposals. We highlight the interconnectivity of quality and environmental concerns, analyze both delivery and design quality aspects, and deduce their impact on the environment. We find that this impact is often not adequately reflected in the costs paid for the quality deviations. To provide a framework for thinking about TEQM's economic significance, we relate these cost categories to external effects. We identify three external effects of poor quality: the pollution, disposal, and excess input effect. We then discuss the challenges of addressing these external effects and delineate the principal means of making TEQM more widespread.

1. Introduction

In recent years, good news about the rise of developing economies have often been accompanied by bad news about environmental neglect or destruction. As the global economy is rapidly integrating and production is conducted in places offering the cheapest input factors, the negative effect of globalization on the environment gets increasingly evident. Reports in the international media play a pivotal role in the process of creating public awareness. Continued media interest and coverage started in the 1980s when audiences in the industrialized West were first confronted with topics ranging from acid rain threatening forests to water pollution responsible for dying fish in rivers and seas. The list of concerns increased since then. In the year 2009, the two most prominent topics are climate change caused by increasing levels of carbon dioxide and reducing biodiversity caused by vanishing natural habitats. The careless abuse of nature as a waste dump and as a mere input for production is increasingly at odds with public opinion and has triggered the establishment of green political parties in many democratic societies. Consequently, companies today are faced with the public demand for a more conscious and efficient use of natural resources.

As important and justified as this demand is, it can somehow be regarded as an amplification of the long-existing concern of the quality movement with the environment. From its very inception, waste prevention has been a major focus of the profession. As early contributors such as Shewhart and Tippett already pointed out in the 1930s, production processes need not be connected to "unavoidable" waste levels. Instead, a profound investigation of the properties of a production technology often leads to the discovery of improvement options that result in reduced waste. This insight has been carried forward into every new quality concept and has been the basis for the continuous improvement philosophy. The widely practiced Six Sigma concept has been especially successful in highlighting that any waste level is connected to high costs, which should be prevented for the direct benefit of the company and the indirect benefit of the environment. The quality movement thereby has had a profound impact on how resources are used in production and thereby positively influenced operational efficiency. The driving force behind the striving for quality is the force of competition in free markets, which trims producers to be as cost-efficient as possible.

Quality today is not only a decision influenced by the individual company. Quality in the new production reality of globalized markets rather is a holistic concept, which needs to be assessed on several distinct levels, including all potential influences from suppliers, customers, and the general public. The same holds true for environmental effects: they might be caused by a variety of actors, or concern a variety of stakeholder groups. What we will be looking at in this chapter is the interconnectivity between quality and environmental effects, and the means of jointly improving both aspects of production. We want to address this topic in the context of *total environmental quality management* (TEQM) by taking a purely economic perspective. As opposed to describing the tools and techniques of achieving TEQM, we will highlight its economic dimensions organized in a simple conceptual framework. The value of this approach rests with the communicability of an economic model. As the language of management is money, using a simple economic framework might help in promoting a correct understanding of the topic and thereby advance its dissemination. An economic assessment also establishes the basis for discussing the topic in political and normative terms, thus providing guidelines of how TEQM might be promoted in future.

Given the growing importance of environmental protection and comprehensive economic assessment of environmentally-friendly production behavior, the purpose of this chapter is to aid practitioners and academics to gain a better understanding of the economic as well as ecologic implications of quality. We will proceed as follows. In Sec. 2, we briefly present the current state of research and outline our approach. Sections 3 and 4 discuss the main environmental and economic aspects of the two complementary dimensions of quality, which are delivery and design quality. In Sec. 5, we focus on the economic concept of external effects to conceptually structure our findings and construct the economic framework. Section 6 then discusses how TEQM can best be promoted to assure widespread implementation. Finally, Sec. 7 provides concluding remarks.

2. Existing Research and New Approach

The combination of total quality management and environmental topics has been a relatively recent focus of researchers. Total quality management has been established on the works of key contributors such as Juran, Deming, and K. Ishikawa and provides companies with a set of tools for detecting and exploring quality improvement potentials. Furthermore, it can be seen as an overarching philosophy guiding companies in their quest for higher operational efficiency. The potential for integrating environmental topics into quality management has first been addressed by Mizuno (1988) and Willig (1994) in the wake of the larger public debate on pollution and use of natural resources. Its conceptual underpinning is the recognition that pollution in all its forms is waste, since pollution incurs costs but no offsetting stream of revenues (Porter, 1991) and should therefore be regarded as a form of both environmental and economic inefficiency (Von Weizsäcker *et al.*, 1997). Highlighting the economic dimension, Hanna *et al.* (2000) argue that minimizing waste leads to a reduction of disposal costs, permit expenditures, and fines. Bhat (1998) coined the term "total quality environmental management" and a collective definition has been proposed by Curkovic and Sroufe (2007) as "an economically driven, system-wide, and integrated approach to the reduction and elimination of all waste streams associated with the design, manufacture, use, and/or disposal of products and materials."

Contributions in the cross section between total quality management and environmental management address a diverse set of topics. Aside from describing the tools and techniques that are mainly based on the traditional body of quality management, authors discuss the motivation of companies to adopt environmental standards (Klassen, 2000; Khanna and Anton, 2002), the ideal phase of integrating environmental thinking in the production process (Ahmed, 2001), or the crucial role of innovation for environmental management systems (Könnölä and Unruh, 2006). Top management has been identified as a crucial and sometimes limiting factor to the success of TEQM. Howard-Grenville *et al.* (2008) find that besides incentive system, culture, identity, and organizational self-monitoring, the company-internal factor "personal commitment and affiliations of top management" is decisive for the adoption of environmental management systems. Previous contributors similarly stress the role of top management for promoting TEQM (Hanna *et al.*, 2000) or the importance of providing top management with guidelines of what TEQM is and how it is beneficial for a company (Bhat, 1998; Revelle, 2000). Economic assessments focus on the detailed accounting for the different costs associated with waste and pollution (Makower, 1994; Steen, 2005) or the importance of a correct and detailed analysis of the costs and benefits of applying environmental standards (Curkovic and

Sroufe, 2007). These two dimensions — economic assessment and top management support — provide the motivation for our approach of constructing a conceptual economic framework highlighting the main interdependencies between quality and environmental concerns and establishing a basis for successful communication.

Deviating from the above definition by Curkovic and Sroufe (2007), we will define TEQM for our purposes as "the totality of economic effects concerning the environment caused by inefficiencies in the production process." There are two basic thoughts underlying this definition. The phrase "totality of economic effects concerning the environment" is used to highlight that in the end, *someone has to pay for the neglect of the environment* — be it the producer, the customer, or third parties, which might directly or indirectly be affected. This corresponds to holistic thinking about the effects of quality, as, for instance, promoted by Taguchi (1986). The phrase "caused by inefficiencies in the production process" is used to highlight that all negative effects on the environment can be traced back *to decisions made in the producing company.* Unless the prospective customer has full information about all environmental effects of his purchase decision, which is rarely the case, only the producing company can be assigned responsibility for these inefficiencies.

Our focus on the "production process" implicitly includes the design phase, the actual production phase, and the final delivery phase. This is important since we will discuss two distinct dimensions of quality commonly used in the standard quality management literature, which is delivery quality and design quality (Juran, 1989). Both have different economic and environmental effects and should thus be treated separately. Design quality is concerned with all aspects leading to the final specifications for a product, i.e., all activities from market research and technological development to the decision on the final blueprint. It is basically a dimension concerned with the product features: do they rightly address the customer needs? For instance, does an e-banking web site contain all the necessary information and analytical tools requested by the customer in easily accessible form? Complementary to that, delivery quality is concerned with all aspects leading to the delivery of the product to the customer, which includes the sourcing of inputs, the design of the production process, the actual production, and the sales and delivery activities. It concerns whether the product is delivered to the customer free of defects. In our e-banking example, this would, for instance, show in a correct and timely quotation of prices of stocks or warrants, which are imperative for making purchase decisions.

Both dimensions of quality are equally important. A company can design the best products available on the market but fail to reap superior profits by disappointing customers with poor delivery quality. *Vice versa*, the best delivery quality might not help a company when the product design is outdated and inferior to the competitor's. To use a simple guideline, we can say that high design quality shows in well-addressed customer needs, whereas high delivery quality shows in well-fulfilled customer expectations. Failure to achieve high quality on both dimensions, or on one of them, might cause severe risks to a company's competitive standing (Freiesleben,

2007). In the same way, environmental effects are applicable to both dimensions, and we have to assess them separately as excelling on one dimension might not necessarily lead to an overall environmental-friendly product.

3. Environmental Delivery Quality

We want to start our analysis by first focusing on the production process and the delivery quality dimension. The most obvious form of environmental neglect is the waste of resources, which happens on the shop floor when input materials, modules, or semi-finished products are made defective. Various reasons might contribute to this: wear of the production machinery, loosening adjustments, insufficient supply of operating fluids, power breakdowns, interruptions of the process flow by inspections, human errors, and the like (Box and Luceno, 1997). As a result, only a certain percentage of the processed volume can be transformed into sellable output. Fundamental to correctly assessing the economic impact of poor delivery quality is the recognition that only good products can generate revenues and thereby cover the incurred cost. Hence, every deviation from quality perfection means the initially planned inputs per unit are exceeded, directly translating into a cost increase. This cost increase is composed of four main cost categories.

3.1. *Wasted inputs*

The input costs per unit of sellable output exponentially increase with the defect rate since the costs for wasted inputs have to be fully borne by the remaining good units (Freiesleben, 2005). Sorting out of defective units additionally causes inspection costs.

3.2. *Disposal of wasted inputs*

Additional costs are incurred for disposal of wasted inputs, e.g., for cleaning, handling, storing, final disposal, recycling, or reworking. The same unit cost logic as in Section 3.2 applies.

Disposal costs inflate production cost, but they often not cover the true environmental impact of the quality problem. While there are serious efforts to correctly estimate long-term environmental costs (German Federal Environment Agency, 2007), disposal costs in most countries fail to sufficiently reflect the disposal's environmental impact. Take, for instance, the common procedure in many parts of the world to dump fluid waste directly into water streams or the sea without previous cleaning, which limits the disposal costs to transporting the waste to the water. Further cost generated by this procedure, such as costs connected to degrading water quality, fish stock, or human health, is excluded. Contrary to the costs of wasted inputs, which can directly be measured, the costs of disposing wasted inputs are in most cases only accounted for with a fraction of its true costs.

3.3. *Resource efficiency of supplier*

Both Sections 3.1 and 3.2 are in principle also applicable to the whole supply chain and can inflate the costs for supplied inputs. When a supplier has high defect rates and sorts out defective units, he will have to charge higher prices due to costs of wasted inputs, inspection, and disposal; when the sorting is done by the producer, his input costs are inflated in a similar fashion by the same cost categories. Whenever suppliers need to be considered, the environmental effects of waste and disposal come into play, and when these suppliers have poor quality levels and come from nations with low environmental standards or lacking legal enforcement, the environmental effect will be more graver than if the producer conducted the respective value-adding activities by himself. In most of today's production settings, a large part of inputs is supplied via global sourcing networks, which often means that the producer has little knowledge about environmental or quality standards of most companies participating in the network. Furthermore, distance means transportation, which has an additional negative environmental effect. However, due to the suppliers' overall production cost advantages and the limited accounting for the environmental impact, the additional costs stemming from poorer quality and transportation might remain largely hidden. The more production is outsourced to suppliers, the greater the likelihood that environmental effects of poor quality are understated.

3.4. *Production technology*

The last category concerns the production set-up, which has two environmental effects. The first is the direct pollution caused by the applied production technology. Here, the difference between the pollution caused by the actual technology and the ideal, minimum-pollution technology constitutes an environmental effect. In this, we will conceptually also include the pollution caused by the supplier, since he can be regarded as a part of the (outsourced) production set-up. The second effect is the consumption of inputs depending on the actual technology. Again, a technology that uses the current minimum of inputs can be seen as more environmental-friendly than an old technology, which is less resource-efficient.

As above, the environmental effects of pollution caused by the applied production technology are economically not adequately reflected in the technology investments conducted by producers. If the use of the environment had an adequate price, the cycles of technological innovation might be shorter and the use of the latest technology might be more widespread as any negative environmental effect would result in increased costs. Sticking to an old technology would then worsen a company's relative cost position due to high pollution fines or expenditures and thereby impede its competitiveness.

A similar argument is valid for the second effect, the use of inputs: a more resource-efficient, new technology would improve a company's relative cost position. However, we can assume that this second effect should in reality only be small as inputs have adequate market prices and the competition would erase large differences in input consumption by pushing resource-inefficient producers out of the market. The perceived need for pollution-reducing technological innovation is therefore likely weaker than the need for input-reducing technological innovation, and both are likely even less pronounced for global suppliers operating out of countries with large relative production cost advantages.

The production technology decision offers an example for a classical trade-off: if there are more resource-efficient technologies available, a company needs to decide whether to invest in it, thereby reducing its variable input costs but increasing its fixed production costs. Alternatively, it can keep the old technology, thereby retaining low fixed costs but relatively high variable input costs. From an environmental point of view, the more resource-efficient technology would always be the top choice since it reduces the input usage. Often, it also causes less pollution. To influence this trade-off in favor of the new technology, it is essential that a correct pricing of environmental effects, especially the accounting for pollution as a significant cost factor, becomes more widespread.

Table 1 depicts the four cost categories connected to delivery quality and its environmental effects. The environmental impact is fully covered when we regard the input-related costs, since inputs have market prices, which are paid by either the producer or the customer. It is, however, not adequately covered when we regard costs related to pollution or disposal. The degrees a, b, and c are conceptual and signify that different degrees of "pricing completeness" exist between these categories. For instance, the disposal costs for the producer can be thought of as being

Table 1. Delivery quality and its impact on costs and the environment.

quality-related effect	related costs	environmental impact accounted for
wasted inputs	costs of wasted inputs	100%
disposal of wasted inputs	costs of disposal	a
efficiency of supplier	costs of wasted inputs	100%
	costs of disposal	b
production technology	costs of excess pollution as compared with minimum-pollution technology	c
	costs of excess input use as compared to minimum-input technology	100%

higher than the disposal costs for a supplier operating out of a country with low environmental standards; hence, $a > b$.

4. Environmental Design Quality

The above discussion has shown that the environmental impact of delivery quality might be considerable, but in certain categories not sufficiently accounted for. We now want to assess whether the same is true for the second dimension, design quality. In the standard economics definition of design quality, high design quality refers to a high level of production features in a product, which often shows in higher production costs and prices, and poor design quality in less or less expensive features. We would like to propose a different definition of design quality as the degree to which products in a certain market segment meet the customer needs in that segment (Freiesleben, 2010). This captures an economically more significant dimension: the possibility that companies deviate from design perfection, which is defined as meeting all needs of the targeted customers. As no product is sold to all customers in the same fashion (even water has different "designs" depending on the targeted customer group), it is decisive for a company to correctly assess the customer needs in the targeted market segment and design the products accordingly. High design quality in this sense signifies that a company has an innate understanding of existing and future customer needs and an ability to translate these into product features. If the design quality is poor, the opposite will be the case.

When we apply this understanding of design quality, we can identify six cost categories and related environmental impacts.

4.1. *Lower sales of poorly designed products*

As a result of poor matching of customer needs, sales of the product will likely be low and cause rising inventories of unsold units, which is equivalent to wasted inputs.

4.2. *Unutilized product features*

Sold products might contain a number of product features of which only a certain percentage is really utilized by the customer. All unutilized features can as well be regarded as waste, since the customer would be equally well-off if they did not exist.

An example for this category can be seen in over-engineered cars with features that only a small percentage of customers would find useful. The materials needed to establish the features could then as well be saved, resulting in less inputs per unit. This also has repercussions on the production process itself, since less features would likely mean a less complex production process. Poor design quality can, of course, also work in the opposite direction: if a company puts too less features into its product, it will sell equally poorly, but it will result in less waste due to saved inputs and in a less complex production process.

4.3. *Input-inefficient establishment of features*

Design quality also concerns the production technology, and with it the question whether certain product features are established using the minimum necessary inputs. Features might be established by several different means, and using the most input-efficient technology can be regarded as high design quality. This is similar to the consideration of production technology for delivery quality, where we asserted that the applied technology has a large influence on the input usage. The design quality dimension comes into effect before, when a company determines the specifications of a certain feature. Take, for instance, a company that needs to decide whether to produce a suspension system using springs or hydraulics. Once it decided on this, it needs to determine the most input-efficient technology, thus influencing its delivery quality level. However, the decision on whether to use springs or hydraulic suspension is a design quality issue — both alternatives clearly derive the same feature (suspension) but one might be more input-efficient than the other. Hence, the most input-efficient feature establishment would constitute high design quality in terms of its environmental impact. As inputs are variable production costs and thereby directly affect the producer's profitability, it is likely that a company naturally strives for lowest possible input costs and therefore the most input-efficient feature establishment, given comparable investments for alternative technology choices.

4.4. *Product's consumption of inputs during use*

Depending on the design, products might have different consumption levels of inputs during use, which have a direct effect on the subsequent pollution. Although the design is determined by the company, this is largely a dimension with impact on the customer as it affects his usage costs. As customers get increasingly alert to the costs of running a product, this might also have repercussions on the sales prospects of a producer.

A popular example is the car engine and its level of gas consumption. As gas prices reached new heights in summer 2008, customers increasingly choose cars with low gas consumption. Producers who are not able to offer these types of cars have to face decreasing market shares. Input consumption during use has to be regarded in relation to the service offered by the product — small town cars offer a different driving comfort than a luxury limousine. This is why, our novel definition of design quality is more meaningful: inside of the same product segment, comparisons are relevant. A large number of products consume inputs during use, for instance, in form of space, energy, water, air, or other physical inputs.

4.5. *Product's reliability and durability*

A cost category easily overlooked is maintenance efforts. The propensity of products to need replacement parts for product maintenance can in fact be regarded as a form of input consumption, since it increases the usage costs for the customer and at the same time has an environmental impact by "wasting" resources, which would not be necessary if the product was more reliable. The reliability can be greatly influenced by the design of the parts used to establish the product.

The product's reliability can also be influenced by poor delivery quality — not every defect immediately imperils the functionality of a product but might do so later. As companies fear negative reputation effects if too many defective products reach the market (Freiesleben, 2006), we can assume that they try to reduce this external defect rate as best as possible — which means that the reliability and durability effect felt by the customer is more likely due to poor design quality than poor delivery quality.

4.6. *Product disposal*

Related to the materials used in a product, we finally have the environmental effect of disposal at the end of a product's lifetime. Depending on the amount of material to be disposed and its hazardousness, the environmental impact might be considerable. Design quality influences this effect since the use of too much material, as, for instance, in the case of too many product features, or the use of hazardous materials is determined in the design phase. We can regard too much material connected to unutilized features or the excessive use of hazardous materials as constituting poor design quality and worsening the environmental impact. Superior design quality reduces the environmental impact of a product to the unavoidable minimum.

More so than delivery quality, the design dimension is subject to constant change as technology progresses, competition strengthens, customers' tastes evolve, and product comparisons are facilitated by the internet. As a result, product life cycles are shortening and the move toward better design quality becomes a prerequisite for staying in business. Although this in itself could have a positive effect on the environment, it could also have a detrimental effect when customers are prematurely inclined to replace old products with newer, better designed ones. This would increase the disposal volume by products that could in principle still work and at the same time increase the consumption of resources. From an environmental viewpoint, this represents a trade-off between keeping a high-impact product and buying a low-impact product at the expense of disposing the old one. This decision has to be made by the individual customer and can only indirectly be influenced by the producer. As it increases their bottom line, most producers would appreciate customers frequently replacing products, and accordingly try to shift customer purchase decisions into this direction.

Analogous to the delivery quality discussion, we need to assess for the design quality dimension whether the environmental impact is adequately accounted for. Lower sales due to poor matching of customer needs result in unsold and stockpiled products, but this form of waste is accounted for since the inputs used in the unsold products have already been paid for by the producer. Unutilized product features in sold products are as well paid for, this time by the customer. The connected opportunity costs of potentially less complex production processes have been paid for by the producer. The direct and indirect costs of the product's input consumption during use are fully paid by the customer. However, the excess pollution caused by the increased resource consumption is not adequately accounted for in terms of its environmental impact. The costs of product maintenance and the costs caused by poor product durability, which are reflected in spare parts, maintenance efforts, or early necessary product replacement, are again fully paid by the customer. Opposed to this, the environmental impact of product disposal is again only partially accounted for.

Table 2 summarizes the discussion. The degrees d and e are again conceptual and signify that different degrees of environmental accounting exist between the quality effects. Comparing Tables 1 and 2, we can see that both design quality and delivery quality induce certain environmental effects, which are not correctly accounted for. Some of the effects concern the use of inputs, which are reflected in their market prices, whereas other effects concern the increased pollution/disposal caused by poor quality, which have no adequate price. This highlights a general challenge when discussing the environmental effects of quality: the difficulty to find correct prices for the use of the environment.

Table 2. Design quality and its impact on costs and the environment.

quality-related effect	related costs	eenvironmental impact accounted for
lower sales of poorly designed products	costs of unsold ("wasted") products	100%
unutilized product features	costs of producing unutilized features	100%
	costs of excess process complexity	100%
input-inefficient establishment of features	costs of excess inputs as compared with minimum-input technology	100%
product's consumption of resources during use	direct costs of consumption	100%
	indirect costs of consumption	100%
	costs of excess pollution as compared with minimum-pollution design	*d*
product's reliability and durability	costs of spare parts/maintenance (additional inputs)	100%
product disposal	costs of disposing products	*e*

5. External Effects

What is the reason for prices not reflecting the use of the environment in adequate ways? For answering this question, we want to refer to an economic concept called "external effects." We speak of an external effect when an economic transaction has an impact on a third party not participating in the transaction. In the case of a negative external effect, certain costs are created, which are "externalized" — that is, shifted onto an uninvolved party. Most often, this third party is not a single individual, but smaller or larger segments of the general public. A prominent example for a negative external effect is the costs incurred by air pollution from cars. Although both gas producers and drivers mutually benefit from the sold gasoline, the residents of main traffic arteries breathe polluted air, potentially resulting in health hazards such as asthma. The costs associated with these illnesses have to be fully borne by the residents (or via rising fees by the subscribers of the affected health insurances) and are not "internalized" by the price paid for the gasoline. In almost all cases of external costs, those who suffer from them do so involuntarily. Although the question of who is negatively affected poses a significant ethical and potentially political problem, from an economic perspective the existence of external costs simply means that *too much* of the respective good is demanded, produced, and consumed. As not all associated costs have to be borne by the user of the good, he pays a too low price; given these costs would be "internalized," a higher price and therefore lower demand would be the result.

Another example for external costs is the long-term costs associated with rainforest logging. Rainforest timber is used as hardwood and woodchips, which is in pure or processed form largely exported out of the countries of origin. As pro-rainforest web sites report, the logging rights for an acre of rainforest are often sold for as low as $2 while the extracted wood can sell for up to thousands of dollars in the importing countries (Raintree, 2008). This large difference between costs of extraction and price paid in the countries of use is in itself a matter of concern; however, it is even more questionable that $2 incorporate the long-term costs of the logging activities. These costs are associated with the likely destruction of the remaining vegetation, mostly done by settlers who follow the loggers' access roads. Large areas of rainforest subjected to logging are subsequently transformed into cattle ranches, soy plantations, or simple farm land. Even if agriculture does not follow, the openings in the otherwise dense treetop layer makes plant species on the ground disappear since they are not adopted to the direct influence of the sun. Hence, virgin rainforest is lost forever to the short-term profits gained by selling logging rights. Virgin rainforest has a tremendous value as a carbon dioxide repository, as an oxygen generator, as a water retainer and climate stabilizer, as a source of potential medical cures developed out of its flora and fauna, and as a habitat for endangered species. Since all of these positive effects have no price (i.e., the costs of *not* having these positive effects are not determined yet), they are not accounted for and not incorporated in the costs for the logging rights.

Applying the logic of external effects to TEQM, all deviations from an optimal price properly reflecting the environmental impact of quality incidents can be explained by the existence of external costs. For this purpose, we group all of the above effects of poor quality into three broad categories. The following list shows all detected effects according to these categories and relates them to external costs.

5.1. *Costs related to excess pollution*

- Excess pollution in production.
- Excess pollution caused by product during use.

The external effects of excess pollution are caused by either production or consumption (use) of the product and are composed of environmental degradation, e.g., in the form of polluted water, air, and soil. The external costs associated with this are composed of direct costs (e.g., the unsuitability of polluted water for human consumption) and indirect costs (e.g., reduction of fish stock in the affected water stream) as well as short-term and long-term effects (e.g., a gradual health degradation of residents using the water body). Essentially, the free use of a finite resource ultimately leads to the resource being over-exploited when demand is unrestricted — a problem called "tragedy of the commons." It exists because "free" resources mean that no property rights are defined and hence there is no one to pay for the use of the resource. The crucial point is to recognize that pollution is a form of using environmental resources since these resources are spoiled and thereby made unfit for use. External effects of excess pollution might in the past not have been assessed correctly because of two perception problems. First, most environmental resources have been perceived as being "infinite," e.g., the seemingly endless air space or the vastness of the oceans. Second, ecosystems have an inherent ability to cope with a certain amount of pollution by decomposing or disposing polluting elements. However, both environmental resources and the ability of ecosystems to contain pollution are finite and getting under increased stress by growing human populations and economic activity, with the effect that resources are getting scarce and ecosystems might degrade to the brink of collapse. A key to reducing these externalities is to internalize the external costs of pollution and thereby charge a (higher) price reflecting the use of environmental resources correctly. Higher prices for pollution directly lead to less pollution and to an earlier adoption of pollution-minimizing technology.

5.2. *Costs related to disposal*

- Disposal of wasted inputs (producer).
- Disposal of wasted inputs (supplier).
- Disposal of product after lifetime.

The external effects of disposal can vary greatly depending on the proficiency with which the input or product disposal is conducted. It ranges from basic dumping, which spoils the environment and may cause similar external costs as pollution, to carefully separating and recycling all materials used, which causes the lowest potential external costs. The decision on disposal proficiency essentially presents an economic trade-off: dumping usually costs less than recycling, and to make recycling attractive the value of the recycled materials must exceed the difference between dumping and recycling costs. A key to reducing these externalities is to influence the trade-off in such a way that environmentally harmful disposal methods get unattractive as compared with environmental-friendly methods. Similar to Section 5.1, this could be achieved by putting a price on dumping, which reflects its environmental impact and hence makes it more costly relative to recycling. It also puts an additional incentive on improving both delivery and design quality and on adopting resource-efficient technologies.

5.3. *Costs related to excess input use*

- Wasted inputs (producer).
- Wasted inputs (supplier).
- Excess input use in production.
- Unsold ("wasted") products.
- Unutilized product features.
- Excess process complexity.
- Excess input use for establishment of features.
- Direct consumption during use.
- Indirect consumption during use.
- Spare parts/maintenance.

Excess input use is the biggest cost category of poor quality. We stated in Sections 3 and 4 that these costs are covered by either the producer via an increase in production costs or the customer via the product's price and its running costs. The strength of either market side generally determines which party pays the majority of these costs. If the producer is a monopolist, he might be able to shift all additional input costs onto the sales price and hence the customer; if he operates in a highly competitive market, he might not be able to shift any except for the input consumption during use.

If either of the two involved parties pays for the extra costs, one could imagine that no external effects exist. However, we want to adopt an alternative viewpoint by stressing that higher input consumption caused by the need for excess inputs has a "market price effect." This effect is caused since the excess demand for inputs increases their market price and thereby negatively affects all other potential customers, who can consume less than in the absence of this effect. An example for this is the surging price for gasoline. On the supply side, depleting natural reserves and limited refining capacity restricts supply growth. On the demand side, economic growth contributes to increasing prices, but excess gas consumption in inefficient cars also does — had cars been more efficient in the past, the total accumulated demand volume would have been significantly lower, and with it the market price for gasoline. Since many car models use more gas than necessary, all car drivers have to cope with higher market prices than had these models been more resource-efficient. This effect is recognizable in all input categories, because all inputs are finite and supply can rarely be enlarged to accommodate surging demand. It is arguable, however, if the "market price effect" could be alleviated by internalization. As the market mechanism is essentially only a spot observation of current market activities, who should estimate future demand and supply patterns and correctly determine a higher price for input wasters? Additionally, excess input use is already a disincentive to either customer or producer in terms of opportunity costs, since lower input use is tantamount to lower costs and hence preferable. Such, it is more challenging to internalize the "market price effect" than the pollution or disposal effect. Its environmental impact, however, can be seen in the excess depletion of natural resources and the damage caused by extraction as well as the excess pollution caused by heightened input consumption, i.e., an amplification of the external effects (Sections 5.1 and 5.2). These environmental costs are, similarly, not internalized in the input prices.

These different forms of external effects related to quality can also be classified according to the concept of efficiency. A widely used definition for production efficiency is the so-called "x-efficiency" by Leibenstein (1966), which states that given a fixed set of inputs, a company is x-efficient when it produces the maximum output possible under these conditions ("x" thereby refers to the amount of forgone output, which should be minimized). In line with most other economic models, x-efficiency presupposes that differences in efficiency are the result of differences in applied production technologies and neglects the possibility of inefficient production behavior (Gonzalez Fidalgo and Carcaba Garcia, 2001). In our context of quality, we focused on the applied production technologies (design quality) as well as the ability of a company to use a given technology in an efficient manner (delivery quality). In business practice, the interplay between the two technological dimensions determines the competitive advantage of a company (Freiesleben, 2007). Using the terminology of Leibenstein, we could call all input-related external effects of quality "x-efficiency effects" since they affect the common input/output ratio. The pollution- and disposal-related external effects could be termed "e-efficiency effects" where e-efficiency means the minimal degree of negative environmental impact caused by the product and its production given today's status of technology.

6. Achieving TEQM

As external effects of poor quality cause environmental degradation and ultimately lead to deteriorating living conditions on the planet, achieving TEQM should be a main focus of attention in the future. Achieving TEQM can best be defined as the combined optimum of x-efficiency and e-efficiency. We already hinted at the main challenge connected to this: the use of nature currently does not have a price correctly accounting for the connected environmental impact — hence, an adequate pricing mechanism has to be found to internalize these external effects. This should ideally be done without running the risk of replacing the current market distortion by too low prices with a distortion by too high prices. Also, as negative environmental effects such as pollution are increasingly "outsourced" to countries with weak environmental laws and protection, a globalization dilemma exists in that national governments can only exert limited influence in the absence of a global governance mechanism (Esty and Ivanova, 2004). In the following, we want to briefly outline potential approaches to achieve TEQM and discuss their practicability.

6.1. *Governmental intervention*

The most direct approach to promote TEQM is through governmental definition and enforcement of regulations, which address environmental concerns in the general framework of business laws. We can distinguish two basic mechanisms for governments to achieve this objective: the first concerns the setting of incentives to promote environmental-friendly behavior, the second concerns punishment for violating legally binding regulations. The incentives can, for instance, be given in the form of tax brakes or direct subsidies for the installation of more resource-efficient and less polluting production technology. Punishment, or disincentives for environmental harmful production, can, for instance, be exerted via the payment of fines or the imposition of high taxes on pollution.

The advantages of governmental intervention are mainly connected to its great assertiveness as a modulator of economic action. As no company in its jurisdiction can escape its reach, compliance to new standards is assured by the government's intervention power. Companies outside its jurisdiction might also be forced to comply with national legislation when they wish to export products to this country. Furthermore, the government employs specialists who can determine the optimal environmental policy for its area of jurisdiction, i.e., have a more comprehensive and holistic perspective than an individual company. This is important when addressing external effects with greater reach. In general, new environmental regulation by governments can be a very effective tool to set the direction toward achieving TEQM.

The disadvantages of governmental intervention are similarly apparent. The power of a government is necessarily limited to the very nation it governs; in the absence of a supra-national governmental body, it can hardly counteract the negative effects of production relocation to countries outside its jurisdiction, except for the imposition of import barriers. Also, a practical question might concern the ability of a government agency to correctly determine the most x- and e-efficient technologies. Technology is the basis for competitive advantage and hence public agencies seldom have detailed insight into the research and development activities of private companies. The government's role would likely be restricted to a disseminator of "best practice" technologies, which might exclude the best available technology since this innovation might be kept secret by its inventors for fear of copying. Also, companies might not employ their latest innovations since this would imperil the amortization of their current production technology investment. Another disadvantage of governmental intervention is the potential for corruption, which might not be widespread in the Western hemisphere but become a serious inhibitor to x- or e-efficiency improvements in other parts of the world. Governmental policy might also be influenced by special interest groups with powerful lobbying organizations.

In conclusion, governmental intervention is an important cornerstone for setting the right incentives for TEQM, but national regulation can in many ways be ineffective. It should ideally be complemented by a supra-national regulation to set global standards for environmental protection and to prevent the evasion of national regulation. However, as long as no supra-national government body with enforcement capabilities exists, a global solution might be difficult to achieve and enforce. Global regulation, which necessarily needs to be a compromise between potentially conflicting interests, can also easily be "watered down" by the veto of countries benefiting from the current system.

6.2. *Political action*

With political action we mean all means of collective action, as opposed to individual action. This can take diverse forms, such as civil tort law class action by groups whose members are harmed by environmental degradation, the use of the political process to influence legislation, the creation of media interest and subsequent public awareness, or other forms of organizing interests to promote specific initiatives. Political action also concerns the exertion of influence in the parliament by opposition parties or by members of the governmental party not directly participating in the government. As indicated in the introduction, the media is of paramount importance to confer credibility and momentum to these political initiatives.

One key advantage of political actions rests in the fact that every concerned group can use this tool to make its claims heard and to support environmental-friendly production behavior. No formalities, as in governmental intervention, need to be adhered to, and interest groups can use their right to voice initiatives without being concerned with implementation feasibility. Hence, in theory this tool is available to anyone with a claim.

The disadvantages of political action are connected to its effectiveness. To create a certain momentum to induce change is a time- and energy-consuming and often also money-consuming activity. The opposing party, using the same means, will likely mobilize its forces for stiff resistance or even counterattack. Groups that start political actions need a profound understanding of the political subsystems, be it the legal, media or parliamentary system, or pay for good advisors. For instance, media reports might be conflicting, influenced by different special interests, or repetitive, requiring a careful media planning to ensure that a message is conveyed effectively. To solve problems with large impact or scale, political action groups should be able to maintain public support and following for extended time periods. As all of these hurdles are limiting political action in democratic societies, in many countries there is not even free speech, free media, or legal access to organized interests.

In conclusion, political action can be a promising tool to push for the adoption of TEQM; however, its outcome is dependent on a variety of parameters, which make predictions about its success difficult. As the means of political action grew over the past decades, mainly due to the easy access and use of the internet as a medium for mass communication, we can expect it to continue to rise in importance. However, solely relying on the change induced by political action might not be enough to make TEQM a widespread industry practice.

6.3. *Market mechanism*

The market mechanism approach subsumes all "non-interventionist" approaches, meaning it solely relies on the forces of the market to solve environmental problems. As stated above, the pricing mechanism of the market does not incorporate costs created by the long-term impact of economic activities, i.e., not addresses the problem of under-priced environmental resources and externalized costs. However, these shortcomings of the pricing mechanism could potentially be alleviated by the two market sides negotiating prices: the producers and the customers. Both market sides could thereby exert individual or collective influence.

6.3.1. *Customer influence*

As the key economic decision maker, the customer can exert tremendous influence on companies by his collective purchase behavior. Availing his influence to the promotion of environmentally-friendly production, he could, for instance, make the TEQM compliance an important criterion in his purchase decision and refuse products that do not meet this criterion. He could aim at internalizing the product's input consumption during use to deduce a price which appropriately reflects his future costs, which would reduce the relative price of the environmental-friendly product. He could furthermore include into the product's life-cycle costs its expected maintenance costs and its durability. Comparing alternative product choices regarding such "full life-cycle costing" could result in a more informed purchase decision of the customer and make the TEQM product the most attractive choice. Recent research in fact suggests that the splitting of costs of environmentally sound products into initial costs and operating costs results in a dissolution of the traditionally assumed dichotomy of "green & expensive" *versus* "brown & cheap" (Känzig and Wüstenhagen, 2008). As our analysis suggests, TEQM products might have lower total costs since poor quality is connected to higher production, operating, and life-cycle costs. The internalization of external costs reinforces this result. Even higher initial costs for TEQM products together with the lower operating cost might result in a more advantageous investment profile compared with a non-TEQM product.

The advantage of customer influence is its great potential power: no other stakeholder group has such a direct and immediate impact on the profitability of a producer. As customers decide on the market success of a new product or a new technology via their purchase decisions, they could similarly influence the adoption of TEQM. This influence works via individual purchase behavior as well as via collective action in the form of customer rights' groups or customer forums. The latter form of influence would classify as political action in our definition from above. It is undeniable that even individual customer purchase behavior has, in aggregated form, a massive influence. Take, for instance, the rising demand for organically-grown products and its effects on agriculture. Ten years ago, these products had a negligible market share and were mostly consumed locally. Today, specialized grocery chains such as Wholefoods or Trader Joes in the US are serving an ever increasing demand for such bio-products nationwide. The same trend can be observed in other Western nations. This has been a direct effect of a changed customer attitude and purchase behavior, which puts a higher value on products with a minimum impact on the environment.

The disadvantage of customer influence is that individual, unorganized customers might often not be able to access information, which could potentially alter their purchase behavior. For instance, most customers have little information about the internal quality level of producers and about their global sourcing activities. Sometimes, they cannot even assess how much resource a product is likely to consume during use, or its likelihood of default and potential maintenance costs. Well-organized customer groups might be able to gain this information and disseminate it inside or even outside the group; however, not all important topics might be covered by this means of information provision. Environmental topics might also be taken on by different groups with different degrees of organization, resulting in conflicting and potentially confusing information. Additionally, not all customer groups might want to share information with the general public because they want to limit "free riding" on costly obtained information.

A key to make customer influence into a more powerful tool for exerting pressure on companies is therefore to enhance information provision to potential customers. As many customers search for purchase-relevant information themselves, the technical information infrastructure, especially the internet, will play a pivotal role in information dissemination. If customers receive TEQM information from organized customer groups, we are again in the realm of political action. We have to be aware that the purchase decision of a customer is subject to many parameters, and making TEQM the decisive parameter might prove difficult. That said increasing awareness for TEQM might over time change the attitude of customers, make them more receptive for information related to it, and create a momentum similar to the bio-product trend.

6.3.2. *Producer influence*

The market side supplying the customer, the individual producer, also has an interest in promoting TEQM. As the assumed dichotomy of "green & expensive" *versus* "brown & cheap" is dissolving for purchase decisions, so is the old misconception that high delivery quality is more costly than poor delivery quality. As Six Sigma companies have repeatedly shown, the opposite is true: less defects in production directly translate into higher profitability (Bisgaard and Freiesleben, 2000). We developed a similar argument for design quality and asserted that given a certain determined product design, poor design quality might be more costly than high design quality because of the effects mentioned in Section 4. Hence, producers have an economic incentive to strive for highest possible x- and e-efficiency and thereby TEQM. Environmentally-friendly producer behavior is congruent with profit maximizing behavior and a high quality attitude.

The advantage of producer influence rests on the business rationale that as TEQM is an economically advantageous concept, it should as enthusiastically be adopted as a new promising technology or a refined business strategy. A key for promoting TEQM is hence to make top management aware of its economic as well as strategic advantageousness. One can assume that despite the existence of opportunistic behavior, a majority of business leaders as globally responsible citizens are concerned about the environmental impact of production decisions. The internalization of external effects, if those effects cannot be avoided in the first place, can also be done by the company causing it. For instance, if a certain degree of air pollution is unavoidable, a company might voluntarily devote financial resources to the future avoidance of the polluting agent and spread the new technology in its business community. The problem of information availability for the customer might be alleviated by a TEQM producer aggressively using the information about his environmental-friendly production in advertising. If comparative advertisements are allowed, environmental-unfriendly behavior of competitors could be exposed. Similarly, rather than finding ways to make customers pay more for TEQM products, the marketing challenge could be re-formulated to lowering customers' perceived initial costs and increasing their awareness of operating costs. On a collective level, all companies in a certain industry could agree on self-imposing environmental standards and lobbying for countermeasures on those companies who do not comply.

There are little disadvantages connected to producer influence. Even if an individual producer cannot convince his competitors to establish collective action to promote TEQM or use comparative advertising to expose his competitors, he can gain profitability by adopting TEQM. It is true that an individual company by itself cannot influence the formation of prices, and for that purpose would have to rely on collective action or governmental intervention. However, as a new product "feature," it can advertise its TEQM behavior, and together with increasing public awareness and potential political action create a momentum that finally leads to the adoption of a common TEQM standard.

Summarizing, there are several different ways of promoting the adoption of TEQM, with direct and indirect influence on the challenges presented above. To address the challenge of a correct pricing of resources, direct influence could be asserted by governments, but also by collective action of producer groups, to set a common standard. Indirectly, this can be influenced by changing purchase behavior of customers and assertive political action. The establishment of a correct pricing automatism, however, remains a challenge since nobody knows with certainty what the long-term costs of using specific resources are. Attempts have been made into this direction, e.g., through the sales of "pollution rights," which put a price on the emission of carbon dioxide, but this is still lacking international consensus. It can, however, be assumed that the efforts into this direction will ultimately lead to a higher pricing of resource use, providing an even stronger incentive for companies to strive for TEQM and be more efficient. For the challenge of "outsourced pollution" to countries with weak environmental laws and protection, both national governments and producer groups can by way of import barriers impose common environmental standards in its area of influence. Still, a supra-national regulation would be a far better approach since import barriers might lead to trade wars or retaliatory action by targeted governments.

From a non-technical perspective, the adoption of TEQM can be promoted by pressure exerted on companies via individual and collective customer influence, individual producer influence, and political action. Most likely, the objective of creating a momentum toward TEQM can best be achieved by a combination of these non-technical, indirect means with the technical, direct means of governmental regulation, and collective producer influence. All of these efforts might work interactively and reinforce each other. For instance, media reports triggered by concerned customer groups might lead to public awareness, and as a result create heightened media interest, political action, and finally governmental regulation or even the establishment of supra-national regulation. A key in this process is, however, the attitude of the individual producer. Ultimately, he is the one adopting and practicing TEQM, and if he does so voluntarily, the momentum might be created more easily that if he only reacts to mounting external pressure. We have found that a producer in fact has a lot of reasons to adopt TEQM voluntarily. From an economic perspective, highest possible x- and e-efficiencies are equal to highest profitability and competitive advantage. From a philosophical perspective, TEQM is much more sophisticated and preferential than wasting resources and harming the environment. Even if a producer has to fight with opportunistic behavior of his competitors, he might address all of the external effects we described — pollution, disposal, and input consumption — in his advertising. By this, he can make customers aware that he offers the best product choice and hence influence the amortization of his investment in TEQM. Such, it might only require a small critical mass of producers to adopt TEQM as first movers to finally make it the best practice in their industries.

7. Conclusion

In the near future, customers might have a more transparent view of producers. TEQM might be common standard in the majority of industries worldwide, and information about the individual producer's attitude toward quality and the environment might be public knowledge, leading to more informed purchase decisions. Resources might be better priced, leading to a more efficient use in production. Pollution might be costly, and companies with high-pollution technologies might have difficulties selling their products profitably. Large inefficiencies in delivery quality as well as design quality might be eradicated, and striving for quality perfection in both dimensions might be a common strategy for competitive companies. As a result, the damage on the environment caused by economic activities might be limited to the unavoidable minimum, and although reconciliation of economy and ecology might never be completely achievable, a workable coexistence might be established.

With customer and producer awareness as well as governmental insight, this illustration of a possible near future might not be an illusion. It is based on the understanding that inefficiencies caused by poor quality are like friction in physical systems, hindering the economic system from working properly, while at the same time being preventable. It is also based on the understanding that the current system, with a market pricing mechanism focusing on spot activities and neglecting long-term implications, needs to be complemented by a focus on sustainability, preferably within a global framework for economic activity. It is finally based on the understanding that a coexistence of economy and ecology is in the common interest of all economic actors.

We have analyzed TEQM as a concept for achieving this possible future. We started by highlighting the interconnectivity of quality and environmental concerns, and showed that quality improvement and environmental protection are congruent objectives. We analyzed both delivery and design quality and deduced their impact on the environment. We found that this environmental impact is often not reflected in the costs paid for the quality deviations. To provide a framework for thinking about TEQM's economic significance, we related these cost categories to external effects. We identified three external effects of poor quality: the pollution, disposal, and excess input effect. We then discussed the challenges of addressing these external effects and delineated the principal means of achieving TEQM. While potentially all of the presented means jointly work toward an adoption of TEQM, we stressed the fundamental role of the individual producer. As obvious as it may seem, businesses have to take responsibility for their actions, and for their roles in the future economic development. If the top management of a company recognizes the economic and philosophical advantageousness of striving for best possible x- and e-efficiencies, the momentum leading to a widespread adoption of TEQM could be triggered by the producers themselves. With this chapter, we hope we have contributed to a correct understanding of these issues and put them into a broader economic perspective.

References

Ahmed, N., 2001. Incorporating environmental concerns into TQM. *Production and Inventory Management Journal* **24**, 25–31.

Bhat, V.N., 1998. *Total Quality Management — An ISA 14000 Approach.* Greenwood Publishing Group, Westport.

Bisgaard, S. and Freiesleben, J., 2000. Economics of six sigma programs. *Quality Engineering* **13**(2), 325–331.

Box, G.E.P. and Luceno, A., 1997. *Statistical Control by Monitoring and Feedback Adjustment.* Wiley, New York.

Curkovic, S. and Sroufe, R., 2007. Total quality environmental management and total cost assessment: An exploratory study. *International Journal of Production Economics* **105**, 560–579.

Esty, D.C. and Ivanova, M.H., 2004. Globalization and environmental protection: A global governance perspective. *Working Paper No. 0402*, Yale Center for Environmental Law and Policy, Yale University.

Freiesleben, J., 2005. The economic effects of quality improvement. *Total Quality Management & Business Excellence* **16**(7), 915–922.

Freiesleben, J., 2006. What are quality reputations worth? *Quality Progress* **39**(2), 35–40.

Freiesleben, J., 2007. Can six sigma claim to be a generic strategy? Reassessing the competitive implications of quality improvement. *International Journal of Six Sigma and Competitive Advantage* **3**(3), 248–265.

Freiesleben, J., 2010. Proposing a new approach to discussing economic effects of design quality. *International Journal of Production Economics*, in press.

German Federal Environmental Agency, 2007. Awareness of external costs — Helps to improve environmental protection. *Internal Research Report*, April.

Gonzalez Fidalgo, E. and Carcaba Garcia, A., 2001. A resource-based interpretation of technical efficiency indexes. *Efficiency Series Paper* 3, Universidad de Oviedo, Departamento de Economia.

Hanna, M.D., Newman, R.W. and Johnson, P., 2000. Linking operational improvement through employee involvement. *International Journal of Operations and Production Management* **20**(2), 148–165.

Howard-Grenville, J., Nash, J. and Coglianese, C., 2008. Constructing the license to operate: Internal factors and their influence on corporate environmental decisions. *Law & Policy* **30**(1), 73–107.

Juran, J.M., 1989. *Juran on Leadership for Quality.* The Free Press, New York.

Känzig, J. and Wüstenhagen, R., 2008. The effect of life-cycle cost information on consumer investment decisions for eco-Innovation. *Proceedings of the Conference of the Sustainable Consumption Research Exchange*, Brussels, Belgium.

Khanna, M. and Anton, W., 2002. What is driving corporate environmentalism: Opportunity of threat? *Corporate Environmental Strategy* **9**(4), 404–413.

Klassen, R.D., 2000. Exploring the linkage between investment in manufacturing and environmental technologies. *International Journal of Operations and Production Management* **20**(2), 127–147.

Könnölä, T. and Unruh, G.C., 2006. Really challenging the course: The limitations of environmental management systems for innovation. *Business Strategy and the Environment* **16**(8), 525–537.

Leibenstein, H., 1966. Allocative efficiency versus X-efficiency. *American Economic Review* **56**, 392–415.

Makower, J., 1994. *Beyond the Bottom Line: Business for Social Responsibility.* Tiden Press, New York.

Mizuno, S., 1988. *Management for Quality: The Seven New QC Tools.* Productivity Press, Cambridge.

Porter, M.E., 1991. America's green strategy. *Scientific American* **4**, 168.

Raintree, 2008. Available at http://www.rain-tree.com/facts.htm.

Revelle, C., 2000. Research challenges in environmental management. *European Journal of Operational Research* **121**(2), 218–231.

Steen, B., 2005. Environmental costs and benefits in life cycle costing. *Management of Environmental Quality* **16**(2), 107–118.

Taguchi, G., 1986. *Introduction to Quality Engineering: Designing Quality into Products and Processes.* Asian Productivity Organization, Tokyo.

Von Weizsäcker, E., Lovins, A.B. and Lovins, L.H., 1997. *Factor Four. Doubling Wealth — Halving Resource Use.* Earthscan, London.

Willig, J., 1994. *Environmental TQM.* McGraw-Hill, New York.

Chapter 13

SUSTAINABLE PLANNING AND DECISION MAKING IN MULTI-STAKEHOLDER CONTEXTS: TRANSBOUNDARY WATERSHED MANAGEMENT

Lanka Thabrew[a], Robert Ries[b] and George Hornberger[a]
[a]*Vanderbilt Institute for Energy and Environment, Vanderbilt University, PMB240, Nashville 37240, TN, USA*
lanka.thabrew@vanderbilt.edu
[b]*M.ERinker School of Building Construction, University of Florida, 311 Rinker Hall, Gainesville, FL 32611, USA*
rries@dcp.ufl.edu

Sustainable watershed management is a multi-stakeholder and a multi-faceted process. A watershed is an essential resource to the existence of human systems as well as ecological systems. Competing water needs of human systems and development activities has led to various conflicts among the users and has imposed a threat to ecological systems. Conflicts are becoming more intense with growing population, rapid urbanization and industrialization, and changing climatic conditions. Finding solutions to the issues needs multi-stakeholder participation at multiple decision-making scales. Finding solutions also requires addressing the integrated nature of multiple dimensions such as changing hydrology, geology, natural cycles, urban development, and supply and demand constraints in the built and natural environment. A framework for stakeholders to participate and identify the system dynamics and the causes and effects of decisions and interventions are needed. This chapter takes the Apalachicola–Chattahoochee–Flint River Basin as a reference example to discuss transboundary water management issues and how a life-cycle-thinking-based framework would be embedded into an integrated planning process for sustainable watershed management.

1. Introduction

Integrated planning brings economic, social, and environmental aspects of development issues and solutions and the cross-sectoral actors responsible to a common planning forum (Allen *et al.*, 2002; Geyer, 2006). The term integrated means (a) to combine parts as a whole; (b) to consider all aspects of issues; (c) to look into all circumstances that may affect a plan in a holistic manner; and (d) to coordinate stakeholders from all respective sectors. The term sustainable development means a process that improves the quality of life of the people and enhances opportunities and choices for growth while safeguarding the earth's life supporting ecosystems (IISD, 2006). Planning helps stakeholders envision the direction of growth and the balance of development with societal and ecological capacities and constraints (Cape Town Gateway, 2007; Geyer, 2006). Hence, integrated planning essentially needs to utilize participatory approaches where the stakeholders become key players in decision making and take ownership of their decisions (Van Asselt and Rijkens-Klomp, 2002; NRC, 2008). Integrated planning processes typically involve stakeholders in a sequence of planning steps to insure that the development interventions are financially viable, realistically achievable, and sustainable. The steps essentially include assessing current status and prioritizing goals, formulating and assessing options for solutions, strategizing, and implementing (Cape Town Gateway, 2007; Wiek *et al.*, 2006; Müller *et al.*, 2005).

In the integrated development planning paradigm, engaging relevant stakeholders belonging to multiple disciplines and sectors is an essential element. The knowledge and experience brought in to the planning process by the stakeholders are diverse, and often, the ways of framing and understanding development-related issues depend on their knowledge, expertise, and experience (Lawrence and Després, 2004). In such a planning exercise, getting the stakeholders to common ground by exchanging and building up on each other's knowledge and expertise becomes a valuable first step toward responsible decision making for collaborative action (Meppem and Gill, 1998; Lee, 2006). The process of exchanging, generating, and unifying knowledge among experts, practitioners, and stakeholders is in a given context which is broadly termed as transdisciplinary learning (Klein, 2004; Klein *et al.*, 2001; Lukas and Siebenhüner, 2007). In a planning process, various assessment methods, frameworks, and tools can be applied to support informed decisions and consensus building toward sustainable action. Holistic system approaches are essential to understanding complex systems (Innes, 1996; Scholz and Marks, 2001; Lee, 2006). Although the life-cycle thinking (LCT) concept has immense relevance in assessing upstream and downstream effects of complex issues and interventions in a systematic manner, its application so far is limited in assessing impacts and integrating economic, social, and environmental aspects within participatory multi-stakeholder planning environments (Thabrew *et al.*, 2008).

2. LCT in Planning

In broader terms, LCT is a framework to assess and understand systems. LCT does not produce easy answers; however, it does provide a framework to recognize and understand complex systems and their interrelationships (Boustead, 1996; Leflar, 2008). In integrated planning, the LCT framework can be a basis for the stakeholders to assess the current system dynamics to understand and agree on the issues at hand and formulate and strategize options considering relevant upstream and downstream requirements and effects. This is to encourage that goals are prioritized acknowledging these effects and viable solutions are negotiated without triggering

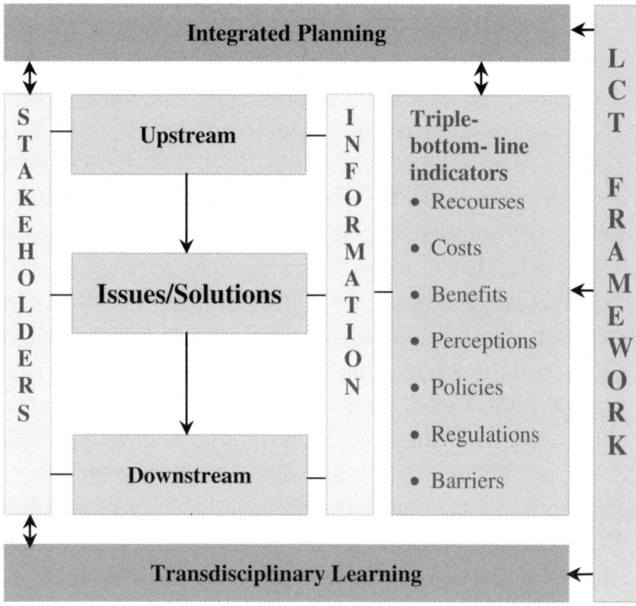

Fig. 1. Conceptual framework of LCT framework embedded in integrated planning.

issues elsewhere. The conceptual basis of the framework is illustrated in Fig. 1. The detailed framework is presented and discussed in Thabrew *et al.*, (2008) and Thabrew and Ries (2009).

At each upstream and downstream phase of an existing issue or a targeted solution, stakeholder responsibilities, behaviors, costs, benefits, and effects are quantified or qualified. In this process, data clarification and verification and stakeholder knowledge integration are important aspects of building trust in the outcomes (Burger *et al.*, 2007; Weik and Walter, 2009). Application steps of the framework include the following:

(1) *Frame of reference for assessment*: Identify current magnitude of issues or alternative solutions.
(2) *Activity-stakeholder mapping*: Chain upstream and downstream of all phases of the issues or solutions and map the stakeholders.
(3) *Indicator development*: Quantify or qualify relevant economic, social, and environmental indicators under each phase.

In the following sections, the application of the framework is discussed with reference to a watershed-scale-integrated planning situation in Apalachicola–Chattahoochee–Flint (ACF) river basin.

3. Background: ACF Watershed

A watershed is defined as the land area where precipitation, i.e., rainfall or snowfall, drains into common surface water bodies. For example, a river system with all its tributaries is one large watershed. Many smaller watersheds, referred to as sub-basins, can be found in large watersheds (USGS, 2008a). Watershed boundaries are natural boundaries and typically cross-political boundaries such as municipal, county, state, or country boundaries. Water is necessary for the proper functioning of ecosystems while agricultural irrigation, electrical power generation, municipal potable water supply, and recreation are some examples of common anthropogenic water uses. Hence, integrated management of watersheds around the world has been a challenging task involving multiple stakeholders. Whereas policies and action should be negotiated and agreed at top political scales, at lower political levels, action should be implemented. In order for the action to be viable and sustainable, the current issues of the watershed should be comprehensively understood and acknowledged by the respective stakeholders and the solution should be developed with due consideration to the interests and priorities of the stakeholders as well as the essential needs of the river systems (Feldman, 2008; Leitman, 2008).

Regarding ACF watershed management, Alabama, Florida, and Georgia have been involved in water negotiations since the early 1970s with some successes and some failures (CRS, 2008; Leitman, 2008; Ruhl, 2005). The three river watershed system that runs in part through each state has become a resource for economic and social development in Georgia, for environmental conservation and preservation in Florida, and for continued agricultural and future industrial and social growth in Alabama (Leitman, 2008). The ACF watershed area covers over 19,800 square miles within the three states. The Apalachicola River is the southernmost river and is formed by the confluence of the Chattahoochee and Flint rivers north of the

Table 1. Basin snapshot.

	Georgia	Alabama	Florida
Population	90%	7%	3%
Basin area	74%	15%	11%
Withdrawals	82%	11%	7%
Area	19,600 sq. mi. or 12.3 million acres		
Population in 1995	4 million		
Estimated population in 2050	7 million		
Land use	6% residential, 2% commercial, 25% agricultural, and 67% undeveloped forest land		
Reservoirs	Hundreds of reservoirs with 5 federal and 11 non-federal dams		
Storage area	W.F. George 45,000 surface acres and Lake Lanier 38,500 surface acres (27% and 30% of total storage area of watershed)		
Storage capacity	W.F. George 123,219 cfs-days and Lake Lanier 583,332 cfs-days (15% and 66.5% of total storage of watershed)		

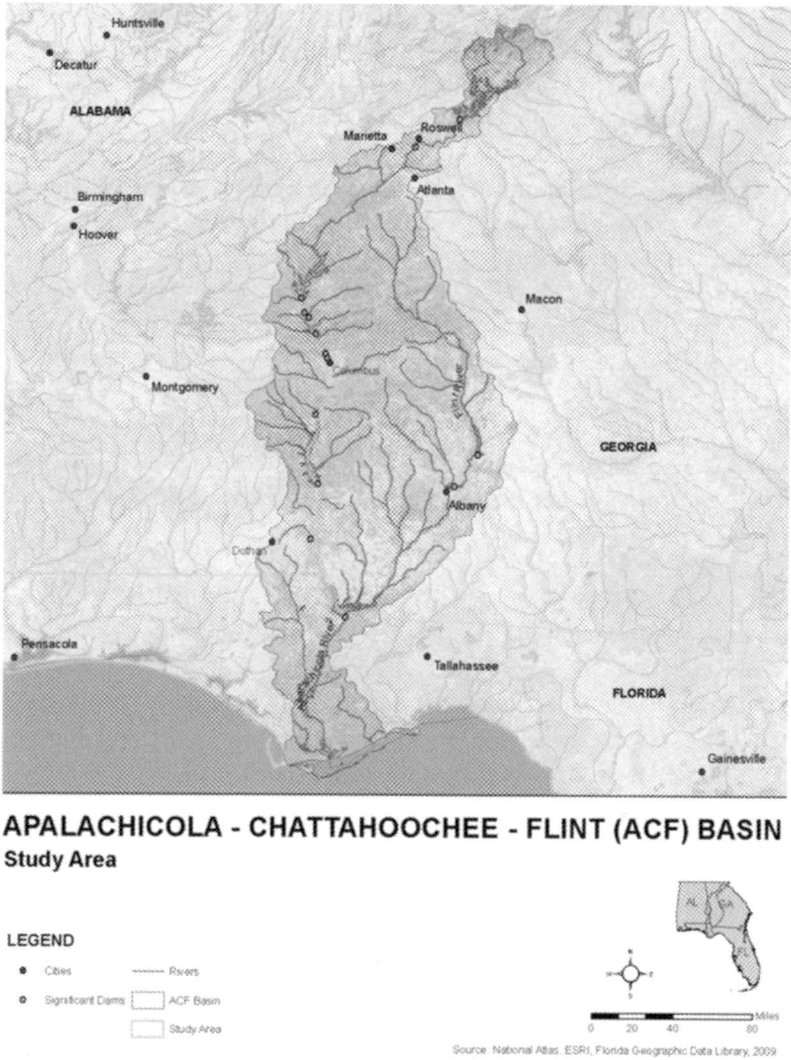

Fig. 2. Apalachicola–Chattahoochee–Flint watershed boundaries and the respective state boundaries.

Florida border. About 75% of the ACF basin resides within the state of Georgia, 10% of the basin is in Alabama and another 10% in Florida. Georgia is responsible for more than 80% of the total withdrawals of the watershed, whereas Alabama and Florida account for 20%. A snapshot of the ACF basin is given in Table 1 (Abram, 2009) and the boundaries of the watershed is given in Fig. 2 (USGS, 2008b).

Water decisions made in Georgia and Alabama have tremendous effects on the quantity and quality of water that reaches the Apalachicola, which are factors considered to be critical to the health of both the Apalachicola River and bay (Eubanks and Greenfield, 1999; Leitman, 2008). Urbanization and industrialization in metropolitan Atlanta have contributed to altering the Chattahoochee's flow and, subsequently, flow in the Apalachicola River. The Apalachicola River supports the highest diversity of freshwater fish species in the state (ARROW, 2008). The shellfish industry around Apalachicola Bay is valued at approximately $10–14 million, which accounts for 90% of Florida's and 10–15% of the nation's supply of shellfish (Florida Department of Agriculture and Consumer Services, 2004). Upper watershed reservoir operations and water withdrawals from the Flint and Chattahoochee Rivers have changed the Apalachicola's natural flow, threatening the survival of native species, and cutting off the state's largest forested floodplain. Rapid development, deforestation, and navigation channel dredging have contributed to river bank erosion and ecological disruption of the Bay area. The dams along the rivers have altered the river flow, disturbing fish movement from important spawning and nursery habitat and seasonal flood patterns essential for maintaining ecological services (Glick and Clough, 2006).

Limited control of ground water pumping and permits for water withdrawals have led to or exacerbated serious low flow conditions, especially along the Flint River (Cummings *et al.*, 2002). Clearing land for increased development and farming opportunities are recognized as serious threats to the Flint River and the surrounding ecosystems. Water quality in the ACF Basin is negatively impacted by Atlanta's older stormwater and sewer systems and inefficient irrigation practices across the states (USGS and US Department of Interior, 1998; Resource Management Institute, 2005). Large built areas with impervious surfaces have disrupted groundwater infiltration and the natural treatment of runoff (IISD, 2006). When the flow is reduced in drought periods, water quality is considerably reduced along the river systems (Eubanks and Greenfield, 1999). Main upriver water uses include municipal and industrial water supply, irrigation, cooling for thermal electricity generation and dams for hydroelectricity, flood control, wastewater dilution, navigation, and recreation. The city of Atlanta, where population has grown from less than 500,000 in 1950 to more than 5 million in 2000, is near the headwaters of the Chattahoochee River. By 2015, population is expected to increase by another 2 million (Giacomini and Hadley, 2005; BEBR, 2008).

Water demand is expected to grow extensively due to this urbanization and industrialization in the greater metropolitan Atlanta area. Georgia's primary concern is to have a sufficient and cost effective water supply to allow for urban and economic growth. Lake Lanier, the primary reservoir and the main resource for water supply in the Atlanta area, maintains storage even during drought periods to safeguard meeting water needs in the region. A higher level of reservoir storage is also important for Georgia to sustain the major recreational activities based around Lake Lanier. Insufficient water releases are a major hindrance to Alabama's interest in having navigation to the Gulf of Mexico, which could potentially contribute to economic growth in the region. Agricultural areas in Alabama and Georgia withdraw a large amount of groundwater for irrigation, which can, in turn, contribute to reducing water flow in the Apalachicola River (Meyer, 2001; Leitman, 2008).

3.1. *Conflict of interests*

The core of the ACF issue has been the conflicting interests of the stakeholders in the three states. So far the states have not been able to agree on how to equitably allocate water resources to meet the current and future needs of the three states. The history of the dispute and the legal procedures are summarized in Ruhl (2005). The comprehensive study conducted upon the courts request by the Army Corps of Engineers (ACE) has been continuously attacked for credibility. The ACE operates 5 of the 11 major dams in the watershed. According to Leitman (2008), some of the key reasons for failure in consensus building among the states include the following: (a) a planning process that extensively involves the respective stakeholders in assessing and compiling data and information had not been conducted; (b) an integrated approach to finding solutions was not used; (c) the states had individually worked out water allocation formulas considering only the individual state agendas; (d) no trust and therefore no agreement were built upon the data used for modeling and analysis by the individual parties (Hawkins, 1997); and (e) the stakeholder working groups developing formulas had not been facilitated by an impartial third party for reaching consensus.

An agreeable water allocation formula that could contribute toward balancing and optimizing human needs as well as the needs of the ecosystems in the ACF and one that comprehensively captures the economic, social, and environmental costs and benefits to the individual states and to the watershed as a whole has not been worked out to date.

3.2. *Recommended action*

Independently, the National Academies' Water Science and Technology Board and National Academy of Environmental Design (NAED) investigated the prevailing issues and potential recommendations for sustainable watershed management in 2009 (UF and NAED, 2009; NRC, 2009). Workshops involving academics and professions from various disciplines who have been working on basin issues investigated how human demands have changed the historic patterns of river flows, water quantity and quality, and life-supporting ecosystems across the ACF basin. The participants discussed and collectively developed broad-scale recommendations on future research and action necessary for sustainable ACF management. Among the recommendations, the value of stakeholder involvement in decision making and operations was recognized as significant aspect; identifying means and processes to more explicitly include stakeholder perspectives and different water management objectives was highlighted as an integral part of the decision making process (NRC, 2009; Richter *et al.*, 2003; Hartmann, 2008). To do so, a working method to connect the scale of the site with the scale of the watershed region, e.g., horizontally and vertically integrated planning methods and implementation strategies, was also discussed as essential.

4. Planning Scales and Stakeholders

According to the discussions and recommendations above, it was clear that integrated planning and mutual learning among the stakeholders at different strategic scales is key to achieving sustainable water management in the ACF. While some issues are seen as clearly important and common sense to one segment of the stakeholders, they are not seen as problematic by other stakeholders. For example, increasing water demand in the upper regions of the basin has a direct effect on the often insufficient and inconsistent water flow, which can lead to salt water intrusion in the Apalachicola at the Gulf of Mexico. This disrupts not only the oyster industry, but also other essential ecosystem functions for life support in the estuary and the mangroves along the Apalachicola River. Similarly, Lake Lanier experienced record low levels of storage during the recent drought cycles and fear of a water crisis in metropolitan areas in Georgia is seen as less relevant by Florida and Alabama (USGS, 2008a). Certain bounded rationalities of the stakeholders on important issues of the basin have led to opportunistic behaviors by various key stakeholders ignoring the real facts about what is appropriate to protect the health of the AFC basin.

Hence, stakeholder-based integrated planning at (1) the watershed scale, (2) the community and the site scale, and (3) the building or unit scale would have immense potential to bring the stakeholders with different interests and priorities to a common understanding of the current watershed issues and formulate options and strategies to solve the issues. The discussions included by scales were as follows:

- *Watershed scale*: Discussions on three rivers, the bioregion, and the human environment over the entire basin.
- *Community and site scale*: Discussions on the elements of specific regions within the basin; for example, the Atlanta metropolitan area and Apalachicola Bay.
- *Building or unit scale*: Discussions on human use including residential, industrial, and agricultural activities.

Creating a transdisciplinary learning process integrating experts', practitioners', and stakeholders' knowledge at the above scales will enhance the ability to develop pragmatic solutions at each scale, which in turn would feed into additional collaborative action (Fig. 3). As described in Meppem and Gill (1997), learning based on data and other relevant information in a transparent manner from the beginning of the planning process would enhance fostering consensus among the stakeholders. Learning from best practices from similar watershed planning problems could also provide useful insights to the stakeholders at the different planning scales of the ACF basin (Fig. 3). The NAED or other potential impartial organizations may be in a position to assume the role of facilitators in the transdisciplinary integrated planning process at appropriate scales. Examples of respective stakeholders at different

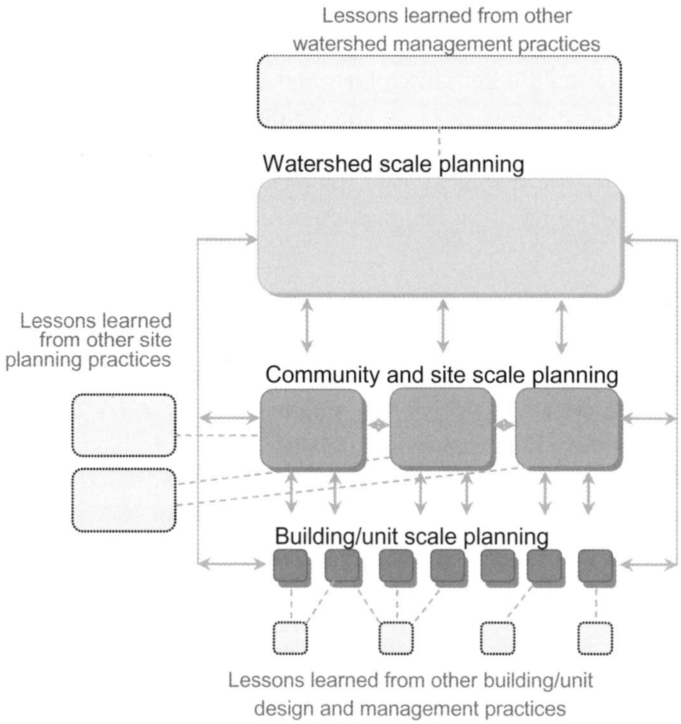

Fig. 3. Integrated water management planning scales.

scales of planning may include the following:

- The watershed scale stakeholders may include State regulatory bodies, state Development Authorities/Councils, state Departments of Watershed Management, ACE, NRC, state Departments of Agriculture and Natural Resources, and the United States Fish and Wildlife Service.
- The site and community scale stakeholders may include Municipalities and local governments, water conservation and floodplain management agencies, representatives from major industries and commercial services, investors, local regulatory bodies, scientists, planners, designers, and community organizations.
- The building or unit scale stakeholders may include Municipalities and local governments, conservation and floodplain management agencies, building or unit owners, planners, designers, contractors, and the community.

5. Application of the Decision Support Framework

LCT can be used at different planning scales and planning steps of an integrated planning process, such as identifying upstream and downstream effects of the current issues and their relationships, and similarly formulating and assessing future scenarios for solutions together with the respective stakeholders. The expert knowledge and information from different disciplines and knowledge and experience from practitioners and stakeholders can be mapped along upstream and downstream events of the issue or solution being analyzed. The information could be quantitative, qualitative, descriptive, or graphical. The complexity of the analysis would depend on the level of detail required for the planning context.

5.1. *Application of the framework in watershed scale*

Watershed scale decisions are ideally made with the broadest perspective possible and consider the scope of the watershed's requirements. Two of the main issues identified in the ACF are water flow (quantity) and water quality. In terms of holistically analyzing the primary issues, it is necessary to identify the current level of the issues. A level would define the reference unit, i.e., functional unit, where the analysis is conducted. In the ACF, the reference unit for the current state analysis could be water flow and the flow regime, which directly controls the flow of the river by changing the levels of storage at reservoirs through dam operations. It may be appropriate to consider the levels of five federal reservoirs, which hold more than 95% of the total storage in the basin. In the option analysis phase, various scenarios could be assessed for alternative flow regimes including extreme scenarios such as no dams or dam operations that better mimic natural drought and flood cycles.

Example: Current state analysis at watershed scale:

- *Step 1*: Identify current levels of water flow related to the operation regime of the dams and reservoirs, specifically,
  - minimum storage at major reservoirs during drought periods;
  - maximum storage at major reservoirs during rainy seasons.
- *Step 2*: Chain the upstream and downstream activity phases relevant to the above levels of flow.
- *Step 3*: Unbundle and assess the components of the upstream and downstream phases in terms of stakeholder involvement, resources, costs, benefits, and impacts using relevant economical, social, environmental indicators.

A graphical representation of the framework applied to assessing the current levels of dam operation is given in Figs. 4–6. Figure 4 shows the primary upstream and downstream phases and related stakeholder mapping for the current levels mentioned above. For both levels, upstream and downstream activity phases and responsible stakeholders are the same, whereas the other components such as costs, benefits, and impacts would be different at each phase. Figures 5 and 6 show the economic, social, and environmental indicators related to the primary activity phases of the two levels. The indicators, which are variable with the seasons, are highlighted in yellow in Figs. 5 and 6.

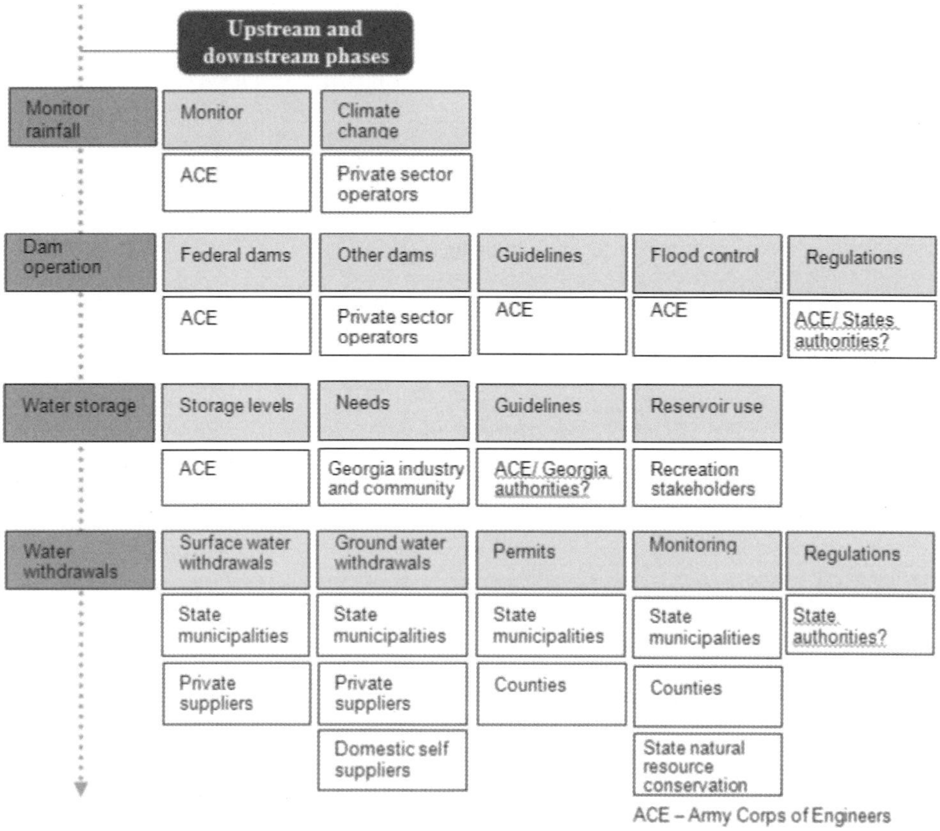

Fig. 4. Activity chaining and unbundling the primary activities mapping the respective stakeholder involvement in dam operation in the ACF watershed.

5.2. *Discussion*

In the above example of a current state analysis exercise using the LCT-based decision-making framework, some of the knowledge missing may have to be generated using stakeholder, practitioner, or expert knowledge in a real planning situation. As facilitators, the experts would gather available information on the issues and solutions in order to have thorough knowledge prior to assessments with the stakeholders, while being open to integrating the knowledge brought into the process by the stakeholders.

Certain information given above could be communicated using different graphical methods. For example, population, industry, urban growth, damage to mangrove areas, and water withdrawals and consumption can be communicated using GIS maps with data validated by the stakeholders. The number of people benefited or affected from certain activities can be graphed or numerically presented. The amount of siltation and evaporation could be communicated using real-time graphics using hydrological modeling software with available and agreeable data demonstrating the uncertainties.

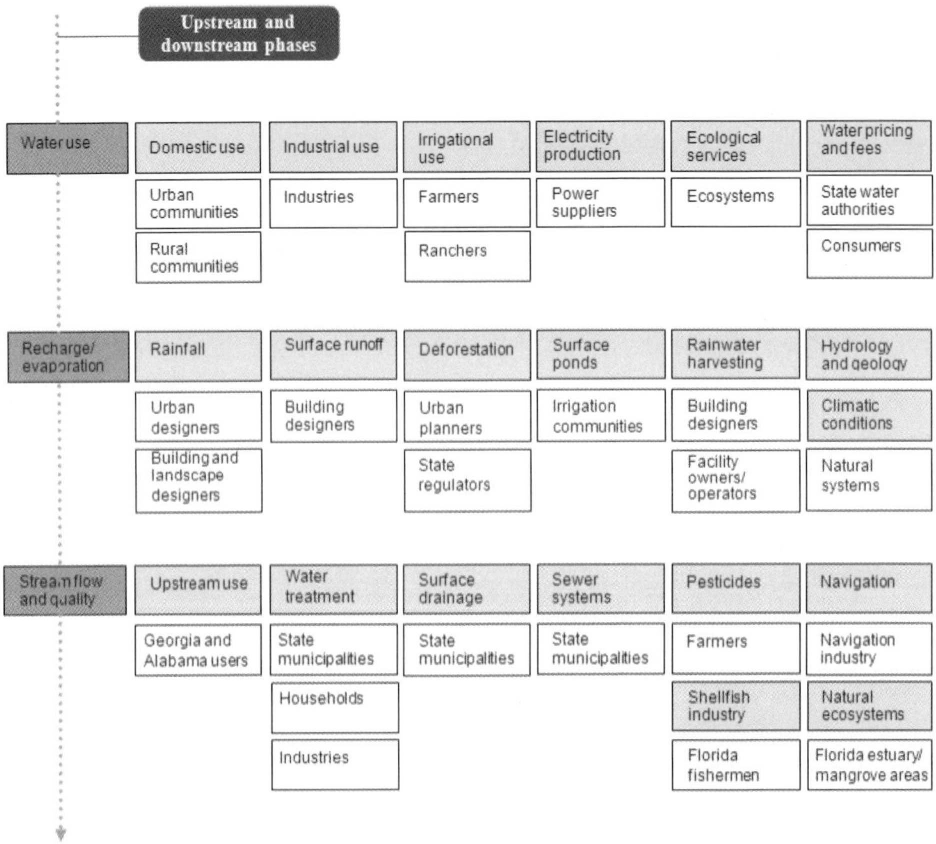

Fig. 4. (*Continued*)

In this application, water flow would be investigated considering the relevant upstream and downstream activity phases. It can be clearly seen that dam operation can have a significant effect on many downstream phases. Analyzing the different levels of operation in the seasons reveals that dam operation does provide benefit to human society. For example, flood control in rainy seasons and essential water supply in drought periods despite the fact that there are significant effects to ecosystems, natural cycles, and species. It also shows that water quality is very much linked to water flow.

It also indentifies the building and landscape issues such as impervious areas and turf grass, which reduce rain water infiltration to the ground and increase evaporation that eventually reduces river recharge. Evaluating the population growth and urbanization patterns could reveal the future trends for these issues and the need for solutions such as better urban, building, and landscape designs.

Limited information on the geology and hydrology for the entire basin could prevent formulating and determining the benefits from options for dam operation,

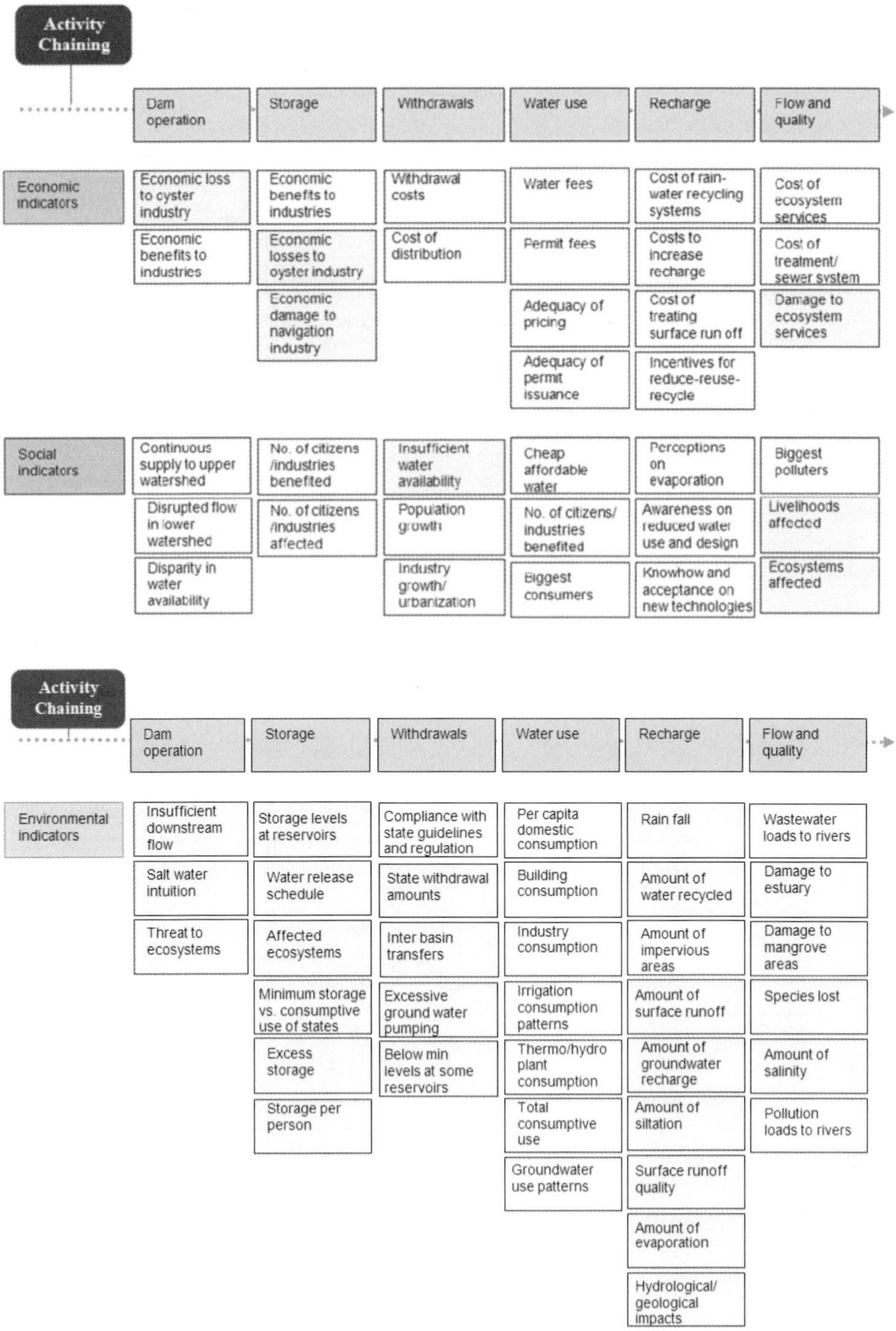

Fig. 5. Unbundling the primary activities and the respective economic, social, and environmental indicators regarding dam operation during drought cycle (minimum storage).

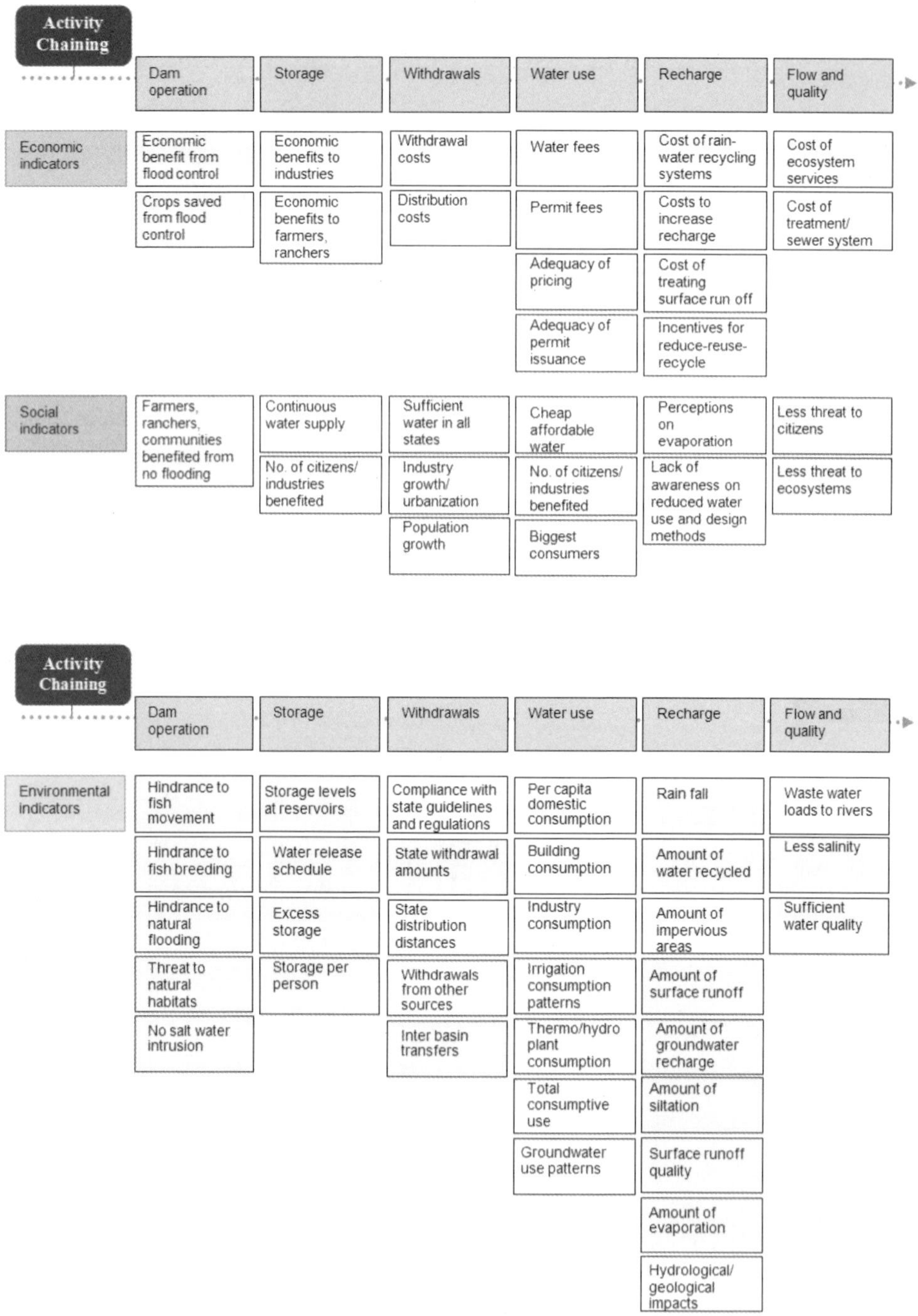

Fig. 6. Unbundling the primary activities and the respective economic, social, and environmental indicators regarding dam operation during flood cycle (maximum storage).

changes in design and construction practices, groundwater withdrawals, and alternative measures to increase recharge to the rivers in the basin.

As is apparent from the above indicators, the benefits to human systems are entangled with negative impacts to the ecological systems. Even though state guidelines and laws are available to safeguard human needs, laws and regulations to protect the needs of the ecosystems are not as prominent in the ACF discussions. Hence, when mapping the stakeholders with the respective upstream and downstream phases, natural ecosystems and species could also be rightfully represented as stakeholders in the watershed system. Finding an ethical balance between the essential human needs and ecological needs with thorough learning regarding system interactions could lead to sound policy negotiations and decisions.

Similarly, alternative scenarios of dam operation and water consumption could be assessed at the watershed scale. Comparative indicators such as minimum storage amounts with consumptive use in the states could be used to examine excess or deficit storage amounts in the reservoirs to understand how the dam operation or water consumption of the states could be optimized. Eliciting the cost of ecological services and respective benefits as well as damage to natural systems and their potential costs and losses would also encourage the stakeholders to evaluate the net benefits in a more comprehensive manner. These discussions would likely bring out the fact that further research needs to be conducted to estimate the cost of ecological services and damage to ecosystems in a manner that could be easily communicated to the stakeholders and responsible policy makers.

6. Conclusions and Outlook

As the methodology suggests, some of the sub-issues can be further analyzed in detail. For example, if excessive water withdrawals and stormwater runoff are sub-issues, the current status of those issues could be unbundled to identify the upstream and downstream activities, resources, causes, effects, and stakeholders. Or, if retrofitting buildings to reduce water consumption and sewer systems to increase water quality are possible scenarios or future options, they can be assessed as described above. When all relevant sub-issues or potential options for solutions are evaluated in the same manner, the not so apparent interconnectivities could be identified. This could lead to prioritizing issues with due consideration to their interconnectivities and the level and type of collaboration necessary. Detailed analysis could be more useful in smaller scales such as site and community scale planning, which is more focused on regional collective action. Then, the issues, scenarios, analyses, indicators, strategies would be more specific to a region.

Along with water management in a region, other related issues such as regional economic and social development programs and activities need to be addressed simultaneously. With such assessments, stakeholders would be able to visualize the costs, benefits, and strategies related to water management activities that would increase the health of the ACF basin, while contributing to the regional development activities in a meaningful manner.

There are various design methods and approaches, which have been developed to reduce water consumption, evaporation, surface water runoff, and siltation at building or unit scale. Improved building designs, efficient irrigation methods, appropriate pricing of water, and incentives for reduced consumption and recycling are some examples included in the discussion. Most methods were designed context specific and applications were limited. They were discussed as successful projects in design, construction, and operational phases, but it was unclear whether the upstream energy, material, and water consumption were less than the savings downstream, i.e., whether significant life-cycle net benefits were achieved. It was also unclear whether the designs would be suitable or acceptable when applied in a broader scale. The impacts from manufacturing, transport, construction, and maintenance are unknown. Furthermore, cost effectiveness, technical know-how and capacities, social and cultural preferences, and stakeholder activities and commitments are essential elements in comprehensively assessing the sustainability of new methods and technologies, which are also unknown in this case.

While improving the flow and quality issues of the ACF, the methods used should not create adverse effects in or outside the ACF system. Hence, at building or unit scale planning, the relevant stakeholders and experts together could use LCT to unify the necessary practical and scientific knowledge in working out comprehensive strategies acceptable to and within the capacities of the stakeholders that would eventually contribute to minimizing the issues in the watershed.

In actual planning situations, once the main issues of the watershed have been identified, stakeholders and experts could choose the appropriate levels and upstream and downstream activity phases that should be included in the assessment. The indicators in the example were introduced in generic form as all the indicators presented were not investigated for their actual values within the first phase of the study. The indicators applicable in this example could be quantitative, semi-quantitative, qualitative, descriptive, or graphical. When the indicators are compared with reference to the respective levels of flow, the effects of the flow regime in various phases could be examined. Especially in the case of comparing alternative levels of dam operation, once the actual values for the full set of economic, social, and environmental indicators are worked out with stakeholder input, they could be organized to further analyze, discuss, and negotiate which levels better meet the stakeholder's needs. Accordingly, sustainable strategies for sound policies and action could be negotiated at the watershed scale.

Acknowledgments

The authors would like to thank Professor Kim Tanzer, president of NAED and Dean of the School of Architecture, University of Virginia; Professor Margaret Carr, David Goldsmith, and Mark Russell, College of Design Construction and Planning,

University of Florida; and Dr. Iris Patten, College of Architecture and Landscape Architecture, University of Arizona for their contributions.

References

Abram, R.H., 2009. *Presentation to USDA-CSREES, National Water Conference*, Savannah, Geogia, 31 January 2007.

Allen, W., Kilvington, M. and Horn C., 2002. Using participatory and learning-based approaches for environmental management to help achieve constructive behaviour change. *Landcare Research Contract Report LC0102/057*. Ministry for the Environment, Lincoln, NZ http://www.landcareresearch.co.nz/research/social/envinfsys.asp (Accessed 3 March 2010).

ARROW (Apalachicola Region Resources on the Web), 2008. *General Information*. Available at http://www.fnai.org/ARROW/almanac/biology/biology_index.cfm (Accessed 20 March 2010).

Boustead, I., 1996. LCA — how it came about: The beginning in the UK. *International Journal of Life Cycle Assessment* **1**, 147–150.

Bureau of Economic and Business Research (BEBR), 2008. *Projections of Florida Population by County, 2007–2035*. Vol. 41, Bulletin 150. Bureau of Economic and Business Research — Florida Population Studies, University of Florida.

Burger, J., Gochfeld, M., Powers, C.W. and Greenberg, M., 2007. Defining an ecological baseline for restoration and natural resource damage assessment of contaminated sites: The case of the Department of Energy. *Journal of Environmental Planning and Management* **50**(4), 553–566.

Cape Town Gateway. 2007. *Integrated Development Planning*. Available at http://www.capegateway.gov.za/eng/pubs/public_info/I/11198 (Accessed 30 June 2008).

CRS (Congressional Research Service), 2008. In: Carter, N.T., Corn, M.L., Abel, A., Kaplan, S.M., Buck, E.H., C.B. and Alexander, K. (Eds), *Apalachicola-Chattahoochee-Flint (ACF) Drought: Federal Water Management Issues. CRS Report for Congress, Order Code RL34326*.

Cummings, R.G., Norton, N.A. and Norton, V.J., 2002. Enhancing in-stream flows in the flint river Basin: Does Georgia have sufficient policy tools? *Water Policy Working Paper #2001-002*, Georgia State University Andrew Young School of Policy Studies, Albany State University Flint River Water Planning and Policy Center.

Eubanks, J.E. and Greenfield, J., 1999. *Proceedings on the 1999 Georgia Water Resource Conference*, University of Georgia, Athens, Georgia, 30–31 March 1999.

Feldman, D.L., 2008. Barriers to adaptive management: Lessons from the Apalachicola–Chattahoochee–Flint compacts. *Society and Natural Science* **21**, 512–525.

Florida Department of Agriculture and Consumer Services, 2004. *Apalachicola*. Available at http://www.fl-seafood.com/apalachicola.htm (Accessed 30 March 2009).

Geyer, Y., 2006. *Integrated Development Planning Handbook Series for Community-based Organizations*. Institute for Democracy in South Africa, Pretoria, South Africa.

Giacomini, R. and Hadley, J., 2005. *Georgia 2015 Population Projections*. Office of Planning and Budget, Atlanta, Georgia.

Glick, P. and Clough J., 2006. An unfavorable tide: Global warming, coastal habitats, and sport fishing in Florida. *Tech. Rep.*, National Wildlife Foundation and Florida Wildlife Federation.

Hartmann, H., 2008. Decision support for water resources management. In: *Uses and Limitations of Observations, Data, Forecasts, and Other Projections in Decision Support for Selected Sectors and Regions. Synthesis and Assessment Product 5.1*, US Climate Change Science Program, pp. 45–55.

Hawkins, D., 1997. Key steps to water allocation formulas for ACT and ACF river basins. In: *Proceedings of the Georgia Water Resources Conference*, March 22–23, University of Georgia, Athens, Georgia.

IISD (International Institute for Sustainable Development), 2006. *What is Sustainable Development.* Available at http://www.iisd.org/sd/ (Accessed 3 March 2010).

Innes, J.E., 1996. Planning through consensus building: A new view of the comprehensive planning ideal. *Journal of the American Planning Association* **62**, 460–472.

Klein, J.T., 2004. Prospects for transdisciplinarity. *Future* **36**, 515–526.

Klein, J.T., Grossenbacher-Mansuy, W., Häberli, R., Bill, A., Scholz, R.W. and Welti, M., 2001. *Transdisciplinarity: Joint Problem Solving among Science, an Effective Way for Managing Complexity.* Birkhäuser Verlag, Basel, Switzerland, 332 p.

Lawrence, R. J. and Després, C., 2004. Futures of transdisciplinarity. *Futures* **36**, 397–405.

Lee, N., 2006. Bridging the gap between theory and practice in integrated assessment. *Environmental Impact Assessment Review* **26**, 57–78.

Leflar, M., 2008. What's the difference between life cycle thinking and assessment? *Converting Magazine.* AllBusiness.com, Inc. Available at http://www.allbusiness.com/environment-natural-resources/pollution-environmental/11710710-1.html (Accessed 30 March 2009).

Leitman, S., 2008. Lessons learned from transboundary management efforts in the Apalachicola–Chattahoochee–Flint Basin, USA. *Transboundary Water Resources: A Foundation for Regional Stability in Central Asia.* Springer, Netherlands, 2008, pp. 195–208.

Luksa, F. and Siebenhüner, B., 2007. Transdisciplinarity for social learning? The contribution of the German socio-ecological research initiative to sustainability governance.

Meppem, T. and Gill R., 1998. Planning for sustainability as a learning concept. *Ecological Economics* **26**, 121–137.

Metropolitan North Georgia Water Planning District, 2009. *Wastewater Management Plan.* Available at http://www.northgeorgiawater.com/html/87.htm (Accessed 1 March 2010).

Meyer, J.L., 2001. *Ecological Impacts of Water Management Decisions for the Apalachicola–Chattahoochee–Flint River Basin in a Changing Climate.* USGS Water Resources Research Grant Proposal by University of Georgia, Athens, GA.

Müller, D.B., Tjallingii, S.P. and Canters, K.J., 2005. A transdisciplinary learning approach to foster convergence of design, science and deliberation in urban and regional planning. *Systems Research and Behavioral Science* **22**, 193–208.

NRC (National Research Council), 2008. Public participation in environmental assessment and decision making research. In: Dietz, T. and Stern, P.C. (Eds), Committee on the Human Dimensions of Global Change, Division of Behavioral and Social Sciences and Education. Panel on Public Participation in Environmental Assessment and Decision Making. The National Academies Press, Washington, DC.

NRC, 2009. *Summary of a Workshop on Water Issues in the Apalachicola–Chattahoochee–Flint and Alabama–Coosa–Tallapoosa (ACF-ACT) River Basins (2009) Water Science and Technology Board (WSTB).* The National Academies Press, Washington, DC.

Resource Management Institute, 2005. *River–Reservoir Water Quality Modeling of the Apalachicola, Chattahoochee, and Flint (ACF) and the Alabama, Coosa, and Tallapoosa (ACT) River Systems, Georgia, Florida, and Alabama.* Available at http://www.rmanet.com/Projects/ACT-ACFRiverSystems.htm (Accessed 1 March 2010).

Richter, B.D., Mathews, R, Harrison, D.L. and Wigington, R., 2003. Ecologically sustainable water management: Managing river flows for ecological integrity. *Ecological Applications* **13**(1), 206–224.

Ruhl, J.B., 2005. Water wars, eastern style: Divvying up the Apalachicola–Chattahoochee–Flint river basin. *Journal of Contemporary Water Research & Education* **131**, 47–54.

Scholz, R.W. and Marks, D., 2001. Learning about transdisciplinarity. In: Klein, J.T., Grossenbacher-Mansuy, W., Häberli, R., Bill, A., Scholz, R.W. and Welti, M. (Eds), *Transdisciplinarity: Joint Problem Solving among Science, an Effective Way for Managing Complexity.* Birkhäuser Verlag, Basel, Switzerland, pp. 236–25.

Thabrew, L., Wiek, A. and Ries, R., 2008. Environmental decision making in multi-stakeholder contexts: Applicability of life cycle thinking in development planning and implementation. *Journal of Cleaner Production* **17**(1), 67–76.

Thabrew, L. and Ries, R., 2009. Application of life cycle thinking in multi-disciplined stakeholder contexts for sustainable cross-sectoral planning and implementation of development projects: A case-study in Sri Lanka. *Integrated Environmental Assessment and Management* **5**(3), 445–460.

USGS (United States Geological Survey), 2008a. *Water Resources Inventory Area 1 Watershed Management: Introduction to Watershed Boundaries.* Available at http://wa.water.usgs.gov/projects/wria01/wb_intro.htm (Accessed 30 March 2009).

USGS, 2008b. *Apalachicola–Chattahoochee–Flint River Basin: The National Water-Quality Assessment Study.* Available at http://ga.water.usgs.gov/nawqa/basinall.html#basin2 (Accessed 11 June 2009).

USGS and US Department of Interior, 1998. In: Frick, E.A., Hippe, D.J., Buell, G.R., Couch C.A., Hopkins, E.H., Wangsness, D.J. and Garrett, J.W. (Eds), *Water Quality of Apalachicola–Chattahoochee–Flint River Basin*, 1992–1995. USGS Circular: 1164.

UF (University of Florida) and NAED (National Academy of Environmental Design), 2009. *Water and Sustainability Symposium: Apalachicola–Chattahoochee–Flint River Basin.* Available at http://naedacf.pbworks.com/ (Accessed 12 May 2009).

Van Asselt, M.B. and Rijkens-Klomp, N., 2002. A look in the mirror: Reflection on participation in integrated assessment from a methodological perspective. *Global Environmental Change* **12**, 167–184.

Wiek, A., Binder, C. and Scholz, R.W., 2006. Functions of scenarios in transition processes. *Futures* **38**, 740–766.

Weik, A. and Walter, A.L., 2009. A transdisciplinary approach for formalized integrated planning and decision-making in complex systems. *European Journal of Operational Research* **197**(1), 360–370.

Chapter 14

ENTREPRENEURSHIP AND SUSTAINABILITY

Abeer Abdel Hamid Salem[a], Robert Anderson[b] and Leo Paul Dana[c,d]
[a]*Education for Sustainability, Prescott College, Arizona, USA*
asalem@prescott.edu
[b]*Hill/Levene Schools of Business, University of Regina, Regina, SK S4S 0A2, Canada*
robert.anderson@uregina
[c]*GSCM — Montpellier Business School, France*
[d]*University of Canterbury, New Zealand*
leo.dana@canterbury.ac.nz

Entrepreneurship is usually thought of as a feature of the capitalist economy and an intrinsic element in its ongoing expansion, and therefore a contributor to decreasing sustainability. As argued by Schumpeter, entrepreneurs do the "creative" destruction as they apply the merciless logic of profit to traditional communities whose lifestyles make no financial sense even if they make cultural sense and are quite sustainable. In this chapter, we argue an alternative and a much more hopeful perspective on entrepreneurship. We view entrepreneurship as a potential force for sustainability. It can bring sustainability not only to the global capitalist economy and capitalist societies, but also to non-capitalist societies. In doing so, entrepreneurship can result in the formation of a complementary set of economies that are mutually compatible and together create a diversified and intricate network giving rise to a sustainable global system. To make this argument, we discuss the concepts of entrepreneurship and sustainability; we propose a sustainability approach to entrepreneurship, and explore Indigenous entrepreneurship as a model that reconciles entrepreneurial and sustainability ideas. We end the chapter with suggestions for alternate, more meaningful forms of wealth and with concluding thoughts on the potential of entrepreneurship to provide a solution for sustainability.

1. Introduction

In the same vein as Kao's (2010) thesis, we view entrepreneurship as a potential force for sustainability. As the late Peter Drucker said, entrepreneurship involves the search for innovated opportunities, the formation venture, and the marshalling of resources to meet the unmet and under satisfied needs of people. And without a doubt, people around the world are desperately seeking development broadly defined (and not as modernization), which is sustainable. Entrepreneurship can bring sustainability not only to the global capitalist economy and capitalist societies, but also to non-capitalist ones. In doing so, entrepreneurship results in the formation of a complementary set of economies that are mutually compatible and together create a diversified and intricate network giving rise to a sustainable global system. To make this argument, we propose a sustainability approach to entrepreneurship, and explore Indigenous entrepreneurship as a model that reconciles entrepreneurial and sustainability ideas.

In this chapter, we present our arguments in support of this view. In the first three sections, we expand on the concepts of entrepreneurship and sustainability. Then, we move to a section that describes a sustainability approach to entrepreneurship including the concepts of connectedness, embeddedness, and contingency. Next, we look at Indigenous entrepreneurship as a demonstrating model of the sustainability approach to entrepreneurship. We end the chapter with suggestions for alternate forms of wealth and with concluding thoughts on the potential of entrepreneurship to provide a solution for sustainability.

2. Entrepreneurship

Broadly defined, entrepreneurship is a process involving the recognition of opportunities in the needs and wants of people and the application of the technology of management to convert such opportunities into viable ventures. These ventures include not only new businesses but also non-profit and government activities (Anderson, 2002).

Because of the widely acknowledged importance of entrepreneurship to the economy, researchers have studied the variables that may inhibit the stock of entrepreneurial talent and related resources to be fully utilized in society, perhaps at some cost to overall economic growth. These barriers include social, individual, geographic, cultural, economic, political, and/or structural disadvantages. Examples of such disadvantages include illiteracy, inability to secure food and shelter, living in isolated areas, racism, complacency to welfare systems, lack of social capital and networking opportunities, and inadequate policies to train and prepare people for self-employment or any kind of entrepreneurial activity (Frederick, 2008).

Based on the social marginalization theory, however, it is hypothesized that the status of disadvantage prompts people to pursue entrepreneurial activity in order to improve their status and living conditions but the research findings testing this hypothesis are inconclusive (Frederick, 2008). It would be interesting if the state of disadvantage is found to be both stimulant and barrier, simultaneously, to entrepreneurial behavior. This would perhaps give prominence to the aspect of human agency and creativity and point to the existence of individual factors intrinsic to the entrepreneurial personality that motivates some people to act on their dire circumstances and pursue entrepreneurial activity, while hindering others who remain in a state of complacency. While these and other factors affect the level of entrepreneurship in a society, entrepreneurial activity is — at the same time — a key definer of social structure, the ecosystems and of the environment.

This depends on the kind of society for this to occur; a society needs to be open and innovative enough to develop new roles as necessary and as needed. Initiating this kind of enabling environment is one of the functions of the entrepreneur (Morrison, 2008).

While entrepreneurial activity is pursued to address certain need(s) of society that are recognized as opportunities by the entrepreneurs, mainstream entrepreneurship is usually perceived mainly as a profit seeking and money-making endeavor. We argue that this understanding of entrepreneurship as seeking merely economic gain is perhaps based on the development paradigm that prevailed since the advent of industrialization. Notions of this development paradigm ruled the world over the past decades and the pressures for economic growth, capitalism, and industrialization that were seen as key to development and modernization dominated the world in tandem. Hornborg conceptualizes the process perceived as development as basically a manifestation of capital accumulation, which has always been an uneven and inequitable process, generating an increased polarization between developed centers and underdeveloped peripheries. He contends that all major issues of global survival including environmental destruction, resource depletion, armament, and world poverty can ultimately be traced to capital accumulation (Hornborg, 2001).

Having regarded economic growth as the ultimate goal for humanity, development indicators have been designed with the assumption that societal development is equivalent to economic growth. Total GNP and average *per capita* stand at the core of indicators, which have been used to measure how developed a given society is, although these indicators are purely quantitative. Development is, however, ultimately about improving human well-being, which is mainly qualitative and although development may include growth as a quantitative measure; much of conventional economic development is indifferent to the qualitative focus on well-being (Calder, 2008). While growth is the quantitative increase in the physical dimensions of the economy, development is the qualitative improvement of society as a whole.

Growth is, therefore, not equivalent to development. However, economic growth is undoubtedly one aspect of the process of economic development — which happens to be the aspect on which traditional development economics and the prevalent development model have concentrated. Traditional development economics can offer immense insights into the field of causation of growth; however, it has been less successful in characterizing social development, which involves expansion of people's capabilities among other qualitative dimensions of development. To reach this kind of development, economic development is only a means, which is oftentimes not the most efficient means either (Sen, 1983).

There is general agreement between the traditional and contemporary theories that economic development must include capital accumulation, introduction of new technology, human resource development, development of institutions and markets, and a permissive international framework (Grilli and Salvatore, 1994). Nevertheless contemporary theory goes a step further to consider the additional requirements for equitable economic development. Hughes (1994) and Goulet (1989) point out that the policy framework adopted and implemented is critical for development. They collectively identify characteristics for equitable sustainable development:

1. economic growth that is oriented toward human progress;
2. equitable distribution of the fruits of growth;
3. optimal participation of non-elite populations in decision making;
4. optimal participation of non-elite populations in implementation;
5. ecological responsibility; and
6. respect for traditional cultures when conducting development activities (Goulet, 1989, p. 5).

This considerably broadened focus emphasizes the increasing social focus of economic development.

The emerging concept "sustainable development" can be conceived as a conceptual counter position to "development" or "modernization" and breaks with the equivalence between development and economic growth that has been taken for granted. It questions the assumption of a continuous, linear and more or less harmonious development for societies along a given track. It strongly negates the idea that there is one and only one way and direction for development to occur. The concept of sustainable development is based on a society-oriented definition of problems (Becker *et al.*, 1997) and definitions of the means to address them.

Despite this enthusiasm about "sustainable development," some still view it as building on the conceptual misgivings of the old development paradigm and it is often interpreted as the continuity of development efforts once initiated in a certain locality, to others, the two concepts "sustainability" and "development" are vague, at best. According to Woodbridge, the concept of sustainable development has been popularized, while "in reality, sustainable development has become the new opiate of our political elites: the rhetoric has become ubiquitous even while the meaning has become ever more oblique (Woodbridge, 2004, p. 8)." The idea of sustainable development, which is introduced in the global development community as a more advanced idea from just plain, old development is quite a controversial issue and not clearly understood.

3. The Concept of Sustainability

Owing to the complexity of the concept of sustainability, there is an inherent difficulty in its definition. Sustainability, when viewed from a global or continental perspective, is very broad and cuts across almost all disciplines, which clearly shows that standard disciplinary approaches cannot address its central issues adequately. There is a risk that sustainability may be narrowly conceptualized as uni-dimensional by specialized disciplines, which negates its overarching and all encompassing nature as a concept (Allen *et al.*, 2003). Moreover, sustainability is generally thought of as a condition of the future, but can also be easily interpreted according to what humans currently value and what is perceived as currently worthy to sustain since predicting what can be valuable to humans' survival in the future is not an easy task.

Acknowledging the centrality of contexts to sustainability and the importance of its active dimension are one useful way of interpreting it. Sustainability is, thus, defined as "maintaining, or fostering the development of, the systemic contexts that produce the goods, services, and amenities that people need or value, at an acceptable cost, for as long as they are needed or valued (Allen *et al.*, 2003, p. 26)." According to this definition, sustainability is more about maintaining an enabling environment that ensures the provision of these crucial elements that secures humans' sustenance.

Deconstructed into its components, the concept can be described as environmental, social, and economic sustainability. Environmental sustainability involves the maintenance of natural capital, which entails maintaining balance between the source and sink functions of nature. Social sustainability requires values such as diversity, love, compassion, humility, and other moral and social capital values. It also requires environmental sustainability as a fundamental pre-requisite. Economic sustainability is the maintenance of capital or keeping a non-declining capital. It is argued that the three types of sustainability — social, environmental, and economic — are clearest when kept separate, although the inevitable overlap and linkages are quite salient (Goodland, 1995; Goodland and Daly, 2006). Along a similar vein, sustainability is interpreted as the three "Es": environment/ecology, economy/employment, and equity/equality. The core innovation of the concept is in its expanded focus that includes economic and social issues alongside the preservation and management of ecology and the environment. This expanded focus enables sustainability to offer the possibility of a broader coalition for positive change both within and between societies (Edwards, 2005). In this sense, sustainability provides a holistic approach to understanding and addressing social, economic, and environmental issues as mutually cross-cutting issues with overlapping cores and boundaries.

While noting that there are some 330 definitions of sustainability, Parajuli proposes an approach to sustainability that focuses on bio-cultural diversity by appreciating the collective memory of nature, and the view that we need a new practice of inter-species partnership. He offers a partnership model for sustainability that includes four kinds of partnerships: intra- and inter-generational, inter-species, inter-cultural, and inter-economic partnerships. Through emphasizing partnership, we can directly relate the concept of sustainability to notions of connectedness and interaction and by emphasizing bio-cultural diversity; sustainability is no longer a "global and continental" issue out there, but a local issue residing in close proximity within people's daily interactions (Parajuli, 2010).

Sustainability is, hence, quite distinct from development. Sustainability is about community, local-based knowledge and intimacy with place. Development, at least the dominant form of it, is about progress and pursuing a foreign ideal created and idealized elsewhere. The value system underlying each of the two notions is incompatible, the values of sustainability are not the values being promoted through the development and modernization ideals. The values of sustainability and those of development represent two opposite poles, so remotely apart and inherently impossible to reconcile.

Sustainability is concerned with encouraging self-sustaining abilities and maintaining wholeness between systemic levels. It respects and appreciates what is already there, conserving and developing inherent creative potential, assisting self-reliance, self-realization, self-sustaining abilities, and resilience (Sterling, 2001). Therefore, as opposed to prevalent development thinking, sustainability is not concerned with continuous progress or modernization, it is rather concerned with modes of coping that are built on understanding, encouraging, and fine tuning what already works and what has worked in the past.

Sustainability emphasizes a holistic approach that includes the economic, socio-cultural, and ecological components and stresses an ecological view in framing and addressing the relevant issues and problems. Sustainability, rather than development or sustainable development, may be the best organizing principle that reflects the fundamental relationships that underlie ecological, economic, and social concerns. Sustainability seeks to create and maintain a fair and just context in which legitimate interests of all parties can be satisfied, within the framework of concern for equity. This means that at the global level, sustainability is oriented toward solutions that "do not doom developing countries to a permanently secondary place in the world economy to address environmental issues (Edwards, 2005, p. 21)." In other words, sustainability provides a framework where developed and developing countries would work together to achieve a sustainable future for all.

Based on this, we can conclude that sustainability is a concept that acknowledges diversity, appreciates the old as well as the new, and incorporates the difficult task of reconciling environmental, economic, and social/cultural well-being, through providing the enabling environment and the contextual conditions that enable healthful relationships, networking, and connectedness to prevail.

4. A Sustainability Approach to Entrepreneurship

Relevant to the concept of entrepreneurship and in alignment with our conviction that what impoverished people mostly aspire for is a decent way to make a living, we discuss what can be described as a sustainability approach to entrepreneurship. Although entrepreneurial activity is often described as having negative consequences on society, resulting in wealth accumulation of the few at the detriment of the majority, we advance a different scenario. We believe that entrepreneurial activity that is undertaken within a framework of certain social, environmental and economic conditions and objectives not only contributes to sustainability, but also can be a most promising solution to needy communities in poverty around the globe.

Sustainability brings to the fore concepts that have been missing in mainstream development discourse. These concepts include social justice, egalitarianism, community, environmental justice, awareness and care for nature, compassion, and diversity, among many other concepts that have been overshadowed by the individualistic inclinations that characterize the dominant development and modernization discourse. In this respect, we highlight three concepts: connectedness, embeddedness, and contingency that underlie the specific social, environmental, and economic framework that constitute the sustainability approach to entrepreneurship.

4.1. *Connectedness*

Connectedness is emphasized as a core element of sustainability. All living beings are connected at some level or another and while functioning separately, their connectedness is a critical factor that sustains their functioning. Connectedness is so intrinsic and essential to our existence that it usually passes unnoticed, and here is where part of the difficulty of recognizing it emerges. It is also inherently difficult to focus on and amplify connectivity and relationships in a modern culture that emphasizes separation (Golley, 1998). However, if we are to take our lessons from nature, we learn that nature is sustained through connections, networks, and communities. Fritjof Capra elaborates on this notion by explaining how the patterns and processes by which nature sustains life emphasize community as a core principle of sustainability, and how nature demonstrates that sustainable systems are possible by creating and nurturing communities. Accordingly, sustainability in nature's language is manifest in communities, in diversified networks and nested systems where cyclical relationships are developed, achieved, and maintained in harmonious flows and dynamic balance (Capra, 2005).

Unlike the modern emphasis on separation, the Indigenous perception philosophically emphasizes interconnectedness of all species in a seamless web of causes and effects, actions and outcomes, behaviors and consequences. "People, animals, plants, natural objects, and supernatural entities are not separate and distinct but are rather linked to each other and to the places where they reside through cultural traditions and interactive, reciprocal relationships" (Turner *et al.*, 2000, p. 1279). The simple fact that Indigenous people are more affiliated with nature than modern people renders them more sustainable through acknowledging and revering their connections with nature and all forms of life.

Given this fine level of ecological intelligence and sensitivity, it is interesting to analyze how Indigenous people approach a business-oriented concept as entrepreneurship. Accordingly, and from a sustainability perspective, it is useful to investigate what features characterize Indigenous entrepreneurial activity and how, from the sustainability perspective, entrepreneurship can be pursued. We base this on the fundamental premise that entrepreneurship is an important skill in society that can not only ensure dignified lives for those who have that skill, but also empower marginalized people and free them from the dependency situation in which they are stuck as a result of successive attempts at modernization and development. It is from this angle and from this base of thinking that our discussion of entrepreneurship and sustainability emerges and ensues.

4.2. *Embeddedness*

In the context of sustainability, Karl Polanyi's (2001) ideas about economic theorizing are very pertinent. We would like to highlight the powerful point Polanyi makes about Embeddedness — how the economy and any theorizing of it should be embedded in society, culture, and nature (Polanyi, 2001). The relevance of this point to our discussion of the relationship between entrepreneurship and sustainability is important since it suggests the contours of a wholesome, well-balanced relationship between the four components of life: the social, cultural, natural, and economic, which make life worth living.

Embedding, as a concept in and of itself, reflects an ecological understanding of the world and the universe: it reflects connectedness and highlights that the relationships among us that we know of and that we do not know of are enhanced by acknowledging them, not by subduing them to an artificial divide. Hence, we argue that Polanyi's concept of embeddedness visualized a non-homogenized pattern of economic development by emphasizing that each economy would reflect, represent, and cater to a certain society belonging to a certain locality that is quite distinctive from others. This visualization is conducive to partnerships, collaboration, and collective relationships, which are core elements in achieving sustainability, rather than competitiveness and domination.

4.3. *Contingency*

As entrepreneurship is perceived as mainly an activity geared to economic gain, it is directly related to society's economic development. The history of development reveals a sequence of uni-dimensional, incomplete perspectives, each mandating conflicting and inevitable futures. The main ones being modernization or dependency, neither benefitting marginalized societies. As a reaction to this, other theories of development emerged that emphasize contingency and human agency. Regulation theory is one of these.

Regulation theory views the economy as an evolving social system created by people. At the core of the theory is the understanding that it is the actions of people that determine the characteristics of the economic system rather than the system that determines the actions of people. Regulation theory analyzes the global economy "in terms of a series of *modes of development* based on combination of the currently ascendant *regime of accumulation* and a variety of *modes of social regulation*" (Hirst and Zeitlin, 1992, pp. 84–85). The regime of accumulation determines the general possibilities for the economy. Scott says it "can be rather simply defined as a historically specific production apparatus ... through which surplus is generated, appropriated, and redeployed" (Scott, 1988, p. 8). Importantly, with respect to geographic scale, the regime of accumulation is a "relationship between production and consumption defined at the level of the international economy as a whole" (Hirst and Zeitlin, 1992, p. 85), and is currently what is commonly called the global capitalist economy.

If the world were Adam Smith's, peopled by the universal perfectly rational "economic man," no regulation of the global economy beyond the "invisible hand" of perfectly functioning markets would be required. However, the world is not Smith's; people are far from perfectly rational and they are driven by many things not economic. Further, they are far from universal in the nature of their variations from the "perfect." As a result, Scott says that stability in the economic system is dependent on the emergence of a further set of social relations that preserve it, for a time at least, from catastrophic internal collisions and breakdowns. These relations constitute a mode of social regulation. They are made up of a series of formal and informal structures of governance and stabilization ranging from the state through business and labor associations to modes of socialization, which create ingrained habits of behavior, and so on (Scott, 1988, p. 9).

Regulation theory does not prescribe the exact nature of a particular mode of social regulation, but each locality may be organized in a multiplicity of ways and since modes of social regulation are based on such things as habits, customs, social norms, and laws among other elements of the system, unique modes can exist at any territorial level from local to global (Anderson, 1997). The importance of the regulation theory lies in its outright consideration of the social dimensions neglected in the discussion of the process of development thus far except as obstacles to be overcome, and the fact that responses to the overall economic system can be shaped by these social dimensions.

The contingency perspective on economic development, advanced by Anderson (1997), emphasizes and elaborates on the notion that development is contingent on human agency and the skill of reproducing social relations. This perspective is also characterized by the focus on the relationship between a developing region and the global economy and the fact that this relationship can exhibit unique characteristics and serve the particular needs of its people. The contingency perspective holds that creation of a successful mode of development requires an active civil sector capable of articulating the needs of the people, a leadership capable of devising the strategies and negotiating the arrangements necessary to address those needs, and a mechanism for the people to modify these arrangements should the outcomes prove to be unacceptable (i.e., feedback from people). The outcome in any particular instance is contingent on many things, not the least of which is the approach taken to participation in the global economy by each community. These approaches can be hegemonic and counter-hegemonic according to the extent to which they consent to the capitalist global economy as it is, dissent from it, or attempt to transform aspects of it (Anderson *et al.*, 2006).

The contingency perspective is one of the alternative perspectives that can be considered the emerging hope for the resurrection of cultural identities, human agency, and creativity and the inherent capacity of humans to strive for their survival. The question remains whether developing nations are going to take on the challenge and realize their own self-designed version of development after decades of suffering the brunt of a development paradigm that was unanimously adopted and that failed to do them justice.

We view Indigenous entrepreneurship as a development related phenomenon and as one of the self-designed models of development. The basic theoretical underpinnings of the field denote how the emergence of Indigenous entrepreneurship reflects a nuanced approach to economic development "from the bottom up" rather than a narrowly focused activity merely targeted to making profits, which is consistent with regulation theory. It represents an alternative approach to attaining sustainable socio-economic benefits that evades the flaws of the dichotomized pattern of dominant development thinking by stressing those variables that were missing from the prevalent discourse: human agency and social responsibility and the concepts of connectedness and embeddedness. While entrepreneurial activity can be economically successful but not relevant to sustainability, our analysis of cases of Indigenous entrepreneurship revealed those elements that make entrepreneurship quite compatible with the concept of sustainability through incorporating the ecological and social/cultural dimensions in addition to the traditional focus on the economic dimension. The ecological and social/cultural focus emanates from an intrinsic belief in embeddedness and the value of connectedness.

5. Indigenous Entrepreneurship: A Model

Based on the above discussion of entrepreneurship, it becomes evident that there is nothing in the nature of entrepreneurship that precludes its successful use by Aboriginal people in pursuit of their economic development goals, as they can successfully identify opportunities in the needs and wants of individuals and apply the tools of management, especially those consistent with traditional values, to the conversion of these opportunities into viable ventures. To Indigenous people, entrepreneurship is perceived of as an "economy building" process (Anderson, 2002, p. 48). And there is nothing preventing these activities from being sustainable; indeed, they are likely to be if they are consistent with traditional values and practices and are observant of and caring for the environment. For example, Hindle and Lansdowne (2005) define Indigenous entrepreneurship as the creation, management and development of new ventures by Indigenous people for the benefit of Indigenous people. They say that this kind of activity can range from the narrow view of economic profit for a single individual to the broad view of multiple social and economic advantages, including sustainability, for entire communities (Hindle and Lansdowne, 2005).

Among the factors that distinguish Indigenous from mainstream entrepreneurship are the management style. The management style of Aboriginal, Indigenous, and Traditional societies across the world is characterized by lower tolerance of power distance than individualistic societies, they are non-hierarchical, and division of labor is based on expertise and responsibility. Leaders are informally recognized by the community not only for their expertise but also for the example they set for others, and decision making is based on consensus. Given these unique characteristics, Anderson advances that this style of management is not conflicting with the mainstream, but it can be argued that they are at the leading edge. It can be safely advanced, hence, that overall, "what is emerging is an Aboriginal approach to management in the same sense as there is a Japanese approach, an American approach, a Spanish approach, and so on" (Anderson, 2002, p. 49).

We are not arguing that there is uniformity or consistency among all instances of Indigenous entrepreneurship. On the contrary, regulation theory implies and in fact, mandates a local and diversified approach to entrepreneurship that is contingent upon the kind of human and environment resources within, and the specific needs and distinctive characteristics of the community. Thus, there will be a diversified pattern of entrepreneurship that characterizes specific localities, despite the common themes and features that can be traced in different cases of Indigenous entrepreneurial activities.

From a sustainability perspective, there are certain features that are salient in the patterns of Indigenous entrepreneurial activity that can clearly relate the field of entrepreneurship to the concept of sustainability. These features are built on the Indigenous beliefs that acknowledge embeddedness and the importance of connectedness, which relate their entrepreneurial activity directly to the three components of sustainability, the social/cultural, the ecological, and the economic components. This demonstrates how Indigenous entrepreneurship can contribute to, and provide, feasible pathways to sustainability.

5.1. *The socio-cultural dimension*

The socio-cultural dimension of Indigenous entrepreneurial activity is salient. The social context directs the objectives, meaning, and development of the entrepreneurial activity, which reflects the importance and weight given to the community where the entrepreneurial activity is being initiated. This strong social basis renders Indigenous entrepreneurial activities different from the classic notion of entrepreneurship in both motivation and process. There is a greater emphasis on collective rather than individual activity, and a greater emphasis on community and cultural preservation than would be expected of purely market-driven entrepreneurial activities (Anderson *et al.*, 2007; Hindle *et al.*, 2005). This gives rise to the emergence of the concept of social entrepreneurship as a defining characteristic of Indigenous entrepreneurship in both purpose and objective.

Social entrepreneurship is defined as that which leads to the establishment of new social enterprises and the continued innovation in existing ones. It is conceptualized as a multi-dimensional construct that involves the expression of virtuous behavior in order to achieve a social mission, a coherent unity of purpose and action in the face of moral complexity, the ability to recognize social value-creating opportunities and decision-making characteristics of innovativeness, proactiveness, and risk taking (Anderson *et al.*, 2006). Profit may play a role in the process but as a means of attaining other broader objectives, and not as an end in itself. And sustainability is certainly one of these broader objectives.

The Aboriginal approach to economic development is grounded on this foundation of social entrepreneurship where social purpose as the principal driver of activity is achieved primarily through entrepreneurship, where there is little if any distribution of profit to individuals as any surplus is reinvested for the long-term benefit of the community, where constituents are democratically involved, and where there is accountability. Social enterprise also includes the creation of jobs and the strengthening of social capital by supporting the integration of marginal people into society at large (Anderson *et al.*, 2006; Dzisi, 2008).

While Aboriginal groups' identification of opportunities and creation of business are clearly entrepreneurial, their reasons for doing so and the organizational forms they have adopted extend far beyond wealth creation for the entrepreneur or owner involved. Any wealth generated from business ventures is used to fund social objectives, broadly defined. Social objectives are also important considerations in the day-to-day activities of these businesses. For instance, the expenditures that are viewed as burdensome expenses to mainstream entrepreneurs such as salaries, training and development, purchases of inputs, and so on, which they seek to minimize to the lowest level possible, are considered as opportunities to deliver socio-economic benefits in the communities by Indigenous entrepreneurs and are used as venues to benefit the community and the people to the maximum extent possible limited by economic feasibility. Similarly, sustainability and profitability are not viewed as trade-offs with more of the later pursued at the sacrifice of the former, if the entrepreneurial activities truly reflect traditional values, as discussed more fully in the following section.

Finally, the social nature of the process is also found in the activity that lies at the heart of entrepreneurship — opportunity identification. A key criterion in the search for and identification of suitable opportunities is the extent of fit with a community's broad objectives and with the capabilities and aspirations (including lifestyle preferences) of community members (Anderson *et al.*, 2006).

5.2. *The ecological dimension*

In the context of sustainability, sensitivity and awareness to the factors that affect the environment and the way humans' deal with it are a core concern. Indigenous people are well aware of their relationship with nature that they are affected by the environment and that they, in turn, affect the life around them. They have resided in a particular locality for a long period of time, depending on the resources of their homelands. Their concept of guardianship over their lands requires careful management and conservation by present generations for the benefit of future ones (Turner *et al.*, 2000).

The ecological sensitivity of Indigenous people is reflected in their perception and application of entrepreneurship. This manifests itself in a variety of ways that can be gleaned from the literature and are summarized as follows: (1) the ecological embeddedness of the entrepreneurial activity is highly acknowledged; (2) an ecological approach to managing the entrepreneurial activity is stressed; (3) valuing the resources that are available in the community; and (4) valuing Traditional Knowledge, Traditional Ecological Knowledge, and Indigenous Knowledge.

In their study on Indigenous entrepreneurship, Hindle and Lansdowne (2005) describe what they perceived as the "individuality *versus* the collectivity" paradox. Based on the assertions gleaned through interviewing Indigenous entrepreneurs, this paradox reflects the relationship with nature as a "collective" value that needs to be respected and protected *versus* values of individual ownership and appropriation. Perceived as a dilemma by the authors, they pose a question that reflects this paradox: "how does deep attachment to the land and the harmonies of nature fit with the drive for profit and success? (Hindle and Lansdowne, 2005, p. 137)." This paradox reflects the concern and reverence Indigenous entrepreneurs have for nature and represents one of the main ecological underpinnings of Indigenous entrepreneurship: the need to balance the collective good with individual benefit. It also emphasizes the reverence for community values, collectivity and the importance of a healthful relationship with nature. The fact that Indigenous entrepreneurs are at least as concerned about these elements as they are about making profits denotes a deep sense of ecological responsibility and sensitivity. These concerns relate Indigenous entrepreneurship most directly to sustainability.

Another study investigated the reaction of the Indigenous peoples' of Canada, the Dene First Nations, to development and globalization (Dana *et al.*, 2008). The study revealed that the Dene First Nations exhibit a very high level of sensitivity and understanding of the negative effects that a petroleum development project can have on their environment, sustenance, and society. Despite promising economic prosperity, livelihoods, and employment opportunities, the development project was seen as temporarily beneficial but detrimental to the environment and the community in the long run. The wisdom inherent in their opinion and decision regarding the development opportunity reflects a deep sense of responsibility toward nature, societal values, and future generations (Dana *et al.*, 2009). The stand taken by Indigenous people regarding this development opportunity resonates with the basic principles of sustainability. However, this should not be taken to mean that all development will be rejected, but rather that development projects will be scrutinized and that Indigenous people have the wisdom to take a stand against forms of development that are degrading and destructive to the natural resources in their bio-region that is the source of their nourishment, sustenance, and livelihood.

Whiteman (2000) introduced the construct of ecological embeddedness to describe the management practices of Indigenous traditional bush managers of a subarctic ecosystem. The managers have "demonstrated sustainable approaches to natural resource management" (Berkes, 1999 quoted in Whiteman, p. 1265) that reflect the extent to which these managers are rooted in the land. In addition to their deep ecological beliefs, traditional managers are genuinely interested in and connected to the land. They "walk the land" (Whiteman, 2000, p. 1271), they prefer experiential learning to gather information and various forms of knowledge, and appreciate the traditional experiential knowledge form the elders. Another important aspect of their wisdom is considering earth as teacher by "listening to the earth and attempting to follow its lessons" (Whiteman, 2000, p. 1274).

The construct of ecological embeddedness obviously relates Indigenous or traditional management to sustainability. The traditional managers' beliefs, knowledge, and practices are strongly anchored in the natural environment. Control over resources is culturally based and defined and is not legally reinforced by specific individual property rights. There is a deep responsibility toward others where distributing the wealth of the land among families is a main commitment. They are also under a strong cultural and social obligation to ensure the productivity as well as the sustainability of their resources and a concern for the well-being of future generations (Whiteman, 2000).

5.3. *The economic dimension*

The relationship between Indigenous entrepreneurial activity and the global market is another defining factor of Indigenous entrepreneurship. This relationship is based on the general understanding that the "interface with the global economy is inevitable" (Hindle and Lansdowne, 2005, p. 134) and based on regulation theory and the contingency perspective (Anderson, 1997), there are different ways to engage (opt-in) or avoid (opt-out) the global economy.

The opt-in and opt-out scenarios illustrate the degree to which a social group or community engages with the global economy and address the nature of this engagement. A further question to consider whether opting in or opting out is the following: "Is the approach to accept the terms of the global economy "as is," or is it to attempt to transform it in some fashion?" (Anderson *et al.*, 2006).

The combination of the two scenarios (Fig. 1) results in four extreme possibilities that can be pursued by the community: (1) passive opting out and choosing not to participate, seeking isolation and even protection from the impact of the global economy, (2) active and aggressive opting out where a group rejects the global economy and seeks to resist it or overthrow it through protest and even "revolution," (3) passive opting in and full participation in the global economy and passively accommodating itself to the requirements of the global economy, and (4) active opting in where participation in the global economy is far from being passive. In the Indigenous case, this often involves an ongoing struggle for land and other rights to allow this participation to be "on their own terms" and the creation of ventures consistent with core traditional values and practices even when these are not a good

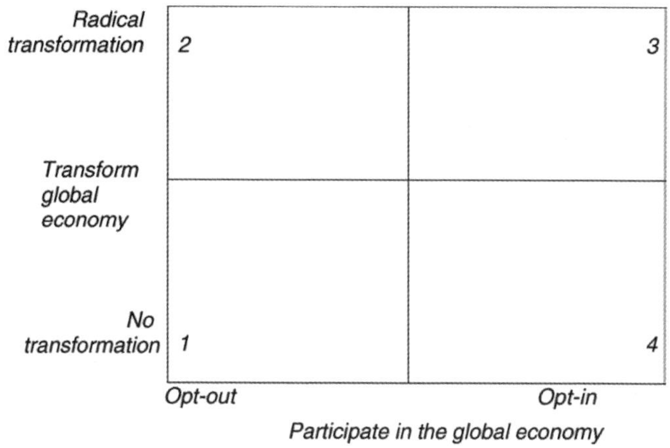

Fig. 1. Community responses to the global economy.
Source: Anderson *et al.*, (2003).

fit with global capitalist expectations (Anderson *et al.*, 2006). One example of the latter is a refusal to accept the sensibility of trading-off environmental damage in exchange for short-term profits.

Aboriginal groups that choose to "opt in" to the global economy are at the beginning of the process, however. To do so successfully, on their own terms or not, they must identify business opportunities and then marshal resources and develop organizations (often as part of a network with other Aboriginal and non-Aboriginal firms) to realize the potential that these opportunities have to satisfy their economic and other development objectives. This is the process of entrepreneurship. Not the entrepreneurship that is narrowly conceived of as a small business operated and/or a new business created by an entrepreneur, but the entrepreneurship that is broadly conceived of as an economy-building process (Anderson *et al.*, 2006), and one that can foster sustainability.

David Newhouse (2003), himself Indigenous, describes an approach to Indigenous people's interaction with the global economy that attempts to balance the Indigenous traditional values and the practice of capitalism. He calls this approach "*red capitalism*" or "capitalism with a red face" (Newhouse, 2003, p. 74). The practice of capitalism within aboriginal society will be affected by their world view and value set incorporating these into any economic activity. These values include kindness, honesty, sharing, strength, bravery, wisdom, and humility, which should be interpreted and translated into community processes, institutions and codes of behavior. Among the factors to be considered is the collectivist orientation of Indigenous society where the needs of the group, whether it is the family, clan or nation, take precedence over the needs of the individual; and the realization that traditional Indigenous people view life as a journey.

It can be safely advanced, hence, that many Indigenous peoples have adopted the fourth variant of the interaction with the global economy, where opting in is coupled with enforcing the traditional values, Indigenous rights, and perception of entrepreneurship and business. Indigenous entrepreneurship has been, in fact, on their own terms. Emphasizing the social and ecological dimensions has demonstrated the will and ability of Indigenous people to enforce their own kind of entrepreneurial activity and perhaps introducing a new model of interaction with the global economy that can inspire similar solutions to engaging in the global economy "on their own terms" for others across the globe.

The three dimensions that reflect the unique characteristics of Indigenous entrepreneurship discussed above readily resonate with the three main components that are advanced as core elements of sustainability. It can be safely advanced that the salience of the social dimension, the ecological underpinnings, and the unique visions of interaction with the global economy that represent salient features of Indigenous entrepreneurship clearly associate it with sustainability and a hopeful future for humanity.

6. Poverty, Destitution, and the Wealth of Community

The situation of poverty, sometimes chronic poverty, is characteristic of marginalized populations worldwide. Manifestations and indicators of poverty vary and include life conditions, income, health and education indicators, among other indicators (Anderson, 2002). We find it quite essential, especially in light of the relationship between entrepreneurship and sustainability to discuss how the concept of poverty became socially constructed post-modernization. We view this link as falling right into the focus of sustainability and ecological intelligence since the way poverty is defined and perceived by people is an important factor in defining the good quality of life, the social elements of success, the measure and value of happiness aspired by all and whether all this is conducive to a kind of living that is sustainable and the role of entrepreneurship in achieving this.

The meaning of poverty has been affected by modernization and its twin mandate of continuous progress and capital accumulation (Bhensdadia and Dana, 2004). From a sustainability perspective, development and modernization created a modern meaning for poverty that is commensurate with the ideals of materialism. While traditional lifestyles enabled people to sustain themselves even with minimal income since community values and nature were there to support the poor, post-modernization poverty means starvation and dire need (Nandy, 2002). Primitive people have few possessions, but they are not poor — they do not feel poor (Shepard, 1998; Sahlins, 1997). In a culture that rests on the ideal of the satisfaction of possessions, only to be again dissatisfied and acquire a new set of newer possessions, poverty represents a relationship between people, a social status and a state of neediness invented by civilization (Sahlins, 1997). Modern poverty as lack of certain kind and amount of possessions is, hence, mistakenly considered synonymous to destitution and deprivation. In doing so, all low-consuming, environment-friendly lifestyles are redefined as being poor and, thus, degrading and unfit for survival in the contemporary, civilized world (Nandy, 2002). When people are considered poor if their circumstances do not meet the levels of income and consumption considered appropriate by the "developed world," what about other broader dimensions of life, other less tangible but more meaningful elements of happiness, and other more sustainability-related criteria?

Poverty should not be interpreted as destitution. Destitution manifests as "an impersonal situation where individualism reigns in the absence of monetary income, one can no longer depend or fall back upon the global commons, either because it is exhausted or depleted, or because it has been taken over by the ubiquitous global market (Nandy, 2002, p. 117)." While community and nature were historically considered a support system for those who are in need, according to modern standards, communities have transformed to a group of individualized citizens who neither expect nor give, and natural global commons are either depleted, privatized or taken over by the global economy.

Community, as mentioned in several parts of this chapter, is at the core of the concept of sustainability. Parajuli notes that a community provides both the "hardware" and "software" for human relationships. As software, the community provides a sense of place, and networks of connection and relations. As hardware, it facilitates arrangements for organizing units such as bio-regions, ecoregions, watersheds, foodsheds, and even food-nations. He outlines the promise of community as the new organizing principle on two fronts. On the one hand, they are the site of remembering, restoring, and regenerating what is lost from the past; whereas, on the other hand, they are a site of constant critique, reform, and transformation of existing relations. He advances that the starting basis for the recovery of community is in preserving and restoring ecosystems and creating zones of self-reliance that contribute to the health of the human members in the communities as well as the communities of all beings (Parajuli, 2009).

If we are to build on the concept of contingency (Anderson, 1997), each community's expectations and goals regarding success, happiness, and a good life are contingent on its tradition, values, practices, and resources and should exceed the current materialistic focus to include the more valuable non-materialistic elements in life. In the context of sustainability, if we all aspire the same material objects to fuel our sense of success, happiness, and good quality of life, the result will be a standardization of dreams and hopes, limiting of human creativity, and decreasing diversity. Ultimately, an inevitable race of humans for material acquisition and accumulation will ensue.

For entrepreneurship to contribute to sustainability, it should be closely congruent to the needs of the community. Entrepreneurship should be an activity that emanates from within the community reflecting its values and practices, is beneficial to the community, and is not imposed. It should also be directly related to the resources and needs of the community and should be ecologically intelligent and sensitive. The features of Indigenous entrepreneurship reflect these means and ends precisely.

Closely relevant to the vital importance and promise of communities based on the contingency perspective that encourages diversified entrepreneurial activities is the concept of cultural commons. Coined by Bowers (2005), the term cultural commons represents

> *"the local traditions of knowledge and patterns of mutual support that enabled communities to be relatively self-sufficient represent what can be called the cultural commons — which are, in turn, dependent upon the environmental commons. While the nature of the cultural and environmental commons are as diverse as the world's cultures and bioregions, what they have in common is that they represent what has not yet been monetized and brought into the industrial approach to markets. In effect, the cultural commons represent sites of resistance to the spread of a money and consumer dependent lifestyle — and thus to the spread of a world monoculture and to the further spread of poverty for those who lack the means to participate in a money economy" (Bowers, 2005).*

Bowers calls for revitalizing of the cultural commons and for reversing the monetization of the cultural commons through industrialization and modernization.

In a similar vein, revitalization of tradition and reviving what qualifies as the cultural commons are equally important elements for Indigenous people and their vision for engaging with the global economy (Newhouse, 2005; Anderson, 1997; Anderson *et al.*, 2005; Hindle *et al.*, 2005). It seems that a unity of thought is emerging among the marginalized of the world, who are equally recognizing their desperate need for finding a form of wealth that they can nourish themselves on their terms and with their own hands and deeds. The helpless feeling of dependence on a wealth being created out there by some other alien source is hampering the genuine development of those people left outside the embrace of modernity and is contributing to their psychic numbing. There are signs that the marginalized are awakening to this reality and they are actively seeking their own solutions to this situation.

A closer look at the collectivity of the different development visions of Indigenous people provides an image of a diversified and rich variety of local patterns that reflect and satisfy the need, resources, and distinctive characteristics of each locality, while emphasizing the importance of partnership and engagement with others in healthful relationships. This image denotes an urge for seeking complementarities among different localities, and inherently values, and even requires networking, dialogue, and active participation for the common good of the collective. In terms of sustainability, such localized approaches that respond to the surrounding ecological and social systems are naturally diverse and are a reflection and manifestation of the bio-diversity of Earth. Hence, it can be safely advanced that the Indigenous pattern of development is conducive to sustainability.

7. Concluding Thoughts

In this chapter, we advanced the idea that entrepreneurship can constitute a variety of solutions for sustainability. Entrepreneurial activity that is driven by social motives, managed in an ecologically sensitive and intelligent way, prioritizes pay back to the community, the needy and the poor first, and strives for the benefit of community and the collective good is one that is conducive to sustainability. In their totality, these different entrepreneurial activities represent a network of sustenance economies. Would it be too much of a stretch to envision that a model of sustenance economies can run as a parallel model to the global economy? Would it be impossible to envision that if it does, it may eventually stand on an equal footing with the current model of the global economy through the choice of the people?

Revitalizing cultural commons is the vital and crucial link in the network of sustenance economies we are envisioning. Revitalizing cultural commons that are specific to each bio-region can be the basis for various possibilities for social entrepreneurial activities that are ecologically sensitive. When each community realizes its own potential and recognizes what resources they can contribute with to help sustain themselves and other communities in a fair and non-competitive network of exchange, this can be conducive to seeding a network of sustenance economies. What we need to do as the first step in this direction? Revitalizing commons in all corners of the globe to truly realize and understand our own potential. Wouldn't this be our first steps toward a more sustainable world?

In conclusion, we are advancing that the promise of Indigenous and local entrepreneurship lies in the creation of a network of various sustenance economies that, together, constitute a parallel to the global economy. Based on this, we are hypothesizing that complementarities, rather than globalization, may well be our way to sustainability since it strikes the balance between local sustenance, conservation, and local well-being, on the one hand; while achieving partnerships, relationships and networking in relating healthfully with the outside world, on the other hand.

References

Allen, T.F.H., Tainter, J.H. and Hoekstra, T.W., 2003. *Supply-side Sustainability.* Columbia University Press, New York.

Anderson, R.B., 2002. Economic development and aboriginal canadians: A case study in economic development. *The Journal of Developmental Entrepreneurship* **7**(1), 45–65.

Anderson, R.B., Camp II, R., Dana, L.P., Honig, B., Nkongolo-Bakenda, J.-M. and Peredo, A.M., 2005. Indigenous land rights in Canada: The foundation for development?' *Int. J. Entrepreneurship and Small Business* **2**(2), 104–133.

Anderson, R., Dana, L. and Dana, T., 2006. Indigenous land rights, entrepreneurship, and economic development in Canada: "Opting-in" to the global economy. *Journal of World Business* **41**, 45–55.

Anderson, R., 1997. Economic development among first nations: A contingency perspective. *Unpublished Dissertation*, University of Saskatchewan, Saskatoon, Canada.

Anderson, R.B., Peredo, A.M., Honing, B., Dana, L.P. and Weir, W., 2007. Indigenous enterprise: The Saskatchewan experience. In: Dana, L.P. and Anderson, R.B. (Eds), *International Handbook of Research on Indigenous Entrepreneurship.* pp. 352–365.

Anderson, R.B., Wingham, D., Gibson, B. and Giberson, R.J., 2007. Indigenous economic development: A tale of two wineries. In: Campbell, G. and Guibert, N. (Eds), *The*

Golden Grape-Wine, Society and Globalization: Multidisciplinary Perspectives on the Wine Industry. Palgrave MacMillan, pp. 201–220.

Bhensdadia, R.R. and Dana, L., 2004. Globalisation and rural poverty. *International Journal of Entrepreneurship and Innovation Management* **4**(5), 458–468.

Bowers, C.A., 2005. *Educating for a Sustainable Future: Mediating between the Commons and Economic Globalization.* Online Article, Retrieved September 15, 2009 from http://www.cabowers.net/CAPress.php.

Bowers, C.A., 2009. Educating students about the political economy of the cultural commons and the nature of sustainable wealth, Chap. 3. In: *Educating for Ecological Intelligence: Practices and Challenges* Online Book. Retrieved September 15, 2009 from http://www.cabowers.net/CAPress.php.

Capra, F., 2005. Speaking nature's language: Principles for sustainability. In: Stone, M. and Barlow, Z. (Eds), Ecological Literacy: Educating Our Children for a Sustainable World. Sierra Club Books, San Francisco, pp. 18–29.

Capra, F., 2004. *The Hidden Connections.* Anchor Books, New York.

Curry, G., 2005. Doing "Business" in Papua New Guinea: The social embeddedness of small business enterprises. *Journal of Small Business and Entrepreneurship* **18**(2), 231–246.

Daly, H.E. and Farley, J., 2004. *Ecological Economics: Principles and Applications.* Island Press, Washington, D.C.

Dana, L.P., Anderson, R.B. and Meis-Mason, A., 2009. A study of the impact of oil & gas development on the dene first nations of the Sahtu (Great Bear Lake) region of the Canadian Northwest Territories (NWT). *Journal of Enterprising Communities: People and Places in the Global Economy* **3**(1), 94–117.

Dzisi, S., 2008. Entrepreneurial activities of Indigenous African women: A case of Ghana. *Journal of Enterprising Communities: People and Places in the Global Economy* **2**(3), 254–264.

Edwards, A., 2005. *The Sustainability Revolution: Portrait of a Paradigm Shift.* New Society Publishers, Gabriola Island, B.C.

Foley, D., 2008. Does culture and social capital impact on the networking attributes of Indigenous entrepreneurs? *Journal of Enterprising Communities: People and Places in the Global Economy* **2**(3), 204–224.

Frederick, H.H., 2008. Introduction to special issue on Indigenous entrepreneurs. *Journal of Enterprising Communities: People and Places in the Global Economy* **2**(3), 185–191.

Golley, F.B., 1998. *A Primer for Environmental Literacy.* Yale University Press, New Haven.

Goodland, R., 1995. The concept of environmental sustainability. *Annual Review of Ecology and Systematics* **26**, 1–24.

Goodland, R. and Daly, H., 1996. Environmental sustainability: Universal and non-negotiable. *Ecological Applications* **6**(4), 1002–1017.

Goulet, D., 1989. Incentives for Development: The Key to Equity. New Horizons Press, New York, NY.

Grilli, E. and Salvatore, D., (Eds.) (1994). *Economic Development Handbook of Comparative Economic Policies*, Vol. 4, Greenwood Press, Westport, CT.

Hindle, K., Anderson, R. Giberson, R. and Kayseas, B., 2005. Relating practice to theory in Indigenous entrepreneurship: A pilot investigation of the Kitsaki partnership portfoli. *American Indian Quarterly* **29**(1 & 2).

Hindle, K. and Lansdowne, M., 2005. Brave spirits on new paths: Toward a globally relevant paradigm of indigenous entrepreneurship research. *Journal of Small Business and Entrepreneurship* **18**(2), 131–41.

Hirst, P., and Zeitlin, J., 1992. Flexible specialization versus post-fordism. In Storper, M. and Scott, A. (Eds.), *Pathways to Industrialization and Regional Development*, Routledge, London.

Hughes, H., 1994. Development policies and development performance. In: Grilli, E. and Salvatore, D. (Eds.), *Economic Development Handbook of Comparative Economic Policies*, Vol. 4. Greenwood Press, Westport, CT.

Kao, R.W.Y., 2010. *Sustainable Economy: Corporate, Social and Environmental Responsibility*. World Scientific, Singapore.

Morrison, K., 2008. Indigenous entrepreneurship in Samoa in the face of neo-colonialism and globalization. *Journal of Enterprising Communities: People and Places in the Global Economy* **2**(3), 240–253.

Nandy, A., 2002. The beautiful, expanding future of poverty: Popular economics as a psychological defense. *International Studies Review* **4**(2), 107–121.

Ndemo, B., 2005. Maasai entrepreneurship and change. *Journal of Small Business and Entrepreneurship* **18**(2), 207–220.

Newhouse, D., 2003. Modern aboriginal economies: Capitalism with a red face. *Journal of Aboriginal Economic Development* **1**(2), 55–61.

Parajuli, P., 2008. At home in Prescott: Ten confluence of streams in my evolution as an interdisciplinary sustainability educator. *Journal of Sustainability Education*. Inaugural Issue, May 2010. Retrieved August 21, 2010 http://www.journalofsustainabilityeducation.org/wordpress/content/at-home-in-prescott-confluence-of-streams-in-my-journey-as-an-interdisciplinary-sustainability-educator_2010_05/

Parajuli, P., 2009. Embedding communities in ecologies: Emergent properties of biological, linguistic and cultural diversities. *Interculture* **157**, 35–54.

Peredo, A., 2003. Emerging strategies against poverty, the road less travelled. *Journal of Management Inquiry* **12**(2), 155–166.

Peredo, A. and Chrisman, J., 2006. Toward a theory of community-based enterprise. *Academy of Management Review* **31**(2), 309–328.

Polanyi, K., 2001. *The Great Transformation*. Beacon Press, Boston.

Sahlins, M., 1997. The original affluent society. In: Rahnema, M. and Bawtree, V. (Eds), *The Post-Development Reader*. Zed Books, New Jersey, pp. 3–21.

Salmon, E., 2000. Kincentric ecology: Indigenous perceptions of the human–nature relationship, *Ecological Applications* **10**(5), 1327–1332.

Satterthwaite, D., 2003. The links between poverty and the environment in urban areas of Africa, Asia and Latin America. *Annals of the American Academy of Political and Social Science* **590**, 73–92.

Skrypietz, I., 2003. *Regulation Theory and the Crisis of Capitalism*. Book review. Capital and Class, Spring.

Sterling, S., 2001. Sustainable education — re-visioning learning and change, Schumacher Society Briefing no. 6, Green Books, Dartington.

Turner, N., Ignace, M. and Ignace, R., 2000. Traditional ecological knowledge and wisdom of aboriginal peoples in British Columbia. *Ecological Applications* **10**(5), 1275–1287.

Whiteman, G., 2004. Why are we talking inside? Reflecting on traditional ecological knowledge (TEK) and management research. *Journal of Management Inquiry* **13**(3), 261–277.

Whiteman, G. and Cooper, W., 2000. Ecological embeddedness. *Academy of Management Journal* **43**(6), 1265–1282.

Zake, P., Kotze, D., Arnab, R., Sihawu, J., Depha, Y., Gaba, M. and Bangani, T. 2005. Poverty alleviation: Intervention, and designing and developing a survey instrument for investigating the effect of entrepreneurial skills on the performance of a small business in South Africa. *Journal of Small Business and Entrepreneurship* **18**(2), 221–230.

Chapter 15

SYSTEM DYNAMICS AND SUSTAINABILITY

Andreas Größler[a] and Jürgen Strohhecker[b]
[a]*Institute for Management Research, Radboud University Nijmegen, P.O. Box 9108, 6500 HK Nijmegen, The Netherlands*
a.groessler@fm.ru.nl
[b]*Frankfurt School of Finance & Management, Sonnemannstraße 9-11, 60314 Frankfurt am Main, Germany*
j.strohhecker@frankfurt-school.de

As the foundation of the field by Forrester in the 1950s, System Dynamics has been applied to sustainability issues of various types. While business was the original research object of Forrester and his group, due to the systems and control theory perspective at its heart, stretching the scope of System Dynamics to other systems happened as a natural next step. We argue that System Dynamics is particularly suitable for sustainability research because it provides (1) a structural theory focusing on feedback, accumulation, delays, and non-linearities within dynamic systems and (2) a methodology to formulate content theories for a broad range of sustainability issues that can rigorously be tested and be analyzed using computer simulations. This statement is substantiated by a selective review of the literature on System-Dynamics-based sustainability research, which shows that the history of the sustainable development movement and System Dynamics is interwoven.

1. Introduction

System Dynamics initially developed from the work of MIT professor Jay W. Forrester, who started the System Dynamics Group at the Sloan School in 1956 and published the seminal article "Industrial Dynamics: A Major Breakthrough for Decision Makers" in Harvard Business Review (1958). System Dynamics is designed for studying complex feedback systems and providing decision support for decision makers by model-based scenario analysis and policy testing. Originally, as the term "industrial dynamics" implies that it was applied to business systems. It aimed at (1) developing "a better intuitive feel for the time-varying behavior of industrial and economic systems," (2) providing "a background showing how the major aspects of a company are related to one another," (3) helping to predict "the future course of an existing organization," and (4) improving "the future prospects of a company" (Forrester, 1958). Very soon, however, it became obvious that these goals and the methodology of industrial dynamics could be extended to all kinds of social systems; applications to systems such as a city (Forrester, 1969) and the whole world (Forrester, 1971) followed; consequentially, the field was renamed System Dynamics.

During these first heydays of System Dynamics, the term and concept of "sustainable development" was not yet commonly used. It was left to the World Commission on Environment and Development in 1987 to put the idea of sustainability into the now famous words: sustainable development ensures that "it meets the need of the present without compromising the ability of future generations to meet their own needs" (World Commission on Environment and Development, 1987). Nevertheless, Forrester's above-stated four goals reveal a relatedness to the sustainability concept. By (1) developing a better intuitive feel for the time-varying behavior of the world system, (2) illustrating how the major aspects of the world are related to one another, (3) helping to predict the future course of the world, and (4) improving the future prospects of the world, a sustainable development of the world seems achievable. Insofar the application of System Dynamics to the issue of providing a sustainable future for humankind appears not merely as a chance event, but rather as corollary.

On the "sustainable development timeline," which is published and regularly updated by the International Institute for Sustainable Development, the probably best-known System Dynamics study "Limits to Growth" (LTG) ranks 15th out of some 100 entries stretching from 1962 to 2009.[1] The non-technical report of a 2-years project initiated by the Club of Rome (Meadows *et al.*, 1972), funded by the Volkswagen Foundation and run by a research group at MIT, was a milestone for both the young field of Systems Dynamics and the even younger movement of sustainable development. A broad, hot, and often emotional debate was triggered off by the rather "simple" and "obvious" statement in the book that "the earth is finite and cannot support exponential growth for very long" (Meadows, 2007).

In the Section 2, we present system dynamics as a structural theory and as a methodology to understand, explain, depict, and influence dynamic systems of any kind. It is discussed that, why System Dynamics is a useful approach when dealing with sustainability issues. In Section 3, we describe and discuss the famous LTG study, a System-Dynamics-based approach to analyze the sustainability of our current and future economic and social conditions. Section 4 summarizes other System Dynamics studies in the field of sustainability; it discusses in more detail with examples from climate change, natural resources development, and sustainable business strategies. This chapter closes with a short summary and an outlook on further potential work in Section 5.

2. Suitability of System Dynamics for Sustainability Research

2.1. *System dynamics as a structural theory and a methodology*

System Dynamics is not only a methodology about how to address and solve complex dynamic problems using a broad variety of mapping, modeling, simulation,

[1] http://www.iisd.org/pdf/2009/sd_timeline_2009.pdf.

and analysis tools. According to Lane ..(1999), System Dynamics can also be seen as a structural theory of dynamic systems, because a set of statements about the core characteristics of dynamic systems lies at its heart. It is based on the main hypotheses that dynamic systems consist of stocks and flows, a feedback structure, accumulation processes, and delays between cause and effect (Forrester, 1961; Forrester, 1968a; Sterman, 2000).

Without engaging in the sprawling philosophical debate on the notion of theories, we pragmatically adopt Popper's (2002a) characterization of scientific theories as universal statements. Seizing on Popper's (2002b, 2003) "searchlight theory of knowledge" and Brunswick's (1952, 1956) "lens model," we understand theories as instruments aiding in the perception of phenomena as well as in their description. Theories are essential for explaining, predicting, and designing (Hempel and Oppenheim, 1948). In this sense "nothing is so practical as a good theory" as Lewin (1945) famously points out.

Theories can be characterized and categorized along a broad variety of criteria. We find the suggestion of Lane (2000) to distinct theories about the content from theories about the structure of systems particularly useful. According to this concept, a structural theory makes principal statements about (1) which elements form a system, (2) how these elements in a system can be configured, and (3) how they causally relate to each other. In contrast, a content theory typically states cause-and-effect relationships between elements of specified systems; one example for such a content theory in the realm of biology would be a set of statements about the impact of increasing water pollution on the fertility of fish.

As a structural theory of dynamic systems, System Dynamics does not offer statements about specific system elements and processes. It does not hypothesize about concrete cause-and-effect relationships. Rather, it states that dynamic systems can be completely described using stock and flow variables. It postulates that dynamic processes function in feedback loops and that the history of systems accumulates over time in state variables. In addition, the accumulated history influences the future development of a system — a process that is often affected by time delays.

Thus, System Dynamics regards feedback loops, accumulation processes, and delays as building blocks of all dynamic systems. The concept of feedback loops emphasizes the fact that actions and results usually are bi-directionally linked: an action leads to an outcome, and the outcome forms the basis for new actions (although the effect of an action may sometimes appear to be remote in time and space). Accumulation processes highlight the notion that the state of a system frequently does not change instantaneously but needs to be gradually developed over time. The focus on delays is dependent on the former two components and makes clear that in each system time lags exist between decisions and actions, or between actions and results: for instance, new information on the state of a system does not immediately lead to new decisions since it needs time to be perceived, considered, and decided on.

Non-linear behavior of systems results from these three building blocks. System Dynamics provides methods for designing formal models and generating their behavior over time, i.e., by simulating the models. In order to do so, a graphical syntax is used, in which flow (rate) and stock (state or level) variables are distinguished. These two types of variables are combined to produce so-called stock and flow diagrams (SFD, Forrester, 1968a; Lane, 2000). The graphical representation of systems using these diagramming techniques proves to be a valuable tool for understanding complex issues. Technically, by the quantification of variables and linkages between variables, a system of differential equations is created that is simulated with the help of numerical algorithms (Sterman, 2000). The validity of the resulting models has to be underpinned by a variety of tests. However, validity can only be judged relative to the purpose of a model (Barlas and Carpenter, 1990).

Another useful classification scheme for theories uses the degree of generality (Popper, 2002b) of its statements to separate "grand theories" from minor theories in the extremes. While a grand theory should be able to cover and explain the complete range of phenomena observed in a system, minor theories focus on a narrow selection of phenomena. Although the development of corroborated grand content theories would be highly desirable and of huge practical relevance, such an endeavor, is rendered unrealistic by the complexity of natural and social systems. Thus, content theories are usually limited in their scope, not claiming to explain complete real-world domains. Nevertheless, grand structural theories exist; for instance, System Dynamics claims to be a theory of the structure of all dynamic systems (Lane, 2000).

Figure 1 summarizes our discussion about the role of System Dynamics in scientific research (as the classification of theories is qualitative only, we depict reasonable ranges only). System Dynamics is a far-reaching theory concerning the structure of dynamic systems. The products of System Dynamics as a methodology (i.e., the models produced) are minor or mid-range theories about a real-world domain. The following section briefly elaborates on this methodological aspect, i.e., on System Dynamics as a method for investigating real processes in a dynamic world, as they are ubiquitous in sustainable development.

2.2. *The process of modeling and simulation with System Dynamics*

Many authors emphasize the importance of the modeling process (and not just that of the resulting model) for gaining insights into the problem (Lane, 1995; Sterman, 1988a). The reason for this emphasis on the process is that System Dynamics models are usually descriptive and not prescriptive, in the sense that they do not provide an "optimal" solution, which has two implications. First, insights from the model can only indirectly be deduced by formulating its equations and running different simulations (in the sense of quantitative scenario analyses). Usually, simulation models do not produce the right answer to solve a problem as a direct output.

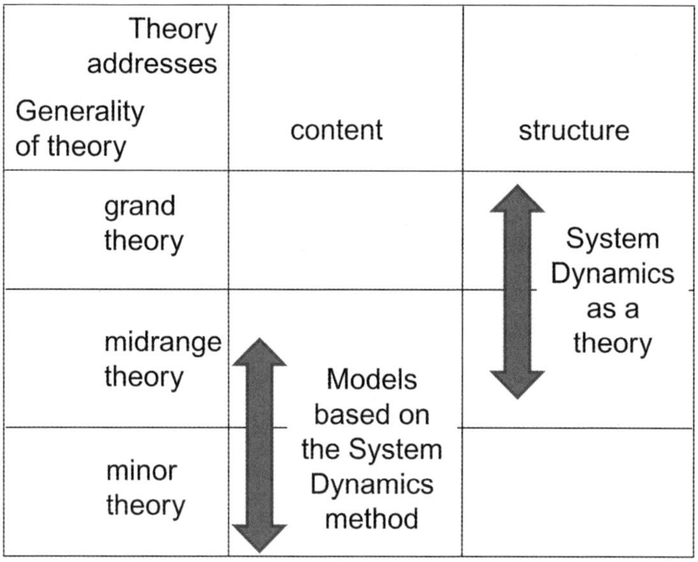

Fig. 1. System Dynamics as theory and methodology.

Second, while being a modeling endeavor, System Dynamics research is substantially empirical: regularly, it does not involve the development of ideal models, but rather the depiction of reality as precisely as possible. Based on a valid picture of what is the case in reality, attempts are made to gain insight into how the real system can be improved. As such, System Dynamics models contain all variables that are deemed important to understand a system, no matter if these variables can be easily measured. The rationale behind this approach is that even a rough estimate of values and linkages is preferable to "leaving out" important aspects, just because they are difficult to measure and/or to operationalize.

The stylized process of System Dynamics modeling and simulation is composed of six steps (Forrester, 1994). Figure 2 depicts this process in graphical form. However, note that the process is not purely sequential as it might occur from the figure: there are various possibilities for jumps back to earlier steps in the process.

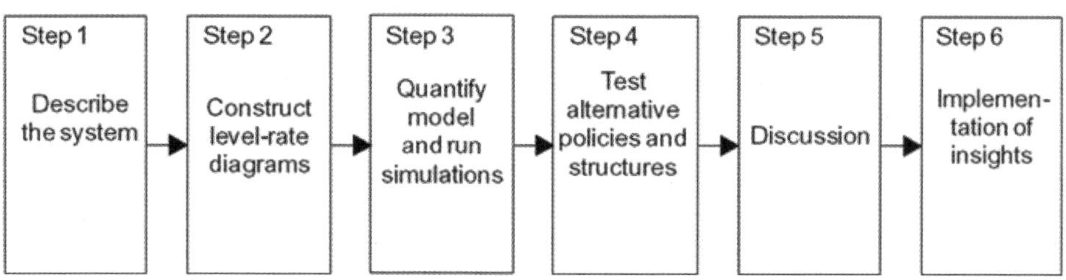

Fig. 2. Stylized System Dynamics process (iterations not depicted).

In the first step, the problem to be modeled is defined and key variables are identified. An important question to answer in this phase concerns the system boundary, i.e., it is determined, which factors should be included and which should not. As a rule of thumb, important feedback loops should not be cut in System Dynamics models, since the assumption is that feedback loops are significantly influencing the behavior of the system. In the second phase, a level-rate diagram is built that represents the system structure as identified before. This diagram is formalized in the next (3rd) step, i.e., mathematical functions are defined that determine how values of a variable are calculated based on the values of other variables. Furthermore, values for constants and initial parameters are laid down. Based on this quantification process, simulation runs are conducted to generate the behavior over time of the model. In the following step (4), additional policies and structures are tested, either to try-out other conceivable givens or to find alternatives for improvement of the system. The fifth step acts as a reflection point: insights from the modeling and simulation activities are discussed and tested for their validity. In the last step, model-based solutions are implemented in order to improve the real-world system. Software packages (such as Vensim, Powersim, or Stella/iThink) support this process of conducting a System Dynamics study.

2.3. *System Dynamics' contribution to sustainable development*

As we have seen in the previous two sub-sections, System Dynamics can contribute to sustainable development in two ways: first, it offers a structural theory that is useful to appreciate and understand sustainability issues; second, it provides a method to analyze and improve those issues. Thus, by applying the "lens" of System Dynamics as a structural theory, we can use System Dynamics methodology to investigate issues in and formulate content theories for sustainable development.

The theoretical contribution to sustainable development must be understood in a way that System Dynamics offers a mindset to better understand dynamic systems and their effects. Since most of the issues sustainable development has to deal with are dynamic by nature, System Dynamics is a framework to address these issues. Many issues in sustainable development can be attributed to feedback relationships, accumulation of material and immaterial goods, or delays between action and reaction within a system. For instance, the actions of a production company that lead to a polluted environment will eventually — maybe delayed by decades — have an effect back on the company:

- its reputation decreases (leading to lower sales and fewer employees wanting to work for them);
- environmental legislations will become stricter (increasing production costs);
- taxes will need to increase to cope with the pollution (again, increasing production costs);
- maybe even the population around the company will move away because of the polluted area (resulting in lower sales on the home market).

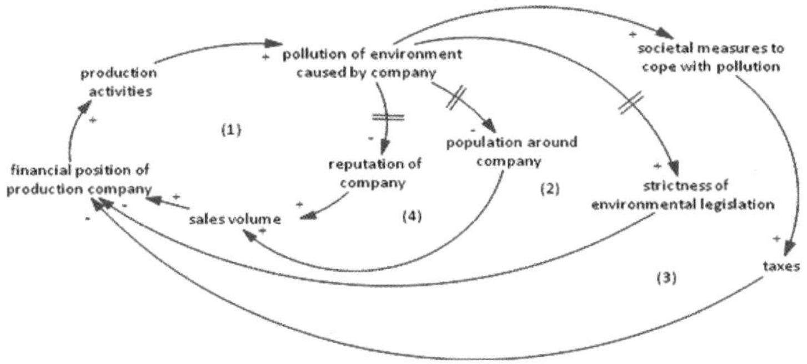

Fig. 3. Exemplary causal-loop diagram.

The theoretically identified characteristics of dynamic systems can be analyzed using System Dynamics methodology. Since — as we have seen in the previous paragraphs — most sustainable development issues are inherently dynamic, System Dynamics can be used to study such issues. In the example given above, the relations that are mentioned could be mapped in a causal-loop diagram (CLD; Maani and Cavana 2007); they could further be quantified in an SFD and simulation experiments could be conducted. The simulations can be used to understand the system's reaction to the company policies and can help to find improvement points for a more sustainable development.

Figure 3 shows a first step toward a formal representation of the production company example in System Dynamics terms. In the CLD, one can see the variables of the system as described above and their interrelation. Note that the linkages between the variables establish feedback loops (numbered as above) and that some linkages are marked with two short stripes indicating delays. The plus and minus signs indicate effects going into the same and the opposite way, respectively. For instance, pollution of environment caused by company has an opposite and delayed effect on reputation of company: if pollution goes up, with some time lag the reputation will go down. While this example is definitely simplified compared with real System Dynamics projects, lacks quantification in its current state, and is depicted here for illustrative purposes only, it should give an idea of how problems are treated when using System Dynamics.

3. The "LTG" Trilogy and Debate

3.1. *A note on the history of the Limits to Growth (LTG) study*[2]

The Club of Rome that initiated and received the LTG study as their first report in 1972 was a rather small group of international professionals from various fields. Founded in 1968 by the Italian industrialist Aurelio Peccei and Scottish scientist Alexander King, the Club of Rome was (and is still) concerned with the definition

[2]This section is mainly bsed on Meadows (2007) and Meadows *et al.* (1972, 2004).

and solution of what it called the "world problematique" later described as "the complex set of the most crucial problems — political, social, economic, technological, environmental, psychological and cultural — facing humanity" (Visser, 2007). In search for order among the list of 66 critical problems that had been compiled by 1970, the Club addressed Forrester, who was known to work on a computer-based approach to solve complex feedback problems. On the return flight from the first meeting in Switzerland, Forrester sketched the first world model on a simple napkin. After some refinements, this model became World1. It started in 1900, showed growth until 1975 and then collapsed.

The negotiations between the Club of Rome and the MIT System Dynamics Group were successful. The Club of Rome agreed to start its Project on the "Predicament of Mankind" using System Dynamics. The Volkswagen Foundation provided the funding, and a team of 17 modelers from seven nations was formed. While Forrester himself continued to improve World1, which lead to the publication of World2 in his book World Dynamics in 1971, the MIT team worked to develop World3. Although similar in their basic structure with five sectors — population, industry, agriculture, natural resources, and the environment — and their basic dynamics, the objective for World3 was to advance the sector's internal structure and strengthen the empirical foundations. Or, as Donella Meadows (2007, p. 192) put it: "Our work made the model more detailed and more consistent with established concepts in demography, economics, geology, ecology, and agriculture."

Early in 1972, a popular version of the results achieved by the MIT team was compiled, handed over to the Club of Rome executive committee, published as a 200-pages paperback (Meadows *et al.*, 1972) and spread among selected policymakers. The book caused a furor. It received an unexpected intense media attention. Overall, it sold more than 12 million copies in 37 languages. Two years later the detailed technical report — Dynamics of Growth in a Finite World — was published (Meadows *et al.*, 1974). It detailed the research and provided the documentation for the World3 model equation by equation.

In 1991, Donella Meadows, Dennis Meadows, and Jørgen Randers conducted a follow-up study based on the World3/91 model. This model introduced only marginal changes. The Dynamo mainframe version was converted to a Stella personal computer model. Seven model constants and table functions were changed to reflect updated data; the way in which technology affected the model coefficients was modernized. "Beyond the Limits," published in 1992, found that "many resource and pollution flows had grown beyond their sustainable limits" (Meadows *et al.*, 1992). In 2004, the authors presented the 30-year update of the LTG study. This study is based on an again only slightly altered model (World3/03) and contains more data that support the hypothesis of the 1992 version that the world is in "overshoot mode" (Meadows *et al.*, 2004). The authors show that the options for implementing a sustainable society have narrowed and actions have to be taken without much delay when overshoot and collapse are to be avoided.

In addition, the Club of Rome had another study conducted with the same research question as for the LTG but another methodology employed (Mesarovic and Pestel, 1974). This work was based on a far more elaborate and regionalized model. It distinguished 10 world regions and consisted of 200,000 equations (compared with 1000 in the LTG model). The basic message of this second report to the Club of Rome was amazingly alike, though. The accelerating growth of industrial production had to be drastically reduced in the rich regions to allow the poor regions some development.

3.2. *Key insights of the study*

The original World3 model consists of roughly 1000 variables that are located in one of five sectors: natural resources, population, environmental pollution, industry, and agriculture. The model is openly available and can be used to generate different scenario runs (for instance, it comes as World3/03 version with the simulation software Vensim, www.vensim.com). Figure 4 shows scenario 1, which has been described in the book "Limits to Growth — The 30-Year Update" (Meadows *et al.*, 2004) to serve as a reference point.[3] It assumes a "business as usual" situation, for instance,

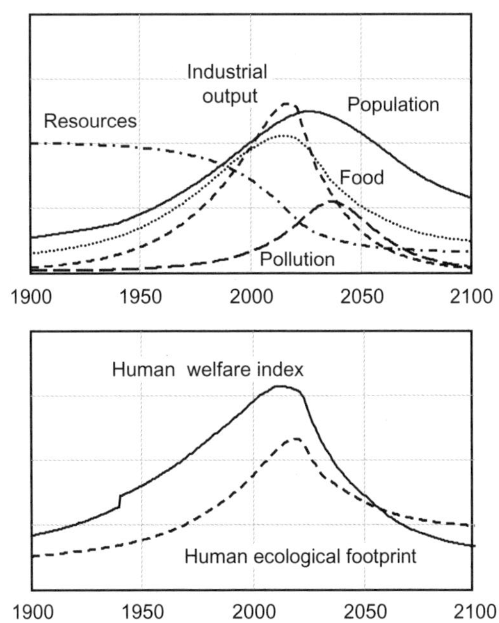

Fig. 4. Scenario 1 as the reference point of the LTG study.

[3]For reasons of simplicity units and scales have been omitted. However, all graphs shown in Figs. 4, 5, and 6 have identical scales to allow comparison. The scenarios were created using the Vensim model world3_03_scenarios.vmf that ships with Vensim 5.8 or newer.

no major alterations are made to the policies employed concerning economic growth. Population, industrial output, and human welfare increase until peaking around the years 2020–2025. From then on, bust follows boom due to resource depletion and pollution initiated by high growth in industry and agriculture that — after a few years delay — affect population.

The structural explanation for this behavior is a fundamental insight: infinite growth cannot exist when resources are limited (Senge, 1990). More and more capital is necessary to solve problems with resource acquisition and environmental pollution. Finally, industrial production comes to a halt and starts shrinking; simultaneously, the production of food and services decreases, which in the end also causes a decline in population.

Assuming that initially twice as much non-renewable resources are available compared with scenario 1, do not alter the boom, and collapse pattern. While industrial output can grow some 20 years longer, the human ecological footprint peaks near 2050 at almost double the level shown in Fig. 4. In this so-called scenario 2, population reaches a maximum of 8 billion and then declines because of food shortages and the negative impact of pollution on health (Meadows *et al.*, 2004).

While many public discussions focused on scenarios 1 and 2 condemning them as prophecies of doom, the positive statements and outlooks in the LTG trilogy were often perceived to a lesser extent. All three books, even the latest one, emphasize that transition to a sustainable society would still be possible and collapse could be avoided, as Fig. 5 indicates. For this, rigorous policies have to be implemented,

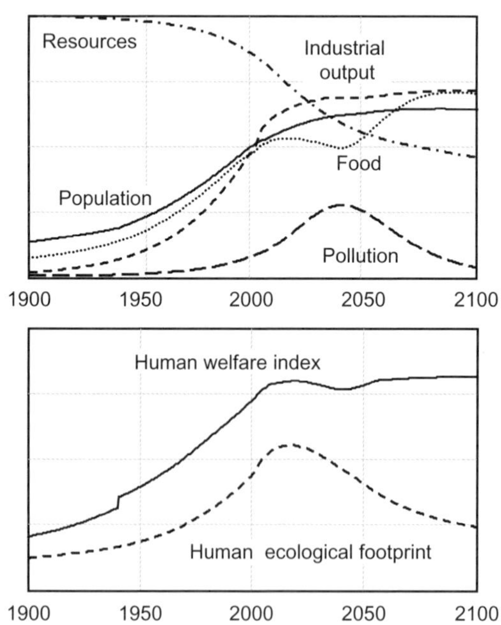

Fig. 5. Scenario 9 — A sustainable future.

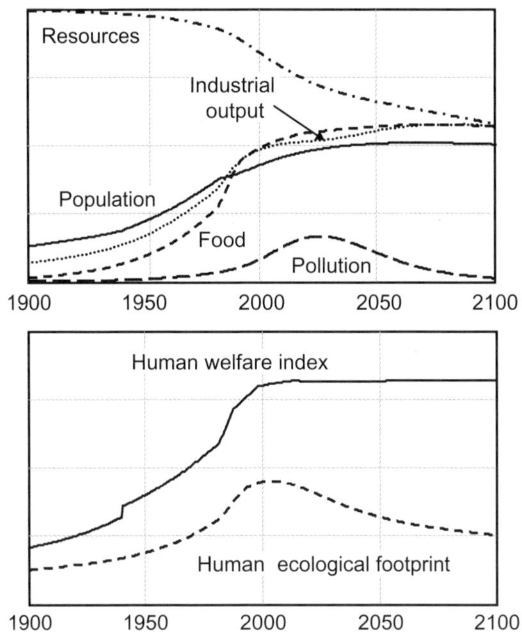

Fig. 6. The difference of 20 years earlier implementation of all scenario 9 policies.

though. Starting in 2002 the desired family size is fixed at two children per couple, perfect birth control is in place and modest limits for industrial production are appointed. Additionally, from 2002 onwards technologies for increasing the efficiency of resource use, decreasing the pollution per industrial output unit, controlling land erosion, and raising land yields are developed and implemented with an average time delay of 20 years at some capital costs.

Although the society in scenario 9 would succeed in avoiding the collapse shown in Fig. 4 and establish a sustainable equilibrium with a population of nearly 8 billion people, a moderate food crisis around 2040, causing the human welfare index to drop, cannot be avoided. Growth had gone on for too long to allow a smooth transition. If a smooth behavior had been intended, the policies from scenario 9 had to be made effective 20 years earlier (Fig. 6).

In addition to the three scenarios shown and described above, the model allows to explore a plethora of other scenarios. However, even when parameters are changed by quantities, the behavior modes of the world system vary in principal between s-shaped growth eventually reaching a sustainable equilibrium and boom and bust with changing peak times. The reason for the similarity of simulation results for various parameter settings lies in the dominance of the system structure over the actual values of the parameters. This explanation is supported by the fact that, for instance, the reference scenario 1 does not change substantially between the three editions of the LTG study, although several parameters and table functions have been updated concerning the latest information at hand in the newer versions.

3.3. *A short synopsis of a hot debate*

The effect of LTG on public and scientific discussions can hardly be overestimated and continuous until today. With some justification can be argued that the development of societal and political movements to protect the environment and international treaties to do so (e.g., the Kyoto protocol) was initiated or amplified by the study or the discussions evoked by it.

Nevertheless, the fundamental results of LTG did provoke criticism of various kinds. Early on such critical accounts are, for instance, Cole *et al.* (1973), Nordhaus (1973), and Beckerman (1974). However, even today, writers are directly linking to and criticizing the Club-of-Rome study, for example, Popper *et al.* (2005), Lomborg (2001), and Stott (2000). The criticism refers to various aspects of LTG:

1. System Dynamics as a method and its usage for scenario analyses (including a principally critical view on modeling and simulation).
2. Assumptions in the model and its boundary (in particular, what has not been included in the model).
3. The results of the study.
4. The consequences and discussions based on these results.
5. The context of the study, for instance, the political and ideological mindset of the modeling group, their assumptions about economic systems.

This chapter argues that System Dynamics is an appropriate tool to model dynamic systems of any kind; thus, it can be used to model the world in total (this statement does not deny the fact that this is of course very complicated). System Dynamics has proven its suitability and usefulness in a wide area of applications. With regard to the general mistrust in modeling and simulation, it seems appropriate to state that modeling and simulation are accepted scientific methods. Therefore, neglecting their principal usefulness is comparable to negating the scientific method as such.

Related to this point is also a criticism that denies the benefit of using System Dynamics modeling over plain mathematics. Indeed, from a mathematical point of view, System Dynamics models are sets of differential equations that are solved numerically. However, the theoretical perspective of System Dynamics as stated above (focus on feedback loops, delays, accumulation, and non-linearities) as well as the mapping techniques provided by System Dynamics (Lane, 2008), make constructing a global model more efficient and effective.

A skeptical perspective regarding the assumptions used in the model is certainly worthwhile, as also Meadows *et al.* (1982) confirm for some of the model's operationalizations. Certainly, a model of the whole world needs to be highly aggregated and as such is a compromise between applicability and complexity of the structures represented. It must be discussed that whether and where this abstraction process (which per definition goes along with model building) has validity. System Dynamics supports this process by making the operationalizations used transparent and open for scrutiny.

As the results of the simulations are generated directly from model structure, criticizing them as such is not adequate. Usually, System Dynamics models are deterministic (as is World3); hence, the scenarios result from the model structure — this structure can be criticized (see the 2nd item) but not its outcomes.

Turner (2008) compared three scenarios from the original 1972 study to environmental data for 1970–2000. He found a surprising match for real data and the behavior shown in the basis scenario of LTG," in particular, if the high level of abstraction within the model and simulation is taken into account. However, more optimistic scenarios showed lower congruency with real-world data. Thus, given the fact that LTG predicts major disruptions not before the middle of the 21st century, it seems too premature to disregard the study as empirically wrong.

Most potential consequences of LTG were not presented in the actual book but put forward from other parties. In the course of the discussion that followed the publication, many mis-interpretations and mis-perceptions can be found of what the study actually says, how the results must be interpreted, and where the limitation of such a study is. For example, information in the book was interpreted in a way that oil reserves would run out in 1992, which obviously did not happen. While this is not really stated in the book, LTG seem to have failed to deliver an understandable and clear message of its results.

This criticism stems from two political camps: left and right. The left camp argues that LTG concentrates too much on physical/materialistic limitations and neglects perhaps more important restrictions of social or political nature. The right camp argues that market mechanisms and technological innovations are capable of mitigating negative effects of human endeavor; such structural changes were not represented in LTG.

4. Sustainability Research with System Dynamics

After reviewing the still most influential contribution of System Dynamics to the sustainability debate (LTG), we are presenting three more up-to-date works that use System Dynamics to study sustainability issues: Climate Change research, eco-system research, and sustainable businesses development. As will be clear in the following sub-sections, the application of System Dynamics in the three case studies has different objectives: (i) the climate case is about informing laypersons about the effects of CO_2 emissions; (ii) the eco-system study aims at consensus building between stakeholders; (iii) the sustainability in business section reports on using System Dynamics by managerial decision makers in order to design sustainable policies for their companies. By concentrating on these newer studies, we do not ignore the important work of other authors using System Dynamics in the field of sustainability, most notably Ford (1978, 1996, 2009), Bossel (1986, 1998, 1999), Fiddaman (2002, 2007), and many others.

4.1. *System-dynamics-based climate change research*

Drawing on the tradition of the LTG project at MIT, recently a new program was launched by the Sustainability Institute involving a broad range of people committed to using innovative approaches to address climate change. The mission of the program *Climate Interactive* is "to develop, extend, and distribute powerful, open-innovation climate simulations for the world to share" (Sustainability Institute, 2010a). As a first step toward accomplishing this mission, the Climate-Rapid Overview and Decision Support Simulator (C-ROADS) was developed advancing and combining previous System Dynamics models and simulators built at MIT (Fiddaman *et al.*, 2009). The purpose of C-ROADS is to "improve public and decision-maker understanding of the long-term implications of possible greenhouse gas emissions futures" (Fiddaman *et al.*, 2009). This purpose shall be achieved by providing a scientifically rigorous computer simulation of the impacts of greenhouse gas emissions and land use on global temperature and sea level rise, which is nevertheless fast and user-friendly to allow frequent iterations and rapid learning.

While a strong scientific consensus exists on the causes and consequences of climate change, the mental models of the public seem to be widely different, blurred, and frequently erroneous. Although a majority at least in the western world is aware of the phenomenon of global warming and prepared to support actions, scientists call for implementation of measures at a decidedly faster pace than the public. One reason for this mismatch in the perceived urgency of actions could be seen in the lack of understanding how the various proposals und pledges affect the climate trajectory and sea level rise. As the world climate is a non-linear, dynamically complex feedback system, it is hardly surprising that most people are overcharged when attempting to foresee the short- and long-term consequences of the different proposals (Sterman, 1989). The problem is even deeper. Recent research shows that even highly educated adults have deficits in understanding the behavior of simple systems consisting of just one stock and one flow, that is, they fail to comprehend the concept of accumulation (Sweeney and Sterman, 2000; Cronin *et al.*, 2009). Because of these flaws, intuitive predictions of the response of the global climate system to emissions' cuts are often wrong; the degree of emissions' reductions needed to stabilize atmospheric carbon dioxide levels is underestimated, as is the lag time between changes in emissions and changes in global mean temperature.

Computer simulations are considered to accelerate the learning process and to deepen the insight in dynamically complex system by providing fast feedback from experimentation with a virtual world (Morecroft and Sterman, 1994; Frensch and Funke, 1995). The existing climate simulations such as MAGICC,[4] GCAM,[5] AIM[6] are based on large, complex, and highly disaggregate models. They are relatively

[4]http://www.cgd.ucar.edu/cas/wigley/magicc/
[5]http://www.globalchange.umd.edu/models/gcam/
[6]http://www-iam.nies.go.jp/aim/index.htm

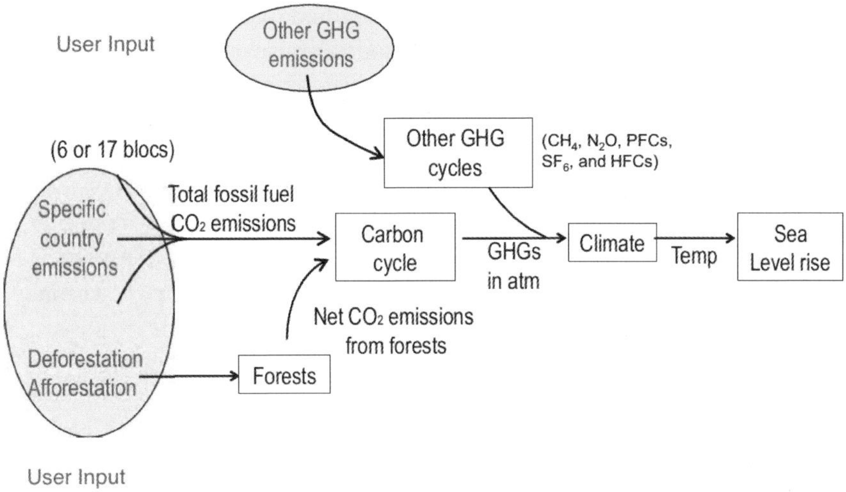

Fig. 7. C-ROADS model sector diagram (Fiddaman *et al.*, 2009).

slow to run and difficult to use for non-experts. C-ROADS and especially the 3-region, simplified version C-Learn are easy to use and fast to run, allowing to test many what-if questions and explore their short- and long-term consequences. By this, they aim at capturing the key insights of larger models and making them available for "rapid policy experimentation" (Fiddaman *et al.*, 2009). The C-ROADS simulator is not seen as a substitute for the larger and more detailed models.

The basic structure of the C-ROADS simulator is outlined in Fig. 7. Total fossil fuel carbon dioxide emissions' scenarios are aggregated from the emissions' scenarios of groups of countries, which can be specified by the user. In addition to, user input is required for land use scenarios implying additional release or uptake of CO_2. Both fossil fuel CO_2 emissions and net CO_2 emissions from forests form the input flows to the carbon cycle sector of the model. For other greenhouse gases (CH_4, N_2O, PFCs, SF_6, and HFCs), explicit cycles are also modeled requiring user input scenarios. The carbon and other greenhouse gas cycles together determine the concentration of greenhouse gases in the atmosphere, the global temperature, and the sea level rise.

First tests of the C-ROADS simulators usefulness were conducted within the course of the UNFCCC negotiations culminating in the COP-15 conference in Copenhagen, Denmark, in December 2009. Sawin *et al.* (2009) report how C-ROADS was used to analyze the expected long-term impacts on the climate of proposals that were put forth by national and regional governments. While the Copenhagen Accord calls for "deep cuts in global emissions ... so as to hold the increase in global temperature below 2 degrees Celsius" (UNFCCC, 2009, p. 5), simulations of the C-ROADS model indicate that this "requires global greenhouse gas emissions to peak by 2020 and then fall 50% below 1990 levels by 2050" (Sustainability Institute, 2010b). This amounts to a cut of approximately 60% below current emissions, which is far beyond the sum of the commitments offered by the individual nations. Confirmed reduction proposals at the time of the Copenhagen conference would raise expected global mean temperature by 3.9 °C — almost double the rise that was agreed on at the conference.

By providing models and simulators like C-ROADS, System Dynamics can help people to develop a better understanding of the causes and consequences of climate change. While more research is needed to increase the scientific knowledge on climate change and more technical innovations may help to achieve the reduction goals, this seems to be not sufficient though. Sterman (2008, p. 533) calls for turning "our attention to the dynamics of social and political change."

4.2. *An example of a System-Dynamics-based study of a natural system*

As an example of a recent System Dynamics project in the area of sustainability of a limited eco-system, we take Videira *et al.* (2009), who report on the usage of a participatory modeling approach for water supply and river basin planning. The modeling project was conducted in Portugal, in the region of the Baixo Guadiana River area. Modeling work actively involved different stakeholder groups.

The dynamic sustainability issue described in the paper is about water management. More precisely, the fact that "[s]ustainable water management is hindered by the lack of adequate water institutions, fragmented institutional structures, upstream and downstream conflicting interests regarding rights and access to water, diversion of public resources for private gain and the unpredictability in the application of laws, regulations and licensing practices" (Videira *et al.* 2009, p. 966). As an issue characterized by high complexity and uncertainty, non-linear dynamics, and significant consequences of any policies taken, participatory modeling following System Dynamics was identified as a method to deal with these characteristics. The aims of the project were to construct CLDs of the most important relationships, generate a simulation model of the river basin, and to agree on objectives and measures for sustainable management of the region in terms of its water resources.

The modeling project was conducted as a sequence of three alternating participatory workshops with stakeholders and "behind the scene" works that, for instance, composed of refining and quantifying models, writing scenario plans, and planning for the implementation of results. With this approach, Videira *et al.* managed to achieve a satisfactory level of commitment and consensus within the group of stakeholders. In particular, participants remarked that the process of building a model and running simulations within a group of stakeholders helped to structure discussions and allowed for a fair process of finding solutions to the problems. Tangible outcomes of the project are, for example, the development of land use plans, the formulation of a policy to reverse population decline, and measures to balance the economic and ecologic requirements for the river basin. These results were achieved despite the fact that the authors report on some problems maintaining the initial group composition.

In the light of this paper, it needs to be re-emphasized that Videira *et al.* (2009) not so much describe a successful sustainable development project based on novel insights from modeling and simulation. Rather, modeling and simulation provided a framework for discussion between stakeholders having differing and conflicting goals (Vennix, 1996; van den Belt, 2004). The structure induced to the process by System Dynamics modeling was perceived to be useful in providing a just, open and targeted platform for discussing the various topics connected to the sustainable management of a water resource.

4.3. *Back to the roots: Sustainable business development*

Sustainable business development can at least have a twofold meaning: First, it can be seen as a request to develop business in an environmentally and ecologically sound way, which may mean, for example, minimizing the use of finite resources, alleviate pollution of all kinds, and behave as eco-sensitive as the business model allows. Second, it can also mean to accept and obey the shareholder's "performance imperative," which dictates to continuously improve "performance over time" (Warren, 2007). Under such a regime not just business performance has to be sustained, the performance growth rate needs to be maintained — meaning that business performance grows exponentially.

As in a finite world growth of any real system has at some point to start leveling off and eventually come to a halt, it would be more realistic for businesses to set themselves s-shaped growth patterns as objectives — at least in the long run. By this, the danger of overexploiting and eroding resources critical for the business model, which results in an ever-present, yet unfavorable boom and bust pattern, could be minimized (Meadows, 2008); exponential growth in the early phases of a company's life cycle is hereby not interdicted.

From the very beginning, System Dynamics was applied to a broad range of business problems. As gaining and sustaining competitive advantage for bringing back or maintaining growth was one of the most frequent issues addressed, the following three references should only be seen as examples. An early case study was published by Forrester (1968b). In "Market Growth as Influenced by Capital Investment" he was pursuing the goal to explain how an "overly cautious capital investment policy" can cause sales stagnation or decline — even in situations where market limits are far afield. The impressive rise and rather sudden fall of People Express airline provided another exemplary case that was taken up by System Dynamics research (Sterman, 1988b; Morecroft, 2007). More recently, Arquitt *et al.* (2005) investigate boom and bust behavior in the shrimp aquaculture industry and suggest improved policies to shift the system toward sustainability.

For connecting System Dynamics and the resource-based view of strategy (Barney, 2007) in striving for an coherent approach to explain and improve business performance over time, Warren (2002, 2007) has to be given significant credit. He promotes the need for rigorous causal explanations of performance and suggests that these have to be searched for predominantly within an organization. Warren

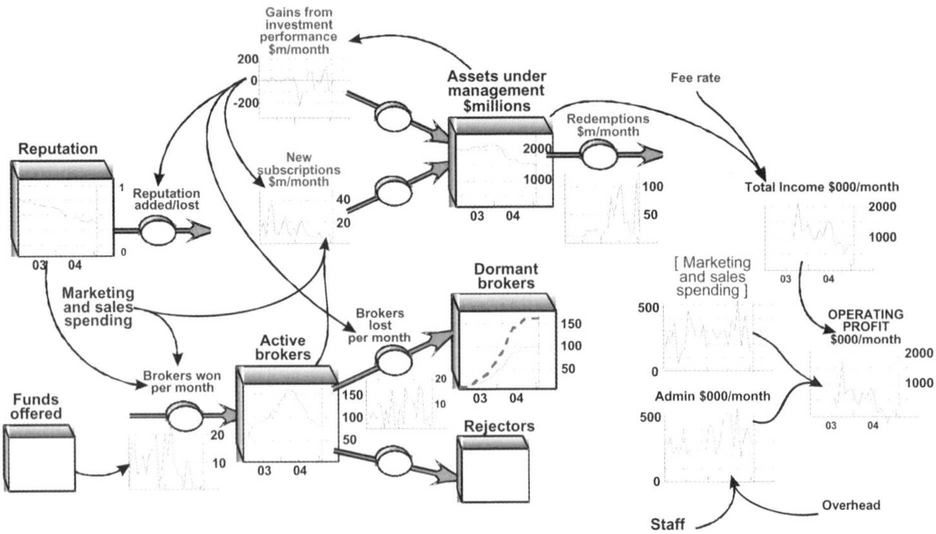

Fig. 8. Resource system for a fund management business as an example.[8]

argues that resources drive performance as they are influencing demand and supply, revenue and costs. Therefore, in order to explain why an organizations performance has shown a specific time path, not only the so-called VRIO[7] resources have to be identified, but all tangible and intangible resources and capabilities that directly or indirectly influence performance. Warren calls for mapping the resource system as a whole to provide an adequate explanation for the observed performance path.

When striving for sustained competitive advantage, managers have to learn how to build and maintain the quantity of each resource in the resource system. As resources are represented as stocks in System Dynamics terminology, they change only by their rates over time and never instantly. To understand that resources accumulate and deplete is critical for managers to identify the appropriate levers for gaining and sustaining competitive advantage.

As the resource system in Fig. 8 for a fund management business shows, the resources *active brokers* or *assets under management* cannot be changed directly. For example, when *active brokers* should be changed the rates *brokers won per month* and *brokers lost per month* have to be adjusted appropriately.

As the example in Fig. 8 illustrates that the eroding operating profit could be traced back to decreasing *assets under management* caused by degrading *reputation* and a peak of *active brokers*. After having found a valid explanation for the problem, the resource system also helps to identify the effective levers to first stabilize and to improve the performance. In the first instance, the erosion in *reputation* had

[7]VRIO stands for valuable, rare, hard to imitate, and supported by other organizational procedures (Barney, 2007).
[8]Figure 8 is reproduced with the permission of the copyright holder Kim Warren.

to be stopped — for the main part by safe investment decisions leading to stable investment performance and gradually rebuilding *reputation*. Only then, intensifying marketing efforts for winning new brokers holds promise. As a third step, launching new investment funds targeting at a different segment of investors may be used to attract additional brokers and re-activate *dormant brokers*.

A System Dynamics model covering the resource system shown in Fig. 8 allows management to test the turn-around strategy outlined above. It can help to explore modifications and alternate tactical and operational decisions based on this strategy. Furthermore, it would allow the management to explore the sensitivity of important assumptions. As System Dynamics models aim for covering all major feedback loops, the danger of ignoring undesirable side effects within the strategy evaluation and selection process is greatly reduced. Thus, the likelihood of really achieving sustained competitive advantage is increased.

5. Conclusion

This chapter demonstrates that System Dynamics is valuable to investigate issues of sustainable development for two reasons. First, as a structural theory System Dynamics offers a lens to understand dynamic phenomena characterized by feedback loops, accumulation effects, delays, and non-linearities — as is frequently the case in sustainable development. Second, System Dynamics as a method provides tools to analyze sustainable development issues due to its diagramming and simulation capabilities. Many applications of System Dynamics in the broader area of sustainability can be identified: starting with the famous and often discussed LTG study to more recent studies of sustainability issues in climate change, natural reserves conservation and management, and sustainable business development.

Of course, having a better cognitive understanding of sustainable development issues — for instance, provided by System Dynamics — does not solve immediately or necessarily such issues (Vennix, 1999; Größler, 2007). As has been demonstrated in the water management case (Section 4.2), System Dynamics can also be used to support the process of consensus finding and commitment. For this purpose, the cognitive, analytical endeavor of modeling and simulation must be augmented by facilitation techniques and methods to come to sustainable decisions in socially conflicting situations.

References

Arquitt, S., Honggang, X. and Johnstone, R., 2005. A system dynamics analysis of boom and bust in the shrimp aquaculture industry. *System Dynamics Review* **21**(4), 305–324.

Barlas, Y. and Carpenter, S., 1990. Philosophical roots of model validation: Two paradigms. *System Dynamics Review* **6**(2), 148–166.

Barney, J.B., 2007. *Gaining and Sustaining Competitive Advantage*. Pearson, Upper Saddle River.

Beckerman, W., 1974. *In Defence of Economic Growth*. Cape, London.

Bossel, H., 1986. Dynamics of forest dieback: Systems analysis and simulation. *Ecological Modelling* **34**(3/4), 259–288.

Bossel, H., 1998. *Earth at a Crossroads: Paths to a Sustainable Future.* Cambridge University, Cambridge.

Bossel, H., 1999. *Indicators for Sustainable Development: Theory, Method, Applications: A Report to the Balaton Group.* International Institute for Sustainable Development, Winipeg.

Brunswik, E., 1952. *The Conceptual Framework of Psychology.* University of Chicago, Chicago.

Brunswik, E., 1956. *Perception and the Representative Design of Psychological Experiments.* University of California, Berkeley & Los Angeles.

Cole, H.S.D., Freeman, C., Jahoda, M. and Pavitt, K.L.R., 1973. *Thinking about the Future: a Critique of the Limits to Growth.* Chatto & Windus, London.

Cronin, M.A., Gonzalez, C. and Sterman, J.D., 2009. Why don't well-educated adults understand accumulation? A challenge to researchers, educators, and citizens. *Organizational Behavior and Human Decision Processes* **108**(1), 116–130.

Fiddaman, T.S., 2002. Exploring policy options with a behavioral climate — economy model. *System Dynamics Review* **18**(2), 243–267.

Fiddaman, T., 2007. Dynamics of climate policy. *System Dynamics Review* **23**(1), 21–34.

Fiddaman, T., Siegel, L.S., Sawin, E., Jones, A.P. and Sterman, J.D., 2009. *C-ROADS Simulator Reference Guide.* Available at http://climateinteractive.org/simulations/C-ROADS/technical/technical-reference/C-ROADS%20reference%20guide%20v27.pdf/view (Accessed 31 March 2010).

Ford, A., 1978. Simulating the impact of regulatory changes on electric utilities. *Simulation* **30**(1), 1–8.

Ford, A., 1996. Testing the snake river explorer. *System Dynamics Review* **12**(4), 305–329.

Ford, A., 2009. *Modeling the Environment.* Island, Washington.

Forrester, J.W., 1958. Industrial dynamics: A major breakthrough for decision makers. *Harvard Business Review* **26**(4), 37–66.

Forrester, J.W., 1961. *Industrial Dynamics.* Productivity, Cambridge.

Forrester, J.W., 1968a. *Principles of Systems.* Productivity, Cambridge.

Forrester, J.W., 1968b. Market Growth as Influenced by Capital Investment. *Industrial Management Review* **9**(2), 83–105.

Forrester, J.W., 1969. *Urban Dynamics.* Productivity, Cambridge.

Forrester, J.W., 1971. *World Dynamics.* Wright-Allen, Cambridge.

Forrester, J.W., 1994. System dynamics, systems thinking, and soft OR. *System Dynamics Review* **10**(2/3), 245–256.

Frensch, P.A. and Funke, J., (Eds.), 1995. *Complex Problem Solving: The European Perspective*, Lawrence Erlbaum, Hillsdale.

Größler, A., 2007. System dynamics projects that failed to make an impact. *System Dynamics Review* **23**(4), 437–452.

Hempel, C.G. and Oppenheim, P., 1948. Studies in the logic of explanation. *Philosophy of Science* **15**(2), 135–175.

Lane, D.C., 1995. On a resurgence of management simulations and games. *Journal of the Operational Research Society* **46**(5), 604–625.

Lane, D.C., 1999. Social Theory and System Dynamics Practice. *European Journal of Operational Research* **113**(3), 501–527.

Lane, D.C., 2000. Should system dynamics be described as a 'hard' or 'deterministic' systems approach? *Systems Research and Behavioral Science* **17**(1), 3–22.

Lane, D.C., 2008. The emergence and use of diagramming in system dynamics: A critical account. *Systems Research and Behavioral Science* **25**(1), 3–23.

Lewin, K., 1945. The research center for group dynamics at Massachusetts Institute of Technology. *Sociometry* **8**(2), 126–136.

Lomborg, B., 2001. *The Skeptical Environmentalist: Measuring the Real State of the World.* Cambridge University Press, Cambridge.

Maani, K.E. and Cavana, R.Y., 2007. *Systems Thinking and Modelling: Understanding Change and Complexity.* Prentice Hall, Auckland.

Meadows, D.H., 2007. The history and conclusions of the limits to growth. *System Dynamics Review* **23**(2–3), 191–197.

Meadows, D.H., 2008. *Thinking in Systems. A Primer.* Chelsea Green, White River Junction.

Meadows, D.H., Meadows, D.L., Randers, J. and Behrens, W.W., 1972. *The Limits to Growth: A Report for the Club of Rome's Project on the Predicament of Mankind.* Universe Books, New York.

Meadows, D.L., Behrens, W.W., Meadows, D.H., Naill, R.F., Randers, J. and Zahn, E.K.O., 1974. *Dynamics of Growth in a Finite World*, Wright-Allen, Cambridge.

Meadows, D.L., Richardson, J. and Bruckmann, G., 1982. *Groping in the Dark: The First Decade of Global Modelling.* Wiley, Chichester.

Meadows, D.H., Meadows, D.L. and Randers, J., 1992. *Beyond the Limits: Confronting Global Collapse, Envisioning a Sustainable Future.* Chelsea Green, Post Mills.

Meadows, D.H., Meadows, D.L. and Randers, J., 2004. *Limits to Growth: The 30-year Update.* Chelsea Green, White River Junction.

Mesarovic, M. and Pestel, E., 1974. *Mankind at the Turning Point: 2nd Report to the Club of Rome.* Dutton, New York.

Morecroft, J.D.W. and Sterman, J.D., (Eds.), 1994. *Modelling for Learning Organizations.* Productivity, Portland.

Morecroft, J.D.W., 2007. *Strategic Modelling and Business Dynamics: A Feedback Systems Approach.* Wiley, Chichester.

Nordhaus, W.D., 1973. World dynamics: Measuring without data. *The Economic Journal* **83**(332), 1156–1183.

Popper, K.R., 2002a. *The Logic of Scientific Discovery.* Routledge, London.

Popper, K.R., 2002b. *Conjectures and Refutations: The Growth of Scientific Knowledge.* Routledge, London.

Popper, K.R., 2003. *The Open Society and Its Enemies.* Routledge, London.

Popper, S.W., Lempert, R.J. and Bankes, S.C., 2005. shaping the future. *Scientific American.* **April**, 66–71.

Senge, P.M., 1990. *The Fifth Discipline — The Art and Practice of the Learning Organization.* Currency Doubleday, New York.

Sawin, E.R., *et al.*, 2009. *C-ROADS — A Simple Computer Simulation of Climate Change — to Support Long-Term Climate Policy Development.* Available at http:// climateinteractive.org / simulations / C-ROADS / simulation-media / publications / C-ROADS%20Analysis%20of%20 Current%20Proposals090311.pdf (Accessed 31 March 2010).

Sterman, J.D., 1988a. A skeptic's guide to computer modeling. In: Grant, L. (Ed.), *Foresight and National Decisions*, Lanham, University Press of America, pp. 133–169.

Sterman, J.D., 1988b. Strategy Dynamics: the Rise and Fall of People Express. *MIT System Dynamics Group, Memo D-3959-1.*

Sterman, J.D., 1989. Misperceptions of feedback in dynamic decision making. *Organizational Behavior and Human Decision Processes* **43**(3), 301–335.

Sterman, J.D., 2000. *Business Dynamics: Systems Thinking and Modeling for a Complex World.* Irwin/McGraw-Hill, Boston.

Sterman, J.D., 2008. Risk communication on climate: Mental models and mass balance. *Science* **322**, 532–533.

Stott, P., 2000. *Political Ecology: Science, Myth and Power.* Arnold, London.

Sustainability Institute, 2010a. *Climate Interactive.* Available at http://www.sustainer.org/?p=42 (Accessed 31 March 2010).

Sustainability Institute, 2010b. *C-ROADS Analysis of the Final Copenhagen Accord.* Available at http://www.sustainer.org/?p=1465 (Accessed 31 March 2010).

Sweeney, L.B. and Sterman, J.D., 2000. Bathtub dynamics: Initial results of a systems thinking inventory. *System Dynamics Review* **16**(4), 249–286.

Turner, G., 2008. A comparison of the limits to growth with thirty years of reality. *CSIRO Working Paper Series*, June 2008, Canberra.

United Nations Framework Convention on Climate Change (UNFCCC), 2009. *Report of the Conference of the Parties on its Fifteenth Session, held in Copenhagen from 7 to 19 December 2009.* Addendum, Part Two: Action taken by the Conference of the Parties at its fifteenth session. Availavle at http://unfccc.int/resource/docs/2009/cop15/eng/11a01.pdf (Accessed 31 March 2010).

van den Belt, M., 2004. *Mediated Modelling: A System Dynamics Approach to Environmental Consensus Building.* Island, Washington.

Vennix, J., 1996. *Group Model-Building: Facilitating Team Learning Using System Dynamics.* Wiley, Chichester.

Vennix, J., 1999. Group model building: Tackling messy problems. *System Dynamics Review* **15**(4), 379–401.

Videira, N., Antunes, P. and Santos, R., 2009. Scoping river basin management issues with participatory modelling: The Baixo guadiana experience. *Ecological Economics* **68**(4), 965–978.

Visser, W., 2007. *The A to Z of Corporate Social Responsibility: A Complete Reference Guide to Concepts, Codes and Organisations.* Wiley, Chichester.

Warren, K., 2002. *Competitive Strategy Dynamics.* Wiley, Chichester.

Warren, K., 2007. *Strategic Management Dynamics.* Wiley, Chichester.

World Commission on Environment and Development, 1987. *Our Common Future.* Oxford University, Oxford.

Chapter 16

SUPPLY CHAINS AND SUSTAINABILITY

Giannis T. Tsoulfas[a] and Costas P. Pappis[b]

[a]*Department of Business Administration (Tutor in Administration of Businesses and Organizations), School of Social Science, Hellenic Open University, Patras, Greece*
tsoulfas@unipi.gr

[b]*Department of Industrial Management, University of Piraeus, 80 Karaoli & Dimitriou str., 185 34 Piraeus, Greece*
pappis@unipi.gr

Over the years environmental and social considerations have been inserted in managerial decision making. Together with the economic dimension, which has always been dominant in business, they are composed of a triad, with which managers are increasingly faced with. Supply chain management (SCM) is an integral part of today's business landscape. In this chapter, we explore recent trends in SCM, which entail sustainability considerations that embrace all its three dimensions, including also the least developed, namely, the social one.

1. Introduction

Logistics management is often identified with supply chain management (SCM). It is important to point out the differentiation of these concepts in the first place. The Council of Supply Chain Management Professionals (CSCMP, 2010) has suggested the following detailed definitions:

- "Logistics management is that part of SCM that plans, implements, and controls the efficient, effective forward and reverse flow and storage of goods, services, and related information between the point of origin and the point of consumption in order to meet customers' requirements. Logistics management activities typically include inbound and outbound transportation management, fleet management, warehousing, materials handling, order fulfillment, logistics network design, inventory management, supply/demand planning, and management of third party logistics services providers. To varying degrees, the logistics function also includes sourcing and procurement, production planning and scheduling, packaging and assembly, and customer service. It is involved in all levels of planning and execution-strategic, operational, and tactical. Logistics management is an integrating function, which coordinates and optimizes all logistics activities, as well as integrates logistics activities with other functions, including marketing, sales, manufacturing, finance, and information technology."

- "SCM encompasses the planning and management of all activities involved in sourcing and procurement, conversion, and all logistics management activities. Importantly, it also includes coordination and collaboration with channel partners, which can be suppliers, intermediaries, third-party service providers, and customers. In essence, SCM integrates supply and demand management within and across companies. SCM is an integrating function with primary responsibility for linking major business functions and business processes within and across companies into a cohesive and high-performing business model. It includes all of the logistics management activities noted above, as well as manufacturing operations, and it drives coordination of processes and activities with and across marketing, sales, product design, finance and information technology."

Thus, logistics refers to activities required to move and position inventory throughout a supply chain. As such, logistics is a subset of the broader framework of a supply chain. Logistics is the process that creates value by timing and positioning inventory. It is the combination of a firm's order management, inventory, transportation, warehousing, materials handling, and packaging, activities that are integrated throughout a facility network. Integration of logistics activities serves to link and coordinate the overall supply chain thus increasing its effectiveness (Pappis, 2010).

As Christopher (2005) aptly mentions, in the past, firms used to establish factories in overseas countries in order to meet local demand. Now, with the significant reduction or even the extinction of trade barriers and the development of transportation infrastructures, fewer factories have such capacities so that they can meet global, rather than local, demand. Indeed, international trade, direct foreign investments, and globalization have had a determinant role in the formation of current logistics reality. Logistics could not be excluded from the introduction of sustainability concepts to traditional business and industry disciplines. On the contrary, since it serves the dominant structure of supply chain, it has provided fertile ground for research during the last two decades. Linton *et al.* (2007) argue that while important contributions have been made in relation to environmental operations and policy, strategy, finance, product design, supplier relations, and post-consumer product management, it is critical to move forward to the systemic issues that exist at the intersection of sustainability and supply chains. Therefore, the connection of logistics with sustainability is dual and the following two dimensions are interrelated, as it will be shown in the sequel:

- As an operational process, logistics is subjected to changes toward its adjustment in the new landscape, particularly the demand for sustainability.

- Owing to its nature, logistics serves as a means to support sustainability purposes (closing materials cycles).

In this, chapter, we present recent trends in SCM, with an emphasis in logistics, which entail sustainability considerations. Reverse logistics is briefly discussed and the emergence of closed-loop supply chains is outlined. Then, the roadmap toward sustainable logistics and sustainable SCM is explored through the incorporation of environmental and social parameters in logistics and SCM. Particularly, several approaches met in the literature are discussed as well as emerging conceptual frameworks with roots in the business world. The chapter ends with some concluding remarks.

2. Reverse Logistics and the Emergence of Closed-loop Supply Chains

A close examination of the situation in the developed countries during the past 20 years indicates that significant attention has been drawn with respect to the post-use treatment of nearly all kinds of products, regardless their size, composition, and initial value. The European Union directives, in particular, which are gradually embodied into national legislations, cover many kinds of end-of-life (EOL) products, e.g., cars, consumer electronics, batteries, packaging materials, and tires. Targets concerning the recovery rates of EOL products are set and networks' requirements are prescribed. Furthermore, specific guidelines are issued regarding the composition of several product categories, which aim at facilitating the recovery of EOL products and the abandonment or proper treatment of hazardous materials. At the same time, the business world has identified certain profit opportunities in the area of used products' recovery, not necessarily in the form of "making more money" but also in the forms of "spending less money" and "adding value to products." In particular, technological advances have made it profitable to use recycled materials, instead of primary ones (e.g., lead and aluminum) or reuse products after some minor treatment, instead of producing new ones (e.g., glass bottles and pallets). As in most managerial innovations, it is true that adapting to the new scenery offered first-time movers a competitive advantage and new ways to add value to core business programs (Hansmann and Claudia, 2001). However, it became evident over the years that meeting such requirements would be a necessary condition for simply doing business.

2.1. *The case of reverse logistics*

The extension of the forward supply chain in order to encompass the activities that occur after the end of a product's useful life cycle has led to the emergence of a new area for research and application, i.e., reverse logistics. According to the European Working Group on Reverse Logistics (RevLog, 1998), reverse logistics is "the process of planning, implementing, and controlling backward flows of raw materials, in process inventory, packaging, and finished goods, from a manufacturing, distribution or use point, to a point of recovery or point of proper disposal." Thus, the reverse flow begins when the end-user returns the product and finishes when the company or a recovery agent has recovered the maximum value out of it. Returns may be distinguished as follows: (Flapper *et al.*, 2005; Kleindorfer *et al.*, 2005):

- production-related returns, which include obsolete materials, packaging materials, production scrap, and defects;
- distribution-related returns, which include consumer (convenience) returns, wrong deliveries, products refused by customers, recalls, and unsold products;
- use-related returns, which include returns in the context of warranties, repair, refurbishment or remanufacturing as well as end-of-lease and end-of-use returns; and
- EOL returns.

All returned items, assuming that are not uncontrollably disposed, are collected and transported to appropriate facilities, where a selection occurs: some of them (or some parts of them) are reprocessed. Of course, certain items can be fed into the market after some minor treatment (reuse). Reprocessing may occur at different levels (De Brito and Dekker, 2004): product level (repair), module level (refurbishing), component level (remanufacturing), selective part level (retrieval), material level (recycling), and energy level (incineration). The waste which is generated during reprocessing is disposed and the reprocessed items are either redistributed and reused preserving their initial usage, or used at material's or component's level for the manufacturing or assembly of different products.

From a logistical perspective, manufacturers have several options with respect to alternative recovery strategies (Toffel, 2003; Östlin *et al.*, 2008). These include the following:

- *Promoting the recycling market*: This strategy mainly refers to the provision of information to end users regarding their options when a product reaches the end of its life cycle, without getting involved in the recovery activities at all.
- *Long-term contract*: A manufacturer signs a long-term contract with a company, which undertakes a part of or the whole reverse logistics activities.
- *Joint venture with a company from a different industry*: Manufacturers may have the option to engage in a joint venture with non-competing companies so that the new company is destined to serve all participants.
- *Joint venture with competitors or establish an industry consortium*: In order to gain economies of scale, manufacturers may co-operate by designating one company responsible to manage EOL product recovery.
- *Vertical Integration*: Manufacturers may decide to undertake themselves reverse logistics activities.

Reverse supply chains differ from forward supply chains in several aspects as follows (Krikke *et al.*, 2002; Tibben-Lembke and Rogers, 2002):

- In addition to economic parameters, there are environmental drivers complicating the design and operation of reverse supply chains.
- There is higher system complexity in reverse supply chains due to increased number of —- and interaction between — goods flows.
- Reverse flows are characterized by uncertainty on the supply (collection) side of the system regarding volumes, quality, composition, and timing. Naturally, this fact directly affects the economic dimension as well.
- There is usually a mismatch between used products supply and demand. "Production" (e.g., processing of used products) is not always coupled with "demand" (e.g., producers' requirements in secondary raw materials).
- There are numerous "suppliers" and few "customers." Indeed, used products are the raw materials for the reverse supply chain. Unlike the forward supply chain, there are a lot more sources of raw materials and they may enter the reverse supply chain at small cost or at no cost at all. However, although often obtained for free, the value of return flows is usually low.
- In several cases, there are unexplored market opportunities. However, sustainability requirements can be the basis of the creation of new markets or result in the reorganization of existing ones for EOL products.

Furthermore, there are several parameters that may affect the efficiency of reverse supply chains. Tsoulfas *et al.* (2009) identified three major categories of such parameters:

- product-dependent, which stem from the product itself;
- organizational, which refer to issues regarding the participants involved in the recovery of used products; and
- social, which refer to the attitudes and values prevailing in societies and may determine practices regarding recovery of used products.

In Table 1, an indicative list of such parameters is given.

2.2. *The emergence of closed-loop supply chains*

Closed-loop supply chains refer to the integration of forward and reverse supply chains. Actually, in most cases the term is used literally since materials seldom return to their origin. However, it is intended to express that once borrowed from the planet, materials are followed and managed during all their transformations. In the literature, the term-extended supply chain is also used in order to address integrated forward and reverse supply chains (Beamon, 1999; Krikke *et al.*, 2002). Post *et al.* (2002) is "the nodal element in a network of actors and interests often referred to as 'stakeholders'" and is characterized by the fact that "the firm creates value through relational, as well as transactional, interactions — both voluntary

Table 1. Parameters affecting the efficiency of reverse supply chains.

Product-dependent
The physical and technical characteristics of EOL products
The remaining value of EOL products
The way used products are replaced by new ones
The recovery options
The capability to change the usage of EOL products or to provide them to different markets
Organizational
The structure of the recovery network
Asset control policies
Marketing
Consumers' involvement
Social
Social habits
Legislation
Social awareness

Source: Tsoulfas *et al.* (2009).

and involuntary — with stakeholders in its social and political environment, as well as those constituting its resource base and industry setting." Mentzer *et al.* (2001) discussed SCM as a set of activities to implement a management philosophy. Given the definition of SCM by the CSCMP, these activities are also relevant to closed-loop supply chains. These are as follows:

- integrated behavior;
- mutually sharing information;
- mutually sharing risks and rewards;
- co-operation;
- the same goal and the same focus on serving customers;
- integration of processes; and
- partners to build and maintain long-term relationships.

According to Corbett and Savaskan (2003), SCM is characterized by three main flows, which exist in both forward and reverse supply chains. These are the following:

- flows of materials;
- flows of information concerning forecasts, inventories, orders, demands, handling, tracking, quality, etc.; and
- financial flows referring to purchase price, lease payments, service payments, disposal costs, etc.

The above flows interact with each other and their coordination is crucial in the supply chain success.

3. The Introduction of Sustainability in Logistics and Supply Chains

In the mid-1990s, "greening" came into the mainstream and was popularized in the business world. Non-business stakeholders (governmental and non-governmental organizations) had their part in this fact by modifying the framework for doing business through issuing environmental laws, regulations, directives and voluntary codes of conduct. Furthermore, environmental management systems such as the ISO 14000 series had a significant role in "greening" business, by cultivating the outstanding success of the preceded quality revolution. Sustainability was the next big thing. Companies started talking about integrating economic, environmental, and social concerns in their operations. The concept of corporate social responsibility (CSR) has been used in order to deal with this extended agenda from a business perspective.

Of course, the fact that companies talk about sustainability issues does not necessarily mean that they are all willing to contribute actively in their efficient integration. In any case, at least theoretically, sustainability has been widely perceived as the suitable pattern for all aspects of human activity. Thus, all kinds of operations should conform to this pattern. Therefore, the incorporation of sustainability considerations in business operations stems from the growing concerns of societies, legislation and relevant regulations as well as various motivations that a company may have, including economic ones. These motivations are briefly presented in Table 2 (Zhu *et al.*, 2005; Tsoulfas *et al.*, 2006; Tsoulfas and Pappis, 2007).

Moreover, companies realize that their stakeholders do not always distinguish between a company and its suppliers (Rao and Holt, 2005). With regard to this, focal companies of supply chains are considered responsible for the sustainability performance of their suppliers, especially for the negative impacts (Seuring and Müller, 2008). Focal companies are those which appear publicly to have a central role in the supply chain, especially by being close to end customers and/or by carrying a widely known brand name. These companies are expected to deal with sustainability issues raised by the entire supply chain, including logistics operations.

Product stewardship is a relevant concept, which was coined in order to describe manufacturers' responsibility with respect to the sustainability issues raised by their

Table 2. Motivations for the incorporation of sustainability considerations.

Internal drivers	External drivers
Strategic management priority	Legislation and regulations
Employee involvement	Market pressure
Costing	Public pressure
Health and safety	Globalization
Marketing issues	
Profit opportunities	

products from "cradle to grave," i.e., throughout their whole life cycle. Therefore, from the perspective of strategic management it is essential that firms consider not only their internal activities but also their supply chain relationships and partnerships in order to identify opportunities of improvement of their sustainability performance, which is inextricably linked with the overall performance of the supply chains in which they participate. Certainly, sustainability performance is not feasible to measure objectively, especially when all sustainability dimensions such as those mentioned above have to be integrated.

It should also be noted that information technology is gaining a central role in the logistics industry with applications that may support all relevant activities. The term e-logistics is used in order to cover the electronic transactions and the computer-aided managerial processes that take place with respect to logistical activities. Sarkis *et al.* (2004) discussed extensively e-logistics and its potential implications in forward and reverse supply chains. Notably the authors emphasized the environmental dimension of e-logistics and analyzed issues with regard to both forward and reverse supply chain. It is also interesting that they implied that the evaluation of the sustainability performance of e-logistics should be done in a long-term horizon by taking into consideration issues such as financial investments and energy consumption of hardware.

Addressing the economic pillar of sustainability is common practice for researchers and practitioners, while during the last two decades the environmental pillar has been considered as well. Notably, the pillar associated with the social dimension of sustainability has not been well-defined and has received only little attention in the literature (Hutchins and Sutherland, 2008). Correspondingly, research in logistics and SCM is dominated by economic and environmental issues. Srivastava (2007) states that green SCM (GSCM) has its roots in both environment management and SCM literature and adding "green" to SCM involves the consideration of the influence and relationships between SCM and the natural environment. Therefore, GSCM is defined as "integrating environmental thinking into SCM, including product design, material sourcing and selection, manufacturing processes, delivery of the final product to the consumers as well as EOL management of the product after its useful life." Hervani *et al.* (2005) use a unique presentation to give a different definition for GSCM:

$$\begin{aligned}\text{Green Supply Chain Management} &= \text{Green Purchasing}\\ &\quad + \text{Green Manufacturing} + \text{Materials Management}\\ &\quad + \text{Green Distribution} + \text{Marketing} + \text{Reverse Logistics.}\end{aligned}$$

In this definition, "reverse" is identified with "green reverse." However, reverse does not entail green by definition. On the contrary, it is possible that reverse logistics and other reverse supply chain activities are responsible for the majority of environmental impacts of a given product system.

It should be noted that GSCM is strongly related to life-cycle thinking and relevant methods, such as life-cycle assessment (LCA). For example, Vermeulen and Ras (2006) cite a sequence of activities that guide greening of supply chains through the application of LCA as follows:

- producers analyze the entire life cycle of their product, identifying major environmental effects;
- they decide on objectives, either self-imposed, aiming at environmental excellence in niche markets, or applying readily available sector-wide standards (such as eco-labeling criteria);
- they analyze at these crucial points the level of reduction of environmental impacts resulting from possible improvements in methods of production, use, and discarding of the product;
- they analyze the economics of suggested improvements and select which improvements are to be implemented;
- they implement the improvements selected within their own production facilities;
- they engage in inter-firm co-operation aiming at implementation of identified improvements on the input side of their product life cycles (procurement, co-operation with main suppliers, co-makerships) and on the output side (organizing recycling, maintenance, information on proper use, *etc.*);
- they develop systems of control, preferably including third parties able to generate legitimacy.

LCA is considered to be the most exhaustive quantitative approach for the appraisal of the environmental performance of product systems. LCA methodology consists of four steps: first, definition of a unit of analysis (the functional unit), definition of the boundaries of the system and scope definition of the study; second, performing inventory analysis of the inputs and outputs of the examined product system; third, conducting impact assessment, which includes classification, characterization, and optionally, normalization and weighting, and where inputs and outputs identified in the previous step are accounted in terms of environmental impacts or damages; and finally, interpretation of the results of the study, which includes a critical review of the previous steps and guidance in decision making.

Several theoretical approaches may be found in the academic literature with regard to the greening of supply chains. For example, Rao and Holt (2005) developed a conceptual framework aiming to cover the major constructs of supply chains, i.e., greening the inbound function; greening production; greening the outbound function; competitiveness; and economic performance. For each of these issues, they suggested several variables. Preuss (2005) proposed a five-step procedure in order to construct a green supply chain, which includes suppliers' compliance with recognized international codes of conduct; greening of the manufacturing processes; assessment of suppliers using environmental criteria; internalization of environmental protection initiatives within the firm; and taking responsibility for the reverse supply chain. Tsoulfas and Pappis (2006) proposed a series of environmental principles applicable to supply chain design and operation, which are classified in six groups corresponding to respective company functions: product design; packaging; collection and transportation; recycling and disposal; greening the internal and external business environment; and other (general) management issues. Linton *et al.* (2007) argue that issues and flows that extend beyond the core of SCM should be integrated: product design; manufacturing by-products; by-products produced during product use; product life extension; product EOL; and recovery processes at EOL. It is noteworthy that although issues such as the above ones are related to sustainability, only the environmental dimension is tackled in these approaches. However, for an integrated approach to sustainability, economic and social issues should be addressed.

Businesses that have realized the significance of environmental issues, including physical impacts on assets, changes in market dynamics for goods and services, escalation of regulation and greater scrutiny from an increasingly sophisticated range of stakeholders, have undertaken initiatives and established monitor and reporting mechanisms with regard to their activities so as to enable the transition toward greener supply chains. A representative example is the Carbon Disclosure Project (CDP) Supply Chain Program. In particular, the mission of the CDP is to facilitate a dialogue between investors and corporations, supported by high quality information from which a rational response to climate change is emerging. CDP focuses on individual companies rather than on nations and brings together institutional investors to focus attention on carbon emissions, energy usage, and reduction wherever companies and assets may be located. CDP provides annual information about the firms that participate in the coalition, covering four principal areas:

- management's views on the risks and opportunities that climate change presents to the business;
- greenhouse gas emissions accounting;
- management's strategy to reduce emissions/minimize risk and capitalize on opportunity; and
- corporate governance with regard to climate change.

The CDP Supply Chain Program is a collaboration of global corporations who have extended their climate change and carbon management strategies beyond their direct corporate boundaries to engage their suppliers via CDP's annual information request. Member companies use CDP's standardized format to communicate with their suppliers in a streamlined, unified, annual request (Carbon Disclosure Project, 2010).

Apart from CDP, several well-known paradigms and initiatives show the policies and engagement of the business world regarding particularly global warming

(Pappis, 2010). Several reports show that many businesses are embracing the new era of climate risk analysis and planning, although serious governance gaps remain, and the work of all businesses to achieve sustainable wealth in a carbon-constrained world is only just beginning. Integrated oil, supermarkets, financial services, insurance, infrastructure and communications companies, and manufacturing groups and intermediate producers are characterized as "sectors moving ahead." Recent and rapid technological innovation is stimulating growth in new and existing industries, as markets receive clearer signals about and draw inferences concerning the long-term growth potential of "low-carbon" products and services. It is important that there are cases, where technological innovations may not only reduce emissions of carbon, but also lead to firms along supply chains becoming more efficient in the use of all inputs, boosting net profit. The following section mentions a number of promising sustainable logistics and supply chains initiatives developed recently.

4. Sustainable Logistics and Sustainable Supply Chains Initiatives

As noted in the previous Section, sustainable logistics and sustainable SCM also include a social dimension, which is, however, much less developed than the environmental one. Academic works facing the crux of the subject, namely, the social dimension of sustainability, are limited and, in some cases, confused and still have not gained wide acceptance neither in the academia, nor among corporations. In the sequel, we mention some examples from the recent literature. Ciliberti *et al.* (2008) focus on logistics social responsibility, where they develop a taxonomy of relevant practices with respect to five areas of interest: purchasing social responsibility; sustainable transportation; sustainable packaging; sustainable warehousing; and reverse logistics. For each of these areas, authors utilize examples from the literature to express the practical implications. Hutchins and Sutherland (2008) review metrics and indicators of social sustainability and CSR and they propose various potential measures of social impacts. Halldórsson *et al.* (2009) identify three main supply chain stages: supply of raw materials and components; production; and distribution and reverse logistics. For each of these stages, the authors identify several relevant dimensions. Sarkis *et al.* (2010) deal with an initial identification of social sustainability issues associated with reverse logistics in order to explain how they can influence organizational strategy and operations.

We should note, though, that logistics and SCM are increasingly linked with CSR and relevant frameworks, which have been discussed in the chapter entitled "Corporate Social Responsibility and Reporting." Some of the instruments referred to in the sequel, which provide business guidance for CSR, have developed specific material for the case of SCM. The reader is invited to refer to the above-mentioned chapter for further insight into the subject.

4.1. *UNEP/SETAC life cycle initiative — Guidelines for social LCA of products*

The UNEP/SETAC Life Cycle Initiative (2009) has recently developed social and socio-economic LCA (S-LCA) guidelines, which aim to complement environmental LCA and life-cycle costing (LCC), so as to contribute to the full assessment of goods and services within the context of sustainable development. Thus, given that supply chains are strongly linked with life cycle thinking, the triad LCC–LCA–S-LCA would be suitable in order to deal with sustainability issues in SCM. Social impacts are perceived as the consequences of positive or negative pressures on social endpoints (i.e., well-being of stakeholders). Social impacts are perceived by the S-LCA guidelines as consequences of interactions, which occur in the context of an activity (production, consumption, or disposal) and/or engendered by it and/or by preventive or reinforcing actions taken by stakeholders. When referring to the causes of social impacts, this generally implies three dimensions, which are interrelated:

- behaviors, where social impacts are those caused by a specific behavior or decision;
- socio-economic processes, where social impacts are the downstream effect of socio-economic decisions; and
- capitals (human, social, cultural), where social impacts may be affected positively or negatively from the attributes possessed by an individual or a group of people.

Subcategories are the basis of an S-LCA assessment because they are the items on which justification of inclusion or exclusion needs to be provided. The subcategories refer to socially significant themes or attributes. They are classified according to stakeholder, impact categories and are assessed by the use of inventory indicators. Inventory indicators and units of measurement may vary depending on the context of the study. Social/socio-economic subcategories may be first classified by stakeholder categories as this might assist with the operationalization as well as ensure the comprehensiveness of the framework. The purpose of the classification into impact categories is to support the identification of stakeholders, to classify subcategory indicators within groups that have the same impacts, and to support further impact assessment and interpretation. The impact categories should preferably reflect internationally recognized categorizations/standards and/or results from a multi-stakeholder process.

4.2. *The United Nations global compact — Supply chain sustainability*

The United Nations Global Compact (UNGC) encourages signatories to engage with their suppliers around the 10 Principles, and thereby to develop more sustainable supply chain practices. The UNGC has been working in order to develop guidance on how to take a more proactive approach to integrate the 10 Principles into SCM practices. This task is closely coordinated with the strategy and work done in the four UNGC issues areas (human rights, labor, environment, and anti-corruption). UNGC has entered into a strategic partnership with Business for Social Responsibility, a global business membership network, and consultancy focused on sustainability, to develop an implementation guide and a learning and assessment tool for signatories (United Nations Global Compact, 2010).

4.3. *The global reporting initiative sector supplement on logistics and transportation*

The Global Reporting Initiative (GRI) G3 Guidelines are already being used by many organizations, including many transport and logistics companies, to promote transparency and accountability. However, GRI recognized that some sectors, such as logistics and transportation, face unique needs that require specialized guidance above and beyond those addressed by the existing G3 Guidelines. Thus, GRI has been working in a relevant sector supplement for logistics and transportation for the following reasons (Global Reporting Initiative, 2010):

- From an environmental point of view, as major consumers of fossil fuels, logistics and transportation companies are expected to minimize the impact of the resultant emissions — which are known to contribute to climate change and urban air pollution.
- From the social side, this sector employs a large number of workers around the world and faces challenges in labor use practices.

This GRI project is still under development and only tentative outputs have been carried out.

4.4. *Future supply chain 2016*

Future Supply Chain 2016 is a study published by the Global Commerce Initiative (GCI) together with the consulting company Capgemini. This ambitious project aims to present a new integrated supply chain model that takes into account sustainability parameters such as reduced energy consumption, better traceability, and reduced traffic congestion, as well as traditional measures such as cost reduction and financial performance. This report is maybe the most integrated effort so far to efficiently link SCM with sustainability.

To address this issue, the GCI and Capgemini have collaborated on the "Future Supply Chain 2016" project, being driven by external trends (e.g., economic, environmental, demographic, technological, and regulatory), which are largely outside the control of industry, but also from key internal trends (e.g., consumer behavior, product flows, information flows). The work has been focused on the following key aspects:

- The challenges ahead that will force companies to change their operations.
- The need for breakthrough change, as the past does not reflect the future the industry will face.

- The innovation that currently exists in the form of new solutions, leading practices, example supply chains, and new ways to calculate the impact of the new parameters on the supply chain.
- Enhanced collaboration, which is essential among all parties in the supply chain as well as between competitors.
- The recognition that now is the time for a step change to a future model, leading to sustainability and new business opportunities.

According to the study, in order to portrait the future supply chain, industry should consider four key components:

- *Solutions areas*: The solution areas address existing challenges as well as those anticipated for the coming decade and are focused on physical supply chain innovation. Seven key solution areas are identified:
  (1) in-store logistics, which includes in-store visibility, shelf-ready products, shopper interaction;
  (2) collaborative physical logistics, which refers to shared transport, shared warehouse, shared infrastructure;
  (3) reverse logistics, which includes product recycling, packaging recycling, returnable assets;
  (4) demand fluctuation management, which involves joint planning, execution, and monitoring;
  (5) identification and labeling;
  (6) efficient assets, which refers to alternative forms of energy, efficient/aerodynamic vehicles, switching modes, green buildings; and
  (7) joint scorecard and business plan.
- *Leading practices*: Examples of existing leading practices are integrated into the model to show how they help address the solutions areas.
- *Application to example supply chains*: Simplified supply chains are used to demonstrate how the model works.
- *New ways to calculate the impact on the supply chain*: These include both macro-level and micro-calculation models.

Integrating these innovative solutions with collaboration concepts into a cohesive model will provide the future supply chain architecture, which will serve sustainability incentives. This analysis demonstrates how the different solutions should be considered in relation to each other, and makes it clear that a big impact on the parameters can be made when the following concepts are merged and implemented:

- information sharing;
- collaborative warehousing;
- collaborative city distribution (including home delivery and pickup);
- collaborative non-urban distribution (including home delivery and pickup).

The total impact of this supply chain redesign could potentially reduce transport costs per pallet, handling costs per pallet, lead time as well as CO_2 emissions per pallet by more than 20%, while also improving shelf availability. Additional energy cost savings can stem from more efficient assets such as green buildings and fuel-efficient/aerodynamic and jumbo trucks. These benefits and others are expected to be achievable when all the elements of the future supply chain are in place. They are expected to provide clear benefits for our society, for the industry, for individual companies, and ultimately for consumers and shoppers.

According to the study, the characteristics of the 2016 Future Supply Chain will be the following:

- The future model will be based on multi-partner information sharing among key stakeholders: consumers (the originators of the demand signal, either from home or from a store), suppliers, manufacturers, logistics service providers, and retailers.
- After production, the products will be shipped to collaborative warehouses in which multiple manufacturers store their products.
- Collaborative transport from the collaborative warehouse will deliver to city hubs and to regional consolidation centers.
- Warehouse locations on the edge of cities will be reshaped to function as hubs where cross-docking will take place for final distribution.
- Non-urban areas will have regional consolidation centers in which products will be cross-docked for final distribution.
- Final distribution to stores, pickup points, and homes in urban and non-urban areas will take place via consolidated deliveries using efficient assets.

5. Concluding Remarks

This chapter has taken a broad look at sustainable SCM and the issues emerging in this field. In particular, sustainable SCM is concerned with balancing economic, environmental, and social issues associated with supply chains by seeking to ensure that decisions made today do not have an adverse impact on future generations.

Linton *et al.* (2007) aptly identify that sustainability has deep roots in both the physical and social sciences. Thus, it is crucial to comprehend the effect and interactions of different activities with the natural environment as well as current and future implications on various issues based on advances in natural sciences. Moreover, the role of social sciences is to enable our interpretation and attitude toward sustainability. As stated by Zaklad *et al.* (2004), the improvement of SCM toward sustainability targets requires focus on business process improvement, enabling technology, and social system transformation. Regarding the latter, in the bottom line, it may be better to modify the problem than solving it. By definition, SCM promotes concepts such as efficiency, collaboration, synergy, and trust among partners. These characteristics are in favor of promoting sustainable development.

The lack of integrated approaches has been a barrier for the enhancement of sustainable SCM. Academics have been mostly engaged to developing principle-based conceptual frameworks. This may be due to the fact that the majority of issues seem self-evident and make common sense. Putting them all together, though, is definitely a contribution. Widely recognized bodies have recently been engaged in linking sustainability to SCM. Their early outputs are very promising and indicate the initiation of rethinking of the business model of supply chain.

References

Beamon, B.M., 1999. Designing the green supply chain. *Logistics Information Management* **12**(4), 332–342.

Carbon Disclosure Project, 2010. *Carbon Disclosure Project Supply Chain Report 2010.* Retrieved April 22, 2010, from https://www.cdproject.net/CDPResults/CDP-Supply-Chain-Report_2010.pdf.

Christopher, M., 2005. *Logistics and Supply Chain Management*, third ed. Pearson Education, Edinburgh, UK.

Ciliberti, F., Pontrandolfo, P. and Scozzi, B., 2008. Logistics social responsibility: Standard adoption and practices in Italian companies. *International Journal of Production Economics* **113**, 88–106.

Corbett, C.J. and Savaskan R.C., 2003. Contracting and coordination in closed-loop supply chains. In: Guide Jr., V.D.R. and Van Wassenhove, L.N. (Eds.), *Business Aspects of Closed-Loop Supply Chains: Exploring the Issues.* Carnegie Mellon University Press, Pittsburgh, PA.

De Brito, M.P. and Dekker, R., 2004. A framework for reverse logistics. In: Dekker, R., Fleischmann, M., Inderfurth, K. and van Wassenhove, L.N., (Eds.), *Reverse Logistics.* Springer.

Flapper, S.D.P., van Nunen, J.A.E.E. and van Wassenhove, L.N., 2005. Introduction. In: Flapper, S.D.P., van Nunen, J.A.E.E. and van Wassenhove, L.N., (Eds.), *Managing Closed-Loop Supply Chains*, Springer.

Global Commerce Initiative and Capgemini, 2008. *Future Supply Chain 2016.* Retrieved April 20, 2010, from http://www.futuresupplychain.com/files/gci_capgemini_future_supply_chain_2016_report.pdf.

Global Reporting Initiative, 2010. *Logistics & Transportation.* Retrieved April 22, 2010, from http://www.globalreporting.org/ReportingFramework/SectorSupplements/LogisticsAndTransportation/LogisticsAndTransportation.htm.

Halldórsson, Á., Kotzab, H. and Skjøtt-Larsen, T., 2009. Supply chain management on the crossroad to sustainability: A blessing or a curse? *Logistics Research* **1**, 83–94.

Hansmann, K.W. and Claudia, K., 2001. Environmental management policies. In: Sarkis, J., (Ed.), *Green Manufacturing and Operations: From Design to Delivery and Back.* Greenleaf Publishing, Sheffield, UK.

Hervani, A.A., Helms, M.M. and Sarkis, J., 2005. Performance measurement for green supply chain management. *Benchmarking: An International Journal* **12**(4), 330–353.

Hutchins, M.J. and Sutherland, J.W., 2008. An exploration of measures of social sustainability and their application to supply chain decisions. *Journal of Cleaner Production* **16**, 1688–1698.

Kleindorfer, P.R., Singhal, K. and Van Wassenhove, L.N., 2005. Sustainable Operations Management. *Production and Operations Management* **14**(4), 482–492.

Krikke, H.R., Pappis, C.P., Tsoulfas, G.T. and Bloemhof-Ruwaard, J., 2002. Extended design principles for closed loop supply chains: Optimising economic, logistic and environmental performance. In: Klose, A., Speranza, M.G. and Van Wassenhove, L.N. (Eds.), *Quantitative Approaches to Distribution Logistics and Supply Chain Management*, Springer-Verlag.

Linton, J.D., Klassen, R. and Jayaraman, V., 2007. Sustainable supply chains: An introduction. *Journal of Operations Management* **25**, 1075–1082.

Mentzer, J.T., DeWitt, W., Keebler, J.S., Min, S., Nix, N.W., Smith, C.D. and Zacharia, Z.G., 2001. Defining supply chain management. *Journal of Business Logistics* **22**(2), 1–25.

Östlin, J., Sundin, E., Björkman, M., 2008. Importance of closed-loop supply chain relationships for product remanufacturing. *International Journal of Production Economics* **115**(2), 336–348.

Pappis, C.P., 2010. *Climate Change, Supply Chain Management and Enterprise Adaptation: Implications of Global Warming on the Economy.* IGI Global. (ISBN 978-1-61692-800-1), In press.

Post, J.E., Preston, L.E. and Sachs, S., 2002. Managing the extended enterprise: The new stakeholder view. *California Management Review* **45**, 6–28.

Preuss, L., 2005. Rhetoric and reality of corporate greening: A view from the supply chain management function. *Business Strategy and the Environment* **14**, 123–139.

RevLog, 1998. *The European Working Group on Reverse Logistics.* Retrieved March 24, 2010, from http://www.fbk.eur.nl/OZ/REVLOG.

Sarkis, J., Meade, L.M. and Talluri, S., 2004. E-logistics and the natural environment. *Supply Chain Management* **9**(4), 303–312.

Sarkis, J., Helms, M.M. and Hervani, A.A., 2010. Reverse logistics and social sustainability. *Corporate Social Responsibility and Environmental Management* DOI: 10.1002/csr.220.

Seuring, S. and Müller, M., 2008. From a literature review to a conceptual framework for sustainable supply chain management. *Journal of Cleaner Production* **16**, 1699–1710.

Srivastava, S.K., 2007. Green supply-chain management: A state-of-the-art literature review. *International Journal of Management Reviews* **9**(1), 53–80.

Tibben-Lembke, R.S. and Rogers, D.S., 2002. Differences between forward and reverse logistics in a retail environment. *Supply Chain Management* **7**(5), 271–282.

Toffel, M.W., 2003. The growing strategic importance of end-of-life product management. *California Management Review* **45**(3), 102–129.

Tsoulfas, G.T. and Pappis, C.P., 2006. Environmental principles applicable to Supply Chain design and operation. *Journal of Cleaner Production* **14**(18), 1593–1602.

Tsoulfas, G.T., Dasaklis, T. and Pappis, C.P., 2007. Identifying success factors for reverse logistic activities. *Proceedings of the International ECO-X Conference*, May 9–11, 2007, Vienna, Austria.

Tsoulfas, G.T., Pappis, C.P. and Karacapilidis, N.I., 2009. Applying web-based collaborative decision-making in reverse logistics: The case of mobile phones. In: Wang, H.-F. (ed.), *Web-Based Green Products Life Cycle Management Systems: Reverse Supply Chain Utilization*, Information Science Reference, London, UK.

UNEP/SETAC Life Cycle Initiative, 2009. *Guidelines for Social Life Cycle Assessment of Products.* Retrieved April 22, 2010, from http://www.estis.net/includes/file.asp?site=lcinit&file=524CEB61-779C-4610-8D5B-8D3B6B336463.

United Nations Global Compact, 2010. *Supply Chain Sustainability.* Retrieved April 22, 2010, from http://www.unglobalcompact.org/Issues/supply_chain/index.html.

Vermeulen, W.J.V. and Ras, P.J., 2006. The challenge of greening global product chains: Meeting both ends. *Sustainable Development* **14**, 245–256.

Zaklad, A., McKnight, R., Kosansky, A. and Piermarini, J., 2004. The social side of the supply chain. *Industrial Engineer* **36**(2), 40–44.

Zhu, Q., Sarkis, J. and Geng, Y., 2004. Green supply chain management in China: Pressures, practices and performance. *International Journal of Operations & Production Management* **25**(5), 449–468.

Chapter 17

ENVIRONMENTAL AUDITING AND COMPLIANCE

Eunjung Shin[a] and Eric W. Welch[b]
Science, Technology and Environment Policy Lab,
Department of Public Administration,
University of Illinois at Chicago,
Chicago, IL 60607, USA
[a]*eshin9@uic.edu*
[b]*ewwelch@uic.edu*

Environmental auditing is a multipurpose activity that provides information that helps organizations meet environmental management goals, ensures compliance with regulations or standards, and communicates these efforts to relevant stakeholders. With the aid of existing literature and multiple examples, the chapter provides a survey and summary of environmental auditing as it has been operationalized over the past 40 years. It seeks to develop a framework for conceptualizing environmental auditing and examines how behavioral and institutional factors matter for their adoption and effectiveness. Lastly, the chapter investigates challenges and tensions related to environmental auditing and identifies consideration for their design and implementation.

1. Introduction

Environmental auditing is a multipurpose activity that satisfies a range of different organizational, regulatory, and verification objectives. Auditing provides information that help an organization assess policies, production processes, and programs in ways that facilitate enhanced triple bottom line performance. Auditing also represents a means of ensuring compliance with regulations or standards as well as a communication device for conveying advances and compliance with environmental regulations and standards to relevant stakeholders. Environmental auditing is a flexible tool that knits together policy and programmatic intentions with demonstration of the accomplishment of outcomes.

Nevertheless, environmental audits are not ubiquitous; not all governments require them, not all environmental management systems (EMSs) incorporate them, and not all organizations adopt them. Partly, this is due to the history of the development and application of environmental audits; mandatory and voluntary approaches have been established to different degrees over time, and the sponsoring organization has varied from government, industry association, and supranational organization. Variation in the application and uptake of environmental auditing is also a function of the institutional context within which the audit is offered and the behavioral assumptions upon which the audit program or policy is established. Institutions and behavioral assumptions also help explain the perceived value and effectiveness of auditing in terms of both performance and compliance.

This chapter has two objectives. First, it provides a survey and summary of environmental auditing as it has been operationalized over the past 40 years. It examines the institutional structures, designs, and objectives that predominate. Second, the chapter provides a simplified framework for understanding and comparing components and effectiveness of institutions. It explores how to think about environmental auditing in a way that integrates the institutional characteristics with the behavioral assumptions to better understand why auditing does or does not accomplish it intended goals. Ultimately, the chapter seeks to provide a guide to understanding the components of auditing practice and theory for managers, policy makers, and academics.

2. A Brief History of Environmental Auditing

The history of environmental auditing is relatively short. In the United States (US), environmental audits were first employed in the early 1970s by a handful companies in order to assess compliance with the Clean Air and Clean Water Acts (Gottlieb *et al.*, 1995). Large oil and chemical manufacturing firms were some of the first adopters, but later that decade and into the 1980s and 1990s, environmental auditing increasingly became commonplace for two main reasons. First, environmental crises and societal demands ushered in a regulatory era that saw the enactment of major new environmental laws and substantial federal enforcement efforts. As that era progressed, EPA increasingly required companies to conduct facility audits as part of specific administrative orders or consent decrees to address, for example, Toxic Substances Control Act non-compliance (Carson and Moulden, 1991; Thomas *et al.*, 2000). Firms also began to conduct environmental audits to avoid legal liability and penalties related to site history (Gottlieb *et al.*, 1995). Hence, the uptake of environmental auditing was partly stimulated by needs for "compliance with regulations and the avoidance of regulatory sanctions" (Tucker and Kasper, 1998).

Second and somewhat later, the US EPA began to encourage broader use of environmental auditing with the issuance of the *Environmental Auditing Policy Statement* (US EPA, 1986). EPA reaffirmed and expanded that policy by issuing *Incentives for Self-Policing* in 1995 in which it identified major economic and legal incentives for companies to voluntarily discover, disclose, and correct environmental violations (Thomas *et al.*, 2000; US EPA, 1995). See Box 1 for a more detailed description. Hence, incentives related to penalty mitigation also encouraged the adoption of environmental auditing in the US.

Box 1. US EPA auditing incentives.

US EPA has encouraged voluntary environmental audits since 1986. EPA's recent self-policing policy clarifies the benefits of self-disclosure which are contingent on conducting an environmental audit (Emery and Watson, 2003). An organization that complies with self-auditing guidelines is eligible for a reduction or elimination of gravity-based penalties, which are those penalties over and above the portion that represents the entity's economic gain from non-compliance (US EPA, 2000). Additionally, voluntarily disclosed audit information is not used for civil prosecution as long as the voluntary audit and self-disclosure were conducted in good faith and an organization adopts a systematic approach to preventing recurrence of the violation. While in most instances, EPA does not request organizations to report audit results (Thomas *et al.*, 2000), disclosure of potential criminal violations may result in a more thorough review (Thomas *et al.*, 2000; US EPA, 2000).

EPA has continued to incorporate incentives for voluntary compliance as part of their 2000 Audit Policy and has developed programs that provide technical resources for auditing of small- and medium-sized manufacturers (Gottlieb *et al.*, 1995; US EPA, 2000). States and other federal agencies have also developed voluntary audit policies. Over the past two decades, a number of states have passed environmental audit immunity legislation as an incentive for facilities to conduct environmental audits (Stafford, 2006).[1]

The emergence of international initiatives and standards in 1990s facilitated the growth of environmental auditing around the world. During that decade, the International Organization for Standardization (ISO) published the ISO 14000 series of standards for environmental management as well as the ISO 19011 auditing principles. In 1995, the European Union also developed the Eco-Management and Audit Scheme (EMAS) (EU, 2010), which integrates environmental audits into EMSs (EU, 2010) (ISO 14001 and EMAS are discussed in greater detail in Boxes 3 and 6 in the chapter.). Meanwhile, the International Organization of Supreme Audit Institutions (INTOSAI) has played the role of international coordinator of auditing through the establishment of the Working Group on Environmental Auditing in 1992 (INTOSAI, 2007).[2] During the last decade or so, international initiatives and standards

[1]In general, environmental audit privilege legislation protects information gathered during the course of an environmental audit from disclosure and prohibits the use of such information in judicial or administrative proceedings. Environmental audit immunity legislation grants limited immunity from civil fines and penalties to companies that voluntarily disclose and remediate violations discovered during the course of an environmental audit. Stafford (2006) shows that 23 states have established privilege legislation and 21 have established immunity legislation.

[2]INTOSAI is the professional organization of supreme audit institutions (SAIs) in countries that belong to the United Nations (INTOSAI, 2010). INTOSAI WGEA encourages international

have increasingly been integrated into firms' management strategies and the environmental audit has been more broadly applied in public, educational, and other not-for-profit organizations.

The evolution of environmental auditing has lead to new approaches and tools. While early environmental auditing focused on compliance with regulation, more recent certificate programs, such as ISO 14000 or EMAS, stress EMSs and beyond compliance performance. Some companies use audits to examine energy efficiency or resource use and reuse (Wang *et al.*, 2000), while others integrate financial and environmental accounting (INTOSAI, 2007) or address health impacts and safety issues (Strasser, 2003). Environmental audits are generally considered to be a key means of assessing progress toward and achieving the objectives set out in EMSs. In addition, several technical tools — software, satellite systems, wireless sensors, and electronic reporting and record keeping — have been developed to facilitate compliance and alleviate operational and managerial burdens associated with environmental auditing (Gilbert, 1999; Jenks, 2002; Lueders, 2005; Rainer, 2004; Huffer, 2000). As environmental auditing becomes more complex, certified professional environmental auditors are in higher demand and multidisciplinary auditing teams are beginning to emerge (Chiang and Lightbody, 2004; De Moor and De Beelde, 2005).

In general, environmental auditing has a short but dynamic history. Established as a means of gaining regulatory compliance, environmental auditing is an increasingly large and diverse area of environmental management that includes multiple actors who use audits for a variety of purposes. To better understand the diversity, we now turn to exploring the modern contours of environmental auditing.

3. The Contours of Environmental Auditing

Environmental auditing is composed of a flexible group of policy and management tools that provides a means of gathering and assessing information relevant to an organization's environmental status. There are two broad characteristics of environmental audits: *process* and *purpose*. In policy documents and prior literature, audits are described as processes that are composed of a systematic, documented, periodic, and objective review that yields information about the status of the environmental facets that matter to the organization (De Moor and De Beelde, 2005; US EPA, 2000). This procedural characteristic differentiates environmental auditing from environmental reporting. Environmental auditing is a whole assessment process that produces documentation that may or may not be disclosed, while environmental reporting concerns integrated documentation and disclosure that could occur as an outcome of an audit.

As for purpose, the two most frequently mentioned objectives of environmental auditing are compliance assessment and performance assessment. Compliance

coordination and sets guidelines for efficient and effective management of internationally funded environmental auditing programs (INTOSAI, 2007; Van Leeuwen, 2004).

assessment concerns the extent to which an organization acts in accordance with the environmental regulations (Cahill, 1998; Shah, 1996; Wei, 2001), whereas performance assessment concerns the extent to which the organization effectively and efficiently meets environmental objectives, which may or may not reflect regulatory criteria or stringencies (Tucker and Kasper, 1998; Whitelaw, 1997; De Moor and De Beelde, 2005; Gottlieb *et al.*, 1995; Maltby, 1995). In performance assessment, environmental audits may be integrated with financial audits by identifying the potential cost savings and profits in addition to checking environmental input–output balances. Performance assessments are intended to be conducted on a systematic, periodic basis and to inform performance improvement. Performance assessment can be used in organizational decision-making process by identifying weak or missing points of a current management system, and identifying priorities for allocating resources in the future plan. In practice, compliance and performance audits are commonly combined (INTOSAI, 2007).

Although environmental audits perform functions similar to other types of audits,[3] they also have several distinguishing factors that are unique to the environmental policy and management arena. In contrast to financial auditing, for example, no generally accepted principles exist for environmental auditing (Coyne, 2006; De Moor and De Beelde, 2005). Environmental audits may include a selection from a menu of measurable financial, environmental, and health criteria. Additionally, environmental audits can be used to assess whether an organization's EMS is designed and implemented to best achieve and maintain compliance, thereby facilitating management control of environmental practices (De Moor and De Beelde, 2005; Gottlieb *et al.*, 1995; Maltby, 1995).

In sum, environmental audits have a wide variety of uses for adopting organizations including

- assessment of an organization's performance against internally established environmental goals or in comparison with external environmental management standards;
- identification of potential cost savings and profits by checking environmental input–output balances;
- identification of weak or missing components of an EMS and assistance in identifying priorities for allocating resources and special emphasis;

[3]INTOSAI (2007) identifies the three broad types of audits used in the public sector — financial, compliance, and performance — which can be used to address environmental, natural resource, and sustainable development issues. Financial audits can assess "whether a government's financial statements reflect its environmental costs and liabilities," whereas compliance audits can assess "compliance with spending authorities, and with environmental laws, treaties, and policies." Financial and compliance audits are known collectively as regularity audits. The third type of audit, performance audits, assesses "whether a government meets its environmental objectives, is effective in producing environmental results, and operates efficiently and economically." It is not uncommon for SAIs to conduct environmental audits that combine some or all aspects of financial, compliance, and performance audits. A combination of two or more of these three types of audits is often described as a comprehensive audit.

- assessment of an organization's compliance with existing environmental regulations;
- and identification of potential environmental risk and assistance in identifying alternative processes or procedure to reduce or eliminate the risk.

Given the variation in definitions and applications, it is important for research, policy, and management to develop a means of conceptualizing and categorizing environmental auditing activities. The next sections provide some guidance in this regard.

4. Framing Environmental Auditing

Surveying the literature, there are two prominent dimensions of environmental auditing — auditor and audience — that help provide a framework for understanding the range of different auditing arrangements and activities. The auditor can be either internal or external to the target organization (Dittenhofer, 1995). An internal auditor constitutes an individual, group, or office that operates within the target organization's boundaries. Additionally, the internal auditor is compensated by the organization to conduct the environmental audit and report on the findings. An external auditor refers to an entity that exists outside the target organization's formal boundaries and is tasked to conduct the audit by either the organization or some other interested party. The second dimension, audience, can also be dichotomized as either external or internal. An internal audience would include a company's board of directors, top managers, or employees who work in the organization that is the subject of the audit.

The environmental audit may be used internally to assess and mitigate environmental risks, improve production inefficiencies, or reduce sustainability gaps. The external audience includes a wide array of potential groups or organizations. An external audience could be a government regulatory body, industry association such as the Chemical Manufacturer Association's Responsible Care Program or the National Biosolids Partnership. Regulatory agencies are major reviewers of environmental audits, especially environmental compliance audits (Fig. 1). In the US, EPA, USDA, and GAO play a major auditing role for several environmental regulations (Gottlieb *et al.*, 1995; INTOSAI, 2007). Other governmental agencies in the state and federal levels are also engaged in environmental audits related to specific regulations. Additionally, environmental compliance audits may be evaluated within the court system when they are submitted during a lawsuit or may be sought by insurance companies during litigation for purposes related to environmental damage claims (Thomas *et al.*, 2000). Citizens and interest groups may also review publicly released audit reports to monitor compliance with environmental regulations and to track the environmental quality of production and management systems.

The two different dimensions of environmental auditing give rise to four quadrants. Quadrant one (internal audience and auditor) depicts the situation in which

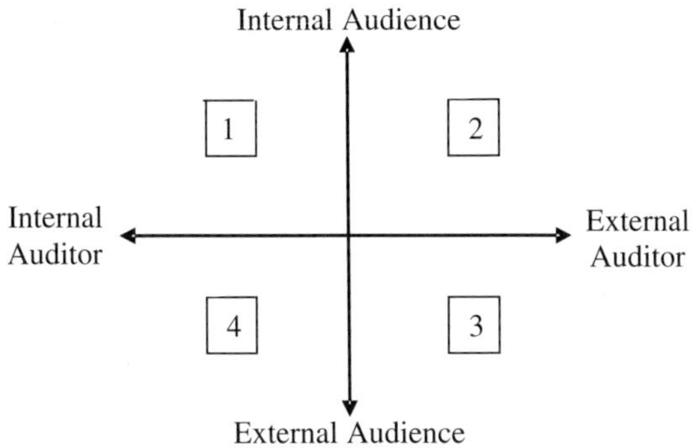

Fig. 1. Auditor and audience dimensions of environmental auditing.

an organization conducts an internal environmental audit solely for its own purposes. Organizations may prefer such audits in situation when a set of pollutants are not regulated but may have pollution risk associated with them. Alternatively, such audits may be conducted to identify material or energy losses that would be the target of efficiency improvements. Quadrant two (internal audience, external auditor) identifies a situation in which an external auditor is hired by an organization to conduct an audit that would only be used for internal management and decision making. External auditors could include consultants or auditing firms that are knowledgeable about environmental auditing practices and the practices of the organization. As with quadrant one, motivation for the audit would be driven by risk, insurance, or economic rationales. Quadrants one and two are most strongly associated with performance objectives of environmental auditing.

Quadrants three and four present situations in which the audit is conducted, at least in part, for an external audience, which may include other supplier or competitor organizations, a government regulatory agency or an international standards organization. These quadrants are most associated with the compliance objectives of auditing. In quadrant three (external auditor and audience), an external government, not-for-profit or private organization conducts the audit and provides the results to at least one interested external party. In quadrant four (internal auditor, external audience), an entity internal to the organization (either an internal department or a contracted organization) conducts the audit and the results are provided to an external agency or organization. These two quadrants are most associated with compliance assessments.

While the dimensionalization of auditing appears to be straightforward, it is important to recognize that two dimensions — auditor and audience — are actually continuums. The relationship between the external auditor and the target organization can vary depending upon the extent to which the target organization has political or economic authority over the auditor. Similarly, the extent to which the audience of a particular audit is internal or external may differ depending upon the marked and regulatory context, and some audits may be broad in scope (numerous regulated and non-regulated environmental aspects), but only a narrow selection of results are made available to external groups with the remainder designed for internal consumption.

Despite the heuristic nature of the framework, the separation of auditor and audience helps set up a discussion about how the design of the audit might relate to the behavior of the organization. To explore the relationship between institutions and behavior as it relates to auditing practices and policy, succeeding sections of this chapter explore existing policy approaches and key theoretical rationales might predict organizational adoption and compliance.

5. Mandatory and Voluntary Environmental Audit Policy

There are two main government policy approaches to environmental compliance audits that are intended for an external audience: mandatory and voluntary. Policy may mandate environmental audits for several reasons including no information on environmental impacts of organizations exists, there is recognition that the environmental aspects of a particular production process carry high risks to human and environmental health, or an organization has been found to be in non-compliance (Macho-Stadler and Perez-Castrillo, 2006; May and Winter, 1999). Mandatory audit policy generally requires an organization to conduct one-time or periodic audit and disclose the audit report. Box 2 provides one example of a mandatory audit scheme.

Box 2. Mandatory government environmental audit.

> Prior to 2009, Indonesian government officials were only able to investigate individuals or organizations when criminal violations of environmental laws had occurred. Among the 517 audits of firms in the mining, oil and gas, manufacturing and pulp and paper sectors in 2008, 40 were found to be non-compliant. None of them were directly brought to the court. In 2009, the Indonesian parliament passed the Environmental Protection and Management Law No. 32, stipulating that organizations whose activities are likely to have a significant impact on the environment or are suspected of non-compliance with environmental regulations must conduct periodic environmental audits. Organizations that do not have an environmental impact statement on file must also conduct an environmental audit by 2011. Moreover, the law requires that civil servant investigation results are submitted directly to prosecutors, thereby strengthening enforcement. When an organization fails to carry out an environmental audit, the Minister for the Environment is authorized to carry out or appoint a third party to carry out the audit (Kandar and Sidharta, 2010; Simamora, 2009, 2010a, 2010b).

In contrast, voluntary auditing policy provides a more flexible approach to gaining improved environmental performance. Proponents of environmental audits argue that this flexibility of voluntary audit policy enables firms to capture of a broad array of environmental factors, many of which regulators may not otherwise detect and may help achieve higher levels of environmental protection without increasing enforcement costs (Innes, 2001; Kaplow and Shavell, 1994). For example, firms may consider a variety of issues — risk reduction, visibility to customers and suppliers, and efficiency gains — that ultimately have positive effects on environmental management and compliance with regulations (Kim, 1997; Mishra *et al.*, 1997; Stanwick and Stanwick, 2001). As noted above, entities such as the ISO, EMAS, Chemical Manufacturers Association provide guidelines for appropriate auditing and establish the processes by which environmental auditing takes place. Boxes 3 and 6 (later in the chapter) describe environmental audit components of the ISO 14001 and EMAS supranational voluntary schemes, respectively.

Box 3. International standards organization voluntary environmental audit.

ISO 14001 is an international standard for environmental management developed by a non-governmental organization, the International Organization for Standardization. Since 1996, the ISO has published the ISO 14000 series of EMSs including eco-labeling and life-cycle assessment. The ISO 14001 out of the series features "a certification (registration) process on establishing, operating, and auditing EMS" (ISO, 2010). The ISO 14001 certification is a voluntary compliance scheme, which is not originally bounded by national or regional-wide regulations, although some countries and industries have integrated ISO standards and certifications into regulatory processes (ISO, 2010).

ISO certification and auditing activities are carried out by the third-party auditors or certification bodies. According to ISO's guidelines on EMS auditing (ISO 14010–14019), a certification process is composed of multiple-stage auditing by third-party auditors or certification bodies. Environmental auditing for an ISO 14001 certificate includes preliminary assessment, document review, interview, on-site audit, examinations and observations of the system. When an organization meets ISO standards, an auditor or a certification body issues an ISO 14001 certificate that is valid for three years. In addition to the primary audit for certification, a certification body conducts a periodic follow-up audit in order to confirm on-going compliance over the three-year period. After three years, the certificate can be extended upon completion of a re-certification audit (ISO, 2010). ISO 14001 does not require an organization to disclose audit information to the public (Potoski and Prakash, 2005a). This feature reflects the current condition that ISO 14001 is primarily an internal management tool used by companies to improve "their systematics, legal security and innovative ability" (Wenk, 2004). Still, an organization may voluntarily disclose its audit information in order to communicate with its external stakeholders or advertise positive environmental performance.

In addition to government policy and international standards associations, industry associations and private corporations develop environmental management standards and systems that incorporate environmental auditing provisions or recommendations. Box 4 provides an example of an on-farm irrigation audit program sponsored by the Ritso Society, a corporation in New Zealand.

Box 4. Private sector environmental auditing initiative.

The Ritso Society is a stand-alone corporation that has the following stated objectives:

- to promote and develop knowledge and understanding of water enhancement schemes on the Central Canterbury Plains between the Waimakariri River to the north and the Rakaia River to the south ("Central Plains") [New Zealand];
- to represent the interests of the Central Plains primary producers on issues and statutory decisions pertaining to the allocation and use of water;
- to raise awareness in Central Plains of the potential that irrigation brings for economic development;
- to promote and develop knowledge and understanding of the issues associated with the use of irrigation and associated land use change (Ritso Society, 2010).

Society members, include Central Plains farmers, business people, and other groups or individuals connected with irrigation and land use change in the Central Plains of Canterbury, New Zealand. Supported through grants from other private and public organizations (MAF, 2010), the Ritso Society collects information on irrigation, water use, and land use to be used as input into a Central Plains Irrigation Scheme Sustainability Code that will ultimately "enable water users to achieve best practices... is practical and economically viable for water users; is dynamic and can be adapted over time to incorporate new technologies, and deal with new issues as they arise (CPWT, 2006)." To implement the irrigation scheme, farmers are required to submit farm plans for approval and monitoring as set out in Ritso guidelines. External audits are conducted to ensure compliance with the farm plan, to inform remediation if necessary, and to help irrigators "achieve high environmental standards and to add credibility to the process (Irrigation New Zealand, 2010)."

The Ritso Society Environmental Farm Plan for Irrigated Land Use Manual outlines how farm plans should be developed and provides information on auditing (Ritso Society, 2009a; Ritso, 2010b). The farm plan guidelines are flexible in the sense that farmers identify objectives and targets for farm practices, while the external auditor determines progress made and level of compliance (fully compliant, compliant with corrective actions, compliant with future actions, non-compliant) and verifies overall compliance before signing off

(*Continued*)

Box 4. (*Continued*)

(Ritso Society, 2009b). Key questions that must be addressed in the audits include:

- Have all the best management practices (BMPs) been used? If not, why not?
- Have any BMPs been rejected in the plan?
- Have any new BMPs been introduced that are not recorded in the plan?
- Do records identify progress/deficiencies in meeting the stated objectives and targets?
- Were the records easily located, understandable, and complete?
- If the objectives/targets have not been met what corrective actions have commenced?
- Plans in place to have a nutrient budget completed for the property within the next 3 months.
- Are these actions working? If not, why not? What further steps are planned?

Sanctions can be assessed for non-compliance, which means that this program voluntary in the sense that the community of users agree to establish an irrigation scheme, but once developed compliance is required.

6. Theoretical Insights to Organizational Auditing Behavior

When audit policies or programs are developed to provide information to external audiences, the design depends upon assumptions and understanding about the behavior of the organization and the institutional norms within which auditing activities take place. Several theoretical approaches help clarify why organizations do or do not conduct environmental audits and how effective a particular audit policy may be in terms of gaining compliance. Rational self-interest theory recognizes that organizations conduct only those activities in which the benefits outweigh the costs: where the reduction in regulatory burden or increase in economic benefits exceeds the resources required to conduct an environmental audit. Principal agent theory helps explain why some institutional structures result in greater levels of compliance, information disclosure, or transparency of the auditing process. Club goods theory shows that by conducting audits and demonstrating compliance with environmental standards, firms and other organizations gain benefits through preferred customer relationships, marketing advantages, and regulatory benefits. Institutional theory shows that not all activities conducted by a firm are strategic; many behaviors occur because the social context within which an organization exists has established norms that are followed by member organizations. Finally, network theory helps understand how ties among organizations, as well as resource flows may determine auditing outcomes. Each of these is discussed in turn.

6.1. *Rational self-interest theory*

According to this approach, organizations are self-interested, willing only to spend time and money to undergo an environmental audit when the expected benefits exceed the anticipated costs (Montero, 1999). Benefits could include economic and managerial advantages realized by both internal firms and external stakeholders. For example, by performing an environmental audit, a company may learn which parts of their production system are vulnerable and they may be able to better anticipate future problems and thereby reducing risk exposure and increasing competitiveness (Kim, 1997; Mishra *et al.*, 1997; Stanwick and Stanwick, 2001). Other benefits of environmental audits may include the development of management decision support data, improved measurement, better environmental compliance, greater awareness or appreciation of employees, and stronger public relations and marketing (Kim, 1997). Basri (1999) notes that environmental audits create transparency that may encourage investor confidence.

The costs of environmental auditing are substantial and include personnel costs, management time, and auditor and consultant fees (Kowal, 1998). When the scope of audit increases, the cost goes up, and in most cases, it is the audited organization that pays. Comprehensive EMS audits require more versatile expertise and more managerial effort to mobilize information and resources needed for systematic diagnosis. Costs can also accrue when documents and other materials recorded during the auditing process result in criminal and civil liability (Arlen, 1994; Hawks, 1998). The production of an audit report can create a paper trail that is usable by governmental agencies and/or private groups in civil or criminal actions against the organization, which may bring about potential legal costs (Arlen, 1997). As organizations disclose their non-compliance, the future costs to correct their violation may also occur. Thornton *et al.* (2008) point out that the costs to voluntarily correct the non-compliance can be higher than the anticipated non-compliance fines, even when considering reduced penalty provisions.

Because the costs associated with auditing generally reduce the motivation to conduct them (Boiral and Sala, 1998; De Moor and De Beelde, 2005), it may be necessary for external entities to offer incentives to encourage voluntary adoption or impose sanctions for non-compliance (Macho-Stadler, 2008; Montero, 1999; Pfaff and Sanchirico, 2000; Stafford, 2006). Incentives offered by government could include reduced or exempted fines for regulatory non-compliance, reduced reporting requirements, technical assistance or protection from litigation for self-disclosure. US EPA and several state governments try to encourage participants to self-audit by offering reduced penalties and technological audit resources (see Box 1). There are several studies on the strategic incentives for self-auditing, including one by Pfaff and Sanchirico (2000), which show how incentives can be properly aligned to firms' auditing efforts. At the same time, several studies illuminate negative effects of incentives and propose stronger policies focused on increasing the costs of non-compliance and the possibility of detection of non-compliance (Macho-Stadler and Perez-Castrillo, 2006; May and Winter, 1999; Thornton *et al.*, 2008).

6.2. *Principal–agent theory*

Principal–agent theory also provides insights about how institutional arrangements affect the behavior of firms and other organizations regarding environmental auditing and compliance. According to this theory, the principal contracts an agent and to conduct an activity that benefits the principal, but is costly for the agent to carry out (Laffont and Mordimort, 2002; Mitnick, 1975; Moe, 2006; Pratt and Zeckhauser, 1991). Using EPA's voluntary self-audit policy as an example, the principal is a regulatory agency or standards organization that delegates the tasks of environmental auditing to firms, which are agents who carry out tasks such as installation of required pollution reduction equipment or implementation of an EMS. In exchange, they may receive permits to operate or certification of compliance. Industry and trade associations may also play a role of the principal requiring member firms to conduct self-audits and self-reports in order to improve the industry's environmental image or fulfill a regulatory demand. Similarly, environmentally-conscious consumers and interest groups may be principals in the sense that they can provide economic or political pressure to bear on firms and thereby encourage environmental behavior.

Because it is difficult to create a fully explicit contract and to observe or monitor the agents and because the goals of the principal and agent are different, several problems arise. For example, when the agent's activity is difficult to observe and monitor, *information asymmetry* — an imbalance in the amount or quality of information held by the two parties — may result (Laffont and Mordimort, 2002; Mitnick, 1975; Moe, 2006). When an agent's behavior is difficult or costly to monitor, principals cannot determine if the agent has conducted the contracted activity. As rational actors, agents resist openness and transparency preferring to minimize costs of compliance through some form of cheating or shirking. Thus, even though an agent may commit to a contract that serves the goals of the principal, information asymmetry provides discretionary power to the agent and enables the agent to pursuit its own interests. For example, because it is difficult for a principal to observe the internal self-auditing activity of a particular manufacturing facility, the facility may prefer not to undergo as detailed an audit process as promised. Because the principal cannot observe the auditing process, information asymmetry exists. Figure 2 provides an illustration. Without including incentives or disincentives that increase the willingness of the agent to be transparent, the agent may not abide by the contract and, consequently, not fulfill the goals of the principal. This problem is called *moral hazard*. Similarly, the problem of *adverse selection* occurs in a market situation in which customers select products that do not contain the expected or advertised qualities. For example, in the absence of verifiable environmental labeling, advertising may dupe customers into purchasing "green" products that are less rather than more sustainable.

Information asymmetry and moral hazard have consequences for the effectiveness of environmental audits (Arlen, 1997). Because a regulatory agency may not be able to predict the level of counterfactual self-audit reports (Montero, 1999), the

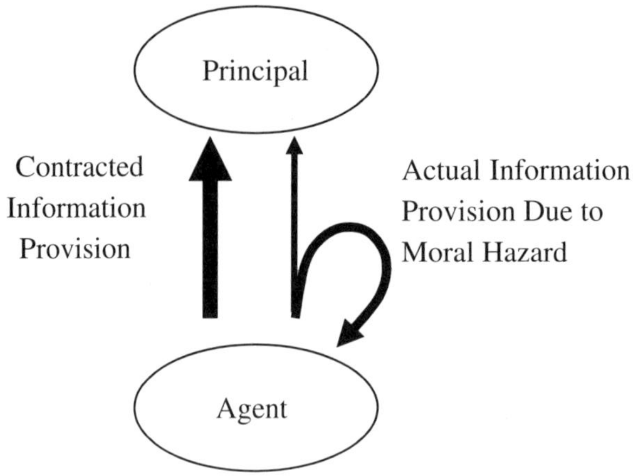

Fig. 2. Information asymmetry.

effectiveness of self-audit policy may be called into question. To respond to this challenge, environmental agencies, trade associations, and international organizations have developed elaborate environmental audit standards, certification programs, and systems of independent third-party auditors. (Two prominent environmental standards and certificates, ISO 14001 and EMAS, are discussed further elsewhere in the chapter.) Third-party auditing systems are designed to simultaneously increase the integrity of the auditing process and provide some level of protection to the firm. Several characteristics of third-party auditors are considered requisite, independence, integrity, objectivity, consistency, and professional knowledge (Gray, 2000, 2001; Karapetrovic and Willborn, 2000).

At its most basic, institutionalization of a third-party auditor is expected to increase the accountability and effectiveness of the environmental auditing system. There are two general forms of third-party auditing, one in which the agent contracts with a professional auditor to conduct an audit and another in which there is a three-way relationship among the third-party auditor, the principal, and the agent. The first form is weaker because the auditor is contracted only by the agent and has no formal relationship with the principal. The two forms are depicted in Fig. 3. Comparing the diagrams on the left- and right-hand sides of the figure, the agent on the left is more likely to retain information when the auditor does not have a formal relationship with the principal.

When the principal and agent have a formal relationship, there is likely to be an increased flow of actual information to the principal. However, even with the right-hand form of third-party auditing, there may still exist a problem of information asymmetry. The auditor, because of customs or laws such as attorney client confidentiality, or because of some unknown relationship with the firm, may agree to remain silent when the firm chooses not to disclose information to the principal. This may be especially true when information disclosure could result in legal compliance

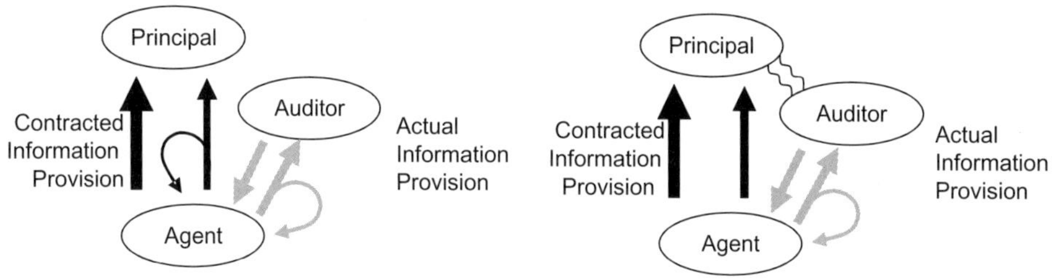

Fig. 3. Information asymmetry with third-party auditor: with and without principal–auditor linkage.

costs or loss of market competitiveness (Delmas, 2002; Gottlieb *et al.*, 1995). Hence, the third-party auditor system may not completely solve principal–agent problems.

6.3. *Other theoretical approaches*

Rational self-interest theory and principal agent theory focus on the behavioral motivations of the organizations and the ways in which incentives, disincentives, and institutional design can solve problems of adoption and compliance. By contrast, club goods, normative institutional and network theories highlight the ways in which organizations are embedded within institutional contexts that foster environmental behavior.

Club goods are a type of impure public good that is generally provided by a non-government entity, is excludable and, depending upon crowding, rival (Buchanan, 1965; Sandler and Tschirhart, 1997). Access to the basket of benefits available through the club depends upon membership, which is either fee-based or requires some cost of entry. Prakash and co-authors have shown that voluntary environmental management standards such as ISO 14000 and EMAS represent clubs in which membership is granted through certification (Kollman and Prakash, 2002; Potoski and Prakash, 2005b). Certification is costly and requires organization to agree to observe established standards. Auditing is a type of standard adherence cost that member firms may be required to undergo.

Research has shown that certification also accrues substantial benefits to members (King *et al.*, 2005; Kollman and Prakash, 2002; Chan and Li, 2001). For example, auditing may provide a signal to other club members that the organization is committed to improvement of environmental performance and that it has met threshold that is accepted by other club members. The conduct of an audit, as well as the disclosure of audit results, may reduce information asymmetries between suppliers and purchaser. Purchasers may prefer establishing contracts with club member because they may be perceived to have lower associated risks, higher environmental quality outputs, or other advantages. Additionally, regulators may perceive club members to be more likely to comply with established regulations and be more willing to either provide them with special treatment, such as reduced monitoring, consolidated reporting, or other forms of reduced oversight and control. Essentially, club goods theory captures the institutional context within which increased benefits or reduced costs associated with environmental auditing — as they are outlined in Sec. 6.1 — can be realized.

Conformity to institutions, including norms, rules, and expectations, represents another set of theoretical rationales for why organization would agree to conduct audits and comply with their commitments. Organizations exist within a social context that carries with it norms of accepted or expected behavior. Conformity with or adaptation to norms may be more important than economic or other rationales. Research in this area indicates that organizations alter structures or behaviors in ways that mimic dominant patterns. Such efforts are designed to increase acceptance and recognition as legitimate actors as well as enhance their own competitive position (DiMaggio and Powell, 1983; Meyer and Rowan, 1977; Powell, 1988). Other contextual factors such as regulatory style, rule formality, societal demands, and expectations have also been shown to affect organizational behavior (Gunningham *et al.*, 2004; LaFollette *et al.*, 1999; Winter and May, 2001).

Social network research has provided a platform for understanding the integration of structure and behavior and the mechanisms for why organizations may adopt and comply with environmental auditing schemes. Organizations are embedded within a network that may provide support or resources. Social exchange theory posits that entities within a network may offer to contribute knowledge, skills, and expertise and in return receive certain benefits (Monge and Contractor, 2003). Exchange requires reciprocal interdependence in which an organization must provide a resource (or have the capacity and intention to provide a resource) in order to receive a resource from another member of the network (Emerson, 1976). Additionally, social support theory suggests that provision of assistance and encouragement meet needs for approval and security, and buffer the impact of stress (House *et al.*, 1988). Organizations may require assistance or obtain support from others in the network depending upon their own circumstances and contextual factors.

These mechanisms — institutional conformity and social exchange and support — provide further pathways to understanding why organizations may undertake environmental auditing. While studies of adoption and compliance with auditing do not generally take advantage of these perspectives, policy examples exist (see Box 5). When organizations embedded within a social context in which environment auditing or public disclosure is the norm, such as when an industry association is a strong proponent or when a large proportion of competitive firms has visibly adopted environmental auditing as a recognized practice, their propensity to adopt the practice may rise. Additionally, when organizations are members of networks that demonstrate traditions of reciprocal exchange of resources or social support, adoption by a few key members who are particularly resource rich or powerful, may induce others to follow. Failure to follow other may compromise norms of reciprocity.

Box 5. Adoption of whole farm plans (WFP) to prevent disease outbreaks.

Skaneateles Lake is located about 20 miles from the city for which it supplies drinking water: Syracuse, New York. The lake is surrounded by approximately 45 small dairy and small crop farms that are located on gradual slopes that lead down to the lake. Farm animals are sources of protozoan parasites, such as giardia and cryptosporidium, that are recognized potential public health threats (AWWA, 1998). Because there is substantial uncertainty about the conditions under which severe outbreaks of these parasites could occur, water suppliers normally install filtration plants to remove the parasites. However, in this case, due to high anticipated costs and willingness of the state to develop a new co-operative regulatory approach to hazard mitigation, the New York Department of Health determined that the city could avoid installation of a filtration plant under the condition that farmers would agree to develop WFP to reduce the risk of cryptosporidium contamination.

Skaneateles Lake Watershed Agricultural Program (SLWAP) is a small interagency pollution prevention program established by NY State in 1994 to encourage new farming practices that would enhance the level of water quality. Geographically located near the farmers, the SLWAP was set up to work with farmers on a cooperative basis to voluntarily assess farming practices, develop WFP that included BMPs and conduct periodic follow up audits to assess progress. Farmers were be responsible for implementation of their WFP, while the SLWAP provided a regulatory buffer between farmers and other regulatory agencies, and ensured that farmers would not incur economic costs resulting from implementation. To build trust, the approach was designed to be open, transparent, and flexible, in which farmer perspectives and interests were visibly integrated into WFPs (Welch and Marc-Aurele, 1999).

Initially, the program received little attention from the farmers, probably due to the complexity of the program, concerns about cost and disclosure, and the history of mistrust between farmers and traditional regulatory agencies. A few farmers, especially those who depended upon farm income and would benefit from lower regulatory oversight and costless improvement on the land, adopted SLWAP's voluntary program early in its history. However, the program stalled and appeared to be in danger of failure. The SLWAP continued to engage the farmers, focusing efforts on gaining the trust of several of the leaders of the Skaneateles farming community who tended to be less economically dependent upon farming and more suspicious of regulators. Over time the SLWAP gained the confidence of a few of the strongest leaders and once they agreed to develop WFP, most of the rest of the community of farmers followed. This case provides some insight into how social networks and institutional context matter for the establishment of a voluntary program that includes environmental auditing.

6.4. *Summary*

The theories described in this section include rational self-interest, principal–agent, club goods, normative, and network approaches to understanding why organizations adopt environmental auditing and comply with their commitments. Self-interest theory illuminates the costs and benefits of environmental auditing as a primary factor to determine organization adoption helps understand the potential effectiveness of environmental audits. Principal–agent theory helps understand the problems caused by authority delegation and information asymmetry in environmental auditing and calls for an attention to transparency and accountability of environmental auditing for better policy and program design. Other theoretical approaches highlight considerations of interdependent relationships and institutional contexts for environmental auditing. While these approaches have mainly been applied by researchers in other areas, including other areas related to environmental policy and management, it is important to recognize their applicability to environmental auditing and to sustainable management more generally. Environmental auditing is likely to continue to be offered as a voluntary option by government, industry, and international organizations. As a result, there is a strong need for policy, and program developers to understand the range of factors that encourage the application of environmental auditing as a constructive tool for improved environmental management. To better understand how auditing may affect management, policy, and governance, the chapter now turns to a final section on challenges and tensions.

7. Challenges and Implications for Auditing

Given the framework and classification of different types of auditing arrangements and activities that occur around the world, it is important to identify some of the key challenges that exist both for the purposes of designing auditing policy and for the development of organizational strategy that implements and uses environmental audits. There are two primary challenges that have been identified in this chapter: integration and tension between flexibility and accountability. The challenge of integration refers to the proliferation of audits, audit standards, and the extent to which environmental audits are integrated with both management practices and other types of audits. It also concerns the extent to which the internal use of an audit to inform management to improve decision making is linked to its external use to inform stakeholders. The second challenge refers to the tension that exists between the flexibility present in voluntary systems and the accountability that actors in society demand when claims of sustainability are made. Both of these challenges have implications for the design and implementation of environmental auditing systems.

7.1. *The management challenge of integration*

There are several different ways in which integration occurs that may affect adoption and implementation of environmental audits. Earlier sections in this chapter often distinguish between the potential internal and external value that environmental audits have for organizations. Internally, environmental audits can play an important role in establishing and maintain EMSs in an organization. Audits enable assessment of the environmental soundness of management systems and allow monitoring of environmental performance. Externally, audits enable an organization to demonstrate compliance with regulations and help persuade other important stakeholders of its commitment to sustainable management. Relatedly, audits can be conceived of as either aids to management or critiques of environmental performance (Maltby, 1995): they may be implemented as either a formative or a summative assessment tool where formative implies the existence of a feedback system for on-going improvement, whereas summative indicates some level of finality. External-internal value and formative–summative approaches to auditing are often driven by different goals — some public and some private; some shorter term and some longer term — and by different internal and external stakeholders. Evidence shows that organizations generally identify a mixture of external and internal motivations for adoption of an EMS, yet little research in this area relates to environmental auditing (see Schucht's work in Box 6). Hence, the challenges for management are to identify ways in which internal and external objectives overlap and to develop auditing systems that maximize external and internal benefits by simultaneously satisfying multiple objectives.

Box 6. EMAS audit requirements.

EMAS is a voluntary supranational standard for environmental management, which is applicable only in Europe (EU, 2010). To be certified, an organization must obtain independent confirmation by the third-party auditor that its EMS is EMAS compliant. The organization is also required to schedule periodic internal checks and independent audits in order to monitor continued compliance with EMAS standards. An organization that participates in EMAS is supposed to conduct internal check every year (Wenk, 2004) and an external audit should be executed at three year intervals, at a minimum (Delmas, 2002, 1998).

EMAS places stronger emphasis on external stakeholders that, which ISO 14001 (Freimann and Walther, 2001; Wenk, 2004). Under EMAS, firms must publish an environmental performance statement after being audited. The statement can be disclosed to the public (Delmas, 2002). There is therefore a possibility that firms participating in EMAS use the environmental statement in order to communicate with stakeholders on their environmental policy (Delmas, 2002).

Firms recognize that joining EMAS results in many different types of benefits that are only available once certification is obtained. For example, a small survey of French EMAS members identified a list of motivations for participation (Schucht, 2000), which are ordered in terms of most to lease important:

(*Continued*)

Box 6. (*Continued*)

improvement of environmental performance of the enterprise; improvement of the company's image; assurance of legal compliance; motivation of employees; improvement in co-operation with public authorities; gain of competitive advantage; cost reduction; gain of preferential treatment from clients (e.g., get more orders); anticipation that the company will be compelled to participate in the future; expectation of simplified administrative procedures; gain of preferential treatment from insurance companies; gain of preferential treatment from banks.

Another way in which integration provides challenges concerns the complexity of the auditing environment. Organizations are increasingly exposed to many different auditing tools, faced with multiple auditing requirements and opportunities (Karapetrovic and Jonker, 2003). Different stakeholders or regulatory agencies may indicate preferences of one type of environmental audit system over another. Different types of audits exist — environmental and financial — and given that environmental performance is not separable from business activities, the integration of financial and environmental audits has the potential to reduce conflicts and bring about synergy effects in management (Karapetrovic and Willborn, 2001). While greater integration of different types of auditing systems would be of benefit for organizations, it is a difficult and complex to conduct at the organization level. Recent research indicates that integration of different EMS audits is most likely to occur among larger organizations that have greater capacity to do so (Bernardo *et al.*, 2010). Hence the challenge for management is to establish auditing systems that effectively integrate multiple auditing activities in ways that combine data and provide insight into different organizational goals — financial, environmental, and social.

7.2. *Flexibility and accountability tension*

As noted several times in this chapter, there exists an underlying tension between the flexibility of an auditing program or policy and the accountability of the auditing system to the broader audience of corporations, not-for-profit organizations, government, and citizens. Flexibility generally refers to the structure of voluntary environmental auditing programs in which organizations follow general auditing guidelines and choose their own methods of implementation. High discretion voluntary audit policies may lead higher level of environmental protection because self-reporting can include a variety of issues beyond the regulatory scheme (Innes, 2001; Kaplow and Shavell, 1994). They may also result in lower costs of policy administration and flexible structures may make implementation more cost effective for the organization.

Simultaneously, environmental audits are expected to be reliable and valid mechanisms for critically assessing the status an organization's environmental performance or compliance. Without such qualities, organizations that undertake audits are not accountable to the stakeholders that offer the voluntary programs. Auditors are expected to be scientifically legitimate and politically independent enough to produce objective audit reports. However, there exists a concern about environmental audits reliability, despite of the advance of audit technology. For example, researchers and others have highlighted limitations of environmental auditing and accounting in the identification and systematic quantification of environmental impacts and risks (Bebbington and Thomson, 2007). Furthermore, there exists a concern about the validity of audits. Because auditors always have more information than the audience, it is challenging for the audience to determine if audit reports are objective and complete. Nevertheless, higher levels of reliability and validity are costly. Greater assurance requires stricter oversight and control or higher expenditures.

The tension between flexibility and accountability leads to several important questions for policy makers and program administrators. For example, are the procedures and instruments of audits reasonable to objectively review the facilities or operation systems? Do cost- or knowledge-constraints negatively affect the precision of the audit? How transparent or confidential should audit information be? How environmentally accountable should business activities be? To what extent should environmental regulations require sustainable practice? All these questions are related to the public values and social norms embedded in a society, which plays an important role to determine audit design. Nonetheless, how they are answered characterizes the underlying social, economic, and environmental values that predominate.

8. Conclusions

Environmental auditing has become increasingly diverse despite a relatively short history. This chapter summarized the different dimensions of environmental auditing that have evolved over the past 40 years. Early environmental auditing focused on compliance with environmental regulations, while more recent approaches stress EMSs. Voluntary audit policies and programs are dominant compared with mandatory approaches, but there exists substantial variation in the way auditing policies integrate incentives, require information disclosure, and take advantage of different institutional and behavioral approaches to persuading organizations to undertake environmental audits.

In order to design effective audit policies or programs, it is important to begin with an understanding the history, character, and components of environmental auditing. The framework introduced in this chapter helps conceptualize the sources of tensions and challenges for environmental auditing policy and practice. Similarly, to implement and carry out effective environmental auditing practices, organizations must traverse multiple challenges and develop ways of integrating multiple goals. Theoretical perspectives provide insight into some of the assumptions and practices related to environmental auditing: costs and benefits, information disclosure, club goods, social networks, and social norms, and the tensions between organization self-interests and social or institutional constraints. It is reasonable to expect that new, innovative approaches to environmental auditing systems will continue to be developed. To meet challenges and address tensions, policy makers, managers, and researchers must consider how environmental auditing systems respond to and fit within the assumptions and practices described in this chapter and elsewhere.

References

American Water Works Association (AWWA), 1998. *Cryptosporidium: Answers to Questions Commonly Asked by Drinking Water Professionals.* American Water Works Association, Denver, CO.

Arlen, J., 1994. The potentially perverse effects of corporate criminal liability. *The Journal of Legal Studies* **23**(2), 833–867.

Arlen, J., 1997. Controlling corporate misconduct: An analysis of corporate liability regimes. *New York University Law Review* **72**, 687–779.

Balikove, H.R. and Jain, A.-S., 1996. The value of EHS auditing in the United States: Solutions for audit protection. *Environmental Quality Management* Winter, 43–45.

Basri, C., 1999. Corporate transparency: The triple bottom line audit. *New York Law Journal.*

Bebbington, J. and Thomson, I., 2007. Social and environmental accounting, auditing, and reporting: A potential source of organisational risk governance? *Environment and Planning C-Government and Policy* **25**(1), 38–55.

Bernardo, M., Casadesus, M., Karapetrovic, K. and Heras, I., 2010. An empirical study on the integration of management system audits, *Journal of Cleaner Production* **18**, 486–495.

Boiral, O. and Sala, J.M., 1998. Environmental management: Should industry adopt ISO 14001. *Business Horizons* **41**, 57–63.

Bovard, J., 1996. The EPA's latest power grab. *Wall Street Journal* A22.

Buchanan, J.M., 1965. An economic theory of clubs. *Economica* **32**(125), 1–14.

Cahill, L.B., 1998. The value-added compliance audit. *The Internal Auditor* **55**, 28–31.

Carson, P. and Moulden, J., 1991. *Green is Gold: Business Talking to Business About the Environmental Revolution.* Harper Collins, Toronto.

Central Plains Water Trust (CPWT), 2006. *Mitigation, Monitoring and Conditions. This Assessment of Environmental Effects Annual Report.* Retrieved on August 20, 2010 from http://www.cpw.org.nz/pdf06/Section9.pdf.

Chan, K.-Y. and Li, X.-D., 2001. A study of the implementation of ISO 14001 environmental management systems in Hong Kong, *Journal of Environmental Planning and Management* **44**(5), 589–602.

Chiang, C. and Lightbody, M., 2004. Financial auditors and environmental auditing in New Zealand. *Managerial Auditing Journal* **19**(2), 224–234.

Coyne, K.L., 2006. Sustainability auditing: Evaluating organizations'progress toward sustainable development. *Environmental Quality Management* Winter, 25–41.

De Moor, P. and De Beelde, I., 2005. Environmental auditing and the role of the accountancy profession: A literature review [Review]. *Environmental Management* **36**(2), 205–219.

Delmas, M.A., 2002. Environmental Management Standards and Globalization.

DiMaggio, P. and Powell, W., 1983. The iron cage revisited: Institutional isomorphism and collective rationality in organizational fields. *American Sociological Review* **48**, 147–160.

Dittenhofer, M., 1995. Environmental accounting and auditing. *Managerial Auditing Journal* **10**(8), 40–51.

Emerson, R.M., 1976. Social exchange theory. *Annual Review of Sociology* **2**(1), 335–362.

Emery, A.R.T. and Watson, M., 2003. Eco-auditing and environmental liability: An international perspective. *Managerial Auditing Journal* **18**(8), 631–636.

European Union (EU), 2010. *Eco-Management Audit Scheme.* Retrieved on August 20, 2010 from http://ec.europa.eu/environment/emas/about/summary_en.htm.

Freimann, J. and Walther, M., 2001. The impacts of corporate environmental management systems: A comparison of EMAS and ISO 14001. *GMI* **36**(Winter), 91–103.

Gilbert, J.B., 1999. Selecting software for environmental compliance. *Chemical Engineering* **106**(10), 127.

Gottlieb, R., Smith, M. and Roque, J., 1995. Greening or greenwashing? The evolution of industry decision-making. In: Gottlieb, R. (Ed), *Reducing Toxics: A New Approach to Policy and Industrial Decision Making.* Island Press, Washington, NW, pp. 170–197.

Gray, R., 2000. Current developments and trends in social and environmental auditing, reporting and attestation: A review and comment. *International Journal of Auditing* **4**, 247–268.

Gray, R., 2001. Thirty years of social accounting, reporting and auditing: What (if anything) have we learnt Business Ethics. *A European Review* **10**, 9–15.

Gunningham, N., Kagan, R.A. and Thornton, D., 2004. Social license and environmental protection: Why businesses go beyond compliance. *Law and Social Inquiry-Journal of the American Bar Foundation* **29**(2), 307–341.

Hawks, R.P., 1998. Environmental self-audit privilege and immunity: Aid to enforcement or polluter protection? *Arizona State Law Journal* **20**, 235–274.

House, J.S., Umberson, D. and Landis, K.R., 1988. Structures and processes of social support. *Annual review of sociology* **14**(1), 293–318.

Huffer, E., 2000. Legal and policy framework for electronic reporting of environmental compliance reports. *Abstracts of Papers of the American Chemical Society* **220**, 124-AGRO.

Innes, R., 2001. Violator avoidance activities and self-reporting in optimal law enforcement. *Journal of Law, Economics & Organization* **17**, 239–256.

International Organization of Supreme Audit Institutions (INTOSAI), 2007. *Evolution and Trends in Environmental Auditing.* Retrieved on August 20, 2010 from http://www.environmental-auditing.org.

International Standards Organization (ISO), 2010. *Environmental management and other management standards.* Retrieved on August 20, 2010 from http://www.iso.org/iso/about/the_iso_story/iso_story_environmental_management.htm.

Irrigation New Zealand, 2010. Retrieved on August 20, 2010 from http://www.irrigationnz.co.nz/assets/Uploads/Ritso-Poster.pdf.

Jenks, R., 2002. Mastering environmental tasks using compliance-management tools. *Chemical Engineering* **109**(11), 83–86.

Kandar, F. and Sidharta, D. (2010). *New Environmental Law — Better Protection or More Legal Hurdles for Industry Asian Legal Business News.* Retrieved on August 20, 2010 from http://asia.legalbusinessonline.com/site-search/new-environmental-law-better-protection-or-more-legal-hurdles-for-industry/46199?keyword=indonesia.

Kaplow, L. and Shavell, S., 1994. Optimal law enforcement with self-reporting of behavior. *Journal of Political Economy* **102**, 583–606.

Karapetrovic, S. and Jonker, J., 2003. Integration of standardized management systems: Searching for a recipe and ingredients. *Total Quality Management and Business Excellence* **14**(4), 451–459.

Karapetrovic, S. and Willborn, W., 2000. Quality assurance and effectiveness of audit systems. *The International Journal of Quality and Reliability Management* **17**, 679–690.

Karapetrovic, S. and Willborn, W., 2001. Audit system: Concepts and practices. *Total Quality Management* **12**, 13–28.

Kim, U., 1997. *Environmental and Safety Auditing: Program Strategies for Legal, International and Financial Issues.* Lewis Publishers, CRC Press, Boca Raton, FL.

King, A.A., Lenox, M.J. and Terlaak, A.N.N., 2005. The strategic use of decentralized institutions: Exploring certification with the ISO 14001 management standard. *Academy of Management Journal* **48**(6), 1091–1106.

Kollman, K. and Prakash, A., 2002. EMS-based environmental regimes as club goods: Examining variations in firm-level adoption of ISO 14001 and EMAS in U.K., U.S. and Germany. *Policy Sciences* **35**, 43–67.

Kowal, S.M., 1998. Corporate compliance programs: A shield against criminal liability. *Food & Drug Law Journal* **53**, 517–525.

Laffont, J.J. and Mordimort, D., 2002. *The Theory of Incentives: The Principal-Agent Model.* Princeton Princeton University Press.

LaFollette, S. *et al.*, 1999. Beyond regulatory compliance — Enhancing environmental health with an education paradigm. *Journal of Environmental Health* **62**(5), 8–13.

Lueders, W., 2005. Wireless sensors for environmental compliance. *Chemical Engineering* **112**(8), 40–43.

Macho-Stadler, I. and Perez-Castrillo, D., 2006. Optimal enforcement policy and firms' emissions and compliance with environmental taxes. *Journal of Environmental Economics and Management* **51**(1), 110–131.

Macho-Stadler, I., 2008. Environmental regulation: Choice of instruments under imperfect compliance. *Spanish Economic Review* **10**(1), 1–21.

Maltby, J., 1995. Environmental audit: Theory and practices. *Managerial Auditing Journal* **10**(8), 15–26.

Maltby, J., 1995. Environmental audit: Theory and practices. *Managerial Auditing Journal* **10**(8), 15–28.

May, P.J. and Winter, S., 1999. Regulatory enforcement and compliance: Examining Danish agro-environmental policy. *Journal of Policy Analysis and Management* **18**(4), 625–651.

Meyer, J. and Rowan, B., 1977. Institutionalized organizations: Formal structure as myth and ceremony. *American Journal of Sociology* **83**, 340–363.

Ministry of Agriculture and Forestry (MAF), 2010. Retrieved on August 20, 2010 from http://www.maf.govt.nz/sff/about-projects/search/05-117/index.htm.

Mishra, B.K., Paul Newman, D. and Stinson, C.H., 1997. Environmental regulations and incentives for compliance audits. *Journal of Accounting and Public Policy* **16**(2), 187–214.

Mitnick, B.M., 1975. The theory of agency: The Policing 'paradox' and regulatory behavior. *Public Choice* **24**, 27–42.

Moe, T.M., 2006. Political control and the power of the agent. *The Journal of Law, Economics and Organization* **22**(1), 1–29.

Monge, P.R. and Contractor, N.S., 2003. *Theories of Communication Networks.* Oxford University Press.

Montero, J.P., 1999. Voluntary compliance with market-based environmental policy: Evidence from the US Acid Rain Program. *Journal of Political Economy* **107**(5), 998–1033.

Pfaff, A.S.P. and Sanchirico, C.W., 2000. Environmental self-auditing: Setting the proper incentives for discovery and correction of environmental harm. *Journal of Law Economics & Organization* **16**(1), 189–208.

Potoski, M. and Prakash, A., 2005a. Covenants with weak swords: ISO 14001 and facilities' environmental performance. *Journal of Policy Analysis and Management* **24**(4), 745–769.

Potoski, M. and Prakash, A., 2005b. Green clubs and voluntary governance: ISO 14001 and firms' regulatory compliance. *American Journal of Political Science* **49**(2), 235–248.

Powell, W., 1988. Institutional effects on organizational structure and performance. In: Zucker, L. (Ed), *Institutional Patterns and Organizations: Culture and Environment.* Cambridge, MA, Ballinger, pp. 115–136.

Pratt, J.W.P. and Zeckhauser, R.J., 1991. *Principals and Agents: The Structure of Business Boston.* Harvard Business School Press.

Rainer, D., 2004. NC State environmental management plan and compliance assistance tools for laboratories. *Abstracts of Papers of the American Chemical Society* **227**, 020-CHAS.

Ritso Society, 2009a. *Preparing Your Environmental Farm Plan for Irrigated Land Use: Audit Report Example Dairy Farm Plan.* Retrieved on August 20, 2001 from http://www.ritso.org.nz/PDFs/Farm%20Plan%20AUDIT%20dairy%20example%20FINAL%20June09.pdf.

Ritso Society, 2009b. *Preparing Your Environmental Farm Plan for Irrigated Land Use: Workshop Manual, An Environmental Management System for Irrigation Schemes in New Zealand.* Retrieved on August 20, 2001 from http://www.ritso.org.nz/PDFs/Farm%20Plan%20Workshop%20Manual%20June09.doc.

Ritso Society, 2010. Retrieved on August 20, 2001 from http://www.ritso.org.nz/index.html.

Sandler, T. and Tschirhart, J., 1997. Club theory: Thirty years later. *Public Choice* **93**, 335–355.

Schucht, S., 2000. *The Implementation of the Environmental Management and Eco-Audit Scheme (EMAS) Regulation in France. Research Paper 2000-B-2*, Cerna. Retrieved on August 20, 2010 from http://portale.unibocconi.it/wps/wcm/connect/resources/file/eb858747912c0b3/WP%20Iraldo%20Rev29.pdf.

Shah, G.C., 1996. Use environmental audits to improve facility-wide compliance. *Hydrocarbon Processing* **75**, 83–86.

Simamora, A.P., 2009. *Polluters to Face a Minimum Sentence of Three Years in Prison. The Jakarta Post.* Retrieved on August 20, 2010 from http://www.thejakartapost.com/news/2009/09/03/polluters-face-a-minimum-sentence-three-years-prison.html.

Simamora, A.P., 2010a. *Firms Given Two Years to Come Clean The Jakarta Post.* Retrieved on August 20, 2010 from http://www.thejakartapost.com/news/2010/03/12/firms-given-two-years-come-clean.html.

Simamora, A.P., 2010b. *Autonomy Exacerbates Damage: Minister. The Jakarta Post.* Retrieved on August 20, 2010 from http://www.thejakartapost.com/news/2010/07/02/autonomy-exacerbates-damage-minister.html.

Stafford, S.L., 2006. State adoption of environmental audit initiatives. *Contemporary Economic Policy* **24**(1), 172–187.

Stanwick, P. and Stanwick, D., 2001. Cut your risks with environmental auditing. *The Journal of Corporate Accounting and Finance* **12**, 11–14.

Strasser, P.B., 2003. Environmental, health, and safety management systems and auditing programs: Part I — The evolution. *Aaohn J* **51**(4), 161–163.

Thomas, W.L., Frey, B.C. and Daves, F.F., 2000. *Crafting Superior Environmental Enforcement Solutions*. Environmental Law Institute, Washington, NW.

Thornton, D., Kagan, R.A. and Gunningham, N., 2008. Compliance costs, regulation, and environmental performance: Controlling truck emissions in the US. *Regulation & Governance* **2**(3), 275–292.

Tucker, R. and Kasper, J., 1998. Pressures for change in environmental auditing and in the role of the internet auditor. *Journal of Managerial Issues* **10**, 340–354.

United Nations Environment Programme (UNEP) Global Reporting Initiative, Unit for Corporate Governance in Africa, 2010. *Carrots and Sticks: Promoting Transparency and Sustainability*.

US EPA, 1986. *Environmental Auditing Policy Statement. 51 Fed. Reg. 25003, 25006* (July 9, 1986).

US EPA, 1995. *Incentives for Self-Policing: Discovery, Disclosure, Correction and Prevention of Violation. 60 Fed. Reg. 66705*, Dec. 22, 1995.

US EPA, 2000. *Incentives for Self-Policing, Discovery, Correction and Prevention of Violations. 65 Fed. Reg. 19618*, Apr. 11, 2000.

Van Leeuwen, S., 2004. Developments in environmental auditing by supreme audit institutions. *Environmental Management* **33**(2), 163–172.

Wang, S., Hu, D. and Yang, J., 2000. A comprehensive environmental auditing tool of products: Life cycle inventory. *Huanjing Kexue* **21**(3), 27–30.

Wei, N., 2001. How to conduct your own environmental audits? *Pollution Engineering* **33**, 39–41.

Welch, E.W. and Marc-Aurele Jr., F.J., 1999. Implementation of environmental policy in Katowice, Poland: Is there cause for concern? *The International Journal of Environmental Studies* **56**, 259–284.

Wenk, M.S., 2004. EU's eco-management and audit scheme. *Environmental Quality Management* Autumn, 59–70.

Whitelaw, K., 1997. *ISO 14001 Environmental Systems Handbook*. Butterworth-Heinemann, London.

Winter, S.C. and May, P.J., 2001. Motivation for compliance with environmental regulations. *Journal of Policy Analysis and Management* **20**(4), 675–698.

Chapter 18

SUSTAINABLE DEVELOPMENT THROUGH GLOBAL LEARNING AND TEACHING

Birgitta Nordén[a] and Elsie Anderberg[b]
[a]*Malmö University, Sweden*
birgitta.norden@mah.se
[b]*Jönköping University, Sweden*
elsie.anderberg@hlk.hj.se

Toward sustainability, the implementation of global learning for sustainable development (GLSD) is crucial. A better understanding of how to — from a global didactic angle — establish globally genuine dialogues forming nuanced conceptions of sustainable development (SD) is necessary. Global teaching and global learning have to identify the challenges in various contexts for transdisciplinary knowledge formation. Aiming to reach established and new target groups, higher education and secondary school, and informal learning situations demands a holistic understanding. Highlighted from a perspective of preventive management strategies for SD, understanding collaboratively could serve as a tool to reach a deeper knowledge formation process through global learning, i.e., GLSD. Notwithstanding, the global perspective has to be integrated in curriculum to achieve a competence-driven global curriculum. Thereby, capabilities through constructive interaction for various (intercultural) qualities of global learning and knowledge formation for SD will be a central part of the outcome.

1. Introduction

Globalization can be seen as entailing a number of problems. However, in the field of education for sustainable development (ESD), imagine if globalization were not the main threat and that lack of globalization is instead the limiting factor? This at a time when a great expansion of education is experienced and digital media equip individuals with sophisticated understanding, through enhanced accessibility and increased diffusion of knowledge sources around the world (Rossi, 2007). For ESD in particular, since the local challenges also occur within global settings, there is a need for a global perspective to enable global learning and teaching toward sustainability. Burbules (2000) highlights the increasing importance of accessible educational communities, since the global character of these issues provides more learning opportunities, which individual users may benefit from. The individual may thus create more nuanced conceptions to cope with increasing societal complexity. In a global context, sustainability issues and other major challenges in contemporary society could be handled easier through global learning and teaching, due to the more sophisticated knowledge formation, greater availability of bodies of knowledge, and by contributing to widespread awareness at a grassroots level (Rossi, 2007). Anderberg *et al.* (2009) highlight the need for more theory-based knowledge of learning and teaching in global settings, since the field is marked by knowledge that is mostly based on policies, while empirical investigations are rare.

According to Burbules (2000), any limits set to the outbid of global education and structural decisions, or modifications in access fees, should be of major concern to educational communities, independently of whether these are local–global, virtual–real, online, or face-to-face. If global learning, for instance, were not accessible for free, because governments or companies restricted resources, this would entail direct consequences and disadvantages for wide segments of society. Educational institutions at all levels, including higher education (HE), could be left outside the globally characterized learning space.

Globalization in learning and teaching depends on a revolution (Burbules, 2000), where global connectivity via the Internet with the speed of an electron works as a "metacommunity." Accordingly, a space is created with instantaneous interaction and communication, bringing

> */.../cultures and societies of enormous variety into immediate contact. Internet is thus both one of the chief manifestations of globalization and one of its primary causes (Burbules, 2000, p. 349).*

It may thus be stated that a global educational community is constituted, or at least facilitated, by the net.

1.1. *The aim of the chapter*

This chapter is not about social policy and how or what tends to promote and reinforce a more global perspective in education. Nor is focus placed on globalization in terms of changes at economic, technological, political, and cultural levels of society. The purpose is instead to highlight some crucial elements of the global dimension in teaching and learning toward sustainability, in the context of preventive sustainable development (SD) management strategies. Within this framework, global learning for sustainable development (GLSD) has been recognized as a productive concept. This approach addresses the challenge of teaching about the complex field of SD in a number of contexts, including HE, secondary school education, informal or life-long learning situations, and in new constellations, such as outreach from university or enterprises, aiming to reach new target groups.

With this chapter, we wish to shed light on the concept of SD from a global didactic angle. Owing to the complexity of SD issues, it is necessary to bring in an open-minded elucidation of the globalization factors actually present already in the foundation of the SD concept. In any educational context and teaching practice, the concept of GLSD continuously needs to be "renegotiated by participants in every concrete learning situation" (Anderberg *et al.*, 2009, p. 371). While Rauch and Steiner (2006) see SD mainly as a narrowing idea, providing a heuristic format for reflection, the complexities of SD also provide a bearing for processes of global research and learning.

The educational perspective of globalization adopted here and limitations in the scope and focus of this chapter are in other words shaped against the background of the ultimate focus on GLSD. Assuming a global didactic angle, at both school and HE levels, gives students and teachers the opportunity to make SD an integral feature of all education. Thereby, both individual and collective self-development and self-determination are emphasized, while learning and teaching practices can be adapted to crucial issues, concerning our planet and its management for SD. In Section 2, we outline the challenges of the field, followed by an introduction to how education of SD has been approached globally. Both the challenges and the character of the introduction are central to how this field has developed. In Section 4, some notes are made concerning the management issues of the field. In Section 5, we present and discuss empirical findings, particularly with regard to an example of learning and teaching globally with SD. In Section 6, some conclusions are formed.

2. Taking Up the Challenge

Global Learning has come to set the scene today, replacing the earlier concepts of global education, development education, and traditional environmental education (EE), according to Anderberg *et al.* (2009) and Brunold (2005).

> *Over the coming years, the concept of global learning will be discussed more and more and developed in the area of developmental and environmental policies and the education of them (Brunold, 2005, p. 297).*

To define global learning, we shall proceed, step by step, examining several interrelated aspects. First, the term *global* is more comprehensive than *international* (Nordén, 2009). International exchange can take place between a few countries, which could even be neighbor countries (e.g., Sweden and Denmark), or limit cooperation to the same continent. In that case, great efforts might not be required to carry out an international learning program, since learning approaches and culture might not differ widely. However, introducing and concerting global meetings, aiming at global learning, when most of the learners come from diverse cultures and nations in different continents — and different parts of these continents — could present a bigger challenge. At the same time, global meetings have the potential of becoming much more enriching, with a wide variety of views exchanged, in a globally extended learning space. Nevertheless, even if examples of rewarding global learning programs are known (e.g., Young Masters Program, YMP), it is a demanding challenge to coordinate learners in different contexts toward a common goal, leading to interactive and fruitful GLSD. Global learning is not about one-way North–South communication. It is also not a "charity" concept, although this has often been the more or less hidden agenda of global education (Anderberg *et al.*, 2009). The diversity of SD issues has to relate further to global perspectives, in the sense of a possible enlargement of the area of contact for the learners involved, if we are indeed committed to support learners' benefits from meetings in a globally extended learning space. These could be achieved by the incorporation of learners' local experiences, reformulated, and exchanged through dialogues within global settings.

In developed and developing countries, as well as in countries under transition, there is a growing need to cooperate and collaborate on GLSD. Gaudelli (2003) has shown that very little is documented on the effectiveness of global education, concerning the development of students' learning. Similar to the notion of "development-policy education," which as a concept lacked explicit didactic references, emphasis on the global didactic angle results in replacing the earlier concept of "global education" by the concept "global learning" (Anderberg *et al.*, 2009; Rauch and Steiner, 2006).

Hartmeyer (2001) quoted in Rauch and Steiner (2006) also emphasizes this as

> *The interfaces between world-wide equity, multicultural society, global environment issues, the peace issue, and the limits of growth in industrialized countries are all in the center of global learning (Hartmeyer, 2001, p. 37).*

A further difficulty is that already since the 1990s, education has been challenged by the emergence of the idea of "global citizenship" and by globalization in everyday life. The pluralistic tradition highlights different conflicts of opinions and a variety of perspectives on sustainability and development issues. It thereby has the advantage of avoiding indoctrination, since the value dimension is considered, Öhman (2008) concludes. To increase the number of attainable solutions to upcoming challenges, plurality and diversity should be shared in both the global and local community (Öhman, 2008). For an overall implementation of ESD, there are also important implications in the fundamental understanding that the challenges for democracy and education are not to create consensus or unity. Based on the pluralistic tradition, school could become an arena in society, where learning SD would include learning democratic practices and values by *participating in* various situations of democratic communication that encourage a deliberative exchange of views (Öhman, 2008). The values of SD would, in this perspective, be continuously reconsidered in an ongoing discussion aiming toward GLSD.

According to Rickinson (2001), until recently, most studies have devoted less attention to learning processes and the students' learning experiences and preferences than to learning outcomes and "factual" knowledge about sustainability. This has serious implications, since

> *Quintessentially, global learning is not about conveying factual knowledge, but is a critical approach to concerns, interests and experiences. Global learning per se cannot serve to create a better world, but encourages self-determination in a global context (Rauch and Steiner, 2006, p. 121).*

More specifically, global learning presupposes competencies, which individuals need to acquire if they want to actively shape the development of world society. These competencies include management skills (Olum, 2004), such as team skills, readiness to compromise and to cooperate, coping with change, creative, and lateral thinking, the ability to deal with insecurity, integrated thinking, and systemic thinking.

In addition to having implications in education, contemporary globalization refers to a set of technological, economic, and cultural changes, reshaping contemporary life (Cole, 2006; Scott, 2000). Prosser and Trigwell (1997) underline that having a limited conception of a subject matter is frequently linked to limited approaches in teaching. This is highly relevant for learning and teaching SD, since this kind of limitations is contrary to the intrinsic complexity of the subject itself. The teaching and learning approaches are widened by a holistic understanding of the concept of SD, and this can only be achieved by professional teachers striving to support understanding of complexities by applying a transdisciplinary approach. In this perspective, disciplines could be intertwined by a facilitating teacher team with complementary competences, who are able to establish connections between the separate aspects of knowledge, perceptions, skills, issues, and values contained in the syllabus, placing them in the overall objective of seeking for and advancing *holistic understanding* (Smuts, 1926; Peirce, 1934; Bateson, 1972) of the issues at stake.

The relationship between theory and practice is balanced through a combination of holism and perception, according to Hansson (2000):

> *A holistic approach requires an interpretation of the meaning of the parts from an interpretation of the whole and vice versa. Holism and relations are important starting points and a relevant epistemological background to the development of knowledge about the environment (Hansson, 2000, p. 143).*

Reinforcing teaching and learning in any knowledge field demands of the individual teacher, not only knowledge of the subject matter, but also the ability to adapt to the learning situation in a responsible and flexible way, depending on the specific circumstances. Sund and Wickman (2008) have pointed out that objects of responsibilities function as anchor points, helping teachers to manage teaching situations that are variable and complex. At the same time, in an ESD context, the role of the teacher is enhanced from only being the expert to "seeking out interesting real-life problems and framing questions around personally discerned needs in collaboration with students." These are crucial management skills required from the teacher and have also been identified by Olum (2004). According to Sund and Wickman (2008), a responsible ESD teacher needs to cooperate with the students as in a team. To support learning of these complex issues, teacher and students need to work together defining problems, analyzing and answering SD questions from a range of different science perspectives, and applying a fundamentally transdisciplinary approach.

Even though in some ways it may feel counterproductive to present the environment as a world of problems, the best starting point to develop an enlightening global awareness could be relating to an acute sense of global problems, for instance, linked to ecological degradation (Rossi, 2007). According to Smyth (2006), scientists and educators should work closely together to successfully produce interpretations that can be acceptable to pupils and society at large based on a high level of understanding of how healthy systems are constituted. One way is to identify benchmarking learning examples to gain an incitement to apply preventive environmental management strategies (PEMS), which is an aspect that is focused in the teaching at the International Institute for Industrial Environmental Economics (IIIEE), at Lund University, Sweden. It is one of the primary educational components in the Institute's M.Sc. in Environmental Management and Policy (the EMP Program). By 2009, graduates from more than 80 countries had been enrolled on the program (Tojo and Lindhqvist, 2009). To further enrich this encounter of students from all over the world, efforts have been made to enroll students with diverse cultural and educational backgrounds preferably with some work experience.

Critique put forward by Rauch and Steiner (2006) underlines that many ESD issues have been transformed into environmental projects. Partly conducted on an interdisciplinary, action-oriented basis, projects "frequently remained on an activist level and failed to reflect on impacts and underlying causes with the result that learning processes remained superficial." Smyth (2006) emphasizes that treating sustainability issues from a holistic point of view demands an equally holistic perspective in education, involving a deep learning approach and including reflection. Rather than merely focusing separate questions of environmental awareness, literacy, responsibility, competence, or citizenship, hopefully educators will instead come to view knowledge more as a necessary means for understanding real-world issues and less as prescriptive lists of topics.

Students in a globalized setting could also achieve intercultural qualities of learning outcomes (Svensson and Wihlborg, 2010) in terms of competencies and capabilities (Bowden, 2004; Anderberg *et al.*, 2007) needed for constructive intercultural encounters and interaction. Svensson and Wihlborg hold that intercultural learning could lead to a development of "global consciousness" and support global citizenship, capabilities, and competencies. On a global level, an "emergent holistic consciousness," through the connection of cultures to a complex collective whole, would form a collective consciousness (Svensson and Wihlborg, 2010). This challenge takes its main point of departure in the particular global perspective formulated above (Svensson, 1986; Hansson, 2000; Anderberg *et al.*, 2009) and concerns ways to see the whole and the parts, on the one hand, whereas, on the other hand, learning how to relate the parts to one another and to the whole, respectively. Because of the complex demands underlying the discourse of GLSD and related topics, a curriculum dimension is also needed for learning and teaching SD in a globalized context. This is what Svensson and Wihlborg recommend, emphasizing that "globalization and the need for curricula change will become the great challenge in HE worldwide in the decades to come" (Svensson and Wihlborg, 2010, p. 15).

3. ESD and SD on the Agenda in the Global Curriculum

This section deals with the development of SD and its relationships to the globalization of the curriculum. SD established its starting point in the *Stockholm Declaration* (UNEP, 1972), where the notion was related above all to environmental issues. ESD was launched by *Our Common Future*, the report of the World Commission on Environment and Development (Brundtland, 1987). The focus altered toward sustainability in a wider sense with the *Talloires Declaration* (University Leaders for Sustainable Future, 1990), in which university administrators declared their commitment to SD. An additional step was taken, when Agenda 21 (UN, 1993) affirmed that an educational focus had the potential of explicitly reorienting education toward SD. ESD was then propelled forward, as concerted efforts were made to transform EE into ESD, benefiting from the impetus of several supplementary conferences. At a global level, the World Summit on Sustainable Development at Johannesburg in 2002 (UN, 2002) identified the need to integrate ideas of SD within curriculum in all disciplines, and as an integral component at all levels of education (Reid and Petocz, 2006).

A global perspective in the curriculum offers students the potential to relate their experiences to a more extensive context. According to Reid and Petocz (2006), this can also contribute to increasing a public awareness of environmental issues, promote environmental training among educators, and improve provision of basic education. Reid and Petocz (2006) highlight that priority must be given to research on innovative educational approaches, with the potential to facilitate real transdisciplinary thinking, and seeking to integrate SD ideas into the curriculum. If SD is indeed supposed to become a central quality of all education, then collective and individual self-development and self-determination also need to be stressed in both a global and local educational context (Anderberg *et al.*, 2009).

Integrating globalization with the concept of SD does not mean that *being-in-the-world* could somehow substitute *being-in-knowledge* (Svensson and Wihlborg, 2010). It is important to integrate the global dimension with the epistemological dimension, concerning the learning process of SD approaches in this type of global perspective. Instead of focusing the content, learners might primarily focus on the global aspects of learning, since "the very situation the learner is a part of provides the immediate context for learning" (Marton and Booth, 1997, p. 141).

> *While the phenomena the learning task relates to might make up the theme of awareness, the thematic field that surrounds it is made up of aspects of a wider, more global world, with roots in the current culture and branches that reach out to the learner's future world (Marton and Booth, 1997, p. 142).*

A global curriculum for learning SD must be thoroughly examined and designed from an educational perspective, rather than simply adapted to globalization as economic or political policy. Globalization should be considered according to its various dimensions and aspects, including cultural/environmental and economic/political issues (Scott, 2000). Scott further distinguishes between globalization as outcome and as hypothesis. In particular, globalization is seen not only as a description of real processes but also as a certain kind of discourse. However, of even greater interest, according to Scott, is the question whether the globalization process is considered to increase diversity by increasing the importance of the "local?" Or, does globalization tend to produce homogenous outcomes? The difference is that if globalization as culture concerns us most, global–local combinations become both more important and more typical, since the global may combine with special local processes in multiple ways — generating "global" effects. Among the implication of such processes is that the empirical indicators of a simple "globalization-as-outcome-thesis" become far from unambiguous. Deeply confusing, in any event, is the fact that the word "globalization" is used to describe both the process and the end-state of that process.

The global call for integration of SD within the university curriculum requires a response. However, any concrete response will ultimately depend on how those responsible for teaching about sustainability understand the notion (Reid and Petocz, 2006). It has been suggested that ESD should be holistic, transdisciplinary, participatory, critical, multi-method, value-driven, and relevant also locally (Robinson and Shallcross, 2006). As a consequence of integrating SD in the curriculum, educational focus must be quick to respond to changes in the global challenges by knowledge and practices that are convincing and appropriate to our times. Acting and reflecting over values become a mandatory foundation for knowledge formation. It follows that objectives of knowledge relate more to the ability to act, rather than to holding knowledge as a "possession" (Hansson, 2000, 2004). Learning for SD also demands that "the learners' experiences are confronted with challenges, on both local and global levels, formulations expressed in a foreign language, abstract inter- and trans-disciplinary scientific concepts, and de-contextualized knowledge formation" (Nordén and Anderberg, 2010, p. 4).

Above all, pedagogical aspects in the process of integrating GLSD and clarifying how it relates to other educational efforts are issues that urgently need to be elaborated. Such a process of implementing GLSD would consist of the interplay between educational change and educational development (Booth and Anderberg, 2005). Challenging overlapping transdisciplinary and transsectorial (Wals, 2009) collaboration will be needed, and teaching must be developed to build on transdisciplinary methods. This is fundamental, in particular, since GLSD as a research field emerges through transdisciplinary reflection on how the approaches could most beneficially be used together:

> *GLSD includes the objective of resolving global problems, involving critical thinking, skills and values, as well as socio-cultural awareness. Determining the characteristics that could guide teaching strategies, inform curriculum design, and serve as a reference for teachers, administrators and policy makers in charge of the practical implementation of GLSD, are all questions which need further research and debate. Substantial empirical research is needed on learning in global settings, in*

> *particular when dealing with complexity in the context of a world society, focusing competencies both as means and outcome. Innovative educational approaches that facilitate real multi-disciplinary and multi-cultural understanding in global settings need to be investigated, and competence-driven ways to integrate SD into curricula must be prioritised (Anderberg et al., 2009, p. 375).*

To achieve these ambitious goals, additional research is needed on net-based learning and its potential contributions to advancing transdisciplinary teaching for SD (Marsella, 2007). Innovative technologies, global modeling, and global communities assist the introduction to understanding how mutually dependent global forces work. These global challenges have to be considered by universities, schools, and other educational institutions. According to Marsella (2007), research on ways to improve GLSD needs to be founded on relevant knowledge, valid methods, and contemporary approaches. Compared with the results of earlier empirical research carried out by Nordén (2008), it appears that collaborative net-based learning develops a supportive global learning perspective, when ESD is enriched by the dimension of globalization (Nordén and Anderberg, 2010).

Even though ESD objectives put forward in national or international documents have generated innovations at schools, many schools do not apply the full scope of professional freedom they dispose of, to build up these ideas. Management initiatives seem to be lacking. Nevertheless, according to Rauch and Steiner (2006), there are examples of thoroughly premeditated support, which have proved successful, with reflexive processes of research and development for school progress through ESD. Brunold (2005) agrees concerning the potential in the deep learning process, considering that "global learning has to be seen as a mediation of a perspective, which assembles connections of nearby observable problems to worldwide processes" and conflicts. The concept of "global learning," with its various pedagogical opportunities, shows possible ways to reach beyond national interests, into a global space. As a result of globalization processes, on the one hand, and through the use of transdisciplinary methods, on the other hand, global learning becomes a strengthening of the horizon of education, in terms of finding better solutions to real-world problems. The very word "global" stands for the idea that content and process are oriented toward the world (Marsella, 2007).

Society, environment, and economy are the three key areas of SD with culture as an underlying dimension (UNESCO, 2004). Svensson and Wihlborg (2010) also emphasize the importance of culture from both the perspective of global and international teaching and learning. The ability to "think globally" involves appreciating cultural flows, and the cultural content in education must be matched in concrete teaching and learning practices:

> *Pedagogical quality assurance in the internationalization of content in higher education demands awareness about intercultural meetings in educational contexts, as well as the management of such meetings (Svensson and Wihlborg, 2010, p. 4).*

According to Tatto (2007), multifaceted globalization leads to far-reaching changes within many sectors in society. Culturally, increasing communication processes challenge the traditions; politically, it makes ideas such as pluralism more acceptable, and in education — in particular up to secondary school levels — decentralization trends can be observed. Other trends that can be noted are added recognition given to the knowledge fields that are connected to the market, for instance, in pedagogies orienting toward problem-solving. Above all, the processes of SD and globalization are connected to a wide range of social, cultural, economic, technological, and ecological factors, and constitute an irreversible worldwide phenomenon (Brunold, 2005).

As identified and argued by the United Nations Decade of Education for Sustainable Development, DESD (UNESCO, 2004), the didactic processes of global learning and teaching must model the values of SD; sustainability issues also thereby need to be reflected in everyday action. Holistic understanding (Peirce, 1934; Bateson, 1972) of the subject matter SD is necessary, due to its complex character, and an understanding of how different teaching and learning practices relate to the subject and goals of SD. Holism and relations are, according to Hansson (2000), important points of departure, and relevant as an epistemological background for knowledge formation about the environment. From this point of view, the need for more knowledge concerning GLSD is obvious. Educational development within GLSD is needed, so that students can learn more effectively in a global context, and teachers can develop a better understanding of what to teach and how to teach SD in a global context (Nordén *et al.*, 2007). Additional motivation and responsibility for developing global learning exists at HE levels and in school, as well as on a parallel individual level (Anderberg *et al.*, 2009; Tatto, 2007). It is suggested that such development can be founded on holistic understanding of the subject matter, through transdisciplinary approaches, research strategies, and interventions for GLSD in the curriculum. There is not much point in attempting to add SD as an extra subject in existing curricula, while maintaining traditional teaching and learning approaches. The need for real integration into the curriculum is in other words critical. Serious attempts to integrate a global perspective in a principle-led curriculum and combine it with a competence-driven curriculum could develop "applied SD knowledge," which would be a step away from the dominant rhetoric, where SD tends to remain at an abstract level of policy alone (Anderberg *et al.*, 2009; Tojo and Lindhqvist, 2009).

Deeper knowledge on how teaching and learning could be best developed in a global context is also needed urgently (Tatto, 2007). Results of academic performance studies such as Trends in International Mathematics and Science Study, Programme for International Student Assessment, or UNESCO's Delors Report 1996 are criticized by Tatto (2007) and the validity of Organization for Economic Cooperation and Development (OECD) indicators. Tatto argues that such evaluations have the effect of validating and legitimizing curriculum reforms, mainly on a basis of "global indicators of quality teaching and learning," with the hidden agenda of promoting a form of "new accountability" to international agencies, such as the OECD. This allows such agencies to impose their own priorities and preferred modalities of education. Teacher-related reforms usually emerge from the administrative level, via regional and global forces. Influenced by societies' views on teachers' opportunities to teach a variety of knowledge and values that are considered important (e.g., YMP during the Decade for ESD, DESD), teachers are expected to impart these values and actually implement them in the classroom (Tatto, 2007).

Formal school is affected by these globalization trends, since they tend to reduce the teachers' autonomy and control over their work. The teachers' professional role diminishes, and instead of developing their knowledge and practice, teachers lose influence over education programs and become "bureaucrats." In other words, they are obliged to reduce teaching to dealing with standards and certification, according to Tatto (2007), to meet these surface knowledge measuring instruments.

4. Global Learning Management for SD — GLSD

Rauch and Steiner (2006, p. 124) emphasize that "learning aims at acquiring a reflective ability to shape the world," rather than adopting a centrally proposed set of action patterns uncritically. A progressive concept of education, the link between SD and global learning, could in this sense be one that stresses self-development — possibly understood as a kind of management — and self-determination of individuals who interact with the world, fellow humans, and their local society. Education would then "refer to the ability to contribute in a reflective and responsible manner to the developing of society for a sustainable future." Interestingly, Rauch and Steiner mention the potential of communities of learners — i.e., teachers, students, pupils, and researchers — who could reflect in shared settings by identifying inter-relations and options for actions. *This progressive view on GLSD is in line with Nordén (2008) and Anderberg et al. (2009), who also point to the need of developing the individual's competencies to act in a community that stresses self-development as shaping part-to-whole relationships between process and content.* In this section, some clarifications will be provided concerning the role of management in this particularly progressive view of GLSD.

Deming has been called the founder of modern quality management (Olum, 2004). Among other assumptions, Deming (2000) *postulated that management builds on effective two-way communication and breaking down barriers, to encourage education and self-improvement for everyone. As an economist,* Deming outlined a system of transformation that is based on cooperation, rather than competition, since he found the latter destructive. Everyone should work together as a system, with the aim for everybody to win. This is one of Deming's points for management, which also includes the following parts: knowledge about variation, theory of knowledge, and psychology. This philosophy suggests abolishing performance examinations for grades in school and at the job. Deming (2000) has described the prevailing management style as a "prison," and instead shows how a style based on cooperation can help individuals develop joy in work and learning.

According to Robinson and Shallcross (2006), the whole institution and the action approaches are evidence for the need of an "extended" role for schools — including not only the current statutory school age range, but at some point also the education of a much broader range of the local population. Widening the educational imperatives to more than those currently dictated by the curriculum is a strategy that could be adopted to extend the variety of learning institutions. The Global Environmental Management Initiative (Cattanach *et al.*, 1995) is an example of the latter. The Initiative shows organizations how to evaluate the state of the art and identify the weakest areas in their environmental program, using an assessment program, to focus future efforts. Included in their Business Charter for SD are — among a total of 16 principles — an emphasis on educating employees and contributing to public education and policy development through openness and listening to employees and the public. Evaluation systems can function as a key to a proactive environmental management system, and many different evaluation systems are available. Companies that go through these quality processes learn to focus on good communication, according to Cattanach *et al.* (1995). *Management commitment means improving quality and the obligation to implement improvements. Referring to "Theory X" by McGregor, Olum (2004) argues that people want to fulfill themselves by seeking self-respect, self-development, and self-fulfillment at work, as in life in general.*

In achieving more sustainable futures through effective (and efficient) education, *the management assumptions* play a significant role. *Therefore, managerial competence needs to be acquired in certain key areas, such as problem solving, administration, human resource management, and organizational leadership* (Olum, 2004). Management can in this context be defined as the art or science of achieving goals through people — leading to "continuous improvement," since managers also supervise, making sure people do what they are supposed to do. Broadly, management is the process of designing and maintaining an environment, in which individuals working together in groups efficiently accomplish selected aims (Koontz and Weihrich, 2007). The aims that interest us here relate to GLSD, as a further development of ESD, giving individuals the opportunity to take command over their learning within this field, and make sense of their own world experiences.

With a global perspective on SD, emphasis is placed on seeking, rather than setting a standard from a pre-defined ESD format. Above all, this implies opening space — space for creating various paths to develop GLSD. Thereby, diverse ways of experiencing the world meet, where learning on the edge may appear through the dissonance that arises from such meetings. While the concept "SD" is hard to grasp, it also offers schools and institutions of HE an opportunity to confront their core values, students' learning, professional educators' teaching, the pedagogy established, and the relationship with the surrounding community (Wals and Jickling, 2002).

It is argued that heuristic structures (Rauch and Steiner, 2006; Jickling and Wals, 2008) for investigation and reflection on learning, teaching, and research conducted on SD can support individual sense-making. These structures serve as means, serving the processes associated with individual sense-making frames for global issues in connection with SD. The ESD concept is "so slippery and open to different interpretations and so potentially complex" (Wals and Jickling, 2002, p. 129). Therefore, explicit heuristic structures could improve the understanding of relationships between epistemic development and the paradigmatic challenge — focusing teaching about and for SD — while assessing the integration of learning and new ways of coming to know, especially in HE (Wals and Jickling, 2002).

Jickling and Wals (2008) view ESD as a threat, in the sense that they see it as "a product and a carrier of globalizing forces." They believe that the globalizing agenda has instrumental and deterministic tendencies that favor "transmissive arrangements for teaching and learning over more transformative ones." In this process, Jickling and Wals fear that traditional EE risks being marginalized. The same goes for individuals and communities, dealing with such questions in a self-determined, relatively autonomous way, within a context (Jickling and Wals, 2008) that engages people in "existential questions about the way human beings and other species live" on our planet. Brunold (2005) instead maintains that differentiation and equality of all questions are significant for upcoming global rescues, trying to look beyond common sense-making of the "one world" dilemma. Not only is rather static and traditional EE included in ESD, traditional approaches to ESD also tend to marginalize the cultural, economic, and social dimensions, which are important to consider for a holistic understanding of the planet's well-being. As a result of globalization processes, the use of transdisciplinary approaches may serve as an amplification of the horizon of teaching SD (Brunold, 2005) — including GLSD.

Approaching these issues from a holistic point of view is the only way forward. All aspects of relationships between human society and nature need to be considered — seeking more, not less diversity of thoughts about the future concerns of the one earth we have available (Hansson, 2000). Scott (2000) argues that globalization will most probably not lead to "cocacolonization" of the world — e.g., more homogeneity — but rather give more prominence to the values of heterogeneity and increasing diversity. The "local" will strongly increase its significance. Thereby, GLSD could become the tool for mirroring the tension and the equilibrium between the global dimensions and how it is interpreted by individuals out of their local SD challenges.

5. The Global Classroom as an Extended Classroom

5.1. *Empirical studies on learning and teaching in the global classroom during the YMP*

In this section, we present some empirical research on a program that is an example of learning and teaching in a global setting. The YMP[1] is an online course about SD and PEMS that has reached out to some 10,000 students in 120 countries in the period 1999–2010. The program was created at the initiative of the IIIEE at Lund University and is offered to upper secondary school students aged 14–18. The YMP has been free of charge and supervised by the IIIEE during the entire period. Young people work across the globe, to learn about SD, exploring a learning environment with ICT-mediated ways of communication. The course includes an important component of global interaction, where young people exchange experiences and develop their own ideas and descriptions. The program adopts a transdisciplinary approach, focusing on social, economic, and environmental dimensions of the students' daily lives (see Nordén, 2005; http://www.iiiee-ymp.org).

Nordén (2005) investigated young people's learning for SD, analyzing their experiences of online learning in a global setting. The study was part of the research project Learning in the ICT-extended University, conducted by the Centre for Learning Lund at Lund University (Booth *et al.*, 2007). The purpose of this study was to describe and analyze how upper secondary school students experienced the YMP, particularly with regard to what and how they learned in a global context. The study investigated their learning process in the field of SD in the ICT-mediated course, which included a number of complex and strategic issues, such as PEMS. Data were collected through a semi-structured online questionnaire, with both closed and open questions, and the analysis took its point of departure in the answers to this questionnaire, provided by participants during the second part of the YMP in the spring semester 2004, ahead of the Global Environmental Youth Convention (GEYC, 2004) that took place at the library of Alexandria in Egypt.

Results show that the YMP (Nordén, 2005) was important to the students as individuals and as members of both local and global society. They felt it had contributed to their knowledge formation by allowing them to explore a global learning space, when interacting globally through new ICT-mediated ways of communication. Participants also expressed the importance of the transdisciplinary approach that encompassed social, economic, ecological, and ethical dimensions. Feedback was

[1] *YMP* information is available at the following web sites:

- *IIEE* at http://www.iiiee.lu.se/site.nsf/AllDocuments/3FF88126AF2B70EDC1256F6B004 89555
- *Lund Calling* at http://www.lucsus.lu.se/lundcalling/
- *YMP* at http://*www.iiiee-ymp.org/*

experienced as essential, and the global meetings seemed to particularly commit the students in their learning process toward a more sustainable direction, based on the concept of PEMS. "Prevention is better than cure" became a mantra among the students, when applying their knowledge in project work that they presented during the GEYC. Young people as stakeholders outside the university had the opportunity to challenge the academic world, and offer a unique opportunity for dialogue about the role of institutions of HE in an overall societal transition toward SD. To increase the understanding of how global learning processes could be designed and carried out as outreach at the university, the YMP might therefore serve as an educational example.

In addition to the various dimensions of ESD, *what* the students experienced that they learned was, above all, teamwork, collaboration, and awareness of challenges in lifestyles. The students recognized that their learning process was rewarding, and felt that *how* they learned was through didactical feedback from one another, through the challenges and problem solving activities that they worked on collaboratively, and through the remote contacts with advising expertise. The participating students also learned via their teamwork with enthusiastic and pioneering teachers, and facilitating mentors from more than 100 countries.

The main *advantages* with GLSD initiated by the YMP online were, according to the participants: having accessible knowledge, which was experienced as a way to come to know more; global citizenship, as if talking and having a dialogue online was similar to "walking around globally," in a kind of meeting and learning from other individuals' points of view; being a part of "culture meeting culture" in a learning context; learning to work with a globalization tool; and finally the flexibility. Some *disadvantages* the participants of the YMP experienced were: insecurity concerning the relevance and the quality of information available via the Internet; control of how adequate references and different sources were; content efficiency and admittance; technology dependency; time scarcity for communication and interaction. Interesting is that students in a global setting wanted more feedback from experts and students in other cultures, particularly, from teachers in other countries. It is obvious that more feedback in general from teachers and mentors would be appreciated.

> *Independently of the role you have in connection to the YMP — if you are a manager or designer of the course, instructor, tutor or teacher giving feedback, a mentor or student facilitating, a parent or a headmaster encouraging the teenagers in the YMP study — the time allocated by the teenagers for this form of the ESD mirrors how important and how seriously engaging they consider the study of the YMP to be (Nordén, 2005, p. 40).*

To establish a common ground in relation to the learning object shared by the teacher and the learner is important for successful learning (Tsui, 2004). According to Laurillard (2002), the student and the teacher should interact and constitute the learning process through a dialogue. The students found it natural to use reflection and critical thinking, due to the experience of process-oriented learning for

SD that the course offered. Dialogue helped reshape and raise their awareness. The *participatory approach* could be seen as a prerequisite for concrete action, and students felt empowered to act for change (Nordén, 2005; Nordén and Anderberg, 2010).

> *Throughout this study the aim has been to get as close as possible to a focus on the learner's experience of learning. And, in order to teach well it is imperative to continuously learn from the learner — to communicate, talk, listen and getting involved in the community of young people as the youth — sooner than we know — will master the world of knowledge (Nordén, 2005, p. 39).*

The students reported that they collaboratively learned from one another. By focusing on the group and noting the meaning captured in words, a process for knowledge formation is developed collaboratively (Anderberg, 2000). The ideas of value to a community could be expected to lead to improvements and most probably would be greater than the sum of the individual contributions. Or, as expressed by one student, commenting on the new educational (GLSD) approach with interaction: "We learned from it the actual meaning of teamwork" (Nordén, 2005). As also stressed by Anderson and Elloumi (2004), the learners clearly experience that learning globally through distance education can be one of the ways to equip one another SD skills and raise awareness of the necessity of sharing resources to combat human suffering, solve socio-economic problems, and fight ecological degradation.

> *... students mean that they raised their awareness of sustainability issues through fruitful dialogues, while reflecting over problems brought up by students in other countries. It started a process of comparing cultures, life and lifestyles, critical thinking and rethinking. A need of deepening their knowledge and learning problem-solving skills through global meetings occurred. Some students mentioned the challenge in optimizing the use and distribution of resources as energy and material, and be able to have a comfortable life and improve our standard of living (Nordén, 2005, p. 42).*

Hansson (2004) has observed that the target group of upper secondary school students is frequently not aware of certain issues, such as the urgency of the distribution of resources on our earth, and various conflicts of interests (UNESCO, 1977). Many of the YMP students globally already embrace the ecological, the economic, and social perspectives of SD (Nordén, 2005). Some of these students experience that thanks to this opportunity to combine learning ESD content in a global setting with learning at their school locally, a process starts that allows them to include various kinds of subject matter more widely and deeply. Teaching about SD is deepened by using reliable and meaningful learning content provided by experts globally. The YMP has also developed a concept that legitimates the schools' work with a more transdisciplinary approach. The forms for GLSD used by the YMP may help many participating nations include economic, ecological, cultural, and social aspects in the future development of GLSD. Exchange of didactical experiences and the pedagogical research within the field of GLSD could also encourage steps to be taken toward developing the integration of GLSD in the curriculum.

> *However, what happens at local level affects and is affected by what happens at global level — broader tensions of globalisation, visible and articulated at world-wide level, find echoes in the problems people face and the solutions they can envisage (UNESCO, 2005, p. 33).*

GLSD benefits from being rooted in local experiences, starting from the grassroots realities, while at the same time providing a global context in which to set these phenomena. Learning examples of how integration of local knowledge and culture would look like within a global approach need to be defined, however, to qualitatively enrich GLSD in the future. The significance of the learning activity for the participants, as individuals and as networked members of both local and global society, is of importance in efforts toward sustainable global development.

Within the GLSD context, teachers motivated by professional development aim at further extending their teaching (Nordén, 2008). As highlighted by Smyth (2006), these teachers head for a focus on SD complexity, developing knowledge from a holistic perspective, in the sense outlined earlier in this chapter. Teachers' experiences of the YMP were investigated by Nordén (2008). The data consisted of interviews undertaken in the spring semester 2006 with the written answers from 26 teachers in Africa, Asia, South America, and Europe. The interviewed teachers worked in a total of 16 countries (Colombia, Denmark, Egypt, Ghana, Greece, India, Jordan, P. R. of China, Lithuania, Mauritius, Poland, Sweden, Tanzania, Turkey, the United Arab Emirates, and Vietnam). The teachers were found to partner up and cooperate with colleagues globally. This is the underlying ground for teachers' community of learning and teaching in a global setting. A transformation affects the teachers when they go from providing education by teaching, to assuming a mentor role instead. The mentor role involves facilitating for young students to unpack, decode, transform, and make knowledge applicable. To enter the global learning space for SD, students further need to train a certain kind of literacy skills. Together they could *co-create* to attempt understanding the complexity code (Nordén, 2008). A "global teacher team," networking online in a "global teacher room," could provide an option for professionally developing teachers and help them acquire the skills needed to reach out in various learning environments in a variety of countries. This professional skill development might include pooling knowledge among colleagues and exchanging ideas for designing teaching in complex knowledge areas of a global character, such as GLSD. A beneficial process of sharing would contribute to teachers' personal and professional development, and thereby enhance the process of implementing GLSD (Nordén, 2008).

Local teacher teamwork could additionally serve as a support for transdisciplinary knowledge formation, since despite their ambitions, teachers frequently experience that they lack full competence in the knowledge field of SD (Nordén and Anderberg 2010). Teachers from various disciplines, teaching different subject matter, could cooperate as a professional local teaching team. Together they could achieve a kind of transdisciplinary integrated competence for additional steps toward the implementation of GLSD in all its complexity. An integration of subject matter from different disciplines would further contribute to developing a holistic approach to SD knowledge formation.

Another empirical study was conducted by Nordén and Anderberg (2010), concerning the implementation of GLSD in the Swedish pilot project *Lund Calling.* The project concerned the intention to facilitate a pilot effort in a number of schools, aiming at implementing the YMP as part of their regular curriculum. In collaboration with the teachers and the headmasters, the objective was to eventually initiate the YMP as an integrated tool in Swedish compulsory schools (years 8–9) and upper secondary schools. It was hoped that the YMP could be progressively offered to all teachers in Lund Municipality at these levels. The focus was partly on the development of a new curriculum, and also included the criteria for marks and the assessment of the integration in upcoming ESD programmes. Results from this study and the literature review carried out in this connection overall showed that competence to act in a global context requires a holistic approach, including both the concept of a global didactic angle and the subject matter of SD. This means that knowledge, skills, and attitudes are closely linked and developed together. The data consisted of semi-structured interviews with students, teachers, and headmasters at schools in Lund Municipality. The main results show that the relevant skills that teachers and pupils needed to develop are of central interest, and that these skills included the ability to analyze conflicts of interest. Jensen and Schnack (1994) have pointed to a similar set of skills as a kind of "grassroots action competence," required for acting globally. The formation of knowledge and attitudes during the implementation process of GLSD, according to findings from researching Lund Calling, depend very much on the extent to which such skills are developed, an aspect which has also been highlighted in earlier research (Anderberg *et al.*, 2009; Nordén, 2008; Booth *et al.*, 2007). Shifting from the notion of "environmental knowledge," to "knowledge for SD" globally, has fundamentally changed the character of learning, from seeing "knowledge as a possession," toward developing "knowledge to act" (Rickinson, 2001; Hansson, 2000). The field of SD is highly complex, and therefore provides plenty of opportunities for various interpretations and actions. Challenges involved in designing a global classroom may therefore include a topic of learning about complexities on an advanced level (Rost, 2004; Scott and Gough, 2003).

Being able to discuss, cooperate, and have a dialogue in public are all essential skills needed in a democratic society. Other important skills include the ability to see challenges, relate different issues to each other, carry out investigations, critically read and write on issues related to their own ways of living, or look for ways of solving problems collaboratively. Another crucial aspect is learning how to present arguments and gain recognition for one's ideas in society. The students need to be listened to when expressing a standpoint. For democracy to function, young people should feel that they are heard, whenever they contribute to society in a respectful way, as citizens (Östman, 2003). By being confirmed at school — and in society — young people can see themselves as a part of the common efforts toward sustainability (SOU, 2004; Öhman, 2008). In view of changing conditions toward sustainability, also globally, democratic skills can be considered as a form of "action competence" for SD, including actions on both the individual and the structural level.

In other words, we observe the need of tools to reach a deeper knowledge formation process, grasping the complexity in a global learning space. Understanding collaboratively is a form of learning that evolves in a *process perspective*. Additionally, this creates the foundation for well-developed competence to act democratically. With GLSD, both necessary skills and a deeper understanding of the content could be gained locally, via global learning activities in global settings.

Organizational implementation tools need to be revised, and in particular various steering documents, such as the curriculum. The question of where to place SD in the curriculum of education is not only about integration, but more an issue of systemic change within educational institutions. Allowing learning for SD presupposes a fundamentally *transformative approach*, which could facilitate implementation processes of GLSD.

6. Conclusions

Globalization, as we have experienced it from the perspective of learning for SD, may not be seen as something radically new. In some sense, globalization could be said to have taken place for millennia, but the discussion here does not deal with this wider notion. In this chapter, we have instead discerned the need for developing a better understanding of the *global didactic angle*, from which GLSD could be recognized. According to Rossi (2007), this approach is in line with the research frontiers connected to the globalization functions and the notion of an ongoing compression–expansion of time/space/meaning. This process appears to be affecting institutions and educational organizations, in their attempts to extend their reach and increase the scope of education offered. Even media and grassroot organizations appear to perceive the call for GLSD, and feel that they may have a task to fulfill in this area. Combining such efforts will hopefully contribute to further developing appropriate practices for GLSD. Thereby, the individual learner will be better equipped to cope with subject matters of great complexity and form nuanced conceptions of SD. One precondition — and maybe the main condition for such processes — is the establishment of a globally genuine dialogue.

References

Anderberg, E., 2000. Word meaning and conceptions. An empirical study of relationships between students' thinking and use of language when reasoning about a problem. *Instructional Science* **28**, 89–113.

Anderson, T. and Elloumi, F., 2004. Introduction. In: *Theory and Practice of Online Learning*. Athabasca University. Web site visited 2010-04-06: http://cde.athabascau.ca/online_book/intro.html

Anderberg, E., Häggström, S. and Nordquist, A., 2007. *Didactics of the Learning Process, Report No. 2007:15.* Department of Civil and environmental Engineering, Chalmers University of technology, Göteborg.

Anderberg, E., Nordén, B. and Hansson, B., 2009. Global learning for sustainable development in higher education: Recent trends and critique. *International Journal of Sustainability in Higher Education* **10**(4), 368–378.

Bateson, G., 1972. *Steps to an Ecology of Mind.* Paladin Fromore, St Albans.

Booth, S., 2007. Learning in the ICT-extended University: Experienced context and constituted meaning in ICT-supported outreach initiatives involving flexibility and diversity. In: Booth, S., *et al.* (Eds.), *Research for Development for Learning in the ICT-extended University. Learning Lund Report No. 3.* Lunds Universitet, Lund.

Bowden, J., 2004. Capabilities-driven curriculum design. In: Baillie, C. and Moore, I. (Eds.), *Effective Teaching & Learning in Engineering.* Routledge Falmer, London, pp. 36–48.

Brundtland, G., 1987. Chairman's foreword. In: *Our Common Future: Report of the World Commission on Environment and Development. Annex to General Assembly document A/42/427*, Development and International Co-operation: Environment. Oxford University, Oxford.

Brunold, A.O., 2005. Global learning and education for sustainable development. *Higher Education in Europe*, **30**(3–4).

Burbules, N.C., 2000. Does the Internet constitute a global educational community? In: Burbules, N.C. and Torres, C.A. (Eds.), *Globalization and Education — Critical Perspectives.* Routledge, New York, pp. 323–356.

Cattanach, R.E., Holdreith, J.M., Reinke, D.P. and Sibik, L.K., 1995. *The Handbook of Environmentally Conscious Manufacturing: From Design & Production to Labeling & Recycling.* Irwin Professional Publishing, Burr Ridge.

Cole, M., 2006. New Labour, globalization and social justice: The role of teacher education. In: Kassem, Mufti and Robinson (Eds.), *Issues and Critical Perspectives.* Education studies.

Deming, W.E., 2000. *The New Economics: For Industry, Government, Education.* MIT Press, Cambridge, Mass.

GEYC 2000. *Global Environmental Youth Convention 2000.* Lund: IIIEE, Lund University. visited 2010-04-06 at http://www.iiiee.lu.se/geyc

GEYC, 2004. *Global Environmental Youth Convention 2004.* IIIEE, Lund University, Lund. Web site visited 2010-04-06 at http://www.birgittanorden.se/geyc2004_Alexandria_Egypt/egypten.htm

Gough, N., 1987. Learning with environments: Towards an ecological paradigm for education. In: Robottom, I. (Ed.), *Environmental Education: Practice and Possibility.* Deakin University, Deakin.

Hansson, B., 2000. *Förutsättningar för gymnasielevers kunskapsbildning och för undervisning inom miljöområdet. Dissertation.* Department of Education, University of Lund, Lund.

Hansson, B., 2004. Formation of environmental knowledge. In: Wickenberg, P., *et al.* (Eds.), *Learning to Change Our World.* Studentlitteratur, Lund.

Hansson, B. and Nordén, B., 2007. Constitution of meaning for "sustainability" in global networked meetings. In: Booth, S., *et al.* (Eds.), *Research for Development for Learning in the ICT-extended University. Learning Lund Report No. 3.* Lunds universitet, Lund.

Hartmeyer, H., 2001. Globales Lernen in Österreich — Erfahrungen, Erwartungen, Perspektiven. In: Halbartschlager, F. (Ed.), *Eine Welt. Beiträge zu globalem Lernen. Symposion globales Lernen.* Südwind Agentur, Vienna, pp. 34–42.

Jensen, B.B. and Schnack, K., 1994. Action competence as an educational challenge. In: Jensen, B.B. and Schnack, K. (Eds.), *Action and Action Competence as Key Concepts in Critical Pedagogy. Studies in Educational Theory and Curriculum*, Vol. 12. Royal Danish School of Educational Studies, Copenhagen, pp. 5–18.

Jickling, B. and Wals, A.E.J., 2008. Globalization and environmental education: Looking beyond sustainable development. *Journal of Curriculum Studies* **40**(1), 1–21.

Koontz, H. and Weihrich, H., 2007. Essentials of Management: An International Perspective, seventh ed. Tata McGraw-Hill, Delhi.

Laurillard, D., 2002. *Rethinking University Teaching: A Framework for the Effective Use of Educational Technology.* Routledge, London.

Marsella, A. J., 2007. Education and training for a global psychology. In: Stevens, M.J. and Gielen, U.P. (Eds.), *Toward a Global Psychology: Theory, Research, Intervention, and Pedagogy.* Lawrence Erlbaum Associates, London.

Marton, F. and Booth, S., 1997. *Learning and Awareness.* LEA, Mahwah, NJ.

Nordén, B., 2005. Youth learning for sustainable development — Analysis of experiences of online learning. *Master thesis, Learning Lund Report No 2.* Lund University, Lund.

Nordén, B., Anderberg, E. and Hansson, B., 2007. Global learning for sustainable development: Identification of the content complexity as a research problem. Paper at the conference "*Learning Environment and Higher Education in a Global Context — from a Sino-Nordic Perspective.*" Nordic Centre, Fudan University, Shanghai, P.R. of China, 8–11 May 2007.

Nordén, B., 2008. *Lärares erfarenheter av förändrad undervisning — lärare möter övergångar mellan det lokala och det globala klassrummet. Lärande Lund Rapport No 3.* Lunds universitet, Lund.

Nordén, B. and Anderberg, E., 2010. *Learning in Global Settings for Sustainable Development: Local Challenges. Learning Lund Report No. 4.* Lund University, Lund.

Öhman, J., 2008. Environmental ethics and democratic responsibility — A pluralistic approach to ESD. In: Öhman, J. (Ed.), *Values and Democracy in Education for Sustainable Development — Contributions from Swedish Research*, Liber, Stockholm. pp. 17–32.

Olum, Y., 2004. Modern management theories and practices: A critical overview. Paper presented at the 15th East African Central Banking Course, 12th July 2004, at Kenya School of Monetary Studies. Web site retrieved 2010–03–04: http://unpan1.un.org/intradoc/groups/public/documents/AAPAM/UNPAN025765.pdf

Östman, L., 2003. *Nationell och internationell miljödidaktisk forskning: En forskningsöversikt [National and International Review of Research in Environmental Education].* Pedagogiska institutionen, Uppsala University, Uppsala. Web site.

Peirce, C.S., 1934. Collected papers V of Charles Sanders Peirce. In: Hartshorne, C. and Weiss, P. (Eds.), Pragmatism and Pragmaticism, Vol. 5 Harward University Press, Cambridge.

Prosser, M. and Trigwell, K., 1997. Perceptions of the teaching environment and its relationship to approaches to teaching. *British Journal of Educational Psychology* **67**, 25–35.

Rauch, F. and Steiner, R., 2006. School development through Education for Sustainable Development in Austria. *Environmental Education Research* **12**(1), 115–127.

Reid, A. and Petocz, P., 2006. University lecturers' understanding of sustainability. *Higher Education* **51**, 105–123.

Rickinson, M., 2001. Learners and learning in environmental education: A critical review of the evidence. *Environmental Education Research* **7**(3), 207–310.

Robinson, J. and Shallcross, T., 2006. Education for sustainable development. In: Kassem, Mufti and Robinson (Eds.), *Education Studies: Issues and Critical Perspectives.*

Rossi, I., 2007. From cosmopolitanism to a global perspective: Paradigmatic discontinuity (Beck, Ritzer, Postmodernism, and Albrow) versus continuity (Alexander and Collins) and emergent conceptualizations (contributors to this volume). In: Rossi (Ed.), *Frontiers of Globalization research: Theoretical and Methodological Approaches*. Springer, New York.

Rost, J., 2004. Competencies for global learning. *The Development Education Journal* **11**(1).

Scheunpflug, A. and Asbrand, B., 2006. Global education and education for sustainability. *Environmental Education Research* **12**(1), 33–46.

Scott, P., 2000. *The Globalization of Higher Education*. Open University Press/SRHE, Buckingham.

Smuts, J.C., 1926. *Holism and Evolution*. The Viking Press, New York.

Smyth, J.C., 2006. Environment and education: A view of changing scene. *Environmental Education Research* **12**(3–4), 247–264.

SOU, 2004. Statens offentliga utredningar (SOU 2004:104). *Att lära för hållbar utveckling*. Web site visited 2010-04-21 at http://www.regeringen.se/sb/d/108/a/34144

Sterling, S., 2004. An analysis of the development of sustainability education internationally: Evolution, interpretation and transformative potential. In: Blewitt, J. and Cullingford, C. *The Sustainability Curriculum: The Challenge for Higher Education*. Earthscan, London.

Sund, P. and Wickman, P.-O., 2008. Teachers' objects of responsibility: Something to care about in education for sustainable development? *Environmental Education Research* **14**(2), 145–163.

Svensson, L., 1986. Färdighet i att lära. In: Marton, F., *et al.* (Eds.), *Hur vi lär*. Rabén & Sjögren, Stockholm.

Svensson, L. and Wihlborg, M., 2010. Internationalising the content of higher education — the need for a curriculum perspective. *Higher Education*, Published online (DOI: 10.1007/s10734-010-9318-6), Springer Netherlands. Web site visited 2010-02-27 at http://www.springerlink.com/content/13170v8121106565/fulltext.pdf

Tatto, M.T., 2007. *Reforming Teaching Globally*. Cambridge University Press, Oxford.

Tojo, N. and Lindhqvist, T., 2009. Teaching fellow students as a way of motivating future decision makers. In: Bommenel, E. and Irhammar, M. (Eds.), *Osynligt och självklart?* Media-Tryck, Lund.

Tsui, A.B.M., 2004. The shared space of learning. In: Marton, F. and Tsui, A.B.M., *Classroom Discourse and the Space of Learning*. Lawrence Erlbaum Associates, Mahwah, NJ.

UN, 1993. *Agenda 21*. Web site visited 2010-04-21 at http://www.un.org/esa/sustdev/documents/agenda21/english/agenda21chapter36.htm

UN, 2002. *Report of the World Summit on Sustainable Development*. Web site visited 2010-04-21 at http://www.un.org/jsummit/html/documents/documents.html

UNEP, 1972. *Stockholm Declaration: Report of the United Nations Conference on the Human Environment*. Web site visited 2010-04-21 at http://www.unep.org/Documents.Multilingual/Default.asp?documentID=97

UNESCO, 1977. *Final Report. Intergovernmental Conference on Environmental Education organized by UNESCO in Co-operation with UNEP*, Tbilisi (USSR) 14–26 October 1977. Web site visited 2010-01-01 at http://www.gdrc.org/uem/ee/EE-Tbilisi_1977.pdf

UNESCO, 2004. *All Different, All Unique — Young people and UNESCO Universal Declaration on Cultural Diversity*. Division of Cultural Policies and Intercultural Dialogue. UNESCO, Paris.

UNESCO, 2005. *United Nations Decade of Education for Sustainable Development 2005–2014. International Implementation Scheme.* Draft January 2005. UNESCO, Paris.

University Leaders for Sustainable Future, 1990. *The Talloires Declaration.* ULSF, Washington.

Wals, A.E.J. and Jickling, B., 2002. "Sustainability" in higher education: From doublethink and newspeak to critical thinking and meaningful learning. *Higher Education Policy* **15**, 121–131.

Chapter 19

ENVIRONMENTAL INFORMATION SYSTEMS

Steven P. Frysinger
Departments of Integrated Science and Technology and Computer Science, James Madison University, MSC 4102 Harrisonburg, VA 22807, USA
frysinsp@jmu.edu

Computer-based information systems are extensively used in support of environmental management and decision making in a variety of contexts. This chapter provides an overview of three different but related technologies of significant interest in environmental management. First, an integrated approach to Environmental Management Information Systems in support of corporate environmental management is detailed. The concept of more highly Distributed Environmental Information Architectures is then introduced, followed by an overview of Geographic Information Systems technology. The chapter culminates with a synthesis of these and other technologies in Environmental Decision Support Systems focused on particular decisions and decision makers and developed through careful interaction design.

1. Introduction

The success of an environmental management system (EMS) depends on the ability of people to make management decisions based on the processes defined by the EMS. This in turn relies on the availability of high-quality information. Therefore, environmental information systems are essential to the success of a modern EMS. As an EMS is an attempt to integrate the management processes that influence the organization's environmental "footprint," integration is likewise the key to successful environmental information management. This chapter considers environmental information integration in three contexts that are of considerable interest to developers of EMS.

First, an object-oriented approach to integrated Environmental Management Information Systems (EMIS) is described, showing how coupling of key information resources within an organization is both necessary and achievable. Then recent advances in the development of Distributed Environmental Information Architectures (DEIAs) to support the discovery of and access to highly distributed information and service resources are introduced. Geographic Information Systems (GIS) are then described. Finally, the development of Environmental Decision Support

Systems (EDSS) for environmental management is explored as a synthesis of these three preceding technologies, among others.

2. Environmental Management Information Systems

Environment, Health, and Safety (EH&S) organizations intending to develop and operate under an effective EMS must organize and manage many disparate processes within their firms. These environmental management processes have evolved over years in response to regulatory events. Process integration into an efficient management system is necessary to support high-quality decisions determining their overall environmental impact, and by facilitating awareness of both risks and opportunities it can significantly improve the effectiveness of the EH&S organization. Sharing of knowledge and effort across regulatory lines can reduce the cost of EH&S operations. For example, a chemical inventory process might serve the needs of both occupational health and hazardous materials' managers, maintaining compliance at a reduced cost. An integrated EMS is therefore essential to an EH&S organization.

However, integrated environmental management cannot be achieved unless environmental *information* management is also integrated. Management arguably depends heavily upon the availability and accessibility of correct and current information, and most EH&S processes are themselves essentially focused on information management in the form of record keeping, reporting, permitting, or training. Therefore, an integrated EMS must stand upon an integrated EMIS.

While integration of diverse software tools will be shown later to be important to environmental decision support, what is described here refers to information rather than software integration. In the past several years, there have been attempts to develop highly-integrated enterprise-wide software packages with the goal of spanning all aspects of a company's operations and integrating data management across organizational boundaries. Unfortunately, EH&S data management is rarely considered in this process because the EH&S organization, as a cost center, is not perceived to add value to the firm and therefore rarely attracts the investment necessary. Considerable and costly disruption is generally required to implement such systems. The EH&S organization is then left to manage its data on its own, even though much of the information on which it depends is in fact generated and maintained by line organizations within the company. Fortunately, it is possible to achieve information integration without attempting the daunting task of software integration across the firm.

2.1. *The need for integration*

Typical EH&S organizations encompass many management processes, which are usually supported by as many diverse information systems, many of them with little or no professional computer support. These information systems have generally evolved in response to individual needs, without regard to interdependencies between processes and their information management needs. In many cases, these include "home made" software systems developed by environmental managers and industrial hygienists using whatever tools are accessible (such as spreadsheet or personal database applications). They are rarely developed according to professional software engineering principles, and they rarely have sufficient documentation to permit their support once the author has left the organization.

Apart from the obvious inefficiencies inherent in such circumstances, this *ad hoc* structure usually results in redundant databases — multiple databases that store the same piece of information. It is a well-known principle of database theory that redundant databases will eventually result in inconsistencies as a value is changed in one of them but not the others. For example, several EH&S information systems may use facilities' lists from different databases, which eventually conflict with one another. This sort of inconsistency ultimately threatens compliance; it is hard to comply with environmental regulations when the list of regulated facilities is in question.

2.2. *An integrated solution*

Fortunately, an organization may resolve this dangerous situation by developing an integrated EMIS. It is important to note that the term "information system," as operationally defined here, is much broader than the computer hardware and software, which might support it. It includes a data model incorporating the structure, definition, and relationships between data elements, and the processes and procedures by which these data are created, modified, used, and destroyed. While much of this can and should be supported by computer systems, this fact has little relevance to the conceptual definition of the information system. Once the EMIS is defined, a software systems engineering activity can readily determine the design and structure of the hardware and software systems to implement it.

2.2.1. *Conceptual framework*

The object-oriented approach described here is predicated on the notion that one can usefully separate data from the management processes that use them. Specifically, most or all data of use to EH&S describe entities, called "objects" in the language of computer science, while the various management processes undertaken by EH&S professionals focus on these objects. For example, this approach to EH&S information might start with the definition of such high-level objects as Employees, Buildings, Vehicles, Services, and Products. Each of these can then be decomposed in a similar fashion, as appropriate, with the terminal objects described by a data structure.

The various EH&S management processes can generally be viewed as operating on the data objects suggested above. For instance, SARA Title III Section 312[1] reporting is focused (by regulation) on buildings and their contents, while OHSA

[1]Legal references in this chapter are made with respect to US law.

training requirements are focused on employees. Such processes would, in general, create, use, modify, and/or delete data objects.

Finally, each process may be supported by one or more software applications. In general, the software applications serving EH&S processes are the agents, which interact with the data required for these processes.

Thus, this approach to integrated environmental information management envisions a clear separation among *data*, *processes*, and *applications*:

- A *datum* may be used by multiple *processes*, e.g., Building Address is used for SARA Title III reporting and for OSHA accident reporting.
- A *process* may be served by multiple *applications*, e.g., one software application might support the SARA Title III inventory maintenance activity by site personnel, whereas another application is used to generate the SARA Title III Tier 2 reports.
- In some instances, *applications* may be used by multiple *processes*, e.g., the software used by site personnel to maintain chemical inventories may serve the purposes of both SARA Title III and Clean Air Act compliance processes.

In essence, this approach addresses our need to understand the relationship between information and processes so that we may ensure the availability of the correct data and the correct software applications to interact with those data. It can be argued that explication of this awareness is one of the primary purposes of EMS.

2.2.1.1 *The path to integration:* There are four essential steps to achieving such an integrated EMIS:

1. develop an integrated data model;
2. map the integrated data model onto corporate databases;
3. define high-level requirements for the EMIS;
4. implement the foundation of the EMIS.

While some of these can be executed concurrently, it is imperative that we recognize the precedence implicit in their ordering. As with any systems engineering activity, in this activity *what* has to lead *how*. It will be advantageous to look ahead to current and future system implementations to help us to achieve an understanding of requirements, but particular discipline must be applied to prevent us from erroneously finding a requirement in what is merely a habit. This discipline will be encouraged by a phased approach, in which we first define an integrated EMIS for the set of processes as they currently exist, admitting that the model will be revisited as a result (and indeed in support) of efforts to reengineer those processes. It is natural to embark on this task as part of the definition of an EMS, since the latter activity already insists upon a high degree of objective introspection.

The first step in the project is the development of an integrated data model, which correctly describes the firm from an EH&S point of view. The initial (baseline) data model must include all data items required by the current set of EH&S processes, but must be orthogonal to these processes so that data objects and fields that are common to multiple processes occur only once in the data model, to be shared by the processes requiring them. This is critical to the identification of shared information and the elimination of redundant databases. Once such a baseline data model has been developed, it can and should be refined and revised as appropriate to reflect the ongoing reengineering of the EH&S organization's structure and processes.

The integrated data model so developed will then be analyzed to determine the appropriate owner for each of the data categories and elements. In many cases, this will be the so-called database of record for the company, and will not be under the control of the EH&S organization. For example, much information about corporate facilities might be maintained by a real-estate organization within the firm, but outside of the EH&S department. Identifying our stake in such external databases is essential, since as customers of these databases we will need to be recognized and have a voice in the implementation and management of the data. There may also be data items of importance to EH&S, which could readily and should be maintained in these external databases; we will want to be in a position to lobby the appropriate organizations for such extensions. Furthermore, interfaces to these data sources must be engineered so that the data truly *will* be shared, rather than simply copied into yet another system, further contributing to data redundancy.

The third step is the definition of high-level requirements for the integrated information system. The integrated data model and analysis described above form the foundation for this. What must be added are the *functional requirements* for the integrated system. For example, if EH&S information must be globally accessible by EH&S leadership, this requirement should be articulated clearly.

The fourth step addresses the implementation of the integrated EMIS. Implementation includes the interaction and negotiation with other organizations whose information assets have been identified as a subset of the EH&S data model in step 2. It also includes the planning and acquisition and/or development of software required to realize the EMIS from the starting position of our existing information management systems. The result of this step is not necessarily a single software system; in fact, this outcome is highly unlikely, given that the software to be used by individuals and groups engaged in the various processes will have to satisfy functional requirements, which may be peculiar to those processes. As long as the ensemble of computer systems finally in use by the EH&S organization (A) implements the integrated data model developed in steps 1 and 2, and (B) satisfies the high-level requirements defined in step 3, then we will have achieved an integrated EMIS, and will reap the benefits thereof. This, perhaps, is the point of departure of this approach from conventional thinking about integration — we seek to achieve the benefits of integrated *information* while valuing diversity of software applications and vendors.

Once these four steps have been executed, the design and implementation of the integrated system using an appropriate combination of existing and new platforms can proceed through conventional project management and software systems engineering activities. In fact, it might be hoped that through effective communication, any ongoing procurement and development activities underway during the execution of these steps can be appropriately guided so as to minimize changes or disruption once they are complete. For example, an early intermediate result will be the identification of data common to the first key processes to be evaluated. This knowledge can surely be used during the procurement of supporting systems to anticipate the results of the integration effort.

The basic steps of the process just elaborated are highly complementary to the development of an EMS. In both cases, thorough identification of environmental management processes and the information used to conduct them must be identified. Therefore, it is logical to consider conducting these developments in tandem.

3. Distributed Environmental Information Architectures

The conceptual framework underlying the corporate architecture described by the foregoing section can also be applied to environmental information management across organizational and even national boundaries. Unlike a corporate EMIS in which shared data are distributed across a relatively limited and closed internal network, more general environmental management will necessarily rely on distributed access to disparate information repositories through a DEIA. In addition, the explosion of the Internet has also made access to digital services widely available. These services, such as graphical rendering tools or computational models of environmental phenomena, may be accessed in a similar fashion to data.

For example, if an environmental manager needs to use a groundwater pollutant transport model with certain characteristics, they might use a software agent to seek out candidates available via a global network. When they had chosen the particular model and negotiated the terms, they could then incorporate this into an integrated system designed to help them resolve the issue at hand.

Recognizing that these repositories can be numerous, heterogeneous, and widely separated, it is apparent that the use of web-enabled networks will be essential. Whether the databases are distributed within or across organizational boundaries, there are often complicated issues associated with access to these data.

3.1. *Discovery*

The first challenge is *discovery*: how do we (or our software agents) find sources for information and services? Services that are available for use (whether free of charge or for a fee) might publish information about their capability, along with terms of use, in formats that are accessible to machine-based search agents. The association of metainformation with available data and services is a partial solution to this problem, and can be augmented by representative ontologies, which help the search tools to discover the semantics associated with distributed resources so that the likelihood of a match to the need can be determined. These help to structure the metainformation about data and services in such a way that differences of expression between authors and host organizations can be overcome.

Mere discoverability on the web is necessary but not sufficient. Somehow, these resources must be accessible by machine, without complicated or onerous human intervention. Once relevant data and/or services have been discovered, there must be transactions to gain access. Access authorization issues need to be managed in a manner, which is independent of particular information structures without compromising the security requirements of the source. An authorization service, acting as an intermediary between the data or service host and the search agent, will generally be necessary to satisfy this requirement. Such support services, whether they perform any actual environmental function or not, are essential to a successful DEIA.

There may also need to be transactional accounting to deal with the potential that the data or services are available only for a fee. An additional intermediary service to support such transactions will therefore be necessary. Even if access to resources is not associated with a fee, hosts may still desire some level of transaction accounting so that more detailed records of use exist (e.g., to document the value of particular data or services by tracking the extent to which they are used).

Finally, given the large number of existing data and service repositories owned and hosted by nearly as large a number of disparate organizations, it is important that an existing repository can be introduced to the network without modification. Among other things, this implies the availability of translation services, which can interpret the metainformation of the various repositories. Such services often take the form of a layer of functionality wrapped around existing repositories without modifying them.

3.2. *Open architectures*

Access to these distributed data and services is predicated on the assumption that they are open to shared use. Proprietary networks do serve a purpose — for example, by allowing a corporate EMIS, as described previously, to share access to various corporate databases within the firm without making those data available outside of the firm. However, in the more general environmental management domain, access to data across organizations and jurisdictions is essential. For example, to conduct flood management planning one must be able to access stream and river monitoring data from various jurisdictions — counties, states, or even countries — because flood waters do not heed borders.

To accomplish this, without requiring modification of existing repositories, it must be possible to create toolsets that can be modified and extended as necessary to suit particular application domains as well as participants. Computer systems exhibiting these properties have been in existence for decades. For example, the UNIX[2] operating system, originally developed by Bell Labs nearly 40 years ago, was built around an open architecture that facilitated the extension of and addition to its toolset. It is important to note that an open architecture need not imply open source code, although the reverse implication is usually true, at least in principle.

In the case of DEIAs, the value of an open architecture approach is that a network may readily evolve as data and service repositories make themselves available through the terms of the architecture. This has the important effect of allowing applications to develop without waiting for all possible resources to subscribe. To continue the previous flood management example, this means that planners in Germany might begin to develop and use their management system even before their neighbors in France have made their data and services available to the network.

Early progress in Europe (Usländer, 2007) has affirmed the value of an open architecture approach, and provided an example of such an architecture based on open system standards. While still in the formative stages, early experience with this sort of DEIA is very encouraging and offers the promise of vastly increased access to data and models essential to sound environmental management.

4. Geographic Information Systems

Some EMS will encompass organizations and/or facilities of significant spatial extent and whose geographical features play an important role in the environmental aspects of concern. In such cases, it may prove valuable to include a class of software technology called GIS in the organization's information technology framework. GIS provides an approach to the representation and analysis of spatial data that were initially developed for use with physical maps, but which has since developed into a highly sophisticated information technology tool. While GIS is an older technology than DEIAs, it is only relatively recently that it has evolved into one which is readily integrated into specialized systems to support environmental management.

4.1. *Map layers*

A core concept within GIS technology is that of the map layer. In analyzing spatial data for the purpose of making environmental decisions, it is often necessary to compare data of different types but which describe the same location. For example, if one is searching for suitable locations for a storm water retention basin, it will be necessary to consider such variables as the elevation of the terrain (e.g., topographic contours), soil type, and vegetation cover. These data can be overlaid, if rendered

[2]UNIX is a registered trademark of The Open Group.

on transparent sheets, allowing the analyst to readily see locations where suitable values for these three variables coexist.

This concept, first employed with physical maps in the environmental planning community, was introduced to computer-based mapping systems, and map layers have become a core feature of GIS. With a GIS implementation, users can decide which of the available data layers should be displayed simultaneously. They can also choose the order in which these should be displayed, which is important because some data layers might occlude others.

Beyond this display capability, these data layers can also be operated on either singly or in combination by a large variety of analytical tools, about which more will be said later.

4.2. *Spatial data structures*

The aforementioned data layers are actually the primary data structure for the spatial data contained within a GIS. These layers are fundamentally two-dimensional in nature, representing a geographic region delimited by the coordinates of the corners of the layer. However, it is possible to partially represent the third spatial dimension, such as through the use of color within the layer. For example, elevation data may be represented by a two-dimensional matrix of cells, the colors of which correspond to elevation values. This so-called "2.5-D" structure supports calculations that consider elevation, and may also be used by visualization tools to create pseudo-three-dimensional renderings.

This type of matrix structure for layers, whether representing the third dimension or some other value, is called a *raster layer*. Each cell within a raster layer represents a geographical subregion within the layer, and the value associated with that cell is normally the value of whatever feature is being represented by the layer aggregated over the subregion of the cell. For example, if a raster layer is being used to represent soil type, then the value of any particular cell might be one that corresponds to the dominant soil type in the subregion it represents. For the elevation layer example mentioned earlier, a cell's value might be the average elevation within the subregion. All cells within a layer are of the same size and therefore represent subregions of the same size. Spatial resolution of this data structure is then, determined by the number of cells into which the layer is divided.

An alternative data structure, also commonly used in GIS, is the *vector layer*. Vector layers use points, line segments, and polygons to represent objects within the geographical space described by the layer. For example, a highway could be stored as a sequence of line segments, each representing a straight stretch of road. The spatial resolution of the endpoints is determined by the number of bits used to store the coordinates, while the resolution of the road segment is arbitrary, limited only by the number of segments used to digitize the road's course. Similarly, a soil layer might consist of a set of polygons bounding regions of uniform soil type. Each polygon is then associated with the corresponding soil type.

In comparing these two data structures, it quickly becomes clear that each has its virtues and drawbacks, depending on the feature being modeled and the analysis which is anticipated. Linear features, such as roads and rivers, can generally be represented with much higher precision in a vector layer than in a raster layer, for in the latter case the representation would be to merely "turn on" a cell if the road or river passes through that cell's subregion. This could only match the precision of a vector layer if there were so many cells within the layer that the dimension of a cell's subregion matched the word size used to represent a vector's coordinates, a massively wasteful use of memory.

On the other hand, raster layers modeling features of area can facilitate inter-layer calculations. For example, to generate a map of those areas that have both a steep slope and pine trees, given raster layers of these two features, one merely has to do a cell-by-cell comparison of the two layers, generating a third layer to contain the results. If a vector layer implementation had been used, complicated algorithms would need to be executed to determine the intersection between the polygons representing forest cover with slopes determined by analyzing elevation contour lines represented by line segments.

While early systems favored one over the other, modern GIS offerings use both vector and raster layers. The display engines make it simple to overlay vector onto raster layers for the purpose of visualization, and many algorithms and tools exist to convert one into the other. For example, kriging is a geostatistical interpolation algorithm that can generate contour lines (vector) from a matrix representing a range of values (raster). Using such a technique, the elevation matrix described earlier can be used to generate a topographic map.

4.3. *Aspatial data structures*

The focus of GIS on spatial data is understandable, but incomplete. The spatial objects modeled within a GIS generally have attributes, which are themselves aspatial. For example, a highway clearly has the spatial features that may be accommodated by a vector layer as previously described. However, this same highway also has many attributes, which should be associated with the spatial representation, such as a route number and/or name. Furthermore, some parts of the highway might have different attribute values than others, such as the number of lanes, surface type, and date of the last repair.

Therefore, GIS implementations include a more conventional relational database, or at least an interface to one, to permit storage of these aspatial data. The data need only be linked to spatial objects stored within layers to facilitate a wide variety of queries.

One can query the relational database to discover spatial entities. For example, a query searching for all sections of Interstate Highway 81 that have at least four lanes would yield the spatial objects (e.g., vector line segments) associated with the query results (generally by highlighting them in the visual display). Alternatively, one can select an area of a spatial layer containing real-estate properties and retrieve a listing of the owners of the selected lots from the relational database. This bidirectional query capability is extremely useful in the management of environmental information.

4.4. *Spatial modeling and analysis*

Thus far, GIS has been described primarily as a class of systems, which can store and display spatial data (and their associated attributes). However, the real strength of a GIS comes from the set of analytical tools that can operate on these data. Early tools focused on the sort of queries, conversions, and comparisons already mentioned. However, modern GIS implementations include many more sophisticated tools.

Some of these are generic in nature, offering geostatistical analyses, alternative renderings, and mathematical calculations using GIS layers or objects as operands. Others are more specific to particular applications, and can include algorithms solving mathematical models using spatial data or attributes as parameters and/or independent variables. For example, most GIS implementations include modeling tools that solve the Universal Soil Loss Equation using appropriate data layers as operands.

The more successful GIS implementations take a building block approach that allows users to extend the functionality of the tool kit by combining existing tools with extensions developed elsewhere. Extending the discussion of open architectures from the previous section, considerable effort has been made in the past two decades to improve interoperability and exchange of GIS data through format standardization, and GIS implementations with both open architectures and open source have been extremely useful in the development of information systems more tightly focused on particular environmental management issues, as illustrated in the following section.

5. Environmental Decision Support Systems

Whether one is considering EMIS, DEIAs, or GIS, the real purpose is to make information and analytical tools available to people to help them make better environmental management decisions. It is a sad fact of history that many of the most important environmental decisions have been made in the absence of the full benefit of existing data and known science regarding them. This is not a criticism of the decision makers so much as of those of us who could and should be making the science more accessible to them.

It is tempting to consider the three types of systems already described in this chapter as decision support systems. Indeed, decision makers addressing environmental management problems do in fact use systems like these. However, they are not well suited to this task because they tend to be generic in nature. That is, they do not address a specific decision and a specific type of decision maker. While general utility might seem like a virtue in a system, in fact it makes these systems difficult to use. As a result, in many cases they are not used, at least not directly by the decision maker who would benefit most from them.

A successful EDSS will certainly incorporate a variety of technologies, possibly including one or more of the three previously described. However, before the composition of an EDSS is determined, it is important to explore the decisions to be made. Such a task-oriented approach is essential to the design of successful interactive systems.

5.1. *Environmental decisions*

Environmental decisions may cover many different topics, but there are some common characteristics that make them particularly challenging to support. In general, environmental decisions are

- spatial;
- multidisciplinary;
- quantitative;
- uncertain;
- procedural;
- political.

Environmental decisions are spatial because the environmental impacts of most human activities are associated with place, and the variables and parameters associated with this place and the impacts upon it are often geographic in nature. In some cases, the spatial nature of the problem may be ignored or reduced by simplifying assumptions; however, ultimately the spatial aspect must be addressed.

Environmental issues cross the boundaries of many seemingly disparate fields of expertise in both the natural and social sciences; hence, it is natural that the decisions associated with these issues must be themselves multidisciplinary. Indeed, Environmental Science, generally defined as the study of the interaction between humans and the natural environment, consists of dozens of fields of inquiry in both the natural and social sciences.

Many of these disciplines are themselves highly quantitative, and the costs and ramifications associated with environmental problems are generally so significant that objective metrics are necessary to help mitigate controversy. Therefore, environmental decisions of any serious nature usually have a significant quantitative character.

The uncertainty in environmental decisions has its roots in a few causes. The most obvious of these is natural variability — it is the nature of nature to be nondeterministic. Added to this, though, is uncertainty due to a lack of empirical data, errors in the data, or the use of imperfect models. This sort of uncertainty can be either biased or unbiased, and is particularly challenging to address.

As many environmental decisions are tied to a regulatory or corporate policy framework, which specifies the steps by which a decision is to be reached, and because the threat of liability dictates a defensible audit trail for the decision process, environmental decisions are often structured according to the policies and/or procedures underlying such a framework. This might guide the actual decision process, or merely impact the *stated* decision process. The difference is critical to the EDSS developer, as they must discover the actual process in order to effectively support it.

Finally, environmental decisions are obviously political, reflecting the fact that environmental management is driven by public policy, influenced by such considerations as economics, social impacts, and public opinion. At its deepest level, this means that individual and collective values will weigh heavily on these decisions, rendering them challenging in the extreme.

Besides the foregoing characteristics of environmental decisions, it is necessary to consider that different categories of decision makers might be involved in making them. For example, we might be trying to support a decision about the location of monitoring wells around a hazardous waste landfill. However, the support of this decision will necessarily depend on whether the decision maker is a geohydrologist, a regulator, or a member of the general public. The needs assessment behind the EDSS design will yield essential differences for each of these types of users. For example, the private citizen (and possibly the regulator) may need assistance in the choice of values for hydrological parameters, which the geohydrologist would not.

These factors make it impractical to contemplate a generic decision framework for environmental management. Even if it were possible to capture all of the elements necessary to address the great variety of decisions to be undertaken, such a system would be virtually unusable. The environmental manager is already confronted with a vastly complex problem space. One of the first jobs of the decision support system is to simplify this space, offering them everything that they need to make the decision at hand — but *only* those things.

5.2. *Interaction design of an EDSS*

Therefore, while EDSSs will depend upon the integration of multiple supporting technologies, we must further restrict this definition to stipulate that EDSSs are focused on a particular decision problem and category of decision maker. Thus, they are not general-purpose tools with which anything can be done — if only you knew how to do it. Rather, they are particularly tailored to the problem facing the analyst and offer a user interface optimized for this problem.

Focusing the EDSS in this way improves the user's interaction with the system, allowing them to concentrate on the problem at hand and the information and tools needed to solve it. It also suggests a software architecture that facilitates the development of sibling systems embracing different decision problems with an essentially common user and data interface (Frysinger, 1995). Such a family of focused EDSS siblings offers user interface simplicity, in that the siblings share interaction style, organization, and fundamental approaches (where appropriate), while maintaining the focus each sibling has on its particular decision and decision maker.

Fortunately, the state of the art of interactive systems design has evolved to support the capture of the needs of a particular category of user facing a particular decision problem. User-centered interaction design allows the EDSS developer to analyze and structure the problem, using such tools as hierarchical task analysis, and then leverage the results through participatory conceptual design in such a way that the resulting system may have a high likelihood of matching the users' mental models of the problem space. Given the complex nature of environmental decisions, as already described, this process is best executed by systems engineers with equal facility in both interactive software systems engineering and environmental management.

5.3. *Components of an EDSS*

The actual composition of an EDSS will necessarily include a combination of technologies, depending on the task analysis previously alluded to. As will be seen, the three technologies previously described might indeed find a home in an EDSS.

5.3.1. *Geographic information systems*

As previously explained, a central feature of many environmental decisions is their spatial context. GIS have already been described as computer software systems, which directly target the management, analysis, and display of spatial information. They may therefore be instrumental in an effective EDSS.

Their analytical capabilities have increasingly permitted GIS technology to be the basis for decision making in many contexts, not the least of these being environmental decision support. In integrating GIS technology into EDSS, we attempt to make the capabilities of the technology accessible to decision makers who are not specialists in the use of GIS products directly. For the subset of GIS-tractable problems that can be approached by the non-GIS specialist, integration into a decision tool addressing their larger problem will bypass the typical batch-oriented, over-the-counter bottleneck, which more often than not results in GIS methods not being used where they might otherwise be put to good effect. Another way to think about this is to consider that an EDSS can bring some elements of GIS to decision makers in such a way that they need not know it is GIS.

5.3.2. *Data access*

While the inclusion of GIS technology within an EDSS framework brings with it the necessary spatial data, along with associated attributes, decisions are naturally based on more general data sets. Indeed, the data that are organized and maintained through the use of an EMIS are valuable precisely because they can be used to help make environmental management decisions.

However, an EMIS itself is much broader than an EDSS. In fact, it is helpful to think about an EDSS as a client of an EMIS. As previously discussed, an integrated EMIS will provide ready access to data throughout the organization in support of environmental management processes. Many of these processes themselves involve decision making. Therefore, an EDSS developed to support such decisions can be seen as one of the software applications previously defined in support of management processes.

As a simple example, a manager who needs to plan a hazardous materials training schedule for employees will benefit greatly from a decision support system that accesses employee data and hazardous materials' inventories within the EMIS in order to help decide which employees need which training.

5.3.3. *Visualization*

While GIS technology goes a long way toward providing spatial information representation capabilities for environmental decision support, it does not satisfy all such needs.

First of all, aspatial data visualization in such forms as plots, charts, and graphs is clearly a staple of any decision support system. Often taken for granted, many conventional data analysis tools provide poor or limited support for such graphical representations.

In addition, GIS displays are overwhelmingly two dimensional in nature, with a strong bias toward representing data in map format, or "plan view." While often useful, such displays are not by themselves adequate. For many environmental decisions, three-dimensional displays are helpful. For example, when evaluating the behavior of a modeled airborne contaminant plume, the analyst should be able to navigate about (and through) the three-dimensional plume in order to get a better feel for its shape and character; contour plots often fail to communicate this information. The technology required to compute and display such volumetric renderings is now readily available, and the algorithms required to efficiently shade and cast virtual light upon three-dimensional objects drawn on a two-dimensional computer screen are the result of considerable research in the field of computer graphics. Many of these algorithms have been known for quite some time; however, the ability to use them to generate very sophisticated volumetric displays in near real time is relatively new, especially on common desktop computing hardware. These tools have begun and will continue to play an important role in environmental decision support, and will be integrated into EDSS platforms with increasing frequency.

Advances in personal computing have also included the development and widespread dissemination of multimedia technology. This category of computer capabilities has added photographic and motion picture display to the more conventional computer graphics world, along with high-fidelity sound generation and reproduction capability within the platform. The ability to include photographs (such as site familiarization photos) and videos (such as a sequence capturing the removal of a well core) has the potential to greatly enhance the information delivery potential of an EDSS. The audio capabilities have an obvious use in delivery of speech (such as in on-line help or co-operative work situations), but also support the use of sound to represent quantitative information, which cannot readily be accommodated by available visual display channels. While not widely used in common desktop data analysis systems, auditory data representation offers an important dimensional extension to an EDSS that is already confined by the two-dimensional display. For example, given a raster representation of soil contamination using color to represent contaminant concentration, sound may be used to represent additional data, which can not otherwise be displayed on the screen, such as confidence in the data at various points.

5.3.4. *Distributed and ubiquitous computing*

The appearance of Distributed Environmental Data Architectures, described earlier in this chapter, portends an increasing ability to greatly extend the capability of EDSS, especially related to the execution of computationally intensive models. Historically, such computations have been relegated to a segment of computer technology called supercomputers. Supercomputers may be defined as computers which are both fast and expensive enough that few of them are in existence. This rather awkward definition is necessary to account for the fact that current personal computers offer a level of computational throughput, which would have been considered supercomputing a few decades ago. It is pointless to attempt to define supercomputers in absolute performance terms, because the technology advances so rapidly as to render such boundaries obsolete in very few years.

Nonetheless, it may be presumed that no matter how fast individual workstations become, there will be still faster computers that are few in number but which may be made available to a wide population. In this work, such supercomputing technology is considered in combination with digital networks because high-bandwidth data networks have made it possible to consider linking supercomputers with personal workstations in such a way that the interactive user need not be aware that computations have been distributed. In this way, computational models might be offered as services by a Distributed Environmental Data Architecture.

In some sense, this sort of approach would represent a distribution of the EDSS architecture across multiple, remotely located machines. This view is especially appropriate if one distributes the data or code repository functions as well. For example, one might keep national meteorological data in a disk farm associated with an NOAA (National Oceanic and Atmospheric Administration) supercomputing facility, which might also store and maintain modeling codes that have been submitted to a quality assurance process. An individual EDSS being used to evaluate potential emissions from a factory might make use of these data and codes, as well as the supercomputer power, to solve a local air-modeling problem. Avoiding the need to distribute the data and codes saves a considerable amount of space (which would have been redundantly consumed on every similar EDSS platform), and also reduces the risk of data (or code) contamination. Alternatively, this application could make use of highly distributed sensors to determine real-time atmospheric conditions without going through NOAA's archives.

In any case, the environmental models currently in use already stretch even high-end workstation capabilities to the point that analysts might wait several days for a single iteration of a model to execute. A particularly clear example of this need may be seen in the support of decisions surrounding global climate change, for which model complexity grows at least as rapidly as the availability of computational horsepower. Although individual workstations can satisfy many environmental management computational requirements, there will be a need for supercomputer access in environmental decision support for the foreseeable future.

5.3.5. *Simulation*

The need for computational brute force in environmental decision support is further magnified by inherent uncertainty within environmental decision problems. In considering the uncertainty of quantitative environmental models, one considers the output of the model to be some function of one or more input coefficients. These coefficients become the parameters of a numerical representation of the model, and the quantitative uncertainty in the modeling solution results from the combined uncertainties of these input parameters. It has been demonstrated (McKone and Bogen, 1991) that using such mono-valued estimates as worst-case values for these parameters introduces "creeping conservatism," which wildly misrepresents the actual likely results.

Stochastic analysis of uncertainty, a viable alternative, is predicated on the ability to articulate the probability distributions of each uncertain parameter and then iteratively solve one or more model equations involving these parameters. To accomplish this, samples are drawn from the parameter distributions, most often employing Monte Carlo or Latin Hypercube sampling methods.

To solve environmental models using such stochastic methods, one solves the model equation iteratively, each time using parameter values drawn from the uncertain parameter distributions by the methods just mentioned. The set of results of these calculations form, themselves, a distribution, which aggregates the uncertainty of each of the parameters, and whose characteristics can be used to describe the model. The moments and upper and lower bounds of such a calculated distribution can be directly employed in decision making based upon the model. While such iterative methods can become computationally intensive, the increased access to distributed supercomputing already described makes this alternative a viable resource for environmental decision support. In fact, the iterative nature of Monte

Carlo simulation makes it a prime candidate for parallel processing via distributed computing resources.

5.3.6. *Expert systems*

Finally, the EDSS developer may incorporate a technology to help capture and deliver the knowledge of experts in particular problem domains.

EDSS is predicated on the notion that *human* intelligence is needed to make responsible environmental management decisions. *Artificial* Intelligence (AI) might therefore seem anachronistic in this work. However, although Expert Systems research has indeed grown out of AI research, the connection stops there. Expert systems offer the possibility of providing advisors to environmental analysts, for example, to help them choose the assumptions and parameters of their conceptual model of the problem. In this sense, the interactive decision maker has the benefit of aggregated advice from many experts who would otherwise be unavailable. However, they still make the decision themselves. This area of research in EDSS is the most prospective, and much work remains to be done before it can be claimed that expert systems technology has contributed substantially. Nonetheless, there is great potential for a productive collaboration.

6. Conclusions

Environmental management is information intensive. Therefore, it lends itself well to assistance from computing technologies developed over the past several decades. Ranging from the fairly traditional EMIS and the increasingly familiar GIS to the groundbreaking work in highly DEIAs, these technologies are ultimately only useful if they and their information are accessible to human decision makers. Therefore, it is imperative to focus on the decision maker when crafting an EMS, and the information systems that will support it.

References

Frysinger, S.P., 1995. *International Journal of Microcomputers in Civil Engineering* **10**(2), 119–126.

McKone, T.E. and Bogen, K.T., 1991. *Environmental Science and Technology* **25**(10), 1674–1681.

Usländer, T., (Ed.), 2007. *Reference Model for the ORCHESTRA Architecture, Open Geospatial Consortium Best Practice OGC-07-097.*

Chapter 20

ADVANCED ENVIRONMENTAL MANAGEMENT AND INNOVATION: A THEORETICAL FRAMEWORK

José Manuel de la Torre-Ruiz,[a] Vera Ferrón-Vílchez,[b]
Javier Aguilera-Caracuel[c] and Rodrigo Martín-Rojas[d]
University of Granada, Business and Management Department, Economics and Business School, Campus Cartuja S/N. 18071, Granada, Spain
[a]*jmtorre@ugr.es*
[b]*vferron@ugr.es*
[c]*javieraguilera@ugr.es*
[d]*rodrigomr@ugr.es*

Environmental impact, a consequence of ever-increasing industrial activities around the globe, is a growing problem in the world. Firms actively engaging in green innovation can not only minimize production waste but also increase their corporate reputation and legitimacy while satisfying stakeholders' demands. In this chapter, we provide a theoretical review of the impact that environmental innovation has on firms' ability to sustain a competitive advantage. We summarize the environmental management literature's most important findings relating to firms' innovation capabilities over the past several decades. In addition, we consider the strategic potential of environmental innovation, explaining how it constitutes a potential source of competitive advantage. Next, we point out a set of important determinants of environmental innovation, focusing on environmental regulations and then analyzing several firm-specific features that may also determine the effectiveness of environmental innovations (a firm's strategic orientation, competencies, and network relations). Finally, we summarize our conclusions, propose future lines of research, and provide managers and policy makers with evidence of the increasing importance of firms' environmental innovation capabilities for protecting the natural environment.

1. Introduction

Environmental problems such as global warming, the depletion of the ozone layer, and air pollution in big cities have increased public concern about environmental issues. Because these problems are usually directly linked to business activity, it is not surprising that, in the past few decades, a large number of companies worldwide have undertaken the organizational initiative of environmental management.

As companies are considered open systems, environmental issues need to be taken into account at all strategic levels of a firm (Aragón-Correa, 1998). At the corporate level, a concern for environmental protection may determine the appeal of an industry or generate new business areas, such as waste treatment or those areas related to renewable energies, but at the same time environmental issues may reduce the attractiveness or even eliminate the profitability of some activities if these activities will have a highly harmful impact on the environment. At the business level, conscientious environmental management may present a great opportunity to develop a competitive advantage, improving the firm's image (Chen, 2008; Christmann, 2004) by increasing the efficiency of production processes and/or by improving the products' quality (Christmann, 2000; Hart, 1995). Conversely, not taking environmental demands into account may have the opposite effect on firms' strategies, with such a lack of concern for the environment becoming a source of competitive disruption.

Thus, it seems clear that many managers currently consider environmental protection as part of their decision-making process. In this regard, innovation may be one of the main drivers by which companies can achieve sustainable growth (Hart, 1995). For this reason, it is necessary to increase our knowledge about how companies' innovative strategies may affect and be affected by environmental issues. Not considering the fact that sometimes innovations must take into account social and environmental pressures, which may convert them from being a source of competitive advantage to being a source of competitive disruption.

In this chapter, we summarize several themes within the environmental management literature from the past several decades relating to the innovation capabilities of firms.

The remainder of the chapter is divided into four sections. Section 2 includes a widely used definition of environmental innovation and attempts to distinguish this concept from a general perspective and offering several examples of eco-innovations at different organizational levels. Section 3 introduces the strategic potential of environmental innovations, explaining how they may have the potential to become a source of competitive advantage for firms. Section 4 offers a set of determinants of environmental innovations, first focusing on environmental regulations and then analyzing several firm-specific features that may also determine the effectiveness of environmental innovations. Specifically, we focus on the strategic orientations of firms, their competencies and their network relations. Finally, Sec. 5 summarizes our main conclusions and suggestions for future lines of research.

2. Defining Environmental Innovation

The concept of innovation is difficult to delimitate, as there are many widely used definitions of the term that have been understood in many ways, according to the different purposes and contexts within which the concept is used (Damanpour and Evan, 1984). For our purposes, innovation can be defined as the adoption of an internally generated or purchased device, system, policy, program, process, product, or service that is new to the adopting organization (Daft, 1982; Damanpour and Evan, 1984; Zaltman *et al.*, 1973). This definition is sufficiently broad to include many different types of innovations pertaining to all parts of organizations and all aspects of their operational development (Damanpour, 1991). Other authors define innovation as the adoption of changes that are new to the environment (Daft and Becker, 1978) or new to the organization and important for the environment (Knight, 1967). Therefore, we can define a firm's innovation capability as the ability of the organization to adopt or implement new ideas, processes, or products successfully (Hurley and Hult, 1998). As the business environment changes, organizations must also change in order to adapt to these new conditions. Consequently, innovation is a means of enacting change within the outputs, structure, or processes of an organization as part of a broader adaptation process.

Accordingly, innovations specifically related to the environment, or eco-innovations, may be defined as "the measures of relevant actors (firms, politician, unions, associations, churches, private households), (1) which develop new ideas, behavior, products, and processes, apply or introduce them, and (2) which contribute to a reduction of environmental burdens or to ecologically specified sustainability targets" (Reenings, 2000, p. 322). In short, eco-innovations are those innovations that attempt to reduce the negative environmental impact of some business activity. Hence, environmental innovation aims to improve both environmental and economic performance within a company, a seemingly win–win situation (Hall and Vredenburg, 2003). However, this perspective also adds complexity to environmental innovation in relation to conventional innovation because eco-innovations now have to consider both business and environmental implications.

In the existing literature, several systems for classifying innovations have been suggested (Zaltman *et al.*, 1973). For instance, Dewar and Dutton (1986) and Ettlie *et al.* (1984) distinguished between radical and incremental innovations. Radical innovations produce fundamental changes in the activities of an organization and represent clear departures from existing practices. Incremental innovations, on the other hand, involve little departure from existing practices (Dewar and Dutton, 1986; Ettlie *et al.*, 1984).

In this chapter, we distinguish between technological and administrative innovations (Daft, 1978; Damanpour, 1987; Eisenhardt and Martin, 2000; Kimberly and Evanisko, 1981). This classification implies a general distinction between social structure and technology within a company (Evan, 1966). Technological innovations pertain to products, services, and processes, and are related to basic work activities (Damanpour and Evan, 1984; Knight, 1967). Technological innovations occur as a result of the use of a new tool, technique, device, or system that produces changes in products or services, or in the way such products and services are produced or rendered (Damanpour, 1987). Computer output microfilm catalogs, *blu-ray* technology for computers, and automatic purchasing systems are just some examples of technological innovations. Administrative innovations, on the other hand, involve developments in an organization's structure and administrative processes. These innovations are indirectly related to the basic work activities of an organization and are more directly related to its management (Damanpour and Evan, 1984; Kimberly and Evanisko, 1981; Knight, 1967). Some examples of administrative innovations are management by objectives, zero-based budgeting, job rotation, staff incentive systems, and flexible timetables.

The distinction between technological and administrative innovations is a widely accepted classification scheme for studies of organizational innovations because (a) this typology portrays important differences in the nature of various innovations, and (b) the two innovation types together can represent changes affecting a wide range of tasks within the organization. Daft (1978) argues that some organizations are structured by a "dual-core model," centered on both a technical core and an administrative core. Each core has its own participants, goals, and problems but also complements the other, and this combination facilitates organizational innovation within the firm.

Additionally, technological innovations can be further classified into two groups: product and process innovations (Damanpour *et al.*, 1989; Romano, 1990). Product innovations are new products or services introduced to satisfy an external user or market need, whereas process innovations are new elements introduced into production operations: input materials, task specifications, work and information flow mechanisms, and equipment used to elaborate a product or render a service (Knight, 1967; Utterback and Abernathy, 1975). The adoption rates of product and process innovations are different during the various stages of a business' development (Utterback and Abernathy, 1975). A realistic assessment of which types of innovations will be successful and how they should be introduced depends upon an understanding of the productive process that will receive them. For instance, during an incipient stage, most technological innovations are applied directly to products that attempt to address simple consumer needs. Few innovations focus on improving processes. Thus, to more realistically assess which types of innovations may be successfully implemented depends on having a better understanding of a firm's productive process. Further, companies may also differ in whether they emphasize either product or process innovation in search of competitive advantages (Ettlie, 1983; Hull *et al.*, 1985).

Consequently, following the aforementioned ways of classifying innovations, there are three differentiated types of eco-innovations within firms: product environmental innovation, process environmental innovation, and organizational environmental innovation.

First, green product innovation performance may be understood as a firm's product innovation activities that are related to environmental innovation, such as energy saving, pollution prevention, waste recycling, avoidance of toxicity, and other green product designs (Chen *et al.*, 2006; Shrivastava, 1995). Thus, product environmental innovation implies an attention to designing products that minimize or eliminate a business' negative impact on the environment not only during the product's use but also when it is thrown away. Therefore, these kinds of innovations mainly focus on reducing energy demands during the use of the product and making the product's disposal clean and easy.

A good example of product environmental innovation might be the prototype of an ecological car fabricated by Volkswagen, called the "Lupo three-liter car." This car was developed in 1999 and was the world's first car to achieve a level of fuel efficiency of only three liters of fuel per one hundred kilometers, thus greatly reducing CO_2 emissions. A more recent example is the Samsung Reclaim (SPH-m560), the first phone in the U.S. made with bio-plastic materials, which are 100% biodegradable. Yet another example is the Ecobioball, a 100% biodegradable, non-toxic golf ball developed by the Spanish company Albus Gold S.L. These balls allow golf to be played near lakes or oceans because they are completely biodegradable, disintegrating and expelling the fish food they contain in less than 48 h.

Second, production processes and the provision of services frequently generate liquid effluents, pollutant emissions, and other inefficiencies that may negatively impact the environment. Process environmental innovation aims to eliminate or at least reduce these negative environmental effects. Therefore, green process innovation performance refers to a firm's performance in process innovation as it connects to energy saving, pollution prevention, or waste recycling (Chen *et al.*, 2006).

For instance, in the coastal province of Málaga (in the southern region of Spain), there is an ecological hotel whose name is "The Ecohotel" and which exemplifies the implementation of environmental innovation in processes. This tourist-focused company utilized several innovative processes both in the hotel's construction and in its daily business. First, during construction, the natural environment where the hotel was to be established was meticulously cared for. Excavations were performed without of the use of dynamite in order to avoid possible environmental alterations. Non-polluting materials were used in the electrical and plumbing systems. Second, most of the utilities (hot water, central heating, air conditioning, and swimming pool heating) are currently based on solar energy. The building is illuminated with low-consumption light bulbs, and their energy usage is further minimized by attaching them to movement sensors. Additionally, used water is recycled for garden irrigation and waste is carefully separated for recycling. All of these efforts were rewarded when the hotel received the *Innova Technology Award*, which is given by the regional government.

Another example of innovation in processes can be seen in the ecologically-friendly gyms of Virgin Active. These gyms, situated in several European and African cities, try to reduce their energy consumption by using boilers and cooling systems at only 65% of their capacity. Additionally, the gyms have systems enabling them to reutilize the heat produced by the air conditioning to heat the swimming pools. Furthermore, they keep track of the number of people in their facilities in order to perfectly calibrate the air conditioning and heating systems.

Finally, organizational environmental innovations are related to organizational structure and administrative processes. The Body Shop, a cosmetics company, clearly exemplifies this type of environmental innovation. More than 30 years ago, Dame Anita Roddick began her business journey with The Body Shop and pioneered the simple idea that "businesses have the power to do good," leading to the birth of the original, ethical cosmetics company. The Body Shop, which sells environmentally-friendly cosmetic products around the world, is a leader in promoting greater corporate transparency. Its five core values are "Support Community Trade," "Defend Human Rights," "Against Animal Testing," "Activate Self-Esteem," and "Protect Our Planet." Thus, by using a completely innovative approach to doing business, this company has demonstrated that environmental protection, innovation, and profitability can work together interdependently.

Therefore, all of these examples highlight how environmental innovations might be successfully implemented at different levels of an organization.

3. Strategic Potential of Environmental Innovations

As is the case with conventional innovations, environmental innovations may be pushed by technological development — for instance, when new eco-efficient technologies emerge within an industry — or they may be pulled by market factors — such as consumer preferences for environmentally-friendly products and images. The main problem is that, unlike in conventional innovation, eco-innovators have to consider both factors simultaneously. Thus, developing an environmental innovation may provide an opportunity for a firm to gain a competitive advantage only if it is able to identify environment-related consumer needs and outpace its competitors in meeting those needs (Noci and Azzone, 1998).

Those organizations that operate in technology-based industries need to innovate in order to achieve a better response from the environment, to improve business performance through expanded capabilities, and to consolidate a sustainable competitive advantage (Zaltman *et al.*, 1973; Bommer and Jalajas, 2004). In this context, a firm's failure to promote innovative projects and activities will negatively affect productivity and business performance because this fact will imply the firm's disadvantaged position with respect to competitors (Lööf and Heshmati, 2002). Adopting innovations may be an effective way for a company to develop several capabilities that reinforce its strategic position and, thus, to obtain greater benefits (Lengnick-Hall, 1992; Lieberman and Montgomery, 1988). In this regard, concern for the natural environment may provide an interesting opportunity for carrying out these innovations within firms. Consumers' increasing demands for companies to lessen their negative environmental impacts sometimes render existing technologies insufficient. Therefore, the growing importance of some areas, such as renewable energy or waste treatment, pushes the development of technological innovations in these fields.

However, as we have stated above, environmental innovation may be not only pushed by technological developments but also demanded by the market. The concern for environmental protection is widely held in society today because the general public has become more conscious of environmental degradation. Environmental problems such as global warming or the depletion of the ozone layer are constantly referenced in the media, with the result that consumers are now more conscious of how firms' activities may have a negative impact on the environment. This awareness has led consumers to demand that firms take into account the impact that their activities have on the environment. As a consequence, nowadays "green consumers" are increasingly demanding environmentally-friendly products and services (Darnall *et al.*, 2007). However this increasing concern is not only a potential opportunity for some firms; it may also represent a potential threat for those that ignore the negative environmental impact of their activities. Consumers may engage in public boycotts against a firm's products or services when they consider that those products or services are not sufficiently eco-friendly (Henriques and Sadorsky, 1999). In the case of such boycotts, the firm's reputation and, consequently, its profitability could be seriously harmed. Hence, for these two reasons, firms must consider the increasing demands of green consumers in order to better satisfy them.

In addition to addressing the needs of green consumers, firms may engage in activities that directly impact the natural environment and, consequently, they may face pressures from secondary stakeholders. These stakeholders' groups may consist of public interest groups as regulators, environmental associations (environmentalists), banks and lenders, community groups (Hoffman, 1999; Etzion, 2007) and professional organizations such as labor unions and trade or industry associations (Etzion, 2007).

Nevertheless, ecologically-conscious products and/or services tend to be offered by innovative firms because their production requires specialized investments, ecological inputs, and other green manufacturing devices. The firms that can pursue such activities usually have consolidated innovations in their productive process specifically and in their management style in general (Christmann, 2000). Thus, eco-innovations must be carried out by those innovative firms that want to integrate environmentally-innovative postures within their internal strategy and consequently satisfy stakeholders' pressures.

In summary, environmental innovations have the potential to become a source of competitive advantage. This kind of innovation may be pursued by introducing or improving products and services to satisfy stakeholders' demands. Additionally, eco-innovations may create wealth or improve regional economic or environmental protections, thus exploiting societal needs (Hall and Vredenburg, 2003).

4. Determinants of Environmental Innovations

4.1. *The role of regulation in environmental innovation*

> *"Regulation creates pressure that motivates innovation and progress. Our broader research on competitiveness highlights the important role of outside pressure in the innovation process, to overcome organizational inertia, foster creative thinking and mitigate agency problems" (Porter and van der Linde, 1995, p. 100).*

Environmental regulation has been considered one of the main factors contributing to environmental innovation (Porter and van der Linde, 1995). Indeed, when an organization does not fulfill environmental regulations, it has to face the threat of legal actions, penalties, and fines (Henriques and Sadorsky, 1996), which could leave firms vulnerable to individual or class action lawsuits (Darnall *et al.*, 2009). Thus, we must consider that environmental innovation may be stimulated by regulators. The influence of a regulatory framework on environmental innovation may be analyzed from two viewpoints.

First, the existence of stringent environmental regulations may force firms to develop and implement several eco-innovations. Complying with environmental regulations may lead to higher costs for firms because of their investments in pollution-preventing technologies and the corresponding organizational changes. At the same time, such high environmental regulatory costs might provide an incentive for organizations to improve their internal efficiency (Porter and van der Linde, 1995). Consequently, eco-innovations may entail a higher and more cost-effective level of compliance with environmental regulations, deriving from higher resource productivity (Porter and van der Linde, 1995). Firms innovate productive processes that reduce their consumption of raw materials or energy, or that reutilize waste. For example, after the introduction of new legal requirements that imposed higher fines for non-compliance, the Molten Metal Technology company developed a catalytic extraction process to treat hazardous waste, subsequently reducing compliance costs but changing little else (Porter and van der Linde, 1995).

Second, environmental regulation may also become a great opportunity for those firms whose environmental commitment goes beyond legal requirements. These firms may advance the development of future environmental regulations, achieving a great advantage over competitors that pay less attention to environmental issues. Table 1 shows several potential eco-innovations based on products and processes that regulators could stimulate within firms.

Table 1. Eco-innovations that can be incentivized by environmental regulations.

Eco-innovations based on product offsets	Eco-innovations based on process offsets
Better performing products	Higher process yield
Higher-quality products	Less downtown through more careful monitoring and maintenance
Safer products	Material savings (substitution, reuse, or recycling of production inputs)
Lower product costs (material substitution or less packaging)	Better utilization of by-products
Products with higher resale or scrap value (ease in recycling or disassembly)	Lower energy consumption
Lower cost of product disposal for users	Reduced material storage and handling costs
	Conversion of waste into valuable forms
	Reduced waste disposal costs
	Safer workplace conditions

Source: Porter and van der Linde (1995).

Additionally, the absence of environmental regulation may even reduce incentives for companies to initiate environmental innovations. Environmental policy, by imposing taxes on polluting firms or by controlling the amount of pollutant emissions, may serve to internalize the external damage produced by some firms. Otherwise, the lack of this kind of environmental regulation may cause competition between environmental and non-environmental innovation to become distorted (Reening, 2000).

4.2. *The strategic orientation of the firm*

Although environmental regulation is an important driver of environmental innovation, other firm-specific features may determine the effectiveness of these innovations. According to Hansen *et al.*, (2003), a firm's environmental innovation capability is dependent on the interaction of its strategic orientation, competencies, and network relations.

Strategic position on the basis of cost reduction pushes firms to innovate technologies and processes that will help them to reduce their operational costs. Research on environmental management, grounded in a resource-based view of the firm (Wernerfelt, 1984; Barney, 1991), has widely contributed to showing the potential that proactive environmental management might hold for generating valuable organizational abilities related to cost reduction (Hart, 1995; Hart and Ahuja, 1996; Nehrt, 1998; Christmann, 2000; Darnall and Edwards, 2006). For instance, Hart's seminal work (1995) argues that environmental strategies related to reducing pollution generate certain organizational capabilities that can help in the minimization of firms' costs. Hart (1995) argues that firms can redesign products in order to eliminate the need for a pollution-generating input in the manufacturing process, or they can redesign products to be less polluting when used by consumers or when thrown out by them. Moreover, the cost-saving approach to pollution prevention helps firms avoid the expense of pollution control technology (Hart, 1995; Sharma and Vredenburg, 1998) and can increase productivity and efficiency because less of the inputs are emitted as waste (Hart and Ahuja, 1996). Pollution prevention activities are not the only way to diminish a firm's costs. Costs can also be minimized by reducing legal sanctions in the form of fines and penalties for non-compliance with environmental regulations (Sharma and Vredenburg, 1998). Then, all of these environmentally-friendly changes to the manufacturing process can decrease firms' costs (Christmann, 2000; Nehrt, 1998) and simultaneously allow firms to reduce their negative environmental impact.

Additionally, strategic positioning in order to differentiate means that firms look for those processes and products that enable them to provide unique or superior products and services to their customers. Developing environmental innovations may be an effective way for firms to differentiate themselves from competitors. This kind of innovation may increase a company's external legitimacy with clients, suppliers, and other economic agents regarding its environmental practices. Such innovative practices may also go a step further, by encouraging durable relationships among a broader array of social agents, such as neighborhood communities, mass media, and public regulators (Bansal and Clelland, 2004). By enhancing their external legitimacy and addressing stakeholders' demands, firms may improve their reputational benefits and increase sales, both of which translate into improved business performance (Hart, 1995; Aragón-Correa, 1998; Rueda-Manzanares *et al.*, 2006).

In sum, the strategic position of a firm and how it seeks competitive advantages may determine its propensity for developing environmental innovations. Accordingly, firms need to integrate environmental management within their general corporate strategy (Aragón-Correa, 1998; Aragón-Correa and Sharma, 2003) in order to develop advanced and innovative environmental practices. Therefore, the development of green innovations is highly connected to a firm's proactive strategy. In fact, firms that follow a proactive environmental strategy may develop competencies that will enable them to implement environmental innovations. Being proactive implies anticipating or initiating changes in a firm's strategic policies rather than simply reacting to events. Moreover, proactive behavior requires the identification and analysis of the possible environmental consequences of a firm's products and services, as well as the establishment of comprehensive management programs (Aragón-Correa, 1998).

4.3. *Competencies*

Additionally, companies will need to develop some internal competencies in order to develop effective environmental innovations. Indeed, the implementation of environmental innovations requires changes in a firm's routines and operations, and in the complex coordination of various human and technical skills and heterogeneous resources (Amit and Schoemaker, 1993; Christmann, 2000).

Internal competencies encompass all of the knowledge, skills, values, and organizational routines that may be necessary to adopt environmental innovations. These requirements are mainly related to the activities of the corporate value chain, the management process, and the company's relationships with stakeholders (Azzone and Noci, 1998).

First, firms that attempt to develop effective environmental innovations must introduce operative principles emphasizing the need to reduce energy consumption and waste in the whole productive process. Such principles should be spread across all stages of the production process, including concerns about the reusing, recycling, or regenerating of the product and/or materials later on in their life cycle. Marketing also plays an important role in the adoption of environmental innovations for two reasons. On the one hand, informing the market about the environmental benefits of the product or process is required; on the other hand, marketing programs will positively contribute to educating customers about environmental issues.

Second, the development of environmental innovation also requires different perspectives within human resource management. Companies' human and organizational resources may be fundamental in helping to develop effective environmental innovations. The implementation of eco-innovations requires not only employees who have enough knowledge and abilities to carry out them but also the development of environmentally-friendly values within the organization (Handfield *et al.*, 2001). Otherwise, managers trying to engage in environmental innovation may find that their obstacles are more human than technological. For these reasons, some firms develop environmental training programs in order to leverage both the environmental knowledge and the environmental motivation of the employees (Jabbour and Santos, 2008). Additionally, organizational support is fundamental in order to create a environment that encourages employees to carry out eco-initiatives (Ramus, 2001). Organizational behaviors such as increasing environmental communication and establishing environmental objectives, responsibilities, and rewards may be effective tools to leverage the environmental commitment of employees.

However, the development of products and/or processes with a reduced environmental impact may not be possible for all companies. Manufacturing these products requires specialized investments, ecological inputs, and green manufacturing devices that are usually held by firms with consolidated innovations in their productive process specifically and in their management style in general. In relation to this, Christmann (2000) finds that firms using pollution prevention technologies without complementary process innovation and implementation capabilities might not save money by adopting proactive environmental practices. In fact, changing a smoothly-running production process without the requisite capacity for process innovation and implementation might make the process less efficient and more risky than it was previously. For this reason, those firms with a pre-existing innovation capability have a better chance of succeeding in environmental innovation.

Finally, managers may pay special attention to the management process as well. Indeed, the implementation of an environmental management system (EMS) implies the integration of environmental protection into firms' operational activities (North, 1992). Firms that adopt an EMS develop a firm-specific environmental policy and implement various processes to assess the environmental impact of their operations throughout their value chain — from raw materials access to the disposition of used products (Fiksel, 1993) — and ensure that their management practices conform to environmental regulations. Thus, the EMS structure may encourage firms to undertake proactive environmental practices that prevent pollution by substituting unregulated for regulated inputs and by eliminating some regulated processes altogether (Darnall and Edwards, 2006). The EMS framework may also promote a greater level of environmental concern throughout the organization. By continually monitoring employees' progress and increasing their knowledge about overall operations, firms may engage employees in environmental issues. All of these actions can also help organizations achieve greater efficiency and create opportunities for firms to improve their strategic value, such as through the promotion of their environmental innovations (Ziegler and Rennings, 2004; Wagner, 2008). In sum, an EMS acts as a coordination mechanism for all the environmental activities of the firm and, consequently, reinforces the firm's capability to carry out environmental innovations.

4.4. *Network relations*

Finally, one of the main drivers facilitating the firm's environmental innovation capability is the establishment of a durable and profitable network of relationships (Hansen *et al.*, 2003). For this reason, it is important to attend all of the agents who interact within the company in order to create this durable network. With regard to this consideration, an accurate knowledge of stakeholder demands allows firms to implement innovations that might serve to satisfy these demands. Stakeholders are defined as "*any group or individual who can affect or is affected by the achievement of an organization's objective*" (Freeman, 1984, p. 46). Stakeholder theory identifies the groups or individuals that affect or are affected by the organization's aims, and it recommends how management can give due regard to the interests of those groups (Mitchell *et al.*, 1997). Additionally, Darnall *et al.* (2010) find that augmented perceptions of stakeholder pressures are associated with a greater number of proactive environmental practices. This finding implies that those companies that are exposed to a great number of pressures from stakeholders are more likely to be willing to implement eco-innovations.

First of all, an effective stakeholders' management requires identifying who are the more salient stakeholders of the firm (Mitchell *et al.*, 1997). Here, we consider three main groups of stakeholders: internal, supply chain and general stakeholders.

First, management employees, non-management employees, headquarters, and shareholders compose the internal stakeholders group. Employees might be capable of pushing for advanced environmental practices in their own companies (Hanna *et al.*, 2000; Ramus and Steger, 2000) and even in their industries (Fineman and Clarke, 1996). Because they are concerned about job security (Fineman and Clarke, 1996), management and non-management employees will attempt to avoid non-compliance with environmental requirements (Henriques and Sadorsky, 1996). Non-management employees often pioneer an organization's proactive environmental practices (Hanna *et al.*, 2000; Ramus and Steger, 2000). Owing to their practical knowledge about the daily manufacturing process, these employees may be able to eliminate inefficiencies while improving environmental protection. However, support and leadership from top-level managers are essential to ensuring environmentally-conscious behavior in organizations (Zutshi and Sohal, 2004). Therefore, by using eco-innovations, management and non-management workers will be motivated to find new, less polluting alternatives that reduce both environmental costs and harmful environmental consequences of the firm's activities. Moreover, corporate leaders and shareholders are very interested in the objective achievements of their operational units because they have an economic stake in the organization (Donaldson and Preston, 1995). These stakeholders have the highest level of responsibility for the company's health, and they are in charge of overseeing its compliance with environmental regulations. Thus, managers might consider eco-innovations essential to satisfying headquarters and shareholder demands. Therefore, due to their concern for the general long-term survival of the company (Henriques and Sadorsky, 2006), managers might use eco-innovative practices to attend to internal stakeholders' demands.

Second, the supply chain stakeholder group is composed of commercial buyers, household consumers, and suppliers. In the context of concerns for the environment, we suggest that this stakeholder group can pressure companies into adopting eco-innovations as follows. Corporate buyers and suppliers may want to protect their reputations for environmental sensitivity and consciousness (Henriques and Sadorsky, 1999) by ceasing delivery of any input or output to a "dirty" organization or by pressuring organizations to use a more environmentally-friendly substitute (Darnall, 2006). For instance, "*U.S. companies that operate in Western Europe are experiencing market pressures from large corporate buyers who request that they provide them with documentation of their environmental procedures*" (Darnall *et al.*, 2007, p. 3). Similarly, as we explained in Section 3, consumers can express their environmental preferences by buying products or services that originate from firms with environmentally-friendly practices (Henriques and Sadorsky, 1999). As a result, eco-innovations help firms to select and be selected by the "greenest" suppliers and corporate buyers. Therefore, the most environmentally-conscious commercial buyers and suppliers, as well as the most environmentally-conscious consumers, are more likely to choose and to be chosen by organizations whose innovations help to protect the environment.

Third and finally, the general stakeholders' group consists of environmental associations, financial institutions, community groups, labor unions, and trade associations. On the one hand, environmental and professional groups have the capacity to mobilize the general public in favor of or in opposition to an organization via public protests, campaigns, consumer boycotts, strikes, and calls for engagement (Henriques and Sadorsky, 1999). These external groups usually hold a higher level of expertise on environmental matters. For instance, professional associations are interested in reducing the uncompensated health effects of harm to the environment. Environmental groups are interested in how organizations seek to reverse the harmful effects of climate change. On the other hand, financial institutions may be reluctant to enter into a business relationship that may transfer environmental liabilities to them in the future (Savage and Jasch, 2004). Banks, financial institutions, and other lenders may identify companies that do not comply with environmental requirements as being too high-risk. As a consequence, environmentally non-compliant companies might face higher charges for financial services and incur higher environmental costs due to their non-compliance. For all of these reasons, general stakeholders can positively influence managers toward using eco-innovations and improve the environmental performance of the firm.

To sum up, all these examples highlight the important roles that various stakeholder groups may play in the development of eco-innovations; companies should take into account each of the objectives of the different stakeholder groups and, consequently, establish relational, long-term, and profitable interactions with them instead of only transactional interactions.

5. Conclusions

The need for environmental innovations has been increasingly emphasized by policy makers and academics as a way to effectively solve environmental problems (Kivimaa, 2008). To date, technological innovations in certain industries have resulted in significantly reduced emissions and increased shares of non-fossil fuels in energy production. Nevertheless, several environmental challenges persist, related to climate change, biodiversity, and chemicals dispersed through consumer products.

Environmental innovation can be defined as hardware or software innovation that is related to green products or processes, including innovations in technologies that are involved in energy saving, pollution prevention, waste recycling, green product designs, or corporate environmental management (Chen *et al.*, 2006). Apart from green product and process innovation, implementing environmental innovation at the organizational level is also required to create the firm core sustainable values that justify the existence of this kind of innovation.

This type of innovation can clearly become a source of competitive advantage. Indeed, environmental innovations have great strategic potential because they may be addressed by introducing or improving products and services that satisfy stakeholders' demands. Moreover, through the implementation of green innovations by the firm, not only will the firm be able to reduce its costs (Christmann, 2000), but also it can increase its transparency, reputation, and legitimacy (Christmann, 2004; Bansal, 2005).

We find four general determinants of a company's propensity for engaging in environmental innovation. First, environmental regulations may encourage firms to develop advanced and innovative environmental practices in order to avoid penalties or legal sanctions (Henriques and Sadorsky, 1996; Darnall *et al.*, 2009). In addition, those firms that self-regulate their environmental management beyond what is required by the law can anticipate the development of future environmental regulations, achieving a greater advantage over those competitors that pay less attention to environmental issues (Porter and van der Linde, 1995).

Second, the strategic position of the firm and how it seeks competitive advantages may determine its likelihood of producing environmental innovations. Strategic positioning on a cost reduction basis and strategic positioning in order to differentiate from competitors are two effective and complementary ways to generate green innovations. Both strategies are connected to firms' adoption of a proactive environmental strategy that allows them to integrate environmental management policies and practices within their general corporate strategy (Aragón-Correa, 1998; Aragón-Correa and Sharma, 2003).

Third, firms' must possess the right internal competencies in order to implement environmental innovations successfully. Firms that attempt to develop effective environmental innovations must introduce operative principles that emphasize the need to reduce energy consumption and waste in the productive process. These principles should be spread over all the stages of the production process and communicated to society. We find that companies' human and organizational resources may be fundamental for developing effective environmental innovations because workers are the ones who need to assimilate and implement environmental innovations. In addition, a firm's capacity for innovation is a key condition that must exist in order for the firm to develop environmental innovations (Christmann, 2000).

Fourth, the establishment of a durable and profitable network of relationships facilitates a firm's environmental innovation capability. Firms that are exposed to a great number of stakeholder pressures are more likely to be willing to implement eco-innovations. There are a variety of stakeholder groups who may impact environmental innovation within a firm, such as employees, shareholders, buyers, consumers, suppliers, environmental associations, financial institutions, community groups, labor unions, and trade associations.

In this chapter, we provide a theoretical review of the increasing importance that environmental innovations hold nowadays. Indeed, firms need to aim their efforts toward achieving successful green innovations and integrating them within corporate strategy. Through the generation of green innovations, firms can not only minimize waste in production and increase productivity but also increase their corporate reputation and legitimacy, thereby enhancing corporate competitiveness in accordance with growing environmental consciousness among consumers and severe international regulations for environmental protection (Chen *et al.*, 2006; Porter and van der Linde, 1995; Shrivastava, 1995). At the same time, we provide evidence to practitioners and policy makers for the increasing need to create incentives and institutional mechanisms (e.g. financial aid and subsidies) to encourage firms to develop valuable and innovative proactive environmental practices that will help protect the natural environment and be socially responsible.

For future research, it would be very relevant to undertake an empirical analysis of the determinants leading firms to successfully generate environmental innovations, including internal factors (e.g. a firm's innovation capability and its absorptive capacity or degree of internationalization) and external conditions (e.g. the impact of different types of stakeholder pressures). Finally, studying the bidirectional relationship between firms' environmental innovation capability and firms' general innovation capability could provide significant value to the environmental management literature.

References

Amit, R. and Schoemaker, P.J., 1993. Strategic assets and organizational rent. *Strategic Management Journal* **14**, 33–45.

Aragón-Correa, J.A. and Sharma, S., 2003. A contingent resource-based view of proactive corporate environmental strategy. *Academy of Management Review* **28**(1), 71–88.

Aragón-Correa, J.A., 1998. Strategic proactivity and firm approach to the natural environment. *Academy of Management Journal* **41**(5), 558–567.

Azzone, G. and Noci, G., 1998. Seeing ecology and "green" innovations as a source of change. *Journal of Organizational Change Management* **11**(2), 94–111.

Bansal, P. and Clelland, I., 2004. Talking trash: Legitimacy, impression management and unsystematic risk in the context of the natural environment. *Academy of Management Journal* **47**(1), 93–103.

Bansal, P., 2005. Evolving sustainably: A longitudinal study of corporate sustainable development. *Strategic Management Journal* **26**(3), 197–218.

Barney, J.B., 1991. Firm resources and sustained competitive advantage. *Journal of Management* **17**(1), 99–120.

Bommer, M. and Jalajas, D.S., 2004. Innovation sources of large and small technology-based firms. *IEEE Transactions on Engineering Management* **51**(1), 13–18.

Chen, Y.S., 2008. The driver of green innovation and green image — Green core competence. *Journal of Business Ethics* **81**, 531–543.

Chen, Y.S., Lai, S.B. and Wen, C.T., 2006. The influence of green innovation performance on Corporate Advantage in Taiwan. *Journal of Business Ethics* **67**(4), 331–339.

Christmann, P., 2000. Effects of "Best Practices" of environmental management on cost advantage: The role of complementary assets. *Academy of Management Journal* **43**, 663–680.

Christmann, P., 2004. Multinational companies and the natural environment: Determinants of global environmental policy standardization. *Academy of Management Journal* **47**(5), 747–760.

Daft, R.L. and Becker, S.W., 1978. *The Innovative Organization*. Elsevier, New York.

Daft, R.L., 1978. A dual-core model of organizational innovation. *Academy of Management Journal* **21**, 193–210.

Daft, R.L., 1982. Bureaucratic versus nonbureaucratic structure and the process of innovation and change. In: Bacharach, S.B. (Ed.), *Research in the Sociology of Organizations*, Vol. 1. JAI Press, Greenwich, CT, pp. 129–166.

Damanpour, F., 1987. The Adoption of technological, administrative, and ancillary innovations: Impact of organizational factors. *Journal of Management* **13**, 675–680.

Damanpour, F., 1991. Organizational innovation: A meta-analysis of effects of determinants and moderators. *Academy of Management Journal* **34**(3), 555–590.

Damanpour, F., Szabat, K.A. and Evan, W.M., 1989. The relationship between types of innovation and organizational performance. *Library and Information Science Research* **14**(4), 465–482.

Darnall, N. and Edwards, Jr., D., 2006. Predicting the cost of environmental management system adoption: The role of capabilities, resources and ownership structure. *Strategic Management Journal* **27**, 301–320.

Darnall, N., 2006. Why firms mandate ISO 14001 certification. *Business and Society* **45**, 354–381.

Darnall, N., Henriques, I. and Sadorsky, P., 2010. Adopting proactive environmental practices: The influence of stakeholders and firm size. *Journal of Management Studies*, forthcoming.

Darnall, N., Jolley, G.J. and Ytterhus, B., 2007. Understanding the relationship between a facility's environmental and financial performance. Edward Elgar Publishing, in association with the Organization for Economic Co-operation and Development (OECD), Northampton, Massachusetts.

Darnall, N., Seol, I. and Sarkis, J., 2009. Perceived stakeholder influences and organizations. Use of environmental audits. *Accounting, Organizations and Society* **34**, 170–187.

Dewar, R.D. and Dutton, J.E., 1986. The adoption of radical and incremental innovations: An empirical analysis. *Management Science* **32**, 1422–1433.

Donaldson, T. and Preston, L.E., 1995. The stakeholder theory of the corporation: Concepts, evidence, and implications. *Academy of Management Review* **20**, 65–91.

Eisenhardt, K.M. and Tabrizi, B.N., 1995. Accelerating adaptive process: Product innovation in the global computer industry. *Administrative Science Quarterly* **40**, 84–110.

Eisenhardt, K.M. and Martin, J.A., 2000. Dynamic capabilities: What are they? *Strategic Management Journal* **21**(10/11), 1105–1121.

Ettlie, J.E., 1983. Organizational policy and innovation among suppliers to the food processing sector. *Academy of Management Journal* **26**, 27–44.

Ettlie, J.E., Bridges, W.P. and O'Keefe, R.D., 1984. Organization strategy and structural differences or radical versus incremental innovation. *Management Science* **30**, 682–695.

Etzion, D., 2007. Research on organizations and the natural environment, 1992-present: A review. *Journal of Management* **33**(4), 637–664.

Evan, W.M., 1966. Organizational lag. *Human Organizations* **25**, 51–53.

Fiksel, J., 1993. Design for environment: The new quality imperative. *Corporate Environmental Strategy* **1**, 49–55.

Fineman, S. and Clarke, K., 1996. Green stakeholders: Industry interpretations and response. *Journal of Management Studies* **33**(6), 715–730.

Freeman, R.E., 1984. *Strategic Management: A Stakeholder Approach.* Pitman, Boston.

Hall, J. and Vredenburg, H., 2003. The challenges of innovating for sustainable development. *MIT Sloan Management Review* **45**(1), 61–68.

Handfield, R.B., Melnyk, S.A, Calantone, R.J. and Curkovic, S., 2001. Integrating environmental concerns into the design process: The gap between theory and practice. *IEEE Transactions on Engineering Management* **48**(2), 189–208.

Hanna, M.D., Newman, W.R. and Johnson, P., 2000. Linking operational and environmental improvement through employee involvement. *International Journal of Operations and Production Management* **20**(2), 148–165.

Hansen, O.E., Søndergaord, B. and Meredith, S., 2002. Environmental innovations in small and medium sized enterprises. *Technology Analysis & Strategic Management* **14**(1), 37–56.

Hart, S.L., 1995. A natural-resource-based view of the firm. *Academy of Management Review* **20**, 986–1014.

Hart, S.L. and Ahuja, G., 1996. Does it pay to be green? An empirical examination of the relationship between pollution prevention and firm performance. *Business Strategy and the Environment* **5**, 30–37.

Henriques, I. and Sadorsky, P., 1996. The determinants of an environmentally responsive firm: An empirical approach. *Journal of Environmental Economics and Management* **30**, 381–395.

Henriques, I. and Sadorsky, P., 1999. The relationship between environmental commitment and managerial perceptions of stakeholder importance. *Academy of Management Journal* **42**, 87–99.

Henriques, I. and Sadorsky, P., 2006. The adoption of environmental management practices in a transition economy. *Comparative Economic Studies* **48**(4), 641–661.

Herbig, P.A. and Day, R.L., 1995. Putting technological innovation into a historial perspective: A quest for better understanding and implementation of the innovation process. *Advances in global high-technology management* **5**(A), 245–274.

Hoffman, A., 1999. Institutional evolution and change: Environmentalism and the U.S. chemical industry. *Academy of Management Journal* **42**(4), 351–372.

Hull, F.M., Hage, J. and Azumi, K., 1985. R&D management strategies: America versus Japan. *IEEE Transactions on Engineering Management* **32**(2), 78–84.

Hurley, R.F. and Hult, G.T.M., 1998. Innovation, market orientation, and organizational learning: An integration and empirical examination. *Journal of Marketing* **62**, 42–54.

Jabbour, C.J.C. and Santos, F.C.A., 2008. Relationship between human resource dimensions and environmental management in companies: Proposal of a model. *Journal Cleaner of Production* **16**, 51–58.

Kimberly, J.R. and Evanisko, M., 1981. Organizational innovation: The influence of individual, organizational, and contextual factors hospital adoption of technological and administrative innovations. *Academy of Management Journal* **24**, 689–713.

Kivimaa, P., 2008. Integrating environment for innovation: Experiences from product development in paper and packaging. *Organization and Environment* **21**, 56–77.

Knight, K.E., 1967. A descriptive model of the intra-firm innovation process. *The Journal of Business* **40**(4), 478–497.

Lööf, H. and Heshmati, A., 2002. Knowledge capital and performance heterogeneity: A firm-level innovation study. *International Journal of Production Economics* **76**, 61–85.

Lengnick-Hall, C.A., 1992. Innovation and competitive advantage: What we know and what we need to learn. *Journal of Management* **18**, 399–429.

Lieberman, M. and Montgomery, D., 1988. First mover advantages. *Strategic Management Journal* **9**, 41–58.

Mitchell, R.K., Agle, B.R. and Wood, D.J., 1997. Toward a theory of stakeholder identification and salience: Defining the principle of who and what really counts. *Academy Management Review* **22**(4), 853–886.

Nehrt, C., 1998. Maintainability of first movers advantages when environmental regulations differ between countries. *Academy of Management Review* **23**(1), 77–97.

Organizational innovation and performance: The problem of organizational lag. *Administrative Science Quarterly* **29**, 392–409.

PDMA, 2004. *The PDMA Glossary for New Product Development.* Product Development & Management Association. available at www.pdma.org.

Porter, M.E. and van der Linde, C., 1995. Toward a new conception of the environment-competitiveness relationship. *Journal of Economic Perspectives* **9**(4), 97–118.

Ramus, C.A. and Steger, U., 2000. The roles of supervisory support behaviors and environmental policy in employee "ecoinitiatives" at leading-edge European companies. *Academy of Management Journal* **43**(6), 605–626.

Ramus, C.A., 2001. Organizational support for employees: Encouraging creative ideas for environmental sustainability. *California Management Review* **43**(3), 85–105.

Reenings, K., 2000. Redefining innovation–eco-innovation research and the contribution from ecological economics. *Ecological Economics* **32**(2), 319–332.

Romano, C.A., 1990. Identifying factors which influence product innovation: A case study approach. *Journal of Management Studies* **27**(1), 75–95.

Rueda-Manzanares, A., Aragón-Correa, J.A. and Martín-Tapia, I., 2006. La medición de las capacidades organizacionales de la empresa: Validación de un instrumento de medida de la gestión medioambiental proactiva. *Revista Española de Financiación y Contabilidad* **35**, 563–582.

Savage, D.E. and Jasch, C., 2004. *International Guidelines on Environmental Management Accounting (EMA).* International Federation of Accountants, New York.

Sharma, S. and Vredenburg, H., 1998. Proactive corporate environmental strategy and the development of competitively valuable organizational capabilities. *Strategic Management Journal* **19**, 729–753.

Shrivastava, P., 1995. Environmental technologies and competitive advantage. *Strategic Management Journal* **16**, 183–200.

Utterback, J.M. and Abernathy, W.J. 1975. A dynamic model of process and product innovation. *The International Journal of Management Science* **3**(6), 639–654.

Wagner, M., 2008. Empirical influence of environmental management on innovation: Evidence from Europe. *Ecological Economics* **66**(2–3), 392–402.

Wernerfelt, B., 1984. A resource-based view of the firm. *Strategic Management Journal* **5**(2), 171–180.

Zaltman, G., Duncan, R. and Holbek, J., 1973. *Innovations and Organizations*. Wiley, New York.

Ziegler, A. and Rennings, K., 2004. Determinants of environmental innovations in Germany: Do organizational measures matter? *Discussion Paper No. 04-30. ZEW*, Mannheim.

Zutshi, A. and Sohal, A.S., 2004. Adoption and maintenance of environmental management systems: Critical success factors. *Management of Environmental Quality: An International Journal* **15**(4), 399–419.

Chapter 21

ENVIRONMENTAL ASSESSMENT TOOLS

Grace K. C. Ding
School of the Built Environment,
Faculty of Design, Architecture and Building,
University of Technology Sydney, Australia
grace.ding@uts.edu.au

Environmental assessment tools are important tools in achieving the goal of sustainable construction and have been used to determine sustainability of the design, construction, and operation of buildings around the world. Since the introduction of the BREEAM in 1990 as the first comprehensive building environmental assessment tool and similar types of environmental assessment tools have been developed in every country as an essential approach in the construction industry to achieve the goal of sustainable development. This chapter discusses the concept and role of sustainability in the construction industry. This chapter also presents an overview of environmental assessment tools, roles, and limitations of the tools in ascertaining building sustainability used in different countries. Finally, the chapter examines the development of the tools in the international context.

1. Introduction

Ecologically sustainable development is the central concern of people from all disciplines (Cole, 1999; Holmes and Hudson, 2000). Construction has been beset with problems ranging from excessive consumption of global resources, both in terms of construction and building operation, to the pollution of the surrounding environment (Spence and Mulligan, 1995; Ahn *et al.*, 2010). The construction industry is closely related to the economic, social, and environmental advancement of society. On the one hand, it employs over 111 million people worldwide and contributes approximately 5–15% of the global domestic product. On the other hand, it also accounts for the world's resource depletion and environmental degradation. It accounts for 20% of the water effluents, 30–40% of the solid waste generation and CO_2 emission, 20% water, 30% raw materials, and 25–40% energy (UNEP SBCI, 2008).

Sustainability is an important consideration in construction and the concept of sustainability in the context of construction is about creating and maintaining a healthy-built environment and at the same time focusing on minimizing resources and energy consumption, thereby reducing damage to the environment, encouraging reuse and recycling, and maximizing protection of the natural environment. The concern also focuses on how to improve construction practices in order to minimize their detrimental effects on the natural environment (Cole, 1999; Holmes and Hudson, 2000). The environmental impact of construction, green buildings, designing for recycling, and eco-labeling of building materials have captured the attention of building professionals worldwide (Crawley and Aho, 1999; Rees, 1999).

Project development in the building sector may readily contribute to achieving the objectives of sustainable development and these objectives may be constructively pursued by thorough assessment at an early stage, as part of the process of choosing the most efficient option among competing alternatives. The choice of the best development depends not only on measures of financial return, but also upon the selection of the most appropriate site to meet environmental criteria and to respect the feelings and views of residents. An eye to future expansion should also be maintained.

Research on green building design and using building materials to minimize environmental impact is also underway. With the increasing complexity of economic, social, and environmental impacts of building on the environment and the environmental issues considering climate change often discussed at the national and international agenda, much attention is paid on building sustainability performance. These discussions inspired the development of tools and methods to assess building sustainability performance in order to influence how buildings are designed, constructed, and operated. The development of environmental assessment tools for building performance has emerged as one of the major issues in sustainable construction (Crawley and Aho, 1999; Cooper, 1999; Holmes and Hudson, 2000). Cole (1998) states the definition of building performance varies according to the different interest of parties involved in building development. A building owner may wish his building to perform well from a financial point-of-view, whereas the occupants may be more concerned about indoor air quality, comfort, health, and safety issues. Therefore, an ideal environmental assessment tool will include all the requirements of the different parties involved in the development. However, using a single method to assess a building's environmental performance and to satisfy all needs of users is not easy task.

The main objectives of this paper are to overview and analyze environmental assessment tools that are used to assess environmental performance of buildings in different countries in terms of their characteristics and limitations in assessing building sustainability. This paper examines the concept of sustainability and sustainable construction development as a foundation of environmental assessment tools. The paper also discusses the role, aims, structure, and limitations of these tools in ascertaining building sustainability. The paper concludes with the challenges for further development of environmental assessment tools and the international collaborative activities in developing a common platform to promote sustainability worldwide.

2. The Concept of Sustainability and Sustainable Construction

2.1. *The concept of sustainable development*

Sustainable development is a catchword that has been interpreted and discussed in all fields (Hill and Bowen, 1997; Ofori *et al.*, 2000). According to Pearce (1998), it is not the difficulty of defining sustainable development, but rather the difficulty of determining ways to achieve the goal. The concept of sustainable development has emerged to describe a new framework for development aimed at achieving economic, social, and environmental balance while maintaining the long-term integrity of ecological systems. This triple bottom line concept is firmly embedded in government policy, legislation and in the environmental policies of private organizations.

According to Goodland and Daly (1995, cited in Herendeen, 1998), sustainability has three levels: weak, strong, and absurdly strong. Weak sustainability requires that the total of the man-made and natural capital does not decline and is close substitutes (Herendeen, 1998; Victor *et al.*, 1998). Strong sustainability is based on a disagreement on the degree of substitution and the natural and man-made capital is not substitutable but complementary in most production functions (Victor *et al.*, 1998). Absurdly strong sustainability tends to stress limits to sustainability. Accordingly, non-renewable resources could never be used at all and renewable resources could only be harvested at the net annual growth (Herendeen, 1998). With the three levels of sustainability, controversy about the meaning and definition of sustainable development is inevitable.

According to Cooper (2002), sustainable means capable of being maintained indefinitely within limits whereas development implies the pursuit of continuous growth. This appears contradictory, as development tends to destroy the ability to sustain. However, Ofori *et al.* (2000) suggest that as long as development is sustained, economic growth will continue and environmental issues will be dealt with through technology. Similarly, Boughey (2000) states that sustainability indicates economic activities, which could continue without long-term damage to the natural environment or general human well-being. This viewpoint indicates that economic growth will continue to thrive while the environment will never be deprived, or even used, at all. However, it is highly unlikely that this will happen, as economic growth requires the consumption of environmental resources to sustain its activities.

The most used definition of sustainable development is derived from the Brundtland Commission on Environment and Development (WCED, 1987, p. 43): "development that meets the needs of present generations without compromising the ability of future generations to meet their needs and aspirations." The four aspects as emphasized in the report are to eliminate poverty and deprivation; to conserve and enhance natural resources; to encapsulate the concept of economic growth, social and cultural variations into a development; and finally, to incorporate economic growth and ecology in decision making. This shows that many factors can contribute to achieving the goal of sustainable development.

Daly (1990) states that the Brundtland definition has made a great contribution by emphasizing the importance of sustainable development, supported by van Pelt *et al.* (1995) who say that the interpretation of sustainable development in the Brundtland report is the clearest. Mitchell *et al.* (1995) describe the definition as being very much about quality of life and ecological integrity. Spence and Mulligan (1995) further state that the idea of sustainable development is well summarized in the report. The report also acknowledges that there are limits imposed by the ability of the biosphere to absorb the effects of human activities and advancements in technology and social organization are needed to improve economic growth.

It is true that sustainable development is about imposing limitations on the use of scarce natural resources in the production and consumption process in order to ensure quality of life of present generations. In that way, sufficient resources may be reserved to allow future generations to have an acceptable level of welfare and quality of life. As the WCED definition appeals to many, it forms the guiding principle for the design of environmentally sound socio-economic policies. It is also clear from the report that unless decisions are taken now to address the deteriorating situation, future generations will not have the ability to correct them. Therefore, there is an immediate need for an action for this crisis situation.

The general acceptance of the Brundtland definition of sustainable means that it is regarded as representing an international consensus of the concept and conflicts between the demand for human development and future protection of environmental systems (Bentivegna *et al.*, 2002). On the other hand, the definition is often considered to be too vague and therefore inadequate in practice (Daly, 1990; Ofori *et al.*, 2000; Bentivegna *et al.*, 2002). Glasby (2002) says that the Brundtland's definition fails to recognize the method and does not provide an agenda to achieve a sustainable society or, most importantly, of how to survive in an unsustainable world. Indeed, the Brundtland definition does not satisfactorily provide guidance on how to define sustainability or be measured in practice. For the principle of sustainable development to be achieved and practiced, a clearer and more precise definition is required for different sectors of society.

Since the publication of the Brundtland definition, over 160 definitions of sustainable development have been developed, used, or interpreted by different groups to suit their own goals (Pearce *et al.*, 1989; Langston and Ding, 2001). According to Pearce *et al.* (1989), sustainable development requires future generations to be left no worse off than present generations, and if sustainable development is to be achieved, depreciation of natural capital must be zero since man-made capital does not provide adequate compensation for a reduction in environmental capital. Similarly, Daly (1990) suggests that an explanation of sustainable development has to satisfy three conditions. First, the rates of use of renewable resources should not exceed the rates of regeneration. Second, the rates of use of non-renewable resources should not exceed the rate at which sustainable renewable substitutes are developed, and finally, the rate of pollution emission should not exceed the assimilative capacity of the environment.

From these discussions, it is clearly shown that the means of achieving sustainable development deals with the concepts of environment, futurity, and equity, with the emphasis that the welfare of future generations should be considered in the decision-making process. On the other hand, The International Institute of Sustainable Development[1] stipulates that sustainable development should also simultaneously consider the improvement of the economy. Stigon (1999) and Rohracher (2001) discuss the concept of sustainable development in the context of considering economic growth in addition to the social and environmental dimensions. Economic growth, with an emphasis on aspects such as financial stability and material welfare creation, is the ultimate goal for every government in order to secure rising standards of living and increase the capacity of providing goods and services to satisfy human needs.

In spite of differing perceptions about the precise meaning and the possible interpretation of the term "sustainable development," it is widely accepted that for a development to be sustainable it must examine environmental, economic, social, and ethical aspects of reality. It also places emphasis on the importance of combining economics and environment in development planning (Spence and Mulligan, 1995; Stigon, 1999). The divergence of opinion relating to the term proves that sustainability is so broad an idea that a single definition cannot adequately capture all meanings of the concept. While there is little consensus about a definition for sustainable development, there are certainly commonly accepted principles that can be used to guide the process of development (du Plessis, 1999). Sustainable development is a continuous process of dynamic balance instead of a fixed destination that must be reached at a certain time (du Plessis, 1999).

In summary, the concept of sustainable development must consist of the examination of economic, social, and environmental aspects of a development. In addition, sustainable development may not be viewed as one-dimensional, but consists of multiple facets of issues that concern people today and in the future. The Brundtland report, although it has been shown to be vague in its own way, provides sufficient explanation of what sustainable development may have meant. To this end, to find a precise definition of sustainable development that satisfies all needs may be difficult. It is more important to find ways to achieve sustainable goals in order to maintain and conserve the environment, so that future generations will not be disadvantaged. It is also difficult to derive a definition that applies to all sectors in the economy; therefore, it is more realistic to define the concept of sustainable development with particular reference to each sector.

2.2. *The concept of sustainable construction*

There is no doubt that the environment is closely linked to economic growth and the continued depletion of environmental assets will be detrimental to the well-being

[1] IISD — http://www.iisd.org.

of mankind. Building development involves complex decisions and the increased significance of external effects has further complicated the situation. Society is not just concerned with economic growth and development but is also conscious of the long-term impacts on living standards for both present and future generations. Sustainable development is now an important issue in project decisions. Ecologically sensitive projects require a different approach to appraisal than most traditional projects. The engagement of a conventional single dimensional evaluation technique in assisting decision making is no longer relevant and much more complicated tools need to be used to handle multi-dimensional arrays of data. Multi-disciplinary appraisal teams and an overall methodology are essential to uphold the goal of a sustainable development.

Sustainable construction is considered as a way for the construction industry to contribute to the effort of achieving sustainable development. It is also a way of portray the construction industry's responsibility toward protecting the environment (Spence and Mulligan, 1995; Hill and Brown, 1997; Ofori *et al.*, 2000; Kaatz *et al.*, 2006). One view considers that the construction industry plays an important role in the world economy through creating man-made capital, a significant contributor to maintaining economic growth and quality of life. It is also a main supporter of economic development by providing infrastructure and buildings. Therefore, continuing and increasing levels of construction activity are essential to all aspects of development. However, the alternate view is that the construction industry has a significant impact on requirements and, therefore, also shares responsibilities of working toward a sustainable future through limiting the environmental impact of building activities. According to Kibert's presentation in the First International Conference on Sustainable Development in 1994, sustainable construction is defined as the creation and responsible maintenance of a healthy-built environment based on resource efficient and ecological principles (cited in Bourdeau, 1999). Bourdeau (1999) states that this has provided a broad definition that forms a starting point for defining sustainable construction. The definition of sustainable construction includes the concepts of construction that prevents environmental degradation and utilizes resources efficiently so that the environmental, economic, and social benefits justify the environmental degradation created throughout the building life cycle (Ofori *et al.*, 2000; Kibert, 2008).

Spence and Mulligan (1995) state that restricting the total amount of construction in order to achieve the principle of sustainable development is not essential. Construction development is interwoven with economic growth, and economic growth is about improving standards of living. du Plessis (1999) suggests that social impacts should have a role in accomplishing sustainable construction. He points out that social achievement through improving the quality of life is the motivation for many actions and it follows that the more development, the more growth, leading to improved standards of living. Therefore, sustainable construction is not just about saving the world, but also about maintaining a comfortable environment for mankind. This view supports the idea that sustainable construction does not imply a complete halt to irreversible change in the natural environment. Some conversion of natural into man-made capital is acceptable providing that the depletion rate of the world's natural capital does not exceeds the rate of accumulation of man-made capital of lasting value (Spence and Mulligan, 1995; Curwell and Cooper, 1998). Unfortunately, this is currently not the case. The consumption rate of the world's natural resources is much faster than the regeneration rate and waste production has already breached the assimilative capacity of many ecosystems (Rees, 1999).

Hill and Bowen (1997) state that sustainable construction starts at the planning stage of a building and continues throughout its life to its eventual deconstruction and recycling of resources to reduce the waste stream associated with demolition. They go on to describe sustainable construction as consisting of four attributes: social, economic, biophysical, and technical. These four attributes form a framework for achieving sustainable development that includes an environmental assessment during the planning and design stages of projects, and the implementation of environmental management systems.

The continuing confusion about concepts and lack of agreement on causes and effects has reduced the ability to provide guidance based on well-accepted and understood concepts for good practice in construction (Bröchner *et al.*, 1999). With regards to the existing definition of sustainable construction, it is important to note that even if the principle of sustainable construction is attained, construction operations would continue to have environmental impacts, but at a reduced rate (Ofori *et al.*, 2000).

Kibert (2008) suggests three ways to achieve sustainable construction: creating a healthy-built environment; restoring damaged and/or polluted environments; and improving barren environments. It is the ultimate goal for the construction industry to restore damaged and polluted environment and to give priority to improve the built environments in order to transform its linear process into cyclic processes. Building green is considered as a way of achieving the goal of sustainable construction and this can be achieved by combined efforts of developers and the design professionals to manage environmental issues. It is important that every development includes environmental protection to the list of project objectives that traditionally include only time, cost, and quality considerations. For building designers, it is important to show environmental consciousness in their design. Bourdeau (1999) suggests that a more integrated approach to design be adopted to consider the fundamentals of sustainable building design and should work together with manufacturers to create new designs, which facilitate material recycling. To achieve the goal of sustainable development in construction, there must be a team effort from different disciplines in the construction industry.

3. Sustainable Construction — The Role of Environmental Assessment Tools

As the problems of natural resource depletion and global environmental degradation become evident, building performance has become a matter of public concern. Most building evaluation methods are concerned with a single criterion such as energy use, indoor comfort, or air quality to indicate the overall performance of a building (Cooper, 1999; Kohler, 1999). As environmental issues become more urgent, a more comprehensive building assessment tool is required to assess building performance across a broader extent of environmental considerations.

Environmental assessment tools for buildings reflect the significance of the concept of sustainability in the context of building design, its subsequent construction work on site, and operation on completion of building works. Designers aim to improve the overall performance of buildings in relation to their effects on both the natural and man-made environments. The primary role of an environmental assessment tools is to provide a comprehensive assessment of the environmental characteristics of a building (Cole, 1999). It is undertaken by providing a common and verifiable set of criteria and targets for building owners and designers to achieve higher environmental standards.

Additionally, the assessment tools help to define the direction for a project and provide information on which to make informed design decisions at all stages and to plan effective environmental design strategies. The development of an environmental assessment tool lays down the fundamental direction for the building industry to move toward environmental protection and achieving the goal of sustainability. It also provides a way of structuring environmental information, an objective assessment of building performance, and measure of progress toward sustainability.

Assessment tools act as a bridge between environmental goals and strategies and building performance during the design and occupancy stages of a building. They are composed of a set of environmental criteria that are relevant to buildings, and are organized and prioritized to reflect the performance of a building. The environmental assessment tools satisfy three major aspects: global, local, and indoor issues. They also include a set of standard guidelines for how individual buildings are assessed and evaluated. They are prepared in order to provide a methodological framework to assess building performance in a broad context of decision making, where environmental issues have a significant role (Cole, 1998; Cooper, 1999; Crawley and Aho, 1999).

Environmental assessment tools do not just provide a methodological framework for assessing building performance but also collect useful information to form guidelines for remedial work in order to meet pre-designed criteria. The collected data can also be used as feedback information for planning future projects of similar design while offering the same level of service and amenity. The accumulated knowledge and expertise of environmental building design contributes to the greater consideration of environmental issues within the decision-making process, thus minimizing the environmental impacts of a building in the long term.

Environmental assessment tools also enhance the environmental awareness of building practices, highlighting concerns about the design and construction of more environmentally oriented projects. Crawley and Aho (1999) state that environmental assessment tools might provide a means for incorporating more holistic environmental performance requirements in national building regulations, which again aim to significantly reduce the environmental impact of new construction. They go on to state that although largely different from each other and designed around different indicators, these systems nevertheless have a positive impact on reducing environmental stress in the short term. However, work is needed to develop a universal life-cycle assessment system based on internationally agreed absolute indicators of environmental performance.

4. Environmental Assessment Tools

4.1. *Overview of tools*

Concern is growing about the impact of building activities on human and environmental health. It is clear that actions are needed to make the built environment and construction activities more sustainable (Holmes and Hudson, 2000; Scheuer *et al.*, 2003; Finnveden and Moberg, 2005). The construction industry and the environment are intrinsically linked and it is inevitable that it has found itself at the center of concerns about environmental impact. Sustainable construction is a way of achieving the goal of sustainable development and a way to demonstrate the construction industry's responsibility toward environmental protection (Spence and Mulligan, 1995; Hill and Bowen, 1997; Bourdeau, 1999; Ofori *et al.*, 2000; Kaatz *et al.*, 2006). According to Kaatz *et al.* (2006), sustainable construction is a shift of practices to deliver a sustainable built environment and leading to sustainable settlement. Zimmerman and Kibert (2007) further state that environmental assessment tools are likely to produce greater and faster progress toward environmental improvement in the built environment because they help altering the mindset about sustainability by providing practical application tools to be used immediately.

Considerable work has gone into developing systems to measure and improve a building's environmental performance and physical facilities over its life cycle. They have been developed to evaluate how successful any development is with regards to balancing energy, environment, and ecology, taking into account both the social and technology aspects of projects (Clements-Croome, 2004). The development of environmental assessment tools is an important step in moving toward improving sustainability in the construction industry.

According to Ness *et al.* (2007), environmental assessment framework can be categorized into three broad areas and these areas include indicators and indices, product-related assessment tools, and integrated assessment tools. Indicators and indices are quantitative measures to allow for the tracking of longer-term sustainability development from a retrospective point of view to allow short-term decisions for the future. Separate indicators or indices based on a single criterion have been developed to monitor aspects of building performance such as air quality and indoor comfort. Product-related assessment tools focus in the production and consumption of material flows and their associated emissions on a life-cycle viewpoint such as life-cycle assessment and life-cycle costing. These tools do not integrate nature–society systems as they are mainly focusing on environmental aspects.

The integrated assessment tools are used for supporting decisions related to a project in a specific region. An integrated assessment tool is essential to provide a thorough evaluation of building performance against a broad spectrum of environmental criteria and this is an environmental assessment tool that has been used widely worldwide to provide guidance for sustainability in buildings. Different environmental assessment tools use different criteria and criteria are characteristics that are considered important and by which the performance of a building is judged. Environmental assessment tools are about assessing how buildings were designed, built, and operated. The early development of environmental assessment tools was to establish a yardstick for benchmarking environmental performance to provide guidelines for best practice and direction for the planning, design, construction, and operation of buildings (Haapio and Viitaniemi, 2008).

The development of environmental assessment tools is to increase the effectiveness in the evaluation of sustainability in buildings and to foster the agenda of sustainable construction. Environmental assessment tools emphasize the life cycle of a building differently. Some tools cover the whole life cycle whereas other tools are more focused on the maintenance and the use of buildings. Environmental assessment tools come in a variety of formats such as LCA-based tools, rating systems, technical guidelines, assessment frameworks, checklists and certificates. In the literature, the terms method, tool, system, and scheme have been used widely and interchangeably. Tool is related to a more practical use, whereas method, system, and scheme may be more relevant to theoretical development. As such the term "tool" is a more relevant and is used widely in this paper.

The release of the BREEAM (Building Research Establishment Environmental Assessment Method) in 1990 was the first comprehensive building environmental assessment tool. It was launched as a credit award system for new office buildings. A certificate of the assessment result is awarded to the individual building based on a single-rating scheme. The purpose of this tool is to set a list of environmental criteria against which building performances are checked and evaluated. This assessment can be carried out as early as at the initial stage of a project. The results of the investigation can be fed into the design development stage of buildings and changes can be made accordingly to satisfy pre-designed criteria. BREEAM is now extended to include assessment of different types of buildings (Saunders, 2008). Crawley and Aho (1999) state that the tool is successfully alerting building owners and professionals to the importance of environmental issues in construction.

BREEAM has made an impact worldwide. Following the introduction of BREEAM, the development of environmental assessment tools has grown tremendously in the past 20 years, with Canada, Australia, Hong Kong, and other countries using the BREEAM methodology in developing their own environmental assessment tools to suit local and national need. Table 1 summarized the environmental assessment tools used in different countries. The driving force behind the development

Table 1. Summary of environmental assessment tools for buildings.

Assessment tools	Origin	Schemes available	Ratings (lowest to highest)	Assessment	References
ABGR (Australian Building Greenhouse Rating)	Australia, 2005	Energy rating	1–5 Stars	Accredited assessor	Seo *et al.*, 2006
AccuRate	Australia, 2006	Residential	1–5 Stars	Accredited assessor	Seo *et al.*, 2006
BASIX (Building Sustainability Index)	Australia, 2004	Residential	Pass 40% energy and water, and pass thermal comfort	Self-assessment	Seo *et al.*, 2006
BEPAC (Building Environmental Performance Assessment Criteria)	Canada, 1993	Office and commercial buildings		Self-assessment	Cole, 1998; Crawley and Aho, 1999
BERDE (Building for Ecologically Responsive Design Excellence)	Philippines, 2009	New and existing buildings	Percentile ranks	Accredited assessor	http://philgbc.org
BREEAM (BRE Environmental Assessment Method)	BRE, 1990	11 Schemes for all types of buildings	Pass, good, very good, excellent, outstanding	Accredited assessor	http://www.breeam.org
CASBEE (Comprehensive Assessment System for Building Environmental Efficiency)	Japan, 2004	8 Schemes for all types of buildings	C, B−, B+, A, S	Accredited assessor	Cole, 2005; Yau *et al.*, 2006; Seo *et al.*, 2006
CEPAS (Comprehensive Environmental Performance Assessment Scheme)	HK, 2001	5 Schemes for all types of buildings	Bronze, silver, gold, platinum	Accredited assessor or self-assessment	Cole, 2005
DQI (Design Quality Indicator)	UK, 2002	New and existing buildings		Facilitator	Clements-Croome, 2004; Cole, 2005
Ecoprofile	Norway, 1999	Existing office	Index	Accredited assessor	Boonstra and Pettersen, 2003
Eco-Quantum	Netherlands, 2002				Yau *et al.*, 2006; Seo *et al.*, 2006

(*Continued*)

Table 1. (*Continued*)

Assessment tools	Origin	Schemes available	Ratings (lowest to highest)	Assessment	References
EEWH (Ecology, Energy saving, Waste reduction and Health)	Taiwan, 1999	New and existing buildings	Diamond, gold, silver, bronze, qualified	Accredited assessor	Yau *et al.*, 2006; ABRI, 2003
Environmental Status (Milijostatus)	Sweden, 1997	Existing buildings	1–5 Points	Self-assessment	Boonstra and Pettersen, 2003
ESGB (Evaluation Standard for Green Building)	China, 2006	New and existing residential, office, commercial, and hotels	1–3 Stars	Accredited assessor	Liu *et al.*, 2005; Yau *et al.*, 2006
DGNB (German Sustainable Building Certification)	Germany, 2008	New office and administration buildings	Bronze, silver, gold	Accredited assessor	http://www.plgc.org
GBI (Green Building Index)	Malaysia, 2009	New and existing buildings	Certified, silver, gold, platinum	Accredited assessor	Tan, 2009
Green Building Certification System	Korea, 2002	New building	Excellent, superior	Accredited assessor	Tae and Shin, 2009
Green Globes	Canada, 2000	4 Schemes for all types of buildings	1–4 Globes	Self-assessment	Boonstra and Pettersen, 2003
Green Mark	Singapore, 2005	New and existing buildings	Certificated, gold, gold plus, platinum	BCA	http://www.bca.gov.sq/GreenMark/
GreenStar	Australia Green Building Council	12 Schemes for all types of buildings	1–6 Stars	Accredited assessor	Seo *et al.*, 2006
GRIHA (Green Rating for Integrated Habitat Assessment)	India	1 Scheme for all types of buildings	1–5 Stars	Accredited assessor	www.grihaindia.org
HQE (High Environmental Quality)	France, 2002	Non-residential buildings	Basic, good, very good	Accredited assessor	Boonstra and Pettersen, 2003

(*Continued*)

Table 1. (*Continued*)

Assessment tools	Origin	Schemes available	Ratings (lowest to highest)	Assessment	References
HKBEAM (Hong Kong Building Environmental Assessment Method)	Hong Kong, 1996	4 Schemes for all building types	Bronze, silver, gold, platinum	Accredited assessor	Yau *et al.*, 2006
LEED (Leadership in Energy & Environmental Design)	USA, 2000	10 Schemes for all types of buildings	Certified, silver, gold, platinum	USGBC	Crawley and Aho, 1999
NABERS (National Australian Building Environmental Rating System)	Australia, 2001	Existing offices, homes, hotels, and retail	1–5 Stars	Accredited assessor	Yau *et al.*, 2006; Seo *et al.*, 2006
LiderA	Portugal, 2005	New buildings	C, B, A, A+, A++	Accredited assessor	http://www/lidera.info
SABA (Green Building Rating System)	Jordan, 2009	New buildings	Not green, green, very green	Self-assessment	Ali and Nsairat, 2009
SBAT (Sustainable Building Assessment Tool)	South Africa, 2004	New and existing commercial and residential buildings	Very poor, poor, average, good, excellent	Self-assessment	Sebake, 2008
SBTool (Green Building Challenge, then GBTool)	International, 1995	New and existing buildings	−1, 0, 1, 3, and 5	Accredited assessor, Self-assessment	http://www.iisbe.org
SBAM (Sustainable Building Assessment Methodology)	Swiss, 2005	New buildings	Rating	Self-assessment	http://www.skat.ch/
VERDE	Span	New buildings	0 to +5	Self-assessment	Sinou and Kyvelou, 2006

was public policy, the growing market demand for environmentally-friendly buildings, and the corporate social responsibilities of major national and international companies.

The development of environmental assessment tools has gained stronger impetus during the 1990s with the release of LEED in the USA and CASBEE in Japan following BREEAM. Many more environmental assessment tools have been developed ever since. Most of the environmental assessment tools are operated voluntarily and market-based by private non-profit making organization such as Green Building Councils. Some of the tools that were developed at the later stage have been mandated such as the Green Mark in Singapore, Evaluation Standard for Green Buildings in China, BASIX in Australia, and Code for Sustainable Homes in the UK. Most tools are either in the form of sustainable design checklist (Wrisberg *et al.*, 2002) or sets of quantitative or qualitative indicators (Ness *et al.*, 2007).

The earlier developed tools have different schemes for different types of buildings such as BREEAM, LEED, and Green Star. However, others may have been developed as "one-size-fits-all" assessment tools, in which the exact same tool is used to assess every type of building from child care center to high-tech buildings such as convention and exhibition centers.

4.2. *Aims and methodology of environmental assessment tools*

Environmental assessment tool is a systematic process designed to identify, analyze, and evaluate the environmental effects of a development. The aim of the process is to identify any foreseeable adverse impacts throughout the life cycle of a development and allows effective integration of environmental considerations and public concerns into decision making.

The environmental assessment tools aim at (Cole, 2005; Kaatz *et al.*, 2006)

- improving sustainability performance of the construction industry;
- contributing to the reduction of environmental footprint of the industry;
- encouraging knowledge and technology transfer so as to improve quality of building design, construction, and management;
- providing a common set of criteria and targets to guide decision making;
- providing a systematic and objective assessment of building environmental performance;
- providing market recognition for buildings that achieve high standards of sustainability;
- providing a blueprint to simulate sustainability and innovation that brings both environmental and financial return; and
- collecting data for future projects.

The use of environmental assessment tools can help in delivering buildings better suited to their physical settings and that impact on their socio-economic and environmental context. Environmental assessment tools can help to enhance the practice of building sustainability assessment by increasing their ability to impact on design and construction practice. They will also help to challenge the existing norms and values of those responsible for the delivery of buildings. They may also provide opportunities for a more conscious use of such a method to influence the quality of building process.

Environmental assessment tools for assessing performance of buildings have been developed and used in many countries. The number and diversity of tools makes it difficult for designers to identify the appropriate tools or information needed for their particular purpose. They are basically in two categories: assessment and rating tools (Seo *et al.*, 2006). Assessment tools provide quantitative performance indicators for design alternatives, whereas rating tools determine the performance level of a building in stars or the like.

According to Trusty (2001, cited in Huovila and Curwell, 2007) environmental assessment tools can be divided into three levels. Level 1 is for product comparison tools and information sources, which are used at the procurement stage and may have a life-cycle analysis perspective. Level 2 is for whole building design, which are used by design team members, and may involve weighting or scoring that provide input data to the Level 3 tools. Level 3 is frameworks for whole building assessment that have a broad coverage with mixed objective and subjective data. They may require trained assessors for the assessment.

According to Cole (1999), environmental assessment tools vary from single dimensional such as energy simulation to comprehensive tools that cover an assessment for the entire building. Comprehensive environmental assessment tools can broadly divided into four key evaluation levels: an assessment level, input level, output level, and presentation of results. The assessment level is the level of environmental assessment that covers the areas of resource use and associated emissions, indoor comfort, indoor environment quality, water and energy efficiency, and transportation. The input level usually requires measurements, audits, calculations, simulations and/or estimations that are used to assess environmental performance of the building. At the output level, assessment results will be presented to provide a comprehensive review of building performance, which is significant to aid decision making. The output results are also allowed for comparison between buildings. The outcomes can also feedback to impact on the design should they be required to re-designing to suit the stated performance. The outputs are usually presented in the forms of graphs, tables, grades, certificates, or reports.

The proliferation of environmental assessment tools usually cover a similar range of issues in broadly similar ways; however, the specific performance measures and priorities assigned to different issues are usually different. Some assessment tools may have weighting procedures such as Green Star. Weighting is particularly important for public projects and the social and environmental issues may have more weight than the economic aspects. However, the situation may be the reverse in a private development as financial return is a crucial driver for private projects. It is only a rare scenario when all criteria carry equal weights.

Whole building tools in general focus on the following eight categories of building design and life-cycle performance:

- sustainable site;
- materials and resources;
- energy;
- water;
- indoor environment quality;
- transport and facilities;
- environmental loadings; and
- biodiversity.

Wrisberg *et al.* (2002) state that environmental assessment tools are not just limited to be used to guide design decision of new buildings. They can be readily used for strategic planning to guide capital investments for new buildings and sustainable upgrading of existing buildings. They are commonly used as a marketing tool to signify the corporate social responsibility for national and international companies for high performance buildings to improve corporate images. Further, environmental reviews of existing buildings can provide a clear direction for sustainability maintenance and operations of buildings by the facility management (Crawley and Aho, 1999).

5. Critique on the Environmental Assessment Tools

As stated by Cole (1998), environmental assessment tools contribute significantly to the understanding of the relationship between buildings and the environment. However, a proliferation of programs and competing systems may slow the market uptake of green building practices. In addition, it may also cause confusion to users to the understanding and responding to environmental issues (Cole, 2006). Sustainability is broad and the interaction between building construction and the environment is still difficult to capture all possible assessment scenarios and contexts of application by using a set of assessment criteria (Kaatz *et al.*, 2006). The environmental assessment tools have limitations that may hamper their usefulness and effectiveness in the context of assessing environmental performance of buildings as discussed below.

5.1. *Usefulness as a design tool*

Environmental assessment tools are usually used to assess performance of a project when its design has completed at a certain stage. However, it may be more useful if any impairments for the pre-design criteria to be assessed and incorporated at the design stage of a development. Incorporating environmental issues can be achieved in the design process, which can minimize environmental damages. Even though these assessment tools are not originally designed to serve as design guidelines, it seems that they are increasingly being used as such (Crawley and Aho, 1999; Cole, 1999). However, Kaatz *et al.* (2006) suggest that the current environmental assessment tool is technical focus, which has little impact on design and construction practice.

The more effective way of achieving sustainability in a project is to consider and to incorporate environmental issues at a stage even before a design is conceptualized. It is important to separate project design and project assessment as building design takes place at an early stage and most of the outcomes of the design have already been established and incorporated into the final design. However, the assessment process works the opposite way around, thus may not be useful as a design tool (Crawley and Aho, 1999; Soebarto and Williamson, 2001; Haapio, 2008). Therefore, in order for environmental assessment tools to be useful as a design tool, they have to be introduced as early as possible to allow for early collaboration between the design and assessment teams.

Some environmental assessment tools may be used to assess existing buildings, such as BREEAM for existing office buildings. However, the usefulness of the environmental assessment tool in this respect is doubtful as the remedial work needed to make a completed building comply with the environmental criteria may be too extensive, too costly, and time consuming (Crawley and Aho, 1999). The remedial work to existing buildings may be impracticable or difficult to facilitate, e.g., replacing an existing ventilation system with a more environmentally-friendly system or installing more windows to allow for natural ventilation. This assessment tool has predominantly been applied to new construction; however, refurbishment and maintenance of existing buildings are also an important part in future construction activities.

5.2. *Project selection*

Environmental assessment tools are less useful for project selection as they are used to evaluate building design against a set of pre-designed environmental criteria. Environmental issues are generally only considered at the design stage of projects where the effect of different development options or locations of development are required to be considered at the feasibility stage.

Project selection starts at an earlier stage and it is at this stage that environmental issues are best considered and evaluated (Lowton, 1997; Ding, 2005). A project may have various development options and choosing the option that minimizes detrimental effects to the environment plays an important role in achieving sustainable goals. Ding (2005) argues that environmental matters are to be considered as early as possible. If they are not dealt with before and during the appraisal stage of a project, later alterations to the brief will cost money and cause annoyance. Sustainability should be considered as early as possible in the selection phase in order to minimize environmental damage, maximize natural resources, and reduce remedial costs. According to Crookes and de Wit (2002), environmental assessment is most efficient during the identification and preparation stages of a proposed project but current environmental assessment tools are designed to evaluate building projects at the (later) design stage to provide an indication of the environmental performance of buildings. However, by this stage it may be too late to consider environmental issues, as changing the design to meet full compliance with the environmental criteria may be costly.

5.3. *Life-cycle economic analysis*

Kibert (2008) states that the existing environmental assessment tools create financial discentives due to lack of life-cycle analysis and use. Environmental assessment tools focus on the evaluation of design against a set of environmental criteria broadly divided into three major categories: global, local, and indoor issues. These tools assess several main issues including resource consumption (such as energy, land, water, and materials), environmental loading, indoor comfort, and longevity. Some assessment tools such as BREEAM, BEPAC, LEED, and HK-BEAM do not include economic analysis in the evaluation framework (Zimmerman and Kibert, 2007). This may contradict the ultimate principle of a development, as financial return is fundamental to all projects because a project may be environmentally sound but very expensive to build. Therefore, the primary aim of a development, which is to have an economic return, may not be fulfilled making the project less attractive to developers even though it may be environment friendly. Environmental issues and financial considerations should both be considered as parts of the evaluation framework when making decisions. As stated in Larsson (1999), the revised Green Building Challenge (GBC) model includes economic issues in the evaluation framework. This is particularly important when the decision-making process starts from the outset at the feasibility stage where alternative options for a development are assessed. As shown, both environmental and financial aspects should be considered when assessing environmental concerns.

5.4. *Regional variations*

Most environmental assessment tools were developed for local use and do not allow for national or regional variations. To a certain extent, weighting systems can offer opportunities to revise the assessment scale to reflect regional variations and criteria order. However, regional, social, and cultural variations are complex and the boundaries are difficult to define (Liu *et al.*, 2005). These variations include differences in climatic conditions, income level, building materials and techniques, building stocks, and appreciation of historic value (Kohler, 1999). Many countries have adapted the BREEAM system for their own use giving rise to new systems such as HK-BEAM and Green Star. Adjustments to customize the system include cultural, environmental, social, and economic considerations. It is unlikely that a set of pre-designed environmental criteria could be prepared for worldwide use without further adjustments. The development of the SBTool (formerly known as GBTool) has been the first international environmental assessment tool to address the issue of regional variations using weighting. However, the SBTool is difficult to use because of the complexity of the framework (Crawley and Aho, 1999).

5.5. *Weighting*

There is insufficient consideration of a weighting system attached to the existing environmental assessment tools. The overall performance score is obtained by a simple aggregation of all the points awarded to each criterion such as LEED. All criteria are assumed to be of equal importance and there is no order of importance for criteria. Cole (1998) states that the main concern is the absence of an agreed theoretical and non-subjective basis for deriving weighting factors. It is currently dependent on the in-depth understanding of the environmental impact of building. The relative importance of performance criteria is an important part of the decision if the stated objectives are to be achieved, for example, the public sector's opinion will definitely differ from that of the private developer. Therefore, weighting environmental criteria should be derived on a project-by-project basis and should reflect the objective of a development. The absence of any readily used methodological framework has hampered existing environmental assessment tools in achieving sustainability goals.

5.6. *Rating scales*

Majority of the environmental assessment tools use different rating scales to determine the performance outcome of assessments such as star rating in Green Star, Gold and Platinum in LEED. The rating scales are based on a point award system and the total score obtained for the evaluation reflects the performance of a building in achieving sustainable goals in the industry. However, there is no clear logical or common basis for the way in which the maximum number of points is awarded to each criterion. Most building environmental assessment tools award their own points to environmental criteria. However, the different rating scales do not facilitates comparable assessment results across countries. Benchmarking the baseline performance for assessment is another difficult area to accurately assess in the existing assessment tools.

Since the introduction of the BREEAM, many tools exist to assess the environmental impact of buildings. The differences between these tools are a barrier to policy makers and organizations developing policies or sustainability measures (Emmann and Elle, 2006; UNEP SBCI, 2009). The proliferation of tools causes confusion to users as they may be developed with different scopes, emphases, functions, underlying assumptions, and limitations (Liu *et al.*, 2006). The ease of accessibility and comparability are crucial to the success of environment assessment tools (Zimmerman and Kibert, 2007).

5.7. *Third-party certification*

Some of the environmental assessment tools based the assessment on the opinion of a trained assessor to validate the achievement of building performance. Not only may the outcome be subjective but also it is only larger projects that can afford external expertise (Crawley and Aho, 1999). In addition, the assessment results are derived from just adding up all the points to get a total score. Even if a building rates poorly in a few key factors such as energy consumption, it can still achieve a high score from meeting other, more marginal criteria.

5.8. *Relative measures against absolute measures*

Common criticisms of environmental assessment tools include the use of checklist or point systems. Cooper (1999) states that most of the environmental assessment tools provide only a relative, not absolute, assessment of a building's performance. Such relative assessments conceal the specific impact of a development on the environment and there is no guarantee that the buildings, which score highly against the framework, are making a substantive contribution to increase environmental sustainability on a global scale. Rees (1999) continues, commenting that such relative assessments do not reveal the global carrying capacity appropriated by the development, and therefore cannot be used to measure progress for sustainability. Owing to the weakness of environmental assessment tools of assessing buildings using relative terms, Cooper (1999) states that the direction for assessing building performance needs to be capable of providing absolute measures. Such absolute measure can reveal the global carrying capacity appropriate by the development and be capable of measuring progress toward sustainability.

5.9. *Public participation*

Most environmental assessment tools primarily a technical assessment undertaken by objective judgments imposed by specialists (Kaatz *et al.*, 2006) and the public participation factor is rarely concerned as part of the decision-making process. This is a significant part of the process as it is the public that will suffer any long-term effects arising from decisions about developments. Indeed, the requirement for public participation is increasing (Kaatz *et al.*, 2006) and is also in line with Principle 10 of the Rio Declaration on Environment and Development (Curwell and Cooper, 1998). According to Kaatz *et al.* (2006), it is important to allow a broader participation by stakeholders to communicate their needs and wants to influence project values and agreeing on a course of action.

6. The Challenges of Environmental Assessment Tools

Sustainability in construction is often a buzz word than an actual practice. Environmental assessment tools for buildings were initially conceived to provide an objective evaluation of the resource use, ecological loadings, and indoor environmental quality of buildings. These environmental assessment tools were largely a framework in organizing existing knowledge and experience into a practical approach for professionals to use. Therefore, these assessment tools are developed to be simple to use, easy to follow, and comprehensive enough to capture social and environmental criteria in a single tool. For the past decade, these tools have made significant contribution to improve building performance toward the sustainability goal. Cole (2006) suggests that the future rating tools will have to be more readily suit the needs of different stakeholders, less complex, and be in a certain degree of consistency in the assessment.

6.1. *Agreeing on the meaning of sustainability*

Definitions of sustainability are many and most definitions center on the triple bottom line approach that integrates economic, social, and environmental dimensions of development. It is important that for assessing sustainability, the economic, social, and environmental aspects have to be turned and checked against each other. There is usually also some consideration of long-term time horizons with respect to giving consideration to future generations. The key challenge for further development of environmental assessment tools is therefore to agreeing on the meaning of sustainability such that causes less confusion to stakeholders and is more readily available to share an understanding of its meaning. At the project level, decision on resource development may deliver environmental protection and enhancement, improvements to human well-being, and economic growth.

6.2. *Affordability assessment*

Nowadays almost every country or region has at least one form of environmental assessment tools to improve sustainability performance of buildings. It is no doubt that these tools help to improve overall environmental awareness among construction professionals toward sustainable practices and achieve the goal of sustainability in the construction industry (Kaatz *et al.*, 2006). As the environmental assessment tools become more mature, Cole (2005) states that environmental assessment tools will need to be evolved in terms of refining performance measures and increasing level of complexity attainable within acceptable costs. With significant costs involved in the certification and the engagement of specialists in evaluation process makes it less attractive to most stakeholders. The documentation and certification costs can be as high as A$ 35,000 for Green Star and A$20,000 for BREEAM (Saunders, 2008). The additional cost and time in the assessment process is the main concern and will be a barrier for the assessment to be more widespread acceptance (Cole, 2006; Kibert, 2008). The web-based development of tools will help to reduce documentation burden and cost.

6.3. *From voluntary to mandatory*

The public sector has become a major driver for green certified buildings as these are only premises to be leased by them. Environmental building assessment tools have moved beyond voluntary market place mechanism as they are now increasingly being specified as performance requirements, are being considered as potential incentives for development approval (Cole, 2005). Some countries or regions have even made environmental assessment of building projects mandatory at some stages of a development, such as BASIX in Australia and EcoHomes in the UK for residential developments, and Green Mark for all types of constructions in Singapore, and ESGB in China (Ding, 2008). They are expected to contribute in reducing environmental impacts, increasing economic viability, and satisfying client's development objects. In addition, the certification process of environmental assessment tools in buildings has been increasingly recognized by banking, financial institutes, and insurance companies to be an acceptable basis for borrowing and investment appraisal (Cole, 2005). This is becoming a trend for the environmental assessment tools to be more effectively implemented through mandatory.

6.4. *Transparency and comparability*

With an increasing in importance existing environmental assessment tools will need further improvements in order to deal with the increasing readiness of its target market for more sophisticated discourse with respect to the understanding of sustainability issues and in facilitating the integration of sustainability consideration in construction decision making (Kaatz *et al.*, 2006; Zimmerman and Kibert, 2007). The role and expectation for building environmental assessment tools are changing and constant improvement is needed to make it more relevant and effective. Saunders (2008) states that further development of environmental assessment tools is essential to encourage more transparency and comparability between schemes. The move toward greater transparency and comparability between schemes may require an international collaboration to develop a set of common minimum standards and indicators to eliminate the effects of local regulations and guidelines in order it can work more efficiently on key issues and give more transparent results (Ness *et al.*, 2007). Such a development will help to promote the sharing of best practices more easily in the design, construction, and operation of the built environment. The growing use of BREEAM and LEED overseas such as Singapore, China, or buildings in the UK obtained a LEED certification has raised an issue of how relevant are these tools to assess buildings since these systems have never been designed to be used across multiple countries. The way of obtaining dual or multi-certification is a waste of resources and effort (Saunders, 2008).

6.5. *Stakeholders participation*

Stakeholders' participation in the assessment process is an area to be encouraged in order to respond effectively to the new challenges and requirements posed by the sustainability agenda. In other words, further improvement in building sustainability assessment may be promoted as collaborative activities among building stakeholders in order to enable a vision of sustainable construction to be of value and to be realizable (Kaatz *et al.*, 2006; Ofori and Ho, 2004; Zimmerman and Kibert, 2007).

6.6. *Assessing building process*

For further development, environmental assessment tools may not be used solely to evaluate the design quality of building performance; however, rather, it should also be used to transform the contents of methods by incorporating the principles of sustainable development directly into the building development process through information exchange and knowledge transfer. The further development will be a way to uphold balanced performance of a building toward economic, environmental, and social objectives; thus, influence the ways the buildings are designed, constructed, used, and demolished. The structure of the building process influences the available opportunities for exploiting economics of scale and the incorporation of sustainability measures into building process is often more effective (Van Bueren and de Jong, 2007). Kaatz *et al.* (2006) further state that environmental assessment tools should be integrated as part of the planning, design, and the decision-making process rather than a single activity so that information and lessons obtained be fed back into the project cycle. Greater emphasis should be placed on the process and transformation that occur within a building system to reflect sustainability values and principles of construction (Kaatz *et al.*, 2006; Shen *et al.*, 2005). Indeed building environmental assessment will need to be considered throughout the entire life span of a development. Through a close integration of building assessment with the building process, sustainability principles can be explicitly integrated with a building's objective and goals.

7. International Context of Environmental Assessment Tools

The absence of a common language between different assessment tools is a major barrier toward the consideration of environmental aspects to allow comparison of buildings internationally and to encourage exchange of best practices. Saunders (2008) states that the contribution of environmental assessment tools of buildings is enormous if the tools can be developed further into international context from which all regional derivatives will be generated and independently calibrated. The rating, then, can be easily understood and recognized worldwide. The development of an international environmental assessment tools is significant, which is flexible and adaptable to local conditions, yet can compare from one country to another.

Since 1995, environmental assessment tools have moved toward an international collaborative effort to develop an environmental assessment tool for international purposes. Known as the GBC, 13 countries collaborated to take a comprehensive look at environmental issues within buildings (Larsson, 1999; Rohracher, 2001). The GBC involves seeking a consensus among a variety of different cultural viewpoints and to reconcile different expectations of an assessment tool by national team members (Dammann and Elle, 2006). GBC's objectives are to establish international benchmarks for building performance and to offer participating countries help in developing regionally sensitive assessment models (Kohler, 1999). Some aspects of BREEAM also served as a model for the GBC framework (Todd *et al.*, 2001). Initially, GBC was difficult to use as a design tool because of its complexity and the data entered into the GBTool were not clearly linked to the scoring system (Todd *et al.*, 2001; Soebarto and Williamson, 2001). In order to simplify the process, GBC was implemented through software called GBTool and is now called SBTool by the iiSBE. The GBC initiative has marked the beginning of international standardization of environmental assessment tools. However, with regional variations and cultural diversities the task of developing international environmental assessment tools is not easy work.

The Built Environment Quality Evaluation for Sustainability through Time (BEQUEST) system was an international project funded by the Research and Technical Development Directorate of the European Commission through the Fourth Framework Programme in 1997. The project's main objective was to identify a common language and framework to assess and implement urban sustainability (Cooper, 1999). The BEQUEST represents a similarly successful international research project to GBC for creating greater collaboration of international partners at a multi-disciplinary level using the BEQUEST Extranet as the communication network. The research project combines the diverse knowledge and expertise of a wide range of environmental researchers, professionals, infrastructure providers, and managers to produce a framework, directory of assessment tools, and set of procurement protocols, which are linked together in the form of a tool kit. The vision of the BEQUEST is to enhance sustainability issues in urban decision making.

The framework of the BEQUEST is different from the existing environmental assessment tools such as BREEAM for assessing individual building sustainability. Instead, the BEQUEST aims at advising users on incorporating sustainability in urban design. As Kohler (2002) suggests, the BEQUEST is a framework for the preparation of projects for the "City of Tomorrow." Indeed, the successful implementation of the BEQUEST will, no doubt, enhance the sustainable development concept in urban development.

The principle of BEQUEST is based on the four principles of sustainable development identified by Mitchell *et al.* (1995): to embrace environment, futurity, equity, and public participation. However, as Cooper (2002) describes, there is a lack of a clear definition of sustainable urban development and its implementation process. Kohler (2002) further states that the problem of valuing the different aspects of sustainable urban development has not been explicitly identified. Thus, no clear basis for future discussions and implementation exists.

Driving force behind the development of an integrated environmental assessment tools on an international perspective correlates to the growth of globalization and the rise of the trans-national business (Saunders, 2008). This is particularly for major international companies to demonstrate their commitment to building in a more environmentally-friendly way, and the prestige speculative developments to attract and satisfy the needs of global tenants. The use of BREEAM and LEED outside the UK and US beyond the places of original has been numerous such as LEED used in assessing buildings in the UK, Singapore, Canada, and China, and BREEAM has also been used to assess building outside the UK as well as the basis for other rating schemes

BREEAM is now extended to establish the BREEAM International to be used to obtain BREEAM certification on non-UK buildings. BREEAM International is an on-size-fit-all assessment tools for all types of new and existing buildings. In June 2009, BRE Global, CSTB, and Certivea have signed a Memorandum of Understanding for the development of a Pan-European building environmental tool to align BREEAM and HQE schemes (BRE, 2009). In addition, a Global Green Rating (GGR) system is being developed and tested in 15 pilot sites in Germany, France, Spain, and the UK. GGR is a scheme that assesses and benchmarks the environmental performance of existing building against a number of measurable indicators (Bureau Veritas, 2010)

Since the development of environmental assessment tools, there have been activities in defining standardized requirements of environmental assessment tools (Cole, 2005; Haapio and Viitaniemi, 2008). The International Organization for Standardization (ISO) has been actively in developing international standards for the requirements and framework in the assessment of sustainable buildings. The aim is to define the process that shall be followed when addressing the economic, environmental, and social impacts of a building using a common framework and a set of indicators. The standards will be generally applicable and relevant for the assessment of integrated performance of building over its life cycle. The standards will describe a harmonized methodology for the assessment of environmental performance of buildings and life-cycle cost performance of buildings, and the quantifiable performance aspects of health and comfort of buildings. The ISO standards concern sustainability in building construction and are composed of the following (ISO, 2006, 2008):

- ISO/TS 21929-1:2006 (Sustainability indicators — Part 1: Framework for development of indicators for buildings) provides a framework, makes recommendations, and gives guidelines for the development and selection of appropriate sustainability indicators for buildings.
- ISO/TS 21931:2006 (Framework for methods of assessment for environmental performance of construction works — Part 1: Building) provides a general framework to improve the quality and comparability of building assessment methods.
- ISO 21930:2007 (Environmental declaration of building products) provides the principles and requirements for type III environmental product declarations (EPD).
- ISO 15392:2008 (General principles) identifies and establishes general principles for sustainability in building construction on a life-cycle perspective.

Furthermore, the European Union has also standardized the requirements and framework for environmental assessment tools in Europe through the European Committee for Standardization (CEN). In March 2004, the European Committee issued Mandate 350 (M350) that directed the CEN to develop a standardized approach for the delivery of environmental performance of construction products and whole buildings. The standards will be voluntary in the expectation that member states will use them as the basis for setting regulation in due course. CEN then set up a Technical Committee (TC350) to be responsible for the development of standardized methods for the assessment of sustainability aspects of new and existing construction works and products. Life-cycle cost analysis will also be included as well as performance of health and comfort of buildings. The standards are developed at the framework level (WG1) for the framework for the assessment of integrated building performance, at building level (WG2) are the calculation methods for the assessment of buildings, and at product level (WG3) for the EPD for the assessment of the environmental performance of buildings and for the assessment of the health and comfort performance of buildings in the framework of the integrated performance of buildings (Haapio and Viitaniemi, 2008).

At an international context, the establishment of the Sustainable Building and Construction Initiative (SBCI) in 2006 is aiming at developing a globally accepted system and definition for sustainable buildings so that it can be served as a platform for countries to establish policies, incentives, or standards in supporting sustainable buildings. The program was launched by the United Nations Environment Program (UNEP) as a partnership between the UN and the building sector stakeholders to develop a metric and protocol to measure and report on the buildings impact worldwide (UNEP SBCI, 2008). The SBCI aims at

- providing a common platform for all buildings and construction stakeholders to collectively address sustainability issues of global significance;
- establishing globally acknowledged baselines for sustainable buildings based on the life-cycle approach;
- developing tools and strategies for adoption of sustainable building practices throughout the world; and
- promoting adoption of the above tools and strategies.

8. Conclusion

Construction is one of the largest end users of environmental resources and one of the largest polluters of man-made and natural environments. The improvement in the performance of buildings with regard to the environment will indeed encourage greater environmental responsibility and place greater value on the welfare of future generations. There is no doubt that environmental assessment tools contribute significantly in achieving the goal of sustainable development within construction. On the one hand, it provides a methodological framework to measure and monitor environmental performance of buildings, whereas, on the other hand, it alerts the building profession to the importance of sustainable development in the building process.

The field of environmental assessment tools is vast, the aim of this study is to overview and analyze the existing tools in the context of sustainable construction. The comparison of the tools is challenging as they have been developed for different needs and purposes and under the influence of different cultural factors and various regulations. This paper summarized the aspects of environmental assessment tools that have been used by many countries to assess environmental performance of both new and existing buildings. The environmental assessment tools have escalated from local programs into an international agenda and from voluntary to mandatory approach. However, existing environmental assessment tools have their limitations as examined in this paper reducing their effectiveness and usefulness. There is a requirement for greater transparency, comparability, interaction, and recognition between members of the design team and various sectors in the industry to promote the popularity of environmental assessment tools. The inflexibility, complexity, and lack of consideration of the weighting system are still major obstacles to the acceptance of environmental assessment tools.

The importance of environmental assessment tools is well documented and recognized as a way for promoting environmental awareness among building professionals in the built environment. However, as discussed in this paper, the usefulness of environmental assessment tools for buildings can be extended from a checklist-type single evaluation tool to a multiple criteria framework that includes physical quantification of criteria to guide how a building is designed, constructed, operated, and demolished in the decision-making process. There is an obvious need for an international standardization on the assessment criteria, processes, and output of results to facilitate a common understand and to ensure consistency. From the development of the GBTool (now SBTool) in 1995 as the first international collaborative activity through to the current international collaboration in defining standardized requirements for the environmental assessment tools such as the International Organization and the European Committee for Standardization, the move toward a greater partnership and coordination among various scheme operators and organizations to improve transparency is inevitable in order that best practice can be shared more easily in the promotion of sustainability worldwide.

References

ABRI, 2003. *Evaluation Manual for Green Building in Taiwan, Ministry of Interior*, 3rd ed. Architecture & Building Research Institute, Taiwan.

Ahn, C., Lee, S.H., Pena-Mora, F. and Abourizk, S., 2010. Toward environmentally sustainable construction processes: The U.S. and Canada's perspective on energy consumption and GHG/CAP emission. *Sustainability* **2**, 354–370.

Ali, H.H. and Nsairat, S.F.A., 2009. Developing a green building assessment tool for developing countries — Case of Jordan. *Building and Environment* **44**(5), 1053–1064.

Bentivegna, V., Curwell, S., Deakin, M., Lombardi, P., Mitchell, G. and Nijkamp, P., 2002. A vision and methodology for integrated sustainable urban development: BEQUEST. *Building Research and Information* **30**(2), 83–94.

Boonstra, C. and Pettersen, T.D., 2003. Tools for environmental assessment of existing buildings, *UNEP Industry and Environment.* April/Sept., pp. 80–83.

Boughey, J., 2000. Environmental valuation, real property and sustainability. *Proceedings of Cutting Edge 2000*, RICS Research Foundation, RICS, London.

Bourdeau, L., 1999. Sustainable development and the future of construction: A comparison of visions from various countries. *Building Research and Information* **27**(6), 354–366.

BRE, 2009. *UK & France Sign MOU to Develop Common Single Rating Tool for Environmental Assessment Certification.* BRE, UK. Available at http://www.breeam.org/newsdetails.jsp?id=565 (Accessed 18 March 2010).

Bröchner, J., Ang, G.K.I. and Fredriksson, G., 1999. Sustainability and the performance concept: Encouraging innovative environmental technology in construction. *Building Research and Information* **27**(6), 367–372.

Bureau Veritas, 2010. *Five Leading International Firms Join Forces to Develop a Global Environmental Rating of Existing Buildings.* Available at http://www.ipd.com/OurProducts/IPDSustainabilityservices/IPDtodevelopenvironmentalbenchmarking/tabid/2030/Default.aspx (Accessed 14/4/2010).

Clements-Croome, D., 2004. *Intelligent Buildings Design, Management and Operation.* Thomas Telford, London.

Cole, R.J., 1998. Emerging trends in building environmental assessment methods. *Building Research and Information* **26**(1), 3–16.

Cole, R.J., 1999. Building environmental assessment methods: Clarifying intentions. *Building Research and Information* **27**(4/5), 230–246.

Cole, R.J., 2005. Building environmental assessment methods: Redefining intentions and roles. *Building Research & Information* **35**(5), 455–467.

Cole, R.J. and Howard, N., 2005. Building environmental assessment tools: Current and future roles. In: *Proceedings of the 2005 World Sustainable Building Conference in Tokyo*, 27–29 September 2005. Available at http://www.sb05.com/homeE.html (Accessed 25/2/2009).

Cole, R.J., 2006. Shared markets: Coexisting building environmental assessment methods. *Building Research & Information* **34**(4), 357–371.

Cooper, I., 1999. Which focus for building assessment methods — environmental performance or sustainability? *Building Research and Information* **27**(4/5), 321–331.

Cooper, I., 2002. Transgressing discipline boundaries: is BEQUEST an example of the new production of knowledge? *Building Research and Information* **30**(2), 116–129.

Crawley, D. and Aho, I., 1999. Building environmental assessment methods: Application and development trends. *Building Research and Information* **27**(4/5), 300–308.

Crookes, D. and de Wit, M., 2002. Environmental economic valuation and its application in environmental assessment: An evaluation of the status quo with reference to South Africa. *Impact Assessment and Project Appraisal* June, 127–134.

Curwell, S. and Cooper, I., 1998. The implications of urban sustainability. *Building Research and Information* **26**(1), 17–28.

Daly, H.E., 1990. Toward some operational principles of sustainable development. *Ecological Economics* **2**(1), 1–6.

Dammann, S. and Elle, M., 2006. Environmental indicators: Establishing a common language for green building. *Building Research & Information* **34**(4), 387–404.

Ding, G.K.C., 2005. Developing a multicriteria approach for the measurement of sustainable performance. *Building Research & Information* **33**(1), 3–16.

Ding, G.K.C., 2008. Sustainable construction — The role of environmental assessment tools. *Journal of Environmental Management* **86**(3), 451–464.

du Plessis, C., 1999. Sustainable development demands dialogue between developed and developing worlds. *Building Research and Information* **27**(6), 378–389.

Finnveden, G. and Moberg, A., 2005. Environmental systems analysis tools — An overview. *Journal of Cleaner Production* **13**, 1165–1193.

Glasby, G.P., 2002. Sustainable development: The need for a new paradigm. *Environment, Development and Sustainability* **3**, 333–345.

Haapio, A. and Viitaniemi, P., 2008. A critical review of building environmental assessment tools. *Environmental Impact Assessment Review* **28**, 469–482.

Herendeen, R.A., 1998. *Ecological Numeracy: Quantitative Analysis of Environmental Issues.* John Wiley, New York.

Hill, R.C. and Bowen, P.A., 1997. Sustainable construction: Principles and a framework for attainment. *Construction Management and Economics* **15**, 223–239.

Holmes, J. and Hudson, G., 2000. An evaluation of the objectives of the BREEAM scheme for offices: A local case study. *Proceedings of Cutting Edge 2000*, RICS Research Foundation, RICS, London.

Huovila, P. and Curwell, S., 2007. Sustainability Assessment of Building Design, Construction and Use. In: *Sustainable Urban Development Volume 2: The Environmental Assessment Methods.* Deakin, M., Mitchell, G. and Nijkamp, P. (Eds.), Taylor & Francis, pp. 329–344.

International Standards Organisation (ISO), 2008. *Sustainability in Building Construction — General Principles*, ISO. Available at http://www.iso.org/iso/iso_ catalogue.htm (Accessed 18 March 2010).

International Standards Organisation (ISO), 2006. *Sustainability in Building Construction — Framework for Methods of Assessment for Environmental Performance of Construction Works*, ISO. Available at http://www.iso.org/iso/iso_ catalogue.htm (Accessed 18 March 2010).

Kaatz, E., Root, D. and Bowen, P., 2005. Broadening project participation through a modified building sustainability assessment. *Building Research & Information* **33**(5), 441–454.

Kaatz, E., Root, D.S., Bowen, P.A. and Hill, R.C., 2006. Advancing key outcomes of sustainability building assessment. *Building Research & Information* **34**(4), 308–320.

Kohler, N., 1999. The relevance of green building challenge: An observer's perspective. *Building Research and Information* **27**(4/5), 309–320.

Kohler, N., 2002. The relevance of BEQUEST: An observer's perspective. *Building Research and Information* **30**(2), 130–138.

Kibert, C.J., 2008. *Sustainable Construction — Green Building Design & Delivery*, 2nd Edn. John Wiley & Sons, Inc, New Jersey.

Larsson, N.K., 1999. Development of a building performance rating and labelling system in Canada. *Building Research & Information* **27**(4/5), 332–341.

Liu, Y., Prasad, D., Li, J., Fu, Y. and Liu, J., 2005. A holistic approach to developing regionally specific framework for green building assessment tools in China. In: *Proceedings of The 2005 World Sustainable Building Conference*, Tokyo, 27–29 September 2005, pp. 1634–1641.

Lowton, R.M., 1997. *Construction and the Natural Environment.* Butterworth Heinemann, Oxford.

Mitchell, G., May, A. and MacDonald, A., 1995. PICABUE: A methodological framework for the development of indicators of sustainable development. *International Journal of Sustainable Development and World Ecology* **2**, 104–123.

Ness, B., Urbel-Piirsalu, E., Anderberg, S. and Olsson, L., 2007. Categorisign tools for sustainability assessment. *Ecological Economics* **60**(3), 498–508.

Ofori, G., Briffett, C., Gang, G. and Ranasinghe, M., 2000. Impact of ISO 14000 on construction enterprises in Singapore. *Construction Management and Economics* **18**, 935–947.

Ofori, G., Gang, G. and Briffett, C., 2002. Implementing environmental management systems in construction: Lessons from quality systems. *Building and Environment* **37**, 1394–1407.

Ofori, G. and Ho, K.K., 2004. Translating Singapore architects' environmental awareness into decision making. *Building Research & Information* **34**(1), 27–37.

Pearce, D., 1998. *Economics and Environment: Essays on Ecological Economics and Sustainable Development.* Edward Elgar, Cheltenham.

Rees, W., 1999. The built environment and the ecosphere: A global perspective. *Building Research and Information* **27**(4), 206–220.

Rohracher, H., 2001. Managing the technological transition to sustainable construction of buildings: A socio-technical perspective. *Technology Analysis and Strategic Management* **13**(1), 137–150.

Saunders, T., 2008. A discussion document comparing international environmental assessment methods for buildings. *BREEAM.*

Scheuer, C., Keoleian, G.A. and Reppe, P., 2003. Life cycle energy and environmental performance of a new university building: Modelling challenges and design implications. *Energy and Buildings* **35**, 1049–1064.

Sebake, T.N., 2008. *A Review of Appropriateness of International Assessment Tools for a Developing Country.* Available at http://researchspace.csir.co.za/dspace/bitstream/10204/3227/1/Sebake5_2008.pdf (Accessed 31/1/2010).

Seo, S., Tucker, S., Ambrose, M., Mitchell, P. and Wang, C.H., 2006. Technical evaluation of environmental assessment rating tools. *Research & Development Corporation, Project No. PN05.1019.*

Shen, L.Y., Wu, Y.Z., Chan, E.H.W. and Hao, J.L., 2005. Application of system dynamics for assessment of sustainable performance of construction projects. *Journal of Zhejiang University Science* **6A**(4), 339–349.

Shi, Q., 2008. Strategies of implementing a green building assessment system in mainland China. *Journal of Sustainable Development* **1**(2), 13–16.

Sinou, M. and Kyvelou, S., 2006. Present and future of building performance assessment tools. *An International Journal of Management of Environmental Quality* **17**(5), 570–586.

Soebarto, V.I. and Williamson, T.J., 2001. Multi-criteria assessment of building performance: Theory and implementation. *Building and Environment* **36**, 681–690.

Stigon, B., 1999. Sustainable development for industry and society. *Building Research and Information* **27**(6), 425–431.

Tae, S. and Shin, S., 2009. Current work and future trends for sustainable buildings in South Korea. *Renewable and Sustainable Energy Reviews* **13**, 1910–1921.

Tan, L.M., 2009. *The Development of GBI Malaysia.* Available at http://www.greenbuildingindex.org/Resources/GBI%20Documents/20090423%20%20The%20Development%20of%20GBI%20Malaysia.pdf (Accessed 31/1/2010).

Todd, J.A., Crawley, D., Geissler, S. and Lindsey, G., 2001. Comparative assessment of environmental performance tools and the role of the green building challenge. *Building Research and Information* **29**(5), 324–335.

UNEP SBCI (United Nations Environment Programme–Sustainable Building & Construction Initiative), 2008. Available at http://earthtrends.wri.org/updates/node/232 (Accessed 26/2/09).

UNEP SBCI (Sustainable Buildings & Construction Initiative), 2009. *Sustainable Buildings & Climate Initiative Stakeholder Round-tables*, October 26–27. Available at http://www.unep.org/sbci/pdfs/SBCI_RoundTables_Singapore09_Summary_final.pdf (Accessed 14/3/2010).

Van Bueren, E. and de Jong, J., 2007. Establishing sustainability: Policy successes and failures. *Building Research & Information* **35**(5), 543–556.

van Pelt, M.J.F., Kuyvenhoven, A. and Nijkamp, P., 1995. Environmental sustainability: Issues of definition and measurement. *International Journal of Environment and Pollution* **5**(2/3), 204–223.

Victor, P. Hanna, S. and Kubursi, A., 1998. How strong is weak sustainability? In: S. Faucheux, M. O'Connor and J. van der Straaten (Eds.), *Sustainable Development: Concepts, Rationalities and Strategies.* Kluwer Academic Publishers, Dordrecht, pp. 195–210.

World Commission on Environment and Development (WCED), 1987. *Our Common Future.* Oxford University Press, Oxford.

Wrisberg, N., Udo de Haes, H.A., Triebswetter, U., Eder, P. and Clift, R., (Eds.), 2002. *Analytical Tools for Environmental Design and Management in a Systems Perspective, The Combined Use of Analytical Tools.* Kluwer Academic Publishers, Dordrecht, The Netherlands.

Yau, R., Cheng, V. and Yin, R., 2006. Building performance assessment in China. In: *Proceedings of the 2nd International Conference on Intelligent Green and Energy Efficient Building & Technologies and Products Expo*, Beijing, 28–30 March 2006.

Zimmerman, A. and Kibert, C.J., 2007. Informing LEED's next generation with the Natural Step. *Building Research & Information* **35**(6), 681–689.

Chapter 22

A HOLISTIC APPROACH TOWARD CORPORATE RESPONSIBILITY

Tarja Ketola
Industrial Management Unit, University of Vaasa,
P.O. Box 700, 65101 Vaasa, Finland
tarja.ketola@uwasa.fi

Corporate responsibility (CR) has developed from the patriarchal care, some large companies took for their employees and local communities at the beginning of the Industrial Age to broad economic, socio-cultural, and environmental responsibilities, all companies are expected to take now in the Information Age. The current challenge is to make the seemingly conflicting three pillars, economic, socio-cultural, and environmental responsibilities, coincide. The purpose of this chapter is to show how these pillars can be first coordinated, and, finally, integrated. A CR portfolio matrix with 5-leap scales is also introduced. Level 5 represents sustainability. Many scales can be used by all companies, but some scales are company-, line-of-business-, or location-specific. The environmental, socio-cultural, and economic responsibilities of a company can be integrated through Fair Business. While Fair Trade is a well-known concept in retail, aiming to secure a fair price for the farmers of developing countries for their socially and environmentally benign products, Fair Business can be adopted by all companies in all countries. Fair Business means Fair Purchases, Fair Production, and Fair Sales, all of which are fair to nature, fair to people, and fair to both the company and society.

1. Introduction

Corporate responsibility (CR) is an old patriarchal concept dating back to the Industrial Revolution at the end of the 19th century when large companies ruled many localities. Companies often took paternal social responsibility for their employees and local community by providing them with housing, health care, children day-care, schools, grocery stores, fire brigades, etc. When the social safety nets of societies were built in the now so-called developed countries during the 20th century, the paternal grip of companies on their personnel and local people loosened, and the idea of corporate social responsibility disappeared into the background — except in developing countries where multinationals continued their patriarchal role (Ketola, 2008).

During the past three decades, as the capitalist market economy has spread all over the world with re-privatization of services and utilities, CR has reappeared in business vocabulary (Goodpaster, 1983; Blowfield and Murray, 2008), along with numerous similar ideas with varying emphases: corporate social performance, corporate social responsibility, and corporate citizenship (Carroll, 1979, 1999; Matten *et al.*, 2003).

Moreover, the concept of CR has widened its scope from social responsibility to two directions: environmental responsibility and economic responsibility. Because of the amplified malignant environmental impacts of industrial operations, such as the effects of air, water and soil pollution on nature, and human health, environmental awareness increased in developed societies in the 1960s and 1970s, leading to demands for corporate environmental responsibility. During the past decade, increasing scientific and public knowledge of climate change and its devastating consequences has accelerated these demands. Financial viability is the cornerstone of capitalist corporations; however, the decreasing acceptance of the dark side of business, namely greed and corruption, has turned mere financial responsibility to shareholders into a demand for broader corporate economic responsibility to all stakeholders. The threesome of environmental, social, and economic responsibility is sometimes called the triple bottom line (Elkington, 1998).

While companies generally accept that they should take at least some kind of social responsibility for their actions, they do not often acknowledge that they should take cultural responsibility as well. Cultural responsibility belongs to all kind of companies, home, and abroad. It is easy to notice the large-scale exploitive operations of oil and gas, companies, pulp and paper corporations, and other multinationals, which destroy indigenous cultures in developing countries, but similar cultural degradation has been taking place also in developed countries. For example, global brands, universal corporate cultures at industrial subsidiaries, and international retail chains may suppress characteristic local cultures and their small-scale activities. Companies can reverse this trend through cultural sensitivity and culturally responsible actions. Hence, the CR areas need to be environmental, socio-cultural, and economic responsibility.

The cultural responsibility notion brings us to the roots of the CR areas. They correspond to the areas of sustainable development (SD). The world-famous Brundtland report by the United Nations World Commission on Environment and Development first mapped them out: environmental, social, and economic sustainability (WCED, 1987). Yet already the United Nations Rio Declaration on Environment and Development in 1992 defined SD as integrating four dimensions: environmental, social, cultural, and environmental (UNCED, 1992). The UNESCO Universal Declaration on Cultural Diversity, Convention for the Safeguarding of the Intangible Cultural Heritage, and Convention on the Protection and Promotion of the Diversity of Cultural Expressions support this view (UNESCO, 2001, 2003, 2005).

2. From Disintegration to Integration of CR Areas

Companies can contribute to the different areas of SD of societies through their environmental, socio-cultural, and economic CR actions. It must be noted, however, that sometimes these three areas of CR may pull in different directions, leading even a well-meaning company to corporate irresponsibility. An environment and health conscious food manufacturer may pack its healthy, organic breakfast cereals in recycled carbon packages — which emit toxic heavy metal residues into the cereals.

Usually, though, environmental and socio-cultural responsibility actions coincide; however, they may conflict with some economic responsibility aspects, mainly financial. Many corporate environmental and socio-cultural responsibility activities incur expenses, which reduces the company's direct profits since in the market economy they are treated as externalities. Should companies aim at maximizing profits the way the market economy expects, or optimizing them at a satisfactory level that covers all expenses and leaves money for future investments? This question explains some of the difference between corporate financial responsibility and corporate economic responsibility: if a company strives for economic responsibility, it needs to set limits to its greed — irrespective of whether their environmental and socio-cultural investments will yield cost reductions, profit increases, image benefits, or partnerships in the long term.

The normal business operations of companies cause many kinds of environmental and socio-cultural damage, which is not internalized in their business costs but remain external expenses that taxpayers must cover. There is no estimation of socio-cultural damage available; however, the United Nations have calculated that the world's 3000 largest firms cause $2200 billion (about 1.600 billion euro) direct environmental damage every year (Jowit, 2010). The amount does not include indirect or long-term effects. This huge sum accounts for 1/3 of these companies' total profits and 6–7% of their total turnover (Jowit, 2010) — less for environmentally benign and considerably more for environmentally malignant companies. If these expenses were internalized, the true financial health of companies would be revealed and the companies would start taking their environmental responsibility seriously. The internalization of socio-cultural damage would require more company-specific calculations, but would lead the companies to fully understand their socio-cultural responsibility.

Research on and practice of environmental, socio-cultural, and economic responsibility have remained disintegrated, mainly due to the different expertise each of these CR areas requires. Economists and lawyers at universities and companies plan economically responsible actions; social scientists and human resource specialists socio-culturally responsible actions; and natural scientists and engineers environmentally responsible actions. Their worldviews are different and they do not have a common language. That is why the CR areas are still considered mostly contradictory and treated as economic *vs* socio-cultural and economic *vs* environmental values.

A holistic approach to CR is needed to first coordinate and finally integrate corporate environmental, socio-cultural, and economic responsibility. The purpose of this chapter is to build a holistic model of CR. This model does not only integrate the three areas of CR but also shows how companies can take major responsibility leaps and, ultimately, reach corporate sustainability (CS).

3. CR with CS at Level 5

Companies talk about their CR in a loose manner, throwing in all kinds of activities that they think will please their stakeholders. Anyone can detect this by studying the web pages and CR reports of different companies. There are some standards that try to make the annual progress of each company transparent and comparisons between companies possible. Most of them focus on one CR area, such as the ISO 14001 standard on environmental responsibility (ISO, 2010), the Social Accountability standard SA8000 on socio-cultural responsibility (SAI, 2010), and the International Financial Reporting Standards on economic responsibility (IASB, 2010). Few standards, such as the Global Reporting Initiative, attempt to cover all three CR areas (GRI, 2010).

The problem with these standards is that they set an acceptable responsibility level to reach but do not encourage, not to mention, challenge companies to become truly sustainable. Even for reaching this acceptable, but still unsustainable level, standards use the concept of continuous improvement, which companies happily adopt.

Genuine CS can be reached by increasing the integrated levels of corporate environmental, socio-cultural, and economic responsibilities — not step by step but — leap by leap. This fast evolutionary alternative suits most currently existing companies to whom revolutionary, one-off change, of course, seems to be impossible and gradual continuous improvement path has been inadequate. The leaps are challenging but not impossible; they simply require creativity. Figure 1 introduces the *CR portfolio matrix*. It measures CR with 5-leap scales — very low, low, moderate, high, and very high levels of responsibility. Level 5 CR represents CS.

Companies using this portfolio matrix to measure their responsibility take on the challenge to reach one responsibility level after another until they reach sustainability at level 5. Even individuals and whole societies can measure their responsibility like this with the ultimate aim of becoming sustainable. We are all parts of the same system: Planet Earth. Economic and socio-cultural sustainabilities are based on the environmental sustainability of the Earth. If the Earth's ecosystems collapse, there is no way (or any need) for individuals, companies, other organizations, or societies to achieve economic or socio-cultural sustainability.

Environmental sustainability can be achieved by staying within the limits of the carrying capacity of the ecosystems. Although we cannot as yet define exactly the carrying capacity of the ecosystems, with amounting data is likely that this will be possible in the near future. As sustainability is a regenerating concept,

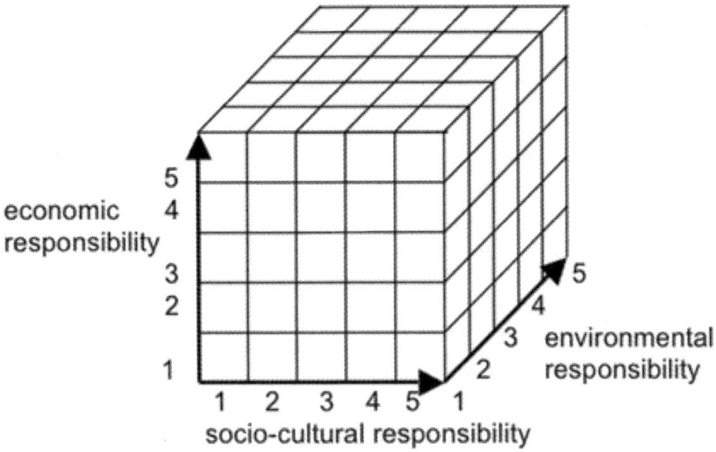

Fig. 1. Corporate responsibility portfolio matrix (Ketola, 2010).

mere abstention of excesses over the carrying capacity of ecosystems is not enough; we should revitalize the ecosystems by protecting their biodiversity and their life-support systems. For companies, zero emissions, use of only renewable energy forms, and use of only recycled or renewable natural resources that are harvested within their capacity of renewal are definite and absolute signs of corporate environmental sustainability. These corporate practices can be preserved to sustain also in the future by investing in the conservation of the Earth's vitality and diversity.

Socio-cultural sustainability can be reached by staying within the limits of the carrying capacity of humans and cultures, and by revitalizing human and cultural diversity and their life-support systems. Although we may not yet have a full list of measurements for socio-cultural sustainability, measures such as human and labor rights as well as gender and ethnic equality can be absolute yardsticks. Companies can also aim at zero occupational and product incidents, as well as at equal pay and opportunities at workplace, in supply chains, company networks, business environments, and different societies.

Economic sustainability incorporates financial and normative sustainability. Economic sustainability can be achieved by staying within the limits of the financial resources and global, regional, and local legal and ethical norms and by advancing them, so that environmentally and socio-culturally good life can be supported both now and in the future. Companies should not only comply with but excel legislation, corporate governance norms, and ethical standards, and give fair rewards to all their stakeholders in exchange of their stakes — pay for their purchases to suppliers, salaries to employees and managers, interests and instalments to financiers, taxes to the states and communities, and reasonable dividends to shareholders. Some money is needed for future payments and plans. Leftover profits can be invested in projects and endeavurs enhancing environmental and socio-cultural sustainability.

Integrating environmental, socio-cultural, and economic sustainability means that all these three areas of sustainability are achieved at the same time. In the following sections, first the ways to leap from responsibility to sustainability in each of the three CR areas are discussed separately and then these measures are integrated into a holistic model of CR leading to CS.

4. Environmental Responsibility with Sustainability at Level 5

Companies have a major role in combating the three environmental problems of the Earth: (a) the limited carrying capacity of the globe, (b) the limited unrenewable natural resources of the globe, and (c) the limited regeneration capacity of renewable natural resources of the globe. Ecological responsibility goals for a company could include

- protecting air water and soil;
- mitigating climate change;
- safeguarding biodiversity;
- using natural resources efficiently and economically.

Table 1 provides examples of possible environmental responsibility scales for companies. The topics and their importance vary between different lines of business.

Apart from using only renewable energy, the amount of energy used should be minimized through energy efficiency measures as any (non-human or non-animal power) energy use burdens the environment. In many countries also, water is a

Table 1. Some environmental responsibility scales (Ketola, 2010).

Level of environmental responsibility	Energy used	Emissions into air, soil and water	Recycling	Environmental incidents	Investments in biodiversity
5 (Very high)	100% Renewables	Zero	100% of raw material and waste	Zero for 10 years	>3% of net sales
4 (High)	75–99% Renewables	301–400% Less than legal limit	90–99% of raw material and waste	Zero for 8 years	2.1–3.0% of net sales
3 (Moderate)	50–74% Renewables	201–300% Less than legal limit	80–89% of raw material and waste	Zero for 6 years	1.1–2.0 of net sales
2 (Low)	25–49% Renewables	101–200% Less than legal limit	70–79% of raw material and waste	Zero for 4 years	0.1–1.0 of net sales
1 (Very low)	0–24% Renewables	≤100% Less than legal limit	<70% of raw material and waste	Zero for 2 years	<0.1 of net sales

scarce resource and its minimization needs to be incorporated in the environmental responsibility scales of companies.

The emissions relevant to the company in question should be detailed and each of them should be scaled: while zero is the ultimate aim at level 5, other levels need appropriate sub-goals. The legal limit concept and its percentages presented in Fig. 1 are probably not at all suitable measurements — they are there only to show that the legal limit is the very minimum level even in countries with stringent environmental legislation.

Recycling at both ends of the production chain links maximization of environmentally benign raw material supply and minimization of waste generation: some side products formerly called waste can be reused in production, others can be sold to other organizations or individuals, and the rest of raw materials needed can be bought recycled or exchanged for the company's own recyclable side products.

Zeronization of environmental incidents should naturally be the goal for all responsible companies; some lines of business are more incident-prone than others, but company-specific variations show that zero is a realistic goal.

Even if the true environmental damage of operations was internalized in the annual profit-and-loss accounts of companies, they should still invest in biodiversity in the long term, to help to safeguard the natural resource base, and, if profitable, behave ethically and donate money for biodiversity initiatives. Few "corporate citizens" do, and their donations are meagre compared with those of individual citizens: many individuals donate per cents of their income, but hardly any company donates even a per mil of their turnover.

Many more environmental responsibility issues than listed in Table 1 are important for different kinds of companies, and their scales should be tailored for each line of business.

5. Socio-cultural Responsibility with Sustainability at Level 5

Socio-cultural responsibility issues to be included in a company's responsibility portfolio may be originated internally from individual employees or managers, or groups of them, and externally from pressures from a variety of external stakeholders. The division of socio-cultural responsibilities between states and companies in different countries is a political question. Some would like companies to take much more socio-cultural responsibility for the communities in which they operate; others find it a threat to individual and community freedom and independence, i.e., a human rights issue. The latter fear that corporate dictatorship replaces citizen democracy. At all events, socio-cultural responsibility goals for a company could include

- protecting human, children's, and labor rights, and traditional cultures;
- mitigating negative and maximizing positive change effects on humans and cultures;
- safeguarding individual and cultural diversity;
- looking after employee, other human and cultural well-being everywhere.

Table 2. Some socio-cultural responsibility scales (Ketola, 2010).

Level of socio-cultural responsibility	Occupational accidents	Layoffs economic reasons	Salary structure	Product incidents	Donations for socio-cultural purposes
5 (Very high)	Zero for ≥10 years	Zero for ≥10 years	Highest salary <4 × lowest salary	Zero for ≥10 years	>3% of net sales
4 (High)	Zero for ≥8 years	Zero for ≥8 years	Highest salary <6 × lowest salary	Zero for ≥8 years	2.1–3.0% of net sales
3 (Moderate)	Zero for ≥6 years	Zero for ≥6 years	Highest salary <8 × lowest salary	Zero for ≥6 years	1.1–2.0% of net sales
2 (Low)	Zero for ≥4 years	Zero for ≥4 years	Highest salary <10 × lowest salary	Zero for ≥4 years	0.1–1.0% of net sales
1 (Very low)	Zero for ≥2 years	Zero for ≥2 years	Highest salary ≥10× lowest salary	Zero for ≥2 years	<0.1 of net sales

Level of socio-cultural responsibility	Gender equality care	Cultural equality care	Disabled employees
5 (Very high)	In global society	In global society	≥10% of work force
4 (High)	In local society	In local society	8–10% of work force
3 (Moderate)	In business environment	In business environment	5–7% of work force
2 (Low)	In production chain	In production chain	2–4% of work force
1 (Very low)	At workplace	At workplace	<1% of work force

Table 2 provides examples of possible socio-cultural responsibility scales for companies. The topics and their importance vary between different lines of business.

The zeronization of occupational incidents and product incidents is consistent with most companies' goal of minimization of them, only taking it to its ultimate end. The last bits of activities aiming to eliminate deaths and injuries may be costly but usually considered ethical even in contemporary societies.

The zeronization of layoffs for economic reasons, on the other hand, is a more controversial goal in the current capitalist market economy. Most companies feel they should have a right to lay off people during recession and laws tend to allow it. However, genuine social responsibility means looking after one's employees under any circumstances.

The salary structure scale presented in Table 2 has been derived from *The Republic* by Plato (2010) who said that, in order to restore peace and harmony, the richest in society should not be more than four times richer than the poorest in society. Should not the same apply to people within companies?

For donations for socio-cultural purposes, the same applies as for donations for biodiversity: corporate citizens should become at least as responsible as individual citizens. The palette of socio-cultural issues to support is wide and all companies can find something of their interest.

Many companies have policies to promote gender equality and/or cultural equality at workplace, but as socially responsible actors they should actively advance them also in their supply and production chain, in their business environment, and in both local and global society.

Another equality issue is the employment and treatment of the disabled. In reality, we are all either physically or mentally disabled to some extent in one way or the other; however, those with severe disabilities also have a right to work. Most of them can contribute as much as anyone else as long as they are treated with the same kindness as others and, in some cases, given special equipment, but even lower work input is valuable. From the global perspective, there are hundreds of millions of severely disabled people, as, in addition to birth defects and illnesses, wars, natural and man-made disasters, accidents, traumatic circumstances and events, and other hardships continuously hit individuals all over the world. The disabled are not a separate group but individuals living and working like anyone. The only reason they are included in Table 2 is that currently most companies are prejudiced against them and hesitate employing them. The exact number of disabled individuals a company employs is not relevant; however, the idea that we all have strengths and weaknesses and yet can and want to work. Hence, larger numbers of disabled employees in a company make them nothing out of the ordinary.

There are numerous other important socio-cultural responsibility issues for companies to observe than those mentioned above and they may vary in different lines of business, in different locations, and at different times.

6. Economic Responsibility with Sustainability at Level 5

A company should not make money at the expense of the well-being of its human stakeholders or the natural environment. Economic responsibility goals for a company could include

- using financial resources efficiently and economically;
- safeguarding financial resources for future investments;
- preventing fraud, bribery, extortion, nepotism, cartels, and other corruption;
- protecting all stakeholders' fair rewards for their stakes.

Table 3 provides examples of possible economic responsibility scales for companies. The topics and their importance vary between different lines of business.

A company should be both profitable and solid. Profits per net sales should not be too low, so that expenses can be covered and investments for future can be secured, but not too high, either, so that greed is not induced in the form of huge bonuses or management stock options and excessive shareholder expectations (Ketola, 2010). Some profits are needed to secure the continuity of the business, give some return on the owners' stake, make environmental and socio-cultural investments, invest in biodiversity, and give donations to charities. Profit percentage of 20 is suggested here as sustainable; however, some may disagree. A well-founded calculation would be needed. Also with return on investment 20% is here estimated sustainable; however, other suggestions are welcome. To measure ROI, companies

Table 3. Some economic responsibility scales (Ketola, 2010).

Level of economic responsibility	Profit, % (= Profit/ net sales)	Return on investment (ROI), %	Equity-to-assets ratio, %	Breeches against laws, fines	Stakeholder reward/stake (r/s) ratio
5 (Very high)	≈20	≈20	>50	Zero for ≥10 years	Highest r/s ≈ lowest r/s
4 (High)	15–19 or 21–25	15–19 or 21–25	40–49	Zero for ≥8 years	Highest r/s max $2 \times$ lowest r/s
3 (Moderate)	10–14 or 26–30	10–14 or 26–30	30–39	Zero for ≥6 years	Highest r/s max $3 \times$ lowest r/s
2 (Low)	5–9 or 31–35	5–9 or 31–35	20–29	Zero for ≥4 years	Highest r/s max $4 \times$ lowest r/s
1 (Very low)	≤4 or ≥36	≤4 or ≥36	<20	Zero for ≥2 years	Highest r/s > $4 \times$ lowest r/s

nowadays use two measurements: return on equity, demonstrating how much profit a company generates with the money shareholders have invested, and return on capital employed, which represents the efficiency with which capital is being utilized to generate revenue.

It is important for a company to be solid; the shareholders' equity should account for much of the total assets. For that reason, the equity-to-assets ratio should be high — the higher the better — and probably more than 50% could be considered sustainable. Zero breeches against laws and zero fines are essential for a sustainable company. Stakeholder reward/stake (r/s) ratio is an idea to be developed further: it could measure the equality between different stakeholders, such as customers, distributors, suppliers, contractors, partners, owners, financiers, the media, local people, NGOs, and the public: does each stakeholder get a fair reward for their stake in the company? The exact measures for this need further contemplation.

In addition to the issues detailed in Table 3, there are other company-specific, line-of-business-specific and location-specific economic responsibility issues for corporations to consider.

The three CR areas — environmental, socio-cultural, and economic — need to be integrated so that they pull in the same direction and turn into holistic CR.

7. A Holistic Model of CR for CS

Environmental, socio-cultural, and economic responsibility can be integrated through the idea of Fair Business, incorporating Fair Purchases, Fair Production, and Fair Sales. Table 4 provides the framework for these holistic responsibility scales for companies. The topics to be included should be chosen separately for each line of business.

Fair Trade is well known in retail, where consumers pay higher prices so that an environmentally certified producer in a developing country can be paid a fair price. These *Fair Purchases* could also be applied to wholesale trade and to business

Table 4. A holistic model of corporate responsibility (Ketola, 2010).

Fair business: Levels of environmental, socio-cultural, economic responsibility	Fair purchases (supplier side): fair to nature, fair to people, fair to company, and society	Fair production (employee side): fair to nature, fair to people, fair to company, and society	Fair sales (customer side): fair to nature, fair to people, fair to company, and society
5 (Very high)	100% of purchases	100% of production	100% of net sales
4 (High)	75–99% of purchases	75–99% of production	75–99% of net sales
3 (Moderate)	50–74% of purchases	50–74% of production	50–74% of net sales
2 (Low)	25–49% of purchases	25–49% of production	25–49% of net sales
1 (Very low)	<25% of purchases	<25% of production	<25% of net sales

contracts within both developing and developed countries (Ketola, 2010). A truly responsible company should bear some of the cost instead of pushing it all to the consumer. A company's purchases can be sustainable only if all of them are Fair Purchases. This will be an achievable goal for most companies when Fair Purchases in wholesale become commonplace in business relations.

Not only supply but also sales can follow fair play rules. Prahalad (2005) introduced the bottom of the pyramid (BOP) approach in which a company participates in solving socio-cultural problems with its products and services, by helping the poor to employ themselves in a way that is profitable also for the company. Such socio-cultural solutions may lead to solutions to environmental problems as well because those who can support themselves sustainably are not forced to exploit their natural environment unsustainably for their families' survival. An example of BOP is affordable renewable energy solutions for poor households and artisans in Asia or Africa. Other customers would also appreciate *Fair Sales*: consumers, craftspeople, and small enterprises all over the world tend to be at a disadvantage in business relations with large companies. A company's sales are sustainable if all of them come from Fair Sales. At the moment, this is a very challenging goal for most companies.

In order to make the whole chain of operations fair, there needs to be *Fair Production* between Fair Purchases and Fair Sales. Fair production treats employees, business partners, and other stakeholders involved in producing the company's goods and/or services on the basis of similar fair play rules as in Fair Purchases and Fair Sales. This means taking their individual socio-cultural circumstances and their environmental impacts into account. The company should make long-term contracts with them with no fear of sudden terminations or layoffs, secure safe and healthy working conditions, pay fair rewards for their ecologically benign inputs in the company, and advance their gender, cultural, and other diversity (Ketola, 2010). A company's production is sustainable if all its operations follow Fair Production rules.

Fair Business through Fair Purchases, Fair Production, and Fair Sales could become a strategic mission for companies: solving with their products and services socio-cultural and environmental problems all over the world. A company is sustainable if all its business is conducted on the basis of Fair Business. The holistic sustainability of the whole chain of corporate operations can be achieved by integrating all the sustainability requirements. In order that Fair Purchases, Fair Production, and Fair Sales are sustainable, they should all meet the level 5 environmental, socio-cultural, and economic sustainability requirements of Tables 1, 2, and 3, respectively, in addition to levels 5 in Table 4.

The holistic approach toward CR through Fair Business is a challenging initiative, which requires multidisciplinary cooperation between academic institutions, companies, non-governmental organizations, governments, and intergovernmental organizations to be developed into a globally responsible economic ideology. Its ingredients already exist in the many concerned quarters of our societies; they just need to be pulled together and committed to by all people, companies, and organizations who care about the future of humankind on the Earth.

References

Blowfield, M. and Murray, A., 2008. *Corporate Responsibility: A Critical Introduction.* Oxford University Press, Oxford.

Carroll, A.B., 1979. A three-dimensional conceptual model of corporate social performance. *Academy of Management Review* **4**, 497–506.

Carroll, A.B., 1999. Corporate social responsibility. Evolution of a definitional construct. *Business and Society* **38**(3), 268–295.

Elkington, J., 1998. *Cannibals with Forks: The Triple Bottom Line of 21st Century Business.* New Society Publishers, Gabriola Island, BC.

Goodpaster, K.E., 1983. The concept of corporate responsibility. *Journal of Business Ethics* **2**(1), 1–22.

GRI, The Global Reporting Initiative, 2010. Global Reporting Initiative, Amsterdam. Available at http://www.globalreporting.org (Accessed 19 February 2010).

IASB, International Accounting Standards Board, 2010. Available at http://www.iasb.org/Home.htm (Accessed 19 February 2010).

ISO, 2010. *ISO 14000 Essentials*. International Organization for Standardization, Geneva. Available at http://www.iso.org/iso/iso_14000_essentials (Accessed 19 February 2010).

Jowit, J., 2010. World's top firms cause $2.2tn of environmental damage, report estimates. *The Guardian*, 18 February 2010. Available at http://www.guardian.co.uk/environment/2010/feb/18/worlds-top-firms-environmental-damage (Accessed 18 February 2010).

Ketola, T., 2008. *From Psychopaths to Responsible Corporations: Waking up the Inner Sleeping Beauty of Companies.* Nova Science Publishers, New York.

Ketola, T., 2010. Five leaps to corporate sustainability through a corporate responsibility portfolio matrix. *Corporate Social Responsibility and Environmental Management* **17**, forthcoming.

Matten, D., Crane, A. and Chapple, W., 2003. Behind the mask: revealing the true face of corporate citizenship. *Journal of Business Ethics* **45**, 109–120.

Plato, 2010. *The Republic (360 B.C.) Translated by Benjamin Jowet.* Available at http://classics.mit.edu/Plato/republic.html (Accessed 20 February).

Prahalad, C.K., 2005. *The Fortune at the Bottom of the Pyramid. Eradicating Poverty through Profits.* Wharton School of Publishing, New York.

SAI, 2010. *Social Accountability International — Human Rights at Work.* Social Accountability International, New York. See http://www.sa-intl.org/ (Accessed 19 February 2010).

UNCED, 1992. Rio declaration on environment and development. In: *The United Nations Conference on Environment and Development,* Rio de Janeiro, 3–14 June 1992. Available at http://www.unep.org/Documents.Multilingual/Default.asp?Document ID=78& ArticleID=1163 (Accessed 19 February 2010).

UNESCO, 2001. Universal declaration on cultural diversity. In: *31st Session of the General Conference of the United Nations Educational, Scientific and Cultural Organization.* UNESCO, Paris, 2 November 2001. Available at http://unesdoc.unesco.org/images/0012/001271/127160m.pdf (Accessed 19 February 2010).

UNESCO, 2003. *Convention for the Safeguarding of the Intangible Cultural Heritage.* United Nations Educational, Scientific and Cultural Organization, Paris, 17 October 2003. Available at http://unesdoc.unesco.org/images/0013/001325/132540e.pdf (Accessed 19 February 2010).

UNESCO, 2005. *Convention on the Protection and Promotion of the Diversity of Cultural Expressions.* United Nations Educational, Scientific and Cultural Organization, Paris, 20 October 2005. Available at www.unesco.org/culture/en/diversity/convention (Accessed 19 February 2010).

WCED, 1987. *Our Common Future. Report of the World Commission on Environment and Development.* Oxford University Press, Oxford.

Chapter 23

ACADEMIA–CORPORATE ON CAMPUS SUSTAINABILITY COLLABORATION: AN EXPLORATION OF THE CONSTRUCT

Phil Cook[a] and Anshuman Khare[b]

[a]*Honeywell*
philip.cook@honeywell.com

[b]*Athabasca University, 301, 22 Sir Winston Churchill Ave., St. Albert AB T8N 1B4, Canada*
anshuman@athabascau.ca

The Academia–Corporate on Campus Sustainability Collaborative (ACCSC) is a construct conceptualized to address many current and urgent issues at Canadian Universities. Among these issues, Accumulated Deferred Maintenance (ADM) at Canadian Universities is of a multi-billion dollar magnitude. Funding for infrastructure refurbishment or replacement has traditionally been difficult to procure and has taken an appropriate secondary position to funding priorities relating to teaching, learning, and research. The ADM issue is not a new issue; however, it is a growing one. The investigation of this construct is general and the research is exploratory in nature. The approach to the chapter is one of novelty that uses Maturana and Varela's (*Autopoiesis and Cognition: The Realization of the Living.* D. Reidel Pub Co., Boston) understanding of the living organization (i.e., Autopoiesis; the theory of self-producing, living systems) as a metaphor for the construct. The construct is disaggregated into its "conceptual elements" and the quality of the relationship between the conceptual elements — the organization of the construct — and the structure — a description of the conceptual elements themselves — is explored. A view on the "structural–environmental coupling" is also explored. A synthesis of the management theory underpinning the construct is also provided. The multiple focus — social, environmental, and economic — nature of sustainably-managed businesses demands an innovative use of management theory. The conclusions of the chapter illustrate a great deal of evidence supporting the positive quality of relationship between the conceptual elements of the construct. It is also concluded that evidence of an accommodating and "demanding" environmental coupling between the ACCSC and the current environmental domain exists.

1. Introduction

This chapter provides an exploration of a construct; namely the Academia–Corporate on Campus Sustainability Collaboration (ACCSC). The authors suggest that the formation and manifestation of the ACCSC would address several distinct issues facing Canadian Universities currently. The infrastructure of universities in

Canada in general has been allowed to deteriorate at an alarming rate and is now a multi-billion dollar issue (A Point of no return,[1] 2000). Sustainable development (SD) is being studied seriously by many great disciplines (Webster and Dautremont-Smith, 2009) and a great deal of interest has been shown toward the inclusion of the concepts of SD in curriculum throughout North America (New Leaders, 2009).

The ACCSC construct brings together Canadian University Administration, Faculty, Students, Energy Services Companies (ESCOs), and provides a means through which many of the failing infrastructure issues can be solved. This is done while simultaneously providing a "living laboratory" setting, in which students and faculty can explore actual SD projects using Action Learning (AL) techniques. The construct also suggests that the projects be financed by the ESCOs and that the ESCOs are paid from monies gained by operation cost savings. The ESCO is a well-tested business model that has been used successfully for many years (Studebaker, 2001).

2. ACCSC Construct: Overview

The ACCSC construct proposes collaboration between Academia, Corporations, Students, and University Staff acted out in and on the "built" environment of the University itself.

At the core of the construct is the act of recommissioning existing campus buildings to optimize energy use, water use, climate control, human comfort, etc. in an effort to bring this environment closer to a sustainable condition. In doing this, the University benefits from reduced costs of operating these buildings and also has an opportunity for students to become involved with a "real-world" sustainability project right on their own campus. Faculty has the opportunity to construct curriculum or courses that incorporate AL opportunities.

Risk to the University is minimized through the use of a Guaranteed Energy Savings Contract (GESC) financed by an ESCO. A point of differentiation is afforded the ESCO who competes for these projects as the offering now has the additional value component of the "AL readiness." The ESCO that demonstrates a superior ability for working with the faculty and students will have a better chance of being awarded the project.

The research approach to this chapter is one of the novelties and as such deserves some clarification of motivation and method. The Oxford Concise Dictionary defines

[1]The info for "A Point of No Return" is a sketch at best. It is adapted from "A Point of No Return: The Urgent Need for Infrastructure Renewal at Canadian Universities" (2000). Canadian Association of University Business Officers ISBN 0-9685895-2-9 available from the National Library of Canada. It mentions in the text that Resources Management Consultants (RMC) Ltd. assisted in the survey, interpreting its findings, and preparing the report.

a construct as "an idea or theory containing various conceptual elements" (Construct, 2009). The analysis of the construct conveyed in this chapter has been carried out in a manner, which envisions any organization, at least metaphorically, as a living entity. As Maturana and Varela (1980) have described the living or "autopoietic" system as a "bounded metabolic network." The network cannot be considered "living" at the disaggregated level of any of its single elements and is only demonstrative of life based on the simultaneous autopoietic quality of the relationship between its singular elements. The significance of the autopoietic thesis for the purposes of the authors is to liken the relationship between the conceptual elements of the construct to those of the relationship between the elements of a living entity. In essence, to reveal the "metabolic network" of the construct by means of confirmation of the positive quality of the relationship between the conceptual elements of the construct and subsequently arrive at a conclusion regarding the feasibility of the construct.

The term "organization" when used in the context of the construct refers to the relationship between the different conceptual elements composing the construct. This is not the traditional definition of organization that refers to "the aggregation of people" rather it refers to the interplay of the conceptual elements toward a common purpose that will eventually define the construct as a whole (Ackoff, 1999). Without empirically-derived evidence to illustrate the "positive quality" (i.e., mutual benefit, need satisfaction, innovation, behavioral intent, sound theoretical foundation, etc.) of these relations, it is posited that the feasibility of the ACCSC be rejected.

The "structure" of the construct refers to the individual conceptual elements and the description of their qualities. It is not imperative that the structure be static for the construct to manifest successfully. It may be that the structure would change as different environmental perturbations arose. What was important to the research was that an initial structure be explored and described and that any structural interaction with the environment (i.e., structural–environmental coupling) (Zeleny, 1997) also be explored.

It is also important to note that the "University as a Built Environment" may technically be considered an environmental element and categorized as such; however, because of the degree of importance to the feasibility of the construct it was deemed that the reader would be better served if the findings and analysis regarding this element be in proximity to the other conceptual elements.

The development of the construct begins with a review of current management theories that converge to form the theoretical foundation of the construct. An observance of the proposed organization and structure of the ACCSC follows with an analysis of both derived from a qualitative gathering of information from academic papers, texts, and online sources. The qualitative information is used to provide evidence illustrative of the quality of relationship between the elements of the construct.

3. Theoretical Foundation

The following review is an illustration of the different contemporary management theories and methodologies that converge to form the theoretical foundation of the ACCSC construct.

3.1. *Collaboration*

Although the different types of "collaboration" are many, the type of collaboration that this chapter is concerned with is that of the "business" collaboration. Welborn and Kasten (2003, p. 1) define "business" collaboration as "the alignment of business activities and processes with another business to create mutual benefit." The imprecision of the definition will suffice as a bucket into which the authors will add several concepts to provide detail and clarity.

Business is perpetually encompassed in uncertainty and instability. The very nature of business carried out in a "free market" suggests complexity and chaos. Evidence of this phenomenon can be witnessed in the volatility of stock markets, the rise and fall of individual organizations, sector defined bubbles, and the reality of nature in general. It is, however, very difficult if not impossible for every corporation to have at its disposal the "core competencies" necessary to satisfy every change in the business environment. It is also not practical to consider that most corporations have the time necessary to develop these competencies given the current rate of change in the environment. One may also posit that organizations that have certain core competencies do not wish to develop other competencies, preserving the "nimble" characteristics of the specialist.

The instability and uncertainty of the environment render traditional strategy tools, which rely on stability as a prerequisite, unusable (Welborn and Kasten, 2003). Sawyer (2007) claims that the "critical instability" of a situation is a catalyst for creativity and emergent innovation. He claims that decades of disaster research reveal that "improvisational groups" are the most effective and fastest solution to rapidly changing conditions caused by natural disaster. Welborn and Kasten (2003) concur that the conditions of the business environment today lend credence to the use of collaboration as a strategic innovation tool. Collaboration, then, can be viewed as a mechanism for dealing with a fast moving chaotic reality. Core competencies can be shared or amalgamated in a synergistic way to provide the collaborators with new customers, new products or services, new market exposures and/or new brand positioning.

This does not imply that planning and strategy should be discontinued. However, both of these processes need to be revisited and viewed in a different way. The strategist must now look much closer on the horizon and devise means through which the team can be best prepared for collaboration. The collaborators must be prepared with an ability to adapt or present a means of communication between the collaborators that is easily understood by two or more parties and amendable to a common understanding quickly. Collaborative software and/or common platforms of dialogue are solutions to this requirement (Svendsen, 1998).

It is put forth by many business theorists that corporations are composed of conversations (Perkins, 2003; Luhmann, 1990). The collaboration, as such, should not be considered any different than a corporation when considering the subjects of conversation and communication. Both progressive and regressive communication can be expected. As one might conclude the success of the collaboration, whether it can be inside the corporation or within the collaboration of many corporations, is contingent to some degree on the "contact architecture" (Perkins, 2003). Contact architecture refers to the various means and contexts of and in which interaction between individuals and groups occurs. If one party is not communicating well with another party or the channel of communication is not open, then serious repercussions are imminent.

Trust is also a prerequisite to a successful collaboration and must be held paramount in the preparation of the collaborative corporation (Kuei and Madu, 2009). Collaborations quite often mean sharing control and so those members of the corporation who will be involved directly with the collaboration must be prepared for what is likely to be a new experience of the loosening of authority. Clarification of roles, goals, objectives, and timelines are all tasks that should be completed in order to begin building a trusting relationship. Conflict resolution and clarification of expectations are also areas of concentration for the corporations preparing for a strategy of collaboration (Svendsen, 1998).

Potential collaborators must also be aware and prepared for differences in cultures, resources, and capabilities. For example, it may come as quite a shock to a corporate collaborator that the not-for-profit has a cooperative and process-oriented culture as opposed to a competitive style.

Planning and preparing for the collaboration are, then, mandatory tasks when using collaboration as a strategy. However, both Svendsen (1998) and Sawyer (2007) imply that the process of planning within the collaboration is iterative. This is to say that the stakeholders involved in the collaboration must be prepared for many cycles of organization and reorganization of parts or the whole of the collaboration during its existence. The nature of collaboration is likely to be much more improvisational than any traditional intra-organizational experience.

3.2. *Action learning*

A unique aspect of the ACCSC is the suggested use of AL to involve students. AL or "active learning" is defined as any strategy that "involves students in doing things and thinking about the things they are doing." It refers to teaching techniques that force students to go beyond listening to lectures and move toward activities requiring them to discuss and reflect on what they are learning (Heriot *et al.*, 2007). The underlying premise of AL is that learning (L) is made up of two elements: programmed knowledge (P) and questioning insight (Q). L = P + Q. P is knowledge that is inherited from traditional instruction. Q is the questioning of the participant inspired by the activity that the participant is involved in. The theory behind such a claim is reliant on the concept that not until one can apply the knowledge learned does one has a legitimate claim to having "learned" anything (Johnson and Spicer, 2006).

AL has many faces and many forms and is not considered, by some, any less effective when used in combination with traditional programmed learning (Raelin and Raelin, 2006). Dependant on where the methodology is applied, the cultural differences and the "readiness" of the potential participants may not be conducive to a full course of AL. It is important to note that many forms of AL are, in general, composed of and involve a group or set of students numbering approximately six, individuals acting on real tasks individually, participating on reflection within the group regarding action, questions as the main help to participants, and the use of a facilitator (Beaty *et al.*, 1993).

As with any other group situation it is the burden of the members to produce the product, which in this case is learning. The facilitator has the responsibility to try to steer the group away from common "group think" problems such as production blocking, social inhibition, and social loafing (Sawyer, 2007); however, it is up to the members of the group to develop themselves with the help of the other group members. Raelin (2006, p. 159) concludes that this method produces students who at the end of the learning process have a strong sense of self-leadership. This is a valuable life skill.

In addition to the stumbling blocks brought about by the group dynamic, Frank (1996) raises a good point regarding validation and evaluation criteria issues. As AL is relatively innovative and non-traditional, the systems through which evaluation of the students' learning (i.e., examinations and written assignments) may also have to be innovative and non-traditional. It is probable that AL methods used in concert with traditional methods are a viable work around for this issue.

Both Ackoff (1999) and Feynman *et al.* (1989) refer to the condition of some teaching systems and methods with a certain degree of dismay. Ackoff (1999), for example, refers to the "knowledge hierarchy" of data, information, knowledge, and wisdom and specifies that not enough teaching of knowledge is carried out in schools. Knowledge in this case is the use of data transformed into information and then applied in reality. Both authors express concern of teaching methods that explain "what" is right and not "why" things are right. A similar sentiment is expressed by Filion (1999) when he suggests the development of "thinking guides" to complement the "how to guide" with "why to" and "for what purpose" publications.

3.3. *Organizational learning and systems thinking*

The term "Organizational Learning" and the term "Learning Organization" are used somewhat synonymously in the literature. Regardless of the term one prefers, each one, to a great extent, implies that the organization displays characteristics other than that of a crude "money-making machine" (Senge, 1990) and more closely related to a "living being." Such a claim is counterintuitive and deserves a digression in order to shed light on cognition, consciousness, and what exactly one can consider to be a "living being."

Life, as a biological phenomenon, can be reduced to the structure of a cell. Everything that is alive on this planet can be said to be composed of at least one cell. Cells display two defining characteristics in their organization: membrane and metabolism. The membrane is pervious, allows the intake of nutrients and the output of waste and forms a boundary for the cell metabolism. Metabolism is the chemical and energy exchange between a cell and its non-living environment. Cells organize into organisms by means of a network of cells. A key concept of systems thinking is the concept that everywhere in the nature of our reality we find networks. The network is a pattern that is common to all life. In addition, a key characteristic of life in the network of cells is the ability for self-generation or Autopoiesis. An autopoietic system is chemically independent in its environment and filters incoming and exiting molecules (Capra, 2002).

The self-generation and subsequent self-perpetuation (Autopoiesis) of cellular networks is stated by the "Santiago Theory of Cognition" to be representative of cognition and therefore of mental activity. Capra (2002, p. 34) paraphrases Varela and Maturana's claim that "Mind — or more accurately mental activity — is immanent in matter at all levels of life." This is a profound defiance of Cartesian theory where mind is the separate "thinking thing." Mind in this case permeates the entire living organism and is in fact part of the structure of the organism. Autopoietic networks undergo two distinct structural changes: self-renewal and developmental changes. Developmental changes are those brought about by a change in the environment in which the network exists or are changes that are brought about by internal dynamics of the system. This ontogenic structural change is particularly useful in the establishment of a system as a living thing. The reaction of a living system to a change in the environment is profoundly different from that of a non-living system to the same change in environment. It is of interest to note at this time that the reaction of the living system to a change in environment is quite likely to be unpredictable as well as different. Continual change in structure in response to environment (i.e., the learning, adaptation, and development cycle) is a hallmark of living systems (Capra, 2002).

Also of note regarding this observation is that throughout the structural adaptation of the living system it maintains its identity. The relationship between the elements remains even though the structure of the elements may change during this process. The structure is subservient to the organization of the complex unity and because of the plastic nature of the structure coupled with the environment the adaptation along the system's ontogeny can be considered "learning" (Maturana, 1975).

The living system is characterized by its autonomy. It stands alone in its non-living environment interacting and adapting to it. In many ways, this typifies the living being. If it is not living, then it is non-living and must be considered because of its complexity to be a machine (de Geus, 1997). Yet the elements of a hard-designed system such as a machine display no autonomous pursuit of purpose that is greater than their physical existence. Too large a hammer does not adapt itself to too small a nail.

Argyris and Schon (1996) suggest that learning in organizations can take one of two forms: single loop or double loop. "Single loop" learning is learning that the organization undergoes in adapting to a situation by correcting a strategy or function to return to previously contrived expectations. "Double loop" learning is learning that requires a change in the organization's values and norms to achieve a desired level of performance previously achieved with a current set of values and norms that are no longer useful for the desired effect under the new current reality. Organizational learning, then, at its most profound can be observed in the organization adapting to its environment through a fundamental structural change. We extrapolate from this observance that the organization changes structurally in response to its environment and therefore begin to see that the organization can be likened to a living being.

Several steps in complexity separate the single cell from the large multinational corporate organization. However, we begin to see that there are similarities and differences that one can note from level to level on the ladder of complexity. Each progression of complexity brings with it a new characteristic whose description has little meaning at the previous level. The cell in the brain has little use for the "concept of thought." The activity of the individual has little meaning in the context of the behavior of the corporation. Each level of complexity brings with it a unique "emergent" characteristic (Checkland, 1981). This is the case when considering that, although the organization or system is cognizant, it is not until the system reaches a certain level of complexity that the system or organization can be said to have consciousness. Consciousness is cognition at a certain level of complexity or can be viewed as an emergent characteristic of complexity.

3.4. *Sustainable development*

For the purposes of this chapter, the authors will use the definition of SD from what has become known as the "Brundtland report." The report states the definition of SD as development which "seeks to meet the needs and aspirations of the present without compromising the ability to meet those of the future" (WCED, 1987). This definition implies that our activities have a social, environmental as well as an economic impact in the present and have an impact on future generations as well. Several terms have been used to describe somewhat similar perspectives and yet there is much confusion over what the terms mean. The authors have selected the term "development" intentionally as it is and must be considered differently from the term growth. In the corporate, world growth is quite often interpreted or cloaked in the idea of development and yet these are two very different terms. Many things can grow, such as garbage piles, that do not develop and many things, such as artists, can develop without growth (Ackoff, 1999). To develop requires an improvement in one's satisfaction of one's needs and the needs of others. Development then implies the increase in capability and competence on behalf of all involved not at the cost of others. For a corporation to be developing is for it to be increasing the quality of reality not only of its employee's but of the society and environment that it is a part of. It is this "virtuous cycle" that is the goal of an SD strategy.

This forces us to revisit our traditional understanding of the economy as an ever expanding cycle of the flow of money between households and firms. Traditionally, we have viewed the economy as a flow of money from households to firms through the purchase of goods and services and subsequently the flow of money from households to firms for investment purposes. The firms used the money to invest in more efficient ways to produce products for the households and the households purchased the goods with the interest paid to them by the firms for the use of their money (Jacobs, 1993). It is important to note that the system is much more complex, because of the interaction with government, imports and exports, etc., but for the purposes of illustrating a new view of the economy the description will suffice. This traditional perspective of the economy still exists; however, an acknowledgement of the imperative of SD introduces the concept that the economy interacts with the biosphere and a much larger and more complex system best described simply through the observance of the laws of thermodynamics.

Thermodynamics allows us to expand the system description of our activity to include the biosphere and to assess the impact of our activities on it. We do not need an exhaustive knowledge of the laws of matter and energy only a brief understanding of the first and second laws of thermodynamics. The first law states that matter and energy can be neither created nor destroyed. This sounds like a marvelous condition to find ourselves in, everything that we make and use is not destroyed and remains with us forever presumably being eventually recycled for our use infinitely. Unfortunately for us the second law of thermodynamics cuts this utopia off at the knees. The second law states that in the use of matter and energy everything in a closed system is flowing from a state of low entropy to a state of high entropy. Entropy can be best understood as a state of disorder or unavailability. The combustion of fossil fuels, for example, renders heat and some atomic byproducts (CO_2, H_2O, NO_x) that are of very high entropy. We will never be able to gain heat from these high entropy byproducts unless we can lower their entropy. In a closed system, then, with no means through which to return materials of high entropy to materials of low entropy we eventually end up with everything at a high state of entropy, or matter and energy unavailable for use.

SD would then be relatively impossible given the parameters of our reality. There is, however, hope. In our reality, we have a source of energy that allows us a means through which we can recycle the high entropy products into low entropy products. The sun is our way out of Newton's vicious cycle of entropy. High entropy products such as CO_2, H_2O, and NO_x can be lowered through the use of sunlight during photosynthesis. Achievement of SD is then not a futile pursuit. Once this fact is established, though we soon come to realize that there has been a shift in the "scarcest resource" from people to the environment (Lovins *et al.*, 2000). The "law of the minimum" is an integral concept for both systems thinking and SD. This often misunderstood concept, which was originally described by von Liebig regarding agronomy, is pertinent in our understanding of SD as well (Meadows and Wright, 2008). Economics that only track labor and capital are quite unhelpful when the scarcest resources are water, clean air, and acceptable forms of energy. The new view of economics then must rely on some old paradigms such as the "law of the minimum" but be exhaustive in its search for the parameters or resources, which are in fact the scarcest. It is very important for practitioners of SD to be aware that measurement of performance of SD be comprehensive. Organizations such as the Global Reporting Initiative (GRI), the Environmental Protection Agency (EPA), and the National Institute of Standards and Technology (NSIT) all provide acceptable frameworks for measurement of parameters not traditionally used for economic performance measurement (Piasecki *et al.*, 1999).

4. Construct Organization and Structure

4.1. *Students*

Across North America, students are displaying a great interest and involvement in SD. A survey of 2418 undergraduate students representing 53 colleges and universities across North America revealed some very interesting statistics regarding their opinions: A total of 73% of the respondents felt that corporate social responsibility courses should be incorporated into all business/management curriculum, 60% of respondents felt that they would focus on securing a job that was socially responsible and 77% felt that they would seek a job that was socially responsible sometime in their career (New Leaders, 2007a). Over half of the respondents to the survey were interested in learning more about international development, environmental sustainability, renewable energy, corporate social responsibility, and social entrepreneurship. A total of 74% of respondents felt that "Universities and colleges should place more emphasis on training socially and environmentally responsible individuals than they currently do" (New Leaders, 2007a).

In two similar surveys of MBA students, one conducted in 2006 and the other in late 2008, some trends of note were revealed. In 2006, the survey reported that 36% of students felt that the for-profit sector should play a role in addressing social and environmental issues. This number increased to 52% in 2008. Also of note was the 2006–2008 increase from 18% to 31% of those who felt that most corporations are working toward the betterment of society. Upto 17%, a 2008 mark of 77% was the student agreement that corporate social responsibility makes good business sense because it leads to financial profits (New Leaders, 2007b, 2009).

A multi-faculty University of British Columbia Study published in April 2009 revealed similar positive responses to the concept of SD. A total of 70.7% of respondents across faculties expressed a ranking of "very interested" when queried about the subjects of clean/renewable energy and energy efficiency. Sustainable transportation and green building design ranked 2nd and 3rd in appeal. Six areas that received a ranking of "very interested" at 40% or above were clean/renewable energy and energy efficiency, sustainable transportation, sustainable community development, sustainable infrastructure (water, waste, and energy), sustainable cities, and climate change. A total 22 other sustainability "areas of interest" were also cited by the respondents in an open comments section of the survey (Sustainability Education at UBC, 2009).

Opinion surveys were not the only indication of student support for SD. The Canadian Federation of Students, representing over 500,000 students in Canada, together with the Dr. David Suzuki Foundation arranged a cross Canada keynote speaking tour that brought speakers such as Dr. Suzuki to 30 campuses (Webster and Dautremont-Smith, 2009).

4.2. *Academia and curriculum*

The Association for the Advancement of Sustainability in Higher Education reported that academic offerings concerning SD increased by 66 programs in 2008. This is up from 27 in 2007, 22 in 2006, and just 3 in 2005. Of the 66, 12 were graduate degrees, 20 were bachelor degrees, and 24 were continuing education programs. Six funding programs were introduced by North American universities to assist faculty in incorporating SD in curriculum (Webster and Dautremont-Smith, 2009).

Class projects in North America throughout 2008 included a biodiesel reactor, a green building, a solar installation, many carbon inventories, a hybrid automobile design, and many more. Faculty redevelopment and course revision are becoming much more common place with many universities holding workshops for faculty specifically focused on SD in curriculum.

Universities are organized by faculties that are very good at vertical pursuit of their particular specialties. Academics are rewarded for research done within these specialties and students are taught within the confines of these specialties. Although not unheard of, multidisciplinary pursuits are rare (Cortese, 2003). Of particular interest is a course offered at Loyola University of Chicago, Center for Urban Environmental Research. The course is interdisciplinary and is named Solutions to Environmental Problems (STEP). The course uses students from over 15 disciplines to address real-world environmental problems and develop solutions to the problems (Varty *et al.*, 2008).

4.3. *The university as a built environment*

The domain of SD emphasizes the need for harmony with our environment. One should not limit one's consideration of the term environment to the vast biosphere, though when one considers the term. There is a vast "built environment" in which we live, work, and learn. Residential and commercial buildings consume 40% of the primary energy supply in the United States. They use 75% of the electricity and 55% of natural gas and contribute more than 33% of the total carbon dioxide emissions (Holness, 2009). A similar situation can be found in most "northern" countries around the globe (Sisson, 2007). Higher education facilities rank third in energy dollars per square foot next to laboratories and hospitals (Mills *et al.*, 2004). Universities are heavy consumers of water as well primarily because of residences (Hopper *et al.*, 2005).

Cortese (2003, p. 17) states that "In many cases, we think of teaching, research, operations and relations with local communities as separate activities; they are not." Human beings learn from their surroundings. This includes the buildings that they live in and learn in. In order to produce graduates that think about sustainability and can compete with common attitudes toward efficiency, speed, and capital, it is important that they make a connection among research, curriculum, and the social and environmental footprint of the institutions from which they have come and to which they are going (Cortese, 2003).

A total of 75–85% of all the buildings that will exist in urban areas in 2030 exist today (Holness, 2009). A great many of these buildings are in poor repair. To understand the magnitude of the current Canadian university situation, a description of the phenomenon of "accumulated deferred maintenance" (ADM) is required. ADM is a phenomenon brought about by the year after year avoidance of maintenance and refurbishment of physical infrastructure. In the case of Canadian Universities, the causes of the situation are many. The average university building in Canada is 32 years old, whereas the average life expectancy of its components is 23 years. Government cutbacks in the 1990s and inflated energy costs in the 1980s left little funding available for capital upgrade. Growth in university programs, enrollment, and research has taken funds that might have been used for refurbishment of facilities. Life-cycle funding mechanisms are not in place at most universities (A Point of No Return, 2000). Thus, no monies from sinking funds (monies accumulated in anticipation of infrastructure replacement) are available when the time for replacement or refurbishment of infrastructure arrives.

The Canadian Association of University Business Officers (CAUBO) estimates the current backlog of infrastructure improvement at Canadian universities to be at \$6.84 billion, of which \$5.1 billion is ADM. They also estimate that of the \$5.1 billion in ADM over half or \$2.4 billon is considered urgent repair or replacement (Deferred Maintenance, 2009). Urgent repair or replacement implies immediate action be taken to avoid a health and safety concern, to avoid escalating deterioration and/or provide protection against costly damage (A Point of No Return, 2000). Not all of the required refurbishment of these buildings represents energy-saving measures; however, almost all of the measures required have SD implications. Anything that has to do with the refurbishment of the built environment implies the use of materials or the design of a process that may be improved or altered to reflect the concerns of the building owners, users, and operators.

4.4. *University administration and operations*

Canadian universities represent a $26 billion enterprise (Trends; Finance, 2008). Naturally, it is important to gain an understanding of how funds move within the university to provide an illustration of where the interplay with the other elements of the construct will occur. The CAUBO describes the method of accounting used by universities as "fund" accounting. Under the system, monies are distributed between the following funds: endowment fund, ancillary fund, capital fund, sponsored research fund, special purpose and trust fund, and the general operating fund.

Endowment funds are funds that are restricted in their use — they are generally donations to be held in perpetuity and spent as per the wishes of the donor or the university. Interest is collected yearly from these donations and applied to the appropriate costs. Ancillary funds are generated from the sale of books, parking, student residences, food services, etc. The capital fund is for the expansion of, or major renovations to, the university infrastructure. Historically, this fund included funds for construction and renovation of research facilities; however, a Federal Government program named the Canadian Foundation for Innovation (CFI) has made for a change in the reporting of these revenues. These revenues, subsequently, are now reported in the sponsored research fund.

Historically, the Sponsored Research fund has been the fund that kept track of all externally funded research monies. Governments, the private sector, and not-for-profit organizations have traditionally been the sources of monies for this fund. Monies from this fund pay for costs directly attributable to research but do not pay for the principle researcher's (faculty) wages. These wages come from the general operating fund.

The general operating fund is the recipient of revenues primarily from the Provincial government and from tuition fees. Faculty and staff salaries and student support services are paid from this fund. In 2006–2007, revenues in the general operating fund totaled $16.9 billion and of this roughly $10 billion went to teaching and learning. The other 41% of revenue went to scholarships and bursaries, health care services, sales of goods and services and miscellaneous revenues, non-credit instruction, institutional support for research sponsored by government, private, and not-for-profit organizations and faculty research time not supported by external research sponsors (Trends: Finance, 2008).

Since 1999 capital expenditures at universities have grown. These expenditures were financed through government grants, private fund raising, and increased debt. Debt is incurred, usually, due to universities having to match a contribution by the government in order to receive a grant for badly needed capital expenditures for facility expansion and refurbishment. Some debt shortfalls have actually been taken out of operating funds leaving funds for teaching and learning short. The interest payments in Ontario alone have increased from $5 million annually to over $50 million annually between 1999 and 2006. This debt load affects how many funds are available in the general operating fund for teaching and learning (Trends: Finance, 2008).

Credit rating companies consider three main categories when rating universities' credit worthiness: academic profile, operating risk profile, and financial risk profile. The academic profile includes such criteria as reputation, where indicators such as entering grades, reputable third-party reviews, and yield rates are considered. Other criteria include program mix, location, competition, size, academic resources, and enrolment trend. The operating risk profile evaluates the university's ability to manage its operations within its means. Criteria include management framework, revenues, expenditures (including operating expenditures, debt servicing, and research), regulatory environment, operating balance, and infrastructure. Concerns regarding infrastructure include the evaluation of existing facilities to meet growth projections and any deferred maintenance issues as based on third-party assessment. The financial risk profile is made up primarily of short-term liquidity measures and an analysis of endowments. Pension plans, capital leases, and law suits are considered at the same time (Beauchemin, 2007).

More and more universities are expected to be a catalyst of innovation and a key contributor to the social and economic well-being of the community (Trends: Finance, 2008). It is because of these expectations that groups such as the American College and University Presidents' Climate Commitment (ACUPCC) and the University and College Presidents' Climate Change Statement of Action for Canada (CCSAC) (University and College Presidents, 2009) have formed. These organizations are a public commitment to the reduction of campus greenhouse gas emissions, the inventorying of carbon sources upstream, on campus and downstream and a schedule for achieving these goals.

In an effort to support their signatories in their goals, the ACUPCC have recently published a reference titled "Leading Profound Change." The resource explains and suggests the use of systems thinking, developing a learning organization, and multidisciplinary approaches to the changes needed to accomplish the commitment statements (Leading Profound Change, 2009).

After 18 years experience with sustainability programs on campuses, Leith Sharp (2009, p. 4) who, in 2000, founded and directed the Harvard University Green Campus Initiative (now the Office of Sustainability) reminds that "[c]ompartmentalization, territorialism, complexity, risk aversion and hidden drivers, to name just a few such dynamics, sometimes conspire to undermine even the most sensible ideas." She goes on to explain that the staffing requirements at her position at Harvard were 24 full-time sustainability employees entrusted to manage change and coordinate on campus activities and how a university with a single sustainability champion entrusted to carry out sustainability initiatives the campus wide will soon be overcome and become ineffective almost immediately. Also of note is her message that the need for ongoing loan funds and the continued disconnect between capital and operating funds must end so that savings from increases in efficiencies in operations be funneled back into further projects to increase efficiencies in operations (Sharp, 2009).

4.5. *ESCOs and GESCs*

ESCOs have been in existence, in one form or another, for over 75 years (Studebaker, 2001). ESCOs provide evaluation, design, implementation, financing, procurement, installation, and measurement services for building owners wishing to retrofit building systems (mechanical, electrical, roof, windows, walls, etc.) to provide increased energy use efficiency. Lately, they have been taking on alternative power generation installation responsibilities as well. Thus, turning buildings into power providers and ultimately rendering buildings in a carbon neutral condition.

The comprehensive approach that ESCOs provide allows for the technical and performance risk to be transferred from the university to the ESCO. The GESC employed by ESCOs ensures that the performance of the measures implemented is met. That is, the ESCO makes up the difference if the energy savings are not what were claimed and the ESCO takes profit if the energy savings are what was claimed or better. The risk in this case lies entirely with the ESCO.

It is important to note that contracts may vary country to country, state to state and province to province. It is beyond the scope of this paper to explore all the possible forms of these contracts. It is suggested that anyone implementing the construct would negotiate a contract most suitable to the situation.

A 2005 US study reported that 72% of projects implemented at Public and Institutional facilities exceed the proposed savings and that only 19% encountered savings shortfalls (Hopper *et al.*, 2005). The same report suggests that the average implementation time for these projects is around six to nine months with the shortest reporting three months and the longest around two years. The average period of the contracts was 9.5 years with contracts ranging up to 26 years. The median costs for projects in this study were around $1.25 million for universities and netted out between 15% and 20% median energy cost savings (Hopper *et al.*, 2005).

It is common for the economics of the projects in question to be explained through a computation of simple payback. Simple payback merely divides the total cost of the project by the predicted savings per year at current costs of energy, construction, etc. For example, the University of Missouri Kansas City engaged with an ESCO in 2008. The scope of the project involved energy conservation measures preformed in 27 university buildings covering over 2 million square ft (University of Missouri, 2008). The contract total cost was pegged at $19.4 million and the savings were predicted and guaranteed at $1.6 million per year. This provided a contract term simple payback of 12.125 years. The net reduction in energy was 27% and the net emissions reduction was 30%.

In general, the financial calculations do not take into account the capital cost avoidance of the projects, the operation and maintenance cost avoidance or the reduction in personnel costs due to the new equipment. Other non-energy costs that are not generally included in these calculations are the increase in productivity and reduction in missed days ill for students and staff that can be directly related to the improved environment, the attractive nature of greener buildings to existing students and staff and potential students and staff, and the increase in building value based on the decrease in building operating costs among others (Khare and Cook, 2009).

4.6. *Structural — environmental coupling*

4.6.1. *Government/legal*

In the most recent Canadian Federal budget, $2 billion dollars was made available for knowledge infrastructure upgrade. Universities in Canada stand to receive $700 million of this in 2009 and $700 million in 2010. The monies are matching funds and the universities or other partners will be responsible for 50% of the cost of the project in order to secure a 50% match from the government. The Federal government has also made incentives available for solar power and heat installations, wind, biomass, geothermal, and hydro-installations at institutional and public buildings. Various provincial programs exist that provide incentives for energy reduction and alternative generation methods as well. Of particular note is the Ontario plan, which sees $0.80 cents per kWh paid to the owner/operators of solar photovoltaic rooftop installations for a period of not less than 20 years (Hamilton, 2009).

Tuition fees have been frozen in Newfoundland and Labrador since 1999. Tuition fees in Manitoba have been frozen since 2001 and Quebec has had a tuition fee freeze for residents since 1998. British Columbia has recently deregulated tuition fees and has seen some increases in tuition fees of over 100%. Ontario and Nova Scotia have never exercised freezes on tuition fees (Campaigns and Lobbying, 2008). It is predicted that Saskatchewan and Manitoba will end their freezes in 2009–2010 joining Alberta that ended its freeze in 2005 (Gillis, 2009). New Brunswick renewed a commitment to frozen tuition fees for 2009–2010 and in PEI tuition fees for the same year rose 4% for domestic students and 9% for foreign students.

Although not legislated at the time of writing carbon taxes or carbon cap and trade schemes are being considered around the globe. The current US administration has been looking seriously at how it can implement a cap and trade scheme as have many European countries. Summits of the G8 and G20 generally include discussion of such instruments. Of specific interest, the government of British Columbia has led the way in Canada with a carbon tax.

4.6.2. *Socio-political*

There is more than compelling evidence of strong social opinion supporting sustainability initiatives. Poll after poll support the rising interest, knowledge of and practice of sustainability. Protocols, such as the Montreal protocol and the Kyoto protocol, display the interest that there is within society for compliance with the biosphere.

The number of university students is expected to grow from the 2006 level of 70,000 to 150,000 by 2016. It is anticipated that a great deal of this growth will be in the graduate and professional programs putting further pressure on universities as these are the most expensive types of programs. These programs also require more student support and more faculty and staff than undergraduate studies (Trends: Finance, 2008).

4.6.3. *Economic*

Having experienced long unprecedented and artificially sustained periods of economic growth for nearly a decade, the world economic bubble burst and suffered a record challenging slow down in the fall of 2008. Canada seemed to have faired better than most countries due to highly regulated and conservative banking cultures. However, a great deal of economic damage was done and a $50 billion plus Canadian budget deficit materialized in 2009. Canada relies heavily on the economic trade it maintains with the US and therefore on how much the US consumer is buying. The US economy seems to be recovering yet the consumer does not seem to be leading the charge and this trend will not help the Canadian economy.

4.6.4. *Technological*

The past 40 years have seen great leaps in technological advance. Building systems and building envelope components (walls, roof, windows, structure, etc.) have followed along with this technological wave of advancement. It is not uncommon now for buildings to be considered intelligent. For the purposes of this paper, "Intelligent buildings are ones in which the building fabric, space, services, and information systems can respond in an effective manner to the demands of the owner, the occupier, and the environment" (Veitch, 2009).

The advancement of technology for buildings does not stop at their front doors, though. In an effort to distribute the generation to the user and produce the energy in a renewable carbon neutral way, commercial, industrial, and institutional solar farms, wind farms, geothermal installations, and tidal generation systems are being constructed. Buildings now have the potential through technology to be zero carbon buildings capable of producing all of their energy needs through renewable means and even sending excess back to the power distribution grid.

5. Analysis of the Construct Organization and Structure

5.1. *Students*

The information suggests that there is a strong positive attitude and perspective of students toward the inclusion of SD in courses of higher education. Year over year, students have increased their agreement with the inclusion of SD concepts in curriculum. Of special interest to the topic at hand was the interest level displayed at the University of British Columbia regarding energy and energy efficiency, green building, and green transportation. The organization of an SD speaking tour by the Canadian Federation of Students is a positive sign in support of the positive quality of relations for this conceptual element.

All of the information collected regarding this conceptual element proved to be suggestive of the positive quality of the relations regarding the concepts of SD, the need for SD in higher education, and the willingness and desire for universities to do more in this regard.

5.2. *Academia and curriculum*

The multidisciplinary nature of SD provides stumbling blocks for the advancement of curriculum development in higher education. There are some preliminary signs that universities are embracing the multidisciplinary requirement for this type of program and this supports a positive relationship between the conceptual elements — students, academia, and curriculum — regarding the concepts of SD. There is also evidence that participative learning is being utilized in curriculum that is concerned with SD.

5.3. *The university as a built environment*

The built environment at universities in Canada is in desperate need of refurbishment. The information would suggest that this problem is growing as well. These buildings are not likely to be torn down and rebuilt considering that monies have not even been made available to repair them over the years.

Although it would be difficult to claim that this conceptual element demonstrates positive relations with the other conceptual elements of the construct it does represent an immediate and growing opportunity for those who might wish to implement such a construct. A built environment that is in such desperate need of refurbishment provides the background upon which such a construct is liable to manifest feasibly.

5.4. *University administration and operations*

The "fund" accounting system is typical of a non-profit accounting system. The system is designed to group revenues with exact expenditures for auditing purposes that are peculiar to corporations that receive funds for specific purposes and endowments. The information suggests that in the past monies intended for teaching and learning have been used for short falls in matching Federal and Provincial funds for capital projects. This precedent occurrence reinforces the favorable relation that the ACCSC would have in producing capital that would be available immediately for much needed projects and would be repaid from operations savings. As well the

ACCSC would provide a means for new funding to be tapped from operations and used for learning and teaching.

The fund system includes the Special Purpose and Trust fund that would be ideally suited for monies arriving from the ACCSC earmarked for AL curriculum that include the concepts of SD. In doing so, the faculty would have a transparent view of the accumulation and expenditure of these monies.

The information would also suggest that the implementation of an ACCSC would increase the credit rating of the university. The credit rating is based on many factors that the ACCSC would improve. The debt incurred for the capital and curriculum funds is debt secured by a third party. Thus, the university has improved its facilities without incurring more debt. The value of the buildings being refurbished increases and the cost of maintenance decreases reducing the deferred maintenance load. All of these factors suggest a positive relationship between the financial environment and the university administration and operations' conceptual elements.

There is a great deal of information suggesting that the university presidents of North America are supportive of the concepts that underlie SD. Commitments have been made toward the reduction of emissions and the move toward sustainable campuses at many universities. This information supports the positive relationship between the university administration, society, and the biosphere. The commitments that the presidents have made and are making everyday are not without challenges. The information suggests that the culture of some universities may make such widespread change very challenging. There is also a great requirement for staff to manage the projects and operations that will lead to a sustainable campus. While the ACCSC introduces another party into the collaboration it does suggest a multidisciplinary actor be added. The ESCO is capable of relieving any group tasked with such projects a great amount of responsibility and risk.

The information suggests that university presidents look toward systems thinking and encouragement of their corporations to become learning organizations. The encouragement of the use of these two tools as well as the suggestion that multidisciplinary solutions to on campus SD problems be used further reinforces the positive relation that is existent between university administration and the other elements of the construct.

5.5. *ESCOs and GESCs*

The research suggests that ESCOs are operating in an industry that is mature (Studebaker, 2001). Any corporation existing in a mature industry will be trying to pursue differentiation. There is no evidence that would suggest that the construct is being used elsewhere currently. Evidence of previous ESCO projects on campuses abounds and suggests that such a project would be easily and effectively carried out today (Hopper *et al.*, 2005).

The information also suggests that the comprehensive nature of the ESCO is especially well suited to the urgent nature of the infrastructure situation at most

Canadian universities (Hopper *et al.*, 2005). A great deal of risk, both technical and financial, is absorbed by the ESCO throughout the length of the contract. Although the most common structure is that of the guaranteed savings contract, the contract structure is flexible and can be adjusted to meet the needs of the situation. Long-term contracts are not without precedent within the industry and therefore should not pose a stumbling point for the feasibility of the ACCSC.

The University of Missouri case study example explored previously states a $19.6 million dollar cost and a $1.6 million dollar per year cost saving. This suggests that if the contract was extended to the average 25 year effective lifespan of the equipment approximately $19.2 million at a rate of $1.6 million per year could be leveraged for teaching and learning purposes. Although every project would be different and many projects would be smaller, one must consider the magnitude of the ADM issue before one discounts the potential of the construct in freeing up monies for teaching and learning. This suggests a very strong positive quality of the relationships among the students, academia, university administration, university as a built environment, and the government.

There are other inherent benefits of the GESC that further the strength of the positive relationship between the previously listed conceptual elements. There is information suggesting the existence of "non-energy" benefits such as increased productivity, marketing collateral, capital cost avoidance, operation and maintenance reduction, and possible personnel reduction or redeployment. All of these benefits further reinforce the positive quality of relationship between the elements of the construct.

5.6. *Structural — environmental coupling*

5.6.1. *Government/legal*

Information regarding government support of infrastructure renewal supports a claim of positive relationship between this conceptual element, the university administration and the university as a built environment (Flaherty, 2009). Although the monies are not available unless matched by another party, it is information that reflects positively on the ACCSC.

The trend toward unregulated and unfrozen tuition fees suggests that universities may well receive some much needed relief for the upcoming anticipated increase in students and the staff and infrastructure needed to support this increase. However, it is likely that the increases in tuition fees will not keep up with the total requirements of the institutions and that the bearing on the feasibility of the ACCSC is at most neutral and at least insignificant.

The probability of a cap and trade system of carbon emission control and/or a carbon tax in the future further reinforces the positive quality of the relationship between this conceptual element and the university administration and operation. The ACCSC would provide a means of mitigating some of the risk brought about should this type of legislation materialize.

5.6.2. *Socio-political*

The general attitude and interest of society in SD is reflected throughout most of the conceptual elements of the construct. There is a great deal of information suggesting that the student, government, university administration, academia, and ESCO elements are supportive of the concepts that underlie SD. This positive outlook and attitude toward SD is at the heart of the construct and therefore must be interpreted as supportive of the feasibility of the construct.

5.6.3. *Economic*

Regardless of the economic times, the construct mitigates a great deal of risk for projects that must be done and provides a further disconnect for the institution from the volatility of the economy. This characteristic of the environment must be viewed as a positive quality of the relationship between the university administration, the government, and the coupling with the environment.

5.6.4. *Technological*

The present state of technology provides for many options including alternate energy production, intelligent buildings, and building envelope modification that complement the construct. Dependent on the geography of the project and the government incentives available, these new technologies have begun to make economic sense as well.

The ACCSC has facility to provide new technologies, which in turn provides an opportunity for students to be exposed first hand in a learning environment to components, materials, and techniques that may give them a competitive edge upon graduation. The ACCSC is therefore supported by the domain of technology and the environmental coupling with it.

6. Conclusions

The ACCSC construct has a deep and rich theoretical foundation. Experts in the SD field reveal the need for the use of Systems Thinking and Organizational Learning (Sharp, 2009; Leading Profound Change, 2009). The construct incorporates the ideas and concepts implied by the term SD in the real-world "built environment." AL has been used in the management domain for many years (Revans, 1980) and is being used in higher education today (Pedler *et al.*, 2005; Johnson and Spicer, 2006; Frank, 1996; Heriot *et al.*, 2007). The ACCSC would be a very good catalyst for an academic interested in involving the student in a "real-world" situation with a firm theoretical foundation. The multidisciplinary nature of SD leads one to the realization that any successful project centered on the concepts implied by the term SD will involve a collaborative mindset. The collaborative nature of these projects may pose one of the most difficult stumbling blocks to overcome in the attempt to engage the construct. It is likely that the successful manifestation of the ACCSC will begin with the stakeholders agreeing to a code of conduct incorporating the improvisational and creative nature implied by the term collaboration as defined and explored in this paper.

There is great interest among students across North America in learning and using the concepts implied by the term SD (New Leaders, 2007a, 2007b, 2009). It appears that this interest may be limited to certain disciplines (Sustainability Education at UBC, 2009) and yet some groups are proposing a time when all higher education will include multidisciplinary core courses centered on the concepts implied by SD (Webster and Dautremont-Smith, 2009).

Interest in SD and AL seems to be alive and well within the world of North American Academia (Webster and Dautremont-Smith, 2009). In the case of SD specifically there is indication of year over year growth in the numbers of those interested in incorporating the concepts of SD in curriculum and of programs and courses that are centered on these concepts (Webster and Dautremont-Smith, 2009).

There is a desperate infrastructure renewal shortfall at Canadian Universities. The shortfall does not seem to be receding and much of it is of an urgent nature (A Point of No Return, 2000). This situation provides a perfect catalyst for the ACCSC construct. In order for there to be monies freed up by the construct, there needs to be a declining infrastructure where improvements will provide savings in operating costs. The current situation at Canadian Universities could not be a better example of this. The ACCSC construct, then, would provide the much needed capital upgrades for infrastructure and also provide funding for academic programs centered on SD projects. This, in turn, would help with the commitments that University Administrations are currently moving toward with regard to sustainable campuses and zero carbon campuses (Leading Profound Change, 2009; University and College Presidents, 2009).

Universities are in desperate need of Sustainability staff who are capable of wide spread coordination and have a great multidisciplinary skill (Sharp, 2009). It is not a great intellectual stretch to conceive that for University Administration staff to address the challenges of a multibillion dollar ADM issue, the production of Sustainable campuses, the multifaceted demands of increasing enrollment, the requirement to become a leader in innovative teaching, the increasing demands for higher education to be an integral part of society, economic instability and rising operating costs that a great deal of resources need to be deployed if any portion of these challenges is to be met with hope of positive outcome. The ACCSC construct implies a collaboration that brings multidisciplinary expertise together to overcome at least some of the aforementioned challenges. There is no need to develop entire departments of sustainability that are able to do the coordination, implementation, design and maintenance of the projects required to meet the commitments and challenges. Risk can also be mitigated through the use of the construct. Campus-centered sustainability champions can be used to coordinate high-level visionary results and lower-level detailed expertise need not be developed *en masse*. This would provide a much more expedient strategy to address an urgent need.

There is information sufficient to forgo the rejection of the feasibility of the ACCSC construct at this point in the research. The environment in which it is proposed that the construct would be set is supportive and conducive to such a construct. The organization of the construct also displays positive qualities further supporting its feasibility. The greatest stumbling block is liable to be the construction of a "learning organization" founded on the concepts of multidisciplinary collaboration and even this is not without precedent (Varty *et al.*, 2008). The time has arrived when the silos separating corporations and higher education need to be deconstructed based on a mutually beneficial opportunity. The time has also arrived when the silos within universities themselves need to be deconstructed to take advantage of this opportunity.

Although the exploration of the construct does not include the university employed "sustainability executive" element it is highly probable that development of information regarding motivation, attitude, and planned behavior of this element would be pertinent. As more and more Canadian Universities hire SD professionals to coordinate on campus SD initiatives, this element will have more and more influence on the construct.

A deeper look at how the Governments would react to the construct is also an area of further research. Many questions, the answers to which may affect the scope of projects and perhaps the possibility of the university receiving benefit on several fronts from the ACCSC, need to be answered.

It is our opinion that great value could be realized with the development of a trial curriculum. Many issues that are thus far unforeseen that relate to the feasibility of the construct may be revealed by this exercise. Logistical details and the details specific to curriculum development that are not explored in this paper may become evident through such an exercise.

Acronyms

ACCSC: Academia–Corporate on Campus Sustainability Collaborative
ACUPCC: American College and University Presidents' Climate Commitment
ADM: Accumulated Deferred Maintenance
AL: Action Learning
ASHRAE: American Society of Heating Refrigeration and Air-Conditioning Engineers
CAUBO: Canadian Association of University Business Officers
CCSAC: University and College Presidents' Climate Change Statement of Action for Canada
CRV: Current Replacement Value
EPA: Environmental Protection Agency
ESCO: Energy Services Company
FCI: Facilities Condition Index
GESC: Guaranteed Energy Savings Contract
GRI: Global Reporting Initiative

HAS: Human Activity Systems
NSIT: National Institute of Standards and Technology
SD: Sustainable development
STEP: Solutions To Environmental Problems

References

A point of No Return, 2000. *The Urgent Need for Infrastructure Renewal at Canadian Universities.* The Canadian Association of University Business Officers, Ottawa, Ontario.

Ackoff, R.L., 1999. *Ackoff's Best: His Classic Writings on Management.* Wiley, New York.

Argyris, C. and Schon, D., 1996. *Organizational Learning II: Theory, Method, and Practice.* Addison-Wesley Pub. Co., Reading, Mass.

Beaty, L., Bourner, T. and Frost, P., 1993. Action learning: Reflections on becoming a set member. *Management Education and Development* **24**(4), 350. Retrieved from http://proquest.umi.com/pqdweb?did=336841&Fmt=7&clientId=12302&RQT=309&VName=PQD.

Beauchemin, E., 2007. *Rating Canadian Universities: Methodology.* DBRS Inc., Toronto, Ontario.

Capra, F., 2002. *The Hidden Connections: Integrating the Biological, Cognitive, and Social Dimensions of Life into a Science of Sustainability*, first ed. Doubleday, New York.

Checkland, P., 1981. *Systems Thinking, Systems Practice.* J. Wiley, Chichester Sussex, New York.

Checkland, P. and Scholes, J., 1990. *Soft Systems Methodology in Action.* Wiley, Chichester, West Sussex, England, New York.

Construct, 2009. ask oxford .com. Retrieved August 8, 2009, from http://www.askoxford.com/concise_oed/construct?view=uk.

Cortese, A., 2003. The critical role of higher education in creating a sustainable future. *Planning for Higher Education* **31**(3), 15–22. Retrieved from http://vnweb.hwwilsonweb.com/hww/jumpstart.jhtml?recid=0bc05f7a67b1790e5ef6c2353165ce95-411126b4a1cf7ce73ec8c12cfba4c76e65b2a1ea00599941&fmt=H PDF.

Deferred Maintenance of University Infrastructure, 2009. *No. Rev. 01-2009.* Association of Universities and Colleges of Canada. Retrieved from http://www.aucc.ca/_pdf/english/reports/2009/deferred_maintenance_fact_sheet_e.pdf.

de Geus, A., 1997. *The Living Company.* Harvard Business School Press, Boston, Mass.

Feynman, R., Leighton, R. and Hutchings, E. (1989). *Surely You're Joking, Mr. Feynman!: Adventures of a Curious Character*, Bantam trade ed. Bantam Books, Toronto, New York.

Filion, L., 1999. Empirical systems modeling methodology: Applications to entrepreneurial actors. *Advances in Qualitative Organization Research* **2**, 201–220.

Flaherty, J., 2009. *Canada's Economic Action Plan: Budget 2009.* National Budget. Department of Finance Canada, Ottawa, Ontario. Retrieved from http://www.budget.gc.ca/2009/pdf/budget-planbugetaire-eng.pdf.

Frank, H., 1996. The use of action learning in British higher education. *Education & Training* **38**(8), 7. Retrieved from http://proquest.umi.com/pqdweb?did=119908240&Fmt=7&clientId=12302&RQT=309&VName=PQD.

Gillis, W., 2009. Saskatchewan joins other western provinces intending tuition freeze. *The Gateway.* Retrieved from http://www.thegatewayonline.ca/articles/news/2009/03/24/saskatchewan-joins-other-western-provinces-ending-tuition-freeze.

Hamilton, T., 2009. *Fixed Prices Proposed for Green Energy Projects.* The Star.Com. Retrieved from http://www.thestar.com/Business/article/601464.

Heriot, K., Cook, R., Matthews, C. and Simpson, L., 2007. Creating active and high-impact learning: Moving out of the classroom with field-based student consulting projects. *Industry and Higher Education* **21**(6), 427–434. Retrieved from http://www.ingentaconnect.com/content/ip/ihe/2007/00000021/00000006/art00006.

Holness, G.V.R., 2009. Sustaining our future by rebuilding our past. *ASHRAE Journal* **51**(8), 16–21.

Hopper, N., Goldman, C., McWilliams, J., Birr, D. and Stoughton, K., 2005. *Public and Institutional Markets for ESCO Services: Comparing Programs, Practices and Services.* Energy Analysis Department Ernest Orlando Lawrence Berkeley National Laboratory, Berkeley, California. Retrieved from http://eetd.lbl.gov/ea/EMS/EMS_pubs.html.

Jacobs, M., 1993. *The Green Economy: Environment, Sustainable Development and the Politics of the Future.* UBC Press, Vancouver.

Johnson, C. and Spicer, D., 2006. A case study of action learning in an MBA program. *Education & Training* **48**(1), 39. Retrieved from http://proquest.umi.com/pqdweb?did=1010989071&Fmt=7&clientId=12302&RQT=309&VName=PQD.

Khare, A. and Cook, P., 2009. The link between continuous commissioning and sustainable growth: A sustainable business model. *Intelligent Buildings International* **1**(2), 142–155.

Kuei, C. and Madu, C.N., 2009. Sustainable operations management. *International Journal of Commerce and Strategy* **1**(1), 2–18.

Leading Profound Change: A Resource for Presidents and Chancellors of the ACUPCC, 2009. No. V 1 American College and University Presidents' Climate Commitment.

Lovins, A., Lovins, L. and Hawkin, P., 2000. *Harvard Business Review on Business and the Environment.* Harvard Business School Press, Boston, MA.

Luhmann, N., 1990. *Essays on Self-reference.* Columbia University Press, New York.

Maturana, H., 1975. The organization of the living: A theory of the living organization. *International Journal of Man–Machine Studies,* **7**(3), 313–332. DOI: 10.1016/S0020-7373(75)80015-0.

Maturana, H. and Varela, F., 1980. *Autopoiesis and Cognition: The Realization of the Living.* D. Reidel Pub. Co., Dordrecht, Holland, Boston.

Meadows, D.H. and Wright, D., 2008. *Thinking in Systems: A Primer.* Chelsea Green Pub., White River Junction, Vt.

Mills, E., Friedman, H., Powell, T., Bourassa, N., Claridge, D., Haasl, T., *et al.*, 2004. The cost effectiveness of commercial buildings commissioning: A meta-analysis of energy and non-energy impacts in existing buildings and new construction in the United States. Retrieved from http://eetd.lbl.gov/emills/PUBS/Cx-Costs-Benefits.html.

New leaders, new perspectives II: A net impact survey of undergraduate student opinions on the relationship between business and Social/Environmental issues, 2007a. Net Impact.

New leaders, new perspectives: A survey of MBA student opinions on the relationship between business and Social/Environmental issues, 2007b. Net Impact.

New leaders, new perspectives: A survey of MBA student opinions on the relationship between business and Social/Environmental issues, 2009. Net Impact.

Pedler, M., Burgoyne, J. and Brook, C., 2005. What has action learning learned to become? *Action Learning: Research & Practice* **2**(1), 49–68.

Perkins, D., 2003. *King Arthur's Round Table: How Collaborative Conversations Create Smart Organizations.* J. Wiley & Sons, Hoboken, N.J.

Piasecki, B., Fletcher, K. and Mendelson, F., 1999. *Environmental Management and Business Strategy: Leadership Skills for the 21st Century.* John Wiley, New York.

Raelin, J., 2006. Does action learning promote collaborative leadership? *Academy of Management Learning & Education* **5**(2), 152–168. Retrieved from http://search.ebscohost.com/login.aspx?direct=true&AuthType=url,ip,uid&db=bth&AN=21253780&site=ehost-live.

Raelin, J. and Raelin, J., 2006. Developmental action learning: Toward collaborative change. *Action Learning: Research & Practice* **3**(1), 45–67. DOI:10.1080/14767330600574615.

Revans, R., 1971. *Developing Effective Managers: A New Approach to Business Education.* Praeger Publishers, New York.

Sawyer, K., 2007. *Group Genius: The Creative Power of Collaboration.* Basic Books, New York.

Senge, P., 1990. *The Fifth Discipline: The Art and Practice of the Learning Organization*, first ed. New York.

Sharp, L., 2009. Higher education: The quest for the sustainable campus. *Sustainability: Science, Practice and Policy* **5**(1), August 29, 2009. Retrieved from http://ejournal.nbii.org/archives/vol5iss1/editorial.sharp.html.

Sisson, B. and Von Aerschot, C., 2007. *Energy Efficiency in Buildings: Business Realities and Opportunities.* World Business Council for Sustainable Development, Geneva, Switzerland. Retrieved from http://www.wbcsd.org/DocRoot/lKDpFci8xSi63cZ5AGxQ/EEB-Facts-and-trends.pdf.

Studebaker, J., 2001. ESCO: *Energy Services Company Handbook.* Penn Well Corporation, Tulsa, Oklahoma. Retrieved from http://books.google.ca/books?id=acK3q0gO5_AC&printsec=frontcover&source=gbs_v2_summary_r&cad=0#v=onepage&q=&f=false.

Sustainability education at UBC: A Student Perspective, 2009. University of British Columbia, British Columbia.

Svendsen, A., 1998. *The Stakeholder Strategy: Profiting from Collaborative Business Relationships* first ed. Berrett-Koehler Publishers, San Francisco.

Trends in Higher Education: Volume 3 Finance, 2008. The Association of Universities and Colleges of Canada, Ottawa, Ontario.

University and College Presidents' Climate Change Statement of Action for Canada, 2009. Retrieved August 29, 2009, from http://www.climatechangeaction.ca/.

University of Missouri Kansas, 2008. Retrieved August 30, 2009, from http://www.naesco.org/resources/casestudies/documents/UMKC.pdf.

Varty, A., Lishawa, S. and Tuchman, N., 2008. Engaging students in campus sustainability initiatives through innovative courses AASHE.

Veitch, J., 2009. Editorial: Whither intelligent buildings? *Intelligent Buildings International* **1**(2), 99–100.

WCED, 1987. *Our Common Future.* Retrieved February 21, 2009, from http://www.are.admin.ch/themen/nachhaltig/00266/00540/00542/index.html?lang=en.

Webster, A. and Dautremont-Smith, J., 2009. *AASHE Digest 2008: A Review of Campus Sustainability News.* The Association for the Advancement of Sustainability in Higher Education, Lexington, Kentucky.

Welborn, R. and Kasten, V., 2003. *The Jericho Principle: How Companies Use Strategic Collaboration to Find New Sources of Value.* J. Wiley, Hoboken, N.J.

Wilson, B., 1984. *Systems, Concepts, Methodologies, and Applications.* Wiley, Chichester, West Sussex, Toronto.

Zeleny, M., 1997. Autopoiesis and self-sustainability in economic systems. *Human Systems Management* **16**, 251–262.

Chapter 24

LOCAL CLIMATE POLICIES IN JAPAN: A CASE STUDY OF NAGOYA CITY'S STRATEGY

Noriko Sugiyama
Graduate School of Environmental Studies, Nagoya University, 464-8061, Furo-cho, Chikusa-ku, Nagoya, Japan
n.sugiyama@nagoya-u.jp

This chapter provides an overview of national and local climate policy in Japan. National climate policy is focused mostly on disseminating information and increasing the efficiency of electric appliances/eco-cars. Japanese local governments are very good at waste management and setting local action plans under national laws, but are weak in the fields of energy and transport policy.

As an advanced case, this chapter features Nagoya City's strategy. In 2009, Nagoya City formulated the Strategy for a Low-Carbon City by 2050. The strategy aims to increase the quality of life, and, as a result, to be a low-carbon city. Finally, several recommendations are made for improving Japanese local climate policy.

1. Introduction

Climate change is a big multilevel challenge from the local to the global level. Therefore, Climate Policy has to act on three levels: the international level (e.g., Kyoto Protocol), national level (e.g., carbon tax), and the local level (e.g., promoting renewable energy). For the reduction of greenhouse gasses (GHG, especially CO_2), activities and measures at the local level such as the management of city infrastructure, transportation, and energy supply are important elements of a comprehensive climate policy.

This chapter shows the National Climate Policy of Japan and the characteristics of local climate policy in Japan, and the Nagoya Strategy for a Low-Carbon City is used as a case study.

2. National Climate Policy of Japan

2.1. *Before COP3 in Kyoto*

It was 1980 that the Japanese government had recognized global warming as a serious environment issue, but it was not until 10 years later that the Japanese government had set up the National Action Plan. Before COP3 (Conference of the

Parties 3) in Kyoto, the Japanese government had set up a committee related to the global environmental issue in 1980; however, Japanese climate policy then stagnated during the 1980s (Sugiyama and Takeuchi, 2007). In 1989, the Japanese government established a council for global environmental issues. Then, after the first report of the Intergovernmental Panel on Climate Change was drafted in 1990, the Action Plan to Arrest Global Warming was set up by the ministerial cabinet of Japan. The action plan aimed to stabilize carbon dioxide (CO_2) emissions at 1990 levels by 2000. However, this action plan did not have any concrete measures or legal binding force.

2.2. *After COP3 in Kyoto*

After COP3 in Kyoto in December 1997, the Japanese government decided to put global warming as their top priority and set up a headquarters to promote measures to prevent global warming immediately. At that time, Japanese emissions of GHG were increasing year by year. In 1997, it seemed difficult to achieve the Japanese obligation under the Kyoto Protocol: a 6% reduction of GHG emissions from 1990 levels by the 2008–2012 period.

In 1998, the Japanese government formulated an outline for the promotion of measures to cope with global warming to achieve the Kyoto target. The basic strategies were good compatibility between the environment and economic activity, a step-by-step policy, the promotion of countermeasures in all areas on every level, and ensuring international co-operation concerning climate change. Also in 1998, the Law Concerning the Promotion of the Measures to cope with Global Warming (Global Warming Law) was enacted, and an act concerning the rational use of energy (Energy Conservation Law) was revised. The Global Warming Law defined the responsibilities of the central government, local governments, businesses, and citizens to promote the prevention measures. The Energy Conservation Law established the Top-Runner Regulation, which requires manufactures to improve the energy efficiency of electrical appliances and cars by adopting, within a given time frame, the standards used by the most energy-efficient appliances.

In 2002, Japan ratified the Kyoto Protocol and amended the Global Warming Law. After the Kyoto Protocol entered into force in 2005, the Japanese government formulated the Kyoto Protocol Target Achievement Plan in accordance with the amended Global Warming Law of 2002. In 2008, the target achievement plan was completely revised (Table 1). It now focuses on formulating low-carbon levels in urban and regional areas and the use of the Kyoto Flexibility Mechanisms (the Clean Development Mechanism) and Joint Implementation.

2.3. *Features of Japanese countermeasures*

It is said that Japanese Climate Policy has two special attributes: one is "breaking through with technology," and the other is "reducing CO_2."

Table 1. Approximate targets of energy-originated carbon dioxide in each sector.

	Base year (FY 1990)	FY2005 level of emissions		Approximate targets in each sector in FY2010		
	A	B	(B – A)/A	C	(C – A)/A	
Estimated results	Million t, CO_2	Million t, CO_2	(Percentage change relative to base year by sector)	Million t, CO_2	(Percentage change relative to base year by sector)	Reference: Differences between FY2010 targets and FY2005 level of emissions
Energy-originated CO_2	1059	1201		1076–1089		
Industrial sector	482	452	(−6.1%)	424–428	(−12.1%)–(−11.3%)	It is estimated that if no measures or policies are taken, emissions will increase through increases in the volume of production resulting from economic growth or other factors. Provisional calculations show that emissions can be reduced by 25–29 million tons from FY2005 level through measures and policies
Commercial and other sector	164	239	(+45.4%)	208–210	(+26.5%)–(+27.9%)	It is estimated that if no measures or policies are taken, emissions will increase through increases in the floor area in buildings or other factors. Provisional calculations show that emissions can be reduced by 29–31 million tons from FY2005 level through measures and policies

(*Continued*)

Table 1. (*Continued*)

Estimated results	Base year (FY 1990)	FY2005 level of emissions		Approximate targets in each sector in FY2010		Reference: Differences between FY2010 targets and FY2005 level of emissions
	A	B	(B − A)/A	C	(C − A)/A	
	Million t, CO_2	Million t, CO_2	(Percentage change relative to base year by sector)	Million t, CO_2	(Percentage change relative to base year by sector)	
Residential sector	127	174	(+36.4%)	138–141	(+8.5%)–(+10.9%)	It is estimated that if no measures or policies are taken, emissions will increase through increases in the number of households and the device ownership rate per household or other factors. Provisional calculations show that emissions can be reduced by 32–35 million tons from FY2005 level through measures and policies
Transport sector	217	257	(+18.1%)	240–243	(+10.3%)–(+11.9%)	It is estimated that if no measures or policies are taken, emissions will increase through increases in the number of automobiles owned or other factors. Provisional calculations show that emissions can be reduced by 14–17 million tons from FY2005 level through measures and policies
Energy conversion sector	68	79	(+16.5%)	66	(−2.3%)	This is self-consumption at power plants, petroleum processing facilities, or the like. Provisional calculations show that emissions can be reduced by 13 million tons from FY2005 level by continuing to steadily promote efficient energy use in these facilities

*In each column, the numbers in all sectors may not add up exactly to the total due to rounding.
Note: The upper and lower limits of approximate targets are provided: the upper limit will be reached if the countermeasures demonstrate their maximum effects, and the lower limit will be reached if they show their minimum effects. Although the Government will certainly aim to maximize their effects, the targets have been set to meet the Kyoto Protocol target even if the countermeasures have their minimum effects.
Source: Kyoto Protocol Target Achievement Plan (Formulated April 28, 2005; Partially revised July 11, 2006, Totally revised March 28, 2008), The Ministry of the Environment.

For "break through with technology," concrete measures are changing to "top runner" (most efficient) electric appliances and cars, increasing the proportion of electric cars, introducing carbon capture and storage to coal-fired power plants, building additional nuclear power-generating plants, and so on. These measures are monolithic systems and focus only on energy efficiency. Introducing these advanced technologies is surely effective in reducing CO_2 emissions. However, there is a "rebound effect" problem. If the number of monolithic systems (electric appliances or cars) is increased, the CO_2 emissions also increase. It is necessary to introduce an efficiency-based social system. Additionally, these measures are for big companies or global companies, not for local businesses, and so it can be said that these measures do not contribute to the local economy and the creation of employment.

For "let us reduce CO_2," concrete measures are based on the dissemination of information to and the edification of citizens. For example, the Japanese government initiated comprehensive campaigns named "Team Minus 6%,"[1] "COOL BIZ[2]/WARM BIZ,"[3] and so on. Team Minus 6% was established in April 2005 as a Japanese international commitment to the Kyoto Protocol and is a national campaign against global warming aimed at cutting GHG emissions by 6%. The team leader is the Prime Minister, and the sub-leader is the Minister of the Environment. This national campaign requires the participation of individuals and organizations against global warming. Team Minus 6% took over "Challenge 25" from January 2010. The number of "Challenge 25" team members is approximately 385,000 individuals and 13,000 businesses and civil groups (as of August 2010).

The Ministry of the Environment of Japan reported that the public recognition of "COOL BIZ" was 96.0%; however, the rate at which the temperature of air conditioning was set higher was 48.1%. And it was estimated that the CO_2 reduction amount was 1,400,000 tons; equivalent to the CO_2 emissions from about 3 million households in one month in FY2007.

Thus, the Japanese government is very good at information dissemination campaigns. The same goes for local governments in Japan; the provision and dissemination of information are easy to implement in a short period of time. Individual awareness and environmental actions are necessary; however, those are not enough for the long-term large reduction.

3. Local Climate Policy in Japan

The Global Warming Law (Article 4) stipulates the responsibilities of local governments to formulate their action plan to promote prevention measures for global

[1] Team-6% Committee & Ministry of the Environment (http://www.team-6.jp/english/index.html).

[2] A business style to deal with global warming, which enables us to feel cool and comfortable when working in offices with the air conditioning set at 28 °C in summer.

[3] A business style which enables us to feel warm and work efficiently in offices with the air conditioning set at 20 °C in winter.

warming in their local region: "Local governments shall promote policies to limit GHG emissions and to enhance sinks in accordance with the natural and social conditions of their areas." As a consequence of the Global Warming Law, a growing number of local governments began formulating local action plans. The Global Warming Law also mandates that local and regional governments formulate reduction plans for government-related activities and that prefectural governors establish a center for climate change action in their prefecture and appoint volunteers to promote climate activities.

In 2002, Japan ratified the Kyoto Protocol and amended the Global Warming Law. The new Article 20 states that "In view of the Kyoto Protocol Target Achievement Plan, prefectural and municipal governments shall endeavor to formulate and to implement comprehensive, plan-based programs for the control of GHG, in accordance with the natural and social conditions of their local areas." The responsibility of regional and local governments to draw up a local climate change action plan was clarified with this amendment. And indeed, as of 2005, all prefectures had formulated a local climate change action plan (Sugiyama, 2008). However, it was not to say that the action plans were in accordance with the natural and social conditions of their local areas, because the reduction targets of their action plans were mostly minus 6%, the same as the national reduction target. Additionally, there were few effective countermeasures in their action plans; most of the concrete measures concerned the dissemination and edification of citizens.

3.1. *The local government climate partnership with USA, Germany, and Japan*

The project "The Local Governments Climate Partnership with USA, Germany, and Japan (LGCP)[4]" was performed by Climate Alliance in Germany and Nagoya University from 2007 to 2009. The LGCP aims to develop comprehensive benchmarks for local climate policy and attempts to apply them to cities in the US, Germany, and Japan. This project also aims to step up each city's local climate policy by identifying their weaknesses and problems, exchanging experiences and good practices, and to activate or reactivate local climate policy partnerships between cities.

Climate Alliance and Nagoya University developed Activity Profiles for local governments based on CLIMATE COMPASS[5] and the different backgrounds of three countries (such as national laws).

The Activity Profiles are used to scan the present state of the local government's climate protection activities. The underlying matrix of activities includes 26 action fields in the categories "climate policy," "energy," transport," and "waste" (Table 2).

---

[4]This project "The Local Governments Climate Partnership with USA, Germany, and Japan (LGCP)" is funded by the German Agency for Environment and The Sumitomo Foundation in Japan.

[5]Climate Alliance (2006). Climate Compass — a roadmap for local climate protection. http://www.climate-compass.net/index.php?id=29&L=0.

Table 2. A total of 26 measures for local climate policy of activity profile.

Action Fields		Local government's climate protection activities
Climate policy (target and plan)	A1	Adopt targets and develop a concept
	A2	CO_2 monitoring (measure, verify, and report performance)
	A3	Institutionalize the climate policy
	A4	Set up and implement the action program (define visions)
	A5	Define the overall planning strategy (urban development)
	A6	Inform about climate change and the climate policy
	A7	Involve local actors
	A8	Open up to the region and initiate common activities
	A9	Compensate carbon emissions
Energy	E1	Bring facilities up to scratch
	E2	Motivate staff to save energy (including economic incentives)
	E3	Become a role model for sustainable energy supply
	E4	Define energy efficiency as a basic principle in urban planning
	E5	Refurbish existing housing stock
	E6	Support the implementation of CHP/district heating in the community
	E7	Support the implementation of renewable energy projects
	E8	Inform and involve citizens
	E9	Work with small- and medium-sized enterprises
	E10	Collaborate with "powerful" partners
	E11	Collaborate with energy supply companies
Waste management	W1	Waste management for CO_2 reduction
Transport	M1	Involve staff in sustainable mobility plans
	M2	Reduce transport demand by sustainable land use planning
	M3	Increase the share of zero carbon transport modes
	M4	Provide convenient public transport
	M5	Restrict traffic flows by managing parking space
	M6	Promote responsible car use

Source: Climate Alliance and Nagoya University.

Each activity has been divided into four levels of ambition from beginner to climate leader (level 1: getting started, level 2: moving forward, level 3: forging ahead, level 4: taking the lead). Therefore, the checklist consists of 104 terms (26 measures ×' 4 levels).

Each participating city has evaluated their activities based on this checklist. The consequence is shown automatically in a Rader Chart. From the Radar Chart, we can see not only the strong points, but also the weak points, of the climate protection activities. In addition, the Activity Profile shows the concrete measures by action fields, items, and four levels; hence, the person in charge of the local government can find what measure to begin, what is the next level, and define their target countermeasures.

In order to find out which types of polices are implemented at the local level, 66 local governments (9 in USA, 18 in Germany, and 39 in Japan) were studied and compared by our research group. The survey and interviews were made between November 2007 and March 2008. The research group consisted of members of the

Climate Alliance, Ifeu (Institute for Energy and Environmental Research Heidelberg GmbH) and the Nagoya University Graduate School of Environmental Studies.

As a result of the research on the 66 cities, we found that there are great differences between the three countries mainly due to different national laws affecting the competencies of local governments. The research results proved to be helpful for the investigated local governments in knowing their strengths and weaknesses.

3.2. *Features of Japanese local climate policy*

Nagoya University surveyed Japanese local governments, and 39 Activity Profiles were collected by March 2008. Figure 1 shows the total number of each measure counted from checklists of activities taken by 39 Japanese local governments.

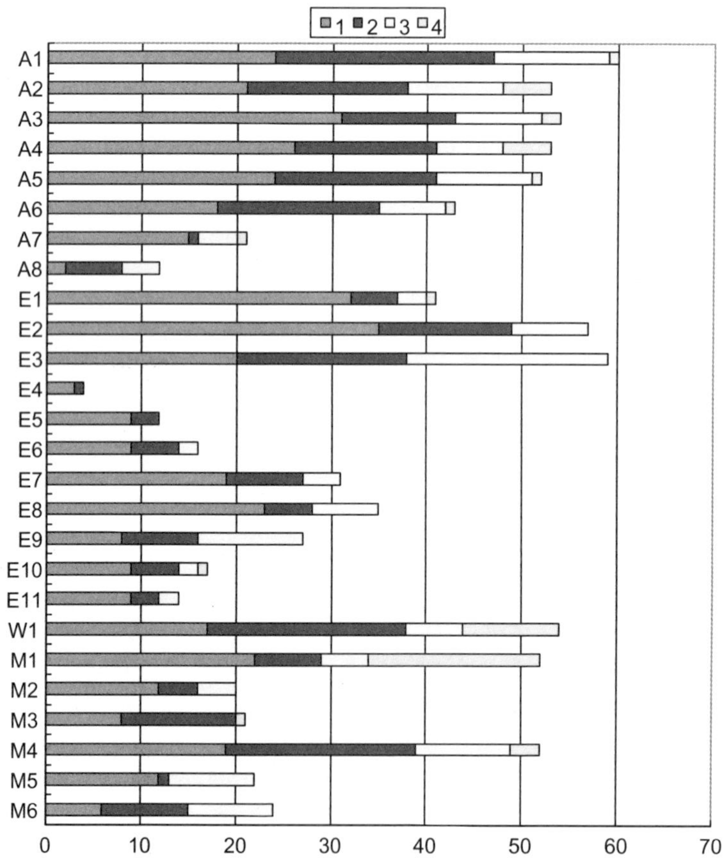

Fig. 1. The total number of each measure taken by 39 Japanese local governments counted from activity profiles (multiple answers, 2007). The number on top of the graph shows the four levels. On the left hand side, letters and numbers refer to the measures shown in Table 2.

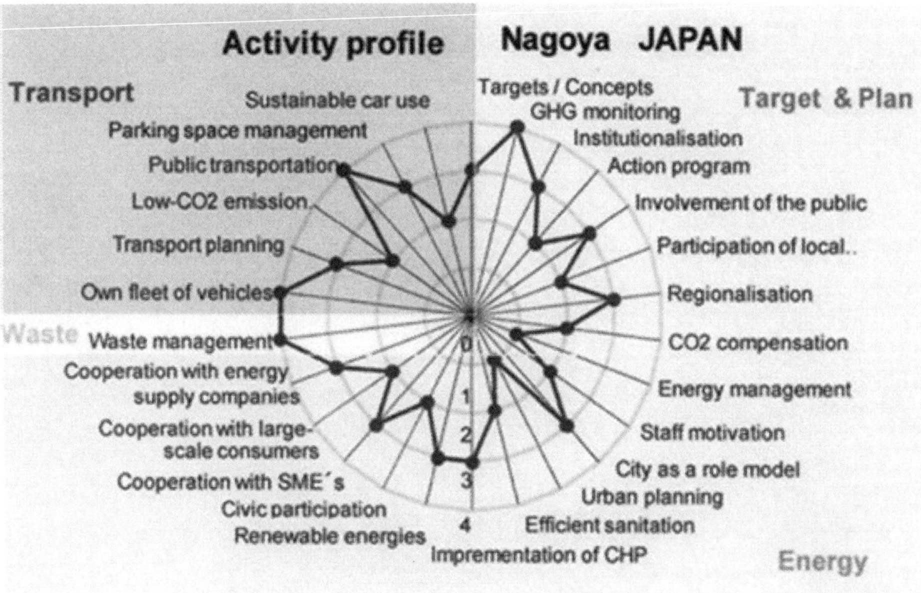

Fig. 2. Rader Chart from the Activity Profile of Nagoya, Japan (2008). The numbers on the Rader Chart show the four levels.

From the results, Japanese local governments can be regarded as being very good at setting targets/planning and waste management; however, energy and transport are not sufficient. On the whole, there are very few Japanese local governments, which had taken measures as high as level 3 or 4, except the measures E3, W1, and M1 (Fig. 1). The E3 — level 3 is "purchase green power to supply a portion of your energy demand," W1 — level 4 is "implement effective waste prevention measures in your community (reduce and reuse)" and M1 — level 4 is "purchase fuel efficient, alternative-fueled, and electric vehicles"

Figure 2 shows the Radar Chart of Nagoya City. Nagoya is the fourth largest city in Japan, with a population of 2.2 million. From this Radar Chart, it is said that one of the strongest points of Nagoya is "waste management" and one of the weakest points is "defining energy efficiency as a basic principle in urban planning."

3.3. *Comparative analysis among USA, Germany, and Japan*

The activity profile results of participating cities from the USA, Germany, and Japan are represented by means of Radar Charts (Figs. 3–5). The Radar Charts represent "no measures" (beginner level) in the center up to "climate leaders" on the outer sphere.

It is said that the US cities are very good at the measures relating to the transport sector and GHG monitoring but not in those relating to energy or target

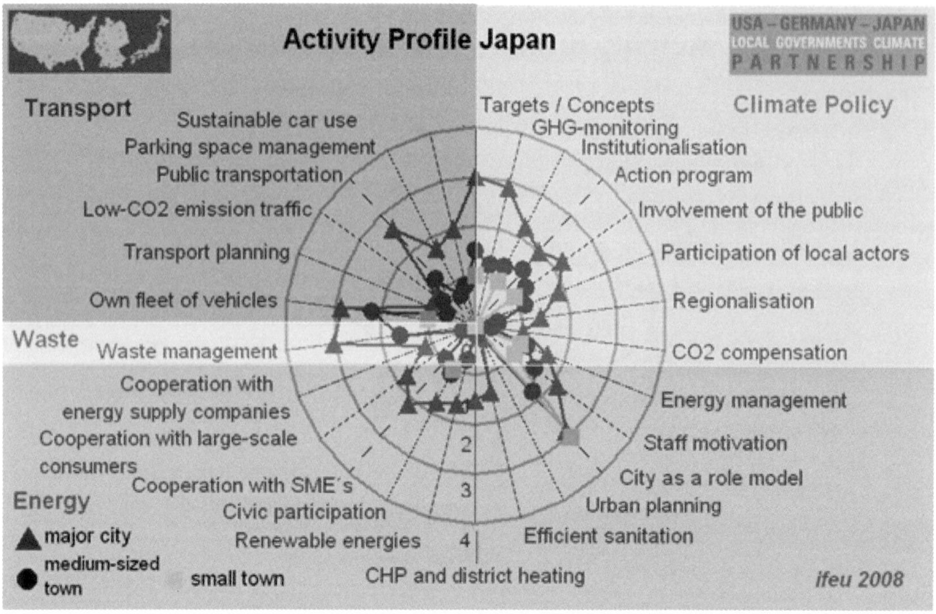

Fig. 3. The Radar Chart of Japan. The numbers on the Rader Chart show the four levels. There are three analyzed levels a city's population: "major city" is more than 1,000,000 residents, "medium-sized town" is 100,000 to 1,000,000, and "small town" is less than 100,000.
Source: ifeu (Institute for Energy and Environmental Research Heidelberg GmbH) and Nagoya University.

setting/planning. German cities, especially large cities, are developed; however, small towns are behind them. Japanese cities are very different from USA and Germany; the Rader Charts show small shapes and sharp edges. One of the reasons for this is that Japanese local governments have very few competencies concerning their energy and transport sectors (Sugiyama and Takeuchi, 2008).

The following are the main results of our research:

(1) The focus of research was on four areas: planning of climate policy, energy, transport, and waste management. We found that there are great differences between the three countries mainly due to different national laws affecting the competencies of local governments.
(2) Many Japanese local governments have implemented the national requirements for planning, information dissemination, and waste management; however, for energy and transport areas, only very few local governments have fulfilled their tasks.
(3) The research results (especially the analytical scheme used) proved to be helpful for the investigated local governments in knowing their strengths and weaknesses. The results were also important for information exchange regarding the best practices between local governments.

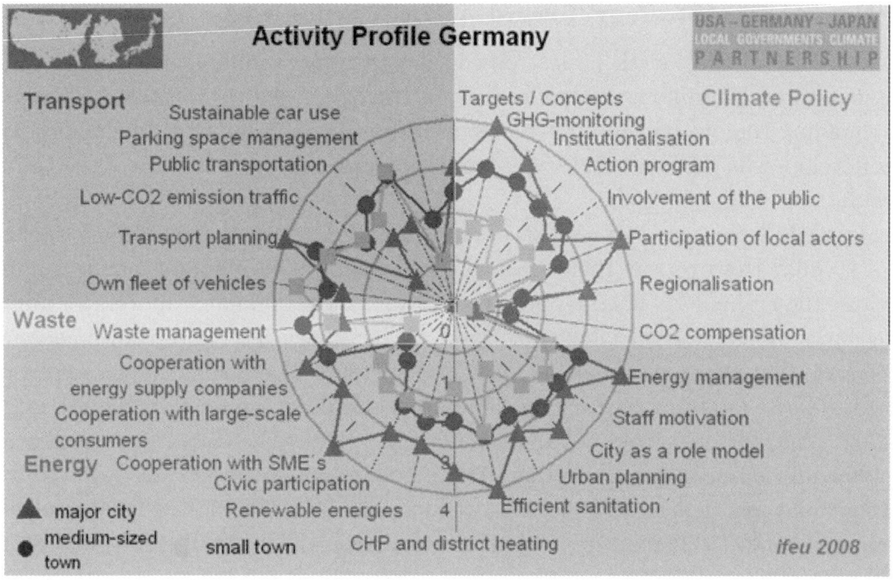

Fig. 4. The Radar Chart of Germany. The numbers on the Rader Chart show the four levels. There are three analyzed levels a city's population: "major city" is more than 1,000,000 residents, "medium-sized town" is 100,000 to 1,000,000, and "small town" is less than 100,000.
Source: ifeu (Institute for Energy and Environmental Research Heidelberg GmbH) and Nagoya University.

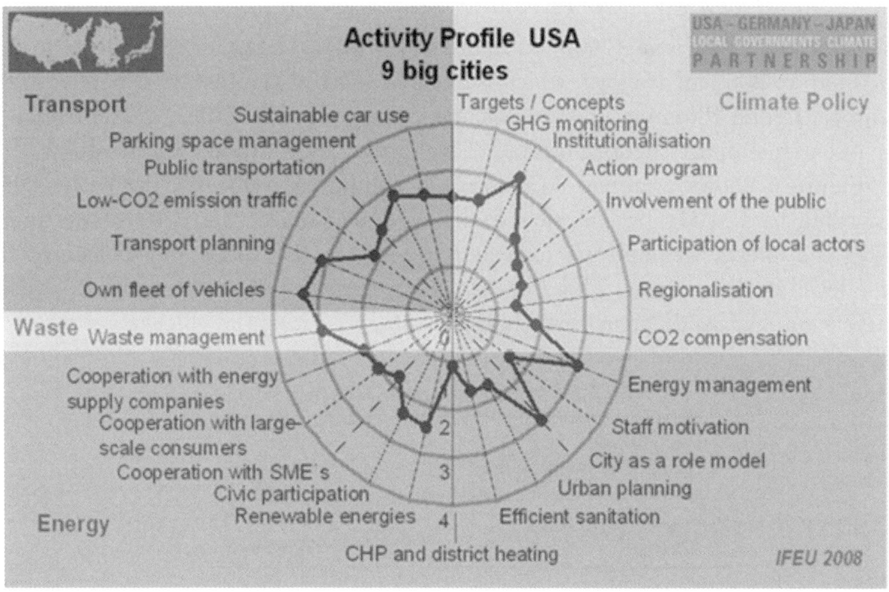

Fig. 5. The Radar Chart of USA. The numbers on the Rader Chart show the four levels. The Rader chart is the average levels of nine big cities.
Source: ifeu (Institute for Energy and Environmental Research Heidelberg GmbH) and Nagoya University.

4. Nagoya City Strategy Toward A Low-Carbon City

Nagoya City is the fourth largest city in Japan, with 1 million households and a population of 2.2 million. A basic outline of Nagoya is shown in Table 3. The area surrounding Nagoya within a 50-kilometer radius has a population of 8 million and has developed as the central city of governance, economy, and culture of the Chubu Region.

From 20 years ago, Nagoya City had been planning to use the Fujimae tideland as a landfill. However, in 1999, Nagoya City gave up the planning of the landfill, because the Fujimae tideland is one of the nation's most popular stopovers for migratory birds. Then, former Mayor of Nagoya Matsubara announced an emergency declaration on waste management in February 1999. The former mayor explained that the waste landfill was reaching critical levels and set the waste reduction target at 20% (200,000 tons) from 1998 (1,020,000 tons). After the emergency declaration in 1999, all citizens worked together. The waste reduction target was achieved, the amount of waste fell to 790,000 tons (reduced by 200,000 tons) in 2000, and was three quarters (750,000 tons) of what is was at the time of the emergency declaration. In 2003, the Fujimae tideland was registered as a Wetland of International Importance under the Ramsar Convention.

Now, on the basis of this successful case, Nagoya City is attempting to reduce GHG emissions by 80% compared with 1990 levels.

4.1. *Proposal of the nagoya minus 80% roadmap*

Takeuchi and Sugiyama (2008) estimated and made a minus 80% Scenario and Roadmap to 2050 for Nagoya, which is one of results of the research "Study on the Strategic Urban Planning and Assessment of Low-Carbon Cities.[6]" The results of our research support the formulation of Nagoya's Low-Carbon City Strategy.

Figure 6 shows a minus 80% scenario compared with 1990 levels by 2050. According to a study by Takeuchi and Sugiyama (2008), the BAU of the population and energy demand are expected to peak in the 2020s and will be at the same level as the present. There are three strategies: "efficiency strategy," "alternative strategy," and "structural change strategy." First, the efficiency strategy is

Table 3. Outline of Nagoya city.

Population	2,253,470
Number of households	1,013,688
Number of people per household	2.22
Population density	6903 persons/km^2
Acreage	326.45 km^2

Source: Nagoya City (April, 2010).

[6]This research is supported by the Global Environment Research Fund by Ministry of the Environment Government of Japan (Hc-086).

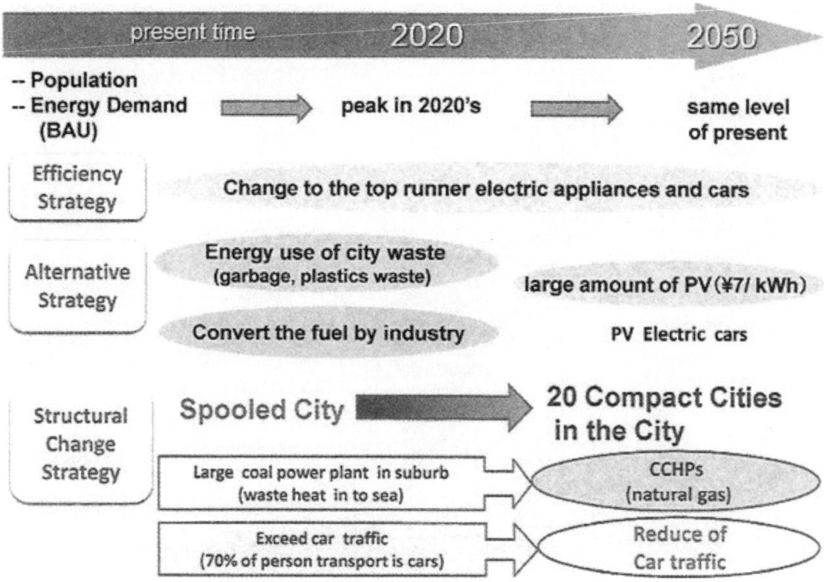

Fig. 6. Nagoya Minus 80% scenario.
Source: Takeuchi and Sugiyama (2008).

Table 4. Profile of Nagoya's CO_2 emissions.

t-CO_2/person/year	Nagoya	Major city average (Tokyo and ordinance-designated cities)	Japanese nationwide
Industry	1.9	2.6	4.9
Transport	2.2	1.5	2.0
Businesses	1.8	1.5	1.9
Households	1.4	1.3	1.4
Total	7.4	6.9	10.1

In each column, the numbers in all sectors may not add up exactly to the total due to rounding.
Source: Nagoya City.

to change to "top runner" electric appliances and cars. Second, the Alternative Strategy is changing fuels, introducing PV (photovoltaic system) and so on. Third, the structural change strategy is to formulate 20 compact cities in Nagoya, introduce CCHP (combined cool, heat, and power), reduce car traffic, etc. We proposed this Minus 80% Roadmap (Fig. 7) to the mayor of Nagoya, the local gas company, the electric company, and so on. According to the Roadmap, 20% of emissions in Nagoya (Table 4) will be reduced due to the top runner change to 2050. An amount 30% of emissions will come from the change from big coal power plants in suburbs to natural gas CCHP in compact cities by 2050.

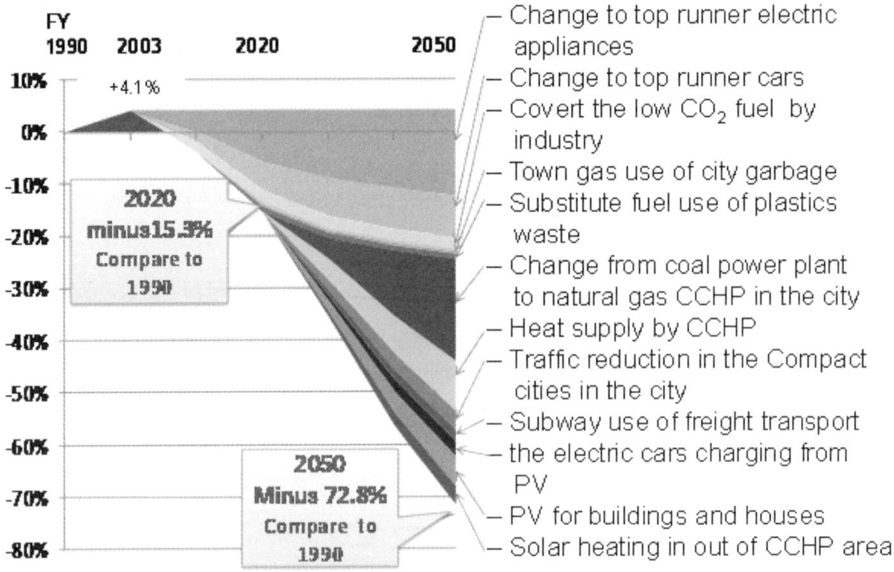

Fig. 7. Proposal of Nagoya Minus 80% roadmap. PV: Photovoltaic system; CCHP: combined cool, heat, and power;
Source: Takeuchi and Sugiyama (2008).

4.2. *Nagoya's low-carbon city strategy*

Nagoya city formulated the "2050 Nagoya Strategy for Low-carbon City"[7] (Nagoya Strategy) in November 2009. The strategy has a subtitle: "Working on transforming Nagoya into a comfortable, Low-Carbon City."

The long-term target of Nagoya Strategy set to reduce GHG emissions to 80% compared with 1990 by 2050 (Fig. 8). The mid-term reduction target is 25% compared with 1990 by 2020.

The visions for the future of this strategy are as follows:

General target: Nagoya as a low-carbon, comfortable city:

- city with small fossil fuel consumption where citizens can move around on foot;
- city with rich nature where air conditioning is not required;
- low-carbon city developed through collaboration of citizens, businesses, and the administration.

Three lifestyles supported by collaborative power of citizens were imagined:

(1) *Life near stations*: Life near public transportation stations where citizens can move around on foot.
(2) *Life with rich nature*: Life in which citizens can enjoy nature in their daily lives.

[7]Nagoya City (2009). *2050 Nagoya Strategy for Low-carbon City* (in Japanese).

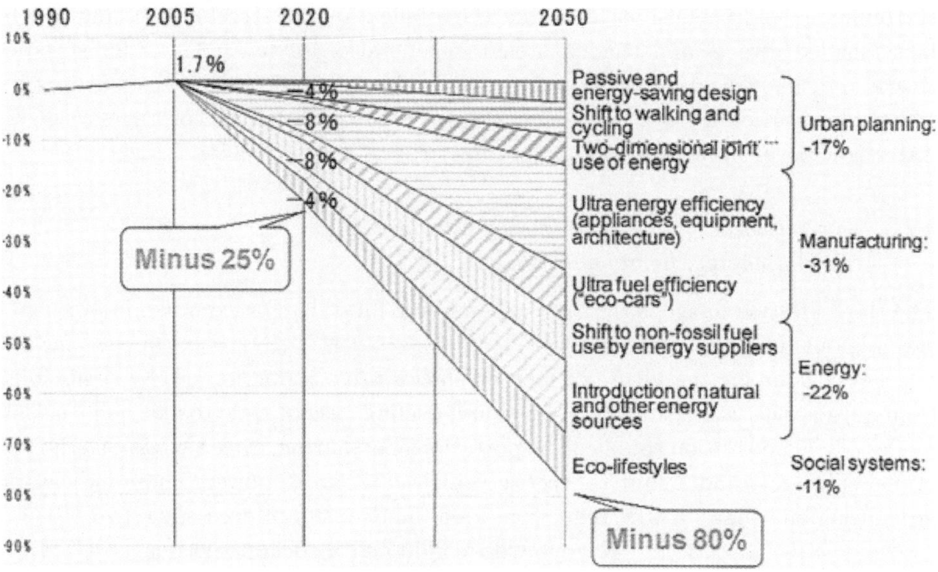

Fig. 8. Estimate of the reduction in energy-derived CO_2 emissions (roadmap of Nagoya). *Source:* Nagoya City.

(3) *Low-carbon lifestyle*: Comfortable low-energy life that harmonizes with nature and super energy-saving instruments:

- cars;
- life and work;
- local energy generation.

4.2.1. *Station-centered lifestyle (EKISOBA Lifestyle)*

One of the three lifestyle images is the station-centered lifestyle (EKISOBA Lifestyle). The vision for the future is redevelopment of the city into walkable living areas near public transportation stations. For example, promoting housing near stations, developing land-use mechanisms (attracting businesses and residents to areas near stations and developing unused land), and implementation of transportation-oriented urban planning (development of new relationships among modes of transportation, urban space, and nature). The indicators are the ratio of the population living near stations: 63–75% (population density near stations: 79–85 people/ha, redevelopment of areas near stations).

4.2.2. *Natural lifestyle (FU-Sui-Rhoku-In Lifestyle)*

The second lifestyle image is the natural lifestyle (FU-Sui-Rhoku-In Lifestyle). Its vision for the future is a lifestyle in which residents can enjoy nearby nature. For example, developing low-impact living arrangements that take advantage of natural surroundings and natural features, the "Greenification" of streets, land, and buildings (shaded streets and shaded areas), regeneration of nature near forests and rivers, and the utilization of natural air conditioning (wind paths and "cool spots": evapotranspiration by trees). The indicators are the ratio of green coverage: 25–40%, water flow permeability: 14–33%, and direct runoff rate: 62–36%.

4.2.3. *Low-carbon living lifestyle (motor vehicles, regional energy, home, and jobs)*

The third lifestyle image is the low-carbon living lifestyle. For example, motor vehicles, regional energy, home, and jobs.

The visions for the future of motor vehicles are a reduction of CO_2 emissions from automobile use; a shift to walking and cycling (use of wide roads), joint use of modes of transportation (public transportation, car sharing, cycle rentals), a tripling of fuel efficiency; and a shift to electric vehicles and green power. Those indicators are automobile share: a 40% reduction, fossil fuel consumption: reduce to 1/10.

Nagoya University and Nagoya City conducted a social experiment regarding the shared use of bicycles, which reused 300 abandoned bicycles or bicycles parked carelessly on sidewalks, from October 20 to December 18, 2009. These community bikes, named "Meichari," were used by a total of 31,000 people in 60 days. As a result, it became obvious that there is high social demand by Nagoya citizens for community bikes.

The other visions for the future of the low-carbon living lifestyle are regional energy, home, and jobs. For example, introducing highly-efficient in-area energy systems, promoting comfortable lifestyles through ultra-energy efficiency and natural air conditioning, using locally produced, locally consumed energy, promoting the use of high R-value insulation, ultra-energy-efficient devices, and natural air conditioning, pushing for the sweeping introduction of sunlight and solar heat use, the effective use of waste products and biomass resources, and two-dimensional joint use. These indicators are residential energy consumption (per capita): a 40% reduction, non-fossil fuel use: increase by a factor of 2.1, and a solar house that uses solar power, green, and wind.

5. Conclusions and Outlook

Japanese local governments still have much more work to do. As a conclusion, the author makes several recommendations to improve the local climate policy in Japan:

(1) It is required to set up an original "Roadmap" to reduce CO_2 by 60–80% compared with 1990.
(2) It is important to change the structure of the city in the long term. The elements of city planning should be set in the local climate policy.

(3) It is necessary for energy supply companies, transportation companies, and waste management companies to incorporate a "Roadmap" or Action Plan of climate change into their business plan/management plan.
(4) The aim should be for a higher quality of life for citizens, in addition to being a low-carbon city.
(5) Japanese local governments should participate in national and international networks of local climate policy. They should then try to raise the level of their local climate policies. Also, it is important that local governments collaborate and lead in national and international climate policy.

Acknowledgments

The project "The Local Governments Climate Partnership with USA, Germany, and Japan (LGCP)" is funded by the German Agency for Environment and The Sumitomo Foundation in Japan. And the research "Study on the Strategic Urban Planning and Assessment of Low-Carbon Cities" is supported by the Global Environment Research Fund by the Ministry of the Environment Government of Japan (Hc-086). The author would like to thank Climate Alliance, Ifeu institute, and the branch of environment of Nagoya City for their cooperation.

The author is thankful to Prof. Chu-hua Kuei and Prof. Christian N. Madu for giving this precious opportunity. The author would also very much like to thank Prof. Tsuneo Takeuchi for valuable comments and Prof. Michio Sugiyama for advice and encouragement.

References

Sugiyama, N. and Takeuchi, T., 2007. Environmental policy study to address the climate policy creation process: From problem perception to decision making. *Journal of Human Environmental Studies* **5**(2), 51–60 (in Japanese).

Sugiyama, N., 2008. A study of planning and measures in regional climate policy. *Planning Administration* **31**(1), 98–105 (in Japanese).

Sugiyama, N. and Takeuchi, T., 2008. Local policies for climate change in Japan. *The Journal of Environment & Development* **17**(4), 424–441.

Takeuchi, T. and Sugiyama, N., 2008. Roadmap to reduce CO_2 emission by 75% by 2050 in Naogya-City. *The Society of Heating, Air-Conditioning and Sanitary Engineers of Japan* **82**(9), 85–90 (in Japanese).

Chapter 25

WASTE MINIMIZATION AND CLEANER PRODUCTION PROGRAMS IN THE ASIA-PACIFIC REGION

Zhang Yi* and Tay Joo Hwa
Department of Environmental Science and Engineering, Fudan University, 220 Handan Road, Yangpu District, Shanghai, 200433 P.R. China
**zhang-yi@fudan.edu.cn*

Cleaner production (CP) and waste minimization are of paramount importance to the preservation of natural environment, the sustainable development of economy, and the overall welfare of human kind. The Asia-Pacific region is the world's most populous area, as well as a region of rapid economy growth and industrialization rate. Therefore, the practice of CP and waste minimization will have significant impact on the local environment and population, and the regional and global economy. In this chapter, it is intended that a brief introduction be given on the CP programs in the Asia-Pacific region, which includes entities of ASEAN+3, Hong Kong, Macau, Taiwan, India, Sri Lanka, Pakistan, and Bangladesh. The scope of discussion was first given, followed by some basic information on the social and economical status of the entities. Owing to the diversity in the sizes, populations, economic development, and industrialization degrees of the entities, various CP collaboration mechanisms and programs exist at several levels. The global and regional programs were briefly introduced, and emphasis was put on the programs at the national level. Six countries in this region have established national centers of CP upon the initiation of UNEP/UNIDO, and each has its unique function and focus. In addition, the other entities in this region have also set up CP centers/programs, or environmental management, waste minimization, and sustainable development plans, incorporating CP concept to various degrees. Furthermore, three CP case studies were presented, each dealing with a major industrial sector in the region, namely, textile, chemical, and electronics manufacturing. The cases were conducted in different countries, of various scales, focuses, approaches, funding schemes, and execution procedures, and were intended as an indication to the diverse scenario of CP and waste minimization programs in the region.

1. Introduction

Cleaner production (CP) is the continuous application of an integrated preventative environmental strategy to processes, products, and services to increase efficiency and reduce risks to humans and the environment (UNEP, 1994). The concept has been continuously replenished and deepened since its initiation, which was to minimize waste and emission from an industrial process. With the development of environment awareness and industrial technology, more scopes have been associated with it, which include waste minimization, pollution prevention, "cradle to grave," life-cycle analysis, source reduction, "cradle to cradle," industrial ecology, green design, energy and resource efficiency, green industry and economy, and green productivity.

Nowadays, the concept of "CP" encompasses much more than merely waste reduction from the "end of the pipe." It emphasizes more on a source reduction strategy, where design and construction can be optimized before the actual production process is put in use, to reduce resource and energy consumption, prevent unnecessary waste generation (including solid, water, and gaseous waste, as well as waste heat and noise), minimize negative impact on both the workers and the environment, and meanwhile maximize product output. "Cleaner" also indicates a continuous effort on the improvement, where all designs, processes, and practices are constantly under scrutiny, and changes are made where and when necessary.

The Asia-Pacific region is the home to several most populous countries in the world, as well as a region experiencing rapid economic growth and industrialization. Innumerous industrial processes are being developed and constructed throughout the region, which, while benefitting the society, also presents potential threats to the public and the environment. Therefore, the practice of CP concept is of tremendous importance in this region, and the respective government-initiated waste minimization and CP programs are to be introduced in this chapter.

2. Scope of Discussion

"Asia-Pacific region" does not have a fixed definition. It can encompass the usual South-East Asia countries such as Vietnam and Indonesia, expand to India and Pakistan of the South Asia and Indian Ocean region, and sometimes even include countries of Australia and Russia. Therefore, in this chapter, the scope of discussion is first identified, as to include the countries and regions of ASEAN+3, Hong Kong, Macau, Taiwan, India, Sri Lanka, Pakistan, and Bangladesh.

In each of the entity, the programs that are in operation can also be very diverse. In terms of funding source, they can be sponsored by multilateral and bilateral organizations, or public and private domestic sectors. In terms of focus, some might be focusing on the traditional area of CP, namely the manufacturing, whereas others have extended into the areas of hotels and fisheries. Moreover, the scale of the programs can range from a single factory to a national level policy change. Not to lose focus, only the international, regional, and national level programs are to be discussed in this chapter.

3. Social and Economical States of the Region

The entities that are to be discussed in this chapter are of various sizes, populations, and economical states. Some of the basic information of these entities found from

Table 1. Social and economical status of the entities in the region.

Category	Entity	Approx. area (km^2)	Approx. population	Total GDP (US$)	GDP per capita (US$)	GDP by sectors (agriculture, industry, service)
ASEAN	Brunei	5800	390,000	11 billion	26,000	1%, 73%, 26%
	Cambodia	180,000	15 million	11 billion	800	30%, 30%, 40%
	Indonesia	1,900,000	230 million	540 billion	2300	14%, 48%, 38%
	Laos	240,000	6 million	6 billion	900	39%, 34%, 27%
	Malaysia	330,000	28 million	190 billion	6900	10%, 42%, 48%
	Myanmar	680,000	50 million	28 billion	460	42%, 20%, 47%
	Philippines	300,000	92 million	160 billion	1700	14%, 30%, 56%
	Singapore	710	5 million	177 billion	37,000	0%, 127%, 73%
	Thailand	500,000	66 million	260 billion	4000	11%, 455, 44%
	Vietnam	330,000	86 million	92 billion	1000	21%, 40%, 39%
South Asia	Bangladesh	150,000	160 million	95 billion	600	30%, 25%, 45%
	India	3,200,000	1.2 billion	1235 billion	1030	28%, 18%, 54%
	Pakistan	800,000	170 million	167 billion	1000	20%, 27%, 53%
	Sri Lanka	66,000	20 million	41 billion	2000	21%, 19%, 60%
East Asia	China	9,600,000	1.3 billion	4920 billion	3600	11%, 47%, 43%
	Japan	380,000	130 million	5000 billion	40,000	2%, 23%, 75%
	Hong Kong	1100	7 million	220 billion	32,000	0.1%, 9%, 91%
	Macau	29	540,000	20 billion	40,000	0.1%, 7%, 93%
	South Korea	100,000	49 million	832 billion	17,000	3%, 40%, 57%
	Taiwan	36,000	23 million	383 billion	17,000	1%, 28%, 71%

Data are from various online sources; therefore, only rough estimates, not precise values.

online sources are listed in Table 1. Nominal Gross Domestic Product (nominal GDP), though not an absolutely accurate criterion, is used as an indicator for economical development progress.

China and India are the two largest countries in the region, whose combined population constitutes more than 1/3 of the world's total, and are in the progress of fast and large-scale industrialization. Japan, while much small in terms of land and population, is the country with the highest degree of industrialization and correspondingly the highest total and per capita GDP. Hong Kong, Singapore, South Korea, and Taiwan have been called the "Four Asian Tigers," due to their high growth rates and rapid industrialization from 1960s to 1990s. These countries have gradually switched from export-driven manufacture industry to more advanced economy, while maintaining their growth. Hong Kong and Singapore became world leading international financial centers, whereas South Korea and Taiwan have turned to information technology.

The Tiger Cub Economies, which includes Indonesia, Malaysia, Philippines, and Thailand, are following the steps of the Four Asian Tigers to pursue an export-driven model of economic development, mainly by manufacturing goods for developed countries. Cambodia, Laos, Myanmar, Vietnam, Bangladesh, and Pakistan have relatively low per capita GDP (<1000 USD) and high percentage of total GDP in the agriculture sector (>20%). However, these countries have adopted various economy development strategies and are showing accelerating economy growth. Sri Lanka, though having higher per capita GDP, suffers from political instability and uncertain economy policy and might also be put into this category. Brunei and Macau are two unique entities, in that their economic structure is less diversified. Brunei relies heavily on its crude oil and natural gas production, which constitutes nearly 50% of its national GDP. Macau' economy is based largely on tourism, and the gaming, tourism, and hospitality industries are estimated to contribute more than 50% of it GDP.

The diversified economical states and structures of the entities in this region determine that the need and the strategy toward CP will also be diversified. Various organizations and collaboration mechanisms have been set up, providing communication, training, project demonstration, technology development, and dissemination programs to the region. The programs at global or regional levels will first be introduced in the following section.

4. Global and Regional CP Organizations and Collaboration Programs

The United Nations Environmental Program (UNEP) has always been a significant driving force in promoting CP, sustainable consumption, and green industry strategies in developing countries. Its Division of Technology, Industry, and Economics and the subordinate Sustainable Consumption and Production Branch (SCP) have been working with governments, industries, and NGOs worldwide for more environmental-friendly practices by all sectors. Asia-Pacific Region is one of the focuses of the program (http://www.unep.fr/scp/cp/).

As aforementioned, Japan started its industrialization the earliest in the region and has been practicing waste minimization and recycling for more than a century. However, in the latter half of 20th century, with the rapid expansion of all industrial sectors and increase is process complexity, Japan has also encountered numerous and often serious environmental and public health problems related to industry development. Out of the 10 most significant pollution events of the 20th century, four happened in Japan (http://countrystudies.us/japan/49.htm).

Since early 1970s, with the increased public awareness, improvement of environmental law, and development of environmental technology, Japan has outgrown the stage of "growth at the expense of pollution," and in the process, accumulated valuable knowledge with regard to pollution prevention, energy and material efficiency and CP. Nowadays, the country is practicing CP on all local and industry levels with various financial tools provided to Small and Medium Enterprises (SMEs), though without a national strategy or program.

Moreover, Japan has actively sought opportunities to disseminate and apply its successful experience with urban environmental management in developing countries. Efforts have been made to attract international environmental organizations, and as a result, the UNEP/DTIE/International Environmental Technology Centre (IETC) was established in 1992 in the Osaka city of Japan (Official site: http://www.unep.or.jp/). The mandate of IETC is to promote the adoption, application, and operation of Environmentally Sound Technologies (ESTs) in developing countries and countries with economies in transition (http://gec.jp/main.nsf/en/AboutGEC-Establishment).

In addition, the Global Environment Center Foundation (GEC), a non-profit organization, was also established in 1992 to provide logistical support to IETC, Osaka, and collaborate with IETC on joint projects. It also acts as a liaison between Japan and the international environmental organizations. GEC has significantly contributed to the promotion of environmental conservation by collecting and disseminating environmental technological information, training programs, seminars, and symposia, as well as research (Official site: http://gec.jp/).

During the years, the Asia-Pacific region has formed some organizations and co-operation mechanisms at the regional level. For example, the Asia-Pacific Roundtable for Cleaner Production was established after a regional meeting of CP workers in Thailand in 1997 (Official site: http://www.aprscp.net/index.htm), and the 9th APRSCP was just held in Sri Lanka, June 2010. The Roundtable is essentially a forum to discuss the latest ideas of CP, promote the acceptation of the new ideas, and share results with the members in the region. The APRSCP network has grown and matured over the years, and now its members, in addition to the entities mentioned in the scope section, also include countries such as Russia, Kazakstan, Uzbekistan, Nepal, Bhutan, Mongolia, Russia, Australia, and Fiji. With the support from international institutes such as UNEP, the members of the organization have co-operated in projects such as monitoring CP activities in the Asia-Pacific region and are prepared to take a more active role in promoting CP.

5. CP Programs at the National Level

In 1994, UNEP and UNIDO (The United Nations Industrial Development Organization) started collaboration to establish and support National Cleaner Production Centers/Programmers (NCPCs/NCPPs) in developing countries and economies in transition. UNIDO primarily focused on technology development and the implementation of CP in enterprises, whereas UNEP worked with various NCPCs to develop new tools in management and policy instruments.

With such initiatives, more than 45 NCPCs have been set up since then, seven of which are situated within the region of discussion. The countries having UNEP/UNIDO-supported NCPCs are Cambodia, China, India, Lao PDR, Republic of Korea, Sri Lanka, and Vietnam (Table 2). The aim of NCPCs is to build local capacity to implement CP and to provide core CP services at the national level.

NCPCs are to provide CP-related services to various stakeholders, e.g., business, government, research, education, and civil societies, and to assist with the

Table 2. List of UNEP/UNIDO-initiated NCPCs in the region (UNIDO, 2010).

Country	Institution	Host	Establish	Official site
Cambodia	Cambodian Cleaner Production Center	Ministry of Industry, Mines, and Energy	2004	http://www.cambodian-cpc.org/en/
China	China National Cleaner Production Center	Ministry of Environmental Protection	1995	http://www.cncpn.org.cn/english/
India	India National Cleaner Production Center	National Productivity Council	1995	http://www.npcindia.org/NationalCleaner.aspx
Lao	Lao Cleaner Production Program	Ministry of Industry and Commerce	2004	http://lao-ncpc.org/index.htm
Republic of Korea	Korea National Cleaner Production Center (KNCPC)	Korea Institute of Industrial Technology	2001	http://www.kncpc.re.kr/eng/
Sri Lanka	National Cleaner Production Center	Small & Medium Enterprise Developers	2001	http://www.ncpcsrilanka.org/
Vietnam	Vietnam Cleaner Production Center (VNCPC)	Institute for Environmental Science and Technology, Hanoi University	1998	http://vncpc.vn/index.php?/

implementation of CP methods, policies, practices, and technologies in their respective countries. In addition, the NCPCs are in an ideal position advocate and promote the acceptance and application of CP, and sometimes the influence can extent outside the country boundary. For example, NPCPs and other institutions in nine Asian countries have collaborated in a three-year project that demonstrated the application of CP methods for achieving energy savings and greenhouse gas (GHG) reductions in five sectors: pulp and paper; cement; iron and steel; chemicals and ceramics (www.energyefficiencyasia.org).

The core services of NCPCs are as follows (http://www.unep.fr/scp/cp/network/):

- Raise awareness of the benefits and advantages of Resource Efficient & CP: Directed to central and local governments, private sector, academia, and national and/or regional financial institutions.
- Provide Training: To deliver training programs to form a cadre of CP professionals in the nation.
- Demonstrate the environmental, financial, and social benefits of RECP through in-plant assessments and demonstration projects: To provide technical support and show the benefit to the industry.
- Help obtain financing for Resource Efficient & CP investments: To support the promotion of RECP investment and to facilitate the transfer of ESTs.

- Provide policy advice to national and local governments: To help governments identify and develop policy tools and economic instruments suitable to their country's context.
- Disseminate technical information: To facilitate sharing of CP information nationally and internationally.

Each NCPC was initially set up as project management units, and hosted and executed by a national host. The host can be a technical institute, a university, or an industry association, and usually accompanied by a government agency as the partner. After establishment, the NCPCs have matured and gained gradual administrative and financial independence from UNIDO and UNEP. In addition, due to the intrinsic diversity in the nations' sizes and economy structures, the characteristics of the respective host institutions, states of environment awareness and technological development, and the individual focuses and aims of the nations, NCPCs have diversified and extended their scope. Areas such as corporate social responsibility, energy efficiency, sustainable consumption and production and environment impact assessment have been included by some NCPCs, and NCPCs have taken various directions of specialization. UNIDO (2008) has identified four such directions, namely, technology specialization, policy specialization, environment-focused diversification, and people-focused diversification, which are not exclusive to each other. The respective service sectors and focus areas of the aforementioned six NCPCs are listed in Table 3.

5.1. *Cambodia*

Recently, Cambodia had been experiencing renewed fast economy growth after the 1997–1998 Asian economy crisis, which calls for a better practice of CP. The

Table 3. List of the services and focuses of NCPCs in the region.

Country	Cambodia	China	India	Lao	Korea	Sri Lanka	Vietnam
Policy and law	○	•	•	○	•	•	•
Financial tool	○	○	○	○	•	•	•
Training and education	•	•	•	•	•	•	•
Information dissemination	•	•	•	•	•	•	•
Technology development	○	•	○	•	•	○	○
Project demonstration	•	•	•	•	○	○	•
Focus and specialization	1, 4	1, 2	2, 4	1, 3, 4	1, 2, 3, 4	1, 3	1, 2, 3, 4

• Active, ○: not so active, 1: technology specialization, 2: policy specialization, 3: environment-focused diversification, 4: people-focused diversification.

Cambodian Cleaner Production Programme started in 2005 to introduce and promote CP concepts and practices in Cambodian manufacturing industry, mainly the export-oriented sectors. The focuses are training, in-plant demonstration programmes, awareness raising, and information dissemination. As the major export goods of Cambodia are clothes and agricultural products, CP demonstration was mainly done in the textile and food processing industries (http://www.cambodian-cpc.org/en/Publication.html).

The Cambodian NCPC has taken such an approach, where the primary focus is set to be the cultivation of a human resource base in enterprises, industry associations, consultant companies, research institutions, and governmental industrial and environmental management agencies, which can later take the concept of CP and develop it organically. Other missions of the Cambodian NCPC have also been identified, which include demonstration of the CP Assessment method to industries, adapting the internationally developed CP approach to Cambodian conditions, assisting policy makers and recommending the proper promotion and legislation for CP, raising awareness among industries and governmental agencies, facilitating CP education in universities, and co-operation with regional and international institutions.

5.2. *China*

China's size of land, population, and economy determine that the Chinese NCPC has a much more heavy and complicated role in promoting CP in China. Different from the Cambodia approach, the Chinese NCPC has taken a much stronger stand in its relationship with the industry and has been focusing on the development of CP-related law and technology (UNIDO, 2008), with personnel training as a secondary emphasis. The Chinese NCPC has lead the legislation process of national CP-related law, which resulted in the adoption of China CP Promotion Law in 2003 and the Circular Economy Promotion Law in 2009. The former created a system of mandatory CP audits for polluting enterprises and sector-specific CP technical guidelines.

In addition to policy and law, Chinese NCPC has also taken an active role in developing various Clean Production audit methodology systems, standards and technical guidelines at the national, local, and sector levels. For example, 55 CP standards have been published for various industries, ranging from chemical production, battery production and recycling, paper and textile manufacturing, food and beverage processing, to even the hotel industry. Pollution prevention and environmental protection technologies have been identified and recommended to industries, as well as lists of obsolete production process, technology, and products to be discouraged.

Owing to the complexity of CP practice in China, different tiers of services have formed to suit various needs. The local level is composed of city Cleaner Production Centres (CPCs) and is mainly engaged in supervising CP audits and administering mandatory CP audit provisions (Mol and Liu, 2005), whereas province level CPCs are more involved in policy planning and evaluation and adapting provincial economy and industry development goals to national CP requirement. Moreover, various industrial associations, e.g., coal, petroleum, and chemical engineering, have also established sector level CPCs. Owing to the insight knowledge of the specific production processes of the industries, sector level CPCs can have tremendous benefit toward improving CP in these areas.

Aside from the policy and technology development, personnel training is also an important service of the Chinese NCPC. Certified CP auditors training program was initiated in 1995, and governmental environmental department employees have been trained at both the national and regional levels. In terms of international collaboration, the NCPC serves as the contact point of China and regional/international CP programs and organizations, facilitates multilateral co-operation projects, and translates and disseminates international CP experience and technologies. In addition, Chinese NCPC provides a platform for CP product/technology trading, and supply environmental management related services to the State Environmental Protection Administration.

5.3. *India*

The Indian NCPC is hosted by the National Productivity Council (NPC), which is a national-level organization established long ago to promote productivity culture in India. The NPC is an autonomous, tri-partite, non-profit organization with equal representation from the government, employers', and workers' organizations, apart from technical and professional institutions and other interests on its governing council (http://www.npcindia.org/AboutUs.aspx).

Possibly due to the characteristics of the host institution, the Indian NCPC acts with far less enforcement as a government agency, compared with the Chinese NCPC case. The CP promotion in the country has taken a somewhat "bottom-up" approach, where local manufacturing capability plays major role in technology upgrades in small-scale industries. Correspondingly, the overall objective of the Indian NCPC is to demonstrate the CP concept to Indian industries, with the focus being small-scale enterprises. The scope of the services includes

- conduct demonstration projects;
- conduct training programmes/workshops;
- disseminate information on CP;
- identify policy level intervention.

As the host institute has a long and established history of providing training and consultancy, as well as undertaking research in productivity technology up-grading, the CP services are essentially integrated into the overall NPC scheme. Various groups of the NPC offer CP-related services, which include the Environmental Management Group, Energy Management Group, and Water Management Group, with an emphasis on resource productivity and energy saving. In deed, the Indian NCPC has been called in to undertake technology studies for the Government of India to define best practices in water and energy saving technologies, whose results have been incorporated into government policy. In 2001, the new Energy Conservation took effect for the aim of promoting efficient use of energy and its conservation. This would in turn foster consideration of CP and encourage better practice in the energy-related aspect thereof.

The Indian NCPC is also very active in training programs and information dissemination. Various posters, video cassettes on quality and productivity, publication and periodicals, guidelines are being published, as well as campaign organized to raise awareness. The posters and video cassettes have been identified as effective tools in disseminate CP-related ideas and technologies (UNIDO, 2008).

5.4. *Lao PDR*

The Lao NCPP aims at raising the awareness of Lao industries and actively promoting CP issues through mechanisms such as CP demonstration project in representative industries, information collection, and experience sharing and dissemination. Positive outcomes to be expected from such practice include increased productivity and profit of the companies, reduced resource consumption in energy, water, and raw materials, reduced generation of toxic waste and lower risk to human and environment. Therefore, it is clear that the scope of Lao NCPP, in addition to waste minimization and CP, also involves industry efficiency development, energy conservation, and environmental protection.

The activities of the Lao NCPP can be concluded as

- *awareness raising*: through various seminars, presentations, publication of case study and annual report, and press involvement;
- *training*: including wastewater treatment design and operation, study tour to other NCPC for experience sharing;
- *in-plant CP assessment to identify possible CP practice*: focusing on sectors such as wood processing, construction, and food processing, consisting of combined class room training and hand-on field experience;
- *education in university*: for 4th year student of the Center for Environment Study and Development in the national university

In addition to these services and activities, Lao's NCPP provides online advice and consultancy to enterprises interested in CP, where a quick scan of the production process can be done to identify possible CP solutions. The Lao CP teams also conduct data collection after the CP assessment and training course, and evaluate relevant ESTs that are techno-economically suitable to the Lao's situation.

5.5. *Republic of Korea*

The Korea National Center of Cleaner Production (KNCPC) is a non-profit organization established in 1999 under the support of the Korean Ministry of Knowledge Economy (http://www.kncpc.re.kr/eng/About/) and joined the UNIDO/UNEP network in 2001. Its main focus is to help Korean industries stay in the international business arena, by enhancing their corporate environmental competitiveness. The scope of KNCPC's activities is comprehensive and encompassing a wide range from policy planning to technology development and dissemination, consultancy and standardization, and also international cooperation. Some of their services are illustrated below:

- Policy development to comply international environmental regulations and multilateral agreements.
- Development of long-term business strategies and industrial plans complying with CP requirement.
- Conducting CP technology research and identification.
- Financing and managing CP technology development projects by individual, collaborative, and collective effort.
- CP assessment service.
- Facilitating local and international technology transfer regarding the tools, infrastructure, methodologies, and technologies available for CP practice.
- Establishment of CP infrastructure.
- Consultancy on SMEs, supply chain, and regional environment management issues.
- Local and regional CP education program.
- Information dissemination on CP standardized methods, guidelines, and indices.
- Portal for regional and international CP collaboration.

The most remarkable feature of the KNCPC is its emphasis on green industry and circular economy, which is understandable considering Korea's state of industrialization. Special attention has been paid to topics such as product life-cycle analysis, where all stages of the product development (designing, manufacture, delivery, use, and disposal) are under the scope of CP; eco-industrial park with internal material recirculation and zero discharge as the aim; and remanufacturing of old products and development of sustainable new products.

5.6. *Sri Lanka*

Though officially established in 2001, the host of the Sri Lanka NCPC, the Small & Medium Enterprise Developers (SMED) had had an active experience of supporting CP activities in the country. As the SMED was established by Federation of Chambers of Commerce and Industry of Sri Lanka and Friedrich Naumann Stiftun, and Ministry of Enterprise Development, Industrial Policy and Investment Promotion of the government of Sri Lanka is the government counter part, the Sri Lanka NCPC has a close collaborating relationship with the nation's industry sectors.

A comprehensive scope of services has been developed by the Sri Lanka NCPC, centering upon the following aspects, with an emphasis on environmental management information and consultancy:

- In-plant assessment of CP and environmental management system: For both CP and waste treatment/minimization.
- Awareness raising: General and case-specific awareness raising, out-reach to children.
- Training: National-level capacity building, new tools for CP and environmental management, annual CP auditor training program.
- Technical information dissemination: Reference library containing publications, CDs, and other sources.
- Promotion of CP investments and financing.
- Policy advise.

In addition to the usual CP services of an NCPC, the Sri Lanka NCPC has participated actively in providing CP financial assistance. As a result, an e-Friends scheme was launched by the National Development Bank as an environmentally-friendly solutions fund for industrial firms interested in CP. Also, the NCPC's effort has lead to the Sri Lankan National Policy and Strategy for Cleaner Production issued by the Ministry and Environment and National Resources in 2005 (http://www.hsbc.lk/1/2/LivingBusiness/NaturalResources/AdditionalResources/SriLankasNationalPolicyandStrategyforCleanerProduction), which places responsibility on all line ministries to develop and implement CP strategies. Nowadays, various important industry sectors including hotel & catering, metal, fiber, food, tea, chemicals, textiles & printing have implemented CP obtained very successful results. The NCPC has also undertaken responsibility of collecting successful cases and sharing the experience.

The Sri Lanka NCPC serves excellently as an anchor point for other international CP-related collaboration programs, sponsored by various organizations, e.g., UNIDO and Asian Development Bank.

5.7. *Vietnam*

The Vietnam Cleaner Production Center is a national focal point for the promotion and implementation of eco-efficient industrial production through CP. It aims at being a "one-stop" service provider for Vietnam's industries' CP needs, by conducting activities of CP assessments, financial engineering, technology advice, training, and information dissemination in an integrated fashion.

Similar to the NCPCs of Sri Lanka and India, the Vietnam NCPC focuses largely on small-scale industries, whether state private owned. Upon the industry's request, the Vietnam NCPC can conduct both preliminary and in-depth in-plant

Table 4. NCPC/NCPP of the other entities in the region.

Entity	National CP center/program	Related institutes /online source	Focus and activity
Bangladesh	NCPC in the planning	http://www.unbconnect.com/component/news/task-show/id-22767	Resource efficiency, energy efficiency, water efficiency, waste minimization, clinical waste management
Brunei	National Environmental Strategy	National Committee on the Environment, Environment Unit of the Ministry of Development	Integrated solid waste management plan
Hong Kong	Cleaner Production Partnership Program	http://www.cleanerproduction.hk/	Awareness raising, on site CP assessment, project demonstration, CP effectiveness verification, information dissemination, focus on HK industries in both, HK and Peal River Delta Area
Indonesia	Indonesian Cleaner Production Center (ICPC)	http://www.ppbn.or.id/en/index.php?option=com_frontpage&Itemid=1	Facilitate interactive dialogue, training and information, dissemination, technical advisory, Awareness raising, demonstration projects, focus on service providers
Macau	NA	NA	NA
Malaysia	Clean Production Virtual Center	Department of Environment (DOE), Standard Industrial Research Institute of Malaysia (SIRIM) http://cp.doe.gov.my/cpvc/	Factory audits, demonstration projects, information dissemination
Myanmar	The National Sustainable Development Strategy (NSDS)	The Ministry of Science and Technology, National Commission for Environmental Affairs	NA
Pakistan	Cleaner Production Program (CPP)	http://www.cpp.org.pk/	Assessment of needs, energy audits, provision of technical assistance to industry in adopting energy efficiency and wastewater recycling techniques, raising awareness

(Continued)

Table 4. (*Continued*)

Entity	National CP center/program	Related institutes/online source	Focus and activity
Philippines	Integrated Program on Cleaner Production Technologies	The Philippines Cleaner Production Technology Center (IPCT) http://cptech.dost.gov.ph/	Capacity building assessment and monitoring information, education, and communication; guidelines and policies for CP technologies; environmental technology verification; environmental management system; research and development
Singapore	Singapore Green Plan	The Singapore Productivity and Standards Board (PSB)	Tax incentive, National Cost of Quality (NCOQ) Program, standards and certificates, technical assistance, financial aid
Taiwan	National Center for Cleaner Production, NCCP Taiwan, R.O.C.	http://www.nccp.org.tw/	CP assessment, technological and management tools, standards and regulations, research and information
Thailand	National Master Plan on Cleaner Production and Cleaner Technology	Thai National Pollution Prevention Roundtable (TNPPR)	Awareness raising, technology development, standards and code of practice, education and training

NA: Not applicable.

CP potential assessment, the scope of which includes energy efficiency, resource consumption, waste generation, and production efficiency. Presently, the CP assessments have been focused on the industry sectors of pulp and paper, metal finishing, food and beverage, textile, and construction materials.

During the assessment, training can be simultaneously carried out, and technical solutions be identified. Upon the completion of the assessment, the NCPC can perform financial engineering for the potential investment on the possible CP practice, should the company requires. Furthermore, the NCPC can assist by linking the company to potential fund providers, e.g., national bank, environmental fund, or international donors. It even helped establish a Green Credit Line for the implementation of higher cost CP.

In addition to the case-specific training that can be conducted simultaneously as the CP assessment, both sector-specific and special skill training at management, supervisor, and worker levels can be arranged in traditional CP and other related areas, e.g., environment management. Information gathering is usually performed after the assessment and training sessions, and dissemination can follow by means of seminars, workshops, publications, and media exposure. By organizing trainings and study tours for governmental authorities, the NCPC also aims at influencing Vietnam's policy makers to the possibility of CP. As a result, the National Cleaner Production Action Plan for the period 2000–2005 was drafted and approved in 2002. Crucial problems, such as a duty and tax system for waste disposal, and a realistic pricing system for raw materials and energy were addressed.

Possibly because of its host institute (Hanoi University) and connection to international expertise, the Vietnam NCPC is strong in the technology aspect. However, instead of directly developing CP technologies, the Vietnam NCPC tends more toward helping the industry form their own expertise. Moreover, the NCPC works with various "intermediaries," i.e., service providers, industrial engineering companies, and financing institutions, and provides support to their activities. CP service providers can receive assistance from NCPC in aspects such as selling their CP assessments, obtaining sector-specific and technical information, and assurance of their CP work quality by the national agency. The Viet Nam Cleaner Production Center Quality Label was specially created for this purpose.

The information available online on the national centers or programs of the other entities in the region is surmised in Table 4. Some of the entities might not have specific programs dedicated to CP and in this case, the closest practice is listed.

6. Case Studies

In this section, three case studies from three Asia-Pacific countries are presented, covering industrial sectors of textile production, electronics manufacturing, and chemical manufacturing, respectively. These sectors are some of the major industries in this area and contribute largely to revenue and tax generation, labor employment, and pollutants production. The three cases were selected based on their various scales, diverse project approaches, and host institutes, and also the various countries in which they occurred, so as to present a clear picture of CP practice in this region.

6.1. *Garment production in Cambodia*

Textile production is a highly active industrial sector in Asia-Pacific. It is estimated that above 70% of the world's production in yarn and cloth is generated in this region (Visvanathan *et al.*, 2000). This industry is also a major polluter. The major process steps leading to the formation of a piece of garment are fiber production and spinning (yarn formation), weaving and knitting (fabrication), dyeing–printing and finishing (wet processing), and garment production. In each step, solid, liquid, gaseous, and noise wastes are generated, and various degrees of energy consumption occur. The spinning and weaving sector consumes the highest portion of electricity, whereas large quantity of steam is needed for wet processing. In addition, water consumption is high for the fabrication, dyeing, printing, and finishing processes, and consequently, wastewaters of various compositions and strengths are produced. These emissions are harmful to the environment and can also have detrimental effect on workers' health. Therefore, various indicators have been developed to evaluate the water/energy efficiency and potential environmental hazard of the processes. These indicators include specific fuel/electricity consumption/cost, specific water consumption, specific wastewater generation, output water quality (biological oxygen demand, chemical oxygen demand, total suspended solid, etc.).

As aforementioned, Cambodia is a country with relatively low degree of industrialization, but is undergoing rapid economy growth, especially in the manufacturing sector. The manufactured products are mainly for exporting purpose, a major one of which is garment. Consequently, the national Cambodian CP program has identified the textile/garment industry as one of its focuses, and conducted several demonstration projects in year 2005. The aim of the CP practice is not only to protect environment and human health, but also to improve the efficiency and competitiveness of the industry; hence, the Cambodian enterprises will be able to maintain and increase its share in the global market.

One of the demonstration projects conducted by the Cambodian national program took place in Roo Sing Garment Factory (Roursey Keo District, Phnom Penh), a large private plant 100% owned by foreign investors. It was employing more than 3700 employees, and had an installed capacity of 5000 tons garment per year, 80% utilized at the time. The main product was denim cotton jeans, and some other garments were also produced from both knitted and woven fabrics. The raw fabric material was imported from Hong Kong and Taiwan, and after garment sewing, the factory conducted several finishing steps before the final products were ready for export to Japan, Europe, USA, and other countries (CCPP, 2005).

A CP training was first arranged for the factory's management staff, followed by the establishment of an in-house CP "task force," which consisted of three

Table 5. Major steps of the wet operation for denim jeans.

Process step	Purpose	CP concerns
Hot washing/desizing	Remove residual sizing chemical; improve fabric wettability; improve adsorption property	Water and electricity consumption; wastewater generation
Soap washing/scouring	Remove superfluous color	Water and electricity consumption; wastewater generation
Bleaching	Improve optical brightness; provide special effects; assist spray-dyeing operation	Steam, water, and electricity consumption; use of bleaching chemicals; wastewater generation
Softening/finishing	Enhance wearing comfort; provide various finishing effects	Water and electricity consumption; use of softener; wastewater generation
Spinning and drying	Dry the garment	Steam and electricity consumption

members from the staff and was supported by national and international experts. The CP team conducted audit for the factory's energy and resource consumption and waste emission, and collected data from the first half of year 2005's operation as the baseline. Based on the systematic study, the wet operation was identified as the focus process, as it had the highest resource consumption, generated significant waste emission, and was responsible for occupational health and safety hazards.

Using denim jeans as the example, the major steps of the wet finishing operation and the respective CP concerns are shown in Table 5.

The CP team carefully inspected the efficiency of the processes and compared them with the benchmark values of similar manufacturers in some other Asian countries. It was discovered that in several areas CP concerns such as high resource consumption and waste emission existed. These concerns and the relative causes are as follows:

- High water consumption:
  - high cloth to water ratio.
- High fuel consumption of boiler to generate steam:
  - poor combustion, low steam generation efficiency;
  - poor steam distribution and utilization.
- High electricity consumption.
- GHG emission high (calculated from fuel consumption).

According to the problems found, 22 CP options were identified, and 18 of them were selected for further feasibility study. These options were distributed among various areas, which included good house keeping (8), input material change (3), process modification/control (5), equipment modification (2), technology change

Table 6. Result of the CP project.

Key process indicator	Result			
	Before	After	Improvement	Economic benefit (US$)
Water consumption (m^3/year)	334,500	280,500	16% reduction	16,000
Electricity consumption (KWh/year)	430,000	380,500	11% reduction	70,000
Fuel oil consumption (US$/year)	750,000	170,000	77% reduction	580,000
Chemical consumption	NQ	NQ	~15% reduction	NQ
Wastewater generation (m^3/ton product)	62	45	27% reduction	NQ
GHG emission (ton/year)	11,680	2,280	80% reduction	NQ
Total investment (US$)			630,000	
Total saving (US$)			666,000	
Payback period			<1 year	

NQ: Not quantified.

(2), and 3R (2). They were then put to implementation based on their technical feasibility, economical viability, and environmental benefit.

Some CP options of the earliest implementation were the improvement on the insulation of steam distribution network, collection of condensate as boiler feed water, recycling of treated wastewater for flue gas scrubbing, optimization of cloth to water ratio, and replacement of three fuel oil fired boiler by one wood fire boiler. All the options were of low cost and easy implementation, except the last one. The result of the CP project is summarized in Table 6.

On the other hand, the limitations and barriers encountered in the project were also identified:

- lack of CP experience of the staff;
- only some of the processes were audited;
- lack of environmental monitoring data, e.g., waste water/gas emission;
- lack of communication between CP experts and the floor staff.

Despite the limitations, the CP demonstration project was quite successful, with significant improvement on the key indicators and short payback period. In addition, one great achievement was made, which was the better understanding of the CP idea by the plant staff. In fact, some of the CP measures were identified by the factory's in-house employees, and more were expected in the subsequent stages. Such achievement would ensure that the CP practice would become a sustainable process, and knowledge and experience would be shared in the future.

6.2. *Fine chemical production in Korea*

The chemical industry is indispensable to the society, as it provides many important materials and products for industry and daily use, such as drugs, dyes, pesticides, fertilizers, and various organic substances and inorganic salts as industrial raw materials. On the other hand, chemical industry is also a big polluter. In addition, being a leading cause of volatile organic carbons emission, which is a major air pollution agent, the production process often generates wastewaters containing various inorganic and organic substances. Some of the substances are resistant toward biological degradation, whereas some are downright toxic to the biosphere. Owing to the complexity and recalcitrance of the waste, and the ever more stringent environment and health-related regulations, waste treatment cost has become a considerable burden on the medium and small companies in this industry.

Instead of trying to treat the highly persistent wastes generated from various chemical production processes, an alternative is to optimize the processes and change the characteristics and minimize the production of the waste. In December 2000, the Korea Institute of Industrial Technology (Chemical Engineering Department, Yonsei University) started an one-year project to establish CP processes in the fine chemical industry (KNCPC, 2004). The project was supervised by LG Environmental Safety Research Institute, and seven Korean leading chemical companies participated. These companies were provided in Table 7.

These companies had diverse product ranges, varied production processes, and hence they faced different CP problems. The problems and the corresponding solutions identified are concluded in Table 8.

The project was fruitful and beneficial both economically and environmentally. The total project cost was US$ 416,000, of which 74,000 was invested on process improvement, with an addition one-time cost of 275,000. However, the payback period was estimated to be lower than one year. The improvement made with Company D alone resulted in nearly USD 260,000 per year, where asphalt from the refinery process was originally outsourced to be treated as waste, but was developed for reuse as water-proofing agent. In addition, the modification also totally eliminated pollution from the asphalt incineration process. The environmental benefit was similarly significant in other companies.

Table 7. Participating companies.

Company	Main products	Scale
Kyung In Synthetic Corp.	Dyestuff	Annual production 20,000 tons
SamWon Industry	Fluorescent brightener	NF
Eyang Chemical	Coagulant, dispersant, decolorant	900 employees; annual production ~10,000 tons for powder-type fluocculant
Jung Woo Fine Chemical	NF	NF
Cheil Petrochemical Co. Ltd.	Polymer	700 employees; annual production 670,000 tons
DuKeun Interline Oil	Recycled base oil	Annual production 8 million gallons
Korea Mirai Industries	NF	NF

NF: Not found

Table 8. The CP problems and solutions of the individual companies.

Company	Problem	Solution	Economical/environmental benefit
A	Wastewater treatment efficiency low, excessive chemicals usage	Replace the existing Fenton oxidation process with chemical aggregation process	Annual cost saving US$ 166,000; 71% reduction of sludge (by 466 tons/year)
	High sludge handling cost	Chemical aggregation reduced sludge volume	
	Lack of denitrification for the wastewater	Intermittent aeration method identified as the best available technology	Investment reduction by 250,000
	Emission of waste solvent and VOC	Diagnosis of the solvent recovery system, optimization of VOC treatment technology	
B	A third refining used to prevent spray dryer erosion, discharging high volume wastewater	Remove copper and inject erosion controller to minimize spray dryer erosion	
	High environmental management cost	Diagnosis of the wastewater treatment facilities, VOC treatment technology optimization	
C	Chrome 6 discharged in the wastewater	Removal and neutralization of Chrome 6	100% removal of 0.53 tons/year chrome 6
	Discharge of cyanogen	Solidify cyanogen and removal	100% removal of 150 tons/year cyanogen
	Discharge of solvent and VOC	Equipment improvement for the solvent recovery process	
	Dust generation from the raw material input system	Automatic input system for powdered raw material	
D	Excessive cost from waste asphalt treatment	Exploitation of asphalt utilization as raw material	Annual cost saving 100,000; additional revenue 158,000; 100% volume reduction of 1450 tons
	Asphalt production high	Process optimization via simulation to increase fuel oil production	
	Production interruption and high repair cost due to plugging and erosion	Material pre-treatment to prevent erosion and plugging	
E	Boiler system efficiency low, low recovery of condensed water	Process and equipment optimization	
	Leakage of solvent *n*-hexane	Diagnosis and prevention of leakage	

(*Continued*)

Table 8. (*Continued*)

Company	Problem	Solution	Economical/environmental benefit
	Improper oil stream discharge	Recommendation of proper handling technology	
F	Macromolecules discharged from the injection procedure	Recommendation of a back filter dust treatment method	
G	Naphthalene odor during unloading and storage	Identification and prevention of leakage	
	Foul odor released due to sampling	Replace sampling with indirect measurement, install gas scrubber	

6.3. *Electronics manufacturing in China*

In the era of information technology, electronic appliances are an integral part of the society. Consequently, electronics manufacturing have become one of the world's biggest industrial sector. In the world's manufacturing chain of electronics, China is the largest manufacturer of the primary components, as well as the most significant consumer of the end products. In the printed circuit board (PCB) manufacturing, a major sector of electronics production, China consists more than 35% of the world's production quantity, with a yearly production value of approximately 16 billion US$ (Okokok, 2010).

However, electronics production is also a heavily polluting industry. Some production processes, e.g., semiconductor manufacturing, can consist of more than 100 steps and are often accompanied by usage and emission of heavy metal, organic solvents, and carcinogen waste streams. In the case of PCB production, the general process is to produce patterns of conductive material on a non-conductive substrate by subtractive or additive processes. The manufacturing steps include cleaning and surface preparation of the base, electroless copper plating, pattern printing and masking, electroplating, and etching. Various gaseous, liquid, and solid wastes are generated from each step, which include ozone depleting organic solvents, acidic vapors, ammonia, polymers, metals (copper, nickel, iron, chromium, tin, lead, palladium, and gold), cyanides, scrap board materials, and various sludge (IFC, 2007).

Unfortunately, though the biggest producer of PCB in the world, the overall CP level of the Chinese manufacturers is not high (Yan, 2009). According to the national regulation and standard, all the manufacturers must install waste treatment facilities, and obtain certain energy and material efficiencies (HJ 450-2008, 2008). However, due to the generally small scale of individual enterprises and the intense competition within the industry, the treatment facilities might not always function at full capacity, and the drive of costly process improvement is not high.

Poor CP status has resulted in serious environmental and health problems, e.g., in 2007 severe water pollution occurred in the Taihu Lake, Suzhou area, which is a center of electronics production.

Various efforts are being made to address the problem by China Printed Circuit Association (CPCA), the major trade association of the PCB industry. One proposed solution is to concentrate various small and medium electronics manufacturers into electronics circuit industry parks, where the wastes generated can be centralized and treated. It is expected to lower the average treatment cost, create industrial ecology within the park, and promote sharing of CP experience. Presently, several such parks are being developed, which include Kunshan Park and Suzhou Huangqiao Park in Zhejiang Province, and several more in Jiangxi, Shantou, Sichuan Provinces. In addition, CPCA is actively supporting CP through R&D measures such as promoting environment-friendly processes and equipments, setting up sample projects and plants, and establishing industry expert database. A college dedicated to the electronics industry is even being planned in Henan Province.

Of all the measures, the sample projects are the closest to the industry practice. Therefore, a company with relatively advanced CP practices is introduced below. Shenzhen Shennan Circuits Co., Ltd. (SCC) is a leading company in PCB production. It was founded in 1984, presently has 2800 employees and an annual production capacity of 550,000 m^2 PCB. SCC has long been engaging in promoting production efficiency and waste minimization. Their efforts have been self-initiated and sustaining, which are focused in the following areas (SCC, 2008):

Education and capacity building:

- Establishing internal CP committee for sustainable CP practice.
- Providing CP training for each new employee.
- Offering at least three environmental trainings/year for all employees.
- Encouraging employees to suggest CP option, rewarding bonus.
- Organizing volunteer work for environment protection.

3R practice:

- Water reuse and management:
  - water usage monitoring at all production points;
  - separation and special treatment for various wastewaters;
  - internal wastewater reuse between various processes;
  - end-point water treatment for industrial and municipal reuse.
- Solid waste management:
  - scraping material reuse for production;
  - outsourcing to certified waste recyclers;
  - improving process to reduce copper usage.

- Energy efficiency:
  - electricity usage monitoring at all points;
  - establishing electricity usage standards;
  - process improvement to phase out low energy efficiency processes;
  - improving air-conditioning efficiency;
  - waste heat reuse by heat-exchange;
  - installation of solar panel for renewable energy generation.

Green supply chain:

- 100% RoHS compliance of the raw materials.
- Green transportation.
- Recycle and reuse of packing materials.

As this is an on-going project, the exact economic investment/return is not calculated. However, in the area of water saving alone, one of the company's plants was able to save 7000 tons water per month, whereas another plant reduced its water consumption by 30,000 tons per month. In terms of energy saving, the reuse of waste heat from the cooling water unit reduced annual electricity consumption by 1.4 million KWh, and the solar panels saved 700,000 KWh per year. The CP practice was also able to achieve significant improvement of wastewater quality, in terms of COD and copper concentrations, etc.

7. Conclusion

Above is a brief introduction of the CP and waste minimization programs in the Asia-Pacific region. As can be seen, the CP opportunities are abundant, and potential is infinite. The entities in this region are conducting various CP projects and programs through diversified mechanisms and funding schemes, with various production scales and project aims, and in all important industrial sectors. Despite the various barriers and difficulties inevitably existing in the region, the future for CP is bright, which will certainly benefit the region and its population, as well as the global environment.

References

CCPP, 2005. *Case Studies: Textile Garment Sector.* Cleaner Production Demonstration Project, Cambodian Cleaner Production Programme.

HJ 450-2008, 2008. China Environment Protection Department, *Cleaner Production Standard PCB Manufacturing.*

IFC, 2007. *Environmental, Health, and Safety Guidelines for Semiconductors and Other Electronics Manufacturing.* International Finance Corporation, World Bank Group.

KNCPC, 2004. *Cleaner Production Case Report 2004-Success Stories.* Korea National Cleaner Production Centre.

Mol, A. and Liu, Y., 2005. Institutionalising cleaner production in China: the cleaner production promotion law. *International Journal on Environment and Sustainable Development* **4**(3), 227–245.

Okokok, 2010. *Global and China PCB Industry Report.*

SCC Shennan Circuit Co., Ltd., 2008. *Company Social Responsibility Report.*

UNEP, 1994. *Government Strategies and Policies for Cleaner Production*, Vol. 32. United Nations Environment Programme, Paris, France.

UNIDO, 2008. *Independent Evaluation of the UNIDO–UNEP Cleaner Production Programme.* United Nations Industrial Development Organisation, Vienna, Austria.

UNIDO, 2010. UNIDO Offices Worldwide — National Cleaner Production Centres (NCPCs) and National Cleaner Production Programmes (NCPPs).

Visvanathan, C., Kumar, S. and Han S., 2000. *Cleaner Production in Textile Sector: Asia Scenario. National Workshop on "Sustainable Industrial Development Through Cleaner Production,"* Colombo, Sri Lanka.

Yan, K., 2009. China Printed Circuit Industry in Changing Period and the Role of the Association. China Printed Circuit Association.

Chapter 26

SUSTAINABLE SOLID WASTE MANAGEMENT FOR DEVELOPING ASIAN COUNTRIES: A CASE STUDY OF PHITSANULOK MUNICIPALITY, THAILAND

Janya Sang-Arun[a] and Magnus Bengtsson[b]
Sustainable Consumption and Production Group Institute for Global Environmental Strategies, Japan
[a]*Policy Researcher*
[b]*Director*

Developing Asia is challenged by rapidly increasing volumes of solid waste and conventional forms of waste management are not coping. The consequences are accumulating piles of uncollected waste in the cities, environmental pollution, increasing cost of waste management, and social conflict over siting of disposal facilities. This chapter describes a sustainable solid waste management approach based on the 3Rs (reduce, reuse, recycle). It provides a case study on implementation of the 3Rs in Phitsanulok Municipality, Thailand, that demonstrates how multiple interventions based on the 3Rs, targeting both the midstream and downstream stages of the waste chain, were effective in reducing waste for final disposal, avoided environmental impacts and greenhouse gas emission, increased resource efficiency, supported the recycling business, generated income for residents, raised the social status of waste pickers, and promoted social acceptance of the practices of reducing, reusing, and recycling. The specific interventions of Phitsanulok Municipality included study of waste composition, institutional capacity, resident behavior and the waste management system; educating residents through a door-to-door campaign and public announcements; establishing a public–private partnership to promote recycling; promoting decentralized composting; and introducing mechanical biological treatment (MBT) prior to landfill. Underlying factors for the success of these interventions included strong and sustained political leadership, a participatory approach, and an integrated approach that combined waste management with economic objectives. The combined impact of these activities was a 80% reduction of waste to be disposed in the landfill, and further reduction could be achieved by using plastic waste fragments as refuse derived fuel, increase composting, and reducing the arsenic contamination of the organic fraction from the MBT.

1. Introduction

Solid waste management is a growing source of concern in developing Asian countries. Both economic development and population increase contribute to the escalating volumes of waste. Local governments — who in most countries are responsible for solid waste management — have insufficient financial resources and qualified personnel to handle the waste in an environmentally sound manner (Ahmed and Ali, 2004; JICA, 2005).

Open dumping in designated areas is, therefore, commonly practiced in the region (UNEP, 2004; Bengtsson *et al.*, 2008). Open burning is sometimes applied, either at the dumpsite or by private land owners, in order to reduce the volume of waste. In parallel, due to inefficient collection and inadequate resources, waste is discarded in the streets, waterways, and empty spaces. These practices have implications for the environment and human health (Ramachandra and Bachamanda, 2007; Puri *et al.*, 2008). Their impacts include giving the city an unpleasant appearance, polluting the water and the air, increasing the risk of flooding, and spreading vermin and insects, some of which are disease carriers.

Some local governments in low-income countries, such as Cambodia, did not have a budget for solid waste management. They, therefore, recruited private companies to take on this task (Sang-Arun, 2010). As they did not receive any compensation from the local governments, these companies anticipated collecting service charges from residents to sustain their operations. However, many residents in urban areas did not want to pay for or participate in solid waste management. They were not aware of the environmental and health impacts of waste and some were of the opinion that waste management is the responsibility of the government. Consequently, some of the waste collection companies went bankrupt. The problem of poor solid waste management in the cities remains largely unsolved.

In more advanced developing Asian countries, some local governments were able to secure funding for upgrading the final disposal sites. However, in many cases they were unable to implement the upgrading projects due to strong and often increasing social resistance from residents living near the sites (Hsu, 2006; Sang-Arun, 2009). Local governments have found themselves facing both increasing waste volumes and decreasing availability of land where waste disposal is both socially acceptable and technically/environmentally suitable (Ahmed and Ali, 2004).

The typical solution to improve solid waste management in developing Asia is an engineered or sanitary landfill. However, from a sustainability perspective, landfill disposal is far from an ideal option for several reasons. First, landfill disposal discourages resource efficiency (Sheehan, 2000; Chen and Greene, 2003). Valuable nutrients and other waste fractions that can be recycled are buried in landfills. Second, organic waste disposed of in landfills can undergo anaerobic degradation and generate methane — a potent greenhouse gas. Third, due to personnel and budget constraints on daily operations, many local governments cannot afford to meet the sanitary landfill criteria, resulting in the leakage of pollutants.

After upgrading to a sanitary landfill, some governments prohibited waste pickers from accessing the site. The waste pickers moved their activities from the landfill to the city, where the working conditions are harsher; they need to walk long distances for many hours to find enough waste to cover their living costs. Landfill access bans can improve the standard of landfill operation; however, they can also harm the economic and social situation of the waste pickers (Medina, 2000).

The mismanagement of solid waste can contribute to global warming. Open burning or incineration of waste derived from fossil resources, such as plastics, synthetic textiles and rubber, liquid solvents and waste oil, releases carbon dioxide into the atmosphere. Furthermore, as mentioned above, the disposal of organic waste in landfills can generate methane, which has 21–25 times higher global warming potential than carbon dioxide (IPCC, 2006).

This chapter outlines an approach for achieving sustainable municipal solid waste management in developing Asian countries. It argues that improved solid waste management has a number of distinct benefits, both for local stakeholders and for the global environment, including enhanced resource efficiency, job creation and reduced poverty, avoided detrimental local and global environmental impacts, and strengthened local economic development.

This chapter first describes a waste management approach based on waste reduction, product reuse, and materials/energy recycling known as the 3Rs, which stands in stark contrast to the waste management systems typically found in cities in developing Asia. Second, it reviews the current regional movement on the 3Rs and how these ideas have been reflected in national legislations. Third, Phitsanulok Municipality, Thailand, is taken as a case study from which lessons are drawn to formulate guidance for other local governments in developing Asia on how to successfully implement the 3Rs for sustainable solid waste management.

The methodologies used for this study were a literature review; analysis of secondary data; semi-structured interviews with officers of Phitsanulok Municipality, selected communities, and private sector actors; and site visits to two communities and the Municipality's disposal facility.

2. Sustainable Solid Waste Management

Sustainable solid waste management can be achieved when the management of solid waste meets the primary goal of human health protection and the secondary goals of promoting environmental quality and sustainability, economic growth, and employment generation (Schübeler *et al.*, 1996). To achieve these goals, solid waste management strategies must go beyond purely technical considerations to political, institutional, social, financial, economic, and environmental concerns.

Local governments in developing Asia lack personnel and financial resources to treat the rapidly increasing volumes of solid waste in their areas of jurisdiction. Therefore, an integrated approach that includes efforts to reduce the volume of waste for final disposal should be considered as the better option to achieve sustainable solid waste management in Asian developing countries.

Reducing the volume of waste delivered to the final disposal site can help reduce the costs incurred by local governments for waste collection, transportation, and disposal. It can also reduce the health hazards and environmental impacts that are generated by landfill and incineration practices at the disposal sites (UNESCAP, 2009). When the quantity of waste for disposal decreases, the likelihood of local governments treating the waste in an environmentally sound manner using their existing budget and personnel increases.

The volumes of waste reaching the final disposal sites can be reduced through the promotion of 3Rs (reduce, reuse, recycle). This approach was traditionally practiced in many countries and was formally brought into international discussions by the Ministry of Environment of Japan (MOEJ) in 2004 (Makiya *et al.*, 2006).

The 3Rs is an approach to enhance resource efficiency by reducing waste generation, reusing materials and products until they can no longer function properly, and recycling materials for further use as long as it is economically feasible to do so, before discarding them at the disposal site (Makiya *et al.*, 2006).

In terms of climate benefits, the 3Rs can directly reduce greenhouse gas emissions from the waste sector by reducing the volume of waste delivered to landfills, especially those where measures to avoid anaerobic conditions are not put in place, and burnt (Fig. 1). The 3Rs can also indirectly contribute to reducing greenhouse gas emissions from non-waste sectors (Sheehan, 2000; Sang-Arun and Bengtsson, 2009). For instance, the composting of organic waste can avoid methane emission from anaerobic landfill practice. Further, the application of compost can reduce the amount of energy used for tilling by reducing soil compaction and increase the

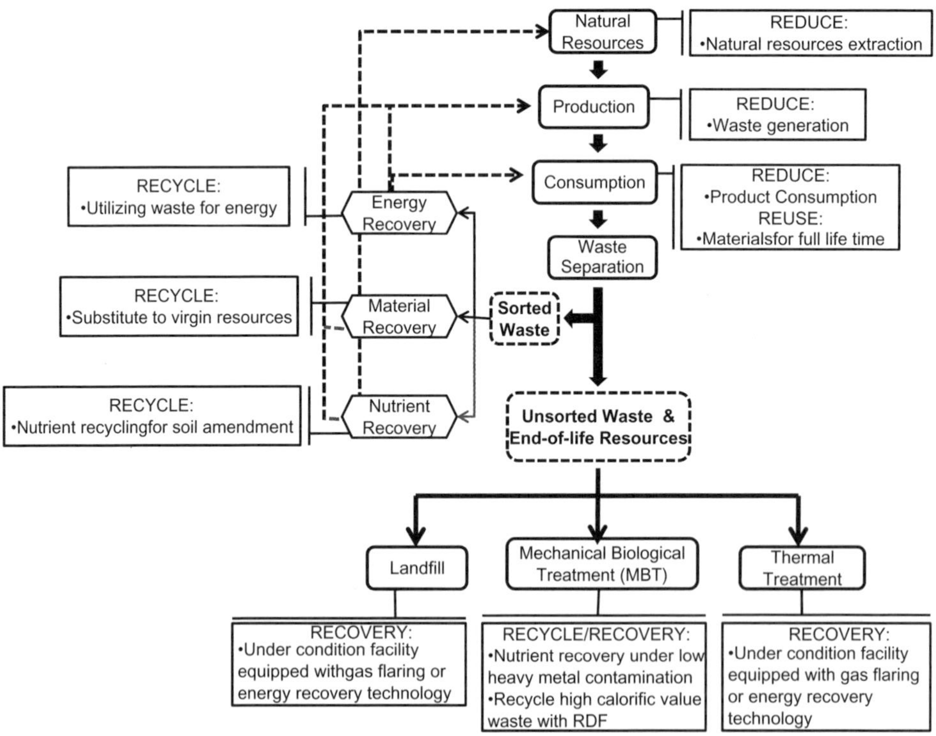

Fig. 1. Resource circulation through the 3R approach (modified from Sang-Arun and Bengtsson, 2009).

Table 1. Linkage of the 3Rs and the MDGs (developed after UNCRD, no date).

S.No	MDGs	Potential contribution of the 3Rs
1	Eradicate extreme poverty and hunger	Waste pickers (very poor people) and residents could generate income through recycling
2	Achieve universal primary education	With their income from recycling, waste pickers could allow their children to attend school, rather than work to support the family
3	Promote gender equality and empower women	Women could generate income by selling recyclable wastes, compost, and crops that were grown using the compost
4 5 6	Reduce child mortality Improve maternal health Combat HIV/AIDS, malaria, and other disease	When local governments improve solid waste management by applying the 3Rs, poor families can enjoy the benefits of clean living environments (land, water, and air) providing flow-on benefits for their health. In addition, proper solid waste management could reduce the food sources and habitats of disease carriers such as mosquitoes and cockroaches
7	Ensure environmental sustainability	The 3Rs could enhance resource efficiency, avoid waste generation, and reduce the volume of waste for disposal; therefore, it can improve environmental quality and sustainability
8	Develop a global partnership for development	For example, through the efforts of the MOEJ and international organizations, a regional 3R Forum was established to stimulate 3Rs implementation for sustainable development in Asia

infiltration rate, which equates with reduced GHG emissions from the energy sector (Shelton, 1991). Furthermore, use of compost can reduce demand for chemical fertilizer, thus reducing nitrous oxide emissions from farmland (Favoino and Hogg, 2008).

At the waste treatment stage, the practice of waste recovery can contribute to food security. Use of compost and nutrients from solid waste can sustain or increase crop yields. In addition, waste recovery can also provide alternative energy through various energy recovery technologies such as anaerobic digestion, mechanical biological treatment (MBT), and gasification.

As shown in Table 1, the 3Rs can also contribute to the United Nations Millennium Development Goals (MDGs; UNCRD, no date). For example, waste pickers and residents could generate income through recycling, which they could use to improve their household's expense (linked to Goal 1) and to enable their children to receive formal education (linked to Goal 2).

3. The 3Rs Movement in Developing Asia

Climate change, one of the most serious environmental issues of our time, is complicated and tangled up with difficult issues such as poverty, economic development, and population growth (UNFCCC, no date). The United Nations Framework

Table 2. The 3Rs in national strategies.

Countries	3Rs in national climate policies	Specific national 3R strategic plan
Bangladesh	No	Yes
Cambodia	No	Yes
China	Yes	No
India	Yes	No
Indonesia	Yes	Yes
Malaysia	No	Yes
Philippines	No	Yes
Thailand	Yes	Yes
VietNam	No	Yes

Note: Update as of July 2010.

Convention on Climate Change, adopted in 1992, and now with 194 Parties, aims to "achieve, in accordance with the relevant provisions of the convention, stabilization of greenhouse gas concentrations in the atmosphere at a level that would prevent dangerous anthropogenic interference with the climate system." The Parties are committed to formulate, implement, publish, and regularly update national programs that list their measures to mitigate climate change. Such policies and measures should be appropriate for the specific conditions of each Party and should be integrated with their national development program (UN, 1992).

Some developing Asian countries, specifically China (NCCCC, 2007), India (PMCCC, 2008), Indonesia (MENLH, 2007), and Thailand (ONEP, 2008), recognize the potential GHG emissions reduction benefits of the 3Rs and have included the 3Rs as a climate change mitigation policy in their national climate change action plan (Table 2).

A review of these national climate change action plans reveals that countries tend to promote the 3Rs for material resource recovery, waste-to-energy conversion, and composting as mitigation actions. China, India, and Thailand clearly indicated the promotion of urban organic waste composting as a mitigation strategy. Further, China and India emphasized the promotion of anaerobic digestion of organic waste for energy generation and utilizing effluent from this treatment for soil improvement (Sang-Arun and Bengtsson, 2009).

3.1. *The 3Rs in national solid waste management strategies*

Some developing Asian countries have witnessed the benefits of the 3Rs for solid waste management and have thus included the 3Rs in their national solid waste management strategies. For instance, India indicated the 3Rs for solid waste management in the National Environment Policy (MOEF India, 2006); the Philippines announced the Ecological Solid Waste Management Act of 2000 to promote the 3Rs (IGES *et al.*, 2009); and China announced the Circular Economy Law in 2008, which entered into force in 2009 — this law clearly notified the 3Rs for economic development, resource utilization, and environmental conservation and rehabilitation.

In addition Bangladesh, Cambodia, Indonesia, Malaysia, the Philippines, Thailand, and Vietnam (Table 2) are now drafting specific national 3Rs strategic plans. The Philippines will focus its national 3Rs strategic plan on improving the working conditions and livelihoods of the informal sector, which makes an important contribution to waste recovery but in doing so endures substandard and unhealthy working conditions (IGES *et al.*, 2009).

This regional progression in the development of national 3Rs strategic plans is a result of ongoing international support provided by the MOEJ, the United Nations Centre for Regional Development (UNCRD), the Institute for Global Environmental Strategies (IGES), and various other international organizations and aid agencies.

3.2. *3Rs implementation at the local level*

The 3Rs were being implemented in many cities in the region before national governments included the 3Rs in their national strategies. Typically, the 3R activities were initiated by NGOs. The replication or extension of these activities was often limited, though when governments provided support scaling up was achieved. Several examples of 3Rs implementation at the local level in developing Asia are presented as follows.

In Bangladesh, an urban composting project was initiated by the NGO Waste Concern. Later on, the composting activity was scaled up with support from both local and national governments. Extension of composting in Bangladesh was successful due to the ongoing efforts of the NGO, an official approval on quality of compost for agriculture, policy support from the Ministry of Agriculture, and with a sales agreement with a fertilizer company (Zurbrügg *et al.*, 2004).

In Cambodia, the 3R activities were initiated by two NGOs: (i) urban composting at Phnom Penh was initiated by the Cambodian Education and Waste Management Organization (COMPED); (ii) recycling of inorganic waste was initiated by the Community Sanitation and Recycling Organization (CSARO). However, the extension of their activities was slow because local governments were unable to provide support because of financial constraints.

In Indonesia, a community-based composting project was initiated by an NGO in a small community of Surabaya city in 2004. Later on, the activity was promoted citywide and resulted in reduction of waste for disposal by 13% in 2007 and 23% in 2008. This successful implementation was due to active co-operation between the NGO, local governments, private companies, and international aid agencies. This practice is being transferred to other cities in Indonesia and also other countries (Maeda, 2009).

In the Philippines, the 3Rs implementation at the local level was driven by the Ecological Solid Waste Management Act (the Republic Act No. 9003) (Congress of the Philippines, 2000), which was enacted in 2001 and entered into force in 2002. This Act requires local governments to establish programs for waste minimization, reusing, recycling, and composting. As a result, 2361 materials' recovery facilities were established serving 6.3% of all barangays (smallest local government unit of the Philippines) in the country. However, recycling still heavily relied on the informal sector that helped to increase the recycling rate from 6% in 1997 to 25% in 2007 (Aguinaldo, 2008).

In Thailand, 3Rs implementation was initiated by various organizations including national government, local governments, NGOs, and universities. Economic incentives are a major driving factor promoting recycling activities in Thailand. Examples include deposition of recyclable wastes at a community waste bank[1], and exchanging recyclable waste for food. A private company, Wongpanit, is actively promoting separation of recyclable waste for sale. They expanded their activity nationwide through a franchise system of waste buyer shops.

The above examples illustrate the fact that at the local level, most 3R activities are the composting of urban organic waste and segregating recyclable waste for sale, regardless of the fact that the preferable order of the 3Rs is reduce before reuse before recycle. The wide promotion of recycling is partly due to the obvious benefits generated from recycling activities, such as income generation for residents and improving the living conditions of waste pickers. In addition, recycling seems not to interrupt business development and economic growth. Another observation is that the 3Rs implementation requires active cooperation from various stakeholders including the project initiator, residents, governments, private enterprise, NGOs, and sometime international aid agencies.

4. Case Study: 3Rs Implementation for Sustainable Solid Waste Management in Phitsanulok Municipality, Thailand

Phitsanulok Municipality is a medium size city located in northern Thailand. The Municipality covers an area of 18.26 square kilometers. Its permanent, official population is approximately 90,000.

Owing to rapid economic growth and the improvement of the waste collection service, the Municipality was confronted with a rapid increase in the volume of waste for disposal. From 1993–1996, waste at the Municipality dump site increased from 50 to 142 ton/day. The Municipality could not handle the waste properly. The waste dumpsites generated a foul smell, were an eyesore, led to housefly outbreaks, and resulted in leachate contamination of the cultivation areas of the local residents in the vicinity. Because of these problems the local residents forced the Municipality to close several dumpsites.

In response to the aforementioned problems, the Municipality gradually upgraded the disposal sites from open dumping and burning to engineered landfill

[1]A waste bank is a recyclable waste storage facility established by schools or communities. Members of the waste bank deposit recyclable wastes and save the money they earn from selling the recyclable waste to the bank. The bank reduces the amount of time the buyers spend on door-to-door waste collection. Students and residents are able to sell recyclable waste regularly.

and then to sanitary landfill. However, after few years the capacity of each dumpsite was exceeded and the Municipality needed to find new and larger locations. The Municipality could not afford to buy land in the vicinity of the town. Therefore, the distance between the disposal site and the Municipality increased, making the transportation of waste expensive.

In 1999, the Municipality aimed to reduce the quantity of waste for disposal by introducing the 3Rs and pretreatment of waste before disposal. The Municipality expected that the 3Rs would reduce solid waste generation at the consumption stage and enhance segregation of reusable and recyclable waste before they are discarded. Therefore, waste transported to the disposal site would decrease and the lifetime of the landfill could be extended. The Municipality also expected that the segregation of recyclable waste could generate income and business opportunities for waste pickers and residents.

The Municipality introduced various activities to promote the 3Rs in both mid-stream (consumption stage — waste generation) and downstream (final disposal site). At the consumption stage, the Municipality aimed to avoid waste generation by promoting use of biodegradable and reusable materials, encouraging waste separation for sale, and supporting composting at household and community levels. At the post-consumption stage, the Municipality aimed to improve final disposal by the pretreatment of waste through MBT prior to landfill.

As a result of these efforts, the quantity of waste transported to the disposal site decreased sharply from 142 ton/day in 1996 to 82 ton/day in 1999, a volume which was maintained thereafter (Phitsanulok Municipality, 2005).

Examples of the 3R activities in Phitsanulok Municipality are explained as follows:

(*a*) *Identify waste composition and waste management problems:* At the beginning of project implementation, the Municipality conducted an investigation into institutional capacities (personnel, budget, communication, etc), waste composition, residents' waste disposal behavior, and the municipal waste management system.

After investigating waste composition, the Municipality divided waste into three categories: organic waste, sellable waste (recyclable waste that has market value), disposal waste (waste that can neither be recycled nor has a market value). It found that the proportion of these categories were 40%, 40%, and 20%, respectively. The Municipality, therefore, aimed to reduce the volumes of organic and recyclable waste transported to the final disposal site by promoting decentralized composting and separation of recyclable waste for sale.

The institutional management of the Municipality shifted to an integrated form of organizational management (Phitsanulok Municipality, 2005) and waste management became a priority policy to support economic development. Several departments were required to accommodate waste management in their action plan: the Department of Social Welfare is now responsible for promoting community-based solid waste management; the Department of Finance is responsible for increasing the coverage of waste fee collection; the Department of Technical Services is responsible for disseminating information on waste and environment to the public; and the Department of Education is responsible for integrating solid waste management into the curriculum, facilitating schools to establish waste banks, and encouraging students to practice waste separation in their households.

The Municipality also reviewed the annual budget for solid waste management. It reduced the budget for the automatic street sweepers, which are much more expensive than labor for the same task. This helped reduce the budget for street sweeping by 70% and provided jobs for 97 residents (Phitsanulok Municipality, 2005).

(*b*) *Educating residents and encouraging self-responsibility for environmental conservation*: The Municipality understood that public participation is very important to reduce the burden of waste treatment. However, most residents did not understand their role in supporting the Municipality in waste management, as was evident by the fact that less than 50% of residents paid for the waste collection service. The Municipality thus introduced an education program on waste and the "polluter pays principle." The Municipality introduced a door-to-door strategy for face-to-face communication and for enhancing two-way communication between the Municipality and residents. University students were involved as volunteers in this activity (Phitsanulok Municipality, 2005).

Several campaigns were conducted to increase the awareness of residents on waste reduction and environmental conservation practices, for instance, reducing plastic waste generation by promoting the use of reusable shopping bags, use of rattan baskets for shopping, and use of tiffin carriers (i.e., a multi-layers Thai style lunch box or food carrier). In addition, a campaign to encourage residents to discard waste in a waste bin rather than on the street was introduced. This campaign reduced the budget for street sweeping by one-third (Phitsanulok Municipality, 2005).

The Municipality, in collaboration with the waste buyer company Wongpanit, provided training to waste pickers on health hazards and environmental problems caused by solid waste. Waste pickers that passed the training program were recognized as "volunteers for environmental conservation" for their contribution to separating recyclable waste from the waste stream. Residents were also encouraged to separate waste for sale or donation. Higher income residents also became involved in this activity as they too realized their role in environmental conservation. This activity helped increase the income, improve working and living conditions, and raise the social status of the waste pickers.

(*c*) *Enhancing waste recycling through public private partnerships*: In order to sustain the activity, the Municipality involved the private sector in the business of recycling. Private businesses played an important role by buying recyclable waste and providing information relevant to waste trading, such as specifications of sellable waste and ways to improve the quality of waste to earn a price premium.

Waste donation events were organized by both the Municipality and the businesses buying waste. The income from these events was used for religious activities or for charity. Such events are quite successful in Thailand where people believed in the virtue of donation. All groups of residents tended to participate in these events.

(*d*) *Promoting community-based solid waste management*: The Municipality employed a participatory approach to reduce the volume of waste for disposal. It provided training workshops on integrated solid waste management for residents, who were encouraged to decide the most suitable approach for their communities, which normally included composting (Fig. 2) and a waste bank (Fig. 3). These

Fig. 2. A community composting at Phitsanulok Municipality.

Fig. 3. A community waste bank at Phitsanulok Municipality.

strategies also helped reduce the Municipality expenses for waste collection, transportation, and treatment at the disposal site.

(*e*) *Promoting decentralized composting*: In order to reduce the contamination of organic waste by other substances, the Municipality encouraged decentralized composting at home or at community level. The Municipality provided technical training, composting equipment, and marketing advice to interested residents and communities. In general, the compost was used for cultivation by the household themselves. In some cases, it was sold to domestic customers.

(*f*) *Improving final disposal*: Owing to the difficulty of landfill siting, the Municipality introduced MBT prior to landfill. Pretreatment by MBT has the potential to efficiently reduce the volume of waste for landfill and to produce a high heating value waste suitable for energy recovery processes (Garg *et al.*, 2007).

After implementation of the MBT Project in Phitsanulok (Fig. 4), an onsite experiment provided evidence of waste reduction after MBT for 5 and 9 months — the quantity of waste treated by MBT decreased by 64% (Fig. 5). The treated waste consisted of 14% small size waste (pieces smaller than 40 mm), 7% medium size waste (pieces from 10–40 mm), and 79% large size waste (pieces greater than 40 mm). The composition and quality of waste after MBT for 5 and 9 months were not so different (Tungtakanpoung, 2006). The analysis indicated that MBT would extend the lifetime of the landfill site of Phitsanulok Municipality from 16 to 44 years.

The small size fraction of the treated waste consisted of almost 100% compost, whereas the medium size fraction consisted of 60–65% compost. These materials could be applied for soil improvement when they contained only low amounts of heavy metal. In a laboratory test, the small and medium size fractions met compost

Fig. 4. Implementation of mechanical biological pretreatment at Phitsanulok Municipality.

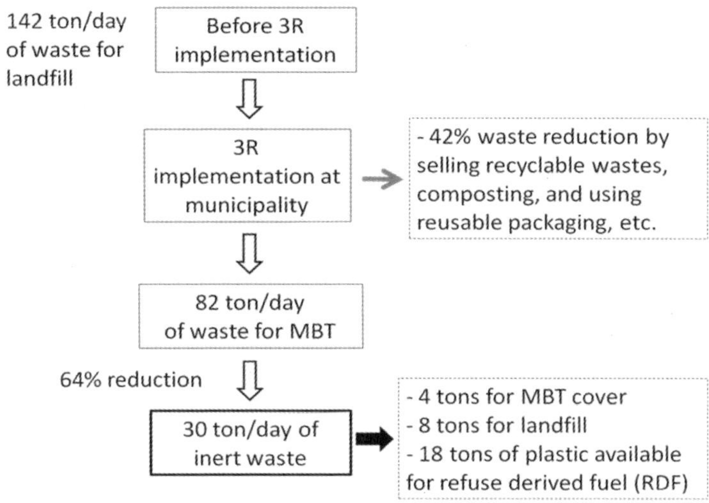

Fig. 5. Waste reduction through 3R practices in Phitsanulok Municipality.

standards for amounts of heavy metals, except for arsenic (As) (Tungtakanpoung, 2006). The material was subsequently used as a top cover for the biological process (composting) of MBT.

The large size fraction of the MBT waste consists of 72–80% plastics. This waste had a high calorific value equivalent to that of diesel (Tungtakanpoung, 2006). Smith *et al.* (2001) state that this type of waste can substitute coal and fossil fuels in energy intensive industries such as power stations and cement kilns. In addition, such waste has potential to be converted to liquid fuel using a technology that was successfully developed in Thailand (Siamrat, 2008).

In addition, the MBT method applied by the Municipality can reduce methane emissions from the landfill as it employs an aerobic windrow composting technique for the biological process. The Intergovernmental Panel on Climate Change (IPCC) 2006 Guidelines for National Greenhouse Gas Inventories, Volume 5: Waste states that methane gas emission from landfill of inert waste after MBT is 95% less than landfill of untreated waste (IPCC, 2006).

From the onsite experiment of MBT by Phitsanulok Municipality, it was estimated that 27 tons of organic waste was delivered for MBT each day, which accounted for 33% of the total waste arriving at the landfill. The greenhouse gas emissions reduction through the MBT practice at Phitsanulok can be estimated using the mass balanced approach explained in the 2006 IPCC Guidelines. We used three assumptions in our calculations. First, we assume that most organic waste is food waste; second, the landfill condition without MBT is a managed anaerobic landfill; third, emission from the biological processes of the MBT is equivalent to the average emission of composting provided by the IPCC. Using these assumptions and the mass balance approach, we estimate that the MBT could avoid emissions of 16.5 ton CO_2 equivalent per day.

5. Lessons Learnt from the Phitsanulok Model

To improve solid waste management, Phitsanulok Municipality introduced various activities including integrated institutional management, public awareness raising, promoting public participation, encouraging waste separation at the source, supporting recycling businesses, upgrading of the status of waste pickers, and improving final disposal practice. Through these activities, the Municipality was able to reduce waste sent to the landfill, by 42%. After arriving at the landfill the volume of waste was further reduced by 64% through MBT. These combined efforts achieved an 80% reduction of waste at the landfill. In addition, if the Municipality is able to sell plastic waste fragments as refuse derived fuel (RDF), a total reduction of 92% could be achieved. There is potential for even greater reduction, by, for example, the Municipality increasing composting and/or reducing the arsenic contamination of the organic fraction from the MBT.

Phitsanulok Municipality succeeded in promoting the 3Rs for sustainable solid waste management. Its achievements appear sustainable with the successes now spanning a decade and observations suggesting that the positive practices and trends will continue. Our analysis found several keys to this long-term success, which included changing the attitude of the Municipality, sustained leadership of the mayor, sharing responsibility and benefits between the Municipality and other stakeholders, and continuous learning by doing including learning from failures.

5.1. *Changing attitudes*

Local governments in the Asia-Pacific region are basically familiar with a top-down approach to project implementation for solid waste management. In addition, their policies typically focus on improving the final disposal site. As a consequence, local governments play a central role in collecting and disposing waste that has been generated by residents.

The Phitsanulok model can be used as positive example to encourage local governments to reconsider this paradigm of solid waste management and to move from focusing on upgrading landfill sites to a midstream-to-downstream waste management approach. Further, the model can encourage local governments to move from a top-down waste management approach to a participatory approach that involve various stakeholders, particularly residents and the private sector, as well as integrated strategies that combine solid waste management with economic development and livelihood creation.

5.2. *Continuous leadership and policy implementation*

The ongoing and strong leadership of the mayor is a driving force that encourages government officials to pay attention to 3Rs implementation. The mayor and her team have the support of the residents and were reelected over the past decade, thereby enabling continued high-level support for the 3Rs policy.

5.3. *Sharing responsibility and benefits*

Generally, local governments in Thailand are not good at business and are not subjected to work for profit. The Phitsanulok case study shows that the involvement of recycling enterprises in the effort to reduce the volume of waste reaching the landfill can sustain recycling activities. Waste buyers benefited from the increase of quantity and quality of recyclable waste and the Municipality benefited through reduced expenses for waste collection and transportation and for the operation of the final disposal site.

The public–private partnership between Phitsanulok Municipality and the large waste buyer, Wongpanit, provides a good example of how solid waste management benefits and responsibilities can be shared. Wongpanit is responsible for providing a health and safe working environment for waste pickers and waste buyers involved in its business. Through this partnership, Phitsanulok Municipality was able to reduce its budget for waste collection, transportation, and disposal because the private sector was now helping to segregate the recyclable waste that contributed to more than 40% of the waste stream. The private sector profited and residents could earn extra income from selling recyclable waste as the numbers of door-to-door waste buyers increased and because of the existence of waste banks in some communities. The unpleasant scene of waste scavenging from public waste bins disappeared.

The co-operation between the Phitsanulok Municipality and the private sector also changed the social attitude about the waste recycling industry. The activities under this cooperation led to an upgrading of the social status of waste pickers and waste buyers from the “dirty poor” to “environmental volunteers.” Moreover, the residents themselves developed the habit of separating waste for sale as part of their social responsibility.

5.4. *Improving practices at the landfill site*

For improvement of the final disposal site, MBT is appropriate for the pretreatment of municipal solid waste that consists of high organic materials. This practice could reduce waste volume by almost two thirds after the biological treatment was completed. It could also avoid methane emission that is generated by degradation of organic waste under anaerobic conditions (Smith *et al.*, 2001). However, this practice is not replicated in other cities of Thailand. A major barrier is that the capital investment and operation costs for MBT over the short term are higher than that of an engineered landfill.

In Thailand, implementation of MBT for pretreatment of waste would be more attractive to municipalities if they could sell the plastic waste that is segregated through the MBT process as an RDF. In the case of Phitsanulok Municipality, negotiation between the Municipality and industry on this issue failed. Even though the plastic waste is used as RDF, the industry proposed to receive it as “waste for treatment.” Therefore, the industry requested the Municipality to pay for the waste treatment service. Another proposal from the industry is that the Municipality offers the plastic waste to the industry without charge and that the Municipality covers the cost of transportation.

5.5. *Learning by doing*

Phitsanulok Municipality received advice from experts dispatched by the German Agency for Technical Cooperation (GTZ) from 1999–2005. The municipal officers learned from these international experts during these six years.

In addition, the Municipality learned from actual implementation. It regularly identified problems and looked for better solutions. Over the decade of the 3Rs implementation, some activities declined after the Municipality terminated the projects. Some activities were continued through the provision of economic incentives and some new activities were introduced to replace unsuccessful ones.

The activities that tend to decline are the community-based composting projects. Some composting groups were able to produce compost for sale in the market but in some cases their composting activities declined due to the changing of the community leader.

Based on this experience of a decline in community-based composting, the Phitsanulok Municipality is now planning a centralized composting system. A private company will be contracted to operate this facility. It is expected that the Municipality will be able to increase the recycling rate of organic waste as well as reduce waste for MBT through this new strategy.

Phitsanulok Municipality has also been able to play an important role as a national expert on 3Rs implementation and integrated solid waste management. The Municipality has shared the lessons that it learned with other cities in Thailand through presentations, workshops, training, and field visits. In turn, the Municipality is also keen to learn from others in seeking solutions to the existing challenges it faces, such as concluding a business agreement on RDF and scaling up the composting activity.

6. Conclusions

Rapid economic development combined with population growth is driving a rapid increase in solid waste generation in developing Asia. Finding sustainable, socially acceptable solutions is critical. The typical solution to solid waste problems is based on end-of-pipe thinking, i.e., landfill disposal. From a sustainability perspective, this downstream solution mindset has serious shortcomings. This chapter argues that effective and sustainable approaches to waste issues must be more holistic and place emphasis on waste reduction, product reuse, materials recycling, and energy recovery. Such measures can be introduced at different stages of the waste chain, including the consumption stage (midstream) and the disposal stage (downstream). Participatory and integrated approaches are needed.

Over the last few years, the importance of the 3Rs (reduce, reuse, and recycle) for sustainability development has gradually become recognized in national policies.

However, in most countries, although there are ambitious objectives laid down in the policy documents and legislations, regulations and supporting mechanisms are largely lacking and implementation is therefore still weak.

The successful implementation of the 3Rs in Phitsanulok Municipality, as described in this chapter, provides evidence that the 3Rs is applicable and affordable to local governments of small- to medium-sized cities throughout developing Asia. This case, together with other success stories, provides some clues to effective implementation. One of the crucial factors is the role taken by the local government. Traditionally, municipalities have either tried to carry out the waste collection and treatment by themselves — through the waste department — or outsourced these tasks to private contractors. However, a more dynamic role is needed for a successful implementation of the 3Rs. In such a role, the local government takes a more integrated approach and assigns key functions — such as education, waste collection, and recycling business promotion — to different parts of the local administration. The government also engages with various stakeholders — including residents, recycling businesses, and waste buyers — and promotes partnerships. This role is different from the one traditionally taken by local authorities, where enforcement of regulations is emphasized.

However, although very significant achievements can be made by local governments through their own efforts, there is also a need for outside support — especially for capacity development and training, and in some cases for financing. Some parts of a sustainable waste management system require substantial investments but cannot generate enough income to be commercially viable. One example is the MBT pretreatment, which was shown to be an important component of the system in Phitsanulok, but where some financial support was needed.

The chapter also pointed out that waste pickers who are among the most vulnerable group of residents in urban areas can significantly contribute to reduce the disposal of recyclable waste at the disposal site. This group can make important contributions toward sustainable waste management, which benefits the whole Municipality. Working with waste pickers — recognizing the value of their work — could help local governments to reduce costs for waste transportation and disposal, as well as increasing the income, and improving the living conditions and social status of the waste pickers.

A paradigm-shift is required to meet the solid waste challenge that Asia now faces. The experience of Phitsanulok Municipality suggests that this will not be easy, but that it is possible.

7. Acknowledgment

The authors are very grateful to Dr. Suthi Hantrakul, deputy mayor of Phitsanulok Municipality, for sharing data and providing insightful comments on this manuscript.

Abbreviation

As: arsenic
COMPED: Cambodian Education and Waste Management Organization
CSARO: Community Sanitation and Recycling Organization
GHG: greenhouse gas
GTZ: German Agency for Technical Cooperation
IGES: Institute for Global Environmental Strategies
IPCC: Intergovernmental Panel on Climate Change
JICA: Japan International Cooperation Agency
MBT: mechanical biological treatment
MDGs: millennium development goals
MOEF: Ministry of Environment and Forests, India
MOEJ: Ministry of Environment, Japan
MENLH: State Ministry for the Environment, Republic of Indonesia
NCCCC: National Coordination Committee on Climate Change, China
NGOs: non-governmental organizations
ONEP: Office of Natural Resources and Environmental Policy and Planning, Thailand
PMCCC: Prime Minister's Council on Climate Change, India
RDF: refuse derived fuel
3Rs: reduce, reuse, recycle
UNCRD: United Nations Centre for Regional Development
UNEP: United Nations Environment Programme
UNESCAP: United Nations Economic and Social Commission for Asia and the Pacific
UNFCCC: United Nations Framework Convention on Climate Change

References

Aguinaldo, E.C., 2008. National and local initiatives on solid waste management and implementation of 3Rs in the Philippines. Presentation at the *ADB Urban Day Conference on "Environment and Livable Cities"* held on 8 September 2008, Manila, 5p.

Ahmed, S.A. and Ali, M., 2004. Partnerships for solid waste management in developing countries: Linking theories to realities. *Habitat International* **28**, 467–479.

Bengtsson, M., Sang-Arun, J., Tsydenova, O. and Maeda, T., 2008. Urban organic waste — from hazard to resource. In: *IGES, Climate Change Policies in the Asia-Pacific: Re-uniting Climate Change and Sustainable Development*, IGES White Paper 2, Hayama, pp. 133–158.

Chen, C. and Greene, N., 2003. *Is Landfill Gas Green Energy*? Natural Resources Defense Council, New York, 72p.

Congress of the Philippines, 2000. *Republic Act No. 9003: An Act Providing for an Ecological Solid Waste Management Program, Creating the Necessary Institutional Mechanisms and Incentives, Declaring Certain Acts Prohibited and Providing Penalties, Appropriating Funds Therefore, and for Other Purposes.* Metro Manila, 24 July 2000, 45p.

Favoino, E. and Hogg, D., 2008. The potential role of compost in reducing greenhouse gases. *Waste Management & Research* **26**(1), 61–69.

Garg, A., Smith, R., Longhurst, P.J., Pollard, S.J.T., Simms, N. and Hill, D., 2007. Comparative evaluation of SRF and RDF co-combustion with coal in a fluidised bed combustor. In: *Proceedings of the Eleventh International Waste Management and Landfill Symposium*, S. Margherita di Pula, Cagliari, Italy, 1–5 October 2007, Paper No. 411.

Hsu, S.H., 2006. NIMBY opposition and solid waste incinerator siting in democratizing Taiwan. *The Social Science Journal* **43**(3), 453–459.

IGES, UNCRD, and AIT/UNEP.RRCAP, 2009. *National 3R Strategy Development* — A Progress Report on Seven Countries in Asia from 2005 to 2009, 132p.

IPCC, 2006. *2006 IPCC Guidelines for National Greenhouse Gas Inventories, Volume 5: Waste.*

JICA, 2005. *Supporting Capacity Development for Solid Waste Management in Developing Countries — Towards Improving Solid Waste Management Capacity of Entire Society*, 218p.

Maeda, T., 2009. *Reducing Waste Through the Promotion of Composting and Active Involvement of Various Stakeholders: Replicating Surabaya's Solid Waste Management Model. IGES Policy Brief No. 9*, 12p.

Makiya, K., Onogawa, K., Mohanty, C.R.C., Shimizu, K., Shibata, T., Kimura, S., Mizushima, T., Coelho, S. and Miura, K., 2006. Cooperation between developed and developing countries in promotion of 3R. *RDD* **27**(1), 152–176.

Medina, M., 2000. Scavenger cooperatives in Asia and Latin America. *Resources, Conservation and Recycling* **31**(1), 51–69.

MENLH, 2007. *National Action Plan Addressing Climate Change.* State Ministry for the Environment, Republic of Indonesia, 101p.

MOEF India, 2006. *National Environment Policy 2006.* Ministry of Environment and Forests, India, 52p.

NCCCC, 2007. *China's National Climate Change Programme.* National Coordination Committee on Climate Change 62p.

ONEP, 2008. *Draft National Strategies for Climate Change Management 2008–2012.* Office of Natural Resources and Environmental Policy and Planning, 51p.

Phitsanulok Municipality, 2005. *Phitsanulok Model for Solid Waste Management.* Phitsanulok, 282p (in Thai).

PMCCC, 2008. *National Action Plan on Climate Change.* Prime Minister's Council on Climate Change, Government of India. 49p.

Puri, A., Kumar, M. and Johal E., 2008. Solid waste management in Jalandhar city and its impact on community health. *Indian Journal of Occupational and Environmental Medicine* **12**(2): 76–81.

Ramachandra, T.V. and Bachamanda, S., 2007. Environmental audit of municipal solid waste management. *Int. J. Environmental Technology and Management*, **7**(3/4), 369–391.

Sang-Arun, J., 2009. Climate benefits of improved organic waste management through the 3Rs approach in developing Asian countries. Presentation at *Inaugural Workshop of Asia Resource Circulation Policy Research* held on 9–10 November 2009, Tokyo. Available online at http://www.iges.or.jp/en/wmr/activity20091109.html.

Sang-Arun, J., 2010. Good practices of the 3Rs for organic waste management in Thailand, Laos and Cambodia. Presentation at the *Second Workshop of "Asia Resource Circulation Policy Research"* held on 15–16 March 2010, Tokyo. Available online at http://www.iges.or.jp/en/wmr/activity20100315.html.

Schübeler, P., Wehrle, K. and Christen, J., 1996. *Conceptual Framework for Municipal Solid Waste Management in Low-Income Countries.* Swiss Centre for Development Cooperation, p. 9–10.

Sheehan, B., 2000. Zero waste, recycling and climate change. *Grass Roots Recycling Network.* Available online at http://www.grrn.org/zerowaste/climate_change.html#4 (Accessed 13 September 2010).

Shelton, J.E., 1991. *Soil Facts: Using Municipal Solid Waste Compost, Publication AG-439-19,* North Carolina Cooperative Extension Service.

Siamrat, 2008. A successful story of Thai's innovation on converting plastic waste to fuel oil [Phumejai khon thai prae rup khya plastic pen namman samrej]. *Siamrat News* 2 July 2008 (in Thai).

Smith, A., Brown, K., Ogilvie, S., Rushton, K. and Bates, J., 2001. Waste management options and climate change. *Final Report to the European Commission,* DG Environment, 205p.

Tungtakanpoung, D., 2006. *Characteristics of Solid Waste after Mechanical Biological Treatment (MBT): Summary Report.* Environmental Research Centre, Naresuan University, 15p.

United Nations, 1992. *United Nations Framework Convention on Climate Change.* FCCC/INFORMAL/84 GE.05-62220 (E) 200705.

UNCRD, no date. 3R Initiative. Available online at http://www.uncrd.or.jp/env/spc/ (Accessed 12 August 2010).

UNEP, 2004. *State of Waste Management in South East Asia.* UNEP IETC and the ASEAN Secretariat, Available online at http://www.unep.or.jp/ietc/publications/spc/State_of_waste_Management/index.asp (Accessed on 12 August 2010).

UNESCAP, 2009. Review of progress, constraints and policy challenges with regard to the implementation of international, regional and national commitments: Waste management (hazardous and solid waste). *Regional Implementation Meeting for Asia and the Pacific Ahead of the Eighteenth Session of the Commission on Sustainable Development,* held on 30 November–1 December 2009, Bangkok, 20p.

UNFCCC, no date. *Essential Background: Feeling the Heat.* Available online at http://unfccc.int/essential_background/feeling_the_heat/items/2918.php (Accessed 13 September 2010).

Zurbrügg, C., Drescher, S., Rytz, I., Sinha, A.H.M.M. and Enayetullah, I., 2004. Decentralised composting in Bangladesh, a win–win situation for all stakeholders. *Resources, Conservation and Recycling* **43**: 281–292.

Chapter 27

SUSTAINABLE MUNICIPAL SOLID WASTE MANAGEMENT AND RECOVERY

Giuseppe D'Alessandro[a], Maria Grazia Gnoni[b] and Giovanni Mummolo[c]
[a]*ASM spa*
[b]*Dept. of Innovation Engineering, University of Salento (lecce), Italy*
mariagrazia.gnoni@unile.it
[c]*Dept. of Mechanical and Management Engineering, Polytechnic of Bari, Italy*

Sustainability analysis of municipal solid waste (MSW) management is usually a complex issue as different type of factors are usually involved in. Source reduction, generation, recovery, and disposal constitute interactive elements of an integrated MSW management systems. The chapter aims to support both technicians and public authorities in order to design a sustainable integrated MSW management system. Starting from literature review about the application of sustainability concepts in MSW management, a comparison between strategies for more sustainable MSW management has been compared in order to point out common features and differences. Collection services have been evaluated as the most critical activity in the both context. Next, integrated MSW management systems that are the operational way for applying sustainability concept in MSW management have been deeply analyzed. Finally, a reference model for designing sustainable collection and recovery services is detailed.

1. Introduction

Sustainability analysis of municipal solid waste (MSW) management is usually a critical issue, which has been analyzed by different points of view: traditionally, sustainability affects three dimensions, i.e., environmental, economic, and social aspects. The design and the management of an MSW system usually heavily affect all these three dimensions (McDougall, 2001; Morrissey and Browne, 2004).

Starting from an analysis based on a legislative approach, sustainability issue represents the fundamental of the European Directive on waste management: the recent new Directive (2008) has confirmed the focus on sustainability of waste plans and policies. By a strategic point of view, the European legislation applies the concept of sustainability by the well-known "waste hierarchy": the hierarchy supplies to Each Member State a strategic guideline for design and managing MSWs. The new hierarchy is defined as follows: prevention represents the first option for a sustainable waste management; next, preparing for reuse, recycling, other recovery (e.g., energy recovery) and disposal represent the subsequent steps, which have to be applied for managing one (i.e., prevention, reuse, recycling/composting, energy recovery, and disposal). The purpose is to supply a common but flexible reference model: therefore, Member States could design their own solutions that deliver the best overall environmental outcome according to their specific context. By an operational point of view, life-cycle thinking represents the basis for evaluating how specific waste streams have to be treated as the overall impacts of waste generation and management have to be analyzed. Depending on the characteristics of the specific cluster or area (such as climate, population density, etc.), MSW plan should address the preferable options to be considered in order to define a local optimal waste hierarchy.

Two main approaches proposed in the European Directive point out the concept of sustainability:

- The "self-sufficiency and proximity" concept: Each member state is responsible to define an integrated and adequate network for both waste disposal and recovery. This concept led to design the so-called Integrated Municipal Solid Waste Management (IMSWM) systems, which represents the current way for applying the European Directive at operational and local level.
- The "Extended Producer Responsibility (EPR)" concept: It involves both the reverse logistics and the design phase of such a product. EPR forces producers to supply information about their products for reuse and recycling; moreover, this concept could promote eco-design procedures for several products. According to this concept, Tanaka (2007) points out the importance of resource productivity as the most effective indicator of sustainability in waste management.

Thus, an IMSWM system will aim to use resources in the most effective way rather than to move up the waste management hierarchy. An IMSWM plan takes an overall approach and manages waste in an environmentally effective, economically affordable and socially acceptable way (McDougall, 2001). This concept also refers to sustainability concept where the three dimensions (economic, environmental, and social) have to be integrated. The strategic decision about the most sustainable set of options for managing waste depends on a variety of factors — such as the type of waste being disposed, transport distances and the availability of specific treatment, the public agreement, the development of secondary materials market, etc. — which are usually contrasting. Therefore, three fundamental concepts have to be evaluated for designing an IMSWM (Greyson, 2007) in accordance with sustainability concept: the definition of effective practices of pre-cycling to prevent waste, the application of the circular economic policy that guarantees a closed loop for each waste stream, and the definition of recycling insurance system to apply EPR by an economic point of view. These concepts are not independent as they are each other influenced; thus, their effects could be amplified by synergies: as an example, the recycling insurance could represent an economic instrument to support the application of circular economy concept.

This brief introduction about sustainability issues in MSW management has highlighted the high complexity, which usually characterize this topic. The aim of the chapter is to provide a critical analysis of these issues in order to supply an effective design of a sustainable MSW management system.

2. MSW Management: Main Issues and Strategies

MSW represents the most complex waste stream, as opposed to more homogeneous waste streams resulting from industrial activities, even if it is associated with the lowest quantity. At the European Level, MSWs represents only a low percentage (about 7% in the year 2006) of waste originated from human activities in the EU, as described in Fig. 1.

Complexity in MSW management is due to several factors, which varies from technical to social points of view. Each waste stream requires a specific treatment, which varies according to different variables (e.g., waste composition, collection services, legislation burdens, etc.). Main waste streams and alternative flows in MSW are depicted in Fig. 2.

According to the European context, the complexity is pointed out by recent data: starting from an average value of 524 kg of municipal waste in 2008 generated by each EU citizen, which is characterized by a quite constant trend, waste streams diverge quantitatively. According to waste treatment, average values show 40% of waste stream was landfilled, 20% incinerated, 23% recycled, and 17% composting. Thus, as previously reported, while sustainable development is the ultimate goal

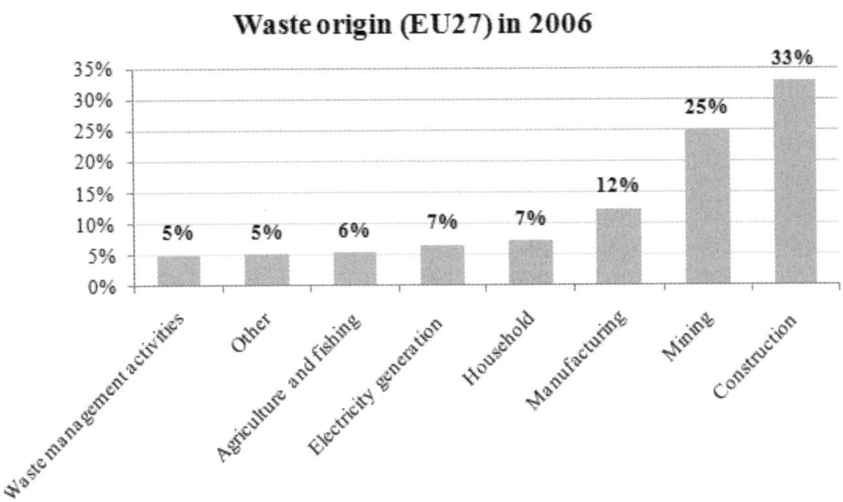

Fig. 1. EU waste composition in the year 2006 (*Data source*: Eurostat).

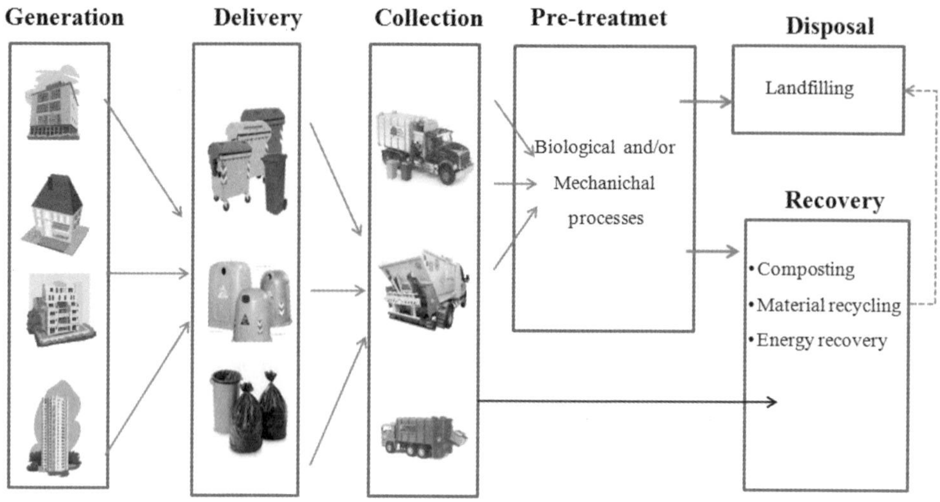

Fig. 2. Waste streams characterizing MSW systems.

of EU waste policy, there is no commonly accepted approach to accomplish this objective.

The problem of sustainability evaluation of MSW management systems represents a critical issue also for developing countries. Differently from developed countries, the main purpose of MSW management is more oriented to solve health problems than to optimize resources (both in the product design and in the end-of-life phase) involved in MSW as European approach. As defined by the European waste hierarchy, a sustainable MSW management system must tend to the recovery option as the most suitable one; integrated approach based on closed-loop concept is also a fundamental.

A similar point of view is spreading in the USA (Kollikkathara *et al.*, 2009): US municipal waste management systems are moving toward more sustainable systems. The analysis shows that progress in waste recycling and the diffusion of the concept of 3Rs (i.e., reduce, reuse, and recycle), cradle-to-cradle approach are supplying interesting results.

As reported in Table 1, Asian and African areas are characterized by quite similar average values of MSW generation rate (i.e., 0.28 and 0.29, respectively); otherwise, European area is characterized by a higher annual average generation rate, i.e., 0.57.

These values could be due not only to the difference in the economic level of these societies, but also to the effect of illegal waste dumping, which is a current practice in developing countries (Ngoc and Schnitzer, 2009; Pasang *et al.*, 2007; Rathi, 2006). Asase *et al.* (2009) point out that in developed counties, sustainability is guaranteed by an approach based on continuous improvement and integration; otherwise, MSW management in developed countries focuses on the optimization of waste management, which is based on resource conservation. This is mainly due

Table 1.

Region	MSW generation rate (tonnes/cap/yr)	MSW landfilled (%)	MSW incinerated (%)	MSW composted (%)	Other MSW management (%)*
Asia					
Eastern Asia	0.37	0.55	0.26	0.01	0.18
South-Central Asia	0.21	0.74	—	0.05	0.21
South-East Asia	0.27	0.59	0.09	0.05	0.27
Africa	0.29	0.69	—	—	0.31
Europe					
Eastern Europe	0.38	0.90	0.04	0.01	0.02
Northern Europe	0.64	0.47	0.24	0.08	0.20
Southern Europe	0.52	0.85	0.05	0.05	0.05
Western Europe	0.56	0.47	0.22	0.15	0.15
America					
Caribbean	0.49	0.83	0.02	—	0.15
Central America	0.21	0.5	—	—	0.50
South America	0.26	0.54	0.01	0.003	0.46
North America	0.65	0.58	0.06	0.06	0.29
Oceania	0.69	0.85	—	—	0.15

Source: IPCC, Guidelines for National Greenhouse Gas Inventories, 2006.
*This category includes data on recycling for some countries.

to the current inadequacy of such a service from collection to disposal. Thus, the focus is heavily on collection service design and optimization as they represent the most scarce phase in MSW management.

Damghani *et al.* (2008) focalize to the social dimensions of sustainability. After reviewing the current conditions of solid waste management in Teheran, they highlight the positive effects related to effective educational policies for improving awareness of both citizens and workers involved in MSW management. McBean *et al.* (2005) also highlight the contribution of the availability of effective waste treatment plants in order to create a more virtuous MSW management systems. El-Hamouz (2008) focuses firstly on economic dimension of sustainability: the author highlights as an optimization of collection activity costs at a social acceptable level could produce several benefits in other dimensions.

Another relevant difference that characterizes MSW management in developing countries is that informal actors are usually involved in. This is due to a lack of capacity in the public sector; consequently, the private sector has been involved to fill the gap. Non-government organizations and community-based organizations are usually the main subjects involved in waste management: they represent informal actors, which supply a structured contribute effectively to the public waste collection service (Fobil *et al.*, 2008).

This cooperation has revealed effective although a partnership is developed for waste treatments (Rathi, 2008) between public and private services. For a more sustainable MSW management system, cooperation among communities, the informal sector, the formal waste collectors, and public authorities is essential factor

(Imam *et al.*, 2008). Usually, private collection systems are mainly based on the improving collection service performance by applying waste separation at source, i.e., citizen deliver waste directly to waste pickers (Parizeau *et al.*, 2006). On the other hand, an effective application of these systems for MSW management also requires a high level of funding resources (Pasang *et al.*, 2007). A sustainable MSW management has to be cost neutral or to provide financial benefit to the community: it has to be noted that different from developed countries, citizen are rewarded directly when they deliver waste to private and/or public service. A common practice in MSW management in developing countries is that citizens are rewarded directly when they deliver waste to public or private systems. Therefore, the cost of the whole service could not fully covered by revenues, which are also affected by the reduced economic level of such a citizen; the application of the well-known "polluter pays" principle could be an effective tool for guarantee economic sustainability of such a service (Abu Qdais, 2007; Sujauddin *et al.*, 2008).

Finally, a common element between developed and developing countries is that funding and affordability remain among the major constraints and challenges in the design and the management of MSWs.

3. A Review Analysis of the Design of IMSWM Systems

Following analyzing the strategies for sustainable MSW management, operational model is described in this section in order to point out how global strategies could be applied in different contexts. It has to be noted that the design of an IMSWM system affects both long- and short-term decisions. Solid waste infrastructures (i.e., treatments and disposal plants) are generally characterized by slow dynamics: this is mainly due to the scale-of-operations, the related capital intensity and the size, and complexity, which affect these industrial infrastructures. Otherwise, these decisions could also influence short-term planning: as an example, collection service evaluation could be influenced by the availability of a recovery plant. Thus, a critical analysis has been carried out.

Several papers are facing with the environmental dimension of sustainability in IMSWM systems. The most widespread approach is based on the application of Life-Cycle Analysis (LCA) for designing and evaluating MSW management performances. A recent study of Del Borghi *et al.* (2009) reviews LCA applications in the waste management context. The analysis starts from models characterizing waste streams and technologies; guidelines and tools for evaluating waste treatment by life-cycle point of view have also been analyzed. LCA is usually applied as decision support tool for designing an IMSWM system by comparing design alternatives at local level (Buttol *et al.*, 2007; De Feo and Malvano, 2009; Diaz and Warith, 2006; Pisoni *et al.*, 2009). Moreover, it could be applied as a more strategic tool in order to evaluate flexible, yet protective, ways to conserve national resources by optimizing resource use and reducing waste (Thorneloe *et al.*, 2007).

As LCA analysis is characterized by high complexity, alternative approaches have been developed. Salvia *et al.*, (2002) and Lei and Wang (2008) proposed a model based on energy (i.e., the embodied energy in a system) analysis for evaluating the sustainability level of an IMSWM system. A model based on strategic environmental assessment is applied by Salhofer *et al.* (2007) for evaluating sustainability of waste management systems.

By a more strategic point of view, Verhoef *et al.* (2006) integrate environmental and engineering issues for a more effective MSW management. Traditional concepts derived from industrial ecology (i.e., closed material cycles, cascaded energy use, and flexible system configuration) have been integrated with technical parameters, which usually characterize recycling processes (i.e., grade and efficiency of recovery activities). This concept of integration of different issues has been deeply analyzed in the scientific literature: strategic guidance to be used in the waste sector should be integrated with quantitative considerations also related to cost and social dimension, as complementary information to the environmental aspects of sustainability in waste management (Mohan *et al.*, 2006).

Multi-criteria decision analysis (MCDA) has been widely applied for designing IMSW plans and systems. Recently, Chang and Davila (2007) describe a decision support system based on a multi-criteria model for designing regional waste plan. The model aims to optimize resources applied for managing wastes.

Morrissey and Browne (2004) propose a review analysis about the application of MCDA in waste management; a critical comparison is proposed between different approaches such as cost benefit analysis and LCA. Results show that main critical steps are the problem formulation and an effective involvement of all relevant stakeholders. Authors also highlight the high computational effort required for applying MCDA in waste management. Thus, an alternative approach aiming to reduce computational complexity is to define performance measurements for different dimensions of sustainability in MSW management. Several studies are focused on generation, delivery, and collection activities. Parfitt *et al.* (2001) propose a set of indicators derived from a comparative cluster analysis based on waste collected and recycled from different districts in UK. Results obtained could support the definition of such a policy for evaluating recycling infrastructure design and the effectiveness of recycling targets defined at National level. In analogous context, Grodzinska-Jurczak (2001) propose strategic indicators to promote sustainable waste management in Poland. Husaini *et al.* (2007) point out different types of technical indicators (e.g., the total quantity of waste reduced and diverted, the time period expected to achieve the set targets, etc.) in order to assess actual IMSWM performances. Cost indicators have not been evaluated due to lack of standardized data. Sufian and Bala (2007) evaluate by a system dynamics model the whole effectiveness of an IMSWM system. The model focuses on evaluating solid waste generation, collection capacity, and energy recovery from solid waste management. Read *et al.* (2009) show the integration of a monitoring system based on performance indicators with a communication tool based on focus group to improve the effectiveness of the whole controlling system.

4. A Proposal for Designing Sustainable Waste Collection and Recycling Services

Thus, according to the two analyses proposed in previous sections, a common result could be outlined: collection and recycling phase usually represent the most critical phases in an IMSWM system according to all three aspects of sustainability. By an economic point of view, collection services represent the highest source of costs (Gomes *et al.*, 2008; Reimer *et al.*, 2006); they also determine the entity of waste flows across each treatment, thus influencing environmental performances of both recovery and disposal processes. By a social point of view, instruments for both reducing waste production and improving recovery rate heavily influence the collection services design. Therefore, a proposal for designing more sustainable IMSWM systems is detailed as follows: the model focuses on collection and recovery services in order to optimize resource involved in MSW management. It has to be noted the analysis proposed in this chapter does not focus specifically on waste minimization: the attention is posed only on recycling and through repair or reuse; waste minimization through point of purchase requires different strategies and approaches (Tonglet *et al.*, 2004). Here are the main pillars:

- first, an *effective waste stream characterization analysis* that supplies reliable information about waste generated in terms of both quantity and type;
- second, an *optimal design of waste delivery and collection services* that has to be evaluated according to specific context conditions;
- finally, the definition of an *efficient control system* for both monitoring actual sustainability level and evaluating follow up policies for continuous improvement. An efficient control system requires the definition of key performance indicators together with economic instruments, such as effective charging system applied in MSW management. This last issue is outlined by Gellynck and Verhelst (2007): wastes rates could be reduced by applying of economic incentives together with an efficient waste collection service. Economic incentives could be applied only if charging systems is working.

Following, each of the three pillars is analyzed in detail.

4.1. *Waste characterization analysis*

MSW collection systems and recycling performances are mainly affected by waste flow composition in terms of weight distribution among their different basic classes of materials, such as organic (or bio-degradable) waste, and recyclable wastes (e.g., paper, glass, plastics, metals, wood). MSW data are essential for solid waste regulation, management, and planning at all levels of government from national

(Dahlén *et al.*, 2007; Magrinho *et al.*, 2006; Sujauddin *et al.*, 2008) to local levels (Hansen *et al.*, 2007). On the other hand, waste characterization often represents a critical activity: a lack of a standardized methodology together with a high local variability of values according to the specific context contributes to increase its complexity (den Boer *et al.*, 2010). Several type of parameters — e.g., from different housing patterns to climate and waste collection schemes — could heavily influence waste generation. An interesting review about different approaches for waste characterization is proposed by Dahlen and Lagerkvist (2010). A model based on intensive application of information and communication technology tools is recently described by Godfrey (2008) and Chowdhury (2009). The aim is to define a national waste information system as an effective tool for supporting different government entities (national, provincial, and local) involved in waste planning.

4.2. *Collection service design*

A high number of parameters and variables usually affect the design phase of the collection service design in IMSWM systems. A first variable that has to be evaluated is the *delivery service* type; two main typologies could be highlighted:

- *Household service or door-to-door services*: In door-to-door collection scheme, combinations of bins, racks, sacks, and bags are used, usually placed outdoors and/or indoors (Dahlen and Lagerkvist, 2010).
- *Drop-off based system*: This typology can be divided in kerbside and neighborhood (or bring systems) collection service. In the first option, a proximity or a low-density collection service could be applied; the first differs on bins dimension. Usually, the planning phase represents a critical activity in kerbside schemes (Woodard *et al.*, 2005): as a example introducing a higher number of collection points could increase the environmental impact (in terms of noise and visual impact); on the other hand, the distance between users and collection points should not increase aiming to reduce citizen collaboration (Bautista and Pereira, 2006; Salhofer *et al.*, 2007). In the second option, wastes have been taken by householders to civic amenity sites ("self-haul" wastes).

Another relevant variable is *the way of collecting waste*: separate or co-collection (or mixed) systems could be applied. In separate collection scheme, citizen have to put recycled and organic wastes into separate bins; mixed system does not require obligatory single-stream collection. Lastly, *the service frequency* represents a critical design variable (Gellynck and Verhelst, 2007; Wilson and Williams, 2007), daily, weekly, fortnightly are the most widespread options for collecting such a waste. A synthetic but complete comparison between three main types of collection schemes is proposed in Table 2: economic and environmental dimensions of sustainability have also been analyzed.

It has to be noted that one single scheme cannot deliver a universal solution for different kinds of wastes generated; the aim of a sustainable collection service

Table 2. Comparison between different types of collection services.

Type	Main features	Delivery system	Collection equipment	Accuracy	Recovery rate	Economic impact	Environmental Impact
Household collection (door-to-door)	Applied for terraced or single-residential house with external or internal stocking areas; high collection frequency required; usually manual bins picking applied	Dedicated waste bins delivered for each user or groups of users	Small household containers; small-sized collection vehicles	Very high	Very high	High	Very low
Kerbside collection Proximity collection	Medium- and high-density buildings; medium- to high-collection frequency; mixed collection (e.g., plastics with paper); manual and automatic bins picking	Street collection bins (one for 50–70 users)	Standard street containers; medium-sized collection vehicles	Medium	Medium	Medium	Medium
Low-density collection	High-density residential buildings; medium to low collection frequency; prevalence of automatic bins picking	Street collection bins (one for 70–200 users)	Large street containers; prevalence of dedicated equipped vehicles	Low	Low	Low	High

is to define the optimal concatenation of material management processes according to specific context features (Wilson and Williams, 2007). Thus, after evaluating critical variables, models proposed for optimizing collection and recycling services are critically analyzed following. Reimer *et al.* (2006) focus on cost analysis of collection services, which usually represents the dominant source. The model aims to optimize truck configurations, as well as optimal compartment sizes for recyclables collection according to different burdens. Gomes *et al.* (2008) compare, according to the economic dimensions, different scenarios for separate collection of organic fraction of MSW, which usually determines the highest impact on service design. Several authors apply analytical models for optimizing the design of integrated collection and recovering services by an engineering point of view.

Lin and Chen (2009) focus on evaluating the optimal number of recycling depots in regions with reduced access to recycling facilities; collection services represent burdens of the proposed model. Teixeira *et al.* (2004) and de Oliveira Simonetto and Borenstein (2007) analyze the vehicle planning problem: the first one proposes an heuristic model for the collection of urban recyclable waste; the second focuses to optimize waste flows in order to reduce land filling. A similar study is proposed by Ghose *et al.* (2008): authors apply GIS-based model to optimize waste collections. Johansson (2006) analyzes the effect of the availability of real-time information derived from waste containers to scheduling and routing policies in collection activities. A simulation-based model has been described by Villeneuve *et al.* (2009) in order to compare performances of integrated collection and recovery systems at a local basin level. The model integrates material balances, energy consumption and production, environmental emissions and costs.

According to socio-economic factors, the design of an integrated collection and recycling service largely depends on household participation (Lin, 2008; Suttibak and Nitivattananonl 2008; Troschinetz and Mihelcic, 2009; Vicente and Reis, 2007) both in developed and in developing countries (Kaseva *et al.*, 2002; Ketlogetswe and Mothudi, 2005). Several studies have demonstrated that different factors derived from social, cultural, and structural contexts could largely influence the effectiveness of recycling system (Clarke and Maantay, 2006; Martin *et al.* 2006; Shaw, 2008). The relationship between recycling attitudes and the design of collection services has been analyzed by Gonzalez-Torre *et al.* (2003), Shaw *et al.* (2007), Sidique *et al.* (2010) and Timlett and Williams (2009). Moreover, in the first study, authors highlight factors — such as the location of collection points, the difficulty of delivery items, and the citizen opinion about the collection service as critical points. Gonzalez-Torre and Adenso-Diaz (2005) study the connection determined by variations in the collection service frequency on the citizen recycling habit by analyzing a real case study from a Spanish region.

4.3. *Efficient control system design*

Different types of *performance indicators* have been proposed in the literature. Wang *et al.* (2008) propose a review analysis of issue about comparison of different organizations of collection services (i.e., formal and informal systems) in developing countries. Karagiannidis *et al.* (2004) point out a set of parameters for evaluating the efficiency and effectiveness of the waste collection process, which have been ranking by a multi-criteria analysis. Wada *et al.* (2009) apply LCA analysis for evaluating performances of collection and transportation alternative in Japan. Calabrò (2009) analyzes, by an environmental point of view, different combinations of separate collection scenarios and different treatments (landfilling and incinerations). Emissions derived from vehicles in collection and transportation activities have been neglected. An economic-based analysis is detailed in McLeod and Cherrett (2008), which compare different operational scenarios for household collections. By focusing on recovery activities, Harder *et al.* (2008) define an innovative approach for monitoring in a more effective way recycling performances achievable by a local authority. The approach is mainly based on evaluating the actual percentage of local waste that could be reasonably recycled. This value depends on the services level provided, the average composition of the local waste, the local population coverage for collection of any materials, and information about how much of each material stream is generally suitable for recycling.

As reported previously, the evaluation of the "optimal" charging system is critical for measuring performances of the MSW management system. Usually, different types could be applied in MSW management: flat, variable (based on weight and volume), or mixed type. The first one is a fixed annual fee; the second type evaluates a unit-pricing by the bag and/or weight-based fees. This option is based on the application of the "pay-as-you-throw (PAYT) mechanism," which could also contribute to reduce individual waste production (Dunne *et al.*, 2008). On the other hand, this type of charging system requires an reliable waste identification and an efficient measurement system of the generated waste and/or corresponding services (Bilitewski, 2008). Thus, if applied correctly, PAYT charge system could be more effective than flat rate (Dahlen and Lagerkvist, 2010; Ku *et al.*, 2009; Puig-Ventosa, 2008). Moreover, a strict relationship has been demonstrated between collection schemes and different types of charging systems according to Dahlén and Lagerkvist (2008).

5. Conclusions

Sustainability analysis of MSW management is usually a critical issue, which has been analyzed by different points of view: traditionally, sustainability affects three dimensions, i.e., environmental, economic, and social aspects. MSWs even if they could not represent the highest quantity of total waste generated require usually the highest computational effort. Thus, the design of sustainable MSW managements systems requires the evaluation of different parameters types starting from the specific features characterizing the local context to economic and technical factors. A context analysis about how sustainability concepts have been applied in MSW management is firstly proposed in the chapter. The analysis has compared strategies for MSW management in developing and developed countries: results have showed common approaches and differences. Results have showed that by an operational point of view, sustainability refers to an integrated MSW management system: each waste stream has to be treated according to a closed-loop approach in order to reduce overall environmental, economic, and social impacts. Next, a proposal for designing more sustainable MSW management system has been detailed based on three main concepts: an effective waste stream characterization analysis; an optimal design of waste delivery and collection services; the definition of an efficient control system for monitoring interventions. The proposed model is largely focused on both collection and recovery services as they are strictly interrelated and they affect heavily the sustainability level of the MSW management systems.

References

Abu Qdais, H.A., 2007. Techno-economic assessment of municipal solid waste management in Jordan. *Waste Management* **27**, 1666.

Asase, M., Yanful, E.K., Mensah, M., Stanford, J. and Amponsah, S., 2009. Comparison of municipal solid waste management systems in Canada and Ghana: A case study of the cities of London, Ontario, and Kumasi, Ghana. *Waste Management* **29**, 2779.

Bautista, J. and Pereira, J., 2006. Modeling the problem of locating collection areas for urban waste management. An application to the metropolitan area of Barcelona. *Omega* **34**, 617.

Bilitewski, A., 2008. From traditional to modern fee systems. *Waste Management* **28**, 2760.

Buttol, P., Masoni, P., Bonoli, A., Goldoni, S., Belladonna, V. and Cavazzuti, C., 2007. LCA of integrated MSW management systems: Case study of the Bologna District. *Waste Management* **27**, 1059.

Calabrò, P.S., 2009. Greenhouse gases emission from municipal waste management: The role of separate collection. *Waste Management* **29**, 2178.

Chowdhury, M., 2009. Searching quality data for municipal solid waste planning. *Waste Management* **29**, 2240.

Chang, N. and Davila, E., 2007. Minimax regret optimization analysis for a regional solid waste management system. *Waste Management* **27**, 820.

Clarke, M.J. and Maantay, J.A., 2006. Optimizing recycling in all of New York City's neighborhoods: Using GIS to develop the REAP index for improved recycling education, awareness, and participation. *Resources, Conservation and Recycling* **46**, 128.

Dahlén, L., Vukicevic, S., Meijer, J. and Lagerkvist, A., 2007. Comparison of different collection systems for sorted household waste in Sweden. *Waste Management* **27**, 1298.

Dahlén, L. and Lagerkvist, A., 2008. Methods for household waste composition studies. *Waste Management* **28**, 1100.

Dahlén, L. and Lagerkvist, A., 2010. Pay as you throw strengths and weaknesses of weight-based billing in household waste collection systems in Sweden. *Waste Management* **30**, 23.

Damghani, A.M., Savarypour, G., Zand, E. and Deihimfard, R., 2008. Municipal solid waste management in Tehran: Current practices, opportunities and challenges. *Waste Management* **28**, 929.

de Oliveira Simonetto and Borenstein, D., 2007. A decision support system for the operational planning of solid waste collection. *Waste Management* **27**, 1286.

Del Borghi, Gallo, M. and Del Borghi, M., 2009. A survey of life cycle approaches in waste management. *International Journal of Life Cycle Assessment* **14**, 597.

De Feo, G. and Malvano, C., 2009. The use of LCA in selecting the best MSW management system. *Waste Management* **29**, 1901.

den Boer, E., Jedrczak, A., Kowalski, Z., Kulczycka, J. and Szpadt, R., 2010. A review of municipal solid waste composition and quantities in Poland. *Waste Management* **30**, 369.

Diaz, R. and Warith, M., 2006. Life-cycle assessment of municipal solid wastes: Development of the WASTED model. *Waste Management* **26**, 886.

Dunne, L., Convery, F.J. and Gallagher, L., 2008. An investigation into waste charges in Ireland, with emphasis on public acceptability. *Waste Management* **28**, 2826.

El-Hamouz, A.M., 2008. Logistical management and private sector involvement in reducing the cost of municipal solid waste collection service in the Tubas are of the West Bank. *Waste Management* **28**, 260.

Fobil, J.N., Armah, N.A., Hogarh, J.N. and Carboo, D., 2008. The influence of institutions and organizations on urban waste collection systems: An analysis of waste collection system in Accra, Ghana (1985–2000). *Journal of Environmental Management* **86**, 262.

Gellynck, X. and Verhelst, P., 2007. Assessing instruments for mixed household solid waste collection services in the Flemish region of Belgium. *Resources, Conservation and Recycling* **49**, 372.

Ghose, M.K., Dikshit, A.K. and Sharma, S.K., 2008. A GIS based transportation model for solid waste disposal — A case study on Asansol municipality. *Waste Management* **26**, 1287.

Godfrey, L., 2008. Facilitating the improved management of waste in South Africa through a national waste information system. *Waste Management* **28**, 1660.

Gomes, A.P., Matos, M.A. and Carvalho, I.C., 2008. Separate collection of the biodegradable fraction of MSW: An economic assessment. *Waste Management* **28**, 1711.

Gonzalez-Torre, P.L. and Adenso-Diaz, B., 2005. Influence of distance on the motivation and frequency of household re cycling. *Waste Management* **25**.

Gonzalez-Torre, P.L., Adenso-Diaz, B. and Ruiz-Torres, A., 2003. Some comparative factors regarding recycling collection systems in regions of the USA and Europe. *Journal of Environmental Management* **69**, 129.

Greyson, J., 2007. An economic instrument for zero waste, economic growth and sustainability. *Journal of Cleaner Production* **15**, 1382.

Grodzinska-Jurczak, M., 2001. Management of industrial and municipal solid wastes in Poland. *Resources, Conservation and Recycling* **32**, 85.

Hansen, T.L., la Cour Jansen, J., Spliid, H., Davidsson, A. and Christensen, T.H., 2007. Composition of source-sorted municipal organic waste collected in Danish cities. *Waste Management* **27**, 510.

Harder, M.K., Stantzos, N., Woodard, R. and Read, A., 2008. Development of a new quality fair access best value performance indicator (BVPI) for recycling services. *Waste Management* **28**, 299.

Husaini, I.G., Garg, A., Kim, K.H., Marchant, J., Pollard, S.J.T. and Smith, R., 2007. European household waste management schemes: Their effectiveness and applicability in England. *Resources, Conservation and Recycling* **51**, 248.

Imam, B., Mohammed, D.C. and Wilson, C.R., Cheeseman, 2008. Solid waste management in Abuja, Nigeria. *Waste Management* **28**, 468.

Johansson, M., 2006. The effect of dynamic scheduling and routing in a solid waste management system. *Waste Management* **26**, 875.

Kaseva, M.E., Mbuligwe, S.E. and Kassenga, G., 2002. Recycling inorganic domestic solid wastes: Results from a pilot study in Dar es Salaam City, Tanzania. *Resources, Conservation and Recycling* **35**, 243–257.

Karagiannidis, A., Xirogiannopoulou, A., Perkoulidis, G. and Moussiopo, N., 2004. ULOS assessing the collection of urban solid wastes: A step towards municipality benchmarking. *Water, Air, and Soil Pollution* **4**, 397.

Ketlogetswe, A. and Mothudi, T.H., 2005. Botswana's environmental policy on recycling. *Resources, Conservation and Recycling* **44**.

Kollikkathara, N., Feng, H. and Stern, E., 2009. A purview of waste management evolution: Special emphasis on USA. *Waste Management* **29**, 974.

Ku, S., Yoo, S. and Kwak, S., 2009. Willingness to pay for improving the residential waste disposal system in Korea: A choice experiment study. *Environmental Management* **44**, 278.

Lei, K. and Wang, Z., 2008. Municipal wastes and their solar transformities: An energy synthesis for Macao. *Waste Management* **28**, 2522.

Lin, A., 2008. A model using home appliance ownership data to evaluate recycling policy performance. *Resources, Conservation and Recycling* **52**, 1322.

Lin, H. and Chen, G., 2009. Regional optimization model for locating supplemental recycling depots. *Waste Management* **29**, 1473.

Magrinho, A., Didelet, F. and Semiao, V., 2006. Municipal solid waste disposal in Portugal. *Waste Management* **26**, 1477.

Martin, M., Williams, I.D. and Clark, M., 2006. Social, cultural and structural influences on household waste recycling: A case study. *Resources, Conservation and Recycling*, **48**, 357.

McBean, E.A., del Rosso, E. and Roversi, F.A., 2005. Improvements in financing for sustainability in solid waste management. *Resources, Conservation and Recycling* **43**, 391–401.

McDougall, F.R., 2001. Life cycle inventory tools: Supporting the development of sustainable solid waste management systems. *Corporate Environmental Strategy* **8**(2).

McLeod, F. and Cherrett, T., 2008. Quantifying the transport impacts of domestic waste collection strategies. *Waste Management* **28**, 2271.

Metin, A., Erozturk, A. and Neyim, C., 2003. Solid waste management practices and review of recovery and recycling operations in Turkey. *Waste Management* **23**, 425.

Mohan, R., Spiby, J., Leonardi, G.S., Robins, A. and Jefferis, S., 2006. Sustainable waste management in the UK: The public health role. *Public Health* **120**, 908.

Morrissey, A.J. and Browne, J., 2004. Waste management models and their application to sustainable waste management. *Waste Management* **24**, 297.

Ngoc, U.N. and Schnitzer, H., 2009. Sustainable solutions for solid waste management in Southeast Asian countries. *Waste Management* **29**, 1982.

Parfitt, J.P., Lovett, A.A. and Sunnenberg, G., 2001. A classification of local authority waste collection and recycling strategies in England and Wales. *Resources, Conservation and Recycling* **32**, 239.

Parizeau, K., Maclaren, V. and Chanthy, L., 2006. Waste characterization as an element of waste management planning: Lessons learned from a study in Siem Reap, Cambodia. *Resources, Conservation and Recycling* **49**, 110.

Pasang, H., Moore, G.A. and Sitorus, G., 2007. Neighbourhood-based waste management: A solution for solid waste problems in Jakarta, Indonesia. *Waste Management* **27**, 1924.

Pisoni, B., Raccanelli, R., Dotelli, G., Botta, D. and Melià, P., 2009. Accounting for transportation impacts in the environmental assessment of waste management plans. *International Journal of Life Cycle Assessment* **14**, 248.

Puig-Ventosa, 2008. Charging systems and PAYT experiences for waste management in Spain. *Waste Management*, **28**, 2767.

Rathi, S., 2006. Alternative approaches for better municipal solid waste management in Mumbai, India. *Waste Management* **26**, 1192.

Read, M., Gregory, M.K. and Phillips, P.S., 2009. An evaluation of four key methods for monitoring household waste prevention campaigns in the UK. *Resources, Conservation and Recycling* **54**, 9.

Reimer, A., Sodhi, M. and Jayaraman, V., 2006. Truck sizing models for recyclables pick-up. *Computers & Industrial Engineering* **51**, 621–636.

Salhofer, S., Wassermann, G. and Binner, E., 2007. Strategic environmental assessment as an approach to assess waste management systems. Experiences from an Austrian case study. *Environmental Modelling & Software* **22**, 610.

Salhofer, S., Schneider, F. and Obersteiner, G., 2007. The ecological relevance of transport in waste disposal systems in Western Europe. *Waste Management* **27**, 47.

Salvia, M., Cosmi, C., Macchiato, M. and Mangiamele, L., 2002. Waste management system optimisation for Southern Italy with MARKAL model. *Resources, Conservation and Recycling* **34**, 91.

Shaw, P.J., Lyasa, J.K., Maynarda, S.J. and van Vugt, M., 2007. On the relationship between set-out rates and participation ratios as a tool for enhancement of kerbside household waste recycling. *Journal of Environmental Management* **83**, 34.

Shaw, P.J., 2008. Nearest neighbour effects in kerbside household waste recycling. *Resources, Conservation and Recycling* **52**, 775.

Sidique, S.F., Lupi, F. and Joshi, S.V., 2010. The effects of behavior and attitudes on drop-off recycling activities. *Resources, Conservation and Recycling* **54**, 163.

Sufian, M.A. and Bala, B.K., 2007. Modeling of urban solid waste management system: The case of Dhaka city. *Waste Management* **27**, 858.

Sujauddin, M., Huda, S.M.S. and Rafiqul Hoque, A.T.M., 2008. Household solid waste characteristics and management in Chittagong, Bangladesh. *Waste Management* **28**, 1688.

Suttibak, S. and Nitivattananon, V., 2008. Assessment of factors influencing the performance of solid waste recycling programs. *Resources, Conservation and Recycling* **53**, 45.

Tanaka, M., 2007. Waste management for a sustainable society. *Journal of Material Cycles and Waste Management* **9**, 2.

Teixeira, J., Antunes, A.P. and de Sousa, J.P., 2004. Recyclable waste collection planning — A case study. *European Journal of Operational Research* **158**, 543.

Timlett, R.E. and Williams, I.D., 2009. The impact of transient populations on recycling behaviour in a densely populated urban environment. *Resources, Conservation and Recycling* **53**, 498.

Thorneloe, S.A., Weitz, K. and Jambeck, J., 2007. Application of the US decision support tool for materials and waste management. *Waste Management* **27**, 1006.

Troschinetz, A.M. and Mihelcic, J.R., 2009. Sustainable recycling of municipal solid waste in developing countries. *Waste Management* **29**, 915.

Tonglet, M., Phillips, P.S. and Bates, M.P., 2004. Determining the drivers for householder pro-environmental behaviour: Waste minimization compared to recycling. *Resources, Conservation and Recycling*, **42** 27.

Verhoef, E.V., van Houwelingen, J.A., Dijkema, G.P.J. and Reuter, M.A., 2006. Industrial ecology and waste infrastructure development: A roadmap for the Dutch waste management system. *Technological Forecasting & Social Change* **73**, 302.

Vicente, P. and Reis, E., 2007. Segmenting households according to recycling attitudes in a Portuguese urban area. *Resources, Conservation and Recycling* **52**, 1.

Villeneuve, J., Michel, P., Fournet, D., Lafon, C., Menard, Y., Wavrer, P. and Guyonnet, D., 2009. Process-based analysis of waste management systems: A case study. *Waste Management* **29**, 2.

Wada, Y., Okumoto, T. and Wada, N., 2009. Evaluating household waste treatment systems with specific examination of collection and transportation processes. *Journal of Material Cycles and Waste Management* **1**, 11.

Wang, J., Han, L. and Li, S., 2008. The collection system for residential recyclables in communities in Haidian district, Beijing: A possible approach for China recycling. *Waste Management* **28**, 1672.

Wilson, C.D.H. and Williams, I.D., 2007. Kerbside collection: A case study from the north-west of England. *Resources, Conservation and Recycling* **52**, 381.

Woodard, R., Harder, M.K., Bench, M. and Philip, M., 2001. Evaluating the performance of a fortnightly collection of household waste separated into compostables, recyclates and refuse in the south of England. *Resources, Conservation and Recycling* **3**, 265.

Woodard, R., Bench, M. and Harder, M.K., 2005. The development of a UK kerbside scheme using known practice. *Journal of Environmental Management* **75**, 115.

Chapter 28

DEVELOPING ENVIRONMENTAL ACTION PLANS TOWARD SUSTAINABILITY: THE TAIWAN EXPERIENCE

P.C. Chiang
Graduate Institute of Environmental Engineering,
National Taiwan University, Taipei,
Taiwan, Republic of China
pcchiang@ntu.edu.tw

The objective of this chapter is intended to present the mission and strategies on three issues: sustainable energy, integrated watershed management, and eco-industrial park. The relative key performance indicators toward sustainability made by the Taiwan Government are also introduced. As suggested by Dr. Ma Ying-Jeou, President of Republic of China, not only we will take the lead in initiating national environmental policies, but also we will demand national policies be focused on the environment. The National Sustainable Development Policy Guidelines were drawn up by the National Council for Sustainable Development based on Agenda 21: Taiwan Sustainable Development Strategic Guidelines. The National Sustainable Development Policy Guidelines include an overall vision, basic principles, concepts, and local policies. Local governments also develop their own environmental action plans toward sustainability based on the National Sustainable Development Policy Guidelines. The best way to implement the above policy for all is to co-ordinate the efforts of all involved stakeholders — manufacturers, academic experts, government officials, environmental protection groups, community organizations, etc — while promoting the concepts of sustainable community.

1. Introduction

The natural environment of Taiwan is a vulnerable and highly isolated island ecosystem. It is vulnerable because its relatively small area and the correspondingly small population size of its native species mean that a single natural disaster could destroy enough of species to cause its extinction. Further, the limited land resources of an island country constitute an innate limitation on its development, so the most important issue for its sustainable development should be to make optimum use of such land resources as it possesses. With its limited land, dense population, shortage of natural resources, frequent natural disasters, and unique international political status, Taiwan's need to pursue sustainable development is more pressing than that of most other countries.

The National Council for Sustainable Development (NCSD) was established by the Executive Yuan on 23 August 1997. NCSD, in its initial stages, first established the Executive Yuan Global Environmental Climate Change Working Group in 1992. In 1994, the team was expanded and renamed the Executive Yuan Global Environmental Climate Change Policy Guidance Working Group. In December 2002, the president promulgated the "Basic Environment Act" through which Article 29 of the Act granted legal status to the NCSD. In addition, the council status has risen, and now the nation's premier and vice-premier serve as Chief Executive Officer (CEO) and vice-CEO of the council. Figure 1 presents the

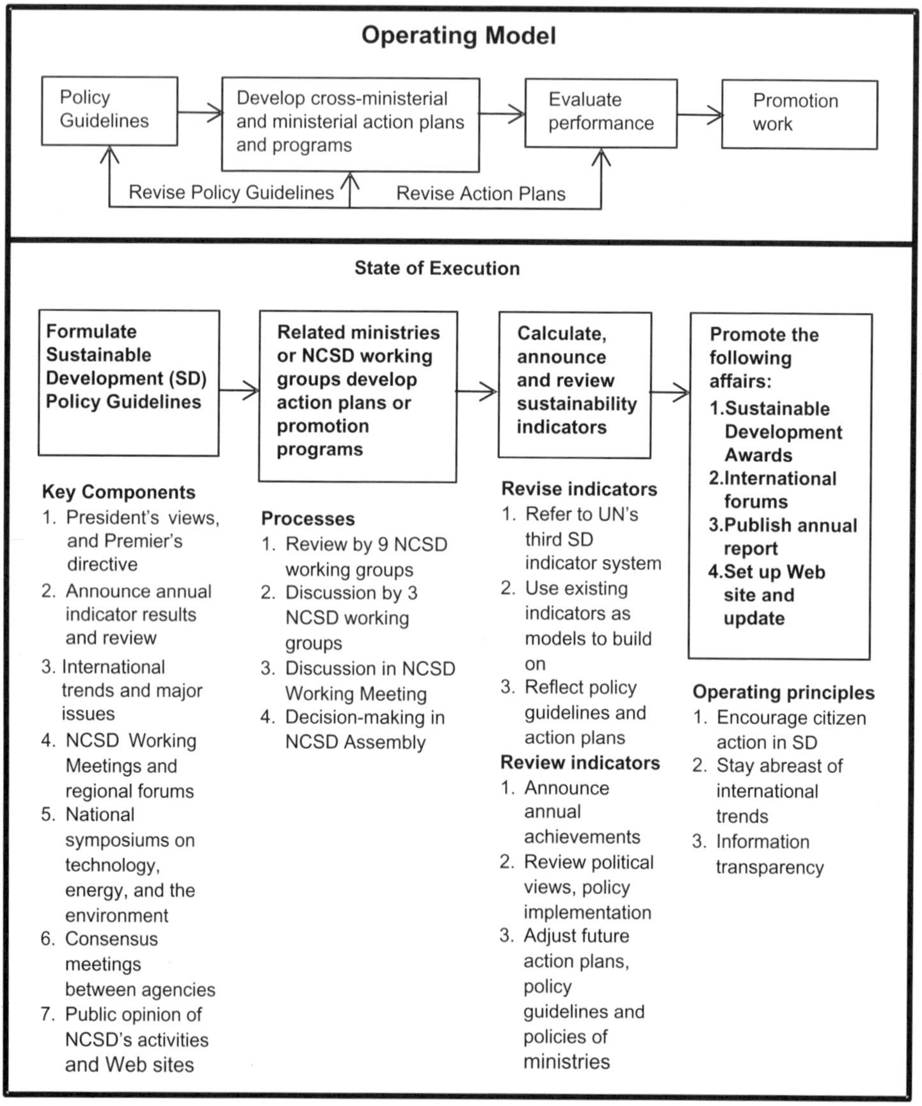

Fig. 1. The framework for NCSD model of operations.

NCSD model of operations, which focuses on the operation model and the state of execution.

Dr. Ma Ying-Jeou, the President of Republic of China, has been responding to international trends by announcing Sustainable Development Policy as follows:

1. Not only we will take the lead in initiating national environmental policies, but also we will demand national policies be focused on the environment.
2. We will consider both the carrying and restoring capacities of the environment while developing our smart growth policies.
3. We will respond to the current and future international treaties and elevate green competitiveness in Taiwan.
4. We will adhere to the spirit of democracy in transparency, involvement, restitution, and practicing environmental justice.
5. We will begin our work at the grass roots and create a clean, high-quality, and sanitary living environment.

Taiwan Agenda 21 proposes the sustainable environment action guidelines, which include nature conservation, pollution prevention and treatment, and environmental planning policies. During the past few years, the government of Taiwan has drawn up its own national Agenda 21 to actively implement the concept of a sustainable Taiwan. Vision and strategies for National Sustainable Development serve as the basic guidelines for all endeavors to keep up with the tide of international development in the 21st century. The conceptual framework for national sustainable development is shown in Fig. 2.

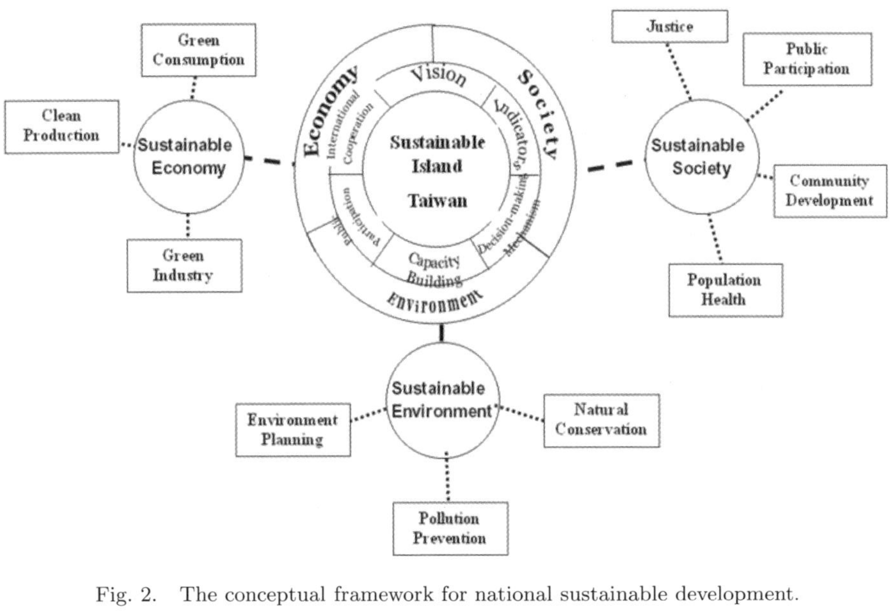

Fig. 2. The conceptual framework for national sustainable development.

One of the salient issues in Taiwan Agenda 21 is to keep the sustainability of energy supply. The integrated watershed management (IWM) developed by Systech Engineering can be used for both point and non-point source pollution management and improvement measures. In addtion to promotion of environmental science and technology park (ESTP) is one of the major missions.

The objective of this paper is intended to present the mission and strategies on the above three issues, i.e., sustainable energy, IWM, and eco-industrial park (EIP). The relative key performance indicators (KPI) toward sustainability made by the Taiwan Government are also introduced.

2. Sustainable Energy

2.1. *Missions*

1. Reduce the national carbon dioxide emission to achieve the following goals:
   (a) The annual CO_2 emissions between the years 2016–2020 will be decreased to and maintained at the same emission rate as of 2008.
   (b) The annual CO_2 emission in 2025 will be decreased to and maintained at the same emission rate as of 2000.
   (c) The annual CO_2 emission in 2050 will be decreased to and maintained at 50% of the emission rate of 2000.
2. Under the framework of clean energy supply, the National Energy Sector should achieve the following goals:
   (a) Develop carbon-free renewable energy, which will account for 15% of the national installed capacity by 2025.
   (b) Increase the utilization of low-carbon natural gas, which will account for more than 25% of the power generated in the Nation by 2025.
3. Reduce the carbon emission intensity from the industrial sector by more than 30% in 2025 via restructuring the industrial sector.
4. Raise the fuel efficiency standard for private vehicles by 25% in 2015.

2.2. *Implementation strategies*

The strategies for sustainable energy contain six categories: (1) low-carbon economy and structural transformation, (2) energy efficiency and sector implementation plans, (3) energy science and technology (EST) and industrial development, (4) energy security and sustainable development, (5) policy instrument and enforcement mechanisms, and (6) energy education and partnership.

(a) *Low-carbon economy and structural transformation*: Reduce the carbon emission intensity via restructuring the industrial sector by developing the best available technology, equipment efficiency standard, energy audit and energy savings counseling, and ecological industrial park. Allocate emission quota and reduce duty to lure industry toward an energy-conserving and emission-reducing production and sales by emission inventory registration and voluntary reduction projects. Assist small- and medium-sized enterprises to improve their emission reduction capacities. Formulate incentive mechanisms and administrative schemes to encourage the application of clean production technologies.

(b) *Energy efficiency and sector implementation plans*: Accelerate the replacement of existing power generating units with more efficient units (coal-fired unit: 42%; gas-fired unit: 52%). Develop green energy industries including energy conserving industries and renewable energy industries, to move toward a clean energy economy. Introduce clean coal technologies and carbon capture and storage technologies through international co-operation.

(c) *Energy S&T and industrial development*: Actively promote the green industries and renewable energy such as the power generations from solar energy (PV), wind mills, and biomass. Promote public use of energy-saving devices including LED illumination and hybrid vehicles, and establish an effective national policy to manage electrical energy. A roadmap for energy S&T should be developed to achieve the objectives of green house gas (GHG) emission reduction. Consequently, the government should increase the national energy research budget from the existing $160 million per year to $320 million per year within the next four-year time period.

(d) *Energy security and sustainable development*: The energy security indicator system and enhancement mechanism should be established for regularly monitoring the linkages among energy security, national security, and industrial competitiveness and impact of structural changes (industry and energy) on energy security.

(e) *Policy instrument and enforcement mechanisms*: The development and enforcement of policy environment impact assessment, choice and integration of economic instruments (energy tax, carbon tax, subsidy, energy pricing, emission trading, etc.), and planning and enforcement of regulatory instruments (goal setting for emission and energy intensity reduction, emission permit allocation, carbon budgeting, etc.) are required to provide the economic instruments and command and control strategies, which could be used to effectively reduce the GHGs.

(f) *Energy education and partnership*: The capacity building and mechanism for public participation, implementation plans for energy education, transparency of information and decision making, and development of international collaboration programs are the main task forces that should be implemented to promote the energy education and partnership program.

3. Local Agenda 21: Environmental Action Plan Toward Sustainability

According to the Sustainable Development Policy Guidelines, the Local Agenda 21: Environmental Action Plan Toward Sustainability was established. Figure 3 presents the Local Agenda 21: Environmental Action Plan Toward Sustainability, which indicates that the (1) sustainable water environment, (2) total emissions

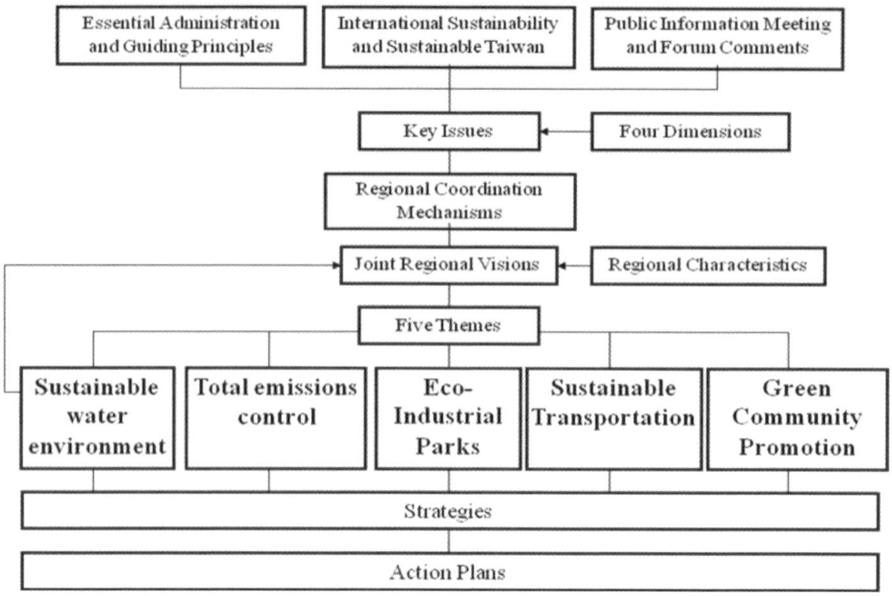

Fig. 3. The framework for local agenda 21: environmental action plan towards sustainability.

control, (3) EIPs, (4) sustainable transportation, and (5) green community promotion are the major five task forces. The detailed missions and implementation strategies on sustainable water environment and EIP, were stated as follows:

3.1. *Establishment of Integrated Watershed Management (IWM) for sustainability*

3.1.1. *Missions*

Although the rainfall in Taiwan is plentiful, it is not evenly distributed in all areas of the island at all times throughout the year. Moreover, it is difficult to preserve water because of the steep terrain and short, swift-flowing rivers. It is also necessary to protect forest areas to reduce mudslides, landslip, and soil erosion. Furthermore, the water resources of lower stream areas should be effectively allocated and water reuse should be strengthened to avoid water shortages.

In response to global climate change issues, the first priority should be to reduce the vulnerability of people and societies to make efforts along hydrometeorological trends and extreme events. The second priority should be to protect and restore ecosystems that provide critical land and water resources and services. The third priority should be to close the gap between water supply and demand. Consequently, the implementation of IWM Plans (IWMP) should include the following:

1. The IWMP must be across political and administrative boundaries by bringing together all interested upstream and downstream parties.

2. The IWMP considers not only technical, but also socio-economic and ecological aspects.
3. The IWMP focuses on complex decision making that involves multi-disciplinary teams.
4. The IWMP addresses specific social and economic needs in addition to the protection of natural resources and ecological health.
5. KPIs for IWMP should be politically accountable, socially acceptable, technically executable, and economically affordable.

3.1.2. *Establishment of the KPIs for IWMP*

The concepts of IWM contain comprehensive management and water, soil, and forest programs collaboration. Programs include environmental protection and monitoring water resources, wetland and marine resources, forestry protection, weather soil and water conservation, and ecologic conservation. It is necessary to develop KPI for outcome assessment. The established KPIs include "water quality and esthetic," "water front," "administration efficiency and effectiveness," "conservation approaches and tools," and "public participation," and are shown in Table 1.

3.1.3. *Implementation strategies*

1. Link KPIs with the Total Maximum Daily Load (TMDL) Program in which the information on water quality and esthetics that form the basis of the TMDL should be gathered and evaluated.

Table 1. KPIs for integrated water management plans.

Aspect	Indicator title
Water quality and esthetic	Criterial contaminants Eutrophication Hazardous contaminants Amenity
Water front	Green infrastructures Ecological engineering practice Access to recreation Landscape conservation
Administration efficiency and effectiveness	Strategic and business plan Point source control Non-point source control Cost-effectiveness analysis
Conservation approaches and tools	Habitat and biodiversity conservation River corridor protection Land protection programs and land use ordinances
Public participation	Stakeholder involvement River patrol and audit Public education Outreach

2. Strengthen pollution reduction programs by constructing sewers and controlling non-point sources pollutants in which the water quality should be assessed by the appropriate model.
3. Clarify the major steps during the design and implementation of national water protection policies including budget issues, outcome assessment tools, i.e., evaluation of ecological and environmental benefits and national sewage fees.
4. Establish roadmaps such as water quality objectives and other quantifiable objectives clearly, including ecological benefits, administrative effectiveness, water front accessibility, and participation by public and stockholders.
5. Enhance involvement to include institutes, commercial entities, individual citizens and "not-for-profits" organizations (NGO). Foster ecological considerations and encourage the participation of local organizations. Strengthen the accessibility to the waterfront, including integrating citizen access and daily life of the public into watershed management programs.

3.2. *Construction of Environmental Science and Technology Park (ESTP) toward sustainability*

3.2.1. *Missions*

One of the measures being implemented as a result of the Sustainable Development Policy Framework is the EIP demonstration plan, which aims to balance the environmental protection and economic development in Taiwan. This plan is a part of the overall effort to promote the national objectives in energy conservation, carbon reduction, and green production. The plan emphasizes the ecological aspects of industrial parks through the establishment of the eco-friendly supply chains so that the parks can be transformed into self-sufficient systems that will produce little pollution and have minimum impacts on the environment.

The concepts of ESTPs are originated from the principles of symbiotic growth, resource recycling, information sharing, and risk reduction, with the focus on technology and eco-industry. The goals of zero waste and carbon reduction based on the concept of sustainable development should be implemented. Different policies and strategies to initiate a recycle-based society have been implemented in different countries. Promotion of EIPs is an important approach to obtain economic and environmental benefits simultaneously. There are five objectives in ESTPs as follows:

1. Establish an integrated structure to embrace economic growth, environmental quality, and social equity.
2. Stimulate investments in the private sector to increase employment opportunities related to resources recycling and to encourage rural and urban community development.
3. Pursue the goal of zero or near-zero emission to build up a recycle-based sustainable society.
4. Manage waste reduction and reuse technologies to gradually achieve the goals of total recovery, and zero waste.

5. Build recycling-based eco-cities and eco-villages to raise resource recycling ratios, and reduce water and energy consumptions, to pursue economic and environmental benefits.

3.2.2. *Establishment of the KPIs for EIP*

The impacts of EPI can be evaluated through three categories: environment, economy, and social. The KPIs for the environmental, economical, and social impacts are provided in Table 2.

3.2.3. *Implementation strategies*

The development of industrial parks is closely linked to the growth of Taiwan's economy and advancement of technology. However, owing to the huge input and output of raw materials and products, the intensive development of industrial parks

Table 2. KPIs for environment, economy, social in EIP.

Aspect	Theme	Indicator title
Environment	Pollution reduction	NOx emissions
		Particles emissions
		VOCs emissions
		SO_2 emissions per unit output value increase
		Wastewater discharge per unit output value increase
		COD emissions per unit output value increase
	Resource recycling	Ratio of reclaimed industrial wastewater
		Water consumption per unit output value
		Material consumption per unit output value
		Waste recycling rate per unit output value
	Improvement of environmental quality	Averaged PSI
		Ratio of green land
		Green building indicators
		Environmental management system
Economic	Cost reduction	Measures for promoting pollution prevention and resource recovery
		Cost reduction of CO_2 emission control by waste recycling
		Ratio of material shipping expense in the total output value
	Profit increase	Output value per recycled waste
	Corporate image promotion	Budget/expenditure of environmental protection
		Total investment for pollution control
Social	Public participation	Number of visitors in open house events per year
		Completeness of message platform
		Writing environmental report
		Public satisfaction of environment
		Public cognition of EIP
	Community development	Interchange plan for public transportation system
		Plan for biking and waling route
	Fairness and justice	Green park area per capital
		Number of pleaded environmental pollution events
		Compliance with laws and regulations

often results in massive energy consumption and the generation of a great amount of waste. Considering the life-cycle assessment and the 3R's (recycling, reducing and reusing), efforts should focus on promoting the development of recycling-based ecological industrial parks. There are six strategies on establishment of an ESTP as follows:

1. Promote protection of ecological environments within industrial parks, upgrade and improve their original environmental qualities, promote the development of research and innovation in environmental technologies, and unify the management of industrial development and administration.
2. Effectively manage the waste produced by industrial parks through reduction and reuse to gradually achieve the goals of total recovery, zero emission, and zero landfill.
3. Build recycling-based eco-cities and eco-villages, expand regional market mechanisms, raise green coverage and resource recycling ratios, and reduce water and energy consumptions, to pursue economic benefits hand-in-hand with environmental benefits.
4. Speed up the promotion of environmental protection sci-tech parks and, with a view in promoting the marketability of regenerated products, assist firms to publicize the sale of such products, vigorously promote green purchasing, and improve the development of marketing channels for recycled and recovered products.
5. Promote eco-industries, improve their conditions for doing business, and promote research and development on green products. Also, expand regional market mechanisms, and provide inducements for domestic and foreign enterprises to develop eco-industries.
6. Review the results of promoting environmental protection sci-tech parks and promote technology diffusion to achieve the environmental benefits of "zero waste" and "recycling and reusing" and create "eco-landscapes" in Taiwan's industrial parks.

4. Major Achievements

Major achievements of the NCSD in 2009 are as follows:

1. *Selection of sustainable development issues relating to presidential campaign platforms for inclusion in the Sustainable Development Action Plan*: Recently, the Council reviewed and selected 41 topics from the "Follow-up List for Implementing Ma Ying-Jeou's Presidential Campaign Platforms" compiled by the Research, Development and Evaluation Commission of the Executive Yuan. These topics were included in the Sustainable Development Action Plan to implement the president's political views on sustainable development. A total of 167 essential tasks were established during the Assembly Committee Meeting

held by the Executive Yuan since the inclusion of the president's views in the NCSD Action Plans.

2. *Evaluation of performance in promoting sustainable development in 2008*: The NCSD held a briefing on the 2008 Sustainable Development Indicators on 17 July 2009 to publicly announce the government's performance in promoting sustainable development in 2008 according to the indicator system. Comparing the 2008 results of the 41 indicators against the previous year (2007), progress has been made in 27 indicators (66%), whereas headway was lost in 11 indicators (27%) and scores remained the same in 3 indicators (7%). Significant attention from all circles recently has focused on the carbon dioxide emissions indicator showing 4.4% overall reduction, equivalent to 4.7% per capita decrease in carbon dioxide emissions.
3. *Establishment of national sustainable development indicators (SDI)*: Sustainable development is a wide-ranging issue that reverberates through all levels of society. In 2009, the new government drew up the Sustainable Development Policy Guidelines that bases the promotion of sustainable development on the expectation that present and future generations will enjoy a quiet environment rich in biodiversity, an open and vibrant economy, and a safe and harmonious society to live in. The new 2009 SDI version includes 12 aspects, 41 themes, and 99 indicators as given in Table 3. Its range and scope are wide, suitable, and sufficient, to demonstrate the effectiveness of our country's sustainable development.

5. Review of the National Land Security and Restoration Plan

The Council for Economic Planning and Development, Executive Yuan, compiled the draft National Land Security and Restoration Plan. The NCSD National Land Resources Working Group and the Urban and Rural Development Working Group held separate working group meetings in co-ordination with four regional forums and three NCSD working meetings. Discussions from these meetings were drawn upon in the review of the National Land Security and Restoration Plan during the NCSD Working Meeting on 31 December 2009.

6. Conclusions and Recommendations

The Sustainable Development Policy Guidelines were drawn up by the NCSD based on Agenda 21: Taiwan Sustainable Development Strategic Guidelines. The Sustainable Development Policy Guidelines include an overall vision, basic principles, concepts, and local policies. Local governments also develop their own environmental action plans toward sustainability based on the Sustainable Development Policy Guidelines.

The current work of planning for sustainable development has to take into account the impact of the extreme weather patterns of recent years on our lives,

Table 3. National sustainable development indicators.

Aspect	Theme	Indicator title
Environment	Air quality	PSI average
		Air pollutant concentration
	Water quality	Water reservoir quality
		Marine environment quality
		Ratio of rivers suffering minor-pollution
		BOD concentration
	Waste	Garbage recycling rate
		Daily per capita garbage volume
		EIA approval rate
	Environmental management	Number of publicly announced toxic substances placed under monitoring
		Ratio of environmental and ecological budget by the central government
		Financial measures in promoting pollution prevention and recycling
Energy conservation and carbon reduction	Greenhouse gas	Per capita CO_2 emissions due to fuel combustion
		Annual increase of CO_2 emissions due to fuel combustion
		Greenhouse gas emissions
	Energy usage	Daily per capita power consumption
		Energy concentration
		Ratio of resource-consumption-based industries to manufacturing industry
	Energy conservation, carbon reduction	Percentage volume of renewable energy
		Energy conserved due to green buildings
		Bicycle path length per 10,000 people
National land resource	Land	Slope variation ratio
		Subsidence land ratio
		Developed land ratio
	Forest	Forest coverage area
	Coasts	Natural coast ratio
		Natural coastline loss ratio
	Water resource	Effective water resource
		Ratio of water usage to production value of the manufacturing industry
		Underground water recharge volume
		Underground water usage volume
	Natural disasters	Total national land area planting betel nuts
		Casualties due to natural disasters
		Economic loss due to natural disasters
Biodiversity	Heredity	Genetic resources and species preservation of biodiversity
	Species	Change in specific wildlife population
		Land area covered by specific exotic plants
		Populations of specific exotic invasive species
	Terrestrial ecosystem	Eco-sensitive area
		Ratio of protected area to total land area
Production	Material usage	Marine protection area

(*Continued*)

Table 3. (*Continued*)

Aspect	Theme	Indicator title
		Ratio of coral ecosystem area to coverage area
		Material strengths used in economic development
		Material strength used in non-manufacturing type of economic development
		Domestic supplies
		Non-manufacturing domestic supplies
		Per capita national gravel production
	Hygiene production	Reuse rate of industrial waste
		Reuse rate of toxic industrial waste
		Reduction rate of low-radioactive solid waste
	Agriculture	Ratio of cultivated land
		Area of organic cultivation
		Fertilizer usage rate per hectare of farmland
		Pesticide usage rate per hectare of farmland
	Fishery	Overfishing
	Labor	Labor production and unit production cost
		Ratio of females receiving salary in non-agricultural sectors
	Overall economic effectiveness	Per capita GDP
		Ratio of gross domestic capital formation to GDP
		Annual increase in consumer price index
	Public finance	Ratio of all levels of government borrowing above one year with outstanding non-selfliquidating debt to GNP
Livelihood	Water usage	Percentage of population with access to suitable drinking water
		Sewage treatment rate
		Daily per capita water consumption
	Transportation	Number of times public transport is utilized
		Domestic energy consumption by transport sector
		Times of tourist visits in Taiwan
		Road: casualties every 10,000 vehicles
		Road maintenance efficiency
	Green consumption	Green procurement amount of public and private sectors
		Number of green marks awarded
Health	Medical care	Percentage of population with access to basic infrastructure
		Infection immunity measures for children's diseases
		Usage rate of preventive health insurance by those 65 and above
	Nutrition	Children nutrition condition
	Health risks	Death rate of standardized cancer
		Infection rate of contagious disease
		Smoking rate of those above 18
		Eating betel nut rate of those above 18

(*Continued*)

Table 3. (*Continued*)

Aspect	Theme	Indicator title
Technology	R&D	Percentage of GDP spent on domestic R&D
	Telecommunications	Ratio of internet users
		Number of people using cell phone for every 100 people
Urban and rural culture	Cultural heritage	The number of ancient monuments and sites appointed
	Community	Number of villages in compliance with SDI
	Urban	Expansion rate of urban areas
		Green area per capita
Welfare	Poverty	Ratio of low-income families
		Accommodation rate
	Average income	Difference in disposable income per household of each division
	Social benefits	Subsidy for the disadvantaged
		Elderly passport and their participation
		Suicide rate
Administration	Crime	Crime rate
	Education	Dropout students
		Adult education participation ratio
Participation	International participation	Condition of Taiwan's participation in UN's international environmental organizations and other MEAs
		International environmental co-operation and assistance to other nations
	Public participation	Civil participation
		Community-based participation of social welfare

particularly the susceptibility of Taiwan to various disasters that rapid climate change is exacerbating. Typhoon Morakot was a good example, and reports produced by a number of experts have recommended that the reconstruction work should cover five major aspects including environmental monitoring and control, disaster assessment and control, reclamation and zone planning of disaster areas, support and guidance for disaster victims, and establishment and assessment of information channels.

In terms of policy planning, promotion of the national energy conservation and carbon reduction campaign must go hand-in-hand with suitable adaptation to real situations. The best way to implement the above policy for all is to co-ordinate the efforts of all involved stakeholders–manufacturers, academic experts, government officials, environmental protection groups, community organizations, etc — while promoting the concepts of sustainable community. Looking to the future, the expected establishment of the Ministry of Environment and Resources will facilitate the integration of drainage area management, taking the drainage area as the administrative unit. This structure should lead to a better overall planning with regard to the use of water resources, water quality maintenance, erosion prevention and flood control in the mountains, protection of coastal areas, and land reclamation.

As suggested by Dr. Ma Ying-Jeou, President of Republic of China, not only we will take the lead in initiating national environmental policies, but also we will demand national policies be focused on the environment.

References

Chiang, P.C., 2010. Development of energy action plans towards sustainability. *Energy Proceedings*, National Taiwan University.

Environmental Protection Administration Executive Yuan, R.O.C., 2010. *Development of Integrated Watershed Management and Performance Indicators.*

Ministry of Economic Affairs Executive Yuan, R.O.C., 2010. *Development of Eco-industrial Parks.*

National Council for Sustainable Development Executive Yuan, R.O.C., 2004. *Local Agenda 21: Environmental Action Plan towards Sustainability.*

National Council for Sustainable Development Executive Yuan, R.O.C., 2009. *Annual Report on National Sustainable Development.*

Chapter 29

SUSTAINABLE SUPPLY CHAIN INITIATIVES IN THE KOREAN AUTOMOTIVE INDUSTRY

Su-Yol Lee[a] and In-Mo Cheong[b]

[a]*College of Business Administration, Chonnam National University, Yongbong, Buk, Gwangju Metropolitan City, South Korea*
leesuyol@chonnam.ac.kr

[b]*Environmental Strategy Planning Team, Hyundai Motor Company, Yangjea, Seocho, Seoul, South Korea*
imcheong@hyundai.com

In response to the emerging and changing external concerns about green and social issues, the sustainable supply chain initiatives in the Korean automotive industry have evolved through three distinct stages. Hyundai, a leading automaker in Korea, has led these initiatives in conjunction with the Korean government. In the first stage, the initiatives focused on fostering the environmental capabilities of suppliers by utilizing two approaches: green procurement and green supplier development. The initiatives were extended to disseminate environmental management practices to second-tier suppliers. Then, the initiatives addressed climate change, an increasingly important topic, in the second stage. In the third stage, the emphasis was placed on transferring best practices in carbon inventory, management, and measurement to supply chain management. These initiatives have focused more on green issues than on social or ethics issues because of contextual differences. However, Hyundai is planning to extend the initiatives to its international supply chain, which is expected to result in the integration of social and ethics issues into the current initiatives.

1. Introduction

The economic importance of supply chain management has been an important research and management topic. The amounts of raw materials and components purchased by US and Korean companies account for approximately 45% and 59% of their total manufacturing expenditure, respectively.[1] Therefore, firms have increasingly relied on their suppliers to achieve and maintain quality, cost, and delivery competitiveness. They have also pressured their suppliers to develop and provide goods that would ensure the adequate environmental and social performance of

[1]The 2004 US data were drawn from www.capsresearch.org, and the 2003 Korean data were from the Korea National Statistical Office, www.kosis.nso.go.kr

their products and processes. Further, more managers have been realizing the presence of increasing numbers of environmental and social risks in their supply chains (Handfield *et al.*, 2005).

The Korean automotive industry has relied heavily on overseas markets. More than 65% of all vehicles produced in Korea are exported.[2] Since the early 2000s, a new array of environmentally and socially driven issues from overseas has been driving the Korean government and automakers to focus on improving green and social performance across the entire supply chain. There have been a number of new global environmental regulations, such as European directives on product take-back and the elimination of certain hazardous substances in products.

As a result, Hyundai Motor Company (herein "Hyundai"), Korea's leading automaker, has implemented new initiatives focusing on sustainable supply chain management in cooperation with the Korean government. Based on the prioritization of key issues, the environment and regional suppliers were addressed first. Later, the initiatives were extended to include climate change issues. Currently, there are plans to expand the initiatives to address social issues and international suppliers.

2. The First Stage: Supply Chain Environmental Management Program (The SCEM Program: 2003–2006)

2.1. *Background*

2.1.1. *External pressures*

Discussions on legislating new types of environmental regulations, such as the European Union's RoHS (restriction of the use of certain hazardous substance), WEEE (waste in electric and electronic equipment), and ELV (end-of-life vehicle) directives, began in the late 1990s. This posed serious challenges to automakers worldwide as they had to address the environment and resource conservation in their business process. The ELV directive was implemented to facilitate the closed-looping of vehicles by improving product design and increasing recycling and re-use. Under the directive, automakers were required to ensure that the components in their vehicles sold in the EU market did not contain major hazardous substances such as mercury, hexavalent chromium, and lead after July 2003 (cadmium after December 2005). In addition, automakers had to take back ELVs at their own expense and increase the rate of re-use and recovery up to 85% per vehicle.

Just before Christmas in 2001, nearly 1.3 million of Sony's best-selling PlayStation game consoles were stopped at the border by Dutch customs agents. A small but legally unacceptable amount of toxic element cadmium was detected in the cables of the consoles. The cables were manufactured by one of Sony's suppliers; however, it was Sony who had to bear the extensive cost for replacing parts, sorting goods, and

[2]The 2005 data were drawn from the web site of the Korea Automotive Manufacturers Association, www.kama.or.kr

then repacking the goods. This Sony case and the infamous Nike sweatshop incident in the mid-1990s have served as an important reminder to large corporations in Korea's automotive and electronic industries that things may go terribly wrong if the supply chain is ignored when addressing newly emerging environmental and social challenges.

2.1.2. *Hyundai's new corporate environmental strategy*

Hyundai became one of the world's 10 largest automakers in 2002, a key milestone in Hyundai's history (it ranked ninth in 2002). Hyundai announced a new and very ambitious corporate strategy; it aimed to become one of the top five global automakers by 2010. In this regard, Hyundai Chairman Chung Mong-koo, who became the CEO in 1999, pushed hard to achieve high brand equity by investing heavily in R&D and quality management. He also globalized the company's operations by establishing new manufacturing plants in the US, China, India, Slovakia, Czech Republic, Russia, and Brazil. After briefly reviewing global trends in the automotive industry, Chairman Chung came to realize that environmental issues were vital to Hyundai's strategic goals. As such, the Environmental Strategy Planning Team was created under Hyundai's Business Strategy Planning Group to identify the relevant environmental issues so that the company could address such issues in advance. By adopting a life-cycle approach, Hyundai focused on three environmental strategies: greening the supply chain, green car technologies, and closing-loop and product recovery (Fig. 1).

2.1.3. *Korea's national initiatives*

The emerging environmental regulations in the EU and the Sony PlayStation incident were issues of major concern to the Korean government. Electric and electrical equipment and automobiles accounted for 38.2% of Korea's total export volume in 2002.[3] The government was concerned that Korean exports might be severely

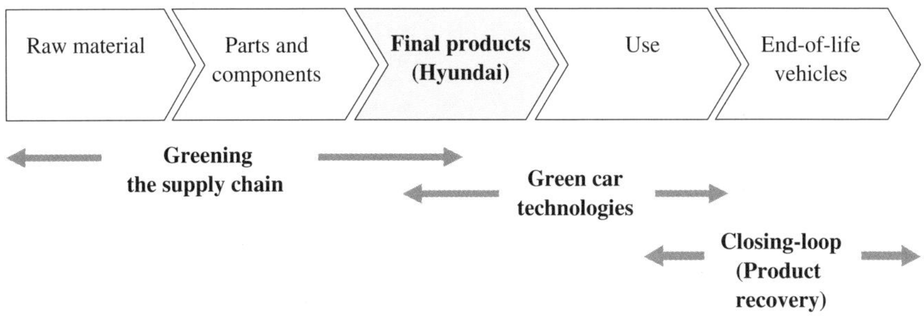

Fig. 1. Hyundai's environmental strategic focus from a life-cycle perspective.

[3]The 2003 data were drawn from the web site of the Korea International Trade Association, www.kita.org

affected by the new regulations because the supply chains of Korea's major industries were relatively unfamiliar with green issues. This concern led the Korean government to implement a national green supply chain initiative. The initiative was designed to foster suppliers' environmental capability by using the relationship between key manufacturers and their suppliers. This national initiative was implemented with funds from the national budget and matching funds from participating large-sized firms. Hyundai participated in this national project by representing the Korean automotive industry. As a result, this was the first stage of the sustainable supply chain initiatives.

2.2. *The SCEM program*

Hyundai's SCEM program incorporates two approaches: green procurement and green supplier development (Fig. 2). First, green procurement changes the supply and supplier assessment criteria and evaluates the environmental performance of incoming goods and the suppliers that provide them. In the process, this approach forces the suppliers to adopt advanced environmental management practices and improve their environmental performance. Second, green supplier development involves providing the suppliers technical and financial support as well as training and education to help them improve their environmental capability and performance.

2.2.1. *Green procurement*

Hyundai's environmental concerns were largely limited to managing its operations in compliance with regulations on manufacturing plants; therefore, very little pressure had been placed on its suppliers regarding environmental issues. Hyundai simply advised its suppliers to comply with appropriate regulations and avoid any disagreement with their local communities.

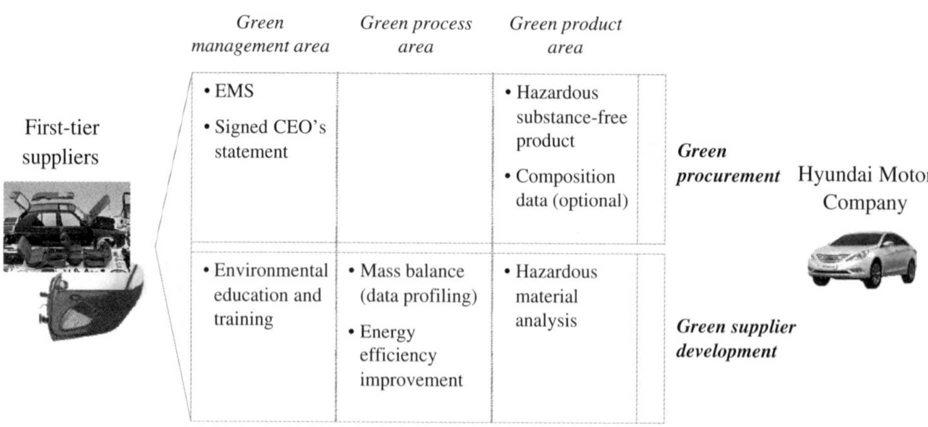

Fig. 2. A systematic approach of the SCEM program.

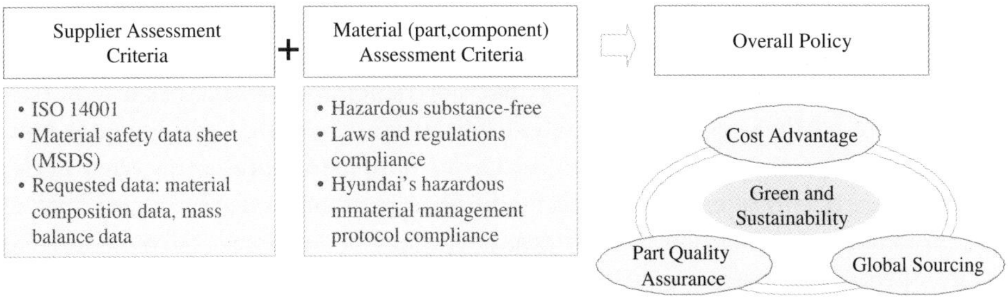

Fig. 3. Hyundai's green procurement policy.

Hyundai instated its green procurement policy in 2003. Hyundai introduced several environmental criteria and integrated them into its conventional procurement process. This newly adopted policy provided for a three-year transition period to give time to the suppliers to make appropriate preparations (Fig. 3).

- First, Hyundai established a hazardous material management protocol and distributed it to its suppliers in 2003. The protocol specified the materials to be banned immediately and to be phased out shortly. It also included test methods. All first-tier suppliers (suppliers that provided goods directly to Hyundai) had to ensure that their products were free of hazardous substances in compliance with the protocol. The suppliers were required to continuously improve the material composition of their products so that they would be safer and more environment-friendly. They were also required to immediately eliminate major hazardous materials including mercury, hexavalent chromium, cadmium, and lead.
- Second, all domestic suppliers were required to establish an environmental management system (EMS) with international certification, such as ISO 14001. In addition, Hyundai asked suppliers to set a mid- or long-term environmental strategic goal and demonstrate its validity with a signed top manager's statement.
- Third, major suppliers providing high-volume materials, parts, and components were requested to provide material composition data for each product and mass balance data for their manufacturing processes. These data were used to conduct life-cycle assessments of vehicles. Hyundai wanted to evaluate and monitor the environmental performance of its automobiles before shipment.

2.2.2. *Green supplier development*

Hyundai initiated a green supplier development program along with its green procurement policy. This support program integrates green issues into supplier development practices. It was expected to enhance the green performance and/or capability of suppliers, allowing them to meet Hyundai's supply needs. Hyundai provided quality and purchasing managers of all domestic suppliers with environmental education programs annually from 2003 to 2005. The programs covered new global environmental regulations, advanced green management practices, and Hyundai's new green procurement policy. The suppliers were also offered advice and assistance on how they could meet Hyundai's new green requirements.

Hyundai chose a small number of first-tier regional suppliers when it participated in the national green supply chain initiative. In-depth environmental management support was given to a total of 15 key suppliers representing the electric, driving gear, decorative, and chassis modules. The environmental support, which was provided in conjunction with external experts, encompassed a diverse range of environmental practices:

- hazardous material analysis of products;
- environmental data profiling of products and processes;
- mass balance management;
- improvements in process and energy efficiency;
- establishment of an EMS;
- implementation of suppliers' own green procurement system; and
- establishment of an environmental reporting system.

2.3. *Consequences*

Hyundai's SCEM program had continued through June 2006. By then, most of Hyundai's first-tier suppliers based in Korea had been exposed to the environmental education program more than once, which effectively enhanced the level of awareness among the suppliers' top management regarding global environmental issues, and fully understood what Hyundai's new green procurement process demanded. Thus, they began eliminating banned substances from the raw materials, parts, and components that they provided. The EMS became widely dispersed among the suppliers and well incorporated into Hyundai's supply chain. By the end of 2006, 361 first-tier domestic suppliers (approximately 84% of the regional supply base) had an operational EMS and achieved an ISO 14001 certificate. The ratio of ISO-certified suppliers rose to 96% and 100% in 2007 and 2008, respectively.

3. The Second Stage: Supply Chain Environmental Partnership Program (The SCEP Program: 2006–2008)

3.1. *Background*

By using the SCEM program, Hyundai had successfully launched its green supply chain initiative; however, it encountered difficulties in completing the entire green supply chain mainly because of second- and third-tier suppliers, that is, small- and medium-sized enterprise (SME) suppliers. The first-tier suppliers were not able to meet Hyundai's supply needs without the cooperation of their own suppliers (i.e., the second-tier suppliers to Hyundai). Most second-tier suppliers to Hyundai, however, were SMEs. In fact, 93.5% of suppliers in the Korean automotive industry were SMEs (Choi, 2003). They had difficulties meeting the emerging environmental standards of their final customer such as Hyundai because of their limited funds and lack of human resource expertise and relevant information. Thus, Hyundai and its first-tier suppliers had to determine how these second-tier suppliers could be made more aware of green issues and improve their environmental capabilities. This required the close involvement of second-tier SME suppliers (and third-tier suppliers wherever possible) in the initiative.

3.2. *The SCEP program*

This SCEP program was an extension of the SCEM and was applied to second-tier suppliers. Some of the major first-tier suppliers to Hyundai provided the very same green supply chain practices that they received from Hyundai to their suppliers. Seven of the 15 first-tier suppliers who experienced the SCEM in the first stage began implementing the SCEP program. Hyundai and its major suppliers took three approaches to green supply chain management: green procurement, green supplier development, and green SC information system (Fig. 4).

3.2.1. *Green procurement: First-tier suppliers*

The seven participating suppliers adopted a green procurement policy by benchmarking Hyundai's policy. They incorporated environmental criteria into their evaluation and selection of their own suppliers; however, the criteria were mostly recommendations than requirements:

- First, the first-tier suppliers requested their suppliers to identify and phase out potentially harmful substances in their products by referring to the hazardous material management protocol of Hyundai.
- Second, the suppliers were requested to develop material safety data sheets (MSDSs) and keep them updated. MSDS information included potential hazards associated with a particular material or product and instructions for safe use; thus, the MSDS was considered as an important component of workplace and product safety.

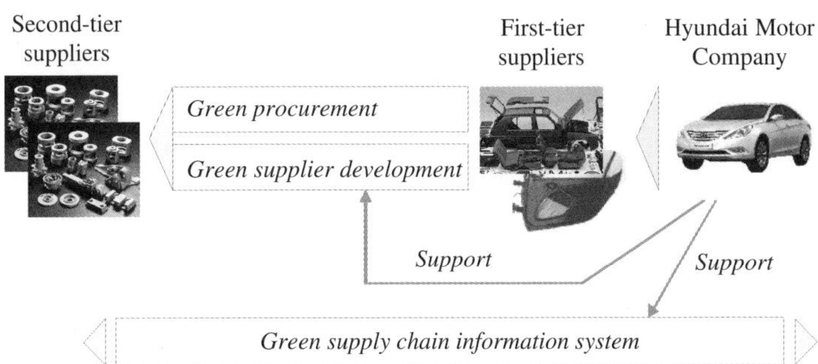

Fig. 4. The cooperative approach of the SCEP program.

- Third, the suppliers were encouraged to establish an EMS for their manufacturing sites and demonstrate its validity by acquiring an ISO 14001 certificate.

3.2.2. *Green supplier development: Second-tier suppliers*

The first-tier suppliers began implementing green supplier development programs in conjunction with Hyundai and external experts. From 2006 to 2008, most suppliers to these seven first-tier suppliers were provided with environmental support and education on the following:

- global environmental regulations;
- green procurement policies and requirements in Hyundai's supply chain;
- Hyundai's hazardous material protocol and advice on compliance;
- MSDS implementation; and
- international material data system (IMDS) utilization.

In addition to this general education and training program, these seven suppliers selected a few of their own suppliers and provided them in-depth environmental management support for demonstration purposes. A total of 17 second-tier suppliers participated in this initiative. Experts specializing in environmental management and energy efficiency from Hyundai, the seven first-tier suppliers and external advisory groups were dispatched to manufacturing sites of the participating second-tier suppliers (Table 1). They conducted a detailed analysis of the suppliers' processes and determined the potential areas for further improvement in the areas of energy and the environment, resulting in specific action plans. By using this supportive consulting process, Hyundai intended to transfer large-sized manufacturing firms' knowledge of energy efficiency, waste reduction, and safety assurance to their upstream suppliers, particularly SME suppliers.

Table 1. The number of second-tier suppliers affected by green procurement and supplier development.

	Suppliers (the first-tier suppliers to Hyundai)						
	A	B	C	D	E	F	G
Green procurement							
Hazardous substance-free goods	All	All	All	All	All	All	All
EMS certification	68	100	3	5	3	n/a	5
MSDS registry	68	100	All	All	Some[b]	10	All
Green supplier development							
Environmental education	38[a]	100	All	All	3	All	5
Hazardous material protocol sharing	38[a]	100	All	All	Some[b]	All	All
MSDS implementation	68	100	10	All	Some[b]	10	All
Green SC Information System	n/a	5	n/a	n/a	13	5	5
Process improvement	2	2	10	4	13	2	2

[a]Only for painting, plating, and injection suppliers
[b]Only for painting and plating suppliers

3.2.3. *Green supply chain information system*

Hyundai wanted to increase its supply chain's environmental visibility, the ability to access or view relevant environmental data or information related to logistics and the supply chain. In this regard, Hyundai began building a green supply chain information system by exploiting its own legacy procurement system and an international network database.

First, Hyundai extended its legacy system for supply chain management, VAATZ, to incorporate environmental information. All information, including that on global environmental regulations, and Hyundai's green procurement policy and requirements were distributed to the suppliers through this system. The suppliers were then required to input their MSDS into VAATZ. Then, these supplier-provided environmental and safety data were incorporated into Hyundai's environment, safety, and health system and greenhouse gases and energy management system, called the I-EHS and the GEMS, respectively. When a supplier failed to submit the required environmental data, the supplier was systematically rejected during Hyundai's supplier selection process. Second, Hyundai joined a group using the IMDS, which is a web-based system that archives all materials and components used in automobile manufacturing. Hyundai's major suppliers providing high-volume materials, parts, and components provided material composition data for each of their product for the IMDS database (Fig. 5).

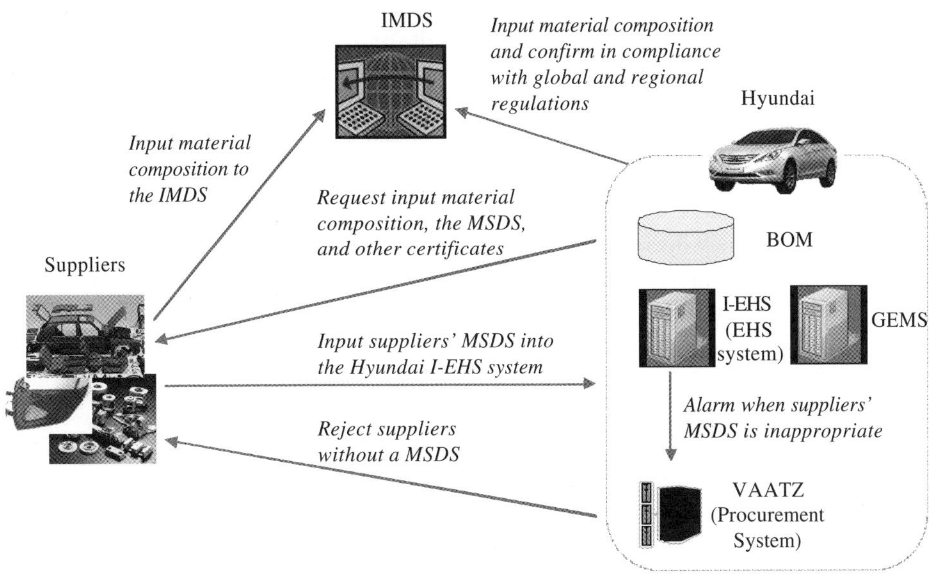

Fig. 5. Network-based green supply chain information system. IMDS: International Material Data System, BOM: Bill of Materials, EHS: Environment, Health, and Safety, MSDS: Material Safety Data Sheet.

By using the information system across the entire supply chain, Hyundai and its suppliers could ensure that they were in compliance with international and national environmental standards and regulations in an easier and more cost-effective way.

3.3. *Consequences*

The second stage of Hyundai's sustainable supply chain initiative ended in March 2008. All of the domestic first-tier suppliers were able to provide evidence that their products were in compliance with international, national, and Hyundai's green standards. Hyundai no longer needed to inspect the environmental performance of inbound goods. Noteworthy is that Hyundai was able to disseminate a higher level of environmental knowledge and environmental management practices to second-tier suppliers. In addition, the SME suppliers that participated in the initiative were able to reduce their energy cost by $1.8 million through the process improvement projects supported by Hyundai and first-tier suppliers.

4. The third Stage: Supply Chain Carbon Management Program (SCCM Program: 2008–2010)

4.1. *Background*

Hyundai's main focus during the first and second stages was on disseminating environmental management practices across the entire supply chain. Since the Kyoto Protocol was entered into force in February 2005, Hyundai considered climate change as one of the most important environmental issues facing the automotive industry. Transport CO_2 emissions account for approximately 20% of Korea's total emissions. Thus, Hyundai anticipated that greater pressure would be placed on automakers to reduce greenhouse gas emissions and be more accountable for global warming. In this regard, Hyundai conceived a plan to assess, monitor, and lessen the carbon footprint of each brand; however, to do this, the involvement and cooperation of its suppliers were necessary. Following Korea's new national vision and development strategy declared in 2008, called the "Low Carbon and Green Growth" strategy, Hyundai implemented the third stage of the sustainable supply chain initiative, which focused mainly on carbon management across the upstream supply chain, in 2008.

4.2. *The SCCM program*

The use-phase CO_2 emissions account for more than 80% of a car's carbon footprint over its entire life cycle. Thus, new technologies and products such as hybrid, electric, and fuel cell cars could be the ultimate answer to carbonabatement. This was the reason why Hyundai has prioritized greener car technology development. In addition, Hyundai determined that the second largest carbon footprint resided upstream (i.e., production and manufacturing). Thus, achieving substantial carbon

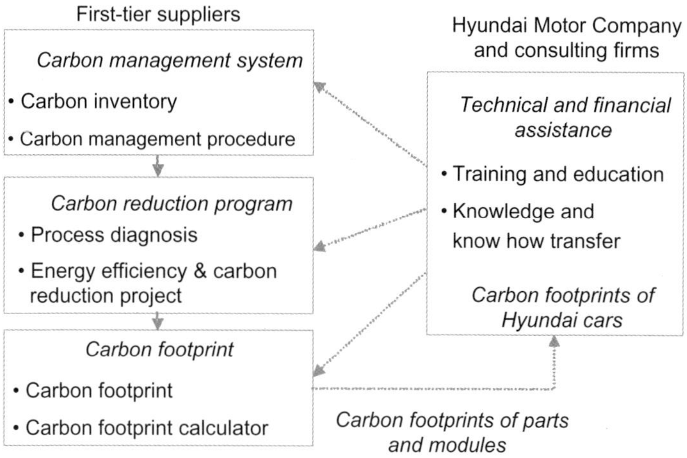

Fig. 6. A systematic and cooperative approach of the SCCM program.

abatement required close collaboration with supply chain partners. In this regard, Hyundai decided to implement a carbon abatement program in its supply chain by taking a supplier development approach (e.g., transferring best practices in energy efficiency and carbon management to key suppliers). A total of 10 major regional suppliers representing the driving gear, chassis, body, decorative, and electric modules participated in the SCCM Program (Fig. 6).

The participating suppliers began to establish their own carbon management system, undertake carbon reduction projects, and measure the carbon footprint of their products. In doing so, Hyundai provided in-depth assistance in conjunction with external carbon experts from consulting firms.

4.2.1. *Carbon management system*

In the first year of the SCCM Program, Hyundai began providing education and training to the 10 participating suppliers (Table 2). General education in a broad range of environmental and carbon issues was provided to senior and middle managers of the suppliers to increase their awareness regarding climate change. In addition, additional in-depth training was implemented to foster the development of the suppliers' internal carbon experts. Hyundai aimed for at least two experts at each supplier (a total of 20 experts for the 10 suppliers), who would play a key role in carbon management within their company after the initiative. In the second year of the SCCM Program, Hyundai extended this education and training program to 15 suppliers.

One of the main purposes of this program was to increase the carbon management capabilities of its suppliers. External carbon experts were engaged to help Hyundai define carbon management capabilities and develop a diagnostic tool for their measurement. Hyundai and its suppliers clarified the capabilities as a set of

Table 2. Demographics of suppliers that participated in the SCCM program.

Supplier	Module	Product	Sales revenue (US$ million)
H	Driving gear	Cylinder head, bracket	133.3
I	Driving gear	Cylinder head, transmission	129.1
E	Chassis	Muffler, dash assembly	304.7
J	Chassis	Brake system	1395.3
K	Body	Door, bumper rail	253.8
L	Body	Bumper assembly	41.6
M	Decorative	Trim, casting	294.1
N	Decorative	Seat	223.3
O	Electric	Connector, wiring part	382.1
P	Electric	Lamp, lever	105.7

Only Supplier E took part in both the SCEP (second-stage) and the SCCM (third-stage) programs.

assets, technologies, and skills that could enable a firm to identify, assess, and manage opportunities and risks, particularly regarding carbon. Operationalizing this definition, they pointed to three specific but interrelated dimensions:

- "Managerial readiness" diagnosed in the company's chief executive officer's awareness and commitment, climate change response strategy, carbon inventory, dedicated team, performance measurement, and external communication (e.g., carbon disclosure project).
- "Risk reduction" diagnosed in energy and resources use and carbon emissions.
- "Opportunity creation" measured in energy-efficient product development and new business analysis.

The suppliers measured and analyzed their carbon capabilities by using a diagnosis tool that Hyundai and its consulting firms developed. They then identified the areas where their capabilities were weak or strong, allowing them to develop their own carbon strategy and action plan.

To set carbon reduction goals and manage carbon footprints effectively, the suppliers needed to take action with the measurement of their emissions by establishing their carbon inventory. Establishing carbon inventory involves quantitative auditing and reporting that cover greenhouse gas emissions from all relevant sources. Carbon experts from Hyundai and environmental consulting firms visited the suppliers and collaborated to define the appropriate emissions scope, identify all relevant sources, measure total emissions, and audit the results in accordance with domestically and internationally accepted methodologies and standards. Through these supportive and cooperative measures, Hyundai was able to transfer its know-how and knowledge of carbon management to its suppliers. As a result, the carbon management system became slowly but surely embedded in the managerial procedures of the suppliers.

4.2.2. *Carbon reduction projects*

Carbon inventory formed the basis for target setting and the formulation of an appropriate carbon reduction strategy. Hyundai set a 5% carbon reduction target for its first-tier suppliers. To achieve this target, the suppliers needed the capacity to precisely diagnose their operating processes, find opportunities for improvements, and take action on energy efficiency and greenhouse gas emissions. In this regard, Hyundai distributed a manual and a guideline showing how the energy efficiency of manufacturing facilities and processes could be analyzed. With this guideline and technical support from external experts, the suppliers were able to identify several opportunities for improvements. A total of 60 projects were carried out by 10 suppliers during the SCCM program period, resulting in a 4744 ton reduction in CO_2 emissions and $900,000 cost savings.

4.2.3. *Carbon footprint*

Managing the carbon footprint of Hyundai's products required close collaboration with suppliers because a car's second largest carbon footprint resides upstream in the supply chain; however, Hyundai's suppliers lacked the skill and knowledge to calculate their own footprints. To help its suppliers calculate and effectively manage their products' carbon footprints, Hyundai supported them in two ways:

- First, 10 suppliers (the SCCM program participants) undertook a carbon footprint project with the support of Hyundai and external consultants. In the first year of this program, they focused on developing a framework of suppliers' carbon footprint by setting proper boundaries and methodologies for calculating emissions. For calculating the carbon intensity of products, the suppliers collected data in scopes 1 and 2 of carbon inventory from their first-tier suppliers, their manufacturing site, and logistics. Following this framework, each supplier estimated the carbon footprint of one major product shipped to Hyundai in the second year of the SCCM Program.
- Second, Hyundai developed and distributed an easy-to-use carbon footprint calculator for use by its suppliers. The calculator was a simplified form of carbon footprinting. It was development based on the carbon footprinting experiences of the 10 suppliers.

Aside from assisting the suppliers to calculate their carbon footprints, Hyundai began to undertake its own exemplary carbon footprinting project in 2008. One year later, Hyundai Sonata YF, a major Hyundai brand, became the first Korean automobile to acquire a certified carbon footprint label from the Korean government.[4] Sonata YF emits 24.9 tons of CO_2 during its life cycle. As expected, approximately

[4]The carbon footprint label data were from the Korea Environmental Industry and Technology Institute web site, http://www.edp.or.kr/carbon/english/main/main.asp

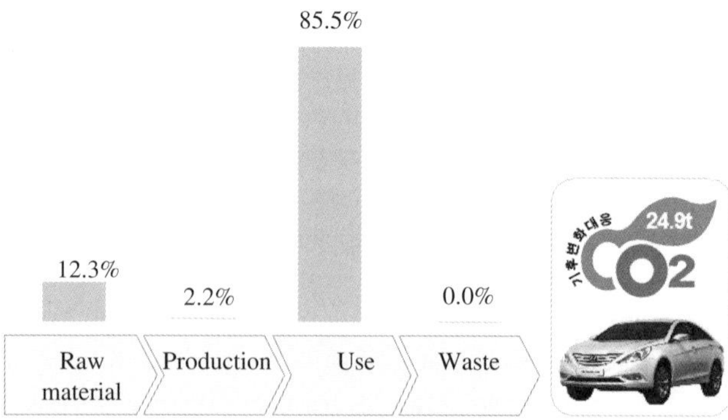

Fig. 7. The carbon footprint of Hyundai Sonata YF across the life cycle. Sonata YF is the first automobile that acquired a carbon footprint label in the Korean automobile industry.

85% and 12% of carbon emissions came from the use-phase and the upstream supply chain, respectively (Fig. 7).

4.3. *Consequences*

Hyundai's third-stage sustainable supply chain initiative program focused on the carbon and climate change issues. Hyundai made efforts to reduce carbon emissions and respond to climate change by managing the carbon footprint of its products across the supply chain. To do this, it took a supply chain intervention and supplier development approach, which was similar to the approach taken during the first and second stages. The third initiative focused on improving suppliers' capabilities in measuring, reducing, and managing carbon emissions as well as on systematically understanding the business risks and opportunities associated with climate change.

When the first year of this program ended in September 2009, a total of 10 major regional suppliers established a carbon management system and carbon inventory. They also completed carbon footprinting of their products. By participating in the program, the suppliers were able to achieve a 4.6% reduction in CO_2 emissions and cost savings of $900,000. In addition, while the Korean government was introducing the carbon label certification program in 2009, Hyundai actively participated in the policymaking process. As a result, Hyundai was able to influence the establishment of a national standard for vehicles' carbon footprint.

5. Conclusion: Evolution of the Sustainable Supply Chain Initiatives in the Korean Automobile Industry

The sustainable supply chain management in the Korean automobile industry changed over time through the dynamic interaction among external stakeholders such as governments, customers, local communities, media, and the public. Internal factors such as leadership and changes in corporate strategies also had considerable influence. The sustainable supply chain initiatives in the Korean automotive industry evolved through three distinctive stages: SCEM, SCEP, and SCCM programs.

Before the late 1990s, the Korean automotive industry was unaware of the sustainability issues in the supply chain. The most pressing issues were complying with regional environmental regulations and acquiring an EMS certificate. A big shift in the sustainable supply chain management in the Korean automotive industry was led by Hyundai, Korea's leading automaker, in the early 2000s (Fig. 8). The change was driven mainly by external factors, particularly global regulations such as the RoHS, WEEE, and ELV directives established by the EU and emerging consumer concerns about green and social issues. A new corporate environmental strategy and the CEO's strategic direction on green management facilitated this change. Hyundai initiated the sustainable supply chain management program by first considering green issues. It shook the supply chain by adding new environmental criteria to its procurement policy. Further, Hyundai made efforts to foster the environmental management capabilities of its major regional suppliers. In this regard, Hyundai employed the green supplier development program, which helped the suppliers enhance their environmental capabilities in the management, process, and product areas.

Once Hyundai's first-tier suppliers became more capable in addressing green issues, the initiatives were extended to include second-tier suppliers, that is, SME suppliers. An absolute majority of Hyundai's suppliers in Korea were SME suppliers. They were viewed as the key bottleneck hampering substantial gains in

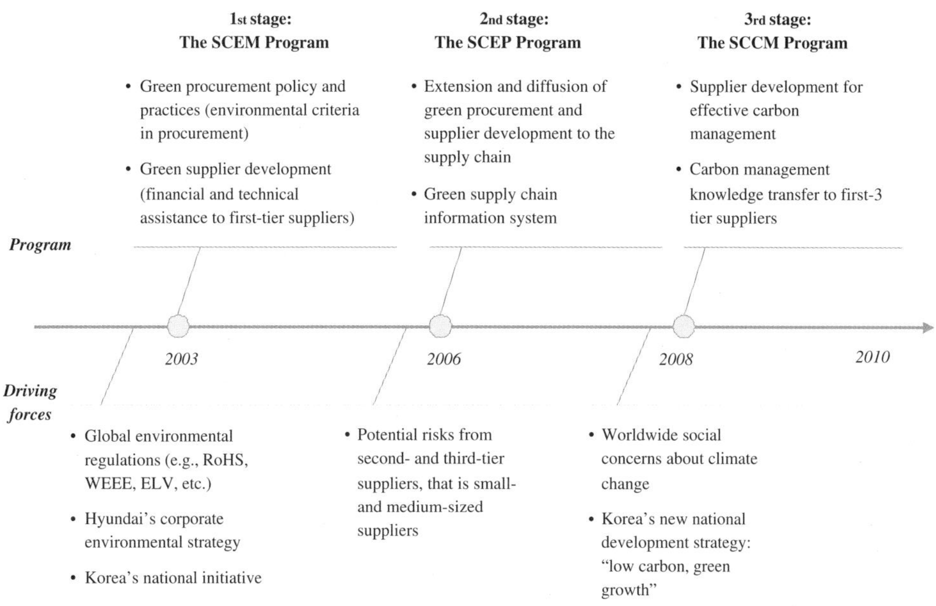

Fig. 8. Evolution of the sustainable supply chain initiative in the Korean automotive industry.

environmental management practices across the entire supply chain. In this stage, the first-tier suppliers played a key role by forcing and supporting their suppliers (i.e., second-tier suppliers) to be more aware of green issues and adopt environmental management practices. IT and network systems were employed to facilitate information and knowledge sharing and increase supply chain environmental visibility across the supply chain.

The sustainable supply chain initiatives evolved into climate change responses. Social concerns about global warming have skyrocketed worldwide. Whereas Japanese and German automakers have thrived with their fuel-efficient cars, American firms have been losing their ground rapidly. Korea's new national strategy of "Low Carbon and Green Growth" and consumers' concerns over energy and global warming have been driving the Korean automotive industry to deliberately translate the importance they place on climate change into action. Hyundai has used this external pressure as an opportunity for supplier development by transferring best practices in carbon inventory management, energy efficiency, and carbon footprinting to its key suppliers. Hyundai and its key suppliers explored the ways to improve operational efficiency and reduce CO_2 emissions in their supply chains.

The sustainable supply chain initiatives have focused mainly on environmental issues; social and ethical issues in the supply chain, such as child and forced labor and diversity, have been left relatively untouched because of contextual differences. In South Korea, because of its highly homogeneous social characteristics and higher regulatory standards on child and forced labor, these social issues have been considered to be less serious. In particular, larger corporations have been in full compliance with appropriate regulations and social expectations. However, Hyundai is planning to extend its sustainable supply chain initiative to include its international supply chain. In doing so, Hyundai may need to incorporate social issues into the initiative because Hyundai has become increasingly reliant on global sourcing from developing and less-developed countries where social and ethical issues have to be addressed. With its successful domestic experience implementing environment- and carbon-focused programs in sustainable supply chain management, the Korean automotive industry is now poised to build a more integrated sustainable supply chain.

References

Choi, J., 2003. *Hana Research Institute Report* (in Korean).

Handfield, R., Sroufe, R. and Walton, S., 2005. *Business Strategy and the Environment* 1–19.

Chapter 30

A EUROPEAN APPROACH OF ENVIRONMENTAL COSTS: A CASE STUDY IN THE SPANISH ROAD FREIGHT TRANSPORT INDUSTRY

Vera Ferrón-Vílchez[a] and Juan Alberto Aragón-Correa[b]
University of Granada, Business and Management Department, Economics and Business School, Campus Cartuja S/N, 18071 Granada, Spain
[a]*vferron@ugr.es*
[b]*jaragon@ugr.es*

Prior literature focused on environmental management has largely studied the external costs caused by business activities. However, less is known about which and how environmental costs are internalized by companies. This work looks for three main objectives: first, to identify and delimitate a clear description of different environmental costs that should be considered by firms; second, to analyze the European regulative context where transport companies establish their business activity; third and finally, to explore connections between the regulatory efforts by policy makers and managers' perceptions and behaviors by means of a case study in the Spanish road freight transport industry. Using a theoretical review of academic papers, institutional agendas, and a case study in the road freight transport industry, we offer our own proposals and suggestions. Our final findings suggest that since managers perceive that environmental issues are relegated to a secondary priority by public authorities, companies might be facing environmental issues not beyond legal requirements.

1. Introduction

The environment is affected by business activity through the consumption of materials and utilities and the generation of wastes and emissions. This consumption of resources and generation of waste represent significant costs that, in some cases, have to be paid by firms (Gibson and Martin, 2004). For instance, GtZ, an international cooperation organization for sustainable development, estimates that non-product-output costs account for between 10% and 30% of total production costs.

As a consequence, the diminishing of these environmental costs has been acquiring an increasing importance in the current business context. This circumstance has highlighted the necessity of a greater concern by firms through pursuing more environmentally proactive behaviors, understanding this proactivity as *the firms' tendency to initiate changes in its various strategies policies rather than to react to events* (Aragón-Correa, 1998, p. 557).

Research on environmental management grounded in resource-based view of the firm (Wernerfelt, 1984; Barney, 1991) has widely contributed to show the potential that a proactive environmental management might have in generating valuable organizational abilities related to cost reduction (Hart, 1995; Hart and Ahuja, 1996; Nehrt, 1998; Christmann, 2000; Darnall and Edwards, 2006). For instance, Christmann (2000) showed that several process-focused practices are essential in order to implement proactive environmental practices and, simultaneously, they might imply cost reduction in a relatively short term. However, little attention is paid about how environmental management could take into account environmental costs and theirs consequences into business strategy in a greater degree. Thus, analyzing what underlies upon the label of environmental costs and how both companies and regulators face them becomes a critical issue.

This study attempts to provide evidence about the following research questions that remain unexplored: is it possible to delimitate different types of environmental costs? and especially, how is environmental cost management related to environmental proactivity of the firm? To address these research questions, our work is focused on the road freight transport industry due to the special features of this economic sector. Prior literature focused on transportation and the environment has largely studied the external costs caused by the transport industry (Mayeres *et al.*, 1996; Eriksen, 2000; Sælensminde, 2004; Quinet, 2004; Haller *et al.*, 2007; Lemp and Kockelman, 2008; Arimura and Iwata, 2008). The "Green Paper" of the European Commission has estimated that external costs caused by the road freight transport sector through noise and pollution account for 0.6% of Communitarian Gross Domestic Product (European Commission, 1995). According to Baumol and Oates (1988), externalities are defined as *spillover effects that occur when an economic transaction affects a third party that does not feature in the considerations of those involved in the transaction* (Hörmandinger and Lucas, 1996, p. 1). In terms of cost, externalities are understood as those costs that are generated as a consequence of business activity but are socially paid. They have been one of the main spotlights of criticism of the transport industry. The road freight transport sector has not been included in those industries affected by emission quotes assigned to diverse European business activities in the framework of Kyoto's protocol. This decision has been the target of numerous disapprovals by environmental activists arguing that public authorities have intentionally forgotten the environmental damage of this industry, whose only purpose is to minimize their operative costs and allow a better profitability. Meanwhile, heads of the industry have frequently argued that, on the one hand, the establishment of those measures would have a harmful effect on European competitiveness and, on the other hand, that actual industry efforts are not being seriously recognized. In this context, since there is a gap regarding how public authorities, practitioners, and ecologists understand the environmental challenge in the transport industry, it seems clear that less is known about which and how these environmental costs are internalized by companies that operate in transport-related industries.

In this study, two main objectives exist covering theoretical and applied issues. First, after reviewing previous literature, our work proposes a typology of different environmental costs that have to be internalized by firms. Our analysis provides a detailed description and some examples to better understand which environment-related costs are. This objective is especially relevant to knowing how firms should face environmental costs effectively and how policymakers might push to avoid externalities, specifically in the transport industry. Identifying the environmental costs is also necessary for the appropriate allocation of these costs. For instance, when environmental costs are not accurately allocated to processes or products that generate them originally, cross-subsidization may occur between processes or products (Graff *et al.*, 1998; Bartolomeo *et al.*, 2000), and this misallocation of environmental costs can send incorrect signals to decision makers (Environmental Protection Agency, 1995; Graff *et al.*, 1998). Second, our study considers those proactive companies that focus their effort on anticipating the appearance of environmental costs beyond legal requirements will be able to develop a rare, inimitable, and valuable capability that ease the achievement of a sustainable competitive advantage. Therefore, the main novelty of this work is attempting to answer whether and how is possible to internalize environmental costs by transport companies. To address this, we analyze the historical and current European regulative policies in transport industry and, more specifically, using the Spanish road freight transport industry as a fieldwork, we also offer some evidence of managers' reactions to these regulative efforts throughout a case study.

The remainder of the study is divided into four sections. Section 2 contains a theoretical classification about environmental costs, bearing in mind the related prior literature. Section 3 introduces a summary about the historical and current situation of the European regulative context for transport industry. Section 4 offers a case study, which highlights how Spanish companies face the challenge of the increasing importance of environmental costs inside the sector. Finally, Section 5 summarizes the main conclusions and future research lines of this work.

2. Environmental Costs under RBV

The RBV of the firm provides a theory to explain competitive advantage as an outcome of the development of valuable organizational capabilities (Wernerfelt, 1984; Barney, 1991). This theoretical approach argues that companies which are able to accumulate resources and capabilities that are rare, valuable, non-substitutable, and imperfectly imitable will achieve an advantage over competitors (Barney, 1991). Related to the environment, several studies focused on the RBV have considered that it is possible to reduce firms' costs by means of the improvements due to the generation of environmentally proactive capabilities (Hart, 1995; Christmann, 2000; Darnall and Edwards, 2006). For instance, the seminal work of Hart (1995) argued that environmental strategies related to activities seeking to prevent pollution generate some organizational capabilities that can be linked to the minimization of firms' costs.[1] Specifically, the cost saving due to pollution prevention can avoid the expense of pollution control technology (Hart, 1995; Sharma and Vredenburg, 1998) and can increase productivity and efficiency because less waste is translated as a better use of inputs (Hart and Ahuja, 1996). However, not only pollution prevention activities are capable to diminish the firm's costs. Costs amount can be minimized by reducing legal sanctions as fines and penalties for the non-compliance with environmental regulations (Sharma and Vredenburg, 1998). Then, all these environmental deeds related to the manufacturing process can lower the amount of firms' costs (Christmann, 2000; Nehrt, 1998) and simultaneously can allow that firms reduce theirs negative environmental impacts.

Additionally, in the current general business context an increasing importance on environment-related costs exists. These environmental costs can account for 5–20% of the total cost of business activities (Ditz *et al.*, 1995). For instance, according to the study of Ditz *et al.* (1995), Amoco Yorktown refinery estimated that its environmental costs were 22% of its total production costs and, the environmental costs of Dupont, the second higher chemical company of the world, were 15.3% of its fixed production costs and 3.8% of its variable production costs. Nonetheless, despite the increasing relevance of environmental costs, there is a conceptual lack about what these environmental costs are (Gibson and Martin, 2004; Environmental Protection Agency, 1995; United Nations Division for Sustainable Development, 2001). Most of the schemes developed internationally note that those costs are clearly related to the control or avoidance of pollution, such as costs to prevent the generation of contaminant emissions, to treat generated waste, and/or for the remediation of polluted sites (Savage and Jash, 2004). In the accounting field, these types of costs are often referred as Environmental Protection Expenditures (EPE). However, our study proposes that environmental costs under environmental management perspective include not only EPEs but also other important environment-related expenses. A comprehensive study about what environmental costs are might suppose a first stage in order to minimize them. In other words, if an accurate understanding of these costs exists by both regulators and practitioners, it will be possible to face them effectively. Thus, it is essential to understand the environmental benefits and economic challenges involved in the analysis of environmental costs that might be internalized and paid by firms.

Based on prior literature related to environmental costs (Environmental Protection Agency, 1995; Graff, *et al.*, 1998; Gale and Stokoe, 2001; Savage and Jash, 2004), the current study proposes a classification that comprehensively delimitates these environmental costs. Our proposal of theoretical classification includes four types

[1]Hart (1995) argued that the firms can redesign products in order to eliminate the need for a pollution-generating input in the manufacturing process or redesign products to be less pollutant for consumers.

of environmental costs that may be present in the firms: (1) internal, (2) external, (3) contingent, and (4) image and relationship costs.

First, internal costs refer to those costs related to the natural environment resulting from the productive process of the firm. These internal costs are composed of direct and indirect costs (Gale and Stokoe, 2001). Direct costs include those costs related to the procurement of raw materials and natural resources (energy, water, etc.). In addition, internal direct costs are composed of expenses caused by the management and elimination of non-product outputs such as solid and/or hazardous waste, wastewater, and air emissions (Savage and Jash, 2004). Both purchasing materials and managing waste are examples of internal direct costs because they can be traced directly to a specific product or activity (Gale and Stokoe, 2001). Unlike direct costs, internal indirect costs are allocated as a whole to cost centers, departments, or facilities (Gale and Stokoe, 2001). For instance, expenses caused by environmental R&D projects or environmental training are internal indirect costs.

Second, external costs are those derived from the environmental damage generated by the firm. Theoretically, this category of costs might be internally monetized in order to delimitate "the maximum amount that people would be willing to pay to avoid the damage or the minimum amount of compensation that they would accept to incur it" (Gale and Stokoe, 2001, p. 122). An illustrative example of this category is the payment of a fine because of non-compliance with environmental regulations.

Third, contingent costs refer to environmental costs that, unlike external costs whose circumstances already exist and must be effectively paid for, are not certain to occur in the future but depend on uncertain future events (Environmental Protection Agency, 1995; Gale and Stokoe, 2001). Budget reserves for remediation costs or for repairing a potential environmental problem or costs related to fit for future compliance of a stricter environmental regulation exemplify this cost category.

Finally, image and relationship costs refer to voluntary expenses incurred for corporate image purposes and/or for maintaining or enhancing relationships with regulators, customers, suppliers, host communities, investors/lenders, and the general public. This category includes costs for advertising or public relations campaigns. On the other hand, it might also be included here that there can be lower earnings due to a negative environmental image of the company. Table 1 summarizes our theoretical classification of environmental costs.

Table 1. Proposal of a classification of environmental costs paid by companies.

Types of environmental costs	Examples
(1) *Internal costs*	
Direct costs	Waste management costs, production and procurement costs
Indirect costs	Environmental R&D projects costs, environmental training costs
(2) External costs	Fines and penalties for non-compliance
(3) Contingent costs	Remediation costs, restoration costs
(4) Image and relationship costs	Environmental advertising expense, lower earnings for a negative environmental image

3. The Road Freight Transport Industry in the European Context

After introducing our proposed classification, this section analyzes how environmental issues of transport industry are addressed in Europe. Focusing on the European transport industry and their environmental impacts is relevant for three main reasons. First, this sector is very connected to environmental issues, not only because of the potential problems generated through contamination by smoke or noise but also because of non-renewable inputs (especially, oil treatment is an important key). Second, the transport industry has to comply with strict and growing regulations in regards to both safety and environmental requirements. Environmental activists often target the industry, offering an increased chance for more regulation and governmental pressure in the future. Third, especially focused on road freight transport sector, the great level of externalities generated by this industry is one of the main spotlights of criticism by activists groups, regulators, and the society, in general. For instance, the "Green Paper" of the European Commission considers that external costs caused by the road freight transport sector through noise and pollution account for 0.6% of communitarian Gross Domestic Product (European Commission, 1995).

In the context of European Union, the transport industry has suffered an important economic liberalizing change in the last decades. The elaboration and subsequent publication of the Commission "White Paper" in 1985, in which the necessity of an establishment of a European common market context for transport services and a gradual deregulation for this sector are definitively admitted (European Commission, 1985), represented a turning point in this sector. Until this report, there was a scarce normative advance related to the economic liberalization process of the transport industry in Europe because of a confrontation between liberalists and other Member States with less liberal tradition (Martínez–Álvaro, 1989). However, the change was started in 1983, when the European Parliament charged Commission with lack of advances in the implementation of a European common policy of transport. The sentence of this charge gave rise to the elaboration of the "White Paper" that encompassed a progressive harmonization of national markets (De-Rus Mendoza and Campos Méndez, 2006). After this milestone, the fulfillment of this liberalizing process took place and, in 1992, the second "White Paper" related to transport industry (European Commission, 1992) was presented. The guiding principle of this document was the opening-up of the transport market, providing an approach toward the elaboration of communitarian policy. Nowadays, the development of the European transport policy is in its last stage of implementation. The last "White Paper" related to transport industry (European Commission, 2001), which is called "European Transport Policy for 2010: Time to Decide," considers that the advance toward liberalization is carried out and new objectives are now proposed for the transport industry: achieving environmental sustainability, ensuring safety, promoting intermodality, among others (De-Rus Mendoza and Campos Méndez, 2006). Figure 1 illustrates the process of liberalization of the European

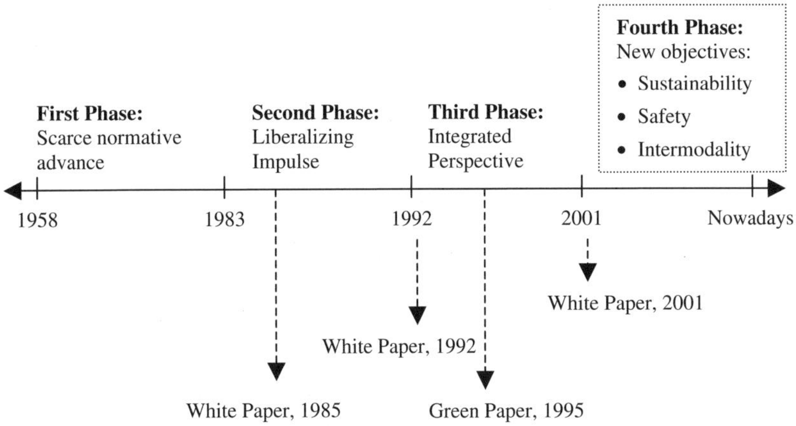

Fig. 1. Four phases in the liberalizing process of European Transport Sector. *Source*: De-Rus Mendoza and Campos Méndez, 2006.

transport policy, which is formed by four main phases that are composed of the current European situation in this economic sector.

According to De-Rus Mendoza and Campos Méndez (2006), the European regulation for transport industry is based on three economic grounds or basics. First, the integration of infrastructures and services that addresses in which degree companies related to building, conservation, and operating of infrastructures (roads, railways, ports, airports, etc.) operate jointly with companies that use these infrastructures in order to produce transport services. Second, the positive effects caused by the growth of efficient transport networks, such as the economic development or the generation of economies of scope. Finally, the special importance of the negative effects that are generated by this sector in terms of caused damages and affected people.

As the main objective of this chapter is to analyze environmental costs upon the management perspective, we will focus on the last economic basic: the special importance of negative externalities in the European transport industry. In terms of environmental costs, externalities are understood as those expenses that are generated as a consequence of negative environmental impacts of business activity but are socially paid. Prior literature focused on transportation and environment has largely studied external costs caused by the transport industry (Mayeres *et al.*, 1996; Eriksen, 2000; Sælensminde, 2004; Quinet, 2004; Haller *et al.*, 2007; Lemp and Kockelman, 2008; Arimura and Iwata, 2008). The meta-analysis about Western European external transport costs made by Quinet (2004) showed that a wide range of measures and specifications exists when researches attempt to study this topic.

Moreover, several studies have classified these external environment-related costs in the transport sector concerning specific characteristics such as vehicle model (Lemp and Kockelman, 2008), or concerning the external cost caused by the use of road transport in cities (Mayeres *et al.*, 1996; Eriksen, 2000; Sælensminde, 2004). For instance, according to Mayeres *et al.* (1996), the marginal external costs of urban transport are congestions, pollution accidents, and noise costs, whereas, according to Lemp and Kockelman (2008), external vehicle costs include emissions of greenhouses and other gases, crash costs, roadway congestion, and space consumption. In general terms, the three main negative externalities generated by transport industry are pollution, accidents, and traffic congestion (De-Rus Mendoza and Campos Méndez, 2006).

Meanwhile accidents and traffic congestion generate social costs — for instance, accidents generate medical costs and traffic congestion generates costs to health and loss of time, pollution generates environmental costs, which are the core of this work. For this reason, we focus on this category of negative externalities related to the transport industry. The first approach to take into account environmental costs associated with transports was during the third stage of the liberalizing process of transport industry — see Fig. 1, phase between 1992 and 2001, more specifically after the appearance of the "Green Paper" in 1995 (European Commission, 1995). In this, document was introduced one proposal for eliminating these environmental costs, which is the internalization of them, understanding that users have to support environmental costs throughout increasing prices of transport services. The introduction of a full tariff-setting for the infrastructures' use was proposed in order to internalize costs of each transport mode (European Commission, 1995, 1998). Through this principle, it is recognized that users had to pay the whole costs of transport infrastructure and services (Nash and Matthews, 2005), forming the general rule for pricing in this industry. Table 2 provides several tools for applying efficient prices according to pollution tariff-setting for road freight transport.

However, nowadays there is no solid Communitarian policy, which homogenizes charges and, in the practical scope, this full tariff-setting is not harmonized in all Member States. For instance, the average charge for a heavy goods vehicle covering 100 kilometers varies between 12 and 24 Euros (European Commission, 2001). Thus, the Communitarian tariff-setting policy is not totally defined. Moreover, related to road freight transport, this sector generates environmental costs higher than charges that currently pay (Proost *et al.*, 2002).

These circumstances highlight the necessity of stricter regulations at Communitarian level that enables the harmonization of charges and taxes among Member

Table 2. Economic tools related to avoid pollution on road freight transport.

Charges in taxes per kilometers
Differentiated taxes concerning polluting features of oil
Global tax for CO_2 emissions
Differentiated taxes concerning geographical areas and schedules
Tariff-setting concerning the infrastructure

Source: European Commission, 1998.

States in order to achieve an efficient market structure. In doing so, this work argues that transport companies play an important role for achieving this objective. The internalization of environmental costs of transport industry must be carried out by companies and these environmental costs management has to become a priority for transport companies. Drawing on RBV, an accurate management of these costs might imply the development of a rare, valuable, and inimitable capability that enables companies to achieve a sustainable competitive advantage. Moreover, as European regulators are still developing a more stringent legal framework for internalizing environmental costs of transport industry, those companies that anticipate the appearance of these costs beyond legal current requirements, which are environmentally proactive companies, will be more able (1) to comply with future stricter regulations, (2) to manage costs in a more effective way, and (3) to protect the environment simultaneously.

4. The Road Freight Transport in Spain

4.1. *Spanish regulation for road freight transport*

We have delimited the geographical scope of our case study in Spain. Although the transport industry is a key sector all around the world, the strategic importance of the transport industry is particularly clear for the Spanish economy as a consequence of the geographic location of the country in a non-central position in its European market, the role of being an intensive supplier of agricultural products to the rest of the European countries, and the importance of the tourism in the country. Additionally, inside the service sector in Spain for 2007, the transport industry accounted for 8.14% of total business volume, occupying fourth place in the non-financial market services sector (EAS, 2008). Four sectors are part of the group of linked-transport services in Spain: passenger transport, freight transport, postal and mailing activities, and annex-transport activities. Road freight transport generates the highest revenue inside this classification. For example, in 2006, more than half of transport activity companies belonged to road freight transport (61.1% of total volume). Moreover, freight transport generated the highest revenue (39.5%) and employed the greatest percentage of people among the related industries (42.3%) (EAS, 2008). Nonetheless, road freight transport is more than three-fourths of the total transported freight (EAS, 2008). The current study is focused on road freight transport. The relevant authority in charge of watching competitiveness and efficiency of this industry for the whole country is the General Department of Road Transport. This agency approved in 2001, with a horizon of five years, a plan called the Strategic Plan for Road Freight Transport (in Spanish, Plan Estratégico para el transporte de mercancías por carretera), also known for its initials in Spanish as PETRA. The main objective of this plan is to build a reference framework where "the interconnected interventions of the whole sector can be delimited in order to reach the modernization of this sector and, through it, a competitive position." Currently, the Strategic Plan for Road Freight

Transport is on its second implantation horizon, which spans from 2007 to 2012, and is called PETRA II since it is merely a continuation of the initial PETRA plan.

Related to environmental issues, PETRA establishes environment protection as one of its cornerstones. With regard to environmental costs that companies have to pay, there are two main factors that economically affect the whole sector. On the one hand, it is investment in "Euro" type vehicles. The "Euro" classification divides vehicles depending on its pollutant power; the "I" category is the most pollutant, while "IV" and "V" categories are less so, as vehicles included in these two levels use urea, a substance that reduces the emissions of pollutant gases. On the other hand, freight operators must pay a tax (€/km) because of their use of infrastructure; the more pollutant the vehicle, the higher the tax amount. To face these two economic factors that affect companies' costs structure, PETRA has launched three programs of measures: (1) encouragement of the investment in ecological vehicles, (2) promotion of safety and public identification of the adaptation to environmental requirements and (3) the establishment of restrictions for vehicles that are not adapted to environmental requirements. Table 3 provides a summary of PETRA's measures and classifications regarding environmental costs, in accordance with the theoretical typology that was introduced in Section 2.

In general, the proposed measures in PETRA plans delimitate a medium-term framework for the environmental objectives. In other words, meanwhile PETRA plans consider that safety, training, and quality in the road freight transport sector are topics of priority importance for improving the competitiveness and efficiency in the transport industry, measures in environmental issues have to be implemented through a medium-term temporal perspective.

Table 3. Summary of Spanish regulations (PETRA Plan) and connections with the classification of environmental cost.

Regulative measure	Objective of the regulation	Type of environmental cost
Encouragement of the investment in ecological vehicles	To favor and to encourage investments in vehicles non-pernicious for the environment, so that external transport costs become minimized at origin	Internal environmental cost, which may be direct or indirect depending on its allocation
Promotion of safety and public identification of the environmental adaptation	To promote the comparison between diverse firms regarding reached achievements in environmental and road safety issues, as a way to boost the adoption of "best practices"	Contingency environmental cost
Establishment of restrictions for vehicles non-adapted to environmental requirements	To favor a quick adoption of ecological measures, which allow reducing external costs at origin	External environmental cost

4.2. *The Spanish case study: Hermanos Aznar S.A.*

This study also highlights what occurs in the practical scope, using a case study based on the company Hermanos Aznar S.A. The SME[2] business orientation and structure of Hermanos Aznar S.A. make this case a good representative of average companies in the Spanish road freight transport industry. The analyzed company is located in Vera, a coastal town in the province of Almeria (Andalucía, Spain). Vera is an important center of road freight transport because of its high level of agricultural exports to the rest of the European Union. Actually, Hermanos Aznar S.A. specializes in the international road transport of agricultural products. It has valuable relationships with commercial agents located in main Central European cities where most of the communitarian trade transactions of this kind of products take place. Hermanos Aznar S.A. had 60 employees, a fleet composed of 57 trucks, and a turnover of more than €3.5 million during 2007.

Related to environmental issues, Hermanos Aznar S.A. has to comply with strict regulations, just like the rest of the industry competitors. The three main environmental issues for the company are (1) the investment in ecological fleets (Euro IV and Euro V vehicles), (2) the payment of environmental fees and taxes related to combustibles, tires, and the use of infrastructure (expressways) and (3) waste management.

First, referring to the investment in an ecological fleet, the current Spanish regulation requires acquiring non-pollutant vehicles. Thus, the "green transformation" of vehicles takes place as companies update their fleets. Hermanos Aznar S.A. has a fleet composed of 57 trucks, 14 of which are type Euro II vehicles, 32 are type Euro III, and 11 of are type Euro IV. This means that approximately 20% of the current fleet of the company is composed of ecological trucks. The acquisition of Euro IV and V vehicles implies environmental cost savings, specifically indirect internal costs savings, because firms profit from tax deductions for each acquisition. However, the renewal of the remaining 80% of the fleet is going to take place over a long period of time. Therefore, because of the mandatory character of the regulation concerning the ecological fleet and the long-term nature of these renewals, the environmental cost savings due to new acquisitions do not entail a priority deal.

Second, the payment of environmental fees and taxes related to combustibles and the use of infrastructure imply an internal and direct environmental cost savings for two reasons. On the one hand, the more the vehicle pollutes, the higher the payment. For instance, Antonio Aznar-Domínguez, manager of Hermanos Aznar S.A, commented that the most expensive tax per traveled kilometer that his company has to pay is for using German expressways. Thus, as this expense is attributed directly and individually to the service, the payment for the use of infrastructure is an internal direct environmental cost. Hermanos Aznar S.A. can reduce this cost by using type IV vehicles when the company operates with German clients. On the

[2]Small- and medium-sized enterprises.

Table 4. Summary of environmental costs of Hermanos Aznar S.A.

Hermanos Aznar S.A.	Theoretical classification	Is it possible to reduce or eliminate it?
Investment in ecological fleet	Internal indirect environmental cost	Yes, by tax deduction for acquisition
Payment of environmental taxes in expressways and combustibles	Internal direct environmental cost	Yes, by accurate planning between freight services and available ecological trucks
Payment of environmental taxes in tires acquisition	Internal indirect environmental cost	No
Waste management costs	Internal indirect environmental cost	No

other hand, the use of urea as a reducer component of emissions implies an increase of the final cost of the service. Concretely, a truck consumes one liter of Ad-Blue product (urea) per 100 kilometers, and one liter L of this component costs approximately 0.5 Euros. According to our theoretical classification of environmental costs, this expense is an internal direct cost because it is attributed directly and individually to the service. However, it is not possible to reduce this environmental cost as ecological trucks (type Euro IV and V vehicles) require this chemical component. Unlike prior examples, the environmental tax that has to be paid by the purchasing of tires is considered an internal indirect environmental cost because it is attributed along with the lifespan of the tire. This cost cannot be reduced.

Third, the company's value chain activity that is most important to the managers of Hermanos Aznar S.A. is waste management. The residue and waste produced by the company's garages (filters, aerosols, dirty clothes, etc.) must be collected by specialized companies in waste withdrawal and treatment. However, this service is not free in all cases. This service is free only for the case of retiring used oil and batteries, and infrastructure and processes needed for this withdrawal are very specific. Therefore, waste management implies a high internal indirect environmental cost that the company has to support in order to comply with environmental regulations.

Table 4 shows a summary of the environmental costs that have been assumed by Hermanos Aznar S.A. and their relationships with the theoretical classification explained in the prior section of this study.

5. Discussion, Implications, and Limitations

5.1. *Discussion*

Drawing on RBV, the main objective of this paper is to answer whether an interest to minimize all types of environmental costs is more related to the implementation of proactive environmental strategies. The analyzed company attempts to face its environmental costs as environmental regulation requirements advance. According to our theoretical classification, not taking into account external, contingent, and image and relationship costs implies that the company is only interested in complying with environmental responsibilities, which are not beyond the limits of legal requirements. Thus, there is an important congruence between the environmental orientation of Hermanos Aznar S.A. and environmentally reactive strategies because the company is interested in environmental issues in a parallel way to regulatory requirements (Roome, 1992).

This result agrees with the case study of Haller *et al.* (2007) about the ecological actions of the Forest Preserve District of Du Page County in Illinois. Haller *et al.* (2007, p. 12) argued that "*the shortfall or delay in assistance and promotion at the federal level limits severely the prospects for voluntary conversion of 'un-mandated' private and public fleets.*" In the same manner, Hermanos Aznar S.A. is not motivated to implement voluntary proactive environmental practices in order to internalize and, subsequently, to minimize its environmental costs because managers perceive that public authorities give a relaxed importance to the environmental issues and do not anticipate serious regulative efforts in the future. The operation manager is clear when he says: "In a global economic bankruptcy, I do not think that governments or managers have real interests in making more serious progresses in the environmental arena; we will have to comply with additional regulations if they appear and we would do it for sure; meanwhile, we will wait for it." Hence, the presented case study highlights the potential presence of an environmentally reactive response pattern within the Spanish transport industry limiting friendly environmental actions as well as legal requirements are advancing.

5.2. *Implications for practitioners and regulator*

In a parallel way to prior objectives previously explained, the current paper analyzes the level of congruence between regulation and practical settings for addressing the environmental costs that have to be paid by companies in the road freight transport industry. By using a case study in Spain, this study highlights that regulators have a very relevant role in promoting environmental costs in the industry through different kinds of planning measures. Our findings suggest that both public authorities and practitioners in the Spanish road freight transport give a secondary role to environmental issues. The responsible organism of the road freight transport sector, the Department of Road Transport, bestows a medium importance on environmental actions in the transport sector. Similarly, nowadays managers of Hermanos Aznar S.A. are focused on obtaining the ISO-9001 quality label; meanwhile, environmental certifications is push into a less important aim. Combining prior arguments, it seems logical that companies are facing the challenge of internalization of environmental costs not beyond the legal environmental requirements. However, it is important to note that the existence of non-internalized negative externalities related to the environment impedes an optimal resources allocation in the transport industry (De-Rus

Mendoza and Campos Méndez, 2006). Thus, both regulators and managers should work together in order to achieve the objectives of modernization, safety, and sustainability within this industry.

Unlike other sectors such as the automobile industry or alimentary industry, whose companies can profit from the benefits of ecological differentiation strategies *via* an increase in sales, managers in the road freight transport industry understand the environmental challenge of an increase in production costs, without an increase in business performance. Moreover, actually in Spain, it looks like public authorities do not contribute enough to change this negative thinking. For all these reasons, it is important to highlight how our study helps to clarify the importance of understanding the different types of environmental costs that are borne by companies. It is especially interesting to see how public regulators do not have a neutral influence on these costs. Furthermore, the selection of regulative measures and their importance in terms of cost have a financial effect not only on firms in the transport industry but probably also on their strategic approaches regarding the natural environment and the vision about how to deal with this issue in the future.

5.3. *Limitations and future works*

There are three main limitations in this study which must be interpreted as future research. First, case study methodology implies a qualitative analysis of circumstances that occur in a specific company; However, this technique does not enjoy benefits of a quantitative analysis, as the extrapolation of results to other companies and/or sectors. Future research may study the presented topic by an empirical approach in which both databases with economic information about companies (cost or expense figures) and statistical techniques (as regression) will be applied. Second, our transport-focused approach limits the generalizability of results and, as a consequence, of the proposed classification of environmental costs. Future research may continue the analysis of this internalization of environmental costs by companies established in other sectors and how public authorities influence this challenge. Finally, future research might study whether a higher amount of environmental costs is correlated to other important strategic variable within companies. For instance, it would be interesting to study whether innovativeness could affect the relationship between environmental costs and proactive environmental strategy.

Acknowledgments

The authors are very grateful to all the executive members and employees of the company Hermanos Aznar S.A. for their willingness to share their experiences and information with us. The authors would especially like to thank Antonio Aznar-Domínguez, current production manager of Hermanos Aznar S.A, who was enthusiastic about the potential of this paper.

References

Aragon-Correa, J.A., 1998. Strategic proactivity and firm approach to the natural environment. *Academy of Management Journal* **41**(5), 556–567.

Arimura, T.H. and Iwata, K., 2008. Economics analysis on motor-vehicle type regulation: Policy evaluation of NOx-PM Law. *Environmental Science* **21**, 103–114.

Barney, J.B., 1991. Firm resources and sustained competitive advantage. *Journal of Management* **17**(1), 99–120.

Bartolomeo, M., Bennett, M., Bouma, J.J., Heydkamp, P., James, P. and Wolters T., 2000. Environmental management accounting in Europe: current practice and future potential. *The European Accounting Review* **9**, 31–52.

Baumol, W.J. and Oates, W.E., 1988. *The Theory of Environmental Policy.* University Press, Cambridge, UK.

Christmann, P., 2000. Effects of "best practices" of environmental management on cost advantage: The role of complementary assets. *Academy of Management Journal* **43**, 663–680.

Darnall, N. and Edwards Jr., D., 2006. Predicting the cost of environmental management system adoption: The role of capabilities, resources and ownership structure. *Strategic Management Journal* **27**, 301–320.

De-Rus Mendoza, G., Campos Méndez, J., 2006. Los fundamentos económicos de la política de transporte europea: un análisis crítico. In: *La política de transporte europea: el papel del análisis económico.* Fundación BBVA, Bilbao.

Ditz, D., Ranganathan, J. and Bank, R.D., 1995. *Green Ledgers: Case Studies in Corporate Environmental Accounting.* World Resources Institute, Washington.

Encuesta Anual de Servicios (EAS), 2008. *Panorámica del Transporte (in English, Transport's Outlook*). National Statistic Institute of Spain (Instituto Nacional de Estadística, in Spanish capitals, INE).

Environmental Protection Agency, 1995. *An Introduction to Environmental Accounting as a Business Management Tool: Key Concepts and Terms.* United States Environmental Protection Agency, Washington, D.C.

Eriksen, K.S., 2000. Calculating external costs of transportation in Norway. *European Journal of Transport and Infrastructure Research*, 9–25.

European Commission, 1985. Completing the internal market. *White Paper COM (1985) 310.*

European Commission, 1992. The development of the future common transport policy. *White Paper.*

European Commission, 1995. Towards fair and efficient pricing in transport policy. Policy options for internalising the external cost of transport in the European Union. *Green Paper. COM (95) 691.*

European Commission, 1998. Fair payment for infrastructure use: A phased approach to a common transport infrastructure charging framework in the EU. *White Paper. COM (1998) 466.*

European Commission, 2001. European transport policy for 2010: Time to decide. *White Paper COM (2001) 370.*

Gale, R.J.P. and Stokoe, P.K., 2001. *Environmental Cost Accounting and Business Strategy.* Kluwer Academics Publishers.

Gibson K.C. and Martin B.A., 2004. Demonstrating value through the use of environmental management accounting. *Environmental Quality Management* **13**, 45–52.

Graff, R.G., Reiskin, E.D., White, A.L. and Bidwell, K., 1998. *Snapshots of Environmental Cost Accounting.* United States Environmental Protection Agency (USEPA).

Haller, M., Welch, E., Lin, J. and Fulla, S., 2007. Economic costs and environmental impacts of alternative fuel vehicle fleets in local government: An interim assessment of a voluntary ten-year fleet conversion plan. *Transportation Research Part D: Transport and Environment* **12**, 219–230.

Hart, S.L., 1995. A natural-resource-based view of the firm. *Academy of Management Review* **20**, 986–1014.

Hart, S.L. and Ahuja, G., 1996. Does it pay to be green? An empirical examination of the relationship between pollution prevention and firm performance. *Business Strategy and the Environment* **5**, 30–37.

Hörmandinger, G. and Lucas, N.J.D., 1996. Is clean enough? The influence of environmental externalities on markets for fuel cells in transport. *Transportation Research Part D: Transport and Environment* **1**, 63–78.

Lemp, J.D. and Kockelman, K.M., 2008. Quantifying the external costs of vehicle use: Evidence from America's top-selling light-duty models. *Transportation Research Part D: Transport and Environment* **13**, 491–504.

Martínez-Álvaro, O., 1989. *Treinta años de política de transportes de la Comunidad Europea.* Centro de Publicaciones del Ministerio de Transportes, Turismo y Comunicaciones, Madrid.

Mayeres, I., Ochelen, S. and Proost, S., 1996. The marginal external costs of urban transport. *Transportation Research Part D: Transport and Environment* **1**, 111–130.

Nash, C.A. and Matthews, B., 2005. *Measuring the Marginal Social Cost of Transport.* Elsevier, Amsterdan.

Nehrt, C., 1998. Maintainability of first movers advantages when environmental regulations differ between countries. *Academy of Management Review* **23**(1), 77–97.

Proost, S., Van Dender, K. and Courcelle, C., 2002. How large is the gap between present and efficient transport prices in Europe. *Transport Policy* **9**, 41–57.

Quinet, E., 2004. A meta-analysis of Western European external costs estimates. *Transportation Research Part D: Transport and Environment* **9**, 465–476.

Roome, N., 1992. Developing environmental management strategies. *Business Strategy and the Environment* **1**, 11–24.

Sælensminde, K., 2004. Cost–benefit analyses of walking and cycling track networks taking into account insecurity, health effects and external costs of motorized traffic. *Transportation Research Part A: Policy and Practice* **38**, 593–606.

Savage, D.E. and Jash, C., 2004. *International Guidelines on Environmental Management Accounting (EMA).* International Federation of Accountants, New York.

Sharma, S. and Vredenburg, H. 1998. Proactive corporate environmental strategy and the development of competitively valuable organizational capabilities. *Strategic Management Journal* **19**, 729–753.

United Nations Division for Sustainable Development, 2001. *Environmental Management Accounting: Procedures and Principles.* United Nations Division for Sustainable Development, New York.

Wernerfelt, B., 1984. A resource-based view of the firm. *Strategic Management Journal* **5**(2), 171–180

Chapter 31

ENERGY PLANNING FOR SUSTAINABLE DEVELOPMENT IN INDIA

Bimlesh Kumar[a] and Rahul B. Hiremath[b]

[a] *Department of Civil Engineering, Indian Institute of Technology Guwahati, Guwahati 781039, India*
bimk@iitg.ernet.in

[b] *Dean, Academic & Research, Walchand Institute of Technology, Solapur 413006, India*
rahulhirem@gmail.com

Decentralized energy planning (DEP) is in the interest of efficient utilization of resources. DEP is one of the options to meet the rural and small-scale energy needs in a reliable, affordable, and environmentally sustainable way. The main aspect of the energy planning at decentralized level would be to prepare an area-based DEP to meet energy needs and development of alternate energy sources at least-cost to the economy and environment. The geographical coverage and scale reflects the level at which the analysis takes place, which is an important factor in determining the structure of models. DEP planning involves multiple objectives and different kinds of constraints. Present work presents the methodology for the DEP. The kinds of objective functions and constraints that have to be included in the DEP have been enumerated in the present work. Decentralized planning involves scaling down energy planning to sub-national or regional scales. Energy planning at village level is the lowest level of the application of decentralized planning principle and district is the upper most level. Present work for the analysis of DEP at panchayat level, which is a cluster of village, has considered different scenarios. Conflicting objectives are considered in the implementation of DEP. DEP is assessed with the help of field studies, available data, and decentralized energy modeling. Through DEP, energy demand at 2020 has been presented for the Yedavani panchayat.

1. Introduction

The energy sector in India has received a high priority in the planning process at macro-level and also in the allocation of public funds for its growth and research and development. Its share of funds has risen from 15% in the Third Five-Year Plan to 27% in the Tenth Five-Year Plan (2002–2007).[1] In spite of continued efforts, the

[1] http://www.teri.in.org (Accessed on 10th February, 2010).

growth of this sector has not been fast enough to keep pace with the growing demand for energy services by the individual consuming sectors.

There is a strong relationship between economic development and energy consumption. The two key socio-economic indicators that drive the pace of growth in energy demand are population and gross domestic product (GDP). The population of India is growing at around 2% annually and GDP is growing at about 8–9% and both are driving demand for energy. India's energy scenario is complex with various sources of energy meeting differing requirements from various sectors. With nearly 70% of over 1.0 billion population being rural, the big challenge is to provide access to modern energy services to the rural population. State Electricity Boards have been aggressively electrifying rural areas in India. However, even though officially 86% of the villages are supposed to be electrified, most of the households away from the center of the village do not have access to the grid. The percentage households electrified in some states such as Bihar and Uttar Pradesh is low at 40% and 55%, respectively, and in Karnataka, it is 90%.[2] Further, the reliability and quality of electricity supply are poor.

The objective of energy planning is to find solutions to match demand and supply of energy sources. Solutions can be demand-oriented (e.g., price policy), supply-oriented (e.g., new supply technologies), or mixtures of supply- and demand-oriented solutions. The framework of rural energy planning has been described earlier by Hulscher (1990). The primary objective of energy planning is to arrive at a set of agreed feasible and consistent targets for energy policy followed by a set of policy actions to implement such a plan (Hiremath *et al.*, 2007).

The energy planning endeavor involves finding a set of sources and conversion devices so as to meet the energy requirements or demands of all the activities in an optimal manner. This could occur at centralized or decentralized level. The current pattern of commercial energy-oriented development, particularly focused on fossil fuels and centralized electricity, has resulted in inequities, external debt, and environmental degradation. For example, large proportions of rural population and urban poor continue to depend on low-quality energy sources and inefficient devices, leading to low quality of life. The current status is largely a result of adoption of centralized energy planning (CEP), which ignores energy needs of rural areas and poor, and has also led to environmental degradation due to fossil fuel consumption and forest degradation. CEP cannot pay attention to the variations in socio-economic and ecological factors of a region, which influence sustained success of any intervention. The regional planning mechanism such as decentralized energy planning (DEP) takes into account various available resources and demands in a region. This implies that the assessment of the demand, supply, and its intervention in the energy system, which may appear desirable due to such exercises, must be at a similar geographic scale. DEP involves scaling down energy planning to sub-national or

[2]http://mop.nic.in (Accessed on 10th February, 2010).

regional scales. This approach allows analysis of the inter-linkages between energy, physical, and biological resources, and the local economic and socio-cultural structures in rural areas. DEP for rural areas should consider all types of energy supply options renewable, fossil fuel, and electricity supplied from the national grid. DEP could lead to the formulation of national policies and programs reflective of the local needs. It is also a very useful approach for environmental analysis, as it provides for detailed analysis of the links between wood energy use and wood production, a prime environmental concern in many regions. Furthermore, it can also allow appropriate analysis of the linkages with agricultural development in both commercial and subsistence farming systems, as these are both consumers and producers of wood for fuel use. Energy planning at the village level is the lowest level of the application of decentralized planning principle. Further, the village level plans must be prepared within the limits set by a panchayat (local administrative unit), block or district level plan, for the sum total of various village plans must correspond to a panchayat, block (or taluka), or district level plans. A panchayat is the lowest administrative unit consisting of a cluster of villages and an elected body to administer developmental activities. Taking into account these features, present work explains the methodology adopted for DEP by estimating the end-use energy requirement and energy resource in the region and developing and analyzing optimized energy plans particularly at panchayat level.

2. Models for DEP

This section deals with formulating a DEP model with multiple objectives and constraints to develop an optimal energy plan at different scales. Energy planning involves finding a set of sources and conversion devices so as to meet the energy requirements of all the activities in an optimal manner. This optimality depends on the objective, such as to minimize the total annual costs of energy or minimization of non-local resources or maximization of system overall efficiency. Factors such as availability of resources in the region, costs, and energy requirement specific activities impose constraints on the regional energy planning exercise. Thus, the DEP turns out to be a constrained optimization problem. Two methods are normally used to solve the DEP problem.

2.1. *Linear programming (LP) model*

This involves a single objective function to be maximized or minimized (subject to constraints). In some situations, there may be more than one competing objective (or goal) and may need to include trade-off objectives against each other. One way of handling problems with multiple objectives is to choose one of the goals as the supreme goal and to treat the others as constraints to ensure that some minimal "satisfying" level of the other goals is achieved. Ramakumar *et al.* (1986) have developed a single objective LP model for the design of Integrated Renewable Energy

System, where energy resource allocation for the minimization of cost was calculated on the basis of system efficiency. Joshi *et al.* (1992) developed an LP model for DEP for three villages in Nepal. The optimization aimed at minimizing cost function considering a mix of energy resources and conversion devices. Sinha and Kandpal (1991) used an LP model for determining an optimal mix of technologies for domestic cooking, lighting, and irrigation sectors in the villages of Kunigal block, Tumkur district, India. Single objective linear programming model for micro-level energy planning was developed for Bangalore North Taluka by Srinivasan and Balachandra (1993), considering different energy sources and their end-use combinations.

2.2. *Goal programming (GP) model*

All the linear programming models developed so far had a single overriding objective, such as maximizing the revenue of the study area or maximizing the biomass energy production or minimizing the total cost of energy sources or maximizing the generation of surplus biomass, etc. However, in reality, achieving such a single objective through mathematically feasible though, the outputs have little utility. Very often optimizing an energy system could involve multiple objectives, namely, minimizing the cost, maximizing use of local energy sources, maximizing employment, reducing emission of pollutants, etc. Thus, an approach or model to optimize multiple objectives for a given set of constraints is necessary. GP is powerful and flexible modeling tool to deal with the above types of multiple criteria decision-making problems in energy planning and management for sustainable development of rural areas. GP provides a way of striving toward several such objectives simultaneously. The basic approach of GP is to establish a specific numeric goal for each of the objectives, formulate an objective function for each objective, and then seek a solution that minimizes the weighted sum of deviation of nine objective functions from their respective goals.

2.3. *The rationale for selecting GP for DEP in the present analysis*

The rationale for selecting GP for DEP in the present analysis is due to the following reasons:

- Multi-criteria objectives have been considered.
- It is less subjective even among techniques based on LP, such as multi-objective LP. It offers a rather straightforward procedure for assigning weightages to the different objectives.
- Best suited for DEP at micro-level.

There have been a few applications of multiple objective models to DEP. GP has been used by Bryson and Joseph (1999) and Ramanathan and Ganesh (1995a) for providing DEP for urban locations. They have incorporated only quantitative criteria in their analysis. Utility theory has been employed for selecting appropriate energy resources by Ahmed and Husseiny (1978), while Nezhad (1990) and Saaty and Mariano (1979) have used the analytic hierarchy process for providing DEP. In the past, several techniques have been used to deal with such a problem. Ramanathan and Ganesh (1995b) used integrated goal programming to optimally match seven energy resources usable for lighting in households against 12 objectives representing the energy–economy–environmental system. Hoog and Hobbs (1993) discussed an integrated resource planning model, which considered several objectives such as cost, emissions, regional economic impact, and net value to customers. More recently, fuzzy dynamic programming has been applied to the resource allocation problem, adding more features to the selection procedure used by decision makers (Hussein and Abo-Sinna, 1995). It can be concluded that the energy sector is an essential part of the whole economic system and that modern energy planning must incorporate social and environmental objectives, leading to a multi-objective optimization problem. The economic objectives consider costs, efficiency, energy conservation, and employment generation. The environmental objectives account for environmental-friendliness factors. The objectives are first quantified, then transformed into mathematical language to obtain a multi-objective allocation model, which will be solved using pre-emptive goal programming techniques. Multi-objective non-pre-emptive goal programming model was developed by Ramanathan and Ganesh (1993). Ghosh *et al.* (1995) have developed a multi-criteria decision making (MCDM) model. MCDM is priority-based goal programming model for optimal allocation of land under cultivation in different seasons. Parikh (1997) employed a goal programming framework to maximize net revenue, minimize emissions, and minimize fertility loss of soil, subject to the energy supply and demand constraints. Deo *et al.* (1991) presented a priority-based goal programming model for village level energy planning with several goals. Kanniappan and Ramachandran (2000) developed goal programming model for Nilakkottai block in Dindigul District, Tamil Nadu. The paper aims to achieve the designed goals through optimum allocation of land area for different crops and allows the block to consume energy generated from locally available energy sources. The following section gives the model description, with different objectives and constraints, and the possible outcomes of the results of different scenarios.

2.4. *Study area*

Yedavani panchayat of Tumkur district in India has been taken a case study in the present work. Map of the study area has been shown in Fig. 1.

Yedavani panchayat of Kunigal block is composed of 12 villages. The total number of households in Yedavani panchayat is 1218. The total population is 7168, with an average household size of 6 per household. According to the questionnaire-based survey findings, cattle to human population ratio are 0.3, and sheep and goat to human population ratio is 0.35. Total dung production is 23.6 tons per day. Total

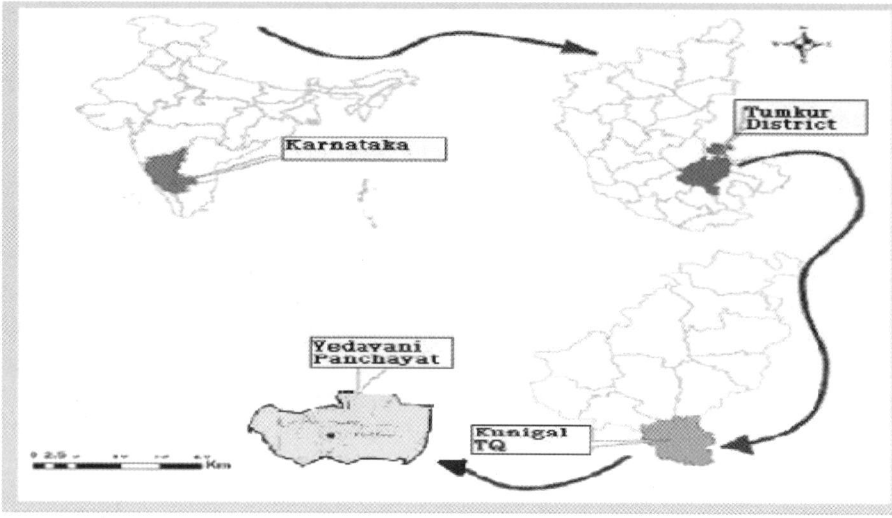

Fig. 1. Location of study area.

geographical area of Yedavani panchayat is 3674 ha. The broad land-use systems include cropland — both cultivated and fallow, wasteland, plantations, settlements and infrastructure. Currently cropland accounts for 1771 ha. The wasteland available is nearly 1050 ha, which accounts for 30% of total geographical area.

2.5. *Energy needs of Yedavani panchayat*

Figure 2 shows the energy needs of Yedavani panchayat. Cooking is the predominant activity (at 7.82 GJ/Capita/year) in Yedavani panchayat accounting for 60% of total energy use. Biomass accounts for 91% of energy for cooking. Traditional wood stoves, LPG stove, and kerosene stove are used for cooking food.

Water heating accounted for 22% of the total energy consumption in Yedavani and all the energy needs for water heating are met by biomass fuels at 2.95 GJ/Capita/year. Kerosene and electricity are the main sources of energy used for lighting in the study area. The total energy consumed in these rural industries is about 11,640 GJ/year.

3. Problem Formulation

The methodology adopted includes implementation of the different steps of DEP in an iterative way, i.e., at different stages of data collection and analysis. This study covers an assessment of the current energy demand and supply situation, the development of several forecasting scenarios (business-as-usual energy scenario, biomass energy intensive scenario, renewable energy scenario (RES), and sustainable development energy scenario), supply-demand balancing, the identification of

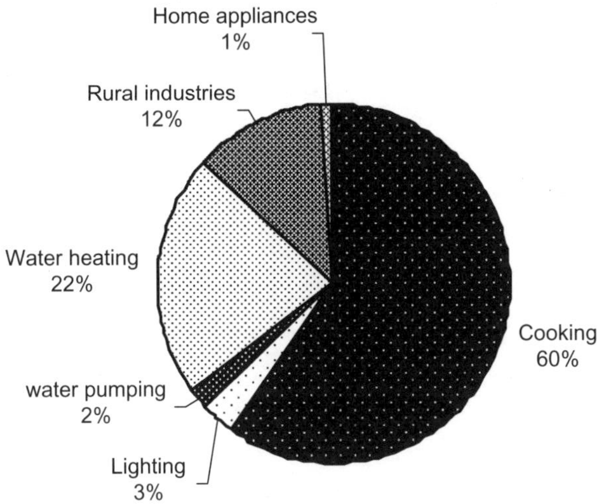

Fig. 2. Share of different activities in total energy use (in %) in Yedavani panchayat for year 2005.

intervention options, and an assessment of impacts of future trends and interventions in economic and environmental terms.

3.1. *Objective function*

The objective of the study is to allocate optimally to each end use, a certain amount of power, X_i, to be supplied by a given resource. The energy resources were selected based on either their present use or potential availability in the study area. The end uses such as cooking, lighting, water pumping for agricultural purposes, water heating, rural industries, and home appliances reflect the existing major applications in the household sector. Altogether, 12 energy resources and six end uses were identified, of which only 56 resource–end use combinations (listed in Table 1) have been selected for the formulation of the current problem based on the feasibility of the combinations. In the proposed energy-allocation problem, seven objective functions are used. First six objectives relate to economic and security considerations, whereas the last objective relates to policies on environmental protection.

3.1.1. *Minimization of cost*

Here, the term "cost" refers to the life-cycle cost (LCC)[3–5] of using an energy resource for satisfying an end use, and is expressed in Indian Rupees (Rs) per kWh (quantum of energy of all types has been converted into a common unit "kWh") of the net energy of the resource allocated. Table 2 provides the unit LCC values used

[3] http://rural.nic.in/eas.htm (Accessed on 10th February, 2010).
[4] http://cgpl.iisc.ernet.in/ (Accessed on 10th February, 2010).
[5] http://planningcommission.nic.in/reports/genrep/ (Accessed on 10th February, 2010).

Table 1. Indices used to identify all possible combinations of energy resources with end users.

Energy resources	Cooking	Lighting	Water pumping	Water heating	Rural industries	Home appliances
Dung cake[a]	1	—	—	29	—	—
Firewood[b]	2	—	—	30	—	—
Agricultural waste (Ravindranath *et al.*, 1990)	3	—	—	31	—	—
Kerosene[c]	4	13	—	32	—	—
LPG[d]	5	—	—	33	—	—
Biogas[e]	6	14	—	34	—	—
Solar thermal[f]	7*	—	—	35*	—	—
Diesel electricity[g]	—	15	22	36	43	50
Bio-diesel electricity[h]	—	16	23	37	44	51
Electricity (grid)[i]	8	17	24	38	45	52
Biomass electricity[j]	9	18	25	39	46	53
PV electricity[k]	10*	19	26	40*	47	54
Biogas electricity[l]	11	20	27	41	48	55
Biomass + diesel electricity (dual-fuel)[m]	12	21	28	42	49	56
Wind energy[n]	NA	NA	NA	NA	NA	NA

*Not considered for the study.

[a]Dung cake has a calorific value of 11.76 MJ/kg and energy equivalent is 22.35 MJ/kg (Ravindranath *et al.*, 1981).

[b]Firewood is classified as hard and soft. Hardwood (dry) has a raw value of 18 MJ/kg, whereas softwood has a raw value of 7.2 MJ/kg (Ravindranath, 1993).

[c]Kerosene fuel has a calorific value of 44 MJ/kg (Rana *et al.*, 1997).

[d]LPG fuel has a calorific value of 46 MJ/kg (Rana *et al.*, 1997).

[e]Biogas is the CH_4/CO_2 gaseous mix made to favor methane production, calorific value of 20.14 MJ/m^3 (Limmeechokchai and Chawana, 2007).

[f]Solar thermal is referred to here for cooking (thermal stoves) and water heating (solar collectors) purposes and not for generation of electricity.

[g]Shaft power from the combustion of diesel fuel. It is mainly used for water pumping and for other industrial purposes. Diesel fuel has a calorific value of 44 MJ/kg (Amulya Kumar *et al.*, 1990).

[h]Bio-diesel: Eco-friendly substitute to diesel. Bio-diesel fuel has a calorific value 37.84 MJ/kg (Mittal and Sharma, 2004).

[i]Electricity (grid): Steam turbines are generating thermal electricity in addition to hydro-power by far the most established resource for electricity generation and commercial investment (Chakrabarti and Chakrabarti, 2002).

[j]Biomass electricity: Engine runs only on producer gas from biomass, single-mode generator.

[k]Photovoltaic-based generation of power (Iniyan *et al.*, 1998).

[l]Biogas electricity: Biogas is converted to electricity.

[m]Biomass + diesel electricity: Generator runs on dual-fuel mode.

[n]Wind energy can provide electricity or pump water at costs below those of other renewable energy technologies, wind potential is not sufficient hence not considered (Ravindranath and Chanakya, 1986).

Table 2. Cost of energy resource associated with different end-use combinations (Rs/kWh).

	Cooking	Home lighting	Water pumping	Water heating	Rural industries	Home appliances
Dung cake[a,b]	0.31	—	—	0.31	—	—
Firewood[a,b]	0.72	—	—	0.72	—	—
Agricultural waste[a,b]	0.5796	—	—	0.5796	—	—
Kerosene[b]	0.82	0.82	—	0.82	—	—
LPG[b]	1.71	—	—	1.71	—	—
Biogas[a,b]	0.27	0.27	—	0.27	—	—
Solar thermal[a,b]	0.37	—	—	0.27	—	—
Diesel electricity[d]	—	15	15	15	15	15
Bio-diesel electricity[b]	—	1.584	1.584	1.584	1.584	1.584
Electricity (grid)[c]	3.96	3.96	1	3.96	3.96	3.96
Biomass electricity[a,b]	4.1688	4.1688	4.1688	4.1688	4.1688	4.1688
PV electricity[b]	15	15	15	15	15	15
Biogas electricity[a,b]	3.888	3.888	3.888	3.888	3.888	3.888
Biomass + diesel electricity (dual-fuel)[a,b]	5.0472	5.0472	5.0472	5.0472	5.0472	5.0472

[a]Planning Commission.[6]
[b]MNRES (Ravindranath and Hall, 1995).
[c]Based on State Electricity Board charges for rural households and irrigation end uses.
[d]Estimated on the basis of cost of diesel.

for analysis. The objective functions can be represented as shown by Eq. (1). This cost minimization objective function is represented as

$$\text{Min} \sum_{i=1}^{6} \left[\sum_{j=1}^{m} \left[\sum_{k=1}^{n} C_{ijk} X_{ijk} \right] \right] \tag{1}$$

where C_{ijk} is the cost coefficient, which is the unit LCC value for combination j given in Table 2.

3.1.2. *Maximization of system efficiency*

Analysis aims to maximize the system efficiency of a particular resource–end-use combination, which is the total efficiency of energy system, including the efficiencies of production, distribution, and end-use utilization. This objective function is represented as

$$\text{Max} \sum_{i=1}^{6} \left[\sum_{j=1}^{m} \left[\sum_{k=1}^{n} \eta_{ijk} X_{ijk} \right] \right], \tag{2}$$

where η_{ijk} represents the efficiency coefficient (see Table 3) for cooking, water pumping, lighting, heating water, and home appliances.

[6]http://mnres.nic.in (Accessed on 10th February, 2010).

Table 3. End-use energy efficiencies (%).

	Cooking	Home lighting	Water pumping	Water heating	Rural industries	Home appliances
Dung cake[a]	9.5	0	0	9.5	0	0
Firewood[a]	14.25	0	0	14.25	0	0
Agricultural waste[a]	9.5	0	0	9.5	0	0
Kerosene[b]	21.6	0.72	0	21.6	0	0
LPG[b]	43.2	0	0	43.2	0	0
Biogas[a]	52.8	0.88	0	52.8	0	0
Solar thermal[d]	20	0	0	20	0	0
Diesel electricity[c]	0	11.52	23.04	21	21	21
Bio-diesel electricity[c]	0	10.72	21.44	16.08	16.08	16.08
Electricity (grid)[c]	18.4	9.2	18.4	13.8	13.8	13.8
Biomass electricity[c]	21.888	10.944	21.888	16.416	16.416	16.416
PV electricity[e]	8	4	8	6	6	6
Biogas electricity[a]	28.16	14.08	28.16	21.12	21.12	21.12
Biomass + diesel electricity (dual-fuel)[c]	22.4	11.2	22.4	16.8	16.8	16.8
Wind electric power[f]	28	14	28	21	21	21

[a]Based on Ravindranath and Hall (1995) and Iiniyan *et al.* (2000).
[b]Based on Domah *et al.* (2006) and Deshmukh *et al.* (2005).
[c]Based on CGPL,[5] MOP.[2]
[d]Solar thermal does not have an external efficiency, as it comes from the Sun.
[e]Based on a 50–5 kW solar photovoltaic system Mezher *et al.* (1998).
[f]Based on Herbert *et al.* (2007) and Iniyan and Jagadeesan (1998).

3.1.3. *Minimization of use of petroleum products*

This objective function is represented as

$$\text{Min} \sum X_i, \tag{3}$$

where $i = 4, 5, 8, 13, 15, 17, 22, 24, 32, 33, 36, 38, 43, 45, 50$, and 52.

3.1.4. *Maximization of use of locally available resources*

The relevant objective functions here can be represented as follows:

$$\text{Max} \sum X_j, \tag{4}$$

where $j = 1, 2, 3, 6, 7, 9, 10, 11, 12, 14, 16, 18, 19, 20, 21, 23, 25, 26, 27, 28, 29, 30, 31, 34, 35, 37, 39, 40, 41, 42, 44, 46, 47, 48, 49, 51, 53, 54, 55$, and 56.

3.1.5. *Maximization of employment generation*

When energy is considered as a sub-system of the overall economic system, it should facilitate attainment of goals such as employment generation (Goldemberg *et al.*, 1998). Table 4 provides details of the number of people employed in developing

Table 4. Gross employment potentials of different energy resources.

Energy resources	No of employees	Consumption (TWh)	No of employees per million kWh (gross)
Coal (including soft coke)	976,660	501	1.22
Petroleum and natural gas	81,125	600	0.135
Grid electricity	—	—	6.900
Non-commercial energy resources	10,806,351	—	250

Source: MOP,[7] Planning Commission Report.[8]

various energy resources, along with the total consumption of energy resources for the year 2004–2005. The corresponding objective functions can be represented as follows:

$$\text{Max} \sum_{i=1}^{6} \left[\sum_{j=1}^{m} \left[\sum_{k=1}^{n} e_{ijk} X_{ijk} \right] \right], \tag{5}$$

where e_{ijk} denotes the number of persons employed per kWh of energy allocated for combination i, as given in Table 4. Job creation can be specified in three ways, namely,

(a) *Direct jobs*: Those jobs resulting directly from the renewable energy project or installation and include the entire production cycle from fuel production and component manufacture to waste management.
(b) *Indirect jobs*: Those jobs that arise in addition to the direct jobs referred to above and include services and inputs to the direct processes.
(c) *Induced jobs*: Those jobs generated through the increased cash flow in the broader society that arise from the wages of those employed in direct and indirect jobs.

Youth employment summit makes use of the factor "Full Time Equivalent" or FTE when expressing jobs: 1 FTE = 1 job (Austin and Worthington, 2003). This varies by country according to the employment regime in the country, namely, the number of hours worked per week, the numbers of days of week, leave per year, etc. In this study, only direct jobs are considered, yet attempt to extend the number of direct jobs created to assess possible indirect jobs based on indicators sourced in the literature is attempted.

3.1.6. *Minimization of CO_2, NO_X and SO_2 emissions*

Three major pollutants are considered to describe the impact of energy generation and use on the air quality. These are carbon oxides (CO and CO_2), sulfur

[7]http://mop.nic.in (Accessed on 10th February, 2010).
[8]http://mnres.nic.in (Accessed on 10th February, 2010).

Table 5. Environmental emissions of combinations for cooking and power generation.

	Carbon emissions (kg/kWh)	Sulfur emissions (kg/kWh)	Nitrogen emissions (kg/kWh)
Cooking	—	—	—
Dung cake	1.0956	0.00822	0.6847476
Firewood	0.73768	0	0.0014316
Agricultural waste	1.10653	0	0.0021474
Kerosene	0.32197	0.00013	0.0003712
LPG	0.14982	3.6E-05	0
Biogas	0.27273	0.00244	0
Solar thermal	0	0	0
Electricity	—	—	—
Diesel electricity	0.26273	0.00121	0.001792
Electricity (grid)	1.2	0.074	0.075
Biomass electricity*	0	0	0.000932
PV electricity	0	0	0
Biogas electricity*	0	0.00457	0
Biomass + diesel electricity (dual-fuel)	0.42059	0.00064	0.0001495
Wind power	0	0	0

Based on sources IPCC reports,[9] and AIT reports (Hiremath *et al.*, 2010).
*Carbon neutral technologies.

oxides, and nitrogen oxides. The pollutant emissions can be computed from these compositions using Stoichiometric equations and other principles of combustion (Deo *et al.*, 1991).[10] The emission of carbon dioxide is expressed in equivalent carbon emissions adjusted by the appropriate system efficiencies, as explained in Sudha and Ravindranath (2000). The emissions of sulfur and nitrogen oxides have been obtained using similar computations and are shown in Table 5. The corresponding objective functions can be written as follows:

$$\text{Min} \sum_{i=1}^{6} \left[\sum_{j=1}^{m} \left[\sum_{k=1}^{n} p_{ijk} X_{ijk} \right] \right], \tag{6}$$

$$\text{Min} \sum_{i=1}^{6} \left[\sum_{j=1}^{m} \left[\sum_{k=1}^{n} q_{ijk} X_{ijk} \right] \right], \tag{7}$$

$$\text{Min} \sum_{i=1}^{6} \left[\sum_{j=1}^{m} \left[\sum_{k=1}^{n} r_{ijk} X_{ijk} \right] \right], \tag{8}$$

where p_{ijk}, q_{ijk}, and r_{ijk} are the emissions of carbon oxides, sulfur oxides, and nitrogen oxides, respectively, for combination j.

[9]http://www.ipcc.ch/ (Accessed on 10th February, 2010).
[10]http://mnres.nic.in (Accessed on 10th February, 2010).

3.1.7. *Maximization of reliability of renewable energy systems*

Reliability is the probability that a device would operate without failure for a given period of time under given operating conditions. If the reliability factor of the system is to be determined, initially the reliability of individual sub-systems or elements has to be estimated. If each component exhibited a constant failure rate, then the reliability factor for each component would be in the form of exponential $(-\lambda t)$. The rth component will have a reliability of $\exp(-\lambda rt)$. Hence, the reliability for the system would be

$$\begin{aligned} R(t) &= \exp(-\lambda_1 t) \times \exp(-\lambda_2 t) \times \cdots \times \exp(-\lambda_s t), \\ &= \exp[-(\text{-}\lambda_1 + \lambda_2 + \cdots + \lambda_s)t] = \exp\left(-\sum \lambda rt\right), \end{aligned}$$

where t is time in hours. The value of reliability $R(t)$ is 1 at $t = 0$, and it decreases continuously thereafter with time. When "t" becomes very large, all the components would fail; thus, $R(t)$ would reach a value of zero:

$$\text{Max} \sum_{i=1}^{6} \left[\sum_{j=1}^{m} \left[\sum_{k=1}^{n} R_{ijk} X_{ijk} \right] \right]. \tag{9}$$

3.2. *Demand constraints*

Cooking energy requirement:

$$\sum_{j=1}^{12} \left[\sum_{k=1}^{n} X_{jk} \right] \geq D_{\text{Cooking}}. \tag{10}$$

Lighting energy requirement:

$$\sum_{j=13}^{21} \left[\sum_{k=1}^{n} X_{jk} \right] \geq D_{\text{Lighting}}, \tag{11}$$

Water pumping energy requirement:

$$\sum_{j=22}^{28} \left[\sum_{k=1}^{n} X_{jk} \right] \geq D_{\text{Water pumping}}. \tag{12}$$

Water heating energy requirement:

$$\sum_{j=29}^{42} \left[\sum_{k=1}^{n} X_{jk} \right] \geq D_{\text{Water heating}}. \tag{13}$$

Rural industries requirement:

$$\sum_{j=43}^{49} \left[\sum_{k=1}^{n} X_{jk} \right] \geq D_{\text{Rural industries}}. \tag{14}$$

Home Appliances energy requirement:

$$\sum_{j=49}^{56}\left[\sum_{k=1}^{n} X_{jk}\right] \geq D_{\text{Home appliances}}. \tag{15}$$

Fuel constraints: Amount of residue used as fuel obtained from irrigated and un-irrigated land area $\geq$ total fuel requirement

$$\sum_{i=1}^{6}\sum_{k=1}^{6}(r_c \times y_c)I_{L_{ik}} + \sum_{i=1}^{6}\sum_{k=1}^{6}(r_c \times y_c)U_{L_{ik}} \leq \mathrm{T_L}. \tag{16}$$

Supply constraints: *Limit for solar thermal usage for cooking*: The solar thermal cookers cannot cook all varieties of food and therefore they are not meeting the total cooking requirement. As such, solar thermal cookers can be used for low-temperature cooking purposes only, which form approximately 20% of the total cooking requirement (Srinivasan and Balachandra, 1993). Therefore, for this reason, the potential limit for the use of solar thermal cookers is considered to be 20% of the total cooking energy requirement. The constraint function is as follows:

$$X_7 \leq 20\% \text{ of total cooking requirement.} \tag{17}$$

Biomass constraints: Quantity of biomass used for different end uses $\leq$ biomass available:

$$\sum_{i=1}^{6}\sum_{k=1}^{6} Q_{BU_{ik}} \leq Q_{\mathrm{BA}}. \tag{18}$$

Dung balance: Quantity of dung used for purpose (dung cake + biogas production) $\leq$ quantity of dung available:

$$\left[\sum_{i=1,4}\sum_{k=1}^{2} Q_{dc_{ik}} + \sum_{i=1}^{6}\sum_{k=1}^{3} Q_{Bg_{ik}}\right] \leq d_b N_b. \tag{19}$$

Fuelwood constraints: Quantity of fuelwood used for (cooking + heating) $\leq$ total quantity of fuel wood available:

$$\sum_{i=1}^{4}\sum_{k=1}^{2} F_{uik} \leq F_T. \tag{20}$$

Biogas balance: Volume of biogas used for different purpose $\leq$ total volume of biogas generated:

$$\sum_{i=1}^{6}\sum_{k=1}^{6} Q_{Bg_{ik}} \leq Q_{BgT}. \tag{21}$$

Biogas constraints: Dung required for biogas generation $\leq$ 50% of total dung available:

$$\sum_{i=1}^{6}\sum_{k=1}^{6} Q_{Bg_{ik}} \leq 50\% T_{\mathrm{DA}}. \tag{22}$$

Dung cake constraint: During the survey, it is observed that a maximum of 50% dung available could potentially be used for making dung cakes. Therefore, it is assumed that 50% of the dung cakes produced are used for cooking and heating applications. Therefore, dung required for making dung cake $\leq$ 50% of total dung available:

$$\sum_{i=1,4}^{4}\sum_{k=1}^{2} Q_{dc_{ik}} \leq 50\% T_{\mathrm{DA}}. \tag{23}$$

Employment constraints: Total number of adult humans available for employment $\leq$ 60% available population of the village or panchayat or block selected:

$$\sum_{i=1}^{6}\left[\sum_{j=1}^{m}\left[\sum_{k=1}^{n} N_{A_{ijk}}\right]\right] \leq 60\% P_{\mathrm{A}}. \tag{24}$$

Reliability constraints: Reliability of different renewable energy resources $\leq$ Potential limit of available resources:

$$\left[\sum_{j=1}^{3}\left[\sum_{k=1}^{n}(1/R_j)\sum X_{ijk}\right]\right] \leq P_j, \tag{25}$$

where i = type of end-use device service; j = type of energy resource available; k = type of energy system used; C_{ijk} = unit cost of energy for i end use, for j-type of energy available by using k type of energy system in Rs/kWh based on life-cycle costing; η_{ijk} = system efficiency coefficient for different end uses; e_{ijk} = employment coefficient (i.e., number of persons employed per kWh of energy allocated for the combination of end use); p_{ijk} = emission rate of carbon oxides of i-type end use of j-type resource and k type of energy system used; q_{ijk} = emission rate of sulfur oxides of i-type end use of j-type resource and k type of energy system used; r_{ijk} = emission rate of nitrogen oxides of i-type end use of j-type resource and k type of energy system used; R_{ijk} = reliability factor of renewable energy systems; r_c = yield to residue ratio/ha of crops c; y_c = yield of crop c/ha; $\mathrm{IRL}_{(\mathrm{fuel})}$ = irrigated area under crop c, the residue obtained for this area is used as fuel; $\mathrm{URL}_{(\mathrm{fuel})}$ = unirrigated area under crop c, the residue obtained for this area is used as fuel; F_{Fuel} = quantity of crop residue of fuel in tons/year; D_{Cooking} = total cooking energy requirement; D_{Lighting} = total lighting energy requirement; $D_{\mathrm{Water\ pumping}}$ = total water pumping energy requirement; $D_{\mathrm{Water\ heating}}$ = total water heating energy requirement; $D_{\text{Rural industries}}$ = total rural industries energy requirement; $D_{\text{Home appliances}}$ = total home appliances energy requirement; Q_{Bg} = volume of biogas used for purpose i (m^3); Q_{BgT} = total volume of biogas produced in (m^3); Q_{B} = total biomass available (tons); Q_{Bu} = quantity of biomass used for purpose i (tons); Q_{BA} = quantity of biomass available for purpose i (tons); Q_{dc} = quantity of dung cake used for purpose i (tons); d_{b} = quantity of dung output from animal type b in tons/year; N_{b} = number of animal type b; F_{U} = fuelwood used as fuel for purpose i (tons); F_{T} = fuelwood available as fuel for purpose i (tons); T_{DA} = total quantity of dung available from animal type b in tons/year; N_{A} = total number of adult human population available for employment; P_{A} = total population of village or panchayat or block; IL = amount of residue used as fuel obtained from irrigated land area (tons); UL = amount of residue used as fuel obtained from unirrigated land area (tons); TL = total fuel requirement (tons); X_{ijk} = quantity of final energy for ith end use, using jth resource by kth end devices; P_j = potential limit of available resources; m = resource for specific end use: cooking $m = 12$ ($j = 1$–12), lighting $m = 21$ ($j = 12$–21), water pumping $m = 28$ ($j = 21$–28), water heating $m = 42$ ($j = 28$–42), rural industries $m = 49$ ($j = 43$–49), home appliances $m = 56$ ($j = 49$–56); n = number of systems used for specific end use: cooking ($n = 8$), lighting ($n = 6$), water pumping ($n = 5$), water heating ($n = 9$), rural industries ($n = 4$) home appliances ($n = 4$).

The basic approach of goal programming is to establish a specific numeric goal for each objective, and then seeks a solution that corresponds to minimum weighted sum of deviation of the objective functions from their respective goals. Further, the pre-emptive goal programming is used, where there is a hierarchy of priority levels for the goals, so that the objective functions are optimized in the order of priority. Based on the objectives and constraints provided in the problem formulation section, a pre-emptive goal programming model has been developed, first by defining goals for each of the objectives, and then by minimizing the sum of the deviation variables. The objective function is solved individually and the optimized value for each objective is used as the corresponding goal. These seven goals may be either *over* ($^+$) or *under* achieved ($^-$).

The goal programming model is described as follows:

The problem formulation:

Minimize:

$$\sum d_j^- + d_j^+$$

Subject to:

$$\left(\sum C_{ijk} \times X_i\right) + w_1 d_{1-} - w_1 d_{1+} = b_1, \quad \text{where } j = 1, 2, \ldots, 56,$$

$$\left(\sum \eta_{ijk} \times X_{ijk}\right) + w_2 d_{2-} - w_2 d_{2+} = b_2, \quad \text{where } j = 1, 2, \ldots, 56,$$

$$\sum X_j + w_3 d_{3-} - w_3 d_{3+} = b,$$

where $j = 4, 5, 8, 13, 15, 17, 22, 24, 32, 33, 36, 38, 43, 45, 50$, and 52.

$$\sum X_j + w_4 d_{4-} - w_4 d_{4+} = b_4$$

where $j = 1, 2, 3, 6, 7, 9, 10, 11, 12, 14, 16, 18, 19, 20, 21, 23, 25, 26, 27, 28, 29, 30, 31, 34, 35, 37, 39, 40, 41, 42, 44, 46, 47, 48, 49, 51, 53, 54, 55$, and 56.

$$\left(\sum e_{ijk} \times X_{ijk}\right) + w_5 d_{5-} - w_5 d_{5+} = b_5, \quad \text{where } j = 1, 2, \ldots, 56,$$

$$\left(\sum p_{ijk} \times X_{ijk}\right) + w_6 d_{6-} - w_6 d_{6+} = b_6, \quad \text{where } j = 1, 2, \ldots, 56,$$

$$\left(\sum q_{ijk} \times X_{ijk}\right) + w_7 d_{7-} - w_7 d_{7+} = b_7, \quad \text{where } j = 1, 2, \ldots, 56,$$

$$\left(\sum r_{ijk} \times X_{ijk}\right) + w_8 d_{8-} - w_8 d_{8+} = b_8, \quad \text{where } j = 1, 2, \ldots, 56,$$

$$\left(\sum R_{ijk} \times X_{ijk}\right) + w_9 d_{9-} - w_9 d_{9+} = b_9,$$

where, $j = 9, 10, 11, 18, 19, 20, 25, 26, 27, 39, 40, 41, 46, 47, 48, 53, 54, 55$.

$\sum X_{jk} \geq$ total cooking energy requirement,
where $j = 1, 2, \ldots, 12$ (cooking energy requirement).

$\sum X_{jk} \geq$ total lighting energy requirement,
where $j = 13, \ldots, 21$ (lighting energy requirement).

$\sum X_{jk} \geq$ total pumping energy requirement,
where $j = 22, \ldots, 28$ (water pumping requirement).

$\sum X_{jk} \geq$ total water heating energy requirement,
where $j = 29, \ldots, 42$ (water heating energy requirement).

$\sum X_{jk} \geq$ total rural industries energy requirement,
where $j = 43, \ldots, 49$ (rural industries energy requirement).

$\sum X_{jk} \geq$ total appliances energy requirement,
where $j = 50, \ldots, 56$ (home appliances energy requirement).

$\sum X_4 \leq 20\%$ of the total cooking energy requirement,
where $j = 4$ (solar thermal constraint).

$\sum X_{jk} \leq$ dung availability, where $j = 1$ and 29 (dung availability limit).

$\sum X_{jk}/\eta_{jk} \leq$ available biogas energy,

where $j = 6, 11, 14, 20, 27, 34, 41, 43, 48$, and 55 (potential limit of biogas energy).

$$\sum X_{jk}/\eta_{jk} \leq \text{available biomass energy},$$

where $j = 9, 12, 18, 21, 25, 28, 39, 42, 49, 53$, and 56 (potential limit of biomass energy); where b_i = goal for the objective j, d_{j+} = over-achievement of goal b_j, d_{j-} = under-achievement of goal b_j, w_i = weighing factor for d_j^- and d_j^+, C_{ijk} = the cost coefficient for combination i, η_{ijk} = the efficiency coefficient for combination i, e_{ijk} = the number of persons employed per kWh of energy allocated for combination, p_{ijk} = emission rate of carbon oxides for combination i, q_{ijk} = emission rate of sulfur oxides for combination i, r_{ijk} = emission rate of nitrogen oxides for combination i, R_{ijk} = reliability factor of renewable energy systems.

3.3. *WINQSB software*

WINQSB software was used to solve the linear and multi-objective optimization problems.[11] The WinQSB software is capable of solving various types of non-linear and linear problems (Amariei *et al.*, 2009). Any types of variables, binary, integer, non-negative, can be defined and solved easily. Problem values are inserted either in table or normal form.

4. Scenarios Considered for Modeling

Energy scenarios provide a framework for exploring future energy perspectives, including various combinations of technology options and their implications.

4.1. *Business as usual (BAU)*

This scenario is based on the direction in which the selected location is headed. Assuming continued moderate economic growth, energy consumption pattern, and modest technological improvement, this scenario leads to adverse environmental impacts, ranging from regional acidification to climate change. Thus, this scenario leads to a higher dependence on carbon-intensive fossil fuels, resulting in high energy-related emissions, and falls short of achieving a transition toward sustainable development. This scenario is sub-divided into two scenarios.

(i) *BAU NP* (no priority): This scenario is developed without assigning any priority to the objective functions.
(ii) *BAU EP* (equal priority): Scenario assumes all the objective functions are taken into account while arriving at energy resource allocation. The objective functions considered are economic, security acceptance, and environmental, this scenario gives equal priority to all the objective functions.

[11] http://www.wiley.com/college/tech/winqsb.htm (Accessed on 10th February, 2010).

4.2. *Economic objective scenario (EOS)*

EOS is given the highest priority. Employment, efficiency, and reliability objectives are given low priority. Government subsidy for agricultural water pumping and grid electricity for rural household are assumed to continue in the year 2020.

4.3. *Renewable energy scenarios*

This scenario is characterized by sustainable development, with a shift toward environmentally benign energy technologies, with a significant role for renewables. Renewable system efficiency is assumed to increase by 25% by 2020; this scenario highlights the implication of renewable energy for future energy supply trends. However, this scenario does not consider the equity aspect aimed at meeting the energy needs of all the population.

- *Emission reduction scenario (ERS)*: Emission reduction objective is ranked higher than the other objectives.
- *Biomass intensive scenario (BIS)*: It would include energy plantations raised on degraded lands.

4.4. *Sustainable energy development scenario (SDS)*

This scenario includes services that promote equity and quality of life based on locally available, convenient, safe, environmentally sound, and sustainable technologies. A considerable change is expected in total energy consumption compared with other scenarios, reflecting varying approaches for meeting the energy needs in the future. Increases in research, development, and deployment efforts for new energy technologies are a prerequisite for realization of the scenario that has the characteristics of sustainable development.

5. Results

Panchayat (local council or government) is the administrative unit in India. Yedavani panchayat consists of 12 villages including Ungra. Results of DEP analysis for different scenarios developed for Yedavani panchayat are presented in this section. A summary of optimized energy and technology mix for different activities under different scenarios is given in Table 6. Optimal energy resource and technology mix for cooking derived from modeling is presented in Table 7 and Fig. 3. Cooking is the dominant energy using activity in Yedavani panchayat under present energy consumption scenarios (PECS) and BAU (NP). Solid biomass is used by majority of households (Fig. 3). Traditional cook stoves with low-thermal efficiency are used for cooking. EOS, RES, and SDS scenarios show that biogas is the option for cooking, based on optimization (Fig. 3).

Water heating practice is similar in all villages of Yedavani panchayat. People take bath twice a week. Water heating demand is too low compared with cooking

Table 6. Optimized energy resource allocation for Yedavani panchayat under different scenarios.

		BAU (2020)		Economic objectives scenario (2020)	RES (2020)		SDS (2020)
Activities	PECS (2005)	No priority	Equal priority	Cost (1),[b] employment (2),[b] efficiency and reliability (3)[b]	ERS: emission (1),[b] economic (2),[b] and security (3)[b]	BIS cost, employment, local and emissions (1)[b]	Economic, security, and emission (1)[b]
Cooking	Biomass (91%), LPG (7%), kerosene (2%)	Biomass (80%), LPG (20%)	Biomass (31%), LPG (31%), biogas (38%)	Biomass using improved cook stoves (3%), biogas (97%)	Biomass using improved cook stoves (3%), biogas (97%)	Biomass using improved cook stoves (7%), biogas (93%)	Biogas (100%)
Home lighting	Kerosene (40%), grid (60%)	Grid (100%)	PV electricity (100%)	Biomass electricity (100%)	PV electricity (100%)	Biomass electricity (100%)	Biomass electricity (100%)
Water pumping	Diesel electricity (5%), Grid (95%)	Grid (100%)	Biomass electricity (100%)	Grid (100%)	PV electricity (100%)	Biomass electricity (100%)	Biomass electricity (100%)
Water heating	Biomass (100%)	Biomass (100%)	Solar thermal (100%)	Biomass using improved cook stoves (100%)	Solar thermal (100%)	Biomass using improved cook stoves (100%)	Biogas (100%)
Rural industries	Diesel (15%), grid (65%), biomass (20%)	Diesel (5%), grid (45%), biomass (50%)	Biomass electricity + diesel electricity (100%)	Biomass electricity (100%)	Biomass electricity (100%)	Biomass electricity (100%)	Biomass electricity (100%)
Home appliances	Grid (100%)	Grid (100%)	Biomass electricity (100%)	Biomass electricity (100%)	PV electricity (100%)	Biomass electricity (100%)	Biomass electricity (100%)

[a]Using improved cook stoves.
[b](1) first priority, (2) second priority, and (3) third priority.

Table 7. CO_2 emissions per capita for different scenarios.

Year	Scenarios	CO_2 emissions (tons/year)	Total population	CO_2/Capita (tons/year)
2005	PECS	8881.61[a]	7168[c]	1.24
2020	BAU (NP)	9266.65[a]	9230[d]	1.00
2020	BAU (EP)	3423.59[a]	9230[d]	0.37
2020	EOS	862.11[a]	9230[d]	0.09
2020	RES	0[b]	9230[d]	0
2020	SDS	0[b]	9230[d]	0

[a]Considering emissions from only unsustainable fuelwood sources, which account for 40% of total emissions for example, total CO_2 emissions for PECS 2005 is 22204 t; thus, 22,204 × 40% = 8881 for CO_2 of PECS scenario).
[b]Carbon neutral energy sources.
[c]2005 population from the survey conducted.
[d]From 1981, 1991, and 2001 census reports population projections for 2020.
[e]CO_2/capita (tons/year) is calculated: example 8881/7168 = 1.24.

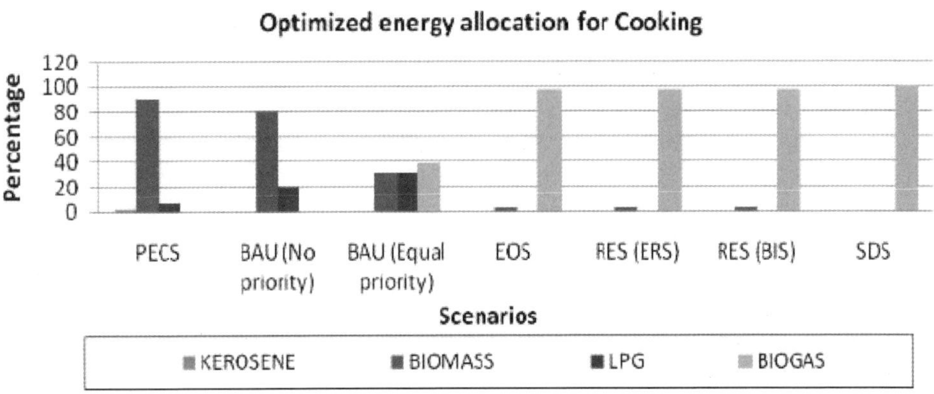

Fig. 3. Optimized energy resource and technology mix for cooking in Yedavani panchayat under different scenarios.

energy demand and this demand can be easily met by biomass using improved bath stoves. Optimized energy resource and technologies for water heating are given in Table 6 and Fig. 4.

Under EOS and RES, efficient wood stove is the technology option for water heating. Solar thermal is an option under RES (ERS) and BAU (EP) scenarios. This is because emissions objective are given higher weightages than other objectives. Biogas from leaf litter and dung is available in surplus under SDS. Cooking and water heating demand can be met together. Hence, biogas is the optimized option in SDS scenario. In the year 2005, 21% of rural households were unelectrified in Yedavani panchayat. Kerosene is used as home lighting fuel in these households. Remaining households depended on grid electricity for home lighting, which is unreliable. If subsidy given to rural households remains at current levels, home lighting is possible through grid electricity under BAU (NP) scenario even in 2020 (Fig. 5).

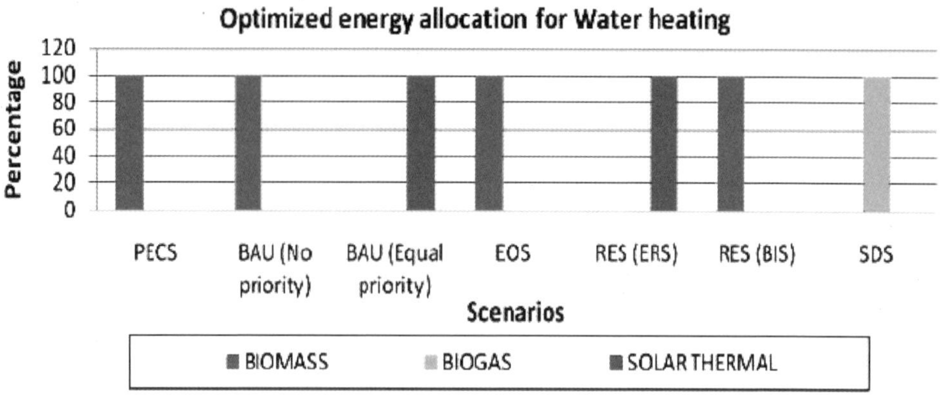

Fig. 4. Optimized energy resource and technology mix for water heating in Yedavani panchayat under different scenarios.

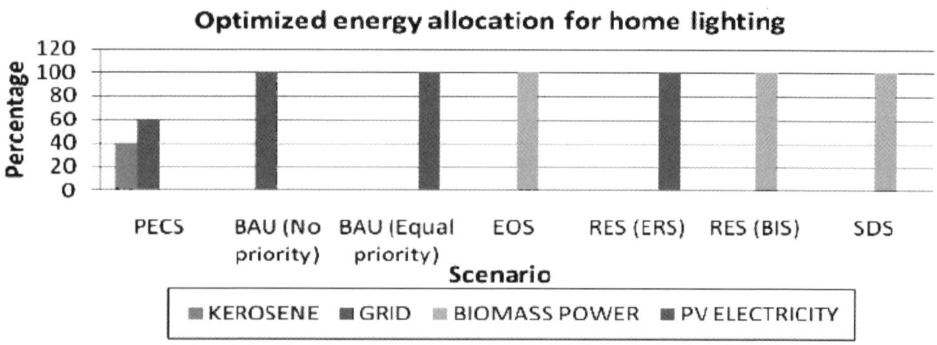

Fig. 5. Optimized energy resource and technology mix for home lighting in Yedavani panchayat under different scenarios.

Current water pumping energy needs are met to the tune of 95% by grid electricity, which is a subsidized source. At a growth rate of 2.1%, 465 pump sets are projected for the year 2020 (Fig. 6). If subsidy continues at current level, 100% of water pumping energy needs can be met by grid electricity.

There are four Jaggery (raw sugar) manufacturing units, which operate for six months. Further, four flourmills, two rice mills, one brick kiln, and one milk production unit operate throughout the year in the panchayat. Diesel (15%), Biomass (20%), and Grid electricity (65%) are the sources of energy for industries in the PECS scenario (Fig. 7).

Considering a 10% growth in energy demand for rural industry, under BAU (no priority) scenario, the optimal mix of energy sources is diesel (5%), biomass (50%), and grid (45%) to meet the projected energy needs for different rural industries in Yedavani panchayat by 2020. BAU with equal priority for different objectives shows that biomass dual-fuel is the optimized option to meet all the power needs.

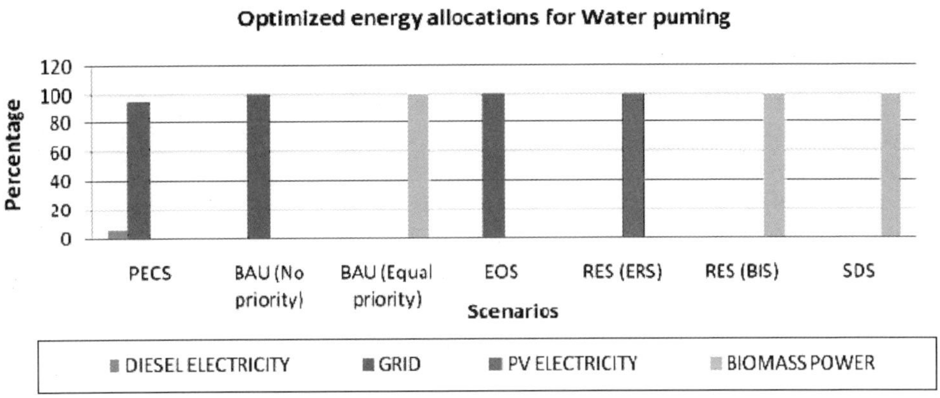

Fig. 6. Optimized energy resource and technology mix for water pumping energy needs.

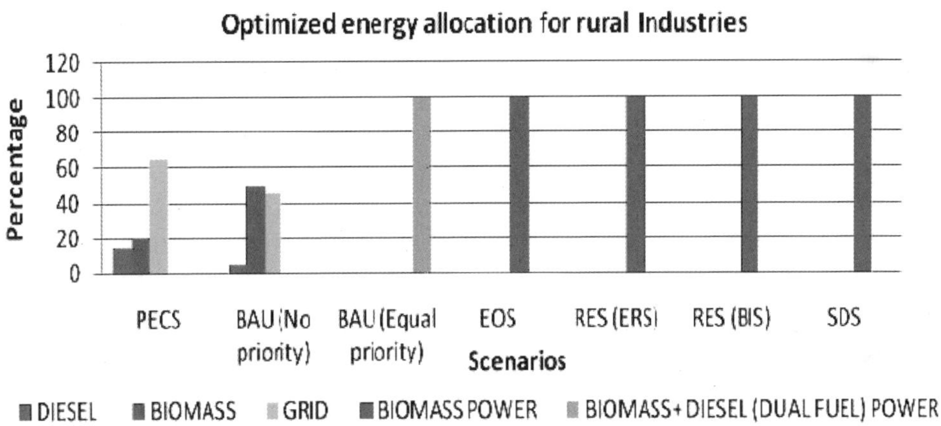

Fig. 7. Optimized energy resource and technology mix for rural industries in Yedavani panchayat for different scenarios.

It can be observed from Table 6 that electricity is the main source of energy for meeting the energy needs of appliances in electrified households. Grid electricity is the energy source for meeting this demand under PECS and BAU (NP) scenarios. Under all other scenarios except RES (ERS), biomass power is the optimized option for meeting the electricity needs of home appliances. PV electricity for home appliances is not a viable option under any scenario. If all households are electrified, biomass power still remains the optimal source due to low cost, reliability, and locally availability. Yedavani panchayat is projected to emit 9266 t CO_2 at about 1 t CO_2/Capita under BAU scenario by 2020. CO_2 emission under SDS and RES scenario is zero as these are carbon neutral scenarios. The associated CO_2 emissions for different scenarios are tabulated for comparing it with the PECS Table 7.

6. Conclusion

Decentralized planning involves scaling down energy planning to sub-national or regional scales. Present work formulates and analyzes DEP result for 2020 for Yedavani panchayat from India. Methodology for making objectives' functions and constraints for DEP has been discussed in this chapter. DEP model is suitably scaled for obtaining the optimal mix of energy resources and technologies using a computer-based goal programming technique. Scenarios considered include PECS, which included current energy consumption pattern. BAU Scenario, EOS, RES, and SDS have been analyzed for the year 2020. Electricity demand of Yedavani panchayat can be met by using biomass power generation systems based on biomass feedstock from raising energy plantation at moderate and high productivity rates (8 t/ha/year and 12 t/ha/year respectively) on the available wastelands under the SDS scenario. Biomass feedstock that can be grown from the energy plantation is adequate to provide electricity through the gasification route. Cooking and water heating energy needs can be met by biogas that is obtained from cattle dung and leaf litter. Cattle dung available in different villages of the panchayat is inadequate to meet the cooking energy needs. Biogas option is not feasible at panchayat level because condensed liquefaction of biogas is not possible for transportation from village to village. Further, transportation of dung and leaf litter is not feasible across villages. Thus, community biogas plants are feasible only at village scale. Biomass power sizing and siting depend on three factors: (a) land availability and energy demand or load, (b) cost of transportation of biomass to the biogasifier plant, and (c) transmission and distribution cost of power. In Yedavani panchayat, Ungra village has high biomass feedstock potential and low-energy demand compared with Keelara village that has high demand and low generation potential for biomass power. Thus, optimal planning for capacity and siting biomass power is essential at multiple village scale in the panchayat. Hence, panchayat or cluster of village planning should be supported by village-level energy planning.

The DEP analysis showed that it is possible to meet the energy requirements of all the services that are necessary to promote development and improve quality of life. Technologies for decentralized electricity generation through biomass gasifier systems and biogas from cattle dung (and plant leaf litter) for cooking are available for meeting the energy needs at decentralized level.

References

Ahmed, S. and Husseiny, A.A., 1978. A multivariate utility approach to selection of energy resources. *Energy* **3**, 669–700.

Amariei, Frunzaverde, D., Popovici, G. and Hamat, C.O., 2009. WinQSB simulation software — A tool for professional development. *Procedia — Social and Behavioral Sciences* **1**, 2786–2790.

Amulya Kumar, N.R., Sumithra, G.D., Balachandra, P. and D'sa, A., 1990. Comparative costs of electricity conservation, centralised and decentralised electricity generation. *EPW* 22, 1201–1216.

Austin, A.W.G. and Worthington, R., 2003. *Employment Potential of Renewable Energy in South Africa.* Denmark.

Bryson, N. and Joseph, A., 1999. Generating consensus priority point vectors: A logarithmic goal programming approach. *Computers and Operations Research* **26**, 637–643.

Chakrabarti, S. and Chakrabarti, S., 2002. Rural electrification programme with solar energy in remote region — A case study in an island. *Energy Policy* **30**, 33–42.

Deo, P., Modak, S. and Shulka, P.R., 1991. *Decentralized Energy Planning.* Oxford and IBH Publishing Co. Pvt. Ltd, India.

Deshmukh, M.K. and Deshmukh, S.S., 2005. Micro-level integrated renewable energy system planning in India. *National Conference on Energy Management in Changing Business Scenario*, BITS Pilani.

Domah, P., 2006. *Technical Efficiency in Electricity Generation — The Impact of Smallness and Isolation of Island Economies.* Queens' College, University of Cambridge, UK.

Ghosh, D., Pal, B.J. and Basu, M., 1995. Determination of optimal solution for MCDM model in agricultural planning a strategy. *International Journal of Management Systems* **11**, 267–283.

Goldemberg, J., Johansson, T.B., Amulya Kumar, N.R. and Williams, R.H., 1998. *Energy for a Sustainable World.* Wiley Eastern Ltd, New Delhi.

Herbert, G.M.J., Iniyan, S., Sreevalsan, E. and Rajapandian, S., 2007. A review of wind energy technologies. *Renew. Sust. Energ. Rev.* **11**, 1117–1145.

Hiremath, R.B., Ravindranath, N.H., Balachandra, P. and Dasappa, S., 2006. Characterization and assessment of selected biomass energy technologies (BETs) in India. *Asian Regional Research Programme in Energy, Environment and Climate — Phase II.* Technical Report. AIT Press.

Hiremath, R.B., Sikha and Ravindranath, N.H., 2007. Decentralized energy planning; modeling and application — A review. *Renew. Sust. Energ. Rev.* **11**, 729–752.

Hoog, D. and Hobbs, B., 1993. An integrated resource planning model considering customer value, emissions, and regional economic impacts. *Energy* **18**, 1153–1159.

Hulscher, W.S., 1990. Energy for rural development. *ITC Journal* **4**, 373–378.

Hussein, M. and Abo-Sinna, M., 1995. A fuzzy dynamic approach to the multi-criterion resource allocation problem. *Fuzzy Sets and Systems* **69**, 115–124.

Iniyan, S. and Jagadeesan, T.R., 1998. Effect of wind energy system performance on optimal renewable energy model — An analysis. *Renew. Sust. Energ. Rev.* **2**, 327–344.

Iniyan, S., Suganthi, L. and Jagadeesan, T.R., 1998. Critical analysis of wind farms for sustainable generation. *Solar Energy* **64**, 141–149.

Iniyan, S., Suganthi, L., Jagadeesan, T.R. and Samuel, A.A., 2000. Reliability based socio economic optimal renewable energy model for India. *Renewable Energy* 291–297.

Joshi, B., Bhatti, T.S. and Bansal, N.K., 1992. Decentralized energy planning model for a typical village in India. *Energy* **17**, 869–876.

Kanniappan, P. and Ramachandran, T., 2000. Goal programming model for sustainable electricity production from biomass. *Int. J. Energy Res.* **24**, 1–18.

Kumar, A., Purohit, P., Rana, S. and Kandpal, T.C., 2002. An approach to the estimation of the value of agricultural residues used as biofuels. *Biomass Bioenergy* **22**, 195–203.

Limmeechokchai, B. and Chawana, S., 2007. Sustainable energy development strategies in the rural Thailand: The case of the improved cooking stove and the small biogas digester. *Renew. Sust. Energ. Rev.* **11**, 818–837.

Mezher, T., Chedid, R. and Zahabi, W., 1998. Energy resource allocation using multi-objective goal programming: The case of Lebanon. *Applied Energy* **61**, 175–192.

Mittal, L.M. and Sharma, C., 2004. *Environmental Impact of Emissions from Thermal Power Generation in India.* The Ohio State University and Ohio Supercomputer Center, USA.

Nezhad, H.G., 1990. World energy scenarios for the year 2010. *Strategic Planning for Energy and the Environment Spring*, 26–48.

Parikh, J., 1997. *Energy Models for 2000 and Beyond.* Tata McGraw Hill, New Delhi.

Ramakumar, R., Shetty, P.S. and Ashenayi, K., 1986. A linear programming approach to the design of integrated renewable energy systems for developing countries. *IEEE Trans Energy Conver* **EC-I 4**, 18–24.

Ramanathan, R. and Ganesh, L.S., 1993. A multi-objective programming approaches to energy resource allocation problems. *Int. J. Energy Res.* **17**, 105–119.

Ramanathan, R. and Ganesh, L.S., 1995a. Energy alternatives for lighting in households: An evaluation using an integrated goal programming — AHP model. *Energy* **20**, 63–72.

Ramanathan, R. and Ganesh, L.S., 1995b. Using AHP for resource allocation problems. *European Journal of Operational Research* **80**, 410–417.

Rana, S., Chandra, R., Singh, S.P. and Sodha, M.S., 1997. Substitution of energy efficient devices for cooking and lighting in rural areas of Madhya Pradesh. *Energy Conversion and Management* **38**, 187–199.

Ravindranath, N.H. and Chanakya, H.N., 1986. Biomass based energy system for a South Indian village. *Biomass* **9**, 215–233.

Ravindranath, N.H. and Hall, D.O., 1995. *Biomass, Energy, and Environment: A Developing Country Perspective from India.* Oxford University Press.

Ravindranath, N.H., 1993. Biomass gasification: An environmentally sound technology for decentralized power generation, a case study of India. *Biomass and Bio-energy* **4**, 49–60.

Ravindranath, N.H., Dattaprasad, H.L. and Somashekar, H.I., 1990. A rural energy system based on energy forest and wood gasifier. *Current Science* **59**, 557–560.

Ravindranath, N.H., Nagaraju, S.M., Somashekar, H.I., Channeswarappa, A., Balakrishna, M., Balachandran, B.N. and Amulya Kumar, N.R., 1981. An Indian village agricultural ecosystem — Case study of Ungra village — Part I: Main observations. *Biomass* **1**, 61–76.

Saaty, T.L. and Mariano, R.S., 1979. Rationing energy to industries: Priorities and input-output dependence. *Energy System Policy* **3**, 85–111.

Sinha, C.S. and Kandpal, T.C., 1991. Decentralized versus grid electricity for rural India: The economic factors. *Energy Policy* **19**, 441–448.

Srinivasan, R. and Balachandra, P., 1993. Micro-level energy planning in India: A case study of Bangalore North Taluka. *Int. J. Energy Res.* **17**, 621–632.

Sudha, P. and Ravindranath, N.H., 2000. A study of Bangalore urban forest. *Landscape and Urban Planning* **47**, 47–63.

Chapter 32

VIETNAM'S SUSTAINABLE DEVELOPMENT POLICIES: VISION VS. IMPLEMENTATION

Debra Lam
Arup, Kowloon Tong, Kowloon, Hong Kong
debra.lam@arup.com

This chapter is divided into two sections. The first section describes the push factors of economic development, environmental degradation, and climate change impacts that drive Vietnam toward sustainable development policies. The second section describes Vietnam's main sustainable development policies, and highlights their institutional, resource, and capacity weaknesses.

1. Introduction

Since the launch of its *Doi Moi* (i.e., renovation) economic policy, Vietnam has made substantial progress in the overall development of the country. Living standards have risen, poverty has decreased, population has increased, and foreign investment has expanded. It is one of the few countries expected to meet most of its UN Millennium Development Goals by 2015.[1] The country recently reached its long-pursued goal of becoming a lower-middle income country and plans to reach industrialized status by 2020.

However, as it strives for this next stage of development with an increased and more affluent urban population, it is finding faults in the traditional path of industrial development, chiefly arising from environmental degradation and climate change impacts. Fossil fuel driven growth, inefficient resource consumption, and land exploitation have severely harmed the environment. Rising carbon emissions have and will cause more severe and frequent climate change impacts, such as typhoons, floods, rising sea level, heat waves, and droughts.

Both environmental degradation and climate change impacts have already resulted in significant socio-economic costs and devastation, and will only increase if unabated. Vietnam's traditional economic growth model worsened these twin

[1] *Vietnam: Country Assistance Plan.* Department for International Development (DFID), February 2008; *Vietnam: Climate Change, Adaptation and Poor People.* Oxfam, October 2008.

challenges, and in turn, these issues have the potential to derail Vietnam's cherished economic goals.

The government has actively pushed for more sustainable development policies and strategies to address environmental degradation and climate change, while still developing the country. Many of the policies are noted for their vision and advancement, especially compared with countries at similar development levels. However, they are plagued by several critical institutional, capacity, and resource problems, including

- weak rule of law;
- high local authority discretion;
- poor government department cooperation;
- lack of resources;
- low stakeholder awareness and engagement.

Failure to address these issues make and will continue to make it difficult for Vietnam's sustainability policies' vision to match its implementation and enforcement. Understanding these issues and addressing its interlinkages can holistically improve the implementation of Vietnam's sustainable policies.

2. Push Factors

2.1. *Economic development*

War and conflict caused Vietnam to adopt market-driven, export-led growth path later than the other Asian tigers. However, its *Doi Moi* economic renovation in 1986, similar to China's open door policy in the late 1970s, helped the country achieve rapid economic growth. Today, Vietnam stands as one of the fastest economies in the world, averaging around 8% GDP growth annually over the past five years, and expanding its foreign trade by close to 20% per year since the mid-1980s. (Baumuller, 2010). Per-capita income grew almost 10-fold — from less than US$100 a year in 1990 to about US$1200 in 2010 (Fig. 1). The poverty rate has also fallen from 58% in 1993 to 14.5% in 2008.[2]

Vietnam successfully transformed itself from an agrarian society to an industrialized economy. From 2001, the manufacturing sector's GDP has surpassed the agricultural sector (Anwar and Nguyen, 2010). Foreign direct investment from the US, European Union, and other Asian tigers has taken advantage of Vietnam's natural resources, cheap labor, and growing consumer base to propel it further onto the international stage. The country became the 150th member of WTO in 2007[3]

[2]Plans on socio-economic development are crucial. *Vietnam News* 19 August 2010. http://vietnamnews.vnagency.com.vn/Economy/202707/Plans-on-socio-economic-development-are-crucial-.html.

[3]*World Trade Organization: Accession Status Vietnam.* http://www.wto.org/english/thewto_e/acc_e/a1_vietnam_e.htm.

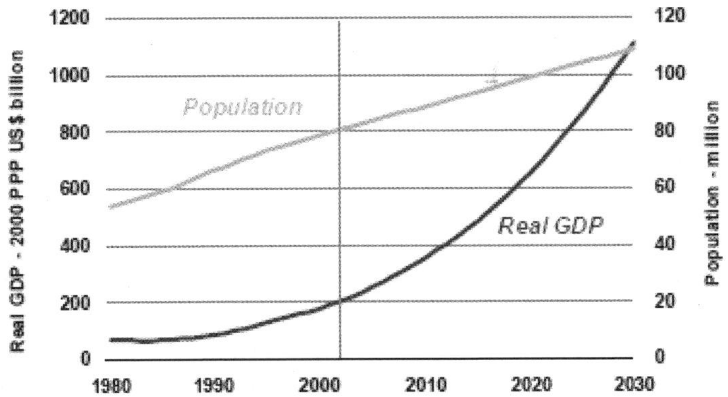

Fig. 1. Vietnam's growth.[5]

and has been active in regional trade with its Southeast Asian neighbors. In fact, its attraction to FDI is attributed to helping it weather the recent global economic recession. As Asian Development Bank Country Director Ayumi Konishi said, "[w]e believe the Government has done extremely well by quickly putting together many measures to minimize adverse impacts from the global economic crisis on Vietnam, and implementing them."[4]

There is still much more to develop, and the country remains ambitious in its economic agenda. Vietnam is actively trying to meet the needs of its growing and more affluent population, projected to reach 109 million by 2030, growing at 1.1%/year from 80.5 million in 2002. At an annual rate of 1.9%, urbanization will hit 43% by 2030.[6] The scale is smaller and slower than in China, which is estimated to have 1% annual growth rate until 2030 at which point 68% of its population will be urbanized.[7] In terms of affordability in housing, goods, and other social services, Vietnam's cities could face similar problems to those encountered by their Chinese counterparts. Poverty has been reduced; however, there remains 16 million classified as poor and 28 million just above the poverty line.[8] Inequality in the socialist country is rising, especially for the rural community, ethnic minorities, and women.

The country remains committed to its economic targets and development. It recently reached its long cherished goal of becoming a lower middle income country.[9]

[4]Economic rebound to accelerate, says ADB. *Vietnam Business News* 13 April 2010. http://vietnambusiness.asia/economic-rebrund-to-accelerate-says-adb/.
[5]APEC Energy Demand and Supply Outlook 2006.
[6]APEC Energy Demand and Supply Outlook 2006.
[7]China's high speed urbanization to end in 2013. *People's Daily* 16 April 2010.
[8]Vietnam: Climate Change, *Adaptation and Poor People*. Oxfam. October 2008.
[9]Under the World Bank's classification, developing countries graduate from low income to lower middle income when per capita gross national income reaches $976. Vietnam stood at $1200 in 2010 *Asian Development Outlook 2010*; Asian Development Bank; Plans for Socio-Economic Development are crucial. *Vietnam News*, 19 August 2010. http://vietnamnews.vnagency.com.vn/Economy/202707/Plans-on-socio-economic-development-are-crucial-.html.

The heart of its next Socio-Economic Development Plan 2010–2015 will be to support its next target of developed country status by 2020. By 2020, GDP per capita is expected to reach about US$3200 and real incomes about three-and-a-half times higher than at present.[10]

Reaching developed country status, however, will not be an easy path. Resources are limited, overused, and unable to meet the growing demand. For example, power shortages are becoming more frequent, longer, and a greater concern to foreign and domestic businesses. Weather, weak investment, poor infrastructure, and low electricity prices are among the reasons given; however, ultimately, it is unlikely to be resolved soon. Do Huu Hao, Deputy Minister of Industry and Trade, admits, "We will always be short of electricity from now through 2020 — not just this year or the next couple of years. We have to pursue rapid economic growth while resources are limited" (Hicky, 2008).

It is estimated that infrastructure investment over the next decade will cost an equivalent of 11% of GDP/year; only half of it met from the budget.[11] The industrial and service sector will continue to drive to economy; however, overall GDP growth is expected to slow down to around 6% annually until 2030.[12] Inefficient state-owned industries continue to burden the economy (Anwar and Nguyen, 2010). Human capital and science and technology remain comparatively low. Most importantly, the traditional development model is severely damaging the environment, bringing devastating socio-economic costs, and threatening to jeopardize Vietnam's economic goals.

2.2. *Environmental degradation*

> *"The major factor for the increased number of asthma patients is the serious environmental pollution" (Head of HCM City Respiration Association Le Thi Tuyet Lan).*[13]

Vietnam's growing and more affluent population and economy have put a severe strain on its environment. There is greater fossil fuel use in electricity and transport, when a rising consumer base buys more electronics, bigger and second homes, and switch to thirstier vehicles. Similarly, industry will move to larger-scale productions as they enter larger and more sophisticated global markets. The effects multiply; for example, open-pit coal mining activities in Quang Ninh province have caused soil erosion and severely polluted nearby Ha long Bay, one of the world's natural heritage sites. Tourism, land use, biodiversity, marine life, and human health can

---

[10]Plans for Socio-Economic Development are crucial. *Vietnam News*, 19 August 2010. http://vietnamnews.vnagency.com.vn / Economy / 202707/Plans-on-socio-economic-development-are-crucial-.html.

[11]*Asian Development Outlook 2010.* Asian Development Bank.

[12]*APEC Energy Demand and Supply Outlook 2006.*

[13]HCM city pollution alarming. *Vietnam News* 1 January 2010. http://vietnamnews.vnagency.com.vn/Environment/195512/HCM-City-pollution-'alarming.html.

all be severely harmed. Rapid urbanization, increased consumption and waste, and resource inefficiency have all contributed to environment degradation.

- *Land*: "In the past 50 years, Vietnam's natural forest cover has shrunk from 43 to 29% of land area, there is an acute shortage of arable land, and habitat loss has led to a rise in the number of threatened species."[14] Raw minerals extraction is rampant and inefficient.[15]
- *Waste*: While there is an active, informal economy of waste recycling and reuse, open dumping remains the principal method of disposing municipal waste. Waste infrastructure and management are not fully funded or implemented, nor have health implications for the (mostly female) informal waste collectors been provided for. Of the 91 disposal sites in the country, only 17 are sanitary landfills and 49 sites have been identified on a national list as hotspots with high environmental and human health risks.[16]
- *Water*: According to Nguyen Viet Anh from the Institute for Environmental Science and Technology under the Hanoi University of Civil Engineering, "[o]nly 6% of wastewater in cities countrywide runs into sewage treatment plants.[17] Increased extraction has caused groundwater levels to dip dangerously low, damaging infrastructure, land, and agriculture. Some groundwater sources have elevated levels of arsenic. "In the Cau and Nhue — Day River sub-basins and the Dong Nai — Sai Gon River basin, the rate of sicknesses related to surface water quality has been high. There is a clear link between the regular use of polluted water and human health effects.[18] As demonstrated by the case of Thai Nguyen (Fig. 2), children are particularly susceptible to polluted water . Weak infrastructure and growing environment and climate issues hinder Vietnam's ability to deliver clean and stable supplies of water to a growing population.
- *Air quality*: Air pollution is getting worse, especially in the urban areas. Dust and pollutants often exceed the permitted levels in 90% of monitoring stations. In Ho Chi Minh City, 60% of the motorbikes do not meet smoke standards and some 80% of the industrial smoke is still untreated.[20] Air pollution will certainly worsen with the completion of 17 new coal-fired power stations by 2020 (Hicky, 2008).

---

[14]Vietnam takes stock of environmental degradation. *Environmental News Services* 19 September 2002. http://www.ens-newswire.com/ens/sep2002/2002-09-19-02.html.

[15]"Sustainable Development is King". *Vietnam Investment Review*. 8 June 2010 http://www.vir.com.vn/Client/VIR/index.asp?url=content.asp&doc=17421

[16]*Vietnam Environment Monitor Solid Waste 2004*. World Bank and MONRE.

[17]Urban pollution getting worse. *Saigon Times* 12 March 2009. http://english.thesaigontimes.vn/Home/business/environment/3560/.

[18]Water quality in Vietnam — with a focus on Cau, Nhue-Day and Dong Nai River Basins. *Vietnam Environment Monitor 2006*. World Bank.

[19]Water quality in Vietnam — with a focus on Cau, Nhue-Day, and Dong Nai River Basins. *Vietnam Environment Monitor 2006*. World Bank.

[20]Air pollution in HCMC getting worse. *The Saigon Times* 19 August 2009. http://english.thesaigontimes.vn/Home/business/investment/5994/.

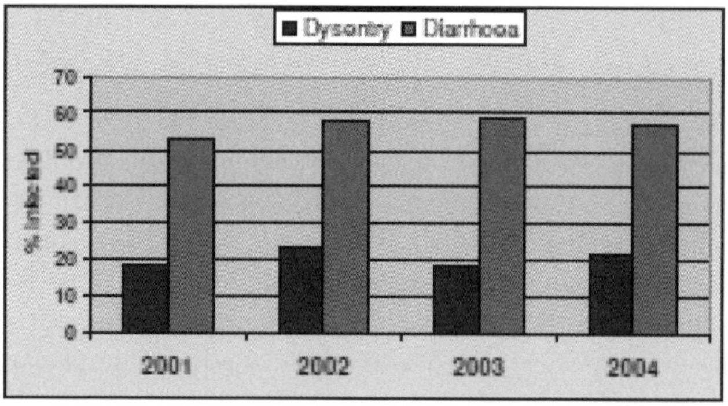

Fig. 2. Percentage of children infected with dysentery and diarrhoea in Thai Nguyen.[19]

The socio-economic costs of Vietnam's environmental degradation are difficult to measure. However, it is growing, causing other indirect impacts, and affecting Vietnam's development. Total environmental damage is estimated to grow to US$19,656 million by 2025, about 7.5% of projected GDP (Nguyen, 2008). Three environmental problems alone — particulate emissions, carbon dioxide damage, and the net loss of Vietnam's forests — reduced gross national income by 2.1% (Bass *et al.* 2010).

Immediate costs to deal with the environmental crisis can be extended into further costs for recovery and prevention. For example, the total financial losses caused by a major oil spill in 2001 were estimated at 250 billion VND; clean up costs for the polluted waters and beaches reached 60 billion VND.[21] The total damage from power generation pollutants in 2005 is assessed at around US$1225 million, equivalent to 2.4% of the nominal GDP. However, those costs do not fully account for the biodiversity and habitat lost, jobs impacted by the destruction, and the need to improve technology and infrastructure. Often the costs are externalities, not fully incorporated into business models or government finances, and therefore escape full accountability.

Moreover, how does one calculate the value of people's livelihoods and health? Over the past four years, six million registered cases of six varieties of water-borne diseases were attributed to increasingly poor water quality. The treatment of cholera, typhoid, dysentery, and malaria accounted for direct costs of at least 400 billion VND.[22] Often it is poor, with weak water and health infrastructure, who are the most susceptible.

Vietnam has one of the highest rates of people dying of diseases caused by outdoor air pollution and indoor air pollution, with up to 230 deaths per million

[21] State of water environmental issues: Vietnam. *Water Environment Partnership in Asia.* http://www.wepa-db.net/policies/state/vietnam/overview.htm.

[22] State of water environmental issues: Vietnam. *Water Environment Partnership in Asia.* http://www.wepa-db.net/policies/state/vietnam/overview.htm.

people/year and up to 400 deaths per million people/year, respectively.[23] Another 24,600–32,800 suffer from pollution-related diseases,[24] affecting work productivity, education, childcare, and overall quality of life. If Vietnam does want to become a developed country, it is essential to build human capital and graduate to a high skilled, educated workforce. This is made increasingly difficult as more and more school-aged children suffer from respiratory and lung diseases, asthma, and ear infections. Successful treatment remains low; only 1% of asthma patients are successfully treated in Vietnam.[25]

Poor environmental quality also detracts top talent from coming and contributing to the country. Hong Kong has started to see this with some of their businesses and expats leaving for greener Singapore (Dissanayake, 2007). It is unknown how many leave or choose not to come to Vietnam because of its environmental degradation. However, as the country aspires to reach developed nation status, it will need to compete with higher sustainability credentials to attract the high skilled workers and investment. Environmental degradation will make it increasingly difficult to maintain further economic development.

2.3. *Climate change*

Vietnam's lower level of development is demonstrated by its ranking as the lowest per capita consumer of electricity in Association of Southeast Asian Nations (ASEAN), and one of the lowest per capita CO_2 emitters in the world (Fig. 3). However, this is rapidly changing. Electricity demand has grown rapidly — even at a faster rate than its GDP (Hicky, 2008). Vietnam's primary energy demand is projected to grow annually at 4.4%, from 42 Mtoe in 2002 to 142 Mtoe in 2030. Vietnam is expected to become a net energy importing economy after 2020, with the energy import dependency projected to reach 15% in 2030.[26] As a result, CO_2 emissions have grown to 1.2 tonnes/capita today compared with 0.3 in 1990. CO_2 emissions are expected to increase to 172 million tons in 2020 (or 1.6 tons per capita) and over 300 million tons in 2050 (or 2.3 tons per capita; Baumuller, 2010). Similarly, global carbon emissions have increased, resulting in climate change and more frequent and severe climate change impacts.

Vietnam is one of the countries most vulnerable to climate change. With major rivers such as the Mekong and the Red, and a coastline that stretches 3260-km along the typhoon-ridden South China Sea, the low-lying country already has a history of tropical storms, droughts, typhoons, floods, and other natural hazards. Climate

---

[23] HCM city pollution alarming. *Vietnam News* 1 January 2010. http://vietnamnews.vnagency.com.vn/Environment/195512/HCM-City-pollution-'alarming.html.
[24] HCM City pollution alarming. *Vietnam News* 1 January 2010. http://vietnamnews.vnagency.com.vn/Environment/195512/HCM-City-pollution-'alarming.html.
[25] HCM City pollution alarming. *Vietnam News* 1 January 2010. http://vietnamnews.vnagency.com.vn/Environment/195512/HCM-City-pollution-'alarming.html.
[26] APEC Energy Demand and Supply Outlook 2006.

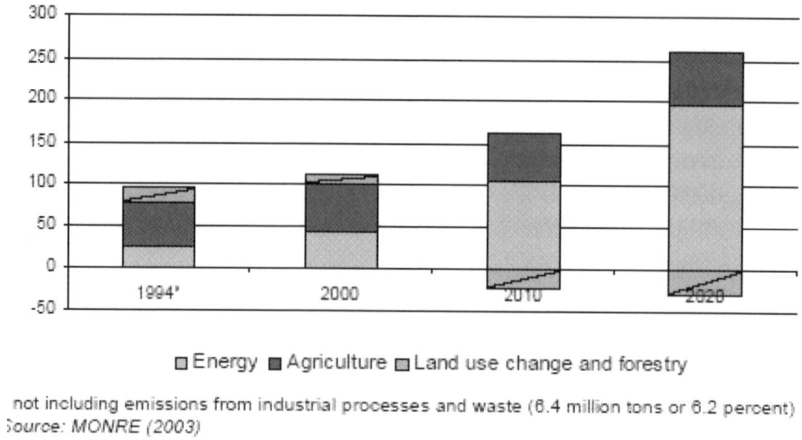

Fig. 3. Estimated GHG emissions by sector (million tonnes $C0_2$; Baumuller, 2010).

change means that those weather conditions will come with greater frequency and severity, and widened regional discrepancies. According to the government's latest climate change data,[27] Vietnam will face the following climate change impacts in the coming years:

- *Temperature Increase*: During the past 50 years (1991–2000), annual average temperature has increased by 0.5–0.7°C. Compared with 1990, temperatures will increase in the range 1.4–1.5°C by 2050 and 2.5–2.8°C by 2100. The number of cold fronts will decrease while the number of days with temperatures higher than 25°C will increase.
- *Typhoons*: Typhoons will be more frequent, occur over a longer period of time, and be more dangerous with higher wind velocity. Typhoons will directly affect 80–90% of the population.
- *Sea level rise*: Data from tidal gauges along the coast show that the sea level rose at a rate of 3 mm/year during 1993–2008. In the past 50 years, the sea level at Hon Dau station rose about 20 cm. The sea level will rise up to 33 cm by 2050, 45 cm by 2070, and 1 m by 2100.[28] Almost half of Ho Chi Minh City will be inundated in water (Carew-Reid, 2007).
- *Precipitation*: About 80–90% of rainfall is concentrated in the rainy season, causing floods and frequent landslides. Rising temperatures will cause increased rainfall during this period with more intense tropical storms. Compared with 1990 levels, annual total rainfall will increase in the range of 2.5–4.8% by 2050 and 4.7–8.8% by 2100. Certain areas will face longer and more severe droughts.

Even without climate change, Vietnam is already one of the most disaster-prone countries in the world. Climate change is expected to further exacerbate these, add

[27] *Climate Change, Sea Level Rise Scenarios for Viet Nam*, MONRE, 2009.
[28] *Initial National Communication to the UNFCCC Regarding Climate Change Impacts in Vietnam*. MONRE, 2003.

additional stresses, and like environmental degradation bring severe socio-economic devastation to the country. The costs of damages associated with increased climate change impacts like flooding have been estimated at US$1.25 billion between 1995 and 2002.[29] Between 1997 and 2006, extreme weather conditions resulted in over 7500 human casualties and economic losses totaling 1.5% of the GDP.

The socio-economic costs will only worsen with more frequent and severe climate change impacts. Key economic sectors including aquaculture, agriculture, manufacturing, tourism, and transport will be severely affected. Tourist and foreign investor confidence will decline. One-meter sea level rise could reduce the nation's GDP by 10% and affect 60% of the provinces, 20% of the communes, six of the eight economic regions, and almost 6 million people (Chaudhry and Ruysschaert, 2007/2008; Carew-Reid, 2007). Other sectors impacted include:

- *Infrastructure*: Major infrastructure such as road networks, power stations, and water and wastewater treatment plants would require substantial resources to support the country's needs, adapt to climate change, and recover from the extreme weather conditions.
- *Health and Safety*: Extreme weather events could lead to an epidemic outbreak. Lack of supporting infrastructure and emergency response plans would increase dangerous conditions, and make it more difficult to quickly reach those in need. Damaged water infrastructure increases toxic effluents and sewage in surface water. Waterborne diseases and insects breed more easily in warmer and more polluted waters. Poverty makes it difficult to buy bottled water or fuel to boil water during emergencies. The Asian Development Bank estimates that as many as 12.2 million people in Vietnam could experience either worsening or new water stress by 2050. Energy shortages or damage upsets critical medical treatment and storage, such as refrigeration.
- *Agriculture and food*: The Mekong Delta, also known as Vietnam's rice bowl, is critical to providing for Vietnam's food security. While its GDP contribution is small and decreasing, about three quarters of the population still live in rural areas and depend on agricultural production. Rising temperatures and heat waves upset water temperatures critical to aquatic species, and increase the water requirements of agriculture and livestock. One-meter sea level rise would impact 7% of agriculture result in heavy losses in livestock, decreased water supplies, etc.
- *Biodiversity*: Changing climate conditions would change the pattern of life for many species and their corresponding food chains. The mix of species in some habitats will shift, and even species lose might occur.

Antithetical to socialist principles, certain groups, such as those living in poverty, ethnic minorities, and women, will be most vulnerable to climate change impacts and have the least resistance against it. Rural residents whose livelihoods

[29] State of water environmental issues: Vietnam. *Water Environment Partnership in Asia.* http://www.wepa-db.net/policies/state/vietnam/overview.htm.

depend on agriculture and aquaculture will see production fluctuations and infrastructure, such as boats, machinery at risk. In many villages, women are hit the hardest by natural disasters. Often they cannot swim, have fewer assets to turn to for alternative livelihoods, and have fewer employment opportunities away from the home.[30]

Both environmental degradation and climate change impacts have already resulted in significant socio-economic costs and devastation and will only increase if unabated. Vietnam's traditional economic growth model worsened these twin challenges, and in turn, these issues have the potential to derail Vietnam's cherished economic goals.

Issues of economic and population growth, environmental degradation, resource inefficiency, and climate change are highly interlinked, with compounded impacts. For example, climate change impacts, such as drought, along with increased water demands from population and economic growth, and more polluted water sources from environmental degradation have threatened the critical hydropower supply at the same time that energy use is increasing. Already the country may face a shortage of almost 1 billion kilowatt-hours as demand is expected to rise as much as 18% for 2010 (Nguyen and Uyen, 2010). Lack of a stable and secure energy supply in turn threatens the country's economic expansion. A sustainable path of development is necessary to curtail environmental degradation, combat climate change impacts, and, most importantly, still allow the country to reach its development goals.

3. Vietnam's Sustainable Development Policies

Fortunately, Vietnam has recognized these twin challenges of environmental degradation and climate change, and enacted major policies to promote sustainable development. Environmental protection has become one of its socialist ideals, promoted in the National Party Congress. Previous Socio-Economic Development Strategies have called for environmental protection, economic growth, and sustainable development. The 5-Year Plan for 2006–2010 calls for "high quality and sustainable economic and human development.[31]" The upcoming Socio-Economic Development Plan for 2011–2020 will echo similar themes. In explaining his vision for the next Socio-Economic Plan, Prime Minister Nguyen Tan Dung said:

> To our country, environmental pollution and the risks posed by climate change and rising sea-levels are enormous challenges. As such, economic development must go hand in hand with environmental protection and improvements. We must be proactive in addressing rising sea-levels, employing energy and material saving technologies to attain the goal of green economic development. It is

[30] *Vietnam: Climate Change, Adaptation and Poor People.* Oxfam, October 2008.

[31] Rapid, sustained growth guide state's development. *Vietnam News* 3 August 2010. http://vietnamnews.vnanet.vn/Economy/202157/Rapid-sustained-growth-guide-States-development.html.

of great importance to sustainable development and must be manifested in all of our socio-economic development plans, from the centre to the sectors and localities.[32]

In 2004, the country promulgated Vietnam Agenda 21 ("Strategic Orientation Strategy on Sustainable Development in Vietnam") to integrate environment and economic objectives and promote sustainable development. To manage the implementation, the National Council of Sustainable Development was established in 2005, with the Vietnam Agenda 21 Office under the Ministry of Planning and Investment (MPI) as its secretariat.

The Council is noted for its high political support, broad inclusivity, and guiding principles. The Council receives the highest political support and leadership. The chair is the Deputy Prime Minister, with the Minister of Planning and Investment serving as Standing Vice Chairman, and the Ministers of Natural Resources and Environment, and of Science and Technology as Vice Chairmen. To ensure broad support, the other 40 members are drawn from the national assembly, party agencies, ministries, sectors, localities, enterprises and social agencies.[33] Eight principles of sustainable development were created to guide the work:

(1) human is the cornerstone of sustainable development;
(2) economic development is the key task of the next stage of development;
(3) protection and improvement of environmental quality development;
(4) development must equitably meet the needs of the present generation and not hamper future generation's life;
(5) science and technology serve as the platform and engine of development;
(6) sustainable development is the cause of the entire Party and agencies of all levels, communities, and people;
(7) consolidate independent economic development with initiatives for international economic integration for national sustainable development; and
(8) incorporate socio-economic-environmental development with defence security maintenance.[34]

However, those qualities are not enough to move the Council into real action and implementation. They seem to be part of the country's noted vision, but sensible rhetoric; little action is done beyond putting it in writing. Though there was support from international agencies such as Asian Development Bank, UN, and the Norwegian Ministry of Foreign Affairs, resources for action seem limited.[35] Aside

[32]Rapid, sustained growth guide state's development. *Vietnam News* 3 August 2010. http://vietnamnews.vnanet.vn/Economy/202157/Rapid-sustained-growth-guide-States-development.html.

[33]Sustainable development implementation in Vietnam. *Ministry of Planning and Investment Vietnam Agenda 21 Office*, 2008.

[34]sustainable development implementation in Vietnam. *Ministry of Planning and Investment Vietnam Agenda 21 Office*, 2008.

[35]*The Vietnam Agenda 21.* Website www.va21.org and email agenda21@va21.org no longer exist.

from the few workshops, forums, and publications, the Council's activities have not been well-known or documented (The Phuong, 2007). Though the membership is politically strong and broad, members rarely meet (Bass *et al.*, 2010). Instead, the broad membership seems more of a disadvantage. As no agency or group claims responsibility, it is treated more as an extracurricular or volunteer activity than an important task to be incorporated into the department's work.

In another example of action failing to match vision is in water management. The Dong Nai and Sai Gong River systems are important water bodies, covering several key cities and provinces in the country. However, as a common good, they have been severely polluted and require greater regional cooperation and administration. In 2005, 12 cities and provinces signed an agreement to have at least 70% of the area's industrial parks establish wastewater treatment systems that met the permissible pollutant discharge standards, and halt environmentally harmful facilities by 2007. As of 2010, only 30% of the centralized wastewater treatment systems had been built or met the required standards.[36]

Policies for interregional cooperation are valuable, especially for borderless, common-good issues, such as water and the environment. It is not enough to have one city impose strict air quality standards when a nearby city is promoting weak industrial standards to capture short-term economic gain. Air pollution will still spread and carbon emissions increase. However, governments have to follow-through with their cooperation agreements or policies, and match their sustainability rhetoric with action.

Additionally, some of the sustainable development policies miss key components, warrant additional considerations, or lack full integration for long-term, comprehensive improvements. In sustainable energy policy, the country is moving aggressively to diversify its energy supplies, engaging in overseas oil and gas exploration,[37] and expanding into nuclear power. Foreign energy dependency, however, poses other geopolitical risks. Specific plans for secured nuclear waste disposal have not been mentioned for the eight recently approved nuclear facilities planned by 2030.[38] The Second National Strategy and Action Plan for Disaster Mitigation and Management 2001–2020 is more designed to "address short-term climate extremes rather than to respond to future climate change, and focuses on emergency response and reconstruction, rather than risk prevention and adaptation" (Chaudhry and Ruysschaert, 2007/2008). Other sustainable development and climate change policies have been criticized for ignoring technological solutions, poverty reduction or addressing vulnerable groups.

[36]Southern rivers suffer heavy pollution. *Vietnam News* 31 July 2010. http://vietnamnews.vnagency.com.vn/Environment/202075/Southern-rivers-suffer-heavy-pollution.html.
[37]APEC Energy Demand and Supply Outlook 2006.
[38]Vietnam plans eight nuclear power plants by 2030. *AFP* 22 June 2010. http://rawstory.com/rs/2010/0622/vietnam-plans-nuclear-power-plants-2030/.

The fact that such policies exist in such an early stage of Vietnam's development, however, shows government recognition of the issues and a greater vision for the future. Vietnam's sustainable development policies are broad, and in some ways more advanced than other more-developed countries. Its renewable energy target of 5% by 2020 is much lower than those of China and the EU, 20%, but still notable when the US has none. Similarly, its National Strategic Program on Energy Saving and Effective Use 2006–2015 has a national target reduction in total energy consumption of 5–8% for 2011–2015 (Baumuller, 2010), while its National Target Programme on Climate Change aims to adapt to climate change impacts and move toward a low-carbon economy (Baumuller, 2010). Although they are not bound by carbon emissions reduction targets, they did ratify the Kyoto Protocol in 2002 and noted the Copenhagen Accord in 2010 (Baumuller, 2010). Regionally, Vietnam assumed the Chairmanship of the ASEAN this year and called for an East Asia Forum on Climate Change.[39]

Yet, it is too early to applaud Vietnam's efforts, especially given its past record. The UNEP and ADB's Vietnam Environmental Performance Assessment Report concurred:

> "Whilst the Government of Viet Nam (GOV) has promulgated new legislation and embarked on programs and projects to mitigate the environmental impact of urban and industrial expansion the effectiveness of these measures has been insufficiently discussed and probed."[40]

No policy is perfect, and the mismatch between Vietnam's stated policies and action reveals deeper issues that must be addressed first before successful implementation. Though not alone, Vietnam's sustainable development policies are plagued with several more critical institutional, capacity, and resource problems:

- weak rule of law;
- high local authority discretion;
- poor government department cooperation;
- lack of resources;
- low stakeholder awareness and engagement.

3.1. *Weak rule of law*

While there are growing environmental and sustainability regulations, they are complex and difficult to interpret and follow. (Tran *et al.*, 2009. Weak criminal regulations complicate the legal situation further. According to Pham Van Loi, director of the Vietnam Environment Administration's Institute of Science for Environmental

[39]PM proposes an East Asia Forum on Climate Change. *lookatvietnam* 30 January 2010. http://www.lookatvietnam.com/2010/01/pm-proposes-an-east-asian-forum-on-climate-change-2.html?utm_source=twitterfeed&utm_medium=twitter.

[40]*Vietnam National Environmental Performance Assessment Report.* UNEP and ADB.

Management, "If any enterprise is discovered causing heavy pollution, the leader of that enterprise could be criminally charged to increase deterrence, but unfortunately, the penal code does not establish criminal responsibility of an individual."[41] Weak criminal regulations lessen the severity of the violation, fear of defiance, or plain ignorance for industries and individuals.

There needs to be stronger rule of law to evoke a sense of seriousness and respect regarding sustainability policies and regulations. Sustainability policies need to be incorporated into the nation's regulations, backed up by a strong enforcement and monitoring framework, and put into development practice. A good starting point could be to ensure all new development follows Vietnam's Sustainable Development Indicators or other key performance measurements and targets. Vietnam's Sustainable Development Indicators are comprehensive and divided into local and national data sets. If used well, they could be an important tool to track progress on sustainable development and identify priority areas to focus on. Governments could provide incentives for developers to follow or surpass the targets, or punish those who fail to adhere to them. Unfortunately, the Indicators now are merely a guide and have little influence on the nation's development progress.

3.2. *High local authority discretion*

National policies and regulations rely on local enforcement and implementation. This is especially true in Vietnam, as the country's history, culture, and geography have promoted a decentralized government. In theory, such systems allow local authorities the flexibility to adapt national policies to best fit their area, minimizing national government burden, optimizing resources, and providing better oversight. However, the benefits discussed materialize only if the local authorities actually implement and enforce the national policies. In reality, laws on environmental safeguards, notably EIA, are not well enforced locally (Bass *et al.*, 2010).

There are a few reasons for this. Economic goals tend to overshadow sustainable development. Sustainability policies can be seen to prevent, conflict, or delay economic goals. There are few incentives or punishments (as seen in the weak rule of law) to encourage local authorities to adhere to sustainable development policies. However, economic gains, even short-term environmentally unsound ones, can bring significant material benefit to the area and prestige for government officials. The Socio-Economic Development Plans and Five Year Plans have prominent economic targets to be reached, and fewer sustainability-related ones. Failure to hit the next GDP target brings much larger implications and attention then failure to hit the green space target.

Economic competition also makes it tough for sustainability cooperation. As discussed, even if the local authorities want to adhere to the sustainability policies,

[41]Heavy polluters still spared from criminal proceedings. *The Saigon Times* 11 May 2010. http://english.thesaigontimes.vn/Home/business/environment/10169/.

they often have to work with other local authorities to make a positive impact. This proves to be difficult if cities are competing for foreign direct investment, skilled workers, and other resources. Finally, weaker communication between local and national governments make it difficult to fully understand the intentions of the policies or how they reflect the realities on the ground (Bass *et al.*, 2010). Local governments need greater guidance and incentives to promote sustainable development policies above economic goals. There needs to be a stronger framework that allows them to communicate better with the national government and with other local governments to best leverage the implementation and enforcement.

3.3. *Poor government department cooperation*

No agency or department is responsibility for sustainable development. As discussed, the National Council for Sustainable Development has yet to fully engage in its coordination and implementation role, or has the resources or mandate to do so. The fact that it falls under the MPI means that it is beholden to that department's interest.

Instead, sustainability responsibility is spread and diluted among the government departments. This results in overlapping responsibilities, conflicting interests, issue gaps, and tougher compliance from industries and citizens. For example, the MPI, is responsible for overseeing environmental issues in the development budget. Environmental policies, however, fall under the Ministry of Natural Resources and the Environment (MONRE), just established in 2002. Energy issues are under the Ministry of Industry and Trade; However, they are not responsible for environmental protection laws, regulations or standards in the energy sector (Do *et al.*, 2008).

It is unclear how the Ministries deal with overlapping and connecting issues. There are 10 ministries that deal with water-related issues, not to mention a number of sub-departments and agencies (Fig. 4).

Departments remain committed to their sectoral issues and short- term gains.

Similar to local authorities, government departments treat sustainability policies as guidelines and fail to fully comprehend how their decisions can result in environmental impacts outside their jurisdiction. For example, more severe and frequent climate change impacts will increase the vulnerability of coastal or low-lying areas. Yet the Ministry of Construction has minimal guidelines for stronger building codes and the MPI has few policies to discourage investment in these high-risk areas. The health implications of environmental degradation and climate change have been discussed; However, health authorities have yet to implement a comprehensive health plan to address long-term issues such as increased asthma rates or disaster prevention and recovery against the rise of water borne diseases in flooding.

[42]Water quality in Vietnam — with a focus on Cau, Nhue-Day, and Dong Nai River Basins. *Vietnam Environment Monitor 2006*. World Bank.

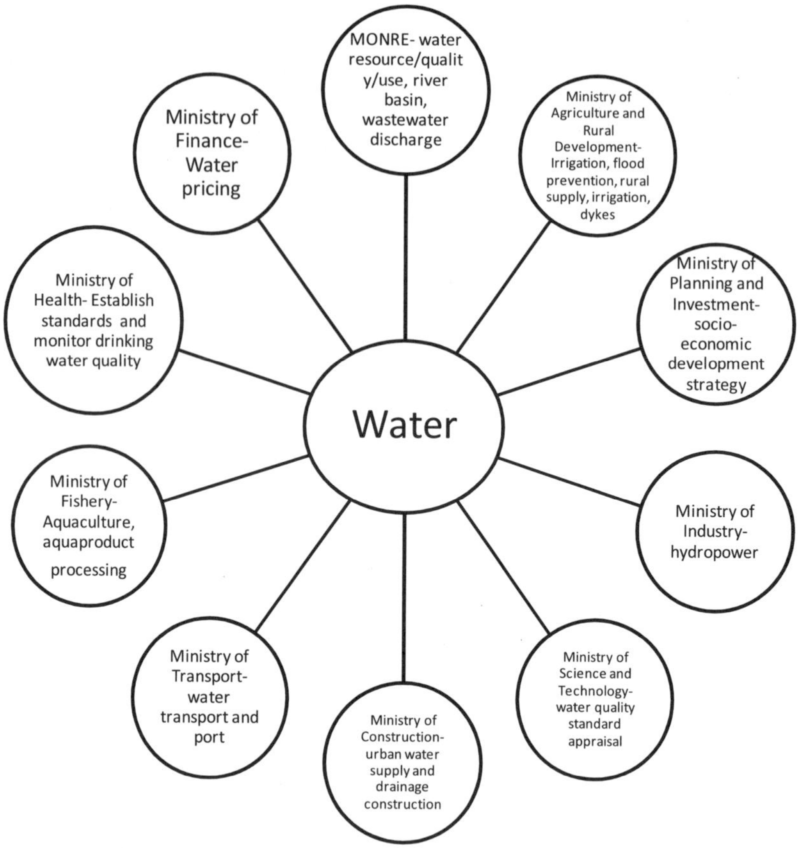

Fig. 4. Water-related responsibilities.[42]

Finally communication and coordination among the departments is insufficient, unstructured, and infrequent. "There is no procedure or established mechanism for organizing regular meetings between MONRE and other agencies to deliberate upon environmental issues."[43] The level of communication varies between departments, personalities, and issues of that time. Greater communication among the departments is one of the first steps to minimize conflict, expose overlaps and gaps, identify sustainability connections beyond traditional jurisdictions, and cooperate on more comprehensive sustainability policies.

3.4. *Lack of resources*

The national government and some provincial governments established an environmental protection fund with a target of 1% of the government budget. The funds are linked to various "polluter-pays" schemes (Bass *et al.*, 2010). Revenue is starting

[43] *Vietnam National Environmental Performance Assessment Report.* UNEP and ADB.

to be generated. For instance, the first half of 2010 has brought the Ho Chi Minh City government VND 530 million in fines from 100 plus enterprises for discharging untreated wastewater and industrial smoke.[44] However, it is unclear how these funds will be spent or prioritized, and their impact on other sources of funding (Bass *et al.*, 2010). Even if the 1% target is met by the national and all provincial governments, it is unknown if it is sufficient to implement, enforce, and monitor sustainability policies.

Current resources are shown to be inadequate in addressing sustainability issues. "Staff and financial resources available for effective implementation of environmental policies have consistently fallen short of what the intent of the policies would have required."[45] Nguyen Thanh Huy, vice head of the agency's Division of Environmental Quality Monitoring and Assessment, said "many targets for reducing the levels of air pollution have been unobtainable because the city does not make appropriate investments for controlling the sources of air pollution."[46] In another example, the proposed EURO-2 standards for vehicle exhaust emissions would prohibit 50% of total vehicles in the country (Do *et al.*, 2008). Such widespread transportation and economic change have not fully been assessed. How would the country deal with the prohibited vehicles? Are there enough finances to replace those vehicles or to supply the public transport infrastructure for the transport mode shift? As targets become higher, with worsening environmental degradation or more severe climate change impacts, it will be harder to provide sufficient resources for successful policy implementation.

Resources also do not just include finances, but also human capital, institutional capacity, technology, and infrastructure. Are environmental officers adequately trained in combating air pollution? Are there additional education and training classes and material on the latest research and development? Do health officers have the data systems to monitor infection and asthma rates? Are decision makers aware of international best practices? Are there enough enforcement officers to track violators? Resource levels vary by cities, provinces, and population groups, making it hard to consistently or comprehensively enact/enforce policy.

If Vietnam does prioritize sustainability as it states in its national policies and strategies, it needs to have the resources to match its commitment. Arguably, it is difficult for a lower-middle income country to do this against other pressing issues, such as education, healthcare, infrastructure, and other services. However, sustainability is often interlinked with these issues and can optimize resources and improve business models and investment decisions.

---

[44]Hotline launched to report pollution, wildlife trafficking. *The Saigon Times* 11 June 2010. http://english.thesaigontimes.vn/Home/business/environment/10800/.

[45] *Vietnam National Environmental Performance Assessment Report.* UNEP and ADB.

[46]Air pollution in HCMC getting worse. *The Saigon Times* 19 August 2009. http://english.thesaigontimes.vn/Home/business/investment/5994/.

Limited resources should not be used to repeatedly address problems with short-term, expensive solutions. Few and Tran's (2010) study of low-income communities in the Central Provinces and the Mekong Delta showed that temporary disaster protection measures were made because of the shortage of funds. Instead, resources should be put into wider livelihood protection. For example, building climate resilient infrastructure means that it does not have to be rebuilt each rainy season, which saves funds for development uses. Boat ownership increases people's ability to reach shelters during floods, limiting government resources to serve other priorities during disasters. It also increases access to other services such as healthcare and education. Similarly, improving air quality reduces the number of days children miss school or employees at work.

The government can start to think about improving or finding alternative ways to improve resources. Are fines high enough to deter pollution and encourage sustainability, or do companies find it easier and cheaper to pay the fines and violate the policies as shown in China? In another case, the National Environmental Action Plan has no provisions to use economic instruments, such as carbon taxes to reduce energy demand (Do *et al.*, 2008).

Resources can also be leveraged jointly with other local and national governments; regional members such as APEC and ASEAN; and international organizations, such as the Asian Development Bank and UN. There are certainly discrepancies between the amount of resources committed and the amount given, and as Vietnam's development status improves, aid will become less. However, other state and non-state actors remain a good place for resources, including ideas, advice, technology transfer, capacity building, funding, etc.

3.5. *Low stakeholder awareness and engagement*

Public, business, media, etc. awareness and engagement in sustainability need to substantially increase. In some cases, groups are ahead of government policies. Some non-profits have been working on sustainable development for decades. Other businesses are forced to deal with the issues because of new European or US legislation. However, there needs to be broader awareness and engagement of these groups. Considering the level of environmental degradation and severity of climate change impacts, public concern is relatively low. In 2007, Nielsen published the results of a global online survey of 25,000 users on the level of concern people had for global warming; out of the 47 countries surveyed, Vietnam came the 36th lowest on the list.[47]

Including more stakeholders in the process makes it easier to implement and enforce policy. Especially with limited resources, stakeholder participation can generate more ideas, incorporate greater understanding of the diverse issues, and increase responsibility towards its success. Indeed, public action is helping with

[47] *Vietnam: Climate Change, Adaptation and Poor People.* Oxfam October 2008.

enforcement and monitoring. With funding from the Danish Embassy, the Environment Police Agency of the Ministry of Public Security launched a public hotline to inform the police of environment pollution or wildlife trafficking violations.[48] This could not only increase compliance, but also increase education among citizens.

Other groups can also support sustainability policy. Academic research can guide government officials on the latest carbon emissions scenarios and forecasting, companies can partner to build and operate new technology. Citizens can decrease their energy and water consumption for step-change. Non-profit groups can run sustainability education and training programs. Vietnam can improve communications channels for feedback and input, increase engagement opportunities, and better seek stakeholder partnerships and resources.

There are five key highly interlinked issues that plague sustainability policies: weak rule of law; high local authority discretion; poor government department cooperation; lack of resources; low stakeholder awareness and engagement. Weak rule of law gives local authorities greater discretion. Lack of resources hinders stakeholder awareness. Poor government department cooperation wastes limited resources. At the same time though, understanding the linkages and combating the issues holistically can provide win–win solutions. Increasing stakeholder engagement can save resources. Strengthening the rule of law increases stakeholder awareness. These issues are not unique to Vietnam and are difficult to solve in the short term. However, acknowledging the limits of Vietnam's sustainability policies will both improve current policy implementation and formulate more effective policies in the future. In time, the aim is to have Vietnam match its sustainability policy vision with implementation.

4. Conclusion

"The challenge is how we are able to drive the economy in an intensive and sustainable manner through application of advanced technology that help improve productivity, resource efficiency, environmental quality and development of renewable energy."

Minister of Planning and Investment and Standing Deputy Chairman of Vietnam National Sustainable Development Council, Vo Hong Phuc[49]

Sustainable development is an iterative process, rather than an end state. Comprehensive policies, regulations, and resources at the national and local levels with broad departmental and stakeholder support are essential toward its successful

[48]Hotline launched to report pollution, wildlife trafficking. *The Saigon Times* 11 June 2010. http://english.thesaigontimes.vn/Home/business/environment/10800/.

[49]Sustainable development is king. *Vietnam Investment Review* 8 June 2010. http://www.vir.com.vn/Client/VIR/index.asp?url=content.asp&doc=17421.

implementation. Vietnam's main priority remains moving the country toward developed status by 2020. However, it realizes that to do so, it must combat environmental degradation and climate change impacts, and enact sustainability policies. Its foresight is much more advanced than many countries at similar stages of economic development. While there are institutional, resource, and capacity challenges in fully implementing the sustainability policies and having action match rhetoric, they are not unique to Vietnam or impossible to confront. Acknowledging these issues and understanding their linkages is the first step. Indeed, Vietnam's sustainable development policies may come to transform the country just as its economic policies have done.

References

Anwar, S. and Nguyen, L.P., 2010. Foreign direct investment and trade: The case of Vietnam. *Res. Int. Business Finance*, doi:10.1016/j.ribaf.2010.05.004.

Baumuller, H., 2010. *Building a Low Carbon Future for Vietnam.* Chatham House, February 2010.

Bass, S. *et al.*, 2010. Integrating environment and development in Vietnam. *IIED, Vietnam/UNDP Poverty Environment Programme.*

Carew-Reid, J., 2007. Rapid assessment of the extent and impact of sea level rise in Vietnam. *Climate Change Discussion Paper 1.* ICEM-International Centre for Environment Management, Brisbane, Australia.

Chaudhry, P. and Ruysschaert, G., 2007/2008. Climate change and human development in Vietnam. *UN Human Development Report 2007/2008.*

Dissanayake, S., 2007. Pollution clouds Hong Kong's Future. *BBC News* 26 June 2007. http://news.bbc.co.uk/2/hi/asia-pacific/6768977.stm.

Do, T.M., Sharma, D. and Nguyen, N.H., 2008. Review of existing energy policies for Vietnam. In: *GMSARN International Conference on Sustainable Development: Issues and Prospects for the GMS 12–14 Nov. 2008.*

Few, R. and Tran, P.G., 2010. Climate hazards, health risks and response in Vietnam: Case studies of social dimensions of vulnerability. *Global Environmental Change* **20**, 529–538.

Hicky, N., 2008. Vietnam's ongoing power shortages. *Energy Tribune*, 16 June 2008. http://www.energytribune.com/articles.cfm?aid=923.

Nguyen, D. and Uyen, T., 2010. Vietnam faces power shortages as drought hurts hydroelectricity. *Bloomberg Businessweek* 6 April 2010. http://www.businessweek.com/news/2010-04-06/vietnam-faces-power-shortage-as-drought-hurts-hydroelectricity.html.

Nguyen, K.Q., 2008. Internalizing externalities into capacity expansion planning: The case of electricity in Vietnam. *Energy* **33**, 740–746.

The Phuong, N., 2007. Sustainable development mechanism in Vietnam. In: *Regional Workshop on National Sustainable Development Strategies in Asia, 7–8 November 2007, Bangkok.* http://www.rrcap.unep.org/nsds/workshop/nsds_RWS_engppt_MPI_VN.pdf.

Tran, T.B., Grafton, R.Q., Kompas, T., 2009. Institutions matter: The case of Vietnam. *The Journal of Socio-Economics* **38**, 1–12.

Chapter 33

SUSTAINABILITY MANAGEMENT IN AGRICULTURE — A SYSTEMS APPROACH

Buddhi Marambe[a] and Pradeepa Silva[b]

[a]*Department of Crop Science, Faculty of Agriculture, University of Peradeniya, Sri Lanka*
bmarambe@pdn.ac.lk

[b]*Department of Animal Science, Faculty of Agriculture, University of Peradeniya, Sri Lanka*
pradeepas@pdn.ac.lk

Success in green revolution was made possible by the introduction of high yielding crop varieties and policy support promoting the use of modern inputs such as chemical fertilizers and pesticides. The incorporation of animals with improved breeds to the crop-driven agricultural system strengthens the approaches in achieving long-term food security. However, the sustainability of such agricultural systems driven by the high external inputs is now being questioned in terms of their sustainability. It is now understood that strategic objectives for agricultural development of a country, especially a developing economy, are to sustain the increases in agricultural production and productivity through an ecosystem approach. Sustainability mainly relates to the efficient and effective use of natural resources, which has become a mandatory for governments, donors, practitioners, and a generalized concern in international spheres during the past three decades. A sustainable agriculture system will incorporate novel technologies or practices that will improve farmer productivity without major negative impacts on the environment and have a positive effect on natural, social, and human capital. In the Sri Lankan context, the development strategy maintained immediately after independence focused on import substitution industries and development of the agriculture sector. Achieving self-sufficiency in rice production has been considered the priority of successive governments during the past six decades to ensure food security, generate employment, and eliminate poverty. The agro-ecosystems of the country, which include crop cultivation, animal husbandry, fisheries, and forestry, collectively provide almost all the food and other needs of the inhabitants. As a result of disintegration owing to structural, technological, and institutional, changes brought about by modem development, the village agro-ecosystem has become de-stabilized. This has led to over-dependence on external inputs and hence, self-sustainability of the village agro-ecosystem has ceased to exist except in some remote areas. Social, economic, and environmental sustainability are the three pillars of the sustainability framework for agriculture and food production. The current high-input-oriented conventional practice of agriculture lacks principles of sustenance where the nutrient balances have become negative for many systems, more marginal lands being cultivated, communal grazing lands have diminished, livestock become more dependent on crop land forages, etc. The development of new and innovative management strategies is thus a need to harness the maximum benefits of integrated systems.

In this regard, the organic methods of food production, while significantly contributing to maintain the equilibrium in the ecosystems, can contribute substantially to feed the current and future human population on the current agricultural land base. Strategic mechanism to promote sustainability management of agricultural systems is required at all levels, with a cognizance from deep-rooted customs and traditions and the time-tested agricultural practices. Development of self-confidence, self-motivation, and empowerment of the farming community would overcome the over-dependence on external suppliers and supplies and would sustain the agricultural development of the country. A new development strategy encouraging pro-poor investments is hence a requirement by review of policies, understanding the suitability and availability of land resources, investing in research, and producing integrated development plans, in order to increase the productivity and profitability of environmentally sustainable practices.

1. Introduction

The world population has been estimated to increase from 6.3 billion in 2008 to 9 billion in 2050. About 90% of the population growth will be observed in developing countries, whereas by 2030, 60% of the population is expected to concentrate in the cities (World Bank, 2007). With this population growth, the food consumption demand (crop-based food) will increase globally by 50%, and with rising income levels, the meat and animal products demand will also increase by 50% (Marambe, 2008). Thus, the total food demand is expected to be doubled in the next 40 years.

The slow pace of commercial imports, food aid, and government market interventions is keeping food prices high in Asia despite falling from 2008 peaks in international markets. This is affecting the food security of large proportions of low-income groups. Rising unemployment and falling income are applying additional pressure on the poor and the vulnerable. More worrying is that once the global economy recovers, the pressure that drove up food prices last year may return. The global hunger has reaches an all-time high, where for the first time in the human history, billion people go to bed hungry (WFP, 2009). The recent global financial crisis has changed face into a food crisis for 583 million people in the Asia and the Pacific region. Studies have indicated that of 19 Asian economies, including India and China, more than 10% of people live on less than 1.25 US$ a day, whereas more than 10% are malnourished. Significant proportions of the South Asia face chronic or transient food insecurity. A recent study carried out by the UNESCAP (2008) has identified 25 countries of food insecurity including Sri Lanka. On top of this, a new report by IFPRI (2008) predicts a gloomy future for developing countries, particularly in South Asia, as climate change may also result in large drops in crop yields by 2050. This can potentially leave 25 million more children malnourished compared with a world without global warming.

All above indicate that there is an enormous challenge in achieving sustainability in food production. The sustainability of food production aims at provision of carbohydrates, fats, and proteins for human consumption. Cereal and root crops and food animals are important components in sustainable agriculture, which are interconnected through resource utilization network. Switching on to more meat consumption as predicted by Marambe (2008) will result in greater demand for grain too, as rearing cattle and poultry are grain-intensive activities. This contradicts the claim for sustainability from animal agriculture point of view, where the animal farming should have the capacity to consume food sources that are complementary to, rather than in competition with, the needs of human beings.

The world economy has faced many problems in the past, but nothing can be or more pressing than the food crisis as well as food price crisis that it faced in the recent past. Governments in the Asia and Pacific region came under enormous pressure due to the rising food prices and dwindling global stocks, as they had to strive hard to provide food required for the most vulnerable and poor in their respective countries. The statistics indicated that over a billion people in the Asia Pacific region were seriously affected by the food price surge, as food expenditure accounts for 60% of the average total expenditure basket. A large number of people live in south Asia experienced the disastrous conditions due to steep food prices as food and energy together account for more than 75% of total spending of the poor in the region (Marambe, 2008). The World's agricultural land is used in different ways (Fig. 1), depending on climatic and soil factors, and also related to cultural and social issues.

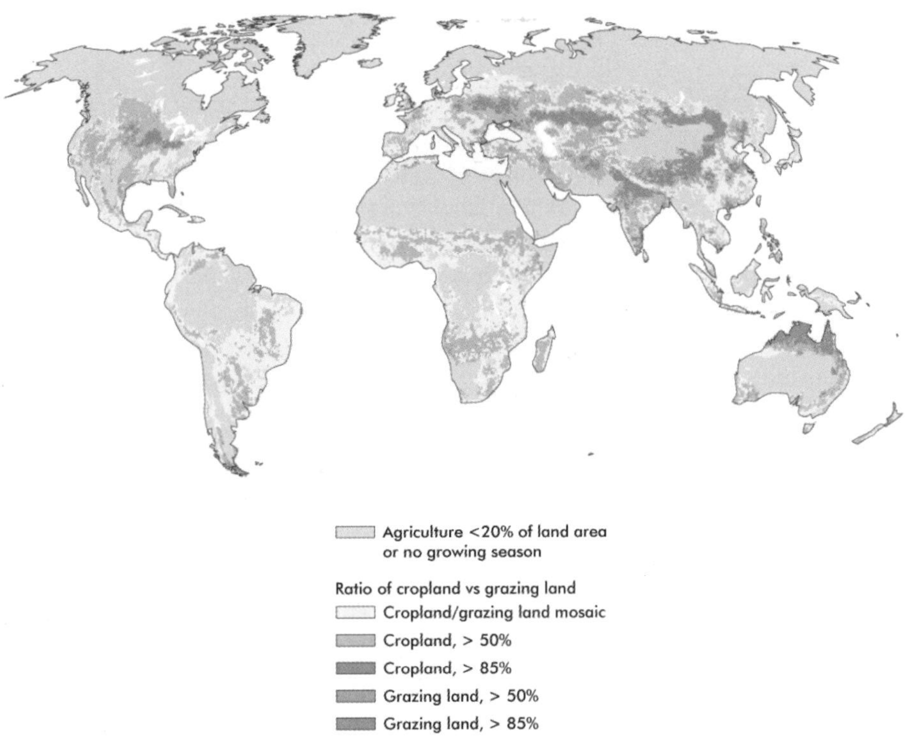

Fig. 1. Agriculture land use distribution — croplands and pasture lands (Author Credit — Hugo Ahelnius, UNEP/GRID-Arendal).

The majority of croplands such as rice, wheat, legumes, and corn — among other crops — are spread out in the Northern Hemisphere, in the temperate zone, and in South and Eastern Asia. Areas where primary livestock is held for agricultural are dominant in Africa, South America, and Australia (http://maps.grida.no/go/graphic/agriculture-land-use-distribution-croplands-and-pasture-land1) where large proportion of lands out of the 35% of the earth's surface suitable for pastoralization is found (Goodwin, 1994).

Climate change has contributed significantly to poor agricultural production at a global scale. In 2007, drought-affected crop and animal production in Australia and the shortfall in the production of staples in Ukraine and Kazakhstan heavily contributed to the global food shortages. Extreme weather conditions have affected the cereal production in China, India, Indonesia, Cambodia, Sri Lanka, etc., in the recent past. In early 2008, the world cereal stocks were adequate to meet 2 months of consumption, the worst scenario experienced since 1970 (Marambe, 2008). In addition, the fuel prices also increased dramatically in 2007–2008 (also due to the weakening US $ against Euro). The effects will be most acute in developing countries, where the increase in demand for consummative goods is expected to be the highest and growing at a rate faster than increases in production (FAO, 2006). These countries are also located in the region where climate change is projected to have its greatest impact. Cline (2007) reported that the agricultural productivity in developing countries is expected to decline by between 9% and 21% due to global warming.

All these developments have lead to food insecurity most parts of the developing world resulting from the core issue of concern "sustainability of agriculture and its operations." South Asia is no exception. The World Summit on Food Security (FAO, 2009) resulted in a declaration highlighting the promotion of new investments to increase sustainable agricultural production and productivity, support increased production and productivity of agriculture, and implementation of sustainable practices, improved resource use, protection of the environment, conservation of the natural resource base, and enhanced use of ecosystem services. The Declaration has also committed to address the sustainable use of land and water, maintaining the health and productivity of all ecosystems, and better management of the biodiversity associated with food and agriculture, thus indicating the need for managing sustainability in agriculture in the long run.

Sustainability, especially in agriculture (including crops, animals, and fisheries), has become an issue close to the heart of many people in the developing world than any other socio-economic issue. Rapid population growth, declining farm size, falling soil fertility, and missed opportunities for income diversification and migration create distress as the powers of agriculture for development remain fallow (World Bank, 2007). In the context of sustainability, two key questions are addressed, namely (a) whether agriculture will be able to meet future global food demands without adversely affecting the resource base, and (b) the optimum approach to enable agriculture to both provide sufficient food and also act as an engine of pro-poor growth despite resource constraints.

2. Sustainability

Sustainability is an important topic in development intervention and discourse, especially related to the use of natural resources. Concern comes from evidence of the depletion and deterioration of resources. This paradigm has become mandatory for governments, donors, practitioners, and a generalized concern in international spheres during the past three decades. At the same time, it has constituted a strong argument to get funds for development programs and projects implemented by governments and non-governmental organizations (NGOs). Within academic spheres, the topic has also been dealt with from different perspectives. It is generally agreed that sustainability is understood as "the need to ensure the preservation of natural resources involved in economic activities over time for future generations."

The sustainable development challenge remains urgent and acute (OECD, 2001) where abject poverty and food insecurity still ravage the lives of a sizeable population in developing countries. For each country, the challenge and the strategy that it chooses to follow in response will be different. All, however, will require deep structural changes, relating to economy, society, and politics. For example, agriculture is sustainable when current and future food demands of a country can be met without unnecessarily compromising economic, ecological, and social/political needs.

3. Sustainability in Agriculture

Environmental health, economic profitability, and social and economic equity are the three main integrated goals of sustainable development, which rests on the principle that the needs of the present should be met without compromising the ability of future generations to meet their own needs (http://www.sarep.ucdavis.edu/concept.htm). The key features of agricultural sustainability include that agricultural strategies be based on more than simple productivity criteria (DFID, 2004). The characteristics of sustainable agriculture include the following:

(a) Replenishing land, water, and air — Maintaining the equilibrium of the environment while working in harmony with the environment, to ensure ecosystem services operate at its potential.
(b) Conservation and use of biodiversity — To maintain biological systems and processes to provide us with food, feed, fiber, and fuel.
(c) Human and animal welfare — To provide healthy and fair employment practices for better yield of food and a stronger community, and minimum standard of cleanliness and space for animals.
(d) Economic viability — Providing employment opportunities to local and regional communities and adoption of fair trade practices.

More than three decades ago, the main concern in agriculture was technological problems to meet a rapidly-growing food demand. The technological development has now allowed reaching unsuspected yield levels by improving different components of the production cycle. Everson and Gollin (2003) attributed 21% increase in yield to the use of plant breeding technologies, 20% to the expansion of the agricultural area, and 59% to the intensive use of agricultural inputs, of which irrigation contributing to a major share. Main concerns related to sustainability of irrigated agriculture tend to look at it only from a technological point of view and thus related to the optimization of inputs, efficiency, and profitability (Khan *et al.*, 2006; Wichelns and Oster, 2006; Haie and Keller, 2008).

4. Sustainability of Agriculture and Livelihood of People Threatened

Challenges posed by external factors due to globalization and open economic policies have directed the country's agriculture to move away from the self-reliance. Competitive export and import opportunities among countries have led to maintain the standard levels of quality and steady levels of production at a lower price in all commodities. This situation demands a firm and perfect policy for country's agriculture. Further, present agriculture has ignored the centuries old wisdom of traditional agriculture and has shown limited sustainability.

Agriculture has experienced significant shocks in recent years due to high fluctuations in oil prices, commodity price spikes, food security fears, and resultant trade restrictions. The anticipated return to global economic growth, rising population, the emerging biofuel markets, and also a higher cost structure are expected to underpin the international commodity markets and prices over the next decade (OECD/FAO, 2010). Developing countries are expected to be the driving force behind the expected growth in agricultural production, consumption, and trade. Despite the recently experienced global-level shocks, agriculture in developing countries such as Sri Lanka has performed remarkably well over the past 50 years, delivering food at progressively low prices by keeping pace with rapid population growth. However, this success is not without the cost of the natural resource base and also a result of overuse of natural resources as inputs. This moves beyond interpretations of sustainability based solely on ecological factors and farming systems. Key features of agricultural sustainability include an acceptance of the fact that agricultural strategies should be based on more than simple productivity criteria, that externalities are of great importance, and that intra- and inter-generational equities are key parameters in assessing agricultural change.

The important issue is that agriculture does not bring a consistent economic gain to the farming community. This aspect is most critically influencing the developing economies. Import–export policies of countries with developing economies do not respond effectively to maintain profitability of farming. Farming without adequate concern on conservation of natural resources such as soil and water has led to deterioration of the agricultural resource base and pollution of the environment. Governments have failed to pay adequate attention, in many instances, to provide farmers with input and marketing facilities in time. Land resource is utilized for various purposes including agriculture without considering its suitability, capability, and vulnerability. Farmers are not much aware of the current trends in agriculture, marketing, and technologies. All these have resulted in developing economies becoming more fragile, exorbitant food prices leading to food and nutrition insecurity, and rising poverty levels. Few examples in this regard from the global scenario are given below.

4.1. *Rising food costs*

The basic staples that feed the world wheat, rice, and corn continue their inexorable rise in cost and scarcity. Rice, an important staple, has already shown signs of short supply. In 2008, the United Nations reported that Nepal, which imports much of its rice, has 2.5 million people in immediate need of assistance and 3.9 million more whose welfare may be compromised by rising prices. In Afghanistan and Pakistan, people have protested to the exorbitant wheat prices where many cannot afford to buy flour to make bread. Recent reports also indicated that India has failed to deliver sufficient food to several hundred million of its people despite its 8% annual growth in its economy. The Director General of FAO, addressing the Conference at the European Parliament in Brussels, has stated that the number of hungry people increased by about 50 million in 2007 as a result of high food prices.

High food prices and shortage of their supply will threaten to increase malnutrition, already an underlying cause of death for over 3.5 million children a year. The world prices of the basic food for human and animal feed have increased dramatically over the past years. Since 2000, the prices of energy, wheat, butter, and milk have tripled, prices of maize, rice, and beef have doubled. In 2008, mirroring price developments of the major agricultural commodities, dairy product prices remain approximately 60% above historical levels. The sudden dairy prices in 2006 took the market by surprise as it started rising faster and sooner than those of any other commodity. Globally, the price of whole milk powder has increased by 100% in the past year and that of skimmed milk has increased by 150% in the same year. Concentrated animal feed price has increased by 50%. Poor families spend up to 80% of their budget on food. Prices are expected to stay high through 2015. FAO (2008) indicated that, from October–December 2007, the prices of whole milk powder increase by 49% in Sri Lanka and 22% in Nepal, whereas during the same period, the prices of fluid milk increase by 25% in Sri Lanka, 27% in Pakistan, and 17% in Nepal.

The year 2009, however, showed a decline trend in increasing food prices in many countries around the globe where countries such as Sri Lanka, Guatemala, and China had relatively small price increase, that is less than 3% when compared with that of 2008 (OECD/FAO, 2010). The OECD/FAO (2010) also reported that

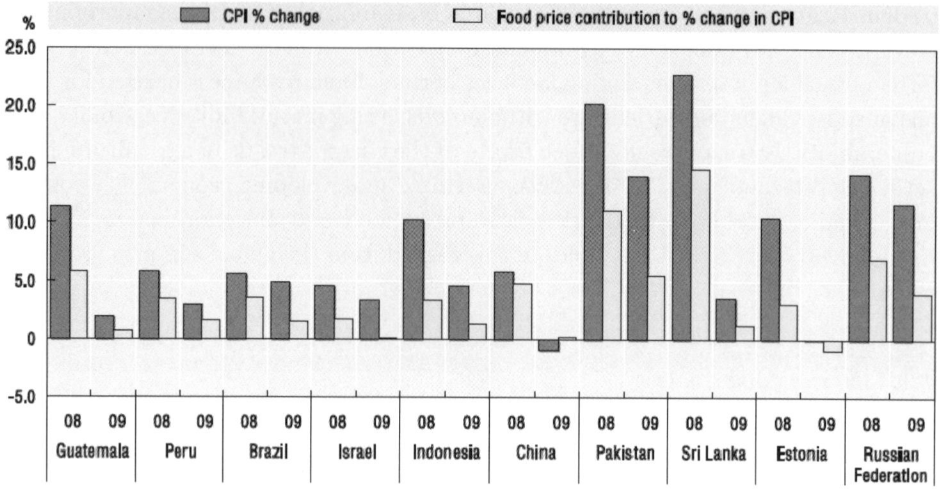

Fig. 2. Contribution of the food price changes to inflation is selected countries.
Source: OECD/FAO, 2010.

in high-income countries, the share of food in the consumer price index (CPI) ranges from less than 10 to 20%; however, in the middle- and low-income countries it is substantially higher, generally in the 30–60% range. For example, the food component accounted for 47% of the CPI in Sri Lanka, 58% in Malawi, 55% in Tanzania, 38% in Peru, and 28% in Brazil. In contrast, in the United States it is, however, 8.2%, 10.4% in Switzerland and Germany, and 11.8% in the United Kingdom. For many middle- and low-income countries, where food expenditures account for a substantial share of household expenditures, their impact on inflation can still be significant even when food prices rise only moderately (Fig. 2). In countries where a large share of household expenditures is devoted to food, the food price inflation has become a crucial economic indicator as higher food prices would result in less money available for other essential items such as housing, transportation, health, and educational services, and affect sustainability in country's economic and social development.

4.2. *Rising poverty levels*

Poverty remains the most pressing issue facing Asia and the Pacific today. Despite some spectacular progress over the past few decades, the region remains home to two-thirds of the world's poor (www.adb.org/poverty/). About 1.2 billion people still live on less than $1 a day. However, 43 countries, with more than 60% of the world's people, have already met or are on track to meet the goal of cutting hunger in half by 2015. The per capita income of the rural people has increased only by 0.54% (CSAPS, 2007). The percentage below national poverty lines of different countries in the SAARC region is illustrated in Fig. 3.

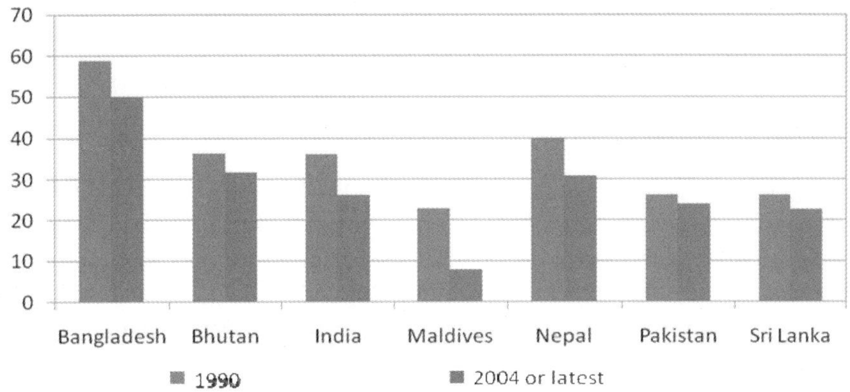

Fig. 3. Percentage of population below national poverty lines.

Large pockets of extreme poverty and hunger persist in Asia, where the global downturn makes it more difficult to achieve UN MDGs goals to reduce the ranks of the poor. The Human Poverty Index for developing countries (HPI-1), which focuses on the proportion of people below a threshold level in the same dimensions of human development as the human development index, living a long and healthy life, having access to education, and a decent standard of living, is 16.9% for Sri Lanka, ranking the country at the 67th position among 135 developing countries for which the index has been calculated.

In 19 Asian economies, including the most populous China and India, more than 10% of people live on less than US $1.25 a day and more than 10% are malnourished. This is despite success over the past 15 years in cutting the number of poor from one in two to one in four. Nepal is the worst off with 55.1% of its population surviving on less than US$1.25 a day. In China and India, 15.9% and 41.6% of the population live below the poverty line, respectively (http://www.nzherald.co.nz/business/news/article.cfm?cid=3&objected=105938 50).[1]

5. Systems Approach

Sustainable development in developing countries takes on important social and economic ramification. Historically, the process of development has heavily relied on the natural resource base of a country, especially to bring people out of poverty. In the recent past, the emerging economies such as Sri Lanka have endorsed the same path to build the country's economy, however, with parallel attempts for conservation and sustainable use of the natural resources. The underlying principle is that the components of an agricultural ecosystems interact, and to maintain "balance" within functional boundaries (Lewis *et al.*, 1997). In an agricultural system, the basic principle for managing variables that are interdependent, especially

[1]Accessed 9 July 2010.

those that are undesirable, is similar to that for other systems, including the social systems. Interdependency demands systems thinking which is a holistic approach with a continuum of activities that ranges from the conceptual models to the technical practice (Richmond, 1991). The system is envisioned in its broadest sense from the individual farm to the local ecosystem *and* to communities affected by this farming system both locally and globally. An emphasis on the system allows a larger and more thorough view of the consequences of farming practices on both human communities and the environment. A systems approach gives us the tools to explore the interconnections between farming and other aspects of our environment. The benefits of a total system approach in any development process are enormous as this approach takes into account the impacts on the natural resources such as the conservation of flora and fauna, energy and non-renewable resources, and improving quality and diversity of landscape. The well-being of people associated with agriculture would be the resulting social benefit in the long run.

Green revolution of 1960s, which is an attempt made in the past for intensification of agriculture to meet the rising food demands, is also recognized as an exercise that yield increases have been at the expense of the environment due to heavy loading of synthetic external inputs, thus making it an unsustainable approach. Many small-scale farmers in the developing nations were unable to reap the actual rewards of scale by embarking on the intensive agricultural practices. The IAASTD (2009) highlighted the need for policies that value, restore, and protect ecosystem services, and address the needs of the world's small-scale and family farmers. The need for a paradigm shift to encourage increased adoption of sustainable ecological agriculture and food systems is thus imperative.

Over the years, the agricultural practices and production systems have evolved, managing the ecosystem services to improve productivity and reduce environmental impact. However, the biological processes resulting from the action of soil-based organisms, pollinators, natural predators, etc. are being complemented as well as replaced by the external inputs such as seeds, fertilizers, pesticides, etc. in the modern day intensive agriculture. However, shifting the production practices to a more nature-based approach will lead to the balance between natural and man-made ecosystems leading to the paradigm of sustainability. The following underlying principles will help achieving in order to assure sustainability in agriculture:

(a) Conservation and sustainable use of plant and animal genetic resources for food and agriculture to ensure food production and meet growing environmental challenges such as climate change.
(b) Conservation agriculture practices to create stable living conditions for micro- and macro-organisms.
(c) Integrated pest management (IPM) to encourage natural predation as a corollary to reducing the overuse of insecticides.
(d) Integrated plant nutrient management to promote the combined use of mineral, organic, and biological resources to balance efficient use of limited/finite

resources and ensure ecosystem sustainability against nutrient mining and degradation of soil and water resources.

(e) Well-integrated crop and livestock systems to increase the diversity and environmental sustainability of smallholder production systems.

(f) Focusing on efficiency and productivity gains by managing the natural resources combining land, water, biodiversity, and ecosystem services to sustain production.

Sustainable agriculture is a dynamic rather than a static concept. What may contribute toward sustainability today may not work as the system evolves, thus requiring a high level of observation and skills that can help adapt to change. Consequently, sustainability is a direction/process and does not by itself result in a final, fixed product (Soni and Salokhe, 2009). The therapeutic approach has been the prevailing strategy for many problems faced by the agriculture sector for over six decades. The ecological disruptions due to unprecedented external use have continued to ensue, and there are renewed appeals for effective, safe, and economically acceptable alternatives. Considerable effort has been directed toward such alternatives, and new technology has been implemented and is still emerging especially in the field of agriculture.

6. Agriculture in Sri Lanka

The development strategy maintained immediately after independence in Sri Lanka focused on import substitution industries and development of the agriculture sector. The development objective was to achieve self-sufficiency in various commodities mainly staple rice and other essential food items. Except for the partial opening of the economy that took place during 1965–1970 with relaxation of import restrictions and encouraging exports, the Sri Lankan economy was characterized as a close economy based on the government policies implemented during 1960–1977.

The efforts of the government during this era helped improve social welfare through multitude of subsidies such as food subsidies, free education (since 1945), health care, etc. that elevated the living standards of Sri Lankan population compared with those in the neighbouring countries. The growth of the economy in real terms averaged 3.6% per annum during 1960–1977 (Central Bank, 1993) and the per capita income grew at 1.55% per annum. The period after 1977 with the change of the government and introduction of more liberal economic policies all price controls and import restrictions removed, tariff structure was simplified and a floating exchange rate system introduced, private sector was encouraged play a vital role in the economy, and foreign investments encouraged. These economic policies were instrumental in creating new activities hitherto undertaken by the public or private entrepreneurs in both agriculture and industrial sectors. This also resulted in the over-dependence on agriculture being replaced by an expanding industrial sector creating more employment and enhanced production.

At the time of independence, more than half of the Sri Lankan population was dependent on agriculture for a livelihood, where the sector directly contributed to about 40% of the national income in 1948. However, there was little interaction between the plantation sector and the domestic non-plantation agricultural sector (Central Bank, 1998). Over the past 62 years, Sri Lanka has made a remarkable progress in terms of food production and availability irrespective its share in the national GDP has been shrinking. For example, in 1953 the country produced only 38% of its requirement of rice for a population that was about 8 million and has reached near self-sufficiency in rice in 2009 with a population of 21.2 million. The total agriculture sector recorded 7.5% growth rate for the year 2008. The third and fourth quarters of the year reported the highest growth rate of 50.5% within past two decades for the paddy production. The other food crops that consist of seasonal crops, cereals, vegetables, and fruits also reported 4.7% growth rate for the year 2008. The fishing sector also reported 9.9% growth rate for the year 2008. This was as a result of government policies and prevailed favorable weather situation in 2008 after a zero growth recorded for the total agriculture in 2004. The country was able to achieve this better performance for agriculture when a food crisis existed in the world. Sri Lanka can be considered as a unique example considering the output features or indicators of economic performance in 2008. Few countries reported positive growth rates for the year 2008 such as China, India, Malaysia Thailand, Philippines, etc., and Sri Lanka is among those countries with a 6.0% GDP growth rate. The first three quarters of the year indicated above 6.0% GDP growth rates. However, the economic activities under agriculture such as tea (−18.7%) and minor export crops (−16.3%) have reported the significant downturn in real terms of value added for fourth quarter of 2008.

7. Sustainable Agricultural Systems: A Case from Sri Lanka

With a civilization dating back to 2500 years, which was based on agriculture determined by the rainfall and suitable soil conditions, Sri Lanka is still a middle-income earning country. The traditional agricultural practices fell in-line with nature and practiced over centuries, and hence practicing agriculture blended with nature, such as organic agriculture, is not a novel concept for Sri Lanka. Despite the progress shown in the growth of the agricultural sector in Sri Lanka, poverty has remained widespread and a significant factor hindering country's development. In addition, the environmental damage resulting from agricultural operations has become more prevalent especially during the past few decades. Agriculture remains the principal livelihood of poor people in developing countries, and particularly the rural poor, and is widely considered to be the major "engine of economic growth" in the majority of developing countries. With the experiences in past contribution of agriculture to poverty reduction and economic growth, it is evident that agriculture has the potential to stimulate further pro-poor growth in the face of existing resource constraints.

The concept of sustainability in agriculture centers on the need to develop technologies and practices that do not have adverse effects on environmental goods and services, leading to improvements in food productivity. Sustainability rests on the principle to meet the needs of the present without compromising the ability of future generations to meet their own needs. Agricultural sustainability in all cases, however, emphasizes the potential benefits that arise from making the best use of both good genotypes of crops and animals and their ecological management (Pretty *et al.*, 2006). Sustainability in agriculture entertains traditional as well as novel technologies such as organic farming, biotechnology, etc., provided they improve productivity for farmers and do not harm the environment (Conway, 1997).

The agro-ecosystem of traditional villages in Sri Lanka included four major components, namely crop cultivation, animal husbandry, fishery, and forestry, collectively providing practically all the food and other needs of the village inhabitants. Thus, the livestock were used to provide draught power for farming, and were fed on crop residues and straw, their dung being used to fertilize the soil. Fishery was mainly on irrigated paddy fields and in the irrigation system, while the fish on the paddy fields consumed harmful insects and worms, and provided fertilizer in their excreta (Ulluwishewa, 1991). The village forest on the catchment area and the trees sparsely grown on the cultivated area and in the irrigation system substantially contributed to maintain the natural productivity, yielding firewood, timber, and various foods. In this manner, the village agro-ecosystem functioned to satisfy the needs of its human inhabitants without impairing its own sustainability.

This village agro-ecosystem is now in the process of disintegration owing to structural, technological, and institutional, changes brought about by modem development. The reason is that these modern development strategies were basically designed to raise the production of the land area rather than to preserve the productivity of the given agro-ecosystem. While farm mechanization displaced the draught animals, the increased population pressure of more and more humans reduced the grazing lands. Both modern development and population pressure had destructive effects on the forest cover, while increased utilization of agro-chemicals adversely affected the fish culture on paddy fields. In this way, livestock, fishery, and forestry, all became insignificant or at least insufficient, and were de-linked from crop cultivation. In consequence, the village agro-ecosystem has become de-stabilized, and dependence on external inputs has increased so greatly that self-sustainability of the village agro-ecosystem has ceased to exist except in some remote areas.

8. Organic Agriculture — A Path to Sustainability in Agriculture

The concept and practice of organic agriculture are based on local resources and a non-chemical practice, and are environmentally friendly, culturally sensitive, socially just and economically viable sustainable agricultural system that maintains an efficient management process. As the market-oriented agriculture started replacing the subsistence agriculture, the agro-chemicals usage in the production systems was intensified. This has particularly occurred after the green revolution. With the introduction of high-input demanding varieties and breeds, the current conventional practice is more market-oriented and commercially driven than the traditional-subsistence-level systems experienced. Organic methods of food production can contribute substantially to feeding the current and future human population on the current agricultural land base, while maintaining soil fertility (Badgley *et al.*, 2007).

The sustainability framework for agriculture and food production consists of three equally important, mutually interacting and reinforcing objectives for social, economic, and environmental sustainability. A simple triangular model may be visualized with social and environmental sustainability at the base and economic sustainability at the apex. A congruence of the objectives of organic agriculture with those of the triangular sustainability framework is obvious.

Under the current commercial and subsistence-level farming practices, the nutrient balances are negative for many agriculture systems with offtake greater than input. The depletion of soil nutrients without adequate replacement would lead to a definite decline of crop yields over time, and the external input or replacement becomes a "must" practice, creating a vicious cycle. According to the present trend, more land is brought under intensive cultivation/husbandry practices in order to maintain production levels, where farmers have to cultivate more marginal areas. In the livestock segment, the communal grazing lands have diminished and livestock have become more dependent on crop land forages, especially in drier areas, where the free grazing was widely practiced in the past. The cycling of nutrients among rangeland, crop land, ruminant livestock, and soils is important to sustain agricultural productivity. However, the nutrient transfer mechanisms are poorly understood by the farmers. Sustaining the productivity of an increasingly fragile ecosystem requires a better understanding of nutrient cycles and the development of new and innovative management strategies in order to utilize the maximum benefit of integrated systems. A well-known linkage between livestock and soil productivity is the cycling of biomass from crop component through animals into excreta (manure, urine) that fertilizes the soil in the integrated systems. The soil health, i.e., soil organic matter, nutrient availability, nutrient exchange, and water holding capacities, is maintained in integrated farming system where the link between crop and animal farming could be exploited for the benefit of sustainability. Moving to stall-feeding of animals could increase nutrient losses if the nutrients excreted by animals are not captured, stored, and spread on cropland.

More than 90% of the agriculture holdings in Sri Lanka come under smallholding category, where the nutrient management is crucial. The sustainability of these smallholder farming operation greatly relies on crop–animal integration, which is the most suitable description for the major integrated production system in Sri Lanka. Crop-ruminant production is carried out under both rainfed and irrigated conditions, especially in the dry zone of Sri Lanka (Silva *et al.*, 2008). The system shows a greater diversity and complexity in the crops grown, the cropping patterns used and in the livestock species/breed raised. As highlighted in many instances the indigenous practice of keeping animals on cropland represents an appropriate manure management practice for the extensive mixed-farming systems.

About 32.2 million ha of agricultural land have been certified according to organic standards (Helga and Kilcher, 2009). The greatest share of the global organic surface area is in Oceania (34.7%), followed by Europe (23.4%) and Latin America (23%). With its vast grazing lands, Australia continues to account for the largest certified organic surface area, with 12 million ha, followed by Argentina (4 million ha) and China (1.9 million ha). The global market for organic products reached a value of over 50 billion US Dollars in 2008, with the vast majority of products being consumed in North America and Europe. These data indicate that the markets worldwide continue to demand more organic produce despite the economic turbulence.

9. Organic Agriculture in Sri Lanka

In Sri Lanka, 17,000 ha of land have been reported under organic management (Helga and Kilcher, 2009), with a share of total agricultural land of 0.72%, and around 4216 organic producers. Sri Lanka is one of the major organic producers in Asia and one of the leading sources of organic tea (http://www.intracen.org/organics/Country-Profile-SriLanka.htm).[2] Most of these organic products are exported to Europe, Japan, and Australia. In the past decade, global trend in organic agriculture has been mainstreamed as a sector in agri-business (http://loamsrilanka.org/srilanka.htm).[3] As a result new market segments have opened-up and product ranges have seen a growth, including marketing of herbal or health care produce and organic cosmetics. The clear indication of the potential for promoting organic production is the demand for many categories of produce, which has often been higher than the supply.

Organic agriculture plays an important role in terms of conserving biodiversity, ensuring nutrient recycling, and tackling climate change related issues. The following section describes the contribution of organic agriculture in such processes showing the path for sustainable development of the country.

10. Biodiversity and Organic Agriculture

Over the past century, population, market pressures, and the development of new agricultural technologies have encouraged patterns of agricultural development tending toward agricultural intensification (i.e., increasing scales of monoculture production, intensive mechanical tillage, irrigation, and the use of synthetic fertilizer, pest control agents, and a restricted diversity of crop and livestock varieties), often leading to natural resources degradation. The majority of the human

[2] Accessed 26 May 2010.
[3] Accessed 14 May 2010.

population increase is expected to take place in the biodiversity-rich developing countries of the tropics such as Sri Lanka.

The convention on biological diversity "encourages the development of technologies and farming practices that not only increase productivity, but also arrest degradation as well as reclaim, rehabilitate, restore, and enhance biological diversity and monitor adverse effects on sustainable agricultural diversity. These could include, *inter alia*, organic farming, IPM, biological control, no-till agriculture, multi-cropping, intercropping, crop rotation, and agricultural forestry" (Decision III/11, 15e). While several agricultural approaches make sustainability claims, organic agriculture is the only well-defined agricultural management system, including recommended and restricted practices that aim at environmental protection and food production. However, the main challenge in protected areas is to conserve biodiversity while providing the basis for the social and economic development of local residents.

Sri Lanka's geographic location, varied climatic conditions, and topography have given rise to its unique biological diversity. Sri Lanka together with Western Ghats has been identified as one of the eight Global Biodiversity Hotspots based on the number of endemic plants and vertebrates, their density, and remaining primary vegetations relative to the original extent. Sri Lanka has the highest species diversity per unit land area of all Asian countries in terms of flowering plants and all vertebrate groups, excluding birds. In the terrestrial ecosystems of Sri Lanka, the wet zone forests in the southwest are especially important as they sustain 75% of the endemic species of flora and fauna.

There are several reasons contributing to the existence of high biodiversity. Among those being an island, high geographical variation and climatic variation are the main contributors. Agro-biodiversity that is a subset of biodiversity is an important concept in the case of sustainable agriculture, where the diversity of the biological and physical resources in a system and their balanced management utilized plays a crucial and deciding role. The level of agro-biodiversity in Sri Lanka is mainly contributed by the 46 agro-ecological zones distributed fairly equally in three main climatic zones: wet zone (>2500 mm rainfall), intermediate zone (1750–2500 mm rainfall), and dry zone (<1750 mm rainfall).

Agriculturalists are now aware of the value of the biodiversity "input" for agriculture. The ecological functions of diverse ecosystems (such as balanced predation, pollination, nutrient cycling, degradation of toxic compounds, carbon sequestration) are today recognized to be central to sustainable food production. Moving away from simplified agricultural systems offers opportunities to produce food while enhancing natural landscapes.

At the same time, biodiversity in all agro-ecosystems is composed of two primary elements, namely, crop and animal biodiversity and systems biodiversity. Crop and animal biodiversity refers to the species present on the farm that will provide direct economic input to the farm. Systems biodiversity is the non-crop/animal component of the biodiversity that is required to sustain the agro-ecosystem. It is both these measures that indicate the state of biodiversity in an agro-ecosystem. There are 11 traditional cultivation/production systems described in Sri Lanka (Pushpakumara and Silva, 2008). These systems have been preserving 2355 crop species and 654 animal species (including fish species). The gain inbiodiversity in an agricultural field is directly proportional to a change in the management regime adopted. Therefore, the challenge is to evolve a system of knowledge that will enable the use of biodiversity data in monitoring and evaluation of organic agriculture, as the present National Standards for Organic Agriculture in Sri Lanka state that "the presence of biodiversity (soil, surface, and aquatic) is an indication of the health of the agroecosystem."

The simplification of agro-ecosystems to monoculture production and the removal of non-crop vegetation from the farm unit (e.g., hedgerows, shelter belts, and field margins) has contributed to the homogeneity of agricultural landscapes by reducing botanical and structural variations. This has resulted in a reduced capacity of agricultural areas to serve as habitat for wild species, and reduced effective internal regulation of populations of pests and disease causing organisms affecting crop productivity has led to a decline in abundance of farm species and diversity across many taxonomic groupings including high rates of wildlife mortality and reduced reproductive success of many species (Gliessman, 1999; Benton *et al.*, 2003). This loss of biodiversity has also resulted in a reduced capacity of agro-ecosystems to perform many essential ecosystem functions such as purification of water, internal regulation of pests and diseases, carbon sequestration, and degradation of toxic compounds (Altieri, 1999). In the past three decades, alternative systems in the form or organic agriculture has been developed in Sri Lanka where the landscapes of farm gardens have been designed to include a variety of annual to both tree crops and animals that provide a host of utility benefits such as food, medicine, timber, fuelwood, fodder, fiber, etc. They are also designed to provide whole forest products and services such as water yield, carbon sequestration, biodiversity conservation, and phytoremediation. Systems of land management in organic agriculture seek to establish ecosystem analogous in architectural structure and ecological function to the original climax or sub-climax vegetation community.

11. Soil Quality and Nutrient Recycling and Organic Agriculture

For the regeneration of vast areas of severely eroded lands, organic farming offers a practical and logical solution. For sustainability of soil productivity, a long-term process of improving soil fertility and soil quality is desired. The term soil quality or soil health refers to an overall soil condition that is highly favorable for plant growth arising from combined effects of physical, chemical, and biological properties of the soil. Long-term organic farming leads to an overall soil rehabilitation because building up the organic matter content of soils causes simultaneous changes in soil properties (physical, chemical, and biological) whose combined effects give rise to soil quality favorable to the plant. For instance, increasing soil organic matter content means concomitant increase in soil fertility, aggregate stability (meaning good soil structure), water holding capacity, ion exchange capacity (meaning capacity to hold nutrients), buffering capacity, and microbial activity (meaning vigorous life in the soil).

Sustainable, well-managed organic farms do not focus on a single species/breed of crop or animal, but involve a diversity of crops and animals that represent multiple nutrient utilization strategies, water uptake requirements, and pest preferences. Carefully planned crop rotations and intercropping systems and crop–animal integrations guard against nutrient movement into waterways by recycling and conserving nutrients within the system. Such practices are most effective and sustainable when they are implemented as part of systems-based approach. Maintaining nutrient balances within fields while minimizing water flows onto fields from off-farm areas, keeping water within fields, and capturing any water that flows away from fields will conserve nutrients on the farm while protecting the environment. Using a diversity of plants as rotation crops, cover crops, intercrops, and integrating with animals enhances soil quality, facilitates nutrient capture, and helps recycle nutrients that would otherwise be leached through the soil. The crops will also provide soil cover, which encourages water infiltration and decreases the potential for nutrient runoff and erosion.

Building up stores of active organic matter and diverse communities of soil organisms will enhance soil storage of nutrient reserves while decreasing the potential for transport of these nutrients to ground or surface waters. Composting organic materials will provide a more uniform nutrient and organic-matter source that is less likely to cause biosecurity risks than fresh manure. During storage, both manure and compost piles should be sited on concrete slabs or soils with a low-leaching potential and with collection or treatment areas for contaminated runoff water. By using practices that conserve nutrients in our crop fields, we are also protecting the environmental quality of nearby streams, lakes, and rivers.

12. Climate Change and Organic Agriculture

Agricultural intensification has had major detrimental impacts on the terrestrial and aquatic ecosystems of the world. The doubling of production during the past 35 years was associated with 6.9 fold increase in nitrogen fertilization, 3.5 fold increase in phosphorus fertilization, and 1.7 fold increase in irrigated land. Agriculture is also affected by climate change. An increase in global warming will shift cultivation zones poleward, plant growth and production being jeopardized by changes in the distribution of rainfall, the increase of UV-B radiation, and changes in the chemical composition of the atmosphere. In regions with continental climate, soils are subject to desiccation, meaning climate change will aggravate problems of salinity, erosion, and desertification. Extreme climatic events will occur more frequently, and pests and diseases favored by a warmer climate will continue to proliferate. All these factors will have negative impacts on productivity of agricultural systems.

According to Hoffman (2010), animal agriculture, in particular, is highlighted as a main actor for both contributes to and is affected by climate change. It has been calculated that the emission of GHG from animals is about 18% of the total emissions or just over the half of the agricultural emission (Seré, 2009), especially, in the form of methane. However, livestock emissions of GHG depend on management system of the livestock and their feed. Grain-fed, factory-farmed cattle in the industrialized countries emit much higher levels of GHG than the grass-fed family-farmed cattle in developing countries such as Sri Lanka (Seré, 2009). Therefore, the concerns of livestock and climate change get a different face in developing countries. The concerns of those countries are more toward the impacts of climate change on livestock and their production than concerns on the extent of GHG emissions from their livestock. The hot and tropical environment in countries such as Sri Lanka usually reduces the productivity of any animal as it has to spend energy on survival. Hotter and more tropical conditions further to the comfort zone for an animal reduce the productivity of the present livestock aggravating the problem of malnutrition in the country. This phenomenon is true for most of the developing countries. Any influence in the livestock production or system of production may affect the livelihood of more than 600 million people in the globe, and around 3.5 million people in Sri Lanka (Ranaweera, 2007) by affecting their economy or by changing the nourishment of their farm families through their livestock products.

Two possible courses of action to alleviate climate change are (a) limiting the greenhouse gas (GHG) emissions and (b) enhancing the removal or uptake of these gases from the atmosphere to stabilize the pools of sediments, trees, and soil organic matter. Organic agriculture not only enables ecosystems to better adjust to the effects of climate change but also offers the potential to reduce the emissions of agricultural GHGs. Moreover, mixed farming and the diversity of organic crop rotations protect the fragile soil surface and may even counteract climate change by restoring the organic matter content. Managing soils to increase the stocks of carbon stored as soil organic matter can be expected to reduce the rate of increase in atmospheric CO_2 and to improve soil quality, thus resulting in additional benefits beyond GHG gas reductions.

Sri Lanka needs to develop better management practices that could increase carbon sinks, energy efficiency improvements, and production of energy from crops and residues. This will result in a further mitigation potential, or cumulative carbon storage. Many long-term experiments in the world support the cognition that organic fertilization (animal manure, green manure, catch, and cover crops) rebuilds soil organic matter. A 20% increase in soil organic matter as a result of organic agriculture would result in an estimated amount of 9 tonnes carbon per ha. There is a considerable potential increase of soil carbon when manure, straw-recycling, minimal tillage, reforestation, and energy-saving plant production are combined. This research demonstrates the need to focus on the use of paddy straw and encourage farmers to engage in the cultivation of green manure crops in Sri Lanka.

13. Biotechnology and Sustainability in Agriculture — Scepticism

The new technological approaches such as biotechnology are a two-stage process. First, there are many applications of gene technology that can be used to improve the management and efficiency of present agricultural practices. Second, there are options for the targeted introduction of transgenic strains, genetically modified for one or more specific traits. Improving the management is the underline theme of sustainable agriculture. The contribution of biotechnology/gene technology to the sustainability in agriculture is already on board in many aspects: (a) better understanding of the functions of crops and animals, (b) targeted selection and breeding to improve the performance and post harvest quality of agriculture products, (c) evaluating genetic resources, (d) efficient diagnostic and management tools for diseases and parasites, and (e) vaccine production for livestock and fish. In the case of transgenic organisms, the strains of various species of crops, trees, livestock, and fish have been developed experimentally, only transgenic crop varieties are in widespread commercial use in agriculture today. As a developing nation, there are concerns in Sri Lanka on the use of biotechnology in agricultural systems sustainably as there is a possibility that it may lead to greater farmer dependence on the providers of the new technology.

14. Managing Sustainability in Agriculture in Sri Lanka — The Strategy

In developing a strategic mechanism to promote an alternative to present agriculture, aiming at sustainable agricultural development, cognizance must be taken from deep-rooted customs and traditions and the time-tested agricultural practices. The dependency of farmers on the external suppliers and supplies that was evolved due to modern agriculture and the government policies dealt with agriculture from time to time should be minimized by developing self-confidence, self-motivation, and empowerment. In this context, it is imperative to develop a new development strategy that encourages the creation of new value in pro-poor investments by increasing the profitability of environmentally sustainable practices.

14.1. *Policy*

Future food production policies in Sri Lanka should be three-pronged: maintain a high degree of self-sufficiency in the major staple, encourage agricultural diversification in potential areas, and develop agro-based industries to ensure household food security, nutritional security, employment, and enhance the income of farmers leading to poverty alleviation and improvement in their quality of life (Mahrouf, 2005). These objectives can be achieved through the proper integration of policies, appropriate technologies, and development of infrastructure facilities in the country.

The National Agriculture Policy of 2007, National Livestock Development Policy of 2007, National Environment Policy of 2003, and *Haritha Lanka* (Green Lanka) program of 2009 are the major policy directives of the government of Sri Lanka to ensure sustainable development in the country. The analysis has revealed that there is an urgent need to strengthen the policies to assure conservation and efficient utilization of soil and water resources in order to implement sustainable agricultural production program in the country. Water being the most previous natural resource that supports livelihood of people and environmental conservation in all directions, policy directives should focus on minimizing water losses from reservoirs and irrigation canals, and improve water use efficiency. Organic agriculture should be encouraged to reduce adverse effects of agro-chemicals and inorganic fertilizer on environment and human health. Creating awareness in general public for consumption of organic products, generating new technology and certifying products, processing and packaging to earn foreign exchange and gain high price to the farmers are current needs. Integrated agricultural practices should be introduced to farmers with crop–animal–fisheries production systems for optimum use of their resources, year round steady income, and effective use of residues.

The policy on integrated/organic agriculture toward sustainable development should be to improve village-level productive farming, discourage synthetic products usage, familiarize livestock farming, and encourage cottage industries to capture foreign and local market with traditional products. A sustainable agriculture policy should stress the importance of conserving natural resources (land, water, forest, atmosphere, etc.), while utilizing them effectively for agricultural production. All land users for agriculture should be encouraged for sustainable use of natural resources by awareness creation, making resource conservation compulsory, generation of improved technologies, safe and efficient management of rainwater, river, tank, and groundwater. Agriculture should be mechanized in all possible ways to reduce the cost of production and improve the quality of production but with no harmful effects on natural resources such as wind and water erosion, increased water and soil nutrient losses, air pollution, etc. National policy on agriculture should emphasize the use of indigenous knowledge in agriculture, which ensures preserving and utilizing traditional crops and varieties, resources conservation practices, medicinal plants, cottage industries, and agricultural heritage of the country.

14.2. *Land resources*

Increasing agricultural productivity should not jeopardize the land, water, and other resources of the country. As many ministries bear the mandate of conserving natural resources, there is a necessity to establish a national advisory board for conservation of natural resources independent of political authority. Lands potentially suitable for agriculture should be given priority for agricultural production to assure the land resource availability for future expansion of agriculture. Primary and secondary forest lands should not be exploited for any purpose other than development of forest vegetations. Decision makers of agricultural production planning should not consider only the national production requirement but also the sustainable production levels of resources including farmers. Thus, food production expectations should aim at national food security, but not always at reducing outflow of foreign exchange for food imports.

14.3. *Research*

Research under the theme of alternative agriculture for self-reliance should be encouraged whereas the most essential knowledge urgently required on (a) measures minimizing the negative impacts of globalization and trade liberalization and to achieve sustainability in agricultural production locally, (b) estimating land and water resources available for agriculture in different parts of the country, (c) methods to develop and allocate such resources for different purposes among competing interests, (d) effective governance framework and institutional mechanisms (policy, legal, and organizational frameworks) to create an enabling environment for cost-effective interventions of sustainable agriculture, (e) relationship between poverty and environmental degradation, (f) poverty alleviation through sustainable agriculture, (f) environmental, social, and economic impacts of land use changes, (g) integrated technological and management strategies required to assure the utilization of natural resources effectively, efficiently, and equitably for agriculture while enhancing environmental services and security, (g) perception of communities in realizing an improving process of sustainability, and (h) decision support information systems to empower the stakeholders in implementing sustainable agricultural development programs. Reorientation of agricultural research agenda from crop or animal-based to resource-productivity-based approach is essential to achieve sustainability. Promotion of endemic fruits, vegetables, and medicinal products and indigenous animal breeds for both local consumption and foreign markets can be initiated through research. Knowledge on conservation of natural resources at present is dispersed and available in various institutions. Gaps need to be identified where further studies are needed and organized by networking them so that any would have the access for utilization.

The reductionist paradigm that gave rise to the technology-focused agriculture in industrial countries has been totally adopted as the research paradigm in developing countries. The world has been producing enough food for all people but the hungry continued to increase to more than 1 billion, and meanwhile, environmental degradation remained unabated. For this reason, research paradigm for developing countries must shift into an encompassing model that addresses people's needs, prevailing state of natural resource base for agriculture, and appropriate technologies to meet those needs while regenerating the environment. This means a shift in research paradigm from reductionist or disciplinary (technology-focused) to trans-disciplinary (integrative approach to technology–people–environment concerns). By integrative paradigm, research is anchored on people's needs and the kinds of resources they have and designed for them to achieve social progress, economic betterment, and environmental regeneration. The integrative paradigm is consistent with the sustainability framework — environmental, social, economic sustainability.

14.4. *Development plan*

A national development plan for agriculture has to be prepared wherein integrated approach of agriculture, livestock, and small agro-based industries is promoted. Any agriculture related activity implemented in the country should be a part of the national agriculture development plan. Committees should be established at provincial, district, divisional, and village level to plan and implement the agricultural production program. Farmers should be protected from adverse effects of free trade policies and globalization.

15. Conclusions

South Asia has the highest number of people who live under the poverty line and who are malnourished. Although the economies are growing at a much faster pace than in the past, food security remains the biggest challenge in the region. The complex and multidimensional nature of food security problem that was initially only viewed in the light of production and availability now also encompasses accessibility, utilization, and vulnerability as its important component. The issues under food security are further aggravated by the low growth of the agriculture sector, restrictive trade policies, and limited social safety nets.

The amount of arable land available to meet increased demand from a burgeoning population is limited. Only about 10% of the world's land surface is suitable for farming, and soil erosion and over-farming are becoming problems in some areas. With no significant room to expand areas of cultivation, increased demand for food in developing countries will have to be met through improved yields of the most widely grown staple crops at a time when the rapid yield growth that characterized the green revolution is beginning to peak.

The demographic profile in most of the developing countries including Sri Lanka reveals an escalating demand for food and employment to cater to the requirements of the ever-increasing population. Deforestation of land for human settlement and cultivation has created environmental degradation based on low productivity in many regions. The lack of arable land can also lead to environmental problems. In many developing countries, for example, biodiversity is threatened because wilderness areas such as tropical rainforests are being converted to farmland to meet the demand of food production. Therefore, farmers need to find new ways to boost production using fewer natural resources.

Sustainability in agriculture is thus important and should be tackled holistically using a systems approach. Some pertinent interlinked constraints, such as poverty of rural farmers, show up immediately, which require integrated strategies to alleviate. Sustainable agriculture does not imply going back to old-fashioned or underdeveloped methods of agriculture. Instead, it seeks to combine the wisdom of past practices, including crop rotation of green manure crops. It also blends in careful use of latest technologies, equipment, new pest-resistant crop varieties, and use of information and communication advancements. The current excessive use of non-renewable resources (because they are relatively cheaper) that will not be available for future generations inhibits the attainment of sustainability. Sustainable agriculture means less emphasis on extractive methods and the use of non-renewable resources, and greater reliance on renewable methods and enhancing the resource base for future generations through exploiting useful biological cycles and thereby saving money on externally purchased inputs.

Appropriateness of natural resource management schemes in agriculture varies with location and time. On the account of small farms and delicate agro-ecosystems, the resources management in the tropical belt is termed under regenerative agriculture, which possesses, and intrinsic mechanism of building and maintaining its productivity through a cyclic farming system. Many studies have been done under tropical climates in developing low-input land and resource management technologies. Organic farming/agriculture with a mixture of crop and animals (including fisheries) has proven to be such an effective and efficient system of integration with sustained land productivity supporting the long-term food production efforts of a country.

References

Altieri, M.A., 1999. The ecological role of biodiversity in agroecosystems. *Agriculture, Ecosystems and Environment* **74**, 19–31.

Badgley, C., Moghtader, J., Quintero, E., Zakem, E., Chappell, M.A., Avilez-Vazquez, K., Samulon, A. and Perfecto, I., 2007. Organic agriculture and the global food supply. *Renewable Agriculture and Food Systems* **22**, 86–108.

Benton, T.G., Vickery, J.A. and Wilson, J.D., 2003. Farmland biodiversity: Is habitat heterogeneity the key? *Trends in Ecology and Evolution* **18**, 182–188.

Central Bank, 1993. *Annual Report.* Central Bank of Sri Lanka.

Central Bank, 1998. *Annual Report.* Central Bank of Sri Lanka.

Cline, W.R., 2007. *Global Warming and Agriculture: Impact Estimates by Country.* Centre for Global Research Development and Peterson Institute for International Economics, 183p.

Conway, G., 1997. *The Doubly Green Revolution: Food for All in the Twenty-First Century.* Comstock Publishing Associates, Ithaca, New York.

CSAPS, 2007. *South Asia — Way Beyond 2015. Reshaping MDG to SDG. Campaign on South Asia Peoples' Solidarity. Briefing paper for Building Opinion.* March 2007.

DFID, 2004. *Agricultural Sustainability.* Department for International Development, United Kingdom. 23p.

Everson, R.E. and Gollin, D., 2003. Assessing the impact of green revolution 1960–2000. *Science* **300**, 758–762.

FAO, 2006. Steinfeld, H., Gerber, P., Wassenaar, T., Castel, V., Rosales, M., de Haan, C. In: Livestock_s Long Shadow — Environmental Issues and Options. (Eds), FAO, Rome, 390p. Available at http://www.fao.org/docrep/010/a0701e/a0701e00.htm.

FAO, 2008. Soaring food prices: Facts, perspectives, impacts and actions required. In: *High Level Conference on World Food Security: The Challenges of Climate Change and Bioenergy.* Food and Agriculture Organization of the United Nations, Rome, 3–5 June, 2008.

FAO, 2009. *Increasing Crop Production Sustainably — The Perspective of Biological Processes.* Food and Agriculture Organization of the United Natinons.

Gliessman, S.R., 1999. *Agroecology: Ecological Processes in Sustainable Agriculture.* Ann Arbor Press.

Goodwin, J., 1994. Sustainable livestock production. *Journal of the Royal Society of Medicine* **87**, 299–301.

Haie, N. and Keller, A., 2008. Effective efficiency as a tool for sustainable water resources management. *J. Am. Water Resour. As. (JAWRA)* **44**, 961–968.

Helga, W. and Kilcher, L., (Eds), 2009. *The World of Organic Agriculture. Statistics and Emerging Trends 2009.* FIBL-IFOAM Report. IFOAM, Bonn, FiBL, Frick, ITC, Geneva, 299p.

http://hdrstats.undp.org /2008/countries/country_fact_sheets/cty_fs_LKA.html.

http://maps.grida.no / go / graphic / agriculture - land - use - distribution - croplands - and - pastureland1 (Accessed on 14 July 2010).

http://www.adb.org/poverty/ (Accessed on 14 June 2010).

http://www.sarep.ucdavis.edu/concept.htm (Accessed on 1 July 2010).

IAASTD, 2009. Agriculture at a crossroads. In: McIntyre, B.D., Herren, H.R., Wakhungu, J. and Watson, R.T., (Eds), *Global Report.* International Assessment of Agricultural Knowledge, Science and Technology for Development.

IFOAM and FiBL, 2006. *The World of Organic Agriculture. Statistics and Emerging Trends.* International Federation of Organic Agriculture Movements (IFOAM), Bonn & Research Institute of Organic Agriculture FiBL, Frick, pp. 108–117.

IFPRI, 2008. *Global Hunger Index — The Challenge of Hunger 2008.* International Food Policy Research Institute, Washington, DC.

Khan, S., Tariq, R., Yuanlai, C. and Blackwell, J., 2006. Can irrigation be sustainable? *Agric. Water Manage.* **80**, 87–99.

Lewis, W.J., van Laneteren, J.C., Phatak, S.C. and Tumlinson, J.H., 1997. A total system approach for to sustainable pest management. *Proc. Nat. Acad. Sci. USA* **94**, 12243–12248.

Marambe, B., 2008. Outcomes of the SAARC towards solving the food crisis in the region. *Economic Review (Special Edition on Food Crisis and Food Security)* **34**, 20–27.

Marambe, B. and Sangakkara, U.R., 1996. Weed control strategies in low-input farming systems. *Commonwealth Agricultural Digest* **5**, 39–67.

Mahrouf, A.R.M., 2005. Enhancing sustainable development of diverse agriculture in Sri Lanka. *UNESCAP-CAPSA Working Paper No. 83*, Indonesia, 116p.

OECD, 2001. Strategies for Sustainable Development — Practical Guidance for Development Cooperation. *Organization for Economic Cooperation and Development*, 7p.

OECD/FAO, 2010. *Agricultural Outlook 2010–2019.* Organization for Economic Cooperation and Development and the Food and Agriculture Organization of the United Nations, 88p.

Pretty, J., Noble, A., Bossio, D., Dixon, J., Hine, R.E., Penning de Vries, P. and Morison, J.I.L., 2006. Resource conserving agriculture increases yields in developing countries. *Environmental Science and Technology* **40**, 1114–1119.

Pushpakumara, D.K.N.G. and Silva, P., 2008. *Agrobiodiversity of Sri Lanka (In Sinhala)*. Biodiversity Secretariat, Ministry of Environment and Natural Resources, Sri Lanka, 80p.

Ranaweera, N.F.C., 2007. *Improved Market Access and Smallholder Dairy Farmer Participation for Sustainable Dairy Development: Lessons Learned, Sri Lanka*. Available at http://www.fao.org/small_holderDairyFAO_milk_report.pdf (Accessed 17/03/2009).

Richmond, H., 1991. *Systems Thinking — Four Key Questions*. High Performance Inc., 8p.

Seré, C., 2009. *No Simple Solution to Livestock and Climate Change*. Available at http://www.Sciencealert.com.au/opinions/20091011-20208.html

Silva, P., Hearth, H.M.S.P. and Chandrasiri, A.D.N., 2008. Indigenous cattle types and production systems in Sri Lanka. *Economic Review*. **34**(5&6), 25–28.

Soni, P. and Salokhe, V.M., 2009: Options for sustainable agriculture in the tropics. *Tech Monitor*, Jan–Feb, 13–20.

Ulluwishewa, R., 1991. Modernization versus sustainability — Disintegrating village agro-ecocomplexes in the dry zone of Sri Lanka. *Environmental Conservation* **18**, 103–109.

UNESCAP, 2008. *Economic and Social Survey of Asia and the Pacific — 2008; Sustaining Growth and Sharing Prosperity*. United Nations Economic and Social Commission for Asia and the Pacific, Thailand.

WFP, 2009. *A Billion for a Billion*. World Food Program. Available at (http://www.wfp.org/1billion).

Wichelns, D. and Oster, J., 2006. Sustainable irrigation is necessary and achievable, but direct costs and environmental impacts can be substantial. *Agricultural Water Management* **86**, 114–127.

World Bank, 2007. *World Development Report*. World Bank, Washington, DC.

Chapter 34

SUSTAINABLE SUPPLY CHAIN INITIATIVES IN THE BRAZILIAN CONSTRUCTION INDUSTRY

K. R. A. Nunes and L. Schebek
Chair Industrial Material Cycle, Institute IWAR,
Darmstadt University of Technology,
Petersenstraße 13, D-64287 Darmstadt, Germany
k.nunes@iwar.tu-darmstadt.de
liselotte.schebek@kit.edu

Brazilian civil construction companies have reached an understanding that socio-environmental responsibility is not just a principle of corporate ethics, but also a business strategy. By taking sustainability into account in their operations, they have established a new business vision that combines economic, environmental, and social aspects, with the goal of assuring companies' survival into the future, as it creates instruments for overcoming crises and gives companies a competitive differential. One of the most significant environmental impacts of the civil construction sector is the quantity of natural resources it uses, recognition of which has made the sustainability of this sector a worldwide priority. For those companies that adopt sustainability policies, the supply chain is an important issue, as some of the impacts are incurred when raw materials are extracted and during suppliers' production processes. The interest in relation to green buildings has also been increasing worldwide and more information about ecological and social aspects of the supply chain companies and their products and services has been required. The aim of this chapter is to give an overview of the process of modernization that Brazil's construction industry is undergoing and the initiatives it has introduced to foster sustainability practices in its supply chain.

1. Civil Construction Industry

1.1. *Introduction*

In 2007, there were 110,000 civil construction companies operating in Brazil, providing jobs for over 1.8 million people in formal employment (around 8.0% of the economically active population). Their total personnel expenses came to US$15.9 billion, of which US$10.8 billion was for wages, labor charges, and other forms of compensation, making the average wage for workers in the sector around US$454.00[1] a month (approximately 2.3 times the Brazilian minimum wage).

[1] *Reference*: US$/R$ exchange rate of 06/30/2004.

In 2009, the sector employed 2.2 million people, a 22% increase vis-à-vis the number employed in 2007 (IBGE, 2007; SINDUSCON-SP, 2009).

The construction work and services undertaken by Brazil's construction companies were worth US$66.5 billion in 2007, a sum that corresponded to 5.1% of the country's GDP. Of this total, US$26.6 billion was for public sector constructions (40.1% of all the constructions built in Brazil) (IBGE, 2007; SINDUSCON-SP, 2009).

The companies from this sector jointly invested around R$5.1 billion in property, plants and equipment, the acquisition of machines and equipment being the biggest investment (44.2% of this total), followed by transportation expenses (23.1%), the acquisition of land and buildings (21.3%) and other acquisitions, such as properties, PCs, and tools (11.4%) (IBGE, 2007). Figure 1 shows the composition of the subsectors of the Brazilian construction industry in 2008.

According to Sinduscon-SP (*Sindicato da Indústria da Construção Civil do Estado de São Paulo*), the São Paulo branch of the Brazilian civil construction trade union, the heightened uncertainty about the global economy toward the end of 2008 had a negative impact on the Brazilian economy (SINDUSCON-SP, 2009).

However, a survey conducted by the same union early in 2009 showed that construction companies were positive about the outlook for economic growth in 2010, thanks to projected rises in the number of mortgages and increases in the amount of capital companies would be raising from the financial markets. There was also an expectation that more new projects would be launched for middle class and low-income families and more investments would be made in infrastructure (CBIC, 2010).

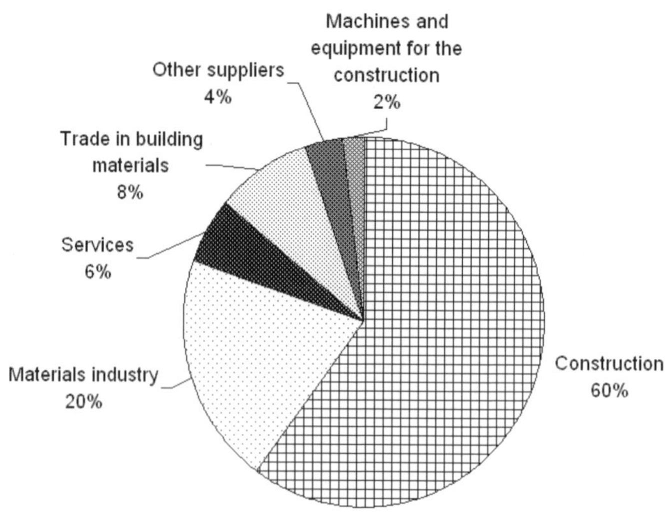

Fig. 1. Composition of the productive chain of the construction industry (CBIC, 2010).

1.2. *Brazilian standards pertaining to civil construction*

Since the 1990s, Brazil's official standards have gained increasing importance in response to the country's new consumer protection code (*Código de Defesa do Consumidor*), quality programs, especially the Brazilian Program for Quality and Productivity in Housing (*Programa Brasileiro da Qualidade e Produtividade do Habitat, PBQP-H*), designed to raise the standards in the civil construction industry, and the opening up of the Brazilian economy.

Consumer protection legislation has been tightened and construction companies have been forced to pay greater attention to the quality of their constructions. With this, it has become imperative to have official standards for the sector, which has led to the drafting and publication of several new standards (CBIC, 2010).

The consumer protection code forbids product and service suppliers from offering any products or services that are not compliant with the standards set by the respective official entities. Even so, despite "the recognition of the importance of standards, the observance of standards is not unanimous across the country, either because they are unknown or because of negligence" (Carvalho, 2004).

Cobracon (*Comitê Brasileiro da Construção Civil*), an entity that is active in the drafting of technical standards for the sector, is engaged in disseminating the use of standards and raising awareness on the market as to the importance of compliance. As a result, buyers can expect to obtain properties of a better quality.

In 2004, Cobracon undertook an analysis of 375 Brazilian standards used in building construction and found that 34% of these had been published prior to 1990, 52% between 1990 and 2000, and just 14% after 2000. They also found that the published standards did not cover all the products and technologies used in civil construction. As a result, the entity also aims to promote a review of published standards and the drafting of new standards, as necessary (Carvalho, 2004).

In Brazil, the entity responsible for technical standards is ABNT (*Associação Brasileira de Normas Técnicas*), a private, not-for-profit entity that generates its own income. The standards are drawn up by study commissions, whose members are volunteers, though their participation is normally sponsored by the companies or institutions they work for. It costs US$3900.00[2] to purchase all 375 standards that apply to buildings, which is a factor that limits access to the standards, especially amongst recently graduated engineers and students of engineering (Carvalho, 2004).

Another point concerning the standards that is highlighted by Carvalho (2010) "has to do with the still widespread culture that compliance with standards will make a construction more expensive. In fact, it is more costly to use standardized systems, but this is offset by the improved performance of constructions, lower post-occupation expenses, and a reduced risk of litigation being brought by consumers" (Carvalho, 2004).

[2] *Reference*: US$/R$ exchange rate of 06/30/2004.

1.3. *Brazilian program for quality and productivity in housing*

It is estimated that around 20–30% of all the material used in Brazil's construction sites is wasted, depending on the construction techniques used (Pinto, 1999). Not only does this have an impact on waste management, but these high levels of waste are one of the factors that raise the cost of buildings.

Part of this waste is the direct result of the poor quality of the materials used in construction processes. According to CBCS (*Conselho Brasileiro de Construção Sustentável, 2009*), the Brazilian Council for Sustainable Construction, "the performance of these products is not satisfactory and a large proportion of them end up being replaced, which generates higher costs and waste. Technical standards are the minimum criteria for quality, and compliance with them is mandatory in Brazil."

In 1991, the federal government created the PBQP-H program (*Programa Brasileiro da Qualidade e Produtividade do Habitat,* Brazilian Program for Quality and Productivity in Housing) with the purpose of organizing the civil construction sector, aiming, in particular, to improve the quality of buildings and modernize the production chain.

Companies' participation in the program is voluntary, and the assessments made by the PBQP-H take into account the differences between the sectors involved (some are less technologically developed than others) and regional inequalities (the south and southeast have the greatest number of and the most modern construction companies and suppliers).

The main agents of the PBQP-H are (a) contractors (mainly components of public sector); (b) sector agents (materials and components manufacturers, and construction and service companies); (c) institutions; and (d) financing and industry funding agents.

A number of measures have been taken to implement the program, including (PBQP-H, 2009)

- appraisals of compliance on the part of construction and service companies;
- improved quality of materials, training and retraining of the workforce;
- creation of technical standards, equipping laboratories and training of laboratory staff, assessment of innovative technologies;
- information for consumers, and fostering of communication between the sectors involved.

The aim of these measures was to enhance the sector's competitiveness, improve the quality of its products and services, reduce costs, and optimize the use of public monies. The main PBQP-H projects are as follows:

(a) *System for assessing service and construction company compliance*: Currently, 1500 construction companies take part in this assessment system. Joining the system is regarded as a market differential.

(b) *System for assessing materials, components, and construction systems*: Since this system was introduced, some significant progress has been noted in the quality of materials and parts used in civil construction. A total of quality programs have been created for the sector, each one focusing on a particular kind of input, such as bricks, mortar, window and door frames, fire extinguishers, etc. These programs, which do not yet cover all the materials available on the market, are being monitored by PBQP-H.

(c) *National System for Technical Assessments*: This system was created to assess the performance of products, whether or not they involve new technologies, for which there are no standards published in Brazil.

(d) The PBQP-H (2010) states that "every year, on average 90% of the technical standards pertaining to civil construction that are published are done so on the back of the PBQP-H sector quality programs. This demonstrates how the PBQP-H has been capable of setting up links with research centers and sector entities to encourage technological research and the development of new quality standards" (PBQP-H, 2010).

Another outcome of the PBQP-H is the heralding of a culture of quality among public administrators. If these entities are to demand higher quality from their contractors, they themselves must have clear internal quality management references. The same applies to private contractors of civil construction works and services (PBQP-H, 2010).

PBQP-H (2010) believes that the program has become a "legal benchmark that validates quality requirements in public tenders, a situation where the State, in the interests of the public, exercises its power of purchase with the aim of inducing improved quality and productivity in the civil construction sector" (PBQP-H, 2010).

1.4. *Supply chain management*

1.4.1. *Recent trends*

With a certain lag behind more developed countries, Brazil has seen its civil construction market go through some major technological developments in recent years. In the view of Vendrametto (2008), "new materials, industrialization, new equipment, modern management and maintenance systems, and the need for a workforce with new skills are breaking down the methods, philosophies, and models traditionally used until the present day."

The economic growth seen in the Brazilian economy in recent years has engendered and accelerated this trend for modernization, with some marked improvements in productivity, quality, costs, and the meeting of deadlines in the construction sector (Vendrametto, 2008).

The modernization seen in the sector, with impacts on its supply chain management, has encompassed the following key aspects (Vendrametto, 2008; Pereira, *et al.*, 2009):

- incorporation of new construction systems;
- management focused on improving the efficiency of production processes by reducing costs, improving the quality of buildings, and enhancing productivity;
- growing dependency on suppliers of materials and construction systems in response to the increasing complexity of the sub-systems;
- acquisition and use of more complex products in constructions (such as parts for modular construction systems), with a view to streamlining construction activities, while also outsourcing a significant proportion of construction work to third parties;
- increasing recognition of the importance of construction quality, with the construction sector drawing on the quality management experiences of other industries and developing its own techniques to address issues specific to their sector.

With these trends, construction companies are establishing more long-standing relationships of cooperation with their supply chains, which brings mutual benefits, particularly of a financial and technological nature. (Pereira *et al.*, 2009).

1.4.2. *Relationship with suppliers*

The relationships to be investigated here are those among construction companies, their materials suppliers, and labor suppliers.

Suppliers of construction materials and systems: In order to foster better and closer ties with their materials suppliers, construction companies have introduced (Pereira *et al.*, 2009):

- procedures for specifying and inspecting materials;
- mechanisms for selecting suppliers;
- techniques for planning, scheduling, and controlling production;
- management of technical and organizational interfaces;
- effective communication systems;
- a vision of supplies that draws on the principles of logistics, such as purchase, receipt, processing, and storage.

The criteria of most concern to Brazilian companies when they select their suppliers are quality, meeting deadlines, punctuality, cost, and supply capacity. Other criteria that are sometimes taken into account include flexibility, reliability, the company's economic status, availability of stock, technological development, and commercial transfer (Pereira *et al.*, 2009).

Labor suppliers: The criteria that are most taken into consideration when subcontractors are selected are price, reliability, references, service quality, production capacity, safety conditions provided, meeting of deadlines, prior experience, provision of a fiscal receipt (*nota fiscal*), and provision of technical support. These criteria reflect the main companies' concern with improving their management of their production systems (Pereira *et al.*, 2009).

2. Sustainability in Civil Construction

2.1. *Corporate sustainability*

Since Rio de Janeiro's Earth Summit (United Nations Conference on Environment and Development) in 1992, the idea of sustainable development has grown increasingly strong. This means that no healthy society or economy can exist in a world with so much poverty and pollution (Nunes, 2010).

The traditional concepts of economic and social development do not consider the real importance of the environment. The aim is to meet people's main requirements and obtain an environment in good conditions as well as a healthy economy for everyone worldwide. Sustainable development covers three dimensions — economic, ecological, and social — and strives to integrate them. The main objective is to preserve environmental and cultural resources for future generations, to respect diversity, and to reduce social inequalities (Nunes, 2010).

The perception of the importance of sustainability has increased steadily over recent years among companies from all sectors of the economy, including civil construction, as investors, financiers, consumers, and the general public have forced companies to take account of and minimize the negative impacts arising from their activities. As Souza (2009) points out that these movements are "consolidating the concept of corporate sustainability as a long-term business view that incorporates social and environmental dimensions into companies' strategies and economic objectives."

The São Paulo stock exchange (BOVESPA — *Bolsa de Valores de São Paulo*) already has a business sustainability index (ISE — *Índice de Sustenta-bilidade Empresarial*), which represents the performance of publicly-traded Brazilian companies according to their sustainability practices. At the end of 2009, 29 companies whose shares were traded on the BOVESPA divulged this index to the public (Geração Sustentável, 2010).

2.2. *Sustainability in the construction sector*

Sustainability in construction has become a worldwide priority, with Brazil taking a leading role in developing sustainable agendas involving different approaches, such as the production of sustainability reports, life-cycle assessments of materials and buildings, and green buildings, involving different kinds of analyses, including energy efficiency, emission reduction, the efficient use of water, use of local and recycled materials, environmental comfort, internal air quality, and waste minimization and management (Souza, 2009).

When it comes to social issues, the sector has started to take a number of initiatives, including social inclusion and social responsibility programs, job and wealth creation, local outreach activities, digital inclusion, literacy and vocational training, work safety and occupational health, culture incentives, and volunteer programs (Souza, 2009).

For those companies that adopt sustainability policies, supply chain management is a key topic. Souza (2009) suggests that construction companies and the entities that contract their services should make demands of an environmental nature, addressing the purchase of materials, products and equipment, as well as investigating the social responsibility practices of suppliers and contractors (service providers) (Souza, 2009).

The CBCS (*Conselho Brasileiro de Construção Sustentável*, 2009), the Brazilian Council for Sustainable Construction, was established on the initiative of business leaders, researchers, consultants, professionals, and opinion formers with the objective of "encouraging the construction sector to use more sustainable practices that will improve the quality of life of users and workers and the environment around buildings." In order to meet this goal, the CBCS "adopts a systemic view of sustainability, focusing on the civil construction sector and its interfaces with the construction materials industry, financial services industry, government, universities, and civil society" (CBCS, 2009).

One of the major initiatives undertaken by the CBCS is the creation and running of six topic-based committees, which discuss guidelines and present them at lectures, symposiums, and other events. The committees are on water, sustainability assessments, economic and financial issues, energy, materials, and projects.

2.3. *Green buildings*

The construction and maintenance of the Brazil's infrastructure is responsible for the consumption of around 75% of all the resources extracted in the country. The quantity of construction waste produced in the country is estimated at 80 million tons a year (CBCS, 2009). In the European Union, the building sector accounts for more than 40% of the final energy demand and 36% of the CO_2 emissions (European Commission, 2010).

In the USA, the building-related impacts are among others (European Commission, 2010; EPA, 2010a,b):

- 39% of total energy use;
- 12% of the total water consumption;
- 68% of the total electricity consumption;
- 38% of the carbon dioxide emissions;
- 70% of all non-industrial secondary material generated in country.

Green building, also known as sustainable building, high performance building, or green construction, is according to EPA (2010a,b) "the practice of maximizing the efficiency with which buildings and their sites use resources — energy, water, and materials — while minimizing building impacts on human health and the environment, throughout the complete building life cycle — from siting, design, and construction to operation, renovation, and reuse" (European Union, 2010).

By adopting the practice of green building, the traditional building design main considerations (economy, utility, durability, and comfort) are expanded to include concerns about the reduction of the overall impact of the built environment on human health and the natural environment (European Commission, 2010; EPA, 2010a).

The potential impacts of the built environmental are resumed in Table 1. Table 2 summarizes some environmental, economic, and social benefits of green buildings.

Recent sustainability design initiatives strive for an integrated and synergistic design to both new constructions and renovations or retrofits of existing structures. According to WBDG (2010), this approach "supports an increased commitment to environmental stewardship and conservation, and results in an optimal balance of cost, environmental, societal, and human benefits while meeting the mission and function of the intended facility or infrastructure (WBDG, 2010)."

Table 1. Impacts of the built environment (EPA, 2010a).

Phases of project	Consumption	Environmental impacts	Ultimate effects
Siting	*Energy*	*Waste*	*Harm to human health*
Design	Water	Air pollution	
Construction	Materials	Water pollution	Environment
Operation	Natural resources	Indoor pollution	Degradation
Maintenance		Heat islands	Loss of resources
Renovation		Stormwater runoff	
Deconstruction		Noise	

Table 2. Some potential benefits of green buildings (EPA, 2010a).

Environmental benefits	Economic benefits	Social benefits
• Enhance and protect biodiversity and ecosystems • Improve air and water quality • Reduce waste streams • Conserve and restore natural resources	• Reduce operating costs • Create, expand, and shape markets for green product and services • Improve occupant productivity • Optimize life-cycle economic performance	• Enhance occupant comfort and health • Heighten aesthetic qualities • Minimize strain on local infrastructure • Improve overall quality of life

The green building practices should consider at least one of the six main principles of sustainable building design that are as follow (WBDG, 2010):

- optimize site/existing structure potential;
- optimize energy use;
- protect and conserve water;
- use environmentally preferable products;
- enhance indoor environmental quality; and
- optimize operational and maintenance practices.

More sustainable building design, construction, operation, and renovation can be achieved through the use of green building rating and certification systems that promote the integration of environmental issues with cost and other decision criteria.

Many countries have developed and published their own rating and certification systems for green building and energy efficiency for buildings. Among them are USA (LEED — Leadership in Energy and Environmental Design), England (BREEAM — Building Research Establishment Environmental Assessment Method), Japan (CAASBEE — Comprehensive Assessment System for Built Environment), South Africa (SBAT — Sustainable Building Assessment Tool), and France (HQE — High Quality Environmental standard). Some of them are adopted and used in other countries (Revista Automatização, 2010).

In Brazil, there are two main building environmental assessment tools in use: LEED Brazil and AQUA (*Alta Qualidade Ambiental do Empreendimento*).

2.4. *Green building in Brazil*

At the beginning of 2010, 400 buildings in the USA have already got the LEED certification and other 3400 were trying to get it. A report of US Green Building Council informs that the regions with the most interest in green buildings are Asia and Middle East, especially the United Arab Emirates, where more than 670 green buildings are ongoing. In the Europe, the interest in green building is growing, mainly in Germany and Italy (IP, 2010).

Brazil is among the first worldwide ranking records of LEED certifications; However, the Brazilian figures are still quite lower than the American records. On the other side, the Brazilian figures reflect the growing interest in green buildings in the country. In 2006, only four projects in Brazil were submitted to get the LEED Certification. In 2007, 39 projects were submitted and in 2008, 56 and in 2009, 36. Hence, 145 projects have tried to achieve the LEED certification and 92% of them are from the State of São Paulo and 13% from the State of Rio de Janeiro (IP, 2010).

Research from consulting companies estimates that the sustainable buildings, in comparison with traditional designed and built projects, besides ensuring a more healthy and productive indoor environment, can reduce up to (IP, 2010):

- 30% the energy consumption;
- 50% the water consumption;
- 35% the carbon dioxide emissions;
- 90% the waste disposal.

According to IP (2010), "the idea of the certification is not to impose limits on the construction market, but invite professionals of this sector to participate adequately in sustainable projects. We are far from having a 100% sustainable building in the country, but moving toward it. Brazil offers 95% of the resources and technology needed for that purpose (IP, 2010)."

3. Considering Sustainability in Supply Chain Management

In the design phase of sustainable buildings, questions about the material and product specifications arise and some of them are as follows (Corcuera, 2010):

- How to specify construction material to green building projects?
- How many certifications exist in Brazil and what do they mean?
- What makes a material a sustainable one?

The existing certifications have not completely covered all the aspects of sustainability; however, some may support by selecting materials and products. For this, it is necessary to know and to understand how information is presented in data sheet or in certifications (Corcuera, 2010).

According to Corcuera (2010), there are in the construction material market three main groups of information levels that are (a) self-declaration and labels; (b) second-party certification; and (c) third-party certification (Instituto Falcão Bauer, 2010).

3.1. *Self-declaration and labels*

The product manufacturer presents data that were not tested or approved by other organizations. An example of self-declaration is the called FISPQ (Safety Data Sheet for Chemical Products) whose model presentation is set by the Brazilian standard NBR 14725/2001 (Corcuera, 2010). The FISPQ should contain 16 sections that are as follows:

- identification of the product and the manufacturing company;
- information and composition of various materials in the product;
- identification of hazardous;
- first aid measures;
- fire-fighting measures;
- control measures for spills or leaks;
- handling and storage;

- exposure control and personal protective equipment;
- physico-chemical properties;
- stability and reactivity;
- toxicological information;
- ecological information;
- considerations about waste treatments and disposal;
- information on transportation;
- related regulations to the product;
- other information.

In some countries, such data sheet is called the material safety data sheet. The FISPQ is mandatory in Brazil for paints and some chemical products used in the construction industry (Corcuera, 2010).

Examples of additional sustainable initiatives in the national industry are the symbol labels for separation of recyclable glas, metals (aluminum and steel), plastics (PET — polyethylene terephthalate, HDP — high density polythylene, LDP — low density polyethylene, PVC — polyvinyl chloride, PP — polypropylene, and PE — polyethylene).

3.2. *Second-party certification*

The second-party certification results from an audit and certification conducted by a trade association or consulting firms. This certification is commissioned by the company that will be evaluated and the evaluation can cover the whole or part of organization, a product or a production process, depending on the interests of the contracting company.

This type of certification provides some patterns; however, there is little assurance about the conflicts of interest. The main problem of this statement is the low transparency of the analyses and results. It has still been not much used in Brazil (Corcuera, 2010).

3.3. *Third-party certification*

Third-party certification results from an audit and certification by an independent institution in a reliable and impartial way. It will be recognized by all the evaluated company's (or product's) customers (Corcuera, 2010).

An example is the Brazilian eco-label of Falcão Bauer released in 2007. It can be applied nationwide and certifies products and sustainable technologies. The eco-label has three categories, which show the impacts and benefits of a sustainable product. The classification will depend on (Instituto Falcão Bauer, 2010):

- the absence of hazardous substances in their composition;
- environmental impact of product and production process less than similar conventional one, according to a carried out LCA (lower consumption of raw material, energy, water, natural resources, packaging, emissions, waste, and an

extended product life and alternatives of disposal after use with lower environmental impacts);
- percentage of renewable raw materials and recyclable material in the product composition;
- actions to improve social and environmental performance of the company.

As already mentioned in Section 2.3, there are different certification systems for green buildings and in Brazil the most important systems are LEED and AQUA. They are certification systems for the whole building and not for single materials or products.

Some products have already been certified by Falcão Bauer and other organizations as ABNT and Forest Stewardship Council, and the demand for such labels should grow in the next year; however, they are presently only few in the construction industry market (Corcuera, 2010; Instituto Falcão Bauer, 2010).

In the absence of more detailed product and producer information, the professionals that are specifying materials and products for sustainable buildings can use the six-step strategy of CBCS.

3.4. *Six-step strategy of CBCS*

CBCS has put forward a six-step strategy for helping architects, engineers, contractors, and users to select suppliers and materials for use in construction work and when a construction is used. While the strategy does not cover all the potential issues involved in a project, it does provide an easy-to-use method for materials purchasers and those responsible for setting specifications for materials and suppliers to make a preliminary assessment of sustainability practices (CBCS, 2009).

The six steps proposed by CBCS for selecting inputs and suppliers based on sustainability criteria are as follows:

- check the supplier's status as a corporate entity;
- check the manufacturing facility's environmental licenses;
- check for compliance with technical standards;
- consult the company's socioenvironmental responsibility profile;
- identify whether "greenwash" (green whitewash) is being used;
- analyze the product's durability.

3.4.1. *Check the supplier's status as a corporate entity*

The formal establishment of a company is important to sustainability. If a company is not registered and has no tax number (CNPJ — *Cadastro Nacional de Pessoa Jurídica*), it is unlikely to have any interest in complying with legislation, be it environmental, labor, or otherwise. A company's CNPJ can be checked on the internal revenue (*Receita Federal*) web site.

If the company's CNPJ is not valid, this either means that it is not paying its taxes or that it is not even registered, meaning that it does not exist as a legal entity. The CNPJ must be included on the bill of lading or receipt that accompanies the product, on the product itself or its packaging (CBCS, 2009; Instituto Falcão Bauer, 2010).

3.4.2. *Check the manufacturing facility's environmental licenses*

In Brazil, industrial activities cannot be legalized unless they have an environmental license, which is issued by the state environmental entity. The fact that a company has such a license does not mean that it is compliant with all the environmental legislation; however, when a company does not have one, it is less likely to be heedful of this legislation.

Companies' environmental licenses can be consulted on the web sites of some state environmental entities, such as INEA (*Instituto Estadual do Ambiente — Estado do Rio de Janeiro* — Environmental Agency of Rio de Janeiro State) and CETESB (*Companhia Ambiental do Estado de São Paulo* — Environmental Agency of São Paulo State); however, in order to do so, the number of the process by which the company applied for the license must be known (CBCS, 2009; CETESB, 2010; INEA, 2010).

3.4.3. *Check for compliance with technical standards*

Before buying a product or service, it can be checked whether the supplier is included in the list of companies that participate in the PBQP-H (companies that operate in compliance with the ABNT technical standards).

There are still some products and services pertaining to the civil construction sector that have not yet been assessed by the PBQP-H. In such cases, one route that buyers can take is to ask the manufacturer to present a certificate issued by a recognized entity that proves the product's quality. When the product in question has only recently entered the Brazilian market and is not subject to any technical standards published in Brazil, an assessment by a third party can be requested, based on Brazilian standard NBR 15.575/2008 (Residential Buildings until five floors — Performance; CBCS, 2009).

3.4.4. *Consult the company's socioenvironmental responsibility profile*

Another point worth considering when making a preliminary assessment of a supplier's sustainability practices is whether they have socioenvironmental responsibility policies in place. This can be proved by certificates they might have received (as ISO 14001 and OSHAS 18001), or sustainability reports or social balance sheets they may produce.

Another method for checking suppliers' socioenvironmental responsibility as put forward by CBCS (2010), aside from checking these certificates and reports, is to analyze a company's web site and product catalogs, checking for

(a) the existence of a clear corporate socioenvironmental responsibility policy;
(b) the breadth of their sustainability practices;

(c) an ongoing review of their socioenvironmental sustainability practices and policy involving the company's employees and partners;
(d) commitment to the occupational health and safety of their employees;
(e) description of their policy for managing the waste produced during the manufacture of their products or delivery of their services and the waste produced after the products have been used;
(f) membership of organizations that promote sustainable construction in the country;
(g) use of life-cycle assessments to improve their products and production processes.

3.4.5. *Identify whether "greenwash" is being used*

Suppliers often present certificates and make claims about the ecological efficiency of their products and processes. Before buying them, consumers should check whether these certificates and claims are in good faith and whether they are recognized by independent organizations. Some ways of identifying greenwash are (CBCS, 2009):

(a) stress on the positive aspects of products while omitting their associated environmental problems or potential limitations;
(b) non-existence of evidence to support the claims of ecological efficiency made about the products or processes;
(c) generic, imprecise information, raising doubts as to the real environmental benefits of products throughout their whole life cycle; and
(d) irrelevance or falsehood: statements about environmental performance that are apparently useful but are actually useless, overstated, or simply untrue.

3.4.6. *Analyze the product's durability*

John (EPA, 2010) states that increasing the durability of construction products is an important topic, and one which has received scant attention in discussions on sustainability. The more durable products are, the less demand there is for replacement products and/or maintenance.

Durability depends on factors such as functional obsolescence, conditions of use, maintenance, and the exposure of buildings or parts of buildings to the elements. Depending on each project, these conditions can change significantly, resulting in different durability performances by the same or similar products (CBCS, 2009).

One of the biggest impacts made by the construction sector arises from the quantity of resources it uses. The extraction of inputs, such as aggregate minerals, iron, and timber, often leads to the destruction of ecosystems and the depletion of deposits and reserves.

Lengthening the useful life of products helps by delaying the generation of post-use waste during building reforms or demolition. Also, the manufacture and assembly of products for new buildings and for replacement parts makes use of energy. The generation of this energy has its own environmental impacts, which in Brazil imply in alterations to ecosystems, the emission of greenhouse gases, and others John (2010).

In economic terms, longer durability explains that John, "improves competitiveness, for the longer the useful life, the lower the cost of the service rendered by a product per unit of time. The lower the investments in replacing structures and fixing worn parts, which reduces maintenance costs." A typical way of estimating annual maintenance costs for buildings is to assume them as being 1% of their construction cost (John, 2010).

4. Tendencies for Suppliers in a Sustainable Construction Industry

According to Coelho (2010), to be sustainable in Brazil is not an easy task. Many consumers doubt the reputation and quality of sustainable products and services, because they confuse sustainability with low quality, rusticity, among other negative aspects. The consumers believe that sustainable products are more expensive and with low market supply, and they often do not know the criteria that become a product in a sustainable one (Coelho, 2010).

Only 29% of the Brazilian companies have developed actions to organize and maintain a net of social responsible suppliers and only 31% have implemented policies to make green purchases (Coelho, 2010).

A total of 10% of the companies that try to obtain the label LEED in Brazil fail to go until the end of the certification process, mainly due to the fact that, besides the final product (the building) itself, its components — materials and products, supplied by other companies — are also analyzed (Coelho, 2010). As already seen in the previous item, in many cases the construction companies still do not have access to basic ecological data (as toxicity, demanded energy, and raw resources by the production) of the purchased products.

The interest in relation to green buildings and corporate sustainability has been increasing worldwide and more information about ecological and social aspects of the supply chain companies and their products and services has been required. Products that own an environmental certification (e.g., labels, second-party certification or third-party certification) may have competitive advantages due to the full or partial compliance with national requirements and, in some cases, even with international requirements.

In Brazil, the segment of products that have volatile organic compounds, such as paints, varnishes, carpets, adhesives and glues was the first to look for environmental certifications, due to issues of toxicity and indoor environmental quality. However, it can already be observed the growing demand for these certifications by building materials with more competitive market such as subfloor void and cleaning materials.

Large Brazilian and multinational companies have already implemented green procurement policies and they give preference to certificated products and companies (Coelho, 2010).

5. Conclusions and Outlook

The Brazilian civil construction sector is going through a period of great dynamism, with a significant volume of business and major investments in modernization. In a competitive market, the adoption of sustainability practices represents an opportunity for companies to gain a competitive edge.

The most pressing of the many impacts incurred by the civil construction industry is its consumption of around 75% of all the natural resources extracted in Brazil. The sector's impacts are not only brought about when a project is built; they happen to a large extent when the natural resources are extracted and during the production activities of materials suppliers. As such, if the construction industry is to effectively implement sustainability policies across the whole gamut of companies that compose it, then not just construction companies, but the companies in their supply chain must also adopt socioenvironmental sustainability practices.

It is estimated that around 20–30% of all the material used in Brazil's construction sites is wasted, depending on the construction techniques used. Part of this waste is the direct result of the poor quality of the materials used in construction processes.

Consumer protection legislation has been tightened and construction companies have been forced to pay greater attention to the quality of their constructions. In the past years, several new standards were published in the country.

Initiatives as CBCS (Brazilian Council for Sustainable Construction), certifications and labels, PBQP-H (Brazilian Program for Quality and Productivity in Housing) have not completely covered all the aspects of sustainability and quality, but make available a support for public administrators and private contractors of civil construction works and services to select material and service suppliers with more sustainable and higher quality practices.

In this chapter, we set out the main aspects at play when addressing the sustainability of supply chains. We also set out an existing methodology for helping professionals from the construction sector to select suppliers and materials. This method does not cover every single aspect of construction projects, but provides guidelines for a preliminary assessment of the sustainability practices of prospective suppliers for building projects.

References

Carvalho, K., 2004. Normas Técnicas. In: *Revista Construção Mercado*, 41, December.

CBCS (Brazilian Council for Sustainable Construction), 2009. *Seis passos para a seleção de insumos e fornecedores com critérios de sustentabilidade*. Available at http://www.cbcs.org.br/hotsite (Accessed 10 February 2010).

CBIC (Brazilian Chamber of the Construction Industry), 2010. *Database*. Available at htto://www.cibicdados.com.br (Accessed 08 June 2010).

CETESB (Environmental Agency of São Paulo State), 2010. *Consulta a Processos*. Available at http://www.cetesb.sp.gov.br/licenciamentoo/cetesb/processo_consulta.asp.

Corcuera, D, 2010. *Certificações para Materiais Sustentáveis*. Available at http://www.casaconsciente.com.br / Textos / Aplicacao%20de%20Estrategias%20Sustentaveis . pdf (Accessed 05.06.2010).

Coelho, L., 2010. Certificação Ambiental. In: *Revista Techne*, São Paulo, February, 2010. Available at http://www.revistatechne.com.br/engenharia-civil/155.

European Commission, 2010. *Energy Efficiency in Buildings*. Available at http://ec.europa.eu/energy/efficiency/buildings/buildings_en.htm (Accessed 08.08.2010).

EPA (Environmental Protection Agency — USA), 2010a. *Basic Information: Green Buildings*. Available at http://www.epa.gov/greenbuilding/pubs/about.htm (Accessed 08.06.2010).

EPA (Environmental Protection Agency — USA), 2010b. *Why Build Green?* Available at http://www.epa.gov/greenbuilding/pubs/whybuild.htm (Accessed 08.06.2010).

Geração Sustentável, 2010. *A Sustentabilidade e o Mercado Financeiro*. Available at http://revistageracaosustentavel.blogspot.com/search?q=ISE, January (Accessed 10 February 2010).

IBGE (Institute of Geography and Statistics), 2007. *Análise de Resultados 2007*. Available at http:// www.ibge.gov.br/home/estatistica/economia/industria/paic/2007/comentario. pdf (Accessed 10 January 2010).

INEA (Environmental Agency of Rio de Janeiro State), 2010. *Licenças disponíveis*. Available at http://www.inea.rj.gov.br/fma/licencas-disponiveis-ca.asp?cat=10.

Instituto Falcão Bauer, 2010. *Certificação de Produto: Selo Ecológico*. Available at http://www.ifbauer.org.br/html/certificacao_prod.asp?produto_escolhido=135 (Accessed 22.06.2010).

IP (Integração Profissional), 2010. *Green Buildings: Brasil segue tendências de edifícios verdes*. Available at http://www.revistaip.com.br/article.php?a=471 (Accessed 07.06.2010).

John, W., 2010. *Por que sustentabilidade?* IBDA (Instituto Brasileiro de Desenvolvimento da Arquitetura). Available at http://www.forumdaconstrucao.com.br/conteudo.hp?a= 23&Cod =195 (Accessed 05. February. 2010).

Nunes, K.R.A., 2010. Business sustainability in Brazil and Germany: Case study of the automotive industry. *International Journal of Energy and Environmental* **1**(2).

Pinto, T., 1999. Metodologia para a Gestão Diferenciada de Resíduos Sólidos da Construção Urbana. *Ph.D. Thesis*, University of São Paulo.

PBQP-H (Brazilian Program of Quality and Productivity in the Buildings), 2009. *Results of PBQP-H*. Available at http://www4.cidades.gov.br/pbqp-h/pbqp_apresentacao.php (Accessed 10 January 2010).

Pereira, S.R., Ohnuma, D.K., Barbosa, A.L.S.F. and Cardoso, F.F., 2009. *Desenvolvimento e Gestao da Cadeia de Fornecedores na Construcao de Edificios*. In: V *International Workshop of Lean Construction*. Instituto de Engenharia, São Paulo, October.

Receita Federal, 2010. *Consulta a Cadastro de Pessoa Jurídica*. Available at http://www.receita.fazenda.gov.br.

Revista Automatização, 2010. Prédios Verdes. *Revista Automatização, Editora do Administrador, São Paulo, Issue 2*, pp. 12–13. Available at http://www. revistaautomatizar.com.br/ (Accessed 10.06.2010).

SINDUSCON-SP (Construction Industry Employer Association Section São Paulo), 2009. *41a. Sondagem Nacional*. Available at http://www.sindusconsp.com.br/msg2.asp?id=3150 (Accessed 20 December 2009).

Souza, R., 2009. *Sustentabilidade nas Empresas do Setor da Construção*. CBCS. Available at: http://www.cbcs.org.br/comitestematicos/avaliacaosustentab/artigos/Sustentabilidade (Accessed 01 February 2010).

Vendrametto, O., 2008. A mudança do perfil do trabalho: Formação versus evolução tecnológica na construção civil. In: *Proceedings of the XXVII National Congress of Industrial Engineering*, October.

WBDG (Whole Building Design Guide), 2010. *Sustainable*. Available at http://www.wbdg.org/design/sustinable.php (Accessed 10.06.2010).

Chapter 35

SUSTAINABLE DEVELOPMENT: A MARKETING PERSPECTIVE

Min H. Lu
Leon Hess Business School,
Monmouth University, W. Long Branch, NJ 07764, USA
mlu@monmouth.edu

To develop a sustainable environment for the future society requires a coordinated effort of many parties of the society. Among them are the major participants in this sustainable endeavor, the businesses, consumers, governments, and social groups. This chapter looks at the recent development of sustainability with particular interest in the marketing perspective. The chapter introduces the model of GAP-ACT to explain the relationships among participants in the society, it expands the model with marketing perspectives to suggest that the achievement of common goals of sustainability requires a thorough understanding of the aims, expectations, strategies, and actions of major participants to ensure the coordinated efforts of businesses, consumers, governments, and social groups such as NGOs to reach their common and respective goals.

1. Introduction

Sustainable development is no longer a new term in today's society. From private companies to government organizations and to consumers, sustainable development means the future for the mankind. The Copenhagen Climate Summit in December of 2009 has clearly demonstrated the urgent need for the whole world to pay close attention to what we as a society do and the consequences of our actions. Such urgency has been seen through the Copenhagen Call to the world leaders at COP15 in Copenhagen. This Copenhagen Call contains the recommendations made by the Copenhagen Climate Council and its partners from private sectors, United Nations agents, and various government agents (Summary Report for Policymakers, 2009). The Copenhagen Call has also demonstrated the readiness of private business sectors around the world, and the need for the governments around the world to come up with the right structure or environment to direct their efforts toward global sustainable development for the future (Summary Report for Policymakers, 2009).

The global attention towards sustainable development has not only demonstrated at the Copenhagen Summit in 2009, the recent trends of businesses and industries have also supported such effort to have a sustainable future for the mankind. However, to reach that destiny it requires serious efforts from businesses, governments, and people (the consumers). As illustrated by Marc Epstein in his recent book, Making Sustainability Work, he identified five reasons for us to pay attention to the global sustainability, regulations, community relations, cost and revenue imperatives, and societal and moral obligations (Epstein, 2008).

His study has found that there are an increasing number of external forces that pressure the business community for a sustainable development of the world economy. Consumers/users, businesses/industries, governments, and third parties or neutral parties such as non-government organizations (NGOs) contribute and benefit as well in their efforts to pursue a sustainable development. The main path toward the sustainability is reflected in the interactions between businesses and consumers, where companies produce products for consumers/buyers to purchase for their consumption and where the governments and neutral parties/NGOs monitor and regulate the interaction between the businesses and consumers. However, the ultimate success of sustainable development of mankind will depend on all these forces to conscientiously participate in this endeavor.

This chapter will look at the recent development in sustainability with a particular eye on the interactions among the businesses, consumers and governments, and also their actions toward this sustainability movement. A general account of marketing concept development will be followed in responding to the market reactions toward sustainable development. The chapter will also introduce a GAP-ACT model to explain the behavioral patterns and relationships among major participants with marketing perspectives in today's society.

2. Sustainable Development

Discussion on sustainable development has generated many different ideas dealing with developing a sustainable society, which would help the general public understand the relationships of humanity with nature and between people. A review by Hopwood *et al.* (2005) provided a broad scope of discussion on the development of sustainability based on a list of over 90 sources from 1970s to 1990s. By using a mapping technique, they categorized these various thoughts on sustainable development into three major views based on the dimensions of socio-economic well-being and equality and the environmental concerns — status quo, reform, and transformation groups. These views reflect the concerns of the nature of the changes necessary to have an optimum relationship between nature and mankind. The status quo views prefer to keep the status of the current structures, the reform views would like to see a fundamental reform but without a major overhaul, and the transformation views suggest a radical transformation as they see the roots of problems are the economic and power structures of society (Reid, 1995).

Recent studies have found that consumers have increased their participation in the effort to help protect our environment in various ways. In a landmark survey done in 2007, the Hartman Group reported that over 90% of the respondents had identified themselves as doing things or willing to do things with respect to sustainable development in their daily lives. Over 70% of the respondents participated regularly in recycling household items, saving water and donating household items for further usage (Hartbeat, 2008).

In a recent survey conducted by Grocery Manufacturers Association (GMA) and Deloitte, they have found that an interest in buying green products spread out across all age, income, and education groups, with 95% of participants open to considering sustainable products, and 67% of them actively looking for these products when shopping (Grocery Manufacturers Association and Deloitte, 2009). However, only 47% of respondents actually found green products and just 22% actually purchased the green products. There has been a huge gap between consumers' behavior and expectation to purchase sustainable products. The survey results indicated that many shoppers concerned about the product performance and the legitimacy of the product's green claims, and cited these as their reasons not to purchase these products.

Another study done by GfK Custom Research North America indicated that in general consumers cited money, time, and lack of information as barriers to green living in this country, and almost a third of the participants in the study agreed that Corporate America is fulfilling its responsibility to the environment. However, 46% of the respondents mentioned that the federal government should lead in the effort toward sustainable development, followed by individual responsibility (39%) and corporate America (32%). A total of 70% of the respondents indicated that companies were not doing their fair share of environmental responsibilities. Hence, there is a major improvement needs to be done by governments, consumers, and business communities (Wong, 2009).

3. Sustainable Development and Marketing

The development of marketing literature has traditionally focused on the exchange relationship between companies and consumers, or demand and supply. Kotler's marketing management theory has been the mainstream of marketing ideas. In recent years, there has been a trend of marketing mainstream moving away from this exchange focus, and shifting to more of relationship focus in the market place among businesses, consumers, and market environments. The sustainable development initiative in the US and other countries have led to a series of reactions from the professional marketing communities in their effort to answer the calls from the market place, fine-tuning the focus of sustainable development in marketing effort, redefining the concept of marketing, and launching an effort in educating the market with sustainability marketing (Belz and Peattie, 2009)

Many new marketing concepts emerged in the market place since the 1970s. In a study done by Cova *et al.* (2006), they listed a total of 78 marketing panaceas drawn from various sources of practical and academic nature to deal with different issues.

The authors grouped them into five groups that are grown out of the mainstream of marketing (the Kotlerian view of the market): *market environment, market niche, subjective experience, consumer competences, and client relationships.* Among these groups, market environment contains such marketing panaceas as ecological marketing, societal marketing, social marketing, green marketing, sustainable marketing, stakeholder marketing, and sustainability marketing.[1] The following paragraphs try to provide a basic understanding of some of the commonly used marketing terms related to environment.

Ecological marketing: Ecological marketing is a marketing approach to highlight products and production methods that improve environmental performance, further ecological causes, or solve environmental problems (Sustainability Dictionary). This concept was developed in the 1990s when businesses marketed their products/services to address concerns about environmental pollutions. Green marketing and environmental marketing may also be used for this purpose.

Societal marketing: Societal marketing emphasizes on marketing practices to go beyond the corporate needs and wants, and seek to meet the customer's needs and societal needs. This allows for more sustainable success rather than short-term accomplishment. The goals are to provide a marketing strategy to better consumers and society's well-being (Kotler and Levy, 1969).

Social marketing: Social marketing is the application of business-oriented marketing principles and practices typically in a non-profit, government, or NGO contexts to change behavior to achieve a social good (Hosfeld, 2009). Social marketing uses the benefits and of doing social good to secure and maintain customer engagement.

Green marketing: Green marketing focuses on the environmental benefits of either the product/service being marketed or the environmental qualities of the process of promotion/advertising. It is primary used in promotion and product design strategies (Hosfeld, 2009). Sometimes, environmental marketing is used for this purpose.

Sustainable marketing: Sustainable marketing accepts the limitations of a market orientation and acknowledges the necessity of regulatory alterations to the market mechanism. Sustainable marketing fosters corporate and collective commitment to necessary alterations of institutional settings and price signals in favor of sustainable development (Belz and Peattie, 2009).

Stakeholder marketing: Stakeholder marketing orients a firm's marketing activities beyond the consideration of the firm's immediate targeted consumers to include others that may be impacted by their marketing activities on a larger base of

[1] Other marketing panaceas dealing with market environment included in the study are cause-related marketing, holistic marketing, macro marketing, maxi marketing, mega marketing, network marketing, non-business marketing, social marketing, societal marketing, symbiotic marketing, and trend marketing (Cova *et al.*, 2006).

constituents, and encourages consideration of the impact of these constituents in planning marketing activities (Hosfeld, 2009; Mish and Scammon, 2010; Chakravorti, 2010).

Sustainability marketing: Sustainability marketing engages in marketing activities in planning, organizing, implementing, and controlling marketing resources and programs to satisfy consumers' wants and needs, and at the same time considering social and environmental well-being and meeting corporate objectives. Sustainability marketing also emphasizes on building and maintaining sustainable relationships between customers, the social environment, and the natural environment, commonly referred to as "triple bottom lines" principle (Belz and Peattie, 2009).

The new marketing concept defined: The mainstream of marketing communities has experienced some recent changes in defining its core concept. A new effort to redefine marketing concept started in the early 2000 almost 20 years after the American Marketing Association (AMA) adopted the 1985 definition of marketing. The AMA came up with a modified definition of marketing in 2004: "*Marketing is an organization's function and a set of processes for creating, communicating, and delivering value to customers and for managing customer relationship in way that benefit the organization and its stakeholders.*" However, this definition still could not break away from its original framework of marketing being considered as a function of an organization and serving a narrow interest between the company and its stakeholders (AMA Definition of Marketing, 2007). Discussion on the focus of marketing concept continued as scholars and researchers presented new ideas regarding new ideas or elements should be considered by the definition for the long term and covering a wider scope of the market place (Sheth and Uslay, 2007; Lusch, 2007).

In 2007, the AMA modified its 2004 definition of marketing to adopt a broad and long-term view of the concept: "*Marketing is the activity, set of institutions, and processes for creating, communicating, delivering, and exchanging offerings that have value for customers, clients, partners, and society at large.*" This latest definition has reflected the new position by the AMA in respond to the changing environment of the market as a whole. It regards marketing as an "activity" not a function of a business and it also positions marketing as providing a long-term value in the market (AMA Definition of Marketing, 2007).

The continuous discussion of current marketing concept has also been reflected in the paper of The New Marketing Myopia by Smith *et al.* (2010) to expand the scope of marketing concept to address multiple stakeholders and engage in value creation for stakeholders involved either directly or indirectly and for the short term and also the long term in the future. Their ideas were also echoed by other marketing scholars and researchers in differentiating the stakeholder marketing orientation from the market orientation, suggesting that firms' businesses characterized by stakeholder orientation or stakeholder marketing were dedicated to learning about and addressing stakeholder issues (Ferrell *et al.*, 2010).

4. The GAP-ACT Model

The GAP-ACT model is based on William T. Powers' (1973, 1989) theory that behavior is the means whereby we control our perceptions. It helps one to understand and improve human performance in the workplace (Nickols, 2008). The basic assumption is that human beings are considered as "living control systems," and we perceive and control our behaviors accomplish what is expected. While we behave and conduct our business, our perception and behavior may also influence others' in their perception and performance. Consequently, it may create differences or gaps as we work to reach our goals or targets. For this study, this model is used to explain that as we are considered as "living control systems" and we interact with other humans in the planet. Our behavior or performance will affect others and as others' will have on ours. It demonstrates a much intertwined relationships of the previously mentioned major forces that influence the global effort to a sustainable development.

It is understandable that a problem exists as a gap or discrepancy between actual and desired conditions or solutions. We do not always achieve what we expect. Therefore, this gap model helps us to understand what we want to do and what we actually can do. In Fig. 1, desired conditions are represented by G (Goal), and perceived actual conditions are represented by P (Perception). The black dot between G and P represents a comparison of the two. If there is a gap or discrepancy between our goal and perception, there will be an action to reduce this gap, which is represented by A (Action). The lower case d represents an action taken as a result of gap between G and P, and lower case i indicates an action takes place to reach the target that will reduce or eliminate the gap. The targeted area or variable we try to affect or control is represented by T (Target). Often times, our action to affect the target is also under the influences of other circumstances that we do not have control. There is a feedback from this action represented by the dotted line of f. The GAP-ACT model suggests that we, as a "living control system," will try to

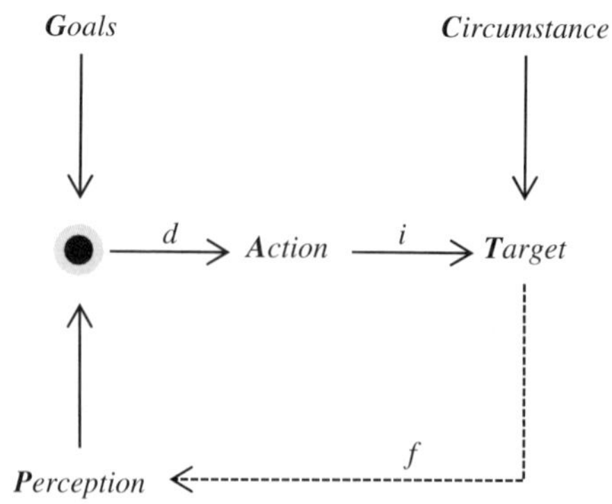

Fig. 1. The GAP-ACT model.

Source: Nickols, 2008.

take actions to adjust ourselves to reach the expected or adjusted target. However, we have to understand that our action is also influenced by other circumstances which also include other "living control systems," i.e., other human beings and their actions.

As we are pursuing a sustainable development path, we consider all human beings and organizations as living control systems. Companies and governments are managed by humans; social groups are also run by human. Therefore, it is safe to say that this model of GAP-ACT can also help us understand the intertwined relationships among the various "living control systems" in achieving a sustainable development of world economy.

As discussed early in the paper, there are many forces that influence the global efforts to achieve a sustainable environment in the future. Governments, businesses, social groups such as NGOs, consumers are the main participants in this endeavor to our planet sustainability. Each participant may have their own goals to achieve, and the actions taken by each participant will have impact on themselves and influencing others. Therefore, when planning for our actions we need to understand how others' actions may affect ours, and ours will affect theirs. The following expanded model of GAP-ACT will illustrate how these four major participants influence each other in a sustainable development path to the future.

5. Expanded GAP-ACT Model and Marketing Perspectives

There are many forces that may influence the effort toward a sustainable development of our planet. However, if we look at the importance of these forces we will see four major ones that have vital roles in developing a sustainable world. Each of these participants is represented in one quarter of the model. Businesses and consumers are in the forefront of producing and consuming products and services that will require the use of natural resources of the planet. Therefore, they are the direct contributors and benefactors of a sustainable world. In an economic sense, they represent the demand and supply curves. Governments and social groups are also important players in that they influence these and other forces in one way or another through regulations, standards, procedures, and public pressures to the businesses, governments and consumers as well.

Governments: Studies have found that more consumers think that governments should hold more responsibilities to regulate the businesses and individuals to follow rules, regulations, and standards to protect our planet (Wong, 2009). Governments should also monitor business doings so that consumers and societies will be protected from unethical business practices.

Businesses: Consumers think that businesses should take the ultimate responsibilities to introduce new and environmental friendly products, and develop technologies to help preserve the natural resources for the future generations.

Social groups: Social groups have played an increasingly important role in today's global society. NGOs have been particularly active in many fronts dealing with many social issues facing the world today. Some of them are well known around the world, such as Green Peace, whereas others are active on the regional or local bases. In many instances, these NGOs' actions could be felt in many regions and countries around.

Consumers: Consumers are the people who are directly participating in economic activities as consumers in most case and also as the initiators for new products and/or technologies. As indicated in recent studies by GfK Roper and Hartman Groups, consumers should hold more responsibility and have increasing their involvement in contributing to the sustainable development of national and global economies (Wong, 2009; Hartbeat, 2008). In today's market place, many companies are compelled to engage in responsible business activities because their consumers or customers want them to do so.

From the basic description of these main participants in the expanded model of GAP-ACT, each of them has an obligation to work toward the target area of the model. Let us say the target is to achieve a sustainable economy. All these main participants will have their respective goals set up to reach this common target area. The following discussion will focus on the marketing implication of companies' activities to create and diver sustainable products and technologies to help the world to reach the target of sustainability.

In Fig. 2, the Expanded GAP-ACT Model, each participant (government, business, social group, and consumer) has their individual models to start with. Each has its own action to take in order to reach the target area. The gap between Goal and Perceived goal will have the participant to adjust its actions to reach the expected target area. That target area represented by "*T*" in this expanded model will be shared by other participants. Let us presume that each of the participants would like to see a sustainable future for this world, which is represented by "*G*" for all the participants. And at the same time, this goal will be different from the perceived goals for the participants, which is represented by "*P*." Let us further presume that government's perceived goal is different and cannot be reached in a short period of time. Hence, the perceived goal may be that government wants to cut CO_2 emission by 30% in 2015. In order to reach this adjusted target, government will take actions, "*A*," such as issuing certain regulations and standards for businesses and individuals to follow. These regulations and standards could be used to encourage new product/technology development by business or penalties if businesses are following the regulations and standards. Therefore, these actions will have impacts, "*C*," on other participants in the model, such as businesses and consumers. Individuals could also be affected by these regulations and standards in their lives. Therefore, based on the original mode of GAP-ACT, one participant or "living control system" will also influence others along its way to reach the target area. Hence, as the government determines to reach the expected target area, it will require its people and organizations to help reach that target.

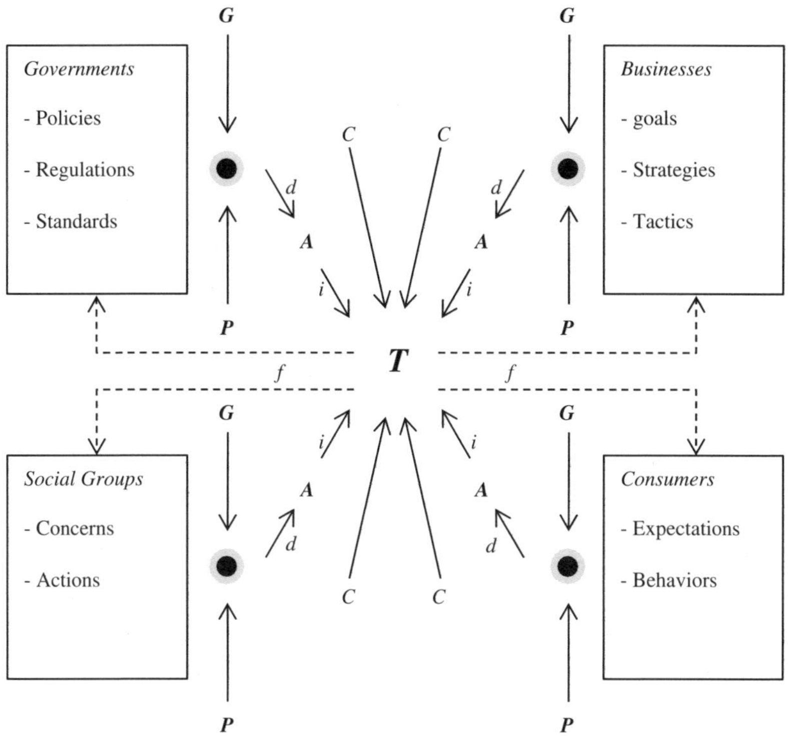

Fig. 2. Expanded model of GAP-ACT.

5.1. *Governments*

Recent developments of governments' effort toward more sustainable development have been reported by the New York Times that US governments and some other foreign governments have required their embassies be more environmentally friendly in climate-conscious times. United States already have three LEED-certified embassies in foreign countries, and the fourth one is coming soon. Some other foreign countries have also joined the effort to make their embassies more green, Finland's LEED-certified embassy in Washington, DC, The Netherlands' LEED-certified embassy in Amman, and the British Embassy in Berlin. In addition to these "green" buildings, these governments also encourage their employees to launch other programs to save water, use public transportation instead of driving, and etc. (Galbraith, 2010).

In the domestic front, US government started a two-prong approach to address the sustainable development issues. On the one end, governments have initiated rebates programs to encourage consumers to use more energy efficient appliances after a successful campaign in the last year's "cash for clunkers" in the auto industry. These programs have met with strong interests from consumers in many states (Zeller and Skipp, 2010). On the other end, governments have started their efforts to reinforce the sustainable policies and regulations to make sure that proper procedures and standards are followed, such as the Department of Agriculture will enforce rules requiring the spot testing of organically grown foods (Neuman, 2010).

5.2. *Businesses*

From the business perspective, companies have long realized that to compete effectively in the national and international market places, they need to design marketing strategies for a sustainable market. Thus, many of them started on their own the ethical and socially responsible standards to guide their business practices. Companies in those countries where they face more urgent need to develop sustainable economy or market due to the lack of sufficient natural resources have long started to look into developing products and technologies to meet the future demands of the world. Businesses as participants in the expanded model of GAP-ACT, they will have their missions or goals to achieve, say, the sustainability of the future market. Their actions as a result of the difference between the desired goals and perceived goals will adjust and lead the businesses to the target area which they share with the governments. Let us presume that if the businesses' goals or missions are to achieve a global sustainability for the future, then the strategies and tactics designed by these businesses are the reflections of the businesses' action leading toward the target area in the model. Their strategies and tactics are also subject to the influences of the governments' regulations and standards. Hence, the businesses and governments become partners working toward the same direction.

As Connie Hedegaard, the Minister of Climate and Energy in Demark, called upon the business communities worldwide: "We, the politicians of the world, have a responsibility to reach a truly global climate change agreement in Copenhagen in December 2009. However, it is the business society that can deliver the tools to turn our vision into reality. Businesses can provide the clever solutions to make it possible to live in a both modern and sustainable society" (Copenhagen Climate Council, 2009). It is quite clear that governments and business communities around the world should work as a partnership to achieve this goal of world sustainability. In many developed and developing countries, companies have started planning and developing sustainable strategies to address the environmental concerns. Companies in many industries, such as textile industry, building industry, organic food industry, apparel manufacturing industry and retailers, and etc., worked with the US government agency, FTC to understand the regulatory side of the requirement to claim products to be green or organic (Rodie, 2008; Yudelson, 2007).

To effectively market products/services in the market place, companies are recommended to fully understand the consumers' concerns and their expectations before their launching of green products in the market. As suggested by Joel Makower (2008, 2009) in his book of Strategies for Green Economy that there are four keys to successfully market green products and services in today's market place: credibility, relevance, effective messaging, and differentiation. Strategies and messages will have to be credible with facts and figures to support the claims. Claims that are vague and general cannot convince consumer to purchase their products or services, and companies of this type will be out market sooner than they expect. Companies without prior reputation in the green industry will have to pay particular attention to their claims in introducing their products. Also, government agents will watch over the companies' claims very carefully to make sure their claims are truthful and accurate. The second key of relevance is that the sustainable strategies will have to create values to companies' stakeholders, especially the key stakeholders. As suggested by the recent broad development of marketing concept to include stakeholders that to ensure long-term success of the companies' business it is advantageous to consider the interests of stakeholders who have keen interests in a company's well-being. Effective messaging requires companies to use right media and distinctive messages to get their information across the company's stakeholders that include company's shareholders, employees, consumers, social groups, government agents, and other people or organizations. As the campaign toward the world of sustainability gets more popular competition will become more intensive and the effectiveness of messaging will be decreased. Therefore, it is important to make sure that the company's voice will be heard through the right channel of communication and at the right time. This leads to the fourth key of differentiation as suggested by Makower (2008). Without a differentiation strategy the market that consumers, governments, and social groups will not notice you and your products/services, or even the claims company has made. Therefore, companies, large or small, will need to use effective communication tactics with hard facts and figure to support, so that to make the company stand out among the competition.

5.3. *Social groups/NGOs*

Since the 1970s the world has felt ever increasing impacts by social activist groups. Some of these groups have become so powerful that companies and governments could not turn a deaf ear to them. In the past couple of decades, large-scale protests have been organized in the major cities around the world in front the international or global summits on various global and international issues. Therefore, as a force influencing national governments and/or multinational corporations, these social groups, also often times referred as NGOs, have played increasingly important roles in help changing global and national economies. Green Peace group and many other powerful ones have already demonstrated their muscles in forcing governments and multinational corporations to change their policies and business practices to ensure equality and social ethics. Their goals and/or actions will have strong influence and also be influenced by governments' policies and businesses' practices.

NGOs have become quite prominent in the field of international development in recent decades. However, the term NGO encompasses a vast category of groups and organizations as the World Bank (Working With NGOs, 1995) defines as "private organizations that pursue activities to relieve suffering, promote the interests of the poor, protect the environment, provide basic social services, or undertake community development." It is suggested that NGOs are a natural outcome of a free democratic and capitalistic society. The idea is that as market forces require more and more relaxations of rules and regulations for their benefit, society will naturally demand social justice to balance out negative aspects of market forces such as exploitation and environmental degradation.

5.4. *Consumers*

Consumers as the last group of the four in the expanded model of GAP-ACT have a fundamental impact on the governments' policies, businesses' goals, and social groups' goals as well. Consumers are the base for everything. They use products/services produced by businesses, they are the market for all the business and social activities. Therefore, consumers' demands for products and services will determine what marketing goals and strategies businesses will create to meet these demands. Governmental policies and regulations will be made to protect consumers of today and tomorrow. Consumers' concerns for the sustainability of future will make businesses, governments, and social groups to consider what appropriate actions they can take to meet consumers' expectations to reach that target area in the model.

Recent studies done by GMA/Deloitte, Hartman Group, and MIT Sloan Management Review (Manget *et al.*, 2009), consumers have become increasingly knowledgeable and more active in seeking sustainable products, and also recession resistant. Some of the common characters found in these studies are as follows:

- Consumers are more aware of sustainable development in the market place and they know more about green products now than before.
- Consumer will spend more money on green products, especially food and healthcare products.
- More consumers are searching for environmental friendly products when shopping, taking advantage of rebates and other incentives.

According to the GMA/Deloitte Green Shopper Study (2009), 95% of the respondents would buy green products, 75% of them know what a green product was, 63% were looking for green product while shopping, 47% of the respondents actually saw green products, and 22% of the surveyed have bought green products. We see that almost everyone would like to buy green products, but only 22% of them actually bought the green products in the store. There is a huge gap between people willing to buy green products and actually purchasing the green products. As illustrated in the expanded model of GAP-ACT, consumers' goals and perceived goals generate a gap and thus leading them to take various actions to reach their expected target area or goals by buying, browsing and searching for more information about green products. This type of behavioral pattern creates a vast opportunity for other major participants to influence the consumers in achieving a sustainable economy.

As we can see so far that all major participants in the expanded model of GAP-ACT will behave and act as their own entity; however, at the same time their actions, strategies, regulations and standards will also influence other participants to achieve this common target area or goals of sustainability of society. To that end, we need a coordinated effort of all participants — consumers, businesses, governments, and social groups so we all can benefit from this endeavor.

6. Conclusions

We will need some sort of cultural revolution in consumption nowadays. As Erik Assadourian, senior researcher at the Worldwatch Institute, said "It's no longer enough to change our light bulbs. We need to change our culture." The bottom line of this sustainable development is that people are the driver behind this movement, people are the influencer and also being influenced.

References

AMA Definition of Marketing, 2007. American Marketing Association. Available at www.marketingpower.com/Community/ARC/Pages/Additional.

Assodourian, E., 2010. *2010 State of the World: Transforming Cultures from Consumerism to Sustainability.* Worldwatch Institute. Available at worldwatch.org.

Belz, F.-M. and Peattie, K., 2009. *Sustainability Marketing: A Global Perspective.* Wiley, USA.

Chakravorti, B., 2010. Stakeholder marketing 2.0. *Journal of Public Policy and Marketing* **29**(1), 97–102.

Cova, Badot and Bucci 2006. *Beyond Marketing: In Praise of Societing.* Available at http://visionarymarketing.com.

Copenhagen Climate Council, 2009. *The Future of Your Business is Being Decided. World Business Summit, Copenhagen, May 24–26, 2009.* Retrieved on April 17, 2010 from www.copenhagenclimatecouncil.com/world-b.

Craig, S.N., Drumwright, M.E. and Gentile, M.C., 2010. The new marketing myopia. *Journal of Public Policy and Marketing* **29**(1), 4–11.

Epstein, M.J., 2008. *Making Sustainability Work.* Berrett-Koehler Publishing, Inc., San Francisco.

Ferrell, O.C., Gonzalez-Padron, T.L., Hult, G.T.M. and Maignan, I., 2010. From market orientation to stakeholder orientation. *Journal of Public Policy and Marketing* **29**(1), 93–96.

Galbraith, K., 2010. Green diplomacy at its most basic level. *The New York Times* April 18, 2010. accessed from www.nytimes.com on April 19, 2010.

Grocery Manufacturers Association and Deloitte, 2009. Finding the green in today's shoppers, sustainability trends and new shopper insights. *GMA/Deloitte Green Shopper Study.* Accessed from http://www.deloitte.com/us/greenshopperstudy09.

Hartbeat, 2008. *The Consumer Side of Sustainability.* Hartman Group, April 22, 2008. Accessed from http://www.hartman-group.com/hartbeat/the-consumer-side-of-sustainability."

Hopwood, B., Mellow, M. and O'Brien, G., 2005. Sustainable development: Mapping Different approaches. *Sustainable Development* **13**, 38–52.

Hosfeld, K., 2009. Finding Your Way Through the Razzle Dazzle, Marketing Jargon, Marjarg, June 18, 2009. Accessed from http://blog.hosfeld.come/category/sustainability-marketing/.

Kotler, P. and Levy, S.J., 1969. Broadening the concept of markting. *Journal of Marketing* **33** (January), 10–15.

Lusch, R.F., 2007. Marketing's evolving identity: Defining our future. *Journal of Public Policy and Marketing* **26**(2), 261–268.

Makower, J., 2009. *Strategies for the Green Economy: Opportunities and Challenges in the New World of Business*. McGraw Hill, New York.

Makower, J., 2008. *The Four Simple Steps to Pitch-Perfect Green Marketing*. Accessed from www.greenbiz.com/print/20778.

Manget, J., Roche, C. and Münnich, F., 2009. *For Real, Not Just for Show. MIT Sloan Management Review. Special Report*, February 25, 2009. Accessed from http://sloanreview.mit.edu/special-report/2009/02/25/for-real-not-just-for-show/.

Mish, J. and Scammon, D.L., 2010. Principle-based stakeholder marketing: Insights from private triple-bottom-line firsm. *Journal of Public Policy and Marketing* **29**(1), 12–26.

Neuman, W., 2010. U.S. plans spot tests of organic products. *The New York Times* March 19, 2010. Accessed from www.nytimes.com.

Nickols, F., 2008. *The GAP-ACT Model, Distance Consulting LLC*. Accessed from www.nickols.us/gapact.pdf.

Powers, W.T., 1973. *Behavior: The Control of Perception*. Aldine de Gruyter, New York.

Powers, W.T., 1989. *Living Control Systems: Selected Papers of William T. Powers*. Control System Group, Gravel Switch, KY.

Reid, D., 1995. *Sustainable Development: An Introductory Guide*. Earthscan, London.

Sheth, J.N. and Uslay, C., 2007. Implications of the revised definition of marketing: From exchange to value creation. *Journal of Public Policy and Marketing* **26**(2), 302–3027.

Sustainability Dictionary. *The Dictionary of Sustainable Management*. Available at www.sustainabilitydictionary.com.

Wong, E., 2009. *Recession Dampens Green Enthusiasm*. Brandweek, April 7, 2009. Accessed on November 22, 2009, from www.brandweek.com/bw/content_display/news-and-features/green-marketing/e3i8...

Working with NGOs, 1995. *World Bank*. Accessed from http://www-wds.worldbank.org/servlet/WDS_IBank_Servlet?pcont=details& eid=000009265_3961219103437.

Summary Report for Policymakers, 2009. *World Business Summit on Climate Change: Shaping the Sustainable Economy*, Copenhagen, 24–26 May, 2009.

Yudelson, J., 2007. *The Keys to Marketing Green — Part 3, Professional Practices, Sustainable Design, Consulting-Specifying Engineer*, July, pp. 31–32, 35–36.

Zeller, T. and Skipp, C., 2010. Appliance discounts, for the Swift. *The New York Times* April 19, Accessed from www.nytimes.com on April 20, 2010.

Chapter 36

ENVIRONMENTAL SERVICE OPERATIONS: CHALLENGES AND OPPORTUNITIES

George I. Kassinis and Andreas C. Soteriou
Department of Public and Business Administration,
University of Cyprus, P.O. Box 20537,
CY 1678, Nicosia Cyprus
kassinis@ucy.ac.cy, basotir@ucy.ac.cy

This chapter discusses the challenges and opportunities surrounding environmental operations in the context of services, the challenge of integrating environmental and quality practices, and how their successful integration can provide customers with legitimate signals regarding a service firm's environmental efforts. We further discuss the relationship among environmental practices, customer satisfaction, and loyalty within the context of an environmental Service Profit Chain framework, and conclude with broad directions for future research in this area.

1. Introduction

Researchers have widely documented the performance implications of the successful implementation of a variety of environmental management practices in both manufacturing and service settings. In the latter, performance gains include cost reduction and resource savings (Goodman 2000; Schendler 2001); however, most importantly the successful adoption of environmental practices has been related to increased customer satisfaction (CS) and loyalty (Kassinis and Soteriou 2003). Such gains notwithstanding, services' distinctive characteristics *vis-à-vis* goods create unique implementation challenges for service firms (Fitzsimmons and Fitzsimmons 2008; Kassinis and Soteriou 2003). One such characteristic is the physical presence of the customer in the system and his role as co-producer, especially in high contact service systems, as well as the resulting simultaneity of service production and consumption (Chase and Tansik 1983; Soteriou and Chase 1998). Often, certain environmental activities are "hidden" from the customer as they take place in the back office. In high contact systems, however, such activities also take place in the front office where an environmentally-conscious customer may not only apply pressure on management to change company policy (Salzman, 2000) but also be involved, as co-producer, in a firm's environmental practices. The challenge and at the same time the opportunity for such high contact service firms are to meet customer demands and manage customer involvement without compromising the quality of services they provide (Goodman, 2000; Schendler, 2001).

Another critical issue that arises in the literature is the importance of combining different organizational capabilities for firm environmental initiatives to be successful. Research has stressed the complexity of successful implementation of environmental management practices (specifically pollution prevention measures; Aragón-Correa and Sharma, 2003) and its dependence on capabilities such as cross-functional integration, stakeholder integration, continuous innovation and improvement, and organizational learning (Aragón-Correa and Sharma, 2003; Eisenhardt and Martin, 2000; Hart, 1995). Moreover, experience from manufacturing shows that quality and environmental problems have many of the same causal factors (including complex interactions and potential conflicts with day-to-day operations) and they are managed with similarly structured reporting systems and information evaluation models. As a result, environmental performance is often positively correlated with quality (King and Lenox, 2001; Pil and Rothenberg, 2003). In his seminal paper, Shrivastava (1995) proposed the integration of quality and environmental considerations by firms in the form of total quality environmental management (TQEM) — a strategic choice that could generate both performance and environmental quality gains for its adopters (Angell, 2001; Klassen and Whybark, 1999a, 1999b; King and Lenox, 2002). While the relationship between the natural environment and quality management has received considerable attention, researchers are only beginning to explore how these programs are most effectively integrated (Angell, 2001; Pil and Rothenberg, 2003).

In this chapter, we discuss (i) environmental operations in the context of services in general, (ii) the relationship between environmental practices and CS and loyalty focusing primarily on the environmental Service Profit Chain (SPC) framework; (iii) the challenge of integrating environmental and quality practices and how their successful integration can provide customers with legitimate signals regarding a service firm's environmental efforts and we conclude with directions for future research in this area.

2. Environmental Operations in the Context of Services

Research shows that the effective integration of environmental management practices into operations leads to lower costs and enhanced efficiencies (Klassen and Whybark, 1999a, 1999b; King and Lenox, 2002; Russo and Fouts, 1997), competitive advantages through product or service differentiation (green products or services), and better servicing of niche markets (that include customers who demand ecologically friendly products or services; Goodman, 2000; Shrivastava, 1995).

Most empirical studies focus on the relationship between environmental performance and financial performance in manufacturing (Dowell *et al.*, 2000; King and Lenox, 2001; Klassen and McLaughlin, 1996) and find a positive relationship between the two. A smaller number of studies focus on the relationship between the implementation of environmental practices and performance and suggest that "green" firms may also be more efficient and innovative (King and Lenox, 2001). Others report mixed results (Christmann, 2000; King and Lenox, 2002) but, nonetheless, point out the benefits of pollution prevention measures for adopting firms.

In services, studies document the positive performance implications of environmental management measures, which include cost reductions and resource savings, customer retention and loyalty, and improved employee morale (Enz and Siguaw, 1999; Goodman, 2000; Grove *et al.*, 1996; Foster *et al.*, 2000; Halme, 2001; Kassinis and Soteriou, 2003; Salzman, 2000; Schendler, 2001). Most such studies rely on case or anecdotal evidence, with a notable exception the paper by Kassinis and Soteriou (2003).

As we argued in the introduction, the successful implementation of environmental practices presents service firms with unique challenges related to the distinctive characteristics of services (Fitzsimmons and Fitzsimmons, 2008; Kassinis and Soteriou, 2003). One such characteristic is the presence of the customer in the system and the resulting simultaneity of service production and consumption. In fact, the impact of customer involvement on the service operating system is arguably one of the most important service idiosyncrasies affecting service performance (Chase, 1981; Chase and Tansik, 1983; Kellogg and Chase, 1995), especially in high contact service systems where customer involvement is high.

All services require some direct or indirect customer involvement (Chase and Tansik, 1983; Lovelock, 1996). However, the physical presence of the customer and the customer's role as co-producer in high contact service systems create numerous challenges for managers (Soteriou and Chase, 1998); it can also influence environmental actions (Foster *et al.*, 2000). Often, certain back office environmental activities are "hidden" from the customer. In restaurants, for example, waste disposal or recycling may take place out of customer view. In high contact systems, however, a number of such activities also take place in the front office. An environmentally-conscious customer may then not only apply pressure on management to change company policy (Salzman, 2000) but be directly involved, as a co-producer, in a firm's environmental practices (for example, in energy and water savings practices in hotels). In practice, therefore, service characteristics will likely influence the environmental technology choices available to a firm and add complexity to the design of its environmental strategy (Kassinis and Soteriou, 2003; Klassen and Whybark, 1999b; Rajaram and Corbett, 2002). More broadly, the challenge for high contact service firms is to meet customer demands and manage customer involvement without compromising the quality of their offering (Foster *et al.*, 2000; Goodman, 2000; Schendler, 2001).

Satisfying the demands of environmentally-conscious customers is critical given the well-documented relationship between CS and performance in the literature. This literature suggests that CS is positively related to the concept of customer loyalty (Heskett *et al.*, 1994, 1997), which, in turn, is linked to increased profits — through enhanced revenues, reduced costs to acquire customers, lower customer–price sensitivity, and decreased costs to serve customers familiar with a firm's service delivery system (Reichheld and Sasser, 1990). The linkages between satisfaction, loyalty, and performance constitute the "external" portion of the SPC framework (Heskett *et al.*, 1994), a framework that proposes a path linking these constructs to performance. An adaptation of the SPC that brings environmental practices into the path is discussed in detail in the following section.

3. CS, Loyalty and the Environmental Service Profit Chain (SPC)

3.1. *CS, loyalty and service performance*

The construct of CS has received considerable attention over the past couple of decades, in different streams of literature, including marketing and operations management. A common thesis in much of this work is that CS results in customer behavior patterns that ultimately have a positive impact on business results (Keiningham *et al.*, 2003). This is particularly true in the case of services, given the increased involvement of the customer in many aspects of the service delivery system and the production of the service. As a result, significant research interest has focused on theoretical and conceptual underpinnings of CS (Oliver, 1997; Rust *et al.*, 1995), its antecedents (Anderson and Sullivan, 1993; Szymanski and Henard, 2001) and outcomes (Mittal *et al.*, 1994; Fornell *et al.*, 1996, Anderson *et al.*, 2004). Interestingly, little attention was paid to the relation of CS to antecedents and outcomes as these related to environmental practices and outcomes, respectively.

Various definitions of CS can be found in the literature. One of the most popular ones positions CS as an evaluation of the perceived discrepancy between prior expectations — or some other norms of performance — and the actual performance as perceived after the consumption of a good or service by a customer (Oliver, 1993). In the marketing domain, cumulative satisfaction, a customer's overall experience to date with a product or service (Johnson and Fornell, 1991; Johnson *et al.*, 1995), is thought to be a fundamental indicator of the firm's past, current, and future performance. It is, in turn, what motivates a firm's investment in CS (Anderson *et al.*, 1994).

The investigation of CS outcomes has recently focused on both financial and customer-related dimensions. While early efforts utilized the PIMS database to establish a relationship among satisfaction, market share, and profitability (Buzzell and Gale, 1987), nowadays many scholars suggest that CS directly impacts the bottom line through a positive impact on financial performance measures, such as return on investment (Anderson *et al.*, 1994) and return on assets (Rust *et al.*, 2002). In turn, customer outcomes remain the focus of a large body of research that specifically examines the relationship of CS with the construct of customer loyalty (Olsen, 2002; Mittal *et al.*, 1994).

Of the various definitions of customer loyalty, we focus on two. According to the first, which defines loyalty as an attitude, different feelings create an individual's overall attachment to a product, service, or organization and define that individual's (purely cognitive) degree of loyalty (Hallowell, 1996). The second considers the behavioral aspects of loyalty. Loyalty behaviors consistent with this second definition include continuing to purchase services from the same supplier and increasing the scale and/or scope of the relationship or the act of recommendation (Yi, 1991). Within the same context, Oliver (1999) extended an earlier definition of loyalty to include the act of consuming. She described it as "... a deeply held commitment to rebuy or repatronize a preferred product/service consistently in the future, thereby causing repetitive same-brand or same brand-set purchasing" despite external influences or marketing efforts that may induce switching behavior (Oliver, 1999, p. 34). Reichheld (1996) and Tellis (1988) also present similar definitions.

Loyalty behaviors, including relationship continuance, increased scale or scope of relationship, and recommendation (word of mouth advertising) are related to customers' beliefs that the value received from one supplier is greater than that from others. Loyalty, in the forms noted above, increases profits through enhanced revenues, reduced costs to acquire customers, lower customer–price sensitivity, and decreased costs to serve customers familiar with a firm's service delivery system (Reichheld and Sasser, 1990). Some studies also examine the cognitive, affective, and conative antecedents of customer loyalty as well as its consequences (Oliver, 1999; Dick and Basu, 1994).

Overall, there is broad consensus in the marketing literature that improvements in satisfaction lead to higher levels of customer loyalty (Bolton and Drew, 1991; Boulding *et al.*, 1993), higher revenues (Rust *et al.*, 1995), and reduced future transaction costs (Reichheld and Sasser, 1990) — all through improved customer loyalty. Similar evidence from the service management literature suggests that CS influences customer loyalty, which in turn affects profitability. Proponents of this theory include Heskett *et al.*, (1994), Reichheld and Sasser (1990), Rust *et al.*, (1995), Schneider and Bowen (1993), and Zeithaml *et al.*, (1990). In this line of research, CS is thought to result from a customer's perception of the value received in a transaction or relationship relative to the value expected from transactions or relationships with competing vendors (Zeithaml *et al.*, 1990).

It is worth noting that despite the increasing attention that environmental practices have been receiving recently, a possible relationship to the bottom line through a mediating effect on CS has not been investigated. We next discuss an adaptation of the SPC that can be used to explore this relationship in the case of services, where employee–customer interaction occurs extensively.

3.2. *The environmental path of the SPC*

The SPC (Heskett *et al.*, 1994), arguably one of the most popular service management conceptual frameworks, which links CS antecedents and outcomes, has recently bridged research in operations and human resource management that focused on organizational effectiveness. Overall, the SPC proposes a pathway linking service operations, employee assessments, and customer assessments to a firm's profitability and growth (Kamakura *et al.*, 2002). SPC builds on the conventional business logic that financial performance is mediated through other constructs and establishes relationships between market performance (revenue growth and profitability), customer loyalty, and CS, on the one hand, and employee satisfaction, loyalty, and productivity, on the other hand (Heskett *et al.*, 1994). Briefly, SPC postulates that (a) the internal quality of the working environment — high-quality support services and policies — drives employee satisfaction, (b) value is created by satisfied employees, (c) CS is driven by the value of services delivered, (d) CS directly impacts customer loyalty, which in turn (e) drives profitability and growth.

A large body of literature that spans a variety of disciplines (from OM to HRM and service management) investigates the proposed SPC relationships in different contexts (Mittal and Kamakura, 2001; Silvestro and Cross, 2000; Soteriou and Zenios, 1999). Interestingly, the proposed relationships are not always supported, which suggests that the exact nature of the relationships maybe complex (Kamakura *et al.*, 2002) and points to the need for further research. For a review and a synthesis of work on SPC, we refer the reader to the work of Yee *et al.*, (2009).

With the notable exception the work of Kassinis and Soteriou (2003), little has been done to expand the conventional SPC to consider the impact of environmental practices on the chain. In process improvements, employee involvement emerges as a key capability associated with a firm's environmental responsiveness. A number of environmental performance improvements (such as waste minimization) may result from employee involvement practices or team projects (Hanna *et al.*, 2000; May and Flannery, 1995). Enz and Siguaw (1999, p. 76) report that all four operations named as environmental best practice champions in a Cornell University study of best practices in the US lodging industry indicated that such practices had a positive impact on employee morale (and thus satisfaction) and enhanced the staff's pride in the hotel. Similar results are reported by Goodman (2000) in the case of Scandic hotels.

Considering the above, Kassinis and Soteriou (2003) expand the SPC's concept of "internal quality" to incorporate environmental practices (see Fig. 1) and suggest that the SPC framework can be used to further investigate the relationship between environmental practice implementation — a component of a firm's operations (Angell and Klassen, 1999; Hanna *et al.*, 2000) — and performance. Environmental practices are an integral part of the "operating strategy and service delivery system" (Heskett *et al.*, 1994) and, as such, are built into service design and might impact CS and firm performance, through customer loyalty.

Their argumentation is consistent with recent literature that argues that the environment must be integrated with management's efforts to address the concerns of all stakeholders involved — with the overarching objective of operations being the improvement of customer value (Angell and Klassen, 1999, p. 594). Environmental

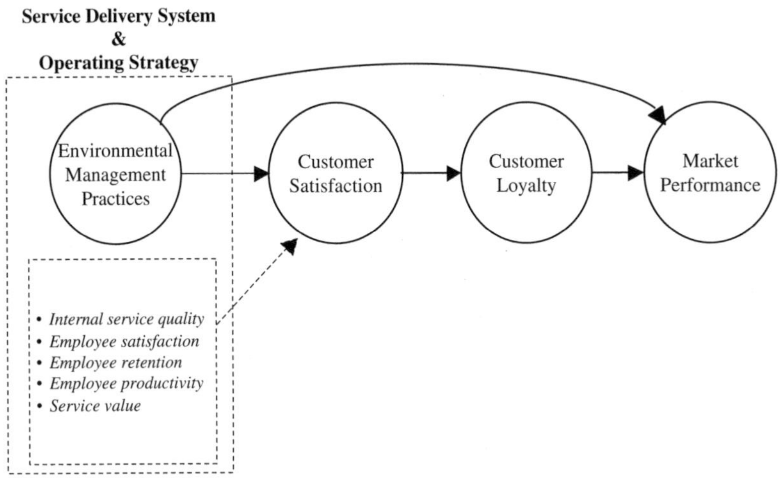

Fig. 1. Environmental management practices and the service profit chain.
Source: Kassinis and Soteriou (2003).

practices are integrated within the service concept and can alter both its structural and managerial elements, including its service delivery, service encounter, quality, and information dimensions (Fitzsimmons and Fitzsimmons, 2008).

It is also consistent with case study evidence. Scandic hotels, for example, whose stated top priority was to communicate their new environmental values and strategy to employees through education and training (Goodman, 2000, pp. 206–207), rolled out an environmental training course to increase customer participation, called the "environmental dialogue." It also invested in new information system infrastructure that monitored and measured key environmental performance indicators and enabled customized reporting and benchmarking, which allowed information sharing and performance comparisons across the chain's hotels.

After conceptually developing the relationship between the adoption of environmental practices and market performance, in the context of services, Kassinis and Soteriou (2003) empirically demonstrated — through the use of structural equation modeling — that the use of such practices is positively related to CS and service performance, through enhanced customer loyalty. Their work utilized survey data from the hotel industry in the European Union where sustainability and environmental management issues have recently been gaining increasing attention. Their work highlighted the importance of environmental management practices and for the first time the related mediating effect of resulting CS on service performance.

4. The Challenge of Integrating Environmental and Quality Practices

As we indicated in the introduction, the successful implementation of environmental management practices — specifically pollution prevention measures — is a complex phenomenon intimately linked to other firm practices aimed at enhancing performance and competitiveness, such as quality practices. Moreover, it depends on dynamic capabilities such as stakeholder integration, cross-functional integration and knowledge sharing, continuous innovation and improvement, and organizational learning (Angell, 2001; Aragón-Correa and Sharma, 2003; Hart, 1995; Pil and Rothenberg, 2003).

Complexity also characterizes efforts to improve service quality. Indicatively, a growing literature links operations and marketing concepts (for example, Roth and van der Velde, 1991; Soteriou and Hadjinicola, 1999; Soteriou and Zenios, 1999) or organizational behavior and marketing concepts (Heskett *et al.*, 1997) with efforts to improve service quality.[1] Moreover, the improvement of service quality is considered one of the fundamental drivers of CS (Oliver, 1993) and lies at the heart of frameworks such as the ones by Heskett *et al.* (1997) and Roth *et al.* (1997). Given the importance of service quality for improving performance, a large literature in marketing (Parasuraman *et al.*, 1985; 1988), operations management (Collier, 1991), and organizational behavior (Schneider and Bowen, 1993) deals with how to deliver high levels of service quality.

Here, we choose to focus on cross-functional integration, one of the dynamic capabilities' service firms need to possess to successfully connect environmental and quality practices (Eisenhardt and Martin, 2000). According to Song and Montoya-Weiss (2001), cross-functional integration involves the interaction and communication, information sharing, coordination, and joint involvement across functions. Furthermore, this capability includes both formal and informal social integration mechanisms for sharing information and knowledge over functional boundaries (Zahra and George, 2002). In his early work, Shrivastava (1995) had in fact proposed that the integration of quality and environmental considerations, in the form of TQEM, could generate both performance and environmental quality gains for implementing such strategic choices. Later, Angell and Klassen (1999) and Corbett and Pan (2002) pointed out that, in fact, total quality management (TQM) principles — such as continuous improvement, training and empowerment of employees, incentive schemes, and quality management systems — apply to environmental improvements as well.

Despite its focus on manufacturing, the concept has relevance for services as well (Angell, 2001). By adopting such a strategic choice, firms could reduce the

[1]One of the most frequently cited definitions of SQ is that of Parasuraman *et al.* (1985) who define service quality as the extent to which the service meets or exceeds consumer expectations. The similarity of this definition to that of customer satisfaction or dissatisfaction (CS/D) — an evaluation of the perceived discrepancy between prior expectations or some other norm of performance and the actual performance as perceived after consumption (Oliver, 1993) — sparked a debate in the literature on the conceptualization of SQ and the causal relationship between service quality and satisfaction (Bolton and Drew, 1991; Cronin and Taylor, 1992; Parasuraman *et al.*, 1994). These differences are further discussed by Rust *et al.* (1994) and by Oliver (1993) who present a theoretical framework integrating the two constructs.

use of energy and conserve resources through increased use of renewable materials, development of ecologically-sensitive purchasing policies and inventory management systems, or even through product redesign. In addition, a focus on production efficiency improvements could minimize waste and reduce costs. Finally, a critical reexamination of product choice and design by firms could minimize life-cycle costs and improve product and service quality (Angell and Klassen, 1999; Shrivastava, 1995). Reporting on the findings of a US Conference Board survey, Angell (2001, p. 307) and Angell and Klassen (1999, p. 584) indicate that linking quality and environmental programs results in "...decreased costs, improved long-term position, increased customer focus, and process simplification."

Clearly, such thinking has influenced the design and development of environmental management systems, with research pointing out that environmental practices have wider performance benefits for the adopting firm (Klassen and Whybark, 1999a and 1999b; King and Lenox, 2002). Significant challenges, however, still hamper the integration of environmental and quality programs. These include poor organizational design, little real implementation of TQM, and insufficient awareness of potential synergies between the two (Angell, 2001; Pil and Rothenberg, 2003). Also, cross-functional implementation teams as opposed to individual project managers are key success factors in the case of successful environmental initiatives (Angell, 2001), a finding that reinforces the importance of strengthening cross-functional capabilities.

4.1. *Service cues, signaling, legitimacy, and CS*

Building on the discussion in the preceding sections, we posit that effectively linking environmental and quality programs is particularly important given the presence of an environmentally-conscious customer in the system and this customer's inability to observe a portion of the service delivery system's back office activities. The intangibility of services (another service characteristic distinguishing them from goods) also increases the level of uncertainty regarding service characteristics, with the customer seeking additional information to reduce the information asymmetry gap (Reimer and Kuehn, 2005). In the absence of such information the customer has to rely on signals or cues regarding the nature of the service firm's environmental orientation. This strengthens the importance of signals that point to a firm's pro-environment orientation and mitigate the information asymmetries that characterize the customer's relationship with the firm.

Cue utilization theory (Baker *et al.*, 2002; Cox, 1967; Reimer and Kuehn, 2005) informs this discussion by pointing out that services (or goods) consist of a series of cues that serve as surrogate indicators of a service's attributes. Many cues are usually available for tangible products, be it physical characteristics or product-related ones that arise from other marketing instruments. The intangibility of services, however, limits the customer to a small number of cues such as price and physical facilities (Reimer and Kuehn, 2005). Physical facilities — also known as servicescape

(Bitner, 1990), a term that describes the physical surroundings of a service company (Reimer and Kuehn 2005, p. 786) — can function as a surrogate quality indicator (Baker *et al.*, 1994, 2002; Parasuraman *et al.*, 1988; Zeithaml *et al.*, 1990). However, only Parasuraman *et al.* (1988, 1991) studied the influence of servicescape (or "tangibles") in conjunction with other determining factors ("intangibles," i.e., reliability, responsiveness, assurance, and empathy) of service quality. Furthermore, empirical research on the effect of servicescape on quality perceptions is rare (Baker *et al.*, 2002; Reimer and Kuehn, 2005) and very little attention has been paid to the question of how the servicescape affects the evaluation of service quality (Reimer and Kuehn 2005, 787). In the context of service environmental operations, front office environmental management measures employed by a service firm may signal its overall environmental orientation and its level of commitment to environmental protection. As such, they can decrease the information asymmetries faced by the interested customer.

Therefore, a service provider has an incentive to provide credible signals to satisfy environmentally-conscious customers, reduce information asymmetries between the firm and its customers, and enhance the legitimacy[2] of the firm's "hidden" environmental efforts (Bansal and Clelland, 2004; King *et al.*, 2005). Of course, customers assess the firm's legitimacy according to their own distinct and diverse norms, cognitive maps, and preferences. Managers, however, can manage customers' impressions and actively shape the way in which they view the firm and its activities. They can do so by conveying information regarding changes in the services offered that demonstrate the firm's commitment to the natural environment, by investing in front office pollution prevention practices, by aligning the firm with symbols or endorsements that convey environmental commitment or other information about otherwise hidden organizational attributes and behaviors (Bansal and Clelland, 2004; King *et al.*, 2005).

Based on the above, we argue that by effectively integrating its quality and environmental programs, a service firm provides a stronger, more legitimate signal of overall excellence, which influences customer perceptions and has a positive effect on CS. The legitimacy of such a signal is supported by literature cited earlier, which suggests that "green" firms are also more efficient and innovative (King and Lenox, 2001).

Recently, in an experimental setting, Kassinis and Soteriou (2009) examined the link between environmental practices and CS in conjunction with service quality practices. They argue that the relationship between environmental practices and CS may be strengthened if environmental practices are linked with service quality practices. They point out that synergies between environmental practices and service quality can increase the level of CS achieved by a service firm — compared with the situation where either environmental or quality practices exist alone.

[2]We follow Suchman (1995) and Bansal and Clelland (2004) and define environmental legitimacy as the generalized perception that the firm's environmental performance is desirable and appropriate.

5. Conclusions and Future Work

One of the themes that emerge from the literature that critically examined the above points is the complexity surrounding environmental service operations. Such complexity manifests itself in a variety of ways that revolve around the distinctive characteristics of services (especially the presence of the customer in the service delivery system and the customer's role as co-producer) as well as the necessary capabilities needed to integrate efforts and exploit synergies across a firm's supply chain, most notably those involving environmental and service quality practices.

As we argued in this chapter, by building the necessary capabilities, service firms can exploit synergies between environmental and service quality practices and increase the level of CS that they would otherwise achieve. The effective linkage of environmental and quality programs is particularly important in services given the presence of the customer in the system, his/her role as co-producer, and the fact that a significant portion of a service firm's environmental operations (and the impacts they are designed to manage) is hidden from the customer. Environmentally-conscious customers depend on signals that point to the environmental orientation of the firm in order to mitigate the information asymmetries that characterize their relationship with the service provider. At the same time, the service firms are motivated to provide such credible signals to legitimize their "hidden" environmental efforts and satisfy environmentally-conscious customers. By effectively integrating quality and environmental programs, service firms deliver stronger signals of overall excellence, something that has positive effects on CS and thus performance.

The argumentation regarding the existence of an interaction effect between service quality and environmental practices (Kassinis and Soteriou, 2009) is broadly consistent with research that stresses that environmental practices need to be integrated within the service concept and alter both its structural and managerial elements, including its service delivery, service encounter, quality, and information dimensions (Kassinis and Soteriou, 2003). Moreover, the "interconnectedness" of service quality and environmental practices' considerations and their potential interactions are in line with the fact that environmental practices — a component of service firms' operations (Angell and Klassen, 1999) — are also a component of a service firm's operating strategy and service delivery system (Heskett, 1994). If then, environmental practices along with quality practices are built into the service design, they could jointly impact CS — hence the interaction effect that is over and above their individual effects on CS. Such integration is in fact advocated in recent literature, which argues that the environment must be integrated with management's efforts to address the concerns of all stakeholders — with the overarching objective of such efforts being the improvement of customer value (Angell and Klassen, 1999).

It is thus important to consider interaction effects when designing environmental management programs in service settings. As mentioned above, although the relationship between the natural environmental and quality management has received considerable attention, researchers are only beginning to explore how environmental and quality programs are most effectively integrated (Angell, 2001; Pil and Rothenberg, 2003). Their integration, in manufacturing and service settings alike, and the challenges associated with it can become the focus of future research.

Methodologically, it is constructive to point out that experimental procedures could be utilized in operations management research to provide additional insights on the relationships between various constructs of interest and their interactions, as Kassinis and Soteriou (2009) have done in exploring interactions between quality and environmental practices within the broader context of service operations. We must note here that a challenge in using such experimental designs is, again, related to the presence of the customer in the system. However, the use of video technology in such experiments, for example, may allow multiple respondents to experience the same scenario and thus maintain consistency, a major requirement of experimental designs. As such, the combination of video technology and experimental designs provides us with a powerful tool for empirical research in service settings.

In this chapter, we also pointed out that the SPC provides a framework within which to test a variety of relationships of interest in environmental service operations. The limited work in this area (see Kassinis and Soteriou, 2003) incorporated environmental practices as an important dimension within the framework but did not fully assess the SPC. We believe that future work needs to fully assess the greening of the SPC — in a variety of service contexts. Such an effort has significant data requirements, however. As Loveman (1998, p. 19) points out "the main impediment to more comprehensive empirical testing" of the service profit chain "... has been the absence of large sample data spanning all (or even most) of the service profit chain components."

More venues for future research include an investigation of the nature of the relationship between environmental practices and CS and loyalty in services using refined measures of the latter. Questions to be addressed may focus on the way or the mechanisms through which environmental practices enhance CS and loyalty. Given the critical role of employees in this context, a full assessment of the SPC needs to establish how employee satisfaction and loyalty are enhanced when firms are environmentally proactive, how employee satisfaction and loyalty impacts CS and loyalty, and through them market performance. In this context, one can look, for example, at employee environmental training or employee involvement in continuous environmental progress.

Overall, a systematic assessment of the effectiveness of service firm environmental practices is needed in future work, that is, an evaluation of their outcomes in terms of reduced environmental loads produced or cost savings achieved. Moreover, the use of finer measures of environmental practices (compared with those used in existing work) that cover the entire value chain of service activities needs to be considered. These may include: first, a distinction between front- and back-office measures; second, an examination of finer measures of resource savings programs;

third, a consideration of environmentally responsible practices in the design and construction of facilities (that influence the servicescape and thus the signals discussed earlier in this chapter).

The complexities surrounding the design and implementation of environmental management practices in service settings remain. Service managers must continuously strive to meet the demands of environmentally-conscious customers and manage their involvement in service production without compromising the quality of services they provide. Further research is thus needed to shed light on how environmental practices can best be integrated into the service design.

References

Anderson, E.W. and Sullivan, M.W., 1993. The antecedents and consequences of customer satisfaction for firms. *Marketing Science* **12**(2), 125–143.

Anderson, E.W., Fornell, C. and Lehmann, D.R., 1994. Customer satisfaction, market share, and profitability: Findings from sweden. *Journal of Marketing* **58**(3), 53–66.

Anderson, E.W., C. Fornell, and Mazvancheryl, S.K., 2004. Customer satisfaction and shareholder value. *Journal of Marketing* **68**(4), 172–186.

Angell, L.C., 2001. Comparing the environmental and quality initiatives of Baldridge Award Winners. *Production and Operations Management* **10**(3), 306–326.

Angell, L.C. and Klassen, R.D., 1999. Integrating environmental issues into the mainstream: An agenda for research in operations management. *Journal of Operations Management* **17**(5), 575–598.

Aragón-Correa, J.A. and Sharma, S., 2003. A contingent resource-based view of proactive corporate environmental strategy. *Academy of Management Review* **28**(1), 71–88.

Baker, J., Parasuraman, A., Grewal, D. and Voss, C., 2002. The influence of multiple store environment cues on perceived merchandise value and patronage intentions. *Journal of Marketing* **66**, 120–141.

Baker, J., Grewal, D. and Parasuraman, A., 1994. The influence of store environment on quality inferences and store image. *Journal of the Academy of Marketing Science* **22**(4), 328–339.

Bansal, T. and Clelland, I., 2004. Talking trash: Legitimacy, impression management, and unsystematic risk in the context of the natural environment. **47**(1), 93–103.

Bitner, J.J., 1990. Evaluating Service Encounters: The effects of physical surroundings and employee responses. *Journal of Marketing* **54**(4), 69–82.

Bolton, R.N. and Drew, J.H., 1991. A multistage model of customers' assessment of service quality and value. *Journal of Consumer Research* **17**(4), 375–384.

Boulding, W., Kalra, A., Staelin, R. and Zeithaml, V.A., 1993. A dynamic process model of service quality: From expectations to behavioral intentions. *Journal of Marketing Research* **30**(1), 7–27.

Buzzell, R.D. and Gale, B.T., 1987. The *PIMS Principles.* The Free Press, New York.

Chase, R.B., 1981. The customer contact approach to services: Theoretical bases and practical extensions. *Operations Research* **21**(4), 698–705.

Chase, R.B. and Tansik, D.A., 1983. The customer conduct model for organizational design. *Management Science* **29**(9), 1037–1050.

Christmann, P., 2000. Effects of "Best Practices" of environmental management on cost advantage: The role of complementary assets. *Academy of Management Journal* **43**(4), 663–680.

Collier, C.D., 1991. The service quality process map for credit card processing. *Decision Sciences* **22**, 406–420.

Corbett, C.J. and Pan, J.N., 2002. Evaluating environmental performance using statistical process control techniques. *European Journal of Operational Research* **139**, 68–83.

Cox, D., 1967. The sorting rule model of the consumer product evaluation process In: Cox, D. (Ed.), *Risk Taking and Information Handling in Consumer Behavior*. Harvard University Press, Boston, MA, 324–369.

Cronin, J.J. and Taylor, T.S., 1992. Measuring service quality: A re-examination and extension. *Journal of Marketing* **56**(3), 55–69.

Dick, A.S. and Basu, K., 1994. Customer loyalty: Toward an integrated conceptual framework. *Journal of the Academy of Marketing Science* **22**(2), 99–114.

Dowell, G., Hart, S. and Yeung, B., 2000. Do corporate global environmental standards create or destroy value?" *Management Science* **46**(8), 1059–1074.

Eisenhardt, K.M. and Martin, J.A., 2000. Dynamic capabilities: What are they? *Strategic Management Journal* 21 (special issue), 1105–1121.

Enz, C.A. and Siguaw, J.A., 1999. Best hotel environmental practices. *Cornell Hotel and Restaurant Administration Quarterly* **40**(5), 72–77.

Fitzsimmons, J.A. and Fitzsimmons, M.J., 2008. *Service Management: Operations, Strategy and Information Technology*, 6th ed. Irwin/McGraw-Hill, Boston.

Fornell, C., Johnson, M.D., Anderson, W.E., Cha, J. and Bryant, B.E., 1996. The American consumer satisfaction index: Nature, purpose, and findings. *Journal of Marketing* **60**, 7–18.

Foster, S., Sampson, S. and Dunn, S., 2000. The impact of customer contact on environmental initiatives for service firms. *International Journal of Operations and Production Management* **20**(2), 187–203.

Goodman, A., 2000. Implementing sustainability in service operations in scandic hotels. *Interfaces* **30**(3), 202–214.

Grove, S., Fisk, R., Pickett, G. and Kangun, N., 1996. Going green in the service sector: Social responsibility issues, implications & implementation. *European Journal of Marketing* **30**(5), 56–66.

Hallowell, R., 1996. The relationships of customer satisfaction, customer loyalty, and profitability: An empirical study. *International Journal of Service Industry Management* **7**(4), 27–42.

Halme, M., 2001. Learning for sustainable development in tourism networks. *Business Strategy and the Environment* **10**(2), 100–114.

Hanna, M.D., Newman, W.R. and Johnson, P., 2000. Linking operational and environmental improvement through employee involvement. *International Journal of Operations and Production Management* **20**(2), 148–165.

Hart, S.L., 1995. A natural resource-based view of the firm. *Academy of Management Review* **20**(4), 986–1014.

Heskett, J., Jones, T., Loveman, G., Sasser Jr., W. and Schlesinger, L., 1994. Putting the service-profit chain to work. *Harvard Business Review* **72**(2), 164–175.

Heskett, J., Sasser, W. and Schlesinger, L., 1997. *The Service Profit-Chain*. The Free Press, New York.

Johnson, M.D. and Fornell, C., 1991. A framework for comparing customer satisfaction across individuals and product categories. *Journal of Economic Psychology* **12**(2), 267–286.

Johnson, M.D., Anderson, E.W. and Fornell, C., 1995. Rational and adaptive performance expectations in a customer satisfaction framework. *Journal of Consumer Research* **21**(4), 695–707.

Kamakura, W.A., Mittal, V., de Rosa, F. and Mazzon, J.A., 2002. Assessing the service-profit chain. *Marketing Science* **21**(3), 294–317.

Kassinis, G. and Soteriou, A., 2003. Greening the service profit chain: The impact of environmental management practices. *Production and Operations Management* **12**(3), 386–403.

Kassinis, G. and Soteriou, A., 2009. Quality and environmental practices in service settings: (Dis)entangling the relationship with customer satisfaction. *Working Paper*, University of Cyprus.

Keiningham, T.L., Munn, T.P. and Evans, H., 2003. The impact of customer satisfaction on share-of-wallet in a business-to-business environment. *Journal of Service Research* **6**(1), 37–50.

Kellogg, D.L. and Chase, R.B., 1995. Constructing an empirically derived measure for customer contact. *Management Science* **41**(11), 1734–1749.

King, A.A. and Lenox, M.J., 2001. Lean and green? An empirical examination of the relationship between lean production and environmental performance. *Production and Operations Management* **10**(3), 244–256.

King, A.A. and Lenox, M.J., 2002. Exploring the locus of profitable pollution reduction. *Management Science* **48**(2), 289–299.

King, A.A., Lenox, M.J. and Terlaak, A., 2005. The strategic use of decentralized institutions: Exploring certification with the ISO 14001 management standard. *Academy of Management Journal* **48**(6), 1091–1106.

Klassen, R.D. and McLaughlin, C.P., 1996. The impact of environmental management on firm performance. *Management Science* **42**(8), 1199–1214.

Klassen, R.D. and Whybark, D.C., 1999a. The impact of environmental technologies on manufacturing performance. *Academy of Management Journal* **42**(6), 599–615.

Klassen, R.D. and Whybark, D.C., 1999b. Environmental management in operations: The selection of environmental technologies. *Decision Sciences* **30**(3), 601–631.

Lovelock, C.H., 1996. *Services Marketing.* Prentice Hall, Upper Saddle River, NJ.

Loveman, G.W. 1998. Employee satisfaction, customer loyalty, and financial performance. *Journal of Service Research* **1**(1), 18–31.

May, D.R. and Flannery, B.L., 1995. Cutting waste with employee involvement teams. *Business Horizons* **38**(5), 28–38.

Mittal, V. and Kamakura, W., 2001. Satisfaction, repurchase intent, and repurchase behavior: Investigating the moderating effect of customer characteristics. *Journal of Marketing Research* **38**, 131–142.

Mittal, V., Ross, W. and Baldasare, P., 1994. The asymmetric impact of negative and positive attribute-level performance on overall satisfaction and repurchase intention. *Journal of Marketing* **62**, 33–47.

Oliver, R.L., 1993. Cognitive, affective, and attribute bases of the satisfaction response. *Journal of Consumer Research* **20**(3), 418–430.

Oliver, R., 1997. *Satisfaction: A Behavioral Perspective on the Consumer.* McGraw Hill, New York.

Oliver, R.L. 1999. Whence consumer loyalty? *Journal of Marketing* **63**(4), 33–44.

Olsen, S.O. 2002. Comparative evaluation and the relationship between quality, satisfaction and repurchase loyalty. *Journal of the Academy of Marketing Science* **30**(3), 240–249.

Parasuraman A., Zeithaml, A.V. and Berry, L.L., 1985. A conceptual model of service quality and its implications for future research. *Journal of Marketing* **49**, 41–50.

Parasuraman A., Zeithaml, A.V. and Berry, L.L., 1988. SERVQUAL: A multiple-item scale for measuring perceptions of service quality. *Journal of Retailing* **64**(1), 12–40.

Parasuraman A., Zeithaml, A.V. and Berry, L.L., 1994. Reassessment of expectations as a comparison standard in measuring service quality: Implications for further research. *Journal of Marketing* **58**(1), 111–124.

Pil, F.K. and Rothenberg, S., 2003. Environmental performance as a driver of superior quality. *Production and Operations Management* **12**(3), 404–415.

Rajaram, K. and Corbett, C.J., 2002. Achieving environmental and productivity improvements through model based process redesign. *Operations Research* **50**(5), 751–763.

Rarasuraman, A., Berry, L. and Zeithaml, V., 1991. Refinement and reassessment of the SERVQUAL scale. *Journal of Retailing* **67**, 420–450.

Reichheld, F.F. and Sasser, W.E., 1990. Zero defections: Quality comes to services. *Harvard Business Review* **68**(5), 105–111.

Reichheld, F.F., 1996. *The Loyalty Effect.* Harvard Business School Press, Boston.

Reimer, A. and Kuehn, R. 2005. The impact of servicescape on quality perception. *European Journal of Marketing* **39**(7/8), 785–808.

Roth, A.V. and van der Velde, M. 1991. Operations as marketing: A competitive service strategy. *Journal of Operations Management* **10**(3), 303–329.

Roth, A.V., Chase, R.B. and Voss, C., 1997. *Service in the U.S.* Seven Trend Plc, UK, Government's Department of Trade and Industry, Department of National Heritage, UK.

Russo, M.V. and Fouts, P.A., 1997. A resource-based perspective on corporate environmental performance and profitability. *Academy of Management Journal* **40**(3), 534–559.

Rust, R.T., Moorman, C. and Dickson, P.R., 2002. Getting return on quality: Revenue expansion, cost reduction, or both? *Journal of Marketing* **65**, 7–24.

Rust, R.T. and Zahorik, A.J. and Keiningham, T.L., 1995. Return on quality (ROQ): Making service quality financially accountable. *Journal of Marketing* **59**(2), 58–70.

Rust, R.T. and Oliver, R.L., 1994. Service Quality: Insights and managerial implications from the frontier. In: rust, R.T. and Oliver, R.L. (Eds.), *Service Quality: New Directions in Theory and Practice.* Sage Publications, London, pp. 1–20.

Salzman, J., 2000. Environmental protection beyond the smokestack: Addressing the impact of the service economy. *Corporate Environmental Strategy* **7**(1), 20–37.

Schendler, A., 2001. Trouble in paradise: The rough road to sustainability in Aspen. *Corporate Environmental Strategy* **8**(4), 293–299.

Schneider, B. and Bowen, D.E., 1993. The service organization: Human management is crucial. *Organizational Dynamics* **21**(4), 39–52.

Shrivastava, P., 1995. The role of corporations in achieving ecological sustainability. *Academy of Management Review* **20**(4), 936–960.

Silvestro, R. and Cross, S., 2000. Applying the service profit chain in a retail environment: Challenging the satisfaction mirror. *International Journal of Service Industry Management* **11**(3), 244–268.

Song, M. and Montoya-Weiss, M.M., 2001. The effects of perceived technological uncertainty on Japanese new product development. *Academy of Management Journal* **44**, 61–80.

Soteriou, A.C. and Hadjinicola, G.C., 1999. Resource allocation to improve service quality perceptions in multistage service systems. *Production and Operations Management* **8**(3), 221–239.

Soteriou, A.C. and Chase, R.B., 1998. Linking the customer contact model to service quality. *Journal of Operations Management* **16**(4), 495–508.

Soteriou, A.C. and Zenios, S.A., 1999. Operations, quality and profitability in the provision of banking services. *Management Science* **45**(9), 1221–1238.

Suchman, M.C., 1995. Managing legitimacy: Strategic and institutional approaches. *Academy of Management Review* **20**(3), 571–610.

Szymanski, D.M. and Henard, D.H., 2001. Customer satisfaction: A meta-analysis of the empirical evidence. *Journal of the Academy of Marketing Science* **29**(1), 16–35.

Tellis, G.J., 1988. Advertising exposure, loyalty and brand purchase: A two stage model of choice. *Journal of Marketing Research* **25**(2), 134–44.

Yee, R.W.Y., Yeung, A.C.L., Cheng, T.C.E. and Lai, K., 2009. The service-profit chain: A review and extension. *Total Quality Management* **20**(6), 617–632.

Yi, Y., 1991. A critical review of customer satisfaction. In: Zeithaml, V.A. (Ed.), *Review of Marketing*. American Marketing Association, Chicago.

Zahra, S.A. and George, G., 2002. Absorptive capacity: A review, reconceptualization, and extension. *Academy of Management Review* **27**, 185–203.

Zeithaml, V.A., Parasuraman, A. and Berry, L.L., 1990. Delivering Quality Service: Balancing Customer Perceptions and Expectations. The Free Press, New York.

Zeithaml, V.A., Berry, L.L. and Parasuraman, A., 1996. The behavioral consequences of service quality. *Journal of Marketing* **60**, 31–46.

Chapter 37

TECHNOLOGICAL TRAJECTORIES AND SUSTAINABILITY: AN INTERSECTOR ANALYSIS OF BRAZILIAN SUBSIDIARIES OF SWEDISH MULTINATIONAL CORPORATIONS

Rita Santos,[a,*] Folke Snickars[a,†] and Evando Mirra[b]
[a]*Royal Institute of Technology KTH,*
Dr Kristinas väg 30, Stockholm SE-100 44, Sweden
[*]*rita.santos@abe.kth.se*
[†]*folke.snickars@abe.kth.se*

[b]*Center for Strategic Studies and Management in Science,*
Technology and Innovation (CGEE), SCN Quadra 2,
Ed. Corporate Financial Center, 70712-900 Brasília, Brazil
emirra@cgee.org.br

Evolutionary theories on technological change are particularly useful for understanding technological regime shifts toward the environment and its sustainability. It is argued that fundamental changes in production processes and consumption patterns underpinned by alternative technological trajectories are required for achieving environmental sustainable direction. Such changes, which go beyond the control of particular pollutants and eco-efficiency improvements, are referred to as technological regime shifts. The focuses of this chapter is to review the relationship between technological trajectories to the issue of environmental technological change. This is illustrated by certain selected sectors of Brazilian subsidiaries of Swedish multinational corporations (MNCs). Technological trajectories are expressed through process innovations: (i) a sequence of steps on incremental innovation; (ii) a sequence of steps oriented toward radical innovation, associated with products to redefine the market, exposing structural change. The chapter seeks to enlarge the understanding of the intricate process of sustainability and technological paths of the subsidiaries of MNCs in newly industrialized countries.

1. Introduction

As a derivation of two Greek words, *techne*, ability or technical, and *logos*, knowledge or science, the technology can be defined as the knowledge of abilities or the technical, or as the science of abilities or the technical. Despite this etymology, the technology has a diversity of interpretations that have been conferred in the literature, for the comprehension of its structure and characteristics. Some metaphors are frequently used to define the technology, for instance, to be compared with a "joint of plans, texts, charts, and equations," supposing that the technology is based on specific and prominent documentation. Another example, the technology is sight as know-how and therefore, it is defined like an assembly of express, the practical ability accumulated to perform a task efficiently.

A common sense on the nature of the technology involves the formal education (or scientific activity) and also the "knowledge not-encoded," incorporated in persons and institutions. Ettlie's definition (2000) suggests that technology is composed of physical objects, equipment and machines, and also by a set of (practical and theoretical) knowledge, know how, methods, procedures, and experiences; a concept that deals with a human labor utilization, especially in a particular area or sector.

This reflection is compatible with the contribution of Dosi (1988) on the generation of technologies involving the use of information obtained by the explicit or formal knowledge codified from the scientific nature; and the use of tacit knowledge constituted by the practical capacities that significantly can be shared by collaborators and colleagues that have a common experience (Dosi, 1988).

Technological capacity, what it is after all? In literature, definitions of technological capacity are found as exemplar concepts such as "a systematic creative effort to obtain new knowledge at the level of production" (Katz, 1976); or in terms of Lall (1982, 1987) technological capacity is an "internal technological effort" to dominate new technologies, to be adapted to local conditions. Completing the concept, Westphal *et al.* (1984) define technological capacity as an "aptitude to effectively use technological knowledge." However, an organization in order to accumulate technological capacity over time depends on the learning process, which affects its dynamics by production performance, functioning, and technical interaction.

All of these definitions are associated with the internal efforts of companies to change or adapt technologies. It involves as proposed by Dosi (1982) the technological paradigm concept as a "model" and a "pattern" of solution of selected technological problems, "based upon selected principles derived from natural sciences and on selected material technologies." Dosi's illustration within a technological paradigm is given from the case of semiconductors, where in order to perform a generic task (switching electrical signals) a material technology is chosen (silicon semiconductors), which uses specific scientific properties, in order to reach some economic maximization of performance (Dosi, 1982). In this sense, "technological paradigm" is seen as a means of coordination among technicians and engineers involved in the generation and administration of a particular technology.

As soon as the "technological paradigm" is established, a "technological trajectory" develops, as a path of technological progress within a given paradigm. It is outlined by perceptions "through market relations, government decisions, institutional pressures and social conflicts, through a variety of economic and social interactions" (Dosi, 1982; Nelson and Winter, 1982).

This chapter emphasizes technology trajectories of subsidiaries of MNCs as organizational units that could consist of an influential factor in technology change, acquisition, and development. As the main focus, the chapter reviews the relationship between technological trajectories to the issue of Brazilian subsidiaries of Swedish multinational corporations (MNCs). Technological trajectories are expressed through process innovation to outline levels of the expected environmental improvement by factors of impact reduction: (i) a sequence of steps on incremental innovation; (ii) a sequence of steps oriented toward radical innovations, associated with products to redefine the market, exposing structural change.

Swedish foreign direct investments (FDI) inflows into Brazil have enhanced an effective bilateral industrial relationship between countries. Two are the distinctive contributions coming out of this chapter: first, it enlarges the understanding of the process of sustainability and technological paths of the subsidiaries of MNCs in newly industrialized countries (NICs). Second, it presents a subsidiary-level analysis regarding Swedish foreign operations in Brazil.

2. Technological Trajectories in Subsidiaries of Multinationals that Operate in NICs

At the early stage of their industrialization, NICs have been characterized by premises such as initial technology acquisition from foreign industrialized advanced countries. Companies that operate in NICs are normally dislocated of the main international markets where they act as suppliers, besides they have been engaged in a trial of learning of technological capacity to become close of the international cross-border of research and development (R&D).

Such companies develop production processes through the acquisition of packaged foreign technology, which includes production know-how, technical personnel, and components and parts. As suggested by Dahlman *et al.* (1987), technological trajectory in NICs is compose of steps of acquisition, assimilation, and improvement, which is in the reverse order of industrialized countries. On the other hand, Katz (1976, 1987) and Lall (1987) argued that NICs cannot be regarded as mere recipients of technology from developed countries.

Acquiring, transferring, and assimilation technology is not a simple matter. It involves a complex process of learning, in which several of its elements are "tacit" or embodied deeply in people and organizations capabilities (information, skills, interactions, and routines). In Nelson and Nelson (2002), tacit elements of new technologies need to be acquired, adapted, and developed locally, *involving time*, costs, and risks.

A transfer of a technology occurs through different mechanisms such as FDI, or licenses, contracts, sales of equipment, joint ventures, and research collaboration between private companies and universities or government agencies. The choice of mechanism for companies depends largely on business strategy, risk tolerance, and available *company's* resources.

Multinational corporate alliances through their overseas subsidiaries use FDIs as a key tool to act as major transferees of technology to NICs, usually centralizing their options for countries with local scientific and technological infrastructure, taking advantage of differences in natural, physical, and human capital.

In respect to the growth in the generation of technology outside the home, Cantwell and Iammarino (2000) suggested that host countries are playing a more important role in the R&D activities of MNCs, as R&D expenditures and patents granted to subsidiaries are increasing. However, as emphasized by Doremus *et al.* (1998) while overseas investments in R&D have risen over the past decades, innovative R&D activities still remain concentrated in their home countries; correspondingly, technology trade flows from parent firms to their foreign subsidiaries remain strong.

Yet, one can affirm that R&D is responsible for technology absorption capacity and apprenticeship. The attribution of R&D units in subsidiaries is composed of activities, among others, of selection and employment of new technologies and applications. In fact, investments in R&D are considered as knowledge generating activities, given that knowledge impact is not bounded within the firm that undertakes such investments, but can spread among firms, industries, and countries (Romer, 1990).

The classification of Hobday (1999) is a useful model to be applied in the levels of transferring and assimilation technology in subsidiaries of multinationals that operate in NICs.

As shown in Fig. 1, the complexity of structure of the technological activities of subsidiaries corresponds to different levels, such as (1) refers to changes in the old stock by means of technological support and engineering services; (2) indicates operational tasks including the equipments performance and installation; (3) means adaptation of products and trials in a short term; (4) production based on Engineering and R&D to levels (5–7); activities of level (7) classified as blue sky R&D in response to near terms product and process. Continuous improvements in the levels (1–4) (Hobday, 1999) are essential for corporate market effectiveness.

In NICs, surveys on subsidiaries performance adopt indicators such as a number of patents and R&D expenses to measure their technological activities. Accordingly, they seek to indicate evidences on advanced R&D composing of only the "top of the pyramid" (levels of 5–7). Such surveys have ignored intermediate levels (2–5) of technological capacity — which prevails in subsidiaries acting in NICs — whose cumulative trial is fundamental to reach advanced technological levels (Dosi, 1988; Ariffin and Bell, 1999; Figueired, 2001; 2003).

From the perspective of interaction with the host country, the subsidiary's R&D role is an important variable in this context. Ssubsidiaries can build R&D exclusively focus on adapting foreign products to the host country market. Or can build up R&D

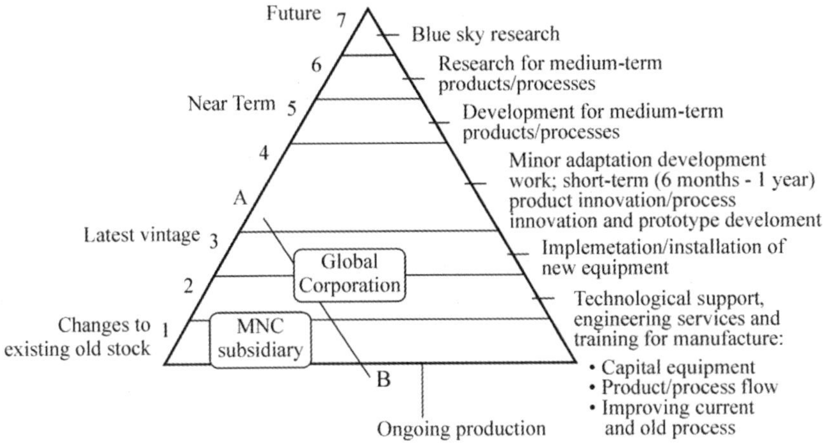

Fig. 1. Technological activities of subsidiaries.
Source: Hobday (1999). A technological profile of a subsidiary is shown in the area A to B, which can progressively move upwards into higher stages of technological activities. Such stages require learning efforts and skills, process and production organization, equipment and new materials.

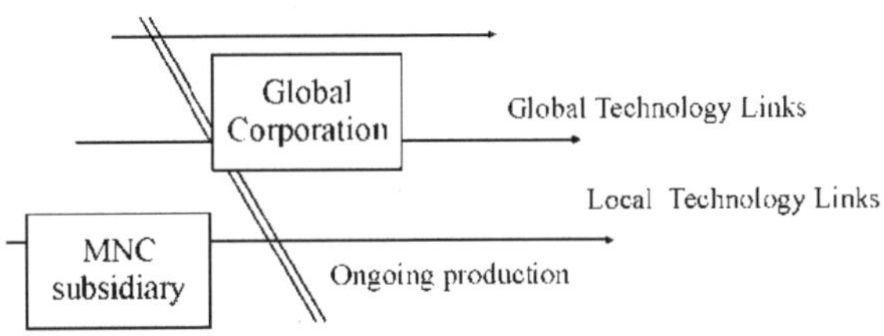

Fig. 2. Technology cooperative links between R&D units of MNC subsidiaries.

units focusing also on the creation of scientific or/and technological links with the host country market.

Figure 2 stems from the concept of Fig. 1 in order to show attributes of R&D roles from an MNC perspective. Local roles encompass activities from adaptation of products to local original production based on Engineering and R&D. Global roles occur when the subsidiary is based on foreign centers of excellence, with internal capacity of knowledge transfer within MNCs, with contribution to the global R&D programs or the development of entire products for the world market.

As seen in Böhe and Zawislak's framework (2004), the cooperative links between R&D units of MNC subsidiaries with host country partner organizations influencing their innovation dynamics. While science links refer to acquisition of scientific knowledge and cooperation with research institutes or universities, technology links refer to acquisition of technological knowledge and cooperation in product development with buyers and suppliers. As noted (Böhe and Zawislak, 2004) subsidiaries able to integrate scientific and technical knowledge under one roof are the one in a position to innovate creating new technologies.

Recently, in NICs technology alliances formed by subsidiaries have been increasing the R&D activities. It involves both science and technology linkages, occurring as a consequence of a managerial action to build up internal R&D capabilities and to establish links with their host country's institutions. Yet, subsidiaries with activities from product adaptation to local original R&D and with focus on the local market (Nobel and Birkinshaw, 1998) tend to be closely connected to host country science and technology system.

3. An Evolutionary View on Sustainability and Technological Trajectories

3.1. *Environmental technological change*

Sustainability is a broad concept, and considering the multitude of its definitions, one can argue that it extends beyond the environment to include dimensions of economic development, society, education, ethics, and technological context. While there is consensus on the multidimensionality of sustainability, the question remains open to a wide variety of interpretations. For instance, by examining trends of decoupling social well-being from the use of biophysical resources, or by a possible gross domestic product (GDP) growth decoupled from material throughput, or simply from the total amount of matter and energy involved at each stage of the economic cycle: extraction, production, use, and disposal.

The capacity to innovate and manage technological change has been shown to be fundamental to achieving economic and social benefits and the internalization of external environmental damage costs. Authors such as Freeman (1995) and Kemp (1997) have been trying to achieve concepts from the evolutionary literature to the conclusion of environmental technological change. It is argued that fundamental changes in production processes and consumption patterns underpinned by alternative technological trajectories are required for achieving environmental sustainable direction. Such changes, which go beyond the control of particular pollutants and eco-efficiency improvements, are referred to as technological regime shifts.

The evolutionary economic literature on technological change makes the important distinction between incremental and radical change in technology systems. Based on Freeman (1995) incremental innovations are relatively minor changes of processes and products that occur more or less continuously; as a continual or gradual improvement of the existing array of products and processes, usually linked to market demands and user needs. They may often occur, not so much as the result of deliberate R&D, but stemming from experiences of engineers in the production process or as a result of initiatives and suggestions by users. Radical innovations, on the other hand, are discontinuous events. They are usually the result of deliberate R&D in enterprises and research activities in university and laboratories. As a decisive factor in structural changes, radical innovations denote a discontinuity in production, distribution, and communication processes (Freeman, 1995).

Most of environmental technological change consists of incremental improvements of existing technologies and the diffusion of technologies integrated in existing production modes. In terms of product improvement, technological change has been focused the distinction between "add-on technologies," often consisting of short-term solutions simply added to existing production systems without modifying them, and "integrated technologies," this may require higher investment and conversion costs concerning substitution, optimization or recycling, which will modify the production system.

Freeman (1996) referred to add-on technologies from the lock-in, which is normally used in pollution control (EOP, end-of-pipe) to transfer the pollutants from one place to another, in contrast to the integrated technologies defined as cleaner processes technologies.

Clean technologies reduce the production of pollutants or inputs of energy and materials by taking into the redesign of all the production process. Furthermore, clean technologies represent a potential strategic resource able to provide firms with unique advantages at each stage of the value chain (Shrivastava, 1995). By contrast, control technologies are EOP additions to production processes. According to Klassen and Whybark (1999), control technologies do not usually reduce the quantity of harmful pollutants, but the risk associated with them. They are operations or equipment added in the final step of an existing production process to capture pollutants and wastes before their discharge, thus leaving the original process and product substantially unchanged. A typical example is that of filters to limit the polluting emissions.

The technologies of pollution control offer advantages to the enterprise, because they gain flexibility in comparison with the clean technologies, which still require adaptation, tacit knowledge, practical capacities, and new institutional arrangements. *Analyzing the factors* on technologies adoption, Murphy and Gouldson (2000) proposed control technologies to be seen as a reactive response to emissions while clean technologies are proactive, because more anticipatory in their nature.

3.2. *Environment policy instruments*

The co-evolution of technology and environment became the focus of attention for policy-makers who face a demand of selection between an existing (dominant) standard lock-in technology and an optimized (changed) clean technology. It has been discussed how alternative types of environmental policy instruments affect rates and directions of technological change (Orr, 1976), or how the effects of environmental protection efforts interfere with the success or failure of new technologies (Kneese and Schultze, 1975). A starting point was provided by Howlett's definition (1991) on environmental policy as the "myriad techniques at the disposal of governments to implement their policy."

As basic types of instruments, regulation is generally concerned as command and control, and economic instruments are market-based. Without financial incentives for businesses to exceed control targets, command-and-control adopts uniform emission standards to force firms to use technology-based by means of controlling pollution, control of equipments, process and products. Market-based policy adopts incentives in a cost-benefit perspective such as pollution taxes, licences, and subsidies to adopt sufficiently low-cost technologies or clean-up process.

Important *questions remain on the effectiveness of environmental policy. First, making* a comparison among taxes, effluents, and technology standards, Magat (1979) analyzed policy instruments as a tool to induce technological change biased toward emissions reduction. Ulph (1998) examined the effects of pollution taxes and control standards on the incentives to undertake R&D. The author observed taxes and regulation as a direct effect of increasing costs, which increases the incentives to invest in R&D to develop cost-saving pollution methods (Ulph, 1998). Katsoulacos and Xepapadea (1996) verified combined effects of R&D subsidy plus emissions tax on the levels of R&D.

A following question refers to the terms of environmental policy as effective inducement for the technological innovation. The Porter hypothesis (Porter, 1996; Porter and van der Linde, 1995a, 1995b) suggested a "win–win" situation — i.e., that all parties could possibly benefit from it — in the sense that environmental policy improves both environment and competitiveness, stimulating innovation, which by enhance productivity and increase firms private's benefits.

In NICs, environmental regulations were enforced much later. From 1980s, after decades of industrialization, such countries are crafting regulatory policies to address environmental problems, relying principally on conventional command-and-control approach. Recently, these countries have adopted economic incentive instruments such as tradable permits, emissions fees, and policies that provide financial incentives for improved environmental performance.

A fact has been the influence of industrialized countries affecting NICs in their choice among alternative environmental policy instruments. In Brazil, for example, where the post-privatization transition period has brought environmental instruments in line with policies of the European Union, the role of technological change became decisive, in the sense that has been a move toward pollution prevention and not solely EOP systems.

4. An Intersector Analysis of Brazilian Subsidiaries of Swedish MNCs

4.1. *Background*

A growing importance of FDI in North–South flows is due to the nature of incentives, enhancing countries attractiveness for oriented investments. Swedish

FDI inflows into Brazil have been attracted mostly by the size of the vast domestic market, and also by favorable Brazilian structural reform from government policies.

Brazil occupies an outstanding position among the NICs mainly for the country's GDP growth rate, ranking among the world's 10 leading economies. The country's GDP is worth 1572 billion dollars or 2.54% of the global economy, according to the World Bank (2010), hosting the fourth most attractive destination for FDI amongst NICs. As outlined in Schneider and Susman (2010), Sweden has one of the most internationally integrated economies in the world and its competitiveness is reflected in large flows of trade and foreign investment. In this respect, Swedish MNCs they have grown markedly, delivering technological breakthroughs, and become world-leading in their fields.

The trading partnership between these countries remains since 1891, when Ericsson supplied the first telephone in Brazil. Currently, more than 200 Swedish MNC companies are locally established with employment over 50,000 people, and their Brazilian operations turn over USD 14 billion per year.

In fact, subsidiaries of Swedish MNCs can play an important role regarding technology diffusion in Brazil as a host country. However, we do not know much on their real impact, which are influenced by some factors such as subsidiary R&D management, costs of local technical and scientific networking, links between companies and research institutes or universities, and time consuming.

Studies in Brazil by Ariffin and Figueiredo (2006) revealed the evolution in capabilities of foreign subsidiaries, despite being limited to some sectors, certain regions of the country and protection and incentive zones. Furthermore, Queiroz *et al.* (2003) observed that MNCs subsidiaries interact more with local universities and research institutes than Brazilian owned companies, and the local market has a strong technological production capacity with high potential to attract R&D. In Sweden, a very high proportion of industrial R&D is performed by a small number of large MNCs (Håkanson and Nobel, 1993).

With the aim to examine Brazilian subsidiaries of Swedish MNCs we chose a sector-based approach from Carlsson *et al.*'s interpretation (2002) that different sectors or industries operate under different and dynamic technological regimes, characterized by particular combinations of opportunities, appropriability conditions, and degrees of cumulativeness of technological knowledge.

A three-stage criteria — value added, energy use (or intensity), technological patterns — was proposed to extend Carlsson *et al.*'s interpretation (2002). The value added is a simple measure of importance to the host country (White and Poynter, 1984) and its scope refers to the range of conditions in which a subsidiary adds value, whether through development, or jobs created including the supply chain, or manufacturing of established products, but also has the capability to develop new products and processes. Sectors were selected on the basis of the classification on the number of employees (OECD/Eurostat, 2005) given direct jobs and jobs generated in the Brazilian chain of production.[1]

Energy intensity refers to a specific scope on the ratio of energy input to industrial output, or the amount of energy used per unit of activity or service, divided into categories of high-energy consumers, high value-added consumers, and low-energy consumers (Manufacturing consumption of Energy, 1991).

Patterns of technology change rely on *Pavitt* (1984) *taxonomy* of industrial *sectors.* In Pavitt's taxonomy industries are clustered along trajectories of technological change according to sources of technology, users, and appropriability regime. It aims to classify innovation modes according to different sectors and knowledge flows between such sectors.

As a sectorial taxonomy, Pavitt's model consists of categories of industrial firms or sectors. *Supplier-dominated* includes traditional manufacturers, representing conventional industries in which R&D intensity is low and innovations are largely developed by suppliers. *Scale-intensive* characterized by mainly large firms producing basic materials and consumer durables. Sources of innovation may be both internal and external to the firm with a medium-level of appropriability. *Specialized-supplier* is composed of specialized firms producing technology to be sold into other firms, e.g., specialized machinery production and high-tech instruments. There is a high level of appropriability due to the tacit nature of the knowledge. *Science-based* characterized by a high degree of R&D intensity and rapid development of underlying sciences. Firms in this sector develop new products or processes and have a high degree of appropriability from patents, secrecy, and tacit know-how.

Falling under value added, energy intensity and different categories of Pavit taxonomy, the following selected sectors[2] are grouped on the basis of the nature of the technological output:

- *Paper and Pulp*: Supplier-dominated classification including firms from mostly traditional manufacturing, which rely on sources of innovation external to the firm.
- *Electrical machinery: Specialized suppliers classification based on capabilities in engineering and frequent interaction with users.*
- *Information and Communication Technologies* (*ICT*): Science-based classification, which rely on R&D from both in-house sources and university research.

Linkages between science-based, specialized suppliers and supplier-dominated industries characterize the core set of industrial sectors that sustained the growth

[1]Number of employees (300–999 Information and Communication Technologies; 1000–4999 Electrical Machinery; 5000 and above Pulp and Paper) composing of direct jobs and jobs generated in the Brazilian chain of production (Report FGV — Fundação Getúlio Vargas, 2007).

[2]The stratification of sectors according to their principal economic activities is based on the International Standard Industrial Classification of all Economic Activities (ISIC Rev. 3) and the Statistical Classification of Economic Activities in the European Community (NACE Rev. 1) respectively, which are appropriate international classifications for this purpose.

of newly industrialized and advanced economies. In a corresponding manner, paper and pulp,[3] electrical machinery,[4] ICT,[5] economically represent privileged sectors in both Brazil and Sweden.

Both countries not only cater to the domestic pulp market but also occupy a prominent position in the international market. Sweden acts as one of the global major machinery equipment countries' sales. Brazil is a potential for electrical machine import. In ICT manufacturing, Sweden has higher than average shares of R&D expenditure in GDP. In Brazil, telecommunications are responsible for the biggest part of the generation of value of the ICT sector, trade, and services activities. Its contribution to the generation of income and employment become the sector as an important economic reference to the country.

4.2. *Implications for technological trajectories and environmental improvements*

It is possible to say that as soon as a "paradigm" is established, the evolution of a given technology is continuous and progressive within a productive unit of a given industrial sector.

However, at some point, this technology may be subject to radical changes causing a disruption in the process of its evolution. From this point of discontinuity, emerges a new set of opportunities or alternatives, some of which can become paths to new technological processes.

Radical changes whether progressive or fundamental do not happen automatically in any technological process; they are results of actions and actors within a given institutional context. Incremental as a continual or gradual improvement of the existing array of products and processes is usually linked to user needs.

This section examines certain sectors of subsidiaries of Swedish MNCs in Brazil. Within a technological framework, Carlsson *et al.*'s interpretation (2002) distinguished at least three levels of analysis composing of technology, products and artifacts, and competence blocs or a set of related firms.

---

[3] *ISIC Rev.3.1 code 21 — Manufacture of paper products*: This division includes the manufacture of pulp, paper, or converted paper products. The manufacturing of these products is grouped together because they constitute a series of vertically connected processes. There are essentially three activities. The manufacturing of pulp involves separating the cellulose fibers from other impurities in wood or used paper. The manufacturing of paper involves matting these fibers into a sheet. Converted paper products are made from paper and other materials by various cutting and shaping techniques, including coating and laminating activities.

[4] *ISIC Rev.3.1 code 31 — Manufacture of electrical machinery and apparatus*: This division includes the manufacture of products that generate, distribute, and store electrical power. Also included is the manufacture of electrical lighting and signaling equipment.

[5] *ISIC Rev.4 ICT — Information and Communication Technologies*: The activities (industries) in the ICT sector can be grouped into ICT manufacturing industries, ICT trade industries, and ICT services industries. For services industries, the products of a candidate industry must be intended to enable the function of information processing and communication by electronic mean (International Standard Classification of activities, ISIC Rev. 3).

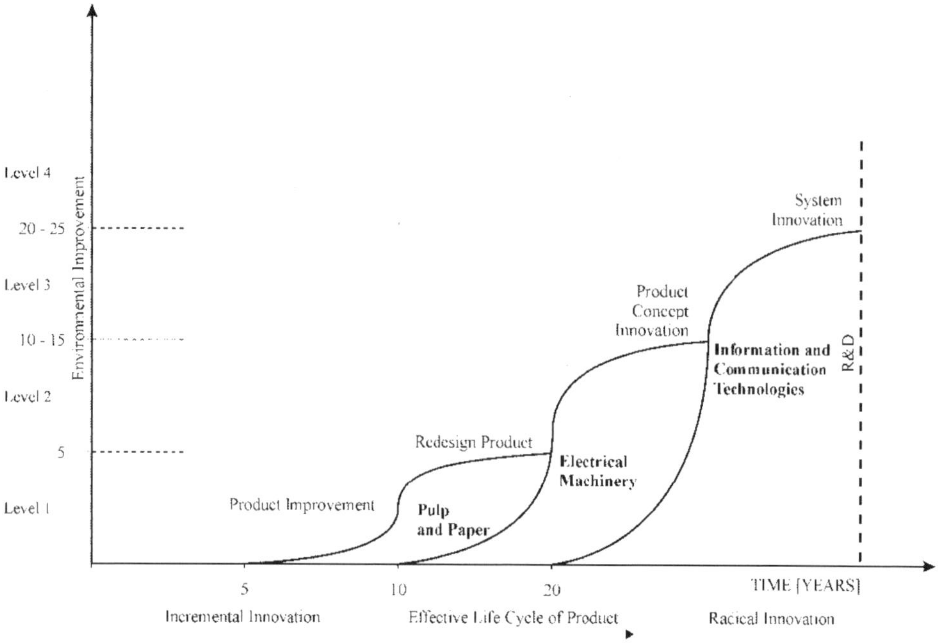

Fig. 3. A sequence of steps on incremental innovation, a sequence of steps oriented toward radical innovations, and environment improvements by sectors or industries.
Source: Adapted from Brezet (1997). Levels 2 and 3 innovations involve a move from product-focused eco-design to transformations to the functions of a product through process innovations — namely the technological trajectory.

We opted for product as a scale of analysis as we are interested in the technological trajectories as a sequence of steps based on incremental and radical innovations, associated with products to the issue of environmental technological change.

Figure 3 is adapted from Brezet's model (1997) as a sequence of s-curves of innovation to outline product, improvement in environment, a sequence of steps on incremental innovation, and a sequence of steps oriented toward radical innovations. It is expressed in factors of impact reduction (vertical axis) and by time scale (horizontal).

In this model, a sequence of steps on incremental innovation, Levels 1 and 2, are environmentally driven, but have different qualities and characteristics: Level 1 represents continual "product improvement" related to resource efficiency with regards to pollution control. In Level 1, prevail add-on technologies (EOP), which toxic substances are treated before their emission to the environment, referred as contamination control.

Level 2 indicates "redesign" by focusing on issues such as minimizing the energy use at several stages in the product life cycle. It is associated with the adoption cleaner technologies, less residuals generation and recycling, improvements in the efficiency by taking into the redesign of all the production process.

Product redesign with entire life-cycle assessment techniques aims to reduce the environmental impact of both products and processes. Level 2 attains factor 5 by further improvements in product environmental performance including all phases of the life cycle. It extends beyond the scope of pollution prevention to include the impact through use and final disposal or recycling of the materials in the product (Owen, 1993).

As a sequence of steps on incremental innovation, Level 3 shows greater environmental improvement involving product concept innovation. It allows the upgrade of product function including considerations of a customer's behavior at an early stage of the design process. Functional innovation, i.e., the function of the product, requires knowledge basis, technical efficiency, and value creation to redirect on how the product function is achieved.

The shift used to characterize "dematerialization" — or the decline over time in weight of the materials used in industrial products — belongs to Level 3. As defined by Hughes and Johnston (2005), dematerialization creates opportunities for decoupling economic growth from the use of limited resources, acting as the key to long-term improvements, by increasing material productivity concerned. Environmental gains in Level 3 refer to the decline in embedded energy in products, also including reuse, renovation, re-manufacturing, and recycling.

Level 4 refers to technological trajectories as a sequence of steps based on "product/systems innovation." It attains positions regarding greater factor × gains through the development of a new product (breakthrough or radical innovation) based on socio-cultural behavior and market needs that are to be fulfilled.

A system of innovation as proposed by Lundvall is constituted of elements, which interact in the production, diffusion, and use of new and economically knowledge in the sphere of societal interactions. As innovation systems stems from a dynamic financial flows, Lundvall (2000) also suggested human flows between universities and firms, as well regulation flows emanating from government agencies toward innovation organizations, and knowledge flows (spillovers) among these institutions.

4.2.1. *Paper and pulp*

As comprised by natural resource-processing firms, the sector is a high-energy consumer converting raw materials into finished goods primarily by chemical (not physical) means. Primary environmental concerns include the use of chlorine-based bleaches and resultant toxic emissions to air, water, and soil. Environmental issues play a key role in driving technology development and related energy use in the pulp and paper industry. It involves the improvement (incremental innovation) of products and the production technique from the perspective of pollution prevention.

In a short time horizon, the environment improvement (Level 1) is considered to be factor 3 in the optimization of existing products. Over a short time horizon, by introducing non-toxic materials, improving distribution, recycling or energy efficiency aims factor 5.

4.2.2. *Electrical machinery*

A high value added energy-consuming sector, which manufacture products that generate, distribute, and use electrical power. It is characterized by dynamic technological regimes and trajectories, closely related to the users and to the scientific area as external sources of knowledge.

The sector attains environmental improvements by product redesign, the product concept remains the same, but parts of the product are improved or replaced, for the purpose of using non-toxic materials, recycling and disassembly, improved distribution, and energy-use reduction with respect to all components over the product life cycle.

Using life-cycle assessments, products and systems that require less material have greater efficiency and consume less energy, which means less harmful emissions — particularly during long operating lifetimes. The environmental gains (Level 2) may reach at a factor 10. In order to achieve factors 15–20 (Level 3), the change of product systems to existing technologies concerns decisions over a medium-term horizon.

4.2.3. *Information and communication technologies*

As presented graphically, the sequence of s-curves shows multi-levels of incremental innovation combined with a sequence of steps oriented toward radical innovations:

- Continuous (evolutionary) innovations in products, resulting in the factor 3 pollution prevention. Product redesign with entire life-cycle assessment techniques, aiming at factor 5.
- In the way in which the product function is changed, ICT provides the desired functions of dematerialization with considerably less materials usage, by decreasing materials' flows in the economy. The environmental performance in fulfilling the desired function is considered to have an improvement potential of factor 10–15.
- Factor 20–25 improvements on environmental impacts to reach sustainability including discontinuous (revolutionary, breakthrough) innovations in entire ICT services. Such factor pertains to long-term decisions, and relates to the impacts both socially and through market expansion. In the science-based pattern, ICT sector has notable sources of R&D activities of firms, based on the rapid development of the underlying sciences in the universities and elsewhere (Pavitt, 1984).

5. Discussion and Conclusions

FDI and technology capacity building in host countries of subsidiaries has been widely discussed in terms of NICs. Regardless, FDI is considered to be an important mechanism for technology and management capacity building, if domestic policies are appropriately tuned (Bell and Pavitt, 1995). There is a need to examine the impact of FDI and its positive side effect on production, including the transfer of environmentally-friendly technology and know-how.

In terms of Brazilian subsidiaries of Swedish MNCs, their dynamic process to the accumulation of technological capacity ("technological catch-up") as an internal technological effort to dominate new technologies seems the way for companies that operate in NICs. It is a process that involves a sequence of accumulation routes, and learning activities composing of tacit knowledge (skills or operational know-how; Agarwal, 2006).

This paper is illustrated by certain selected sectors of subsidiaries of Swedish MNCs hosted in Brazil, with the aim to verify levels of technological accumulation capacity and factors of environment impact reduction by time scale. Such factors vary significantly from sector to sector. As well as knowledge processing for the domain of a technology has a relevant variable between sectors.

If sectors such as paper and pulp or electrical depend of tacit knowledge applied on the production performance, the science-based industries such as ICT require considerable encoded scientific knowledge to foster their absorptive capacity of innovation.

On capability building of pulp and paper sector, the pathways followed by firms in their accumulation of innovation capability involve basically incremental innovation by product improvement. It is based on a departure from the established technological trajectory at an early stage in the development of the firms' capability.

Electrical machinery sector has attained progressively higher levels of *technology accumulation* on firm level. It has more rapidly developed a close cooperation with the specialized suppliers of precision instruments and machinery in order to undertake increasingly complex, but firm-centered innovation efforts based on in-house R&D facilities.

ICT differ in terms of their technological capabilities, especially their innovative trajectories tends to be based on a mixed of continuous improvement of product and process associated with breakthrough (or discontinuous) innovations. The sector differs by a greater ability to create new technological knowledge internally, having innovation processes close to the scientific advances achieved by universities and research institutes. ICT firms play a central position in the knowledge chain, as they receive technological inputs from advanced knowledge providers, while in turn providing technologically advanced products. By increasing the knowledge flows and societal interactions, this sector plays a pivotal role in the dynamic of the innovation system.

Correspondingly, these sectors constitute interfaces between Brazil as the host country system innovation and system innovation of Sweden as the country where MNC subsidiaries are located.

System innovation for sustainability tends to create a new system. Accepting the premise that this study opted for product as scale of analysis for technology trajectories and environment improvements by sectors, a system of innovation shift toward higher sustainability requires, notably

- a multi-stakeholder structure, composing of efficiency across product chain regularly assessed to identify potential energy savings — during production, distribution, and supply, as well as in installation and operation;
- a legal or a merger environmental policy and technology policy to combine entire technological system (product, production chain and institutional structure) with infrastructure entails to achieve breakthrough innovations taking into account environmental aspects.

Moreover, a particular focus on knowledge synthesized within the socio-technical spheres for action transmitted to the turning ideas into the new products related to the sustainability target.

Acknowledgment

The authors would like to thank the organization *Swedish–Brazilian Chamber* of Commerce (Sweden) for the dialogues that have informed much of this work.

References

Agarwal, S.P., 2006. *Foreign Investment in R&D in India.* International Business Summit, FIIE, New Delhi, 31st October–2nd November, 2006.

Ariffin, N. and Bell, M., 1999. Firms, politics and political economy: Patterns of subsidiary-parent linkages and technological capability-building in electronics TNC subsidiaries in Malaysia. In: Jomo, K.S., Rasiah, R. and Felker, G. (orgs.), *Industrial Technology Development in Malaysia.* Rutledge, London.

Ariffin, N. and Figueiredo, P.N., 2006. Globalization of innovative capabilities: Evidence from local and foreign ?rms in the electronics industry in Malaysia and Brazil. *Science, Technology and Society* **11**(11), 191–227.

Bell, M. and Pavitt, K., 1995. The development of technological capabilities. *Trade, Technology and International Competitiveness* 69–10.

Böhe, D.M. and Zawislak, P.A., 2004. R&D roles in subsidiaries of multinational companies: when does the institutional environment matter? Paper presented at the *DRUID Summer Conference Industrial Dynamics, Innovation and Development*, Elsinore, Denmark, June 14–16, 2004. Available at http://www.druid.dk/uploads/txpicturedb/ds 2004-1312.pdf.

Brezet, H., 1997. Dynamics in ecodesign practice. *UNEP Industry and Environment* **20**(1–2), 21–24.

Cantwell, J. and Iammarino, S., 2000. Multinational corporations and the location of technological innovation in the UK regions. *Taylor and Francis Journals* **34**(4), 317–332.

Carlsson, B., Jacobsson, S., Holmén, M. and Rickne, A., 2002. Innovation systems: Analytical and methodological issues. *Research Policy* **31**, 233–245.

Dahlman, C., Ross-Larson, B. and Westphal, L.E., 1987. Managing technological development: Lessons from the newly industrializing countries. *World Development* **15**(6), 759–775.

Doremus, P.N., Keller, W.W., Pauly, L.W. and Reich, S., 1998. *The Myth of the Global Corporation.* Princeton University Press, Princeton.

Dosi, G., 1982. Technological paradigms and technological trajectories: A suggested interpretation of the determinants and directions of technological change. *Research Policy* **11**, 147–162.

Dosi, G., 1988. Sources, procedures, and microeconomic effects of innovation. *Journal of Economic Literature* **26**, 1.120–1.171.

Ettlie, J.E., 2000. *Managing Technological Innovation.* John Wiley and Sons, New York.

Figueiredo, P.N., 2001. *Technological Learning and Competitive Performance.* Edward Elgar Publishing, Cheltenham, UK, Northampton, MA, USA.

Figueiredo, P.N., 2003. Aprendizagem tecnológica e performance competitiva. *Rio de Janeiro: FGV.*

Freeman, C., 1995. The national system of innovation in historical perspective. *Cambridge Journal of Economics* **19**(01), 5–25.

Freeman, C., 1996. The greening of technology and models of innovation. *Technological Forecasting and Social Change* **53**, 27–39.

Håkanson, L. and Nobel., 1993. Foreign research and development in Swedish multinationals. *Research Policy* **22**(5–6), 373–396.

Hobday, 1999. Understanding innovation in electronics in Malaysia. In: Jomo, K.S., Felker, G. and Rasiah, R. (orgs.), *Industrial Technology Development in Malaysia: Industry and Firm Studies*, Routledge, UK.

Howlett, M., 1991. Policy instruments, policy styles and policy implementation. *Policy Studies Journal* **19**(2), 1–21.

Hughes, B.B. and Johnston, P.D., 2005. Sustainable futures: Policies for global development. *Futures* **37**(8), 813–831.

Katsoulacos, Y. and Xepapadea, A., 1996. Environmental innovation, spillovers and optimal policy rules. In: Carraro, C., *et al.* (Eds.), *Environmental Policy and Market Structure.* Kluwer Academic Publishers, Dordrecht, 143–150.

Katz, J., 1976. *Importación de Tecnologia, Aprendizage y Industrialización Dependiente.* Fondo de Cultura Economica, México.

Katz, J., 1987. Domestic technology generation in LDCs: A review of research findings. In: Katz, J. (Ed.), *Technology Generation in Latin American Manufacturing Industries.* St. Martin's Press, New York.

Kemp, R., 1997. *Environmental Policy and Technical Change. A Comparison of the Technologica Impact of Policy Instruments.* Edward Elgar, Cheltenham, UK.

Klassen, R.D. and Whybark, D.C., 1999. The impact of environmental technologies on manufacturing performance. *Academy of Management Journal* **42**(6), 599–615.

Kneese, A. and Schultze, C., 1975. *Pollution, Prices, and Public Policy.* Brookings Institution, Washington, DC.

Lall, S., 1982. Technological learning in the third world: Some implications of technology exports. In: Stewart, F., James, J. (orgs.), *The Economics of New Technology in Developing Countries.* Frances Pinter, Londres.

Lall, S., 1987. *Learning to Industrialise: The Acquisition of Technological Capability by India*, Macmillam, Londres.

Lundvall, B.A., 1992. *National Systems of Innovation: Towards a Theory of Innovation and Interactive Learning.* Pinter Publisher, London.

Lundvall, B.A., 2000. Introduction. In: Edquist, C. and Maureen, M. (Eds), *Systems of Innovation: Growth, Competitiveness and Employment, An Elgar Reference Collection.* Edward Elgar Publishing, UK.

Magat, W.A., 1979. The effects of environmental regulation on innovation. *Law and Contemporary Problems* **43**, 3–25.

Manufacturing Consumption of Energy, 1991. DOE/EIA-0512(91), December 1994.

Murphy, J. and Gouldson, A., 2000. Environmental policy and industrial innovation: Integrating environment and economy through ecological modernisation. *Geoforum* **31**, 33–44.

Nelson, R. and Winter, S., 1982. *An Evolutionary Theory of Economic Change*. Belknap Press, Cambridge.

Nelson, R.R. and Nelson, K., 2002. Technology, institutions and innovation systems. *Research Policy* **31**(2), 265–272.

Nobel, R. and Birkinshaw, J., 1998. Innovation in multinational corporations: Control and communication patterns in international R&D operations. *Strategic Management Journal* **19**, 479–496.

Orr, L., 1976. Incentive for innovation as the basis for effluent charge strategy. *American Economic Review* **66**, 441–447.

OECD/Eurostat, 2005. Oslo Manual — Guidelines for Collecting and Interpreting Innovation Data, third ed. OECD, Paris, France.

Owen, J., 1993. Environmentally conscious manufacturing. *Manufacturing Engineering* **October**, 44–55.

Pavitt, K., 1984. Sectoral patterns of technical change: Towards a taxonomy and a theory. *Research Policy* **13**, 343–373.

Porter, M.E. and van der Linde, C., 1995a. Toward a new conception of the environment competitiveness relationship. *Journal of Economic Perspectives* **9**, 97–118.

Porter, M.E. and van der Linde, 1995b. Green and competitive. *Harvard Business Review* **73**(5), 120–134.

Porter, M., 1996. What is strategy. *Harvard Business Review* **Nov/Dec**.

Queiroz, S., Zanatta, M. and Andrade, C., 2003. Internationalization of MNCs' technological activities: What role for Brazilian subsidiaries? Paper presented at *SPRU Conference in Honour of Keith Pavitt*, Brighton, 13–15 November 2003. Available at ww.sussex.ac.uk/Units/spru/events/KP_Conf_03/documents/Queiroz.pdf.

Romer, P.M., 1990. Endogenous technological change. *Journal of Political Economy* **98**(2), 71–102.

Shrivastava, P., 1995. The role of corporations in achieving ecological sustainability. *Academy of Management Review* **20**(4), 936–960.

Schneider, G. and Susman, P., 2010. Uneven development and grounded comparative institutional advantage: Lessons from Sweden and Mondragon. *Forum for Social Economics, New York, NY* **39**(1), 1–11.

Ulph, D., 1998. Environmental policy and technological innovation. In: Carraro, C. and Siniscalaco, D. (Eds.), *Frontiers of Environmental Economics*. Edward Elgar, Cheltenhsam, UK.

Westphal, L.E., Kim, L. and Dahlman, C.J., 1984. *Reflections of Korea's Acquisition of Technological Capability*. World Bank Research Department, Economics and Research Staff, Washington, DC.

White, R. and Poynter, T., 1984. Strategies for foreign owned subsidiaries in Canada. *Business Quarterly* **48**(4), 59–69.

World Bank, 2010. *World Development Indicators database*. World Bank.

Chapter 38

AN ANP MODELING APPROACH FOR CARBON MANAGEMENT OF SUPPLIER SELECTION IN GREEN SUPPLY CHAINS

Chia-Wei Hsu[a] and Allen H. Hu[b]

[a] *Department of Environmental Management, Tungnan University, 152, Sec. 3, Peishen Rd., Shenkeng, Taipei 22202, Taiwan, R.O.C.*
jcwhsu@mail.tnu.edu.tw

[b] *Institute of Environmental Engineering & Management, National Taipei University of Technology, 1, Sec. 3, Chung-Hsiao E. Rd., Taipei 10643, Taiwan, R.O.C.*
allenhu@ntut.edu.tw

With the increased consciousness of climate changes, the capabilities of carbon management has been incorporated into green supplier selection in green supply chain in order to mitigate risks associated with global warming. This chapter presents an analytic network process (ANP) approach to incorporate the issue of carbon management into supplier selection. In this chapter, identification of criteria of carbon management competence is categorized into three dimensions. An illustrative example in an electronics company is presented to demonstrate how to select the most appropriate supplier in accordance with the requirements of carbon management.

1. Introduction

One of the biggest challenges our planet faces is climate change, which is now recognized as the most serious environmental threat in sustainable development (IPCC, 2007). Now is the time for companies to move to the next-generation carbon emissions management. Therefore, companies must address the carbon impact of their supply chains, as 50% of a product's value and often above 70% is typically derived from suppliers (CDP, 2010). With increased consciousness in climate change, the capabilities of carbon management have been incorporated into green supplier selection in the green supply chain management (GSCM) to mitigate the risks associated with global warming. GSCM is generally understood to involve screening suppliers based on their environmental performance and doing business only with those that meet certain environmental regulations or standards (Rao, 2002). Empirical investigations by Hu and Hsu (2006) and Seuring and Müller (2008) have identified supplier management as the most important dimension in implementing GSCM or sustainable supply chain management. An increasing number of authors have addressed supplier selection issues in green supply chain viewed from environmental perspectives (Noci, 1997; Enarsson, 1998; Walton *et al.*, 1998; Zhu and Geng, 2001; Handfield *et al.*, 2002; Humphreys *et al.*, 2003a, 2003b; Rao, 2005; Wu *et al.*, 2007; Hsu and Hu, 2009; Lee *et al.*, 2009; Bai and Sarkis, 2010). According to the 2010 supply chain report from the Carbon Disclosure Project (CDP), more than half of its members surveyed said that in the future, they would cease doing business with suppliers that do not manage their carbon. More recently, some CDP members have attempted to develop a way to address the impact of green supply chain in electronics companies on climate change. These companies include Acer, Dell, Fujitsu, HP, IBM, and Sony Corporation. Apparently, companies are increasingly requiring their suppliers to manage their greenhouse gas (GHG) emissions as a condition for doing business with them.

In view of the importance of carbon management in green supply chains in addressing the issue of the emerging worldwide climate change, it implies to be an urgent need for developing a green-supply-chain' based operational methodology to systematically evaluate supplier competence with respect to carbon management. As pointed out by the supply chain report of CDP (2010), supplier relationship management processes are now becoming more important than ever, with an emphasis on green criteria for supplier selection. Specifically, suppliers need to perform detailed assessments of their current capabilities in climate change and carbon emission management, and set ambitious targets to reduce their emissions. Dell Corporation, for instance, requires primary suppliers to measure and publicly report their GHG emissions, and also asks them to set improvement goals of their own and set expectations for their suppliers, including a lot of electronics manufacturers from Taiwan. If suppliers fail to meet these requirements with Dell, suppliers can be impacted on the ranking and may be diminished potentially on ability to compete for Dell's business (Dell, 2009). Many of Dell's suppliers are from Taiwan, which is one of the most industrialized countries in the Asia-Pacific region and home to a large number of electrical and electronics manufacturers involved in Original Equipment Manufacturing (OEM) and Original Design Manufacturing (ODM) (Hsu and Hu, 2009). The main risks and pressures OEMs and ODMs face with their suppliers include carbon management with regard to climate change in the green supply chain. Therefore, both types of manufacturers must select suppliers capable of delivering both high-quality products and competent carbon management.

Nevertheless, to the best of our knowledge, green supplier selection specifically considering carbon management competence in the electronics industry has never been found in previous literature. In addition, most of the early literature may be limited to exploring the broad environmental criteria of either quantitative or qualitative property with regard to environmental cost, production process, product, and management system. Some typical supplier selection models are illustrated below.

By incorporating green competence, environmental efficiency, green image, and life cycle cost into the supplier selection, the framework proposed in Noci (1997) designed green vendor rating systems for the assessment of a supplier's environmental performance. As later pointed out in Zhu and Geng (2001), environmental consideration of supplier selection is a key competitive issue for large- and medium-sized enterprises, and thus it should be taken into account to maintain the long-term relationships with their suppliers. Similarly, considering the corresponding evaluation factors of environmental performances, Handfield *et al.* (2002) proposed an environmentally conscious purchasing decision tool to assist managers in understanding the trade-offs between environmental dimensions using the analytic hierarchy process (AHP). Moreover, Humphreys *et al.* (2003b) proposed a knowledge-based system to evaluate the supplier environmental performance, which is recognized into several categories of environmental costs, management competencies, green image, green design, environmental management system, and environmental competencies. Presently, Rao (2005) pointed out that companies are embracing the concept of greening of suppliers in the South East Asian region, aiming to provide an insight into the extent of greening that has been implemented and the underlying reason for Asian companies to adopt increasingly. As further pointed out in Wu *et al.* (2007), environmental principles applicable to green supplier evaluation have been proposed using the AHP and fuzzy logic. Their study considers the complete environmental impact of a product during its entire life cycle. By incorporating the issue of hazardous substances into green supplier selection, Hu and Hsu (2009) utilized the ANP method to construct an evaluation framework of supplier selection in a Taiwanese electronics company, including five main criteria, namely, procurement management, R&D management, process management, incoming quality control, and management system. Recently, Lee *et al.* (2009) have proposed an integrated model that adopts environmental and non-environmental criteria for selecting a green supplier in the high-tech industry. These criteria include quality, technology capability, pollution control, environmental management, green product, and green competencies. Similarly, Bai and Sarkis (2010) integrated a number of sustainability factors into the model of supplier selection that include economic, environmental, and social issues, particularly in the field of social dimension, where employment practices, health and safety, local communities influence, contractual stakeholders influence, and other stakeholders influence are covered. However, these previous studies may be still limited to either the broad environmental criteria or integrating criteria, as they do not consider the carbon management issue in the operation of the corresponding green supplier selection.

Supplier selection is a multi-criteria decision problem (Akarte *et al.*, 2001; Liaoa and Rittscherb, 2007; Huang and Keskar, 2007), for which the related literature has proposed several supplier selection methodologies. Some familiar examples of systematic analysis for supplier selection include the AHP (Akarte *et al.*, 2001; Tam and Tummala, 2001; Chan, 2003), fuzzy AHP (FAHP) (Wu *et al.*, 2007; Kahraman *et al.*, 2003; Zaim *et al.*, 2003; Chan and Kumar, 2007), ANP (Bayazit, 2006; Gencer and Gürpinar, 2007; Hsu and Hu, 2009), case-based reasoning systems (Choy *et al.*, 2003; Humphreys *et al.*, 2003b), and multiple objective programming (Ghodsypour and Obrien, 2001; Feng *et al.*, 2001; Zhu, 2004). The application of the mathematical programming model to supplier selection may have problems including qualitative criteria, particularly for supplier partnership policies (Ghodsypour and O'Brien, 1998). Furthermore, the computational complexities inherent in multiple objective programming frequently prohibit consideration of many crucial attributes for supplier selection (Chan and Kumar, 2007). Either the weighting model of the AHP or the ANP can be used since it is more useful for treating qualitative factors than other models such as mathematical programming models (Lee *et al.*, 2001). More recently, various studies (Bayazit, 2006; Gencer and Gürpinar, 2007; Hsu and Hu, 2009) utilized ANP in supplier selection to provide good insights in terms of feedback systematic and interdependencies property. ANP can capture the interdependencies between the criteria under consideration, thus allowing for more systemic analysis (Ravi *et al.*, 2005). Additionally, ANP can be used as a decision analysis tool to solve multi-criteria supplier selection problems that contain interdependencies (Bayazit, 2006) and provide systematic feedback (Gencer and Gürpinar, 2007). ANP modeling thus better fits the problem examined in this study and offers the advantage of providing a systematic approach to supplier selection in terms of carbon management.

2. Criteria to Carbon Management

Based on integrating the categories and criteria identified from the literature, a selection framework was designed for incorporating environmental criteria for the competency of carbon management into supplier selection in GSCM. A total of 13 criteria were determined and categorized into three main clusters as follows (see Table 1).

2.1. *Planning dimension (D_1)*

2.1.1. *Carbon governance (C_{11})*

Companies that integrate climate change into their board and executive structures, as well as their public reporting mechanisms, are far more likely to maintain the long-term commitment and comprehensive approaches needed to address climate change risks and opportunities effectively across their entire business structure (Cogan *et al.*, 2008). Over 90% of the CDP members have tasked either a board committee or another executive body with the overall responsibility for climate change management to ensure that the strategy is turned into action effectively (CDP, 2010).

2.1.2. *Carbon policy (C_{12})*

Aside from disclosing their own climate change strategies, companies also need to be transparent about their views on climate change regulatory actions, policies, and regulations they support (Cogan *et al.*, 2008). The CDP (2010) pointed out that its

Table 1. Criteria for supplier selection in carbon management.

Dimensions	Criteria	Reference
Planning (D_1)	Carbon governance (C_{11})	Cogan *et al.*, 2008; CDP, 2010
	Carbon policy (C_{12})	Cogan *et al.*, 2008; CDP, 2010
	Carbon reduction targets (C_{13})	Weinhofer and Hoffmann, 2008; Cogan *et al.*, 2008
	Carbon risk assessment (C_{14})	Skjaerseth and Skodvin, 2003; Okereke, 2007; CDP, 2010
	Training-related carbon management (C_{15})	Cogan *et al.*, 2008
	Life-cycle cost management (C_{16})	Kolk and Pinkse, 2004
Implementation (D_2)	Measures of carbon management (C_{21})	Okereke, 2007; Kolk and Pinkse, 2004; Schultz and Williamson, 2005; Boiral, 2006; Weinhofer and Hoffmann, 2008
	Involvement in initiatives for carbon management (C_{22})	CDP, 2010; Cogan *et al.*, 2008
	Management systems of carbon information (C_{23})	Sullivan, 2009
	Supplier collaboration (C_{24})	Cogan *et al.*, 2008; Kolk and Pinkse, 2004
Management (D_3)	Carbon accounting and inventory (C_{31})	Kolk and Pinkse, 2004; Cogan *et al.*, 2008
	Carbon verification (C_{32})	Cogan *et al.*, 2008
	Carbon disclosure and report (C_{33})	Southworth, 2009; Barnard, 2006

members have integrated carbon policies into their procurement departments and that a large majority of these companies (90%) have a carbon emissions reduction plan in place. Thus, the company can facilitate carbon management practices by establishing a carbon policy as a manifestation of its position regarding carbon emissions disclosure, carbon reduction targets, and carbon emissions certification, among others.

2.1.3. *Carbon reduction targets (C_{13})*

With respect to the mitigation of climate change, Weinhofer and Hoffmann (2008) revealed that GHG reduction targets reflect a long-term need to decrease emissions. Setting targets to reduce GHG emissions has become the norm for corporate climate change strategies that include quantitative emission reduction targets for their Scopes 1 and 2, and occasionally even Scope 3 GHG emissions (Cogan *et al.*, 2008). A company must set its carbon reduction target sufficiently high to enable real and measurable progress in addressing climate change.

2.1.4. *Carbon risk assessment (C_{14})*

Perception of risk from physical damages associated with climate change is also an important factor influencing corporate climate strategy (Skjaerseth and Skodvin,

2003). Some companies have adopted climate change activities due to fear of risks from the direct impact of climate change (Okereke, 2007). Companies also recognize that current and upcoming regulations aimed at reducing carbon emissions will put pressure on their suppliers and ultimately on them. Increasing their knowledge in carbon emissions within their supply bases (chains) can help these companies to judge the risks presented by suppliers better and to collaborate more efficiently with suppliers to mitigate and turn these risks into opportunities (CDP, 2010).

2.1.5. *Training related to carbon management (C_{15})*

Companies initially encounter challenges when implementing carbon management initiatives because such practices are generally very complex, and most employees are unaware of carbon management. To promote awareness, IBM recently held an internal innovation that matters video contest wherein employees were encouraged to submitted videos related to energy and environment practices with clients, at work, and in their homes (Cogan *et al.*, 2008). In view of the importance of consciousness of carbon management for employees, education and training associated with carbon issues need to be launched to promote environmental consciousness. For example, regulations related to the emerging standards, PAS 2050 or ISO 14064, are used to collect and report carbon data from either product or organizational scope.

2.1.6. *Life-cycle cost management (C_{16})*

In the empirical investigation conducted by Kolk and Pinkse (2004), respondents emphasized the business opportunities related to climate change rather than the risks. Cost increases due to rising energy prices pose a risk for energy-intensive companies. Meanwhile, emissions' reductions are also associated with cost reductions because of less energy use and are then seen as potential opportunities (Kolk and Pinkse, 2004). While incorporating life-cycle cost management into GHG emissions mitigation, companies can obtain insightful analysis in carbon management in terms of the cost-effective perspective.

2.2. *Implementation dimension (D_2)*

2.2.1. *Measures of mitigation of carbon emissions (C_{21})*

To mitigate carbon emissions, many companies no longer concentrate solely on influencing policy debates. Instead they have begun to pursue various firm-specific practical actions against climate change within the framework of a corporate climate strategy (Okereke, 2007). Companies can take internal and external measures with regard to their carbon dioxide emissions (Kolk and Pinkse, 2004; Schultz and Williamson, 2005; Boiral, 2006). Internal measures are usually defined as activities within the business operations of the company, whereas external measures represent emission compensation measures (Weinhofer and Hoffmann, 2008).

2.2.2. *Involvement of initiatives for carbon management (C_{22})*

As a well-known initiative for the mitigation for GHG emissions, CDP works with key stakeholders to encourage companies all over the world to measure, manage, disclose, and ultimately reduce their GHG emissions and related climate change risks (CDP, 2010). By incorporating the low-carbon principle into product design, electronics companies have organized specialized product design teams to work on environmental initiatives and collaborate with peers and governments in developing product standards. Members of the non-profit Standard Performance Evaluation Corporation (SPEC), including Dell, Intel, Hewlett-Packard, IBM, and Sun Microsystems, have helped established a benchmark for server-class computers known as SPECPower, which was released in December 2007 (Cogan *et al.*, 2008).

2.2.3. *Management systems of carbon information (C_{23})*

Based on the survey results of the management of GHG emissions in large European companies, Sullivan (2009) argued that the majority of large companies have established the basic management systems and processes necessary to manage their emissions and related business risks effectively. In order to disclose and manage GHG data, suppliers need to establish a management system as a platform to collect carbon emissions data from their organizational or product scope. This is the first step in managing carbon emissions.

2.2.4. *Supplier collaboration (C_{24})*

Based on the 2008 survey on corporate governance and climate change in consumer and technology companies, Cogan *et al.* (2008) pointed out that Sun Microsystems launched the Sun Eco Advantage Program that provides partner companies with tools and training to build their eco-IT practices. These tools include datacenter efficiency training, technical assessment services, assistance in modeling investment returns and carbon savings scenarios, and implementation methodologies. Furthermore, climate change is not a signal issue that can be tackled by one company or even one sector. BP is facing up to the impact of product emissions through both short- and long-term partnerships aimed at developing innovative, cost-effective technology (Kolk and Pinkse, 2004).

2.3. *Management dimension (D_3)*

2.3.1. *Carbon accounting and inventory (C_{31})*

Carbon accounting and inventory is an essential step in developing strategies and evaluating progress for controlling GHG emissions in the operations of a company, product, and supply chain because, to start with, companies need to know their current situation (Kolk and Pinkse, 2004). Cogan *et al.* (2008) found that more than 60% of the companies evaluated conducted a GHG emissions inventory. Recently, most companies have applied various standards, such as ISO 14064-Parts I and II and PAS 2050, to inventory and to account GHG emissions.

2.3.2. *Carbon verification (C_{32})*

External verification of inventories is becoming increasingly important. The ISO has launched a new standard, ISO 14065:2007, for use in accrediting or recognizing verified GHG. The International Auditing and Assurance Standards Board is also working on a standard for assurance of carbon emissions information (Cogan *et al.*, 2008). Sullivan (2009) argued that most companies have had their corporate social responsibility reports verified, whereas other companies use a separate verification process for their emissions data.

2.3.3. *Carbon disclosure and report (C_{33})*

Industry reporting of GHG emissions is an important first step to a successful climate change strategy. Many multinational corporations have chosen to report voluntarily to various emission accounting projects, such as the Chicago Climate Exchange, CDP, Global Reporting Initiative, Climate Change Registry, and the US DOE 1605(b) Voluntary Reporting Database (Southworth, 2009). These projects may lend credibility to corporate voluntary reporting of emissions by providing open access to industry estimates and add a degree of regularity to the reporting process (Barnard, 2006).

3. The Analytic Network Process

ANP is the general form of the AHP, which has been used in multi-criteria decision making (MCDM) to release restrictions associated with hierarchical structure (Huang *et al.*, 2005). AHP can integrate qualitative information and quantitative values (Meade and Sarkis, 1998) and handle MCDM problems (Saaty, 1980). Nevertheless, AHP suffers the disadvantages of not sufficiently considering interdependencies (Chung *et al.*, 2005) and not allowing for integrated dynamic modeling of the environment (Meade and Sarkis, 1998). As a result, Saaty (1996) introduced a super-matrix approach for dealing with the interdependencies among clusters. This approach is what today is referred to as the ANP method. The advantages of ANP include the ability to incorporate dependencies and feedback using a hierarchical decision network, and representing and analyzing interactions, and synthesizing their mutual effects by a single logical procedure (Sarkis and Sundarraj, 2002). ANP is currently widely applied in different decision-making processes, such as supplier selection (Bayazit, 2006; Gencer and Gürpinar, 2007; Hsu and Hu, 2009), alternative fuels for residential heating (Erdogmus *et al.*, 2006), selection of logistics service providers (Jharkharia and Shankar, 2007), multidimensional data in multidimensional scaling (Huang *et al.*, 2005), strategic analysis of logistics and supply chain management (Meade and Sarkis, 1998), enterprise partner selection for vocational education (Chen *et al.*, 2004), financial crisis forecasting (Niemira and Saaty, 2004), logistics strategy selection (Meade and Sarkis, 1998), and product mix planning in semiconductor fabricator (Chung *et al.*, 2005). The ANP methodology and its application to case companies in an MCDM environment are illustrated in the following section.

3.1. *An illustrative example for a electronic manufacturer*

GSCM is a significant issue in the Taiwanese electronics industry, as recent studies have shown that most of the world's manufacturing will be relocated to Asia within the next two decades (US-AEP, 1999). In adherence to the awareness of climate change in the green supply chain, the consideration of supplier selection has been changed from broad environmental issues into the specific issue of carbon management. Taiwan is one of the most industrialized countries in the Asia-Pacific region, with numerous electrical and electronics manufacturers involved in OEM and ODM. Notably, the main risks and pressures OEMs receive from their suppliers include carbon management in GSCM practice. Therefore, OEMs must select suppliers capable of both delivering high-quality components and competent carbon management.

The case company is a leading provider of innovative products for both global and domestic markets, including imaging products, enclosures, power supplies, and light-emitting diode (LEDs). Its primary objective is to be number one in Taiwan and within the top three in the global market for all its product lines. The case company began in the Taiwanese LED industry, where it has ranked number one for the past 30 years. It is also the world's largest notebook adapter manufacturer, holding a 60% global market share. The company's other major products are projector power supplies, phone camera modules, keyboards, and enclosures, all of which have large global market shares.

The case company in this study is interested in incorporating carbon management into supplier evaluation and selection for GSCM practice because it suffers pressure from buyers and it has become a CDP member in carbon management in the green supply chain. As regards to the increasing environmental regulations on climate change initiatives, the case company would like to implement a systematic method of selecting appropriate suppliers based on competency of carbon management.

The proposed decision model for supplier selection in green supply chain was implemented for 13 criteria under three main criteria clusters. The relative importance of criteria for supplier selection is determined based on expert opinion as determined by sampling the company's GSCM team. The expert group consists of three people from the case company who are responsible for the planning and management of green initiatives, evaluating the suppliers' performance and maintaining the list of the approved suppliers with respect to the capability of carbon management. The case experience provided assistance in understanding how to establish

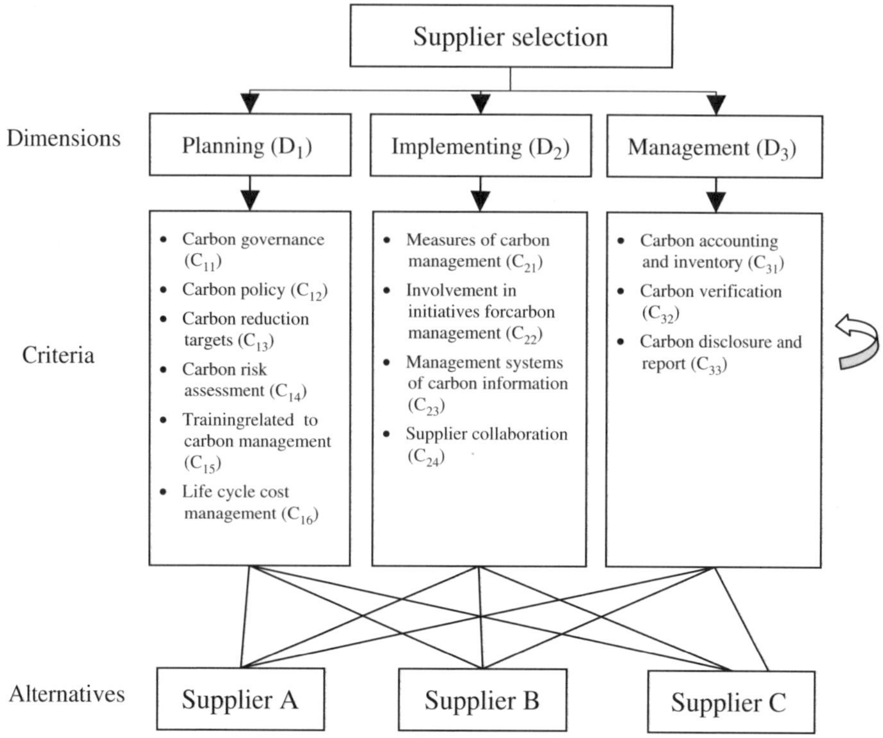

Fig. 1. ANP-based model for selecting supplier to carbon management.

the decision model for supplier selection and selecting appropriate suppliers in green supply chain. The application and analysis of ANP methodology is presented in the following steps.

3.1.1. *Step 1: Model construction and problem formation*

The first step in ANP method implementation is to construct the decision structure of the supplier selection problem and identifying the relevant criteria and alternatives developed based on the literature. This model has four levels (see Fig. 1). The second level consists of three main criteria clusters or dimensions; they are planning (D_1), implementing (D_2), and management (D_3). There are 13 criteria under the above-mentioned three dimensions. The fourth level is the alternatives they are supplier A, supplier B, and supplier C in the illustrated case.

3.1.2. *Step 2: Pair-wise comparison matrix of dimensions*

During this step, the decision maker is asked to respond to the relative weighting of each dimension via a pair-wise comparison matrix. A scale of 1–9 is used to compare the two components in this comparison. A score of 1 indicates that the two components have equal importance whereas a score of 9 indicates the overwhelming dominance of the considered component (row component) over the comparison component (column component). If the impact of one component is weaker than that of its comparison component, it will be scored from 1 to 1/9, with 1 indicating indifference and 1/9 indicating overwhelming dominance of the column component over the row component. To make a reverse comparison between already compared components, a reciprocal value is automatically assigned within the matrix, so that in a matrix $a_{ij}a_{ji} = 1$. This is complied with the method recommended by Saaty (1980). In the formation of a pair-wise comparison matrix, group decision making may be applied to avoid decision-maker bias toward particular providers. The authors thus adopted the geometric mean of the individual judgments based on the recommendations of Dyer and Forman (1992).

After the pair-wise comparison matrices are developed, a vector of priorities (i.e., vector or eigenvector) in each matrix is calculated and is then normalized to sum to 1.0 or 100%. This study utilized a two-stage algorithm to calculate the e-vector, which first adds the value in each column of the matrix and then separates each entry in each column by the total of that column; the normalized matrix is acquired through meaningful comparison among components. Subsequently, row sums are calculated and normalized to get the e-vectors of these relative importance weights as shown in Table 2. And these e-vectors are used as Pj in Table 8 to calculate the desirability indices for alternatives. In this case, the "planning dimension (D_1)" criterion was found to have the highest priority (0.5502) in selecting supplier, followed by "management dimension (D_3)" (0.2337), and "Implementing dimension (D_2)" (0.2161).

3.1.3. *Step 3: Pair-wise comparison matrices between dimensions/criteria*

In this step, the decision maker is asked to respond to a number of pair-wise comparisons in which two components are compared at a time in terms of an upper lever control criteria. The pair-wise comparisons of the components at each level are performed regarding their relative influence toward the control criterion. Pair-wise comparison is thus performed between the applicable criteria within a given dimension cluster. In this model, three such pair-wise comparison matrices are formed and the e-vectors acquired from these matrices are imported from Table 8 asA^D_{kja}. One such pair-wise comparison matrix for planning dimension is listed in Table 3. For the pair-wise matrix, the question asked to the decision maker is "what is the relative impact on planning dimension by criterion a when compared with criterion b."

Table 2. Pair-wise comparison of dimensions.

Dimension	D_1	D_2	D_3	e-Vector
D_1	1.0000	1.8171	3.6342	0.5502
D_2	0.5503	1.0000	0.6300	0.2161
D_3	0.2752	1.5874	1.0000	0.2337

Table 3. Pair-wise comparison for planning dimension.

	C_{11}	C_{12}	C_{13}	C_{14}	C_{15}	C_{16}	e-Vector
C_{11}	1.000	1.738	3.037	5.192	2.410	3.271	0.3532
C_{12}	0.575	1.000	2.466	0.843	3.000	0.523	0.1682
C_{13}	0.329	0.405	1.000	0.523	2.714	1.419	0.1225
C_{14}	0.193	1.186	1.913	1.000	1.710	1.040	0.1378
C_{15}	0.415	0.333	0.368	0.585	1.000	0.485	0.0735
C_{16}	0.306	1.913	0.705	0.961	2.061	1.000	0.1447

Table 4. Pair-wise comparison for carbon governance under planning dimension.

	C_{12}	C_{13}	C_{14}	C_{15}	C_{16}	e-Vector
C_{12}	1.0000	3.0366	1.6134	4.8203	2.2894	0.3758
C_{13}	0.3293	1.0000	0.3057	1.0000	1.1968	0.1122
C_{14}	0.6198	3.2711	1.0000	1.9129	2.9240	0.2817
C_{15}	0.2075	1.0000	0.5228	1.0000	0.4750	0.0960
C_{16}	0.4368	0.8355	0.3420	2.1055	1.0000	0.1343

In Table 3, the relative importance of "carbon governance (C_{11})" when compared with "carbon policy (C_{12})" with respect to planning dimension is five. Table 3 also reveals that the criterion "carbon governance (C_{11})" has the maximum influence (0.3532) and "training related to carbon management (C_{12})" has the minimum influence (0.0735) on the planning dimension. It implies that "carbon governance (C_{11})" has the most influence on the dimension of planning, with a priority in selecting best supplier, which followed by "carbon policy (D_{12})" (0.1682) and "life-cycle cost management (C_{16})."

3.1.4. *Step 4: Pair-wise comparison matrices for criteria*

This step involves performing pair-wise comparisons to consider the interdependencies among the criteria (third level). There are 13 such pair-wise comparison matrices in this stage. Table 4 lists one such comparison that represents the result of the planning dimension with "carbon governance" as the control criterion over other criteria. The decision maker is asked to answer the question for evaluating the interdependencies related to "when considering 'carbon governance' with respect to increasing planning, what is the relative impact of criterion a compared with criterion b." Table 4 reveals that "carbon policy (C_{12})" has the strongest impact (0.3758) on the planning dimension, with "carbon governance (C_{11})" as the control criterion over others. Furthermore, "training related to carbon management (C_{15})" has the weakest impact (0.0960). It shows that carbon policy has the most influence on the carbon governance for the dimension of planning. The e-vector from these matrices is thus used to form super matrices. The e-vector from Table 4 has been used in the 13 column of the super matrix in Table 6.

Table 5. Pair-wise comparison of importance of carbon governance on the alternatives.

	Supplier A	Supplier B	Supplier C	e-Vector
Supplier A	1.0000	1.4142	7.4833	0.5272
Supplier B	0.7071	1.0000	6.9282	0.4082
Supplier C	0.1336	0.1443	1.0000	0.0646

3.1.5. *Step 5: Evaluation of alternatives*

The final set of pair-wise comparisons is made for the relative impact of each of the alternatives on the criteria influencing the dimensions. The number of such pair-wise comparison matrices is determined based on the number of criteria included in each dimension. The present case involves three alternatives (supplier A, B, and C) and 13 criteria; hence, 13 such pair-wise comparison matrices were constructed in this step.

Table 5 lists one example of such a pair-wise comparison matrix, in which the impacts of alternatives are evaluated based on the criterion "carbon governance (C_{11})." In comparing the three suppliers, we asked which supplier is more preferable with respect to determining the best alternative under the carbon governance criterion. Supplier A appears superior to the other two alternatives. The e-vector from Table 5 reveals the relative importance of the criteria on the alternatives, which will be used in the second row of columns 6, 7, and 8 of the desirability indices in Table 8.

3.1.6. *Step 6: Super-matrix formation*

The super matrix permits a resolution of the interdependencies that exist among the components of a system. It is a partitioned matrix where each sub-matrix is composed of a set of relationships between and within the levels, as represented by the decision-maker model. The super matrix M, as shown in Table 6, presents the results of the relative importance measure for each of the criteria for each dimension of the best supplier. The components of super matrix have been imported from the pair-wise comparison matrices of interdependencies (Table 4). As there are 13 such pair-wise comparisons matrices, one for each interdependent criterion, the super matrix contains 13 non-zero columns. Each of the non-zero values in a column expresses the relative importance weight associated with the interdependent pair-wise comparison matrices.

In the next step, the super matrix M is made to converge to acquire a long-term stable set of weights, as listed in Table 7. For convergence to occur, the super matrix needs to be column stochastic, which means the sum of each column of the super matrix must be one. Raising the super matrix M to the power 2^{k+1}, where k is an arbitrarily large number, achieves the convergence of the interdependent relationship.

Table 6. Super-matrix M before convergence.

		D_1						D_2				D_3		
		C_{11}	C_{12}	C_{13}	C_{14}	C_{15}	C_{16}	C_{21}	C_{22}	C_{23}	C_{24}	C_{31}	C_{32}	C_{33}
D_1	C_{11}	0.0000	0.3133	0.3061	0.2100	0.2702	0.2490							
	C_{12}	0.3758	0.0000	0.2301	0.1710	0.1141	0.1682							
	C_{13}	0.1122	0.3024	0.0000	0.3358	0.0952	0.2079							
	C_{14}	0.2817	0.2241	0.2318	0.0000	0.3525	0.3005							
	C_{15}	0.0960	0.0558	0.0614	0.0623	0.0000	0.0744							
	C_{16}	0.1343	0.1044	0.1706	0.2209	0.1680	0.0000							
D_2	C_{21}							0.0000	0.3610	0.3649	0.4352			
	C_{22}							0.1094	0.0000	0.1738	0.2482			
	C_{23}							0.3551	0.1565	0.0000	0.3166			
	C_{24}							0.5355	0.4825	0.4613	0.0000			
D_3	C_{31}											0.0000	0.5866	0.4609
	C_{32}											0.3548	0.0000	0.5391
	C_{33}											0.6452	0.4134	0.0000

Table 7. Super matrix M after convergence 14.

		D_1						D_2				D_3		
		C_{11}	C_{12}	C_{13}	C_{14}	C_{15}	C_{16}	C_{21}	C_{22}	C_{23}	C_{24}	C_{31}	C_{32}	C_{33}
D_1	C_{11}	0.2122	0.2122	0.2122	0.2122	0.2122	0.2122							
	C_{12}	0.1890	0.1890	0.1890	0.1890	0.1890	0.1890							
	C_{13}	0.1862	0.1862	0.1862	0.1862	0.1862	0.1862							
	C_{14}	0.2097	0.2097	0.2097	0.2097	0.2097	0.2097							
	C_{15}	0.0656	0.0656	0.0656	0.0656	0.0656	0.0656							
	C_{16}	0.1373	0.1373	0.1373	0.1373	0.1373	0.1373							
D_2	C_{21}							0.2840	0.2840	0.2840	0.2840			
	C_{22}							0.1535	0.1535	0.1535	0.1535			
	C_{23}							0.2301	0.2301	0.2301	0.2301			
	C_{24}							0.3323	0.3323	0.3323	0.3323			
D_3	C_{31}											0.3421	0.3421	0.3421
	C_{32}											0.3093	0.3093	0.3093
	C_{33}											0.3486	0.3486	0.3486

Table 8. Desirability indices.

Dimensions	P_j	Criteria	A^D_{kj}	A^I_{kj}	S_{1kj}	S_{2kj}	S_{3kj}	Supplier A	Supplier B	Supplier C
D_1	0.5502	C_{11}	0.3532	0.2122	0.5272	0.4082	0.0646	0.0217	0.0168	0.0027
	0.5502	C_{12}	0.1682	0.1890	0.5662	0.3656	0.0682	0.0099	0.0064	0.0012
	0.5502	C_{13}	0.1225	0.1862	0.6358	0.2936	0.0707	0.0080	0.0037	0.0009
	0.5502	C_{14}	0.1378	0.2097	0.3772	0.5539	0.0689	0.0060	0.0088	0.0011
	0.5502	C_{15}	0.0735	0.0656	0.4606	0.4438	0.0957	0.0012	0.0012	0.0003
	0.5502	C_{16}	0.1447	0.1373	0.6241	0.2830	0.0929	0.0068	0.0031	0.0010
D_2	0.2161	D_{21}	0.3312	0.2840	0.6532	0.2846	0.6532	0.0133	0.0058	0.0133
	0.2161	D_{22}	0.0714	0.1535	0.4568	0.4568	0.0863	0.0011	0.0011	0.0002
	0.2161	D_{23}	0.1948	0.2301	0.5824	0.3143	0.1034	0.0056	0.0030	0.0010
	0.2161	D_{24}	0.4026	0.3323	0.4687	0.4687	0.0626	0.0136	0.0136	0.0018
D_3	0.2337	D_{31}	0.2382	0.3421	0.6128	0.3150	0.0722	0.0117	0.0060	0.0014
	0.2337	D_{32}	0.3332	0.3093	0.4865	0.4058	0.1077	0.0117	0.0098	0.0026
	0.2337	D_{33}	0.4286	0.3486	0.4706	0.4706	0.0588	0.0164	0.0164	0.0021
Desirability indices D_i								0.1270	0.0957	0.0296
Normalized desirability indices D_{iN}								0.5034	0.3793	0.1173

3.1.7. *Step 7: Selection of the best alternative*

The selection of the best supplier is determined based on the value of the normalized desirability indices. The desirability indices D_i for alternative i is defined as

$$D_i = \sum_{j=1}^{J} P_j \; A_{kj}^D \; A_{kj}^I \; S_{ikj},$$

Where P_j denotes the relative importance weight of the dimension of best supplier; A_{kj}^D represents the relative importance weight of best supplier enabler k of the dimension of best supplier j for the dependency (D) relationships between component levels; A_{kj}^I is the stabilized relative importance weight (determined by the super matrix) for supplier attribute enabler k of the dimension of supplier j for the interdependency (I) relationships within the supplier attitude enablers' component level; S_{ikj} denotes the relative impact of best supplier alternative i on best supplier attribute enabler k of the dimension of best supplier j; and J represents the index set for the dimension of best supplier alternative.

Table 8 lists the desirability indices (D_i) and their normalized values (D_{iN}) for selecting the best supplier. These indices are based on the relative weights obtained from the pair-wise comparison of alternatives, dimensions, and weights of criteria from the converged super matrix. In Table 8, the values of the second column indicate the relative importance of the dimensions imported from Table 2. Furthermore, the values in the fourth column denote the relative importance of enablers through their respective dimensions, which also imported from Table 3. Moreover, the values in the fifth column of Table 8 are the stable independent weights of enablers acquired via a converged super matrix (Table 7). Additionally, the next three columns (the sixth, seventh, and eighth) contain the pair-wise comparison matrices, which show the relative impact of each of the alternatives on the enablers. The final three columns represent the weighted values of the alternatives $(P_j \times A_{kj}^D \times A_{kj}^I \times S_{ikj})$ for each of the enablers. An illustrative example, the value of supplier A corresponding to the carbon governance is 0.0217 $(0.5502 \times 0.3532 \times 0.2122 \times 0.5272)$. Summing these values for each of the alternatives provides the value of desirability indices, D_i, and their normalized desirability indices, D_{iN}, are shown in the last two row of Table 8. Supplier A (0.5034) thus received the highest ranking, implying that it was the best supplier.

4. Conclusion and Further Research

The GSCM-based conceptual framework and operational model to incorporate carbon management into supplier selection have been presented. By identifying the related criteria of carbon management activities for the proposed framework, an ANP methodology was applied to an electronics company.

Compared with the previous investigations, the proposed method may have following contributions. First, a new model for selecting suppliers with emphasis on carbon management issues has been developed. Such a framework has never being found in the previous literature. And from the illustrated example, this model shows its potential advantage in selecting suitable suppliers in terms of carbon management. Second, ANP methodology was applied in supplier selection and it is rarely found from the previous studies. ANP can capture both quantitative and qualitative criteria and reflect more realistic results among decision attributes and alternatives owing to the existence of interdependent relationships in the real supplier selection and evaluation environment. Therefore, ANP modeling can serve as a new method and offer insights to managers in selecting suppliers systematically.

If a company wants to incorporate carbon management into supplier selection and evaluation in green supply chain, a company can adopt the presented model, which includes all the criteria for understanding the competence of its suppliers and prioritizing the suppliers. In addition, the results of weights determined in the case study can also be adopted as a reference. Moreover, if the company is not that familiar with ANP technique the weights can be easily determined by an expert group who are responsible for tasks associated with carbon management in the company. Otherwise, the weights of criteria can be re-determined by an expert group and ANP technique.

Although the results obtained from this study are satisfactory, this case study can still be improved. The outcome of the ANP model conducted in this study is determined by managers of the case company exclusively; and since carbon management includes different tasks and thus the criteria involved in supplier selection are complex; a cross-functional team is required to participate in supplier selection for future research.

Acknowledgment

The authors would like to thank the National Science Council of Taiwan, for financially supporting this research under grant NSC 99-2815-C-236-005-H.

References

Akarte, M.M., Surenda, N.V., Ravi, B. and Rangaraj, N., 2001. Web based casting supplier evaluation using analytical hierarchy process *Journal of the Operational Research Society* **52**, 511–522.

Bai, C. and Sarkis, J., 2010. Green supplier development: Analytical evaluation using rough set theory. *Journal of Cleaner Production* **18**(12), 1200–1210.

Barnard, J.W., 2006. Corporate boards and the new environmentalism. *William and Mary Law Review* **31**, 291–295.

Bayazit, O., 2006. Use of analytic network process in vendor selection decisions. *Benchmarking: An International Journal* **13**(5), 566–579.

Boiral, O., 2006. Global warming: Should companies adopt a proactive strategy? *Long Range Planning* **39**(3), 315–330.

Carbon Disclosure Project, 2010. *Supply Chain Report.*

Chan, F.T.S., 2003. Performance measurement in a supply chain. *International Journal of Advanced Manufacturing Technology* **21**, 534–548.

Chan, T.S. and Kumar, N., 2007. Global supplier development considering risk factors using fuzzy extended AHP-based approach. *Omega* **35**(4), 417–431.

Chen, S.H., Lin, H.T. and Lee, H.T., 2004. Enterprise partner selection for vocational education: Analytical network process approach. *International Journal of Manpower* **25**(7/8), 643–655.

Choy, K.L., Lee, W.B. and Lo, V., 2003. Design of a case based intelligent supplier relationship management system — The integration of supplier rating system and product coding system. *Expert Systems with Applications* **25**, 87–100.

Chung, S.H., Lee, A.H.I. and Pearn, W.L., 2005. Analytic network process (ANP) approach for product mix planning in semiconductor fabricator. *International Journal of Production Economics* **96**, 15–36.

Cogan, D., Good, M., Kantor, G. and McAteer, E., 2009. *Corporate Governance and Climate Change.*

Dell, 2009. *Inspire and Innovation — Achievements in Corporate Responsibility.* Available at http://i.dell.com/sites/content/corporate/corp-comm/en/Documents/Dell_CR_Summary_Report_FINAL.pdf.

Enarsson, L., 1998. Evaluation of suppliers: How to consider the environment. *International Journal of Physical Distribution and Logistics Management* **28**(1), 5–17.

Erdogmus, S., Aras, H. and Koc, E., 2006. Evaluation of alternative fuels for residential heating in Turkey using analytic network process (ANP) with group decision-making. *Renewable and Sustainable Energy Reviews* **10**, 269–279.

Feng, C.X., Wang, J. and Wang, J.S., 2001. An optimization model for concurrent selection of tolerances and suppliers. *Computers & Industrial Engineering* **40**, 15–33.

Gencer, C. and Gürpinar, D., 2007. Analytic network process in supplier selection: A case study in an electronic firm. *Applied Mathematical Modelling* **31**(11), 2475–2486.

Ghodsypour, S.H. and O'Brien, C., 1998. A decision support system for supplier selection using an integrated analytic hierarchy process and linear programming. *International Journal of Production Economics* **56–57**, 199–212.

Ghodsypour, S.H. and Obrien, C., 2001. The total cost of logistics in supplier, under conditions of multiple sourcing, multiple criteria and capacity constraint. *International Journal of Production Economics* **73**, 15–27.

Handfield, R., Walton, S., Sroufe, R. and Melnyk, S., 2002. Applying environmental criteria to supplier assessment: A study in the application of the analytical hierarchy process. *European Journal of Operational Research* **141**, 70–87.

Hsu, C.W. and Hu, A.H., 2009. Applying hazardous substance management to supplier selection using analytic network process. *Journal of Cleaner Production* **17**(2), 255–264.

Hu, A.H. and Hsu, C.W., 2006. Empirical study in the critical factors of green supply chain management (GSCM) practice in the Taiwanese electrical and electronics industries. In: *Proceeding of 2006 IEEE International Conference on Management of Innovation and Technology*, Singapore, pp. 853–857.

Huang, J.J., Tzeng, G.H. and Ong, C.S., 2005. Multidimensional data in multidimensional scaling using the analytic network process. *Pattern Recognition Letters* **26**, 755–767.

Huang, S. and Keskar, H., 2007. Comprehensive and configurable metrics for supplier selection. *International Journal of Production Economics* **105**(2), 510–523.

Humphreys, P., McIvor, R. and Chan, F., 2003b. Using case-based reasoning to evaluate supplier environmental management performance. *Expert Systems with Applications* **25**, 141–153.

Humphreys, P.K., Wong, Y.K. and Chen, F.T.S., 2003a. Integrating environmental criteria into the supplier selection process. *Journal of Materials Processing Technology* **138**, 349–356.

IPCC (Intergovernmental Panel on Climate Change), 2007. *Climate Change 2007: The Physical Science Basis — Summary for Policymakers*. IPCC, Geneva.

Jharkharia, S. and Shankar, R., 2007. Selection of logistics service provider: An analytic network process (ANP) approach. *Omega* **35**(3), 274–289.

Kahraman, C., Cebeci, U. and Ulukan, Z., 2003. Multi-criteria supplier selection using fuzzy AHP. *Logistics Information Management* **16**(6), 382–394.

Kolk, A. and Pinkse, J., 2004. Market strategies for climate change. *European Management Journal* **22**(3), 304–314.

Lee, E.K., Ha, S. and Kim, S.K., 2001. Supplier selection and management system considering relationships in supply chain management. *IEEE Transactions on Engineering Management* **48**(3), 307–318.

Lee, H.I., Kang, H.Y., Hsu, C.F. and Hung, H.C., 2009. A green supplier selection model for high-tech industry. *Expert Systems with Applications* **36**(4), 7917–7927.

Liaoa, Z. and Rittscherb, J., 2007. A multi-objective supplier selection model under stochastic demand conditions. *International Journal of Production Economics* **105**(1), 150–159.

Meade, L. and Sarkis, J., 1998. Strategic analysis of logistics and supply chain management systems using the analytical network process. *Transportation Research Part E* **34**(3), 201–215.

Niemira, M.P. and Saaty, T.L., 2004. An analytic network process model for financial crisis forecasting. *International Journal of Forecasting* **20**, 573–587.

Noci, G., 1997. Designing 'green' vendor rating systems for the assessment of a supplier's environmental performance. *European Journal of Purchasing & Supply Management* **3**(2), 103–114.

Okereke, C., 2007. An exploration of motivations, drivers and barriers to carbon management: The UK FTSE 100. *European Management Journal* **25**(6), 475–486.

Rao, P., 2002. Greening the supply chain a new initiative in south East Asia. *International Journal of Operations & Production Management* **22**(6), 632–655.

Rao, P., 2005. The greening of suppliers in the South East Asian context. *Journal of Cleaner Production* **13**, 935–945.

Ravi, V., Shankar, R. and Tiwari, M.K., 2005. Analyzing alternatives in reverse logistics for end-of-life computers: ANP and balanced scorecard approach. *Computers & Industrial Engineering* **48**, 327–356.

Saaty, T.L., 1996. *Decision Making with Dependence and Feedback: The Analytic Network Process*. RWS Publications, Pittsburgh, PA.

Saaty, T.L., 1980. *The Analytic Hierarchy Process*. McGraw-Hill, New York.

Sarkis, J. and Sundarraj, R.P., 2002. Hub location at digital equipment corporation: A comprehensive analysis of qualitative and quantitative factors. *European Journal of Operational Research* **137**, 336–347.

Schultz, K. and Williamson, P., 2005. Gaining competitive advantage in a carbon-constrained world: Strategies for European business. *European Management Journal* **23**(4), 383–391.

Seuring, S. and Müller, M., 2008. Core issues in sustainable supply chain management — A Delphi study. *Business Strategy and the Environment* **17**, 455–466.

Skjaerseth, J. and Skodvin, T., 2003. *Climate Change and Oil Industry: Common Problems Varying Strategies*. Manchester University Press, Manchester.

Southworth, K., 2009. Corporate voluntary action: A valuable but incomplete solution to climate change and energy security challenges. *Policy and Society* **27**, 329–350.

Sullivan, R., 2009. The management of greenhouse gas emissions in large European companies. *Corporate Social Responsibility and Environmental Management* **16**, 301–309.

Tam, M.C.Y. and Tummala, V.M.R., 2001. An application of the AHP in vendor selection of a telecommunications system. *Omega* **29**, 171–182.

US-AEP, 1999 *Supply Chain Environmental Management-lessons for Leader in the Electronic Industry.* Clean Technology Environmental Management (CTEM) Program, US–Asia Environmental Partnership.

Walton, S.V., Handfield, R.B. and Melnyk, S.A., 1998. The green supply chain: Integrating suppliers into environmental management processes. *International Journal of Purchasing and Materials Management* **34**(2), 2–11.

Weinhofer, G. and Hoffmann, V.H., 2008. Mitigating climate change — How do corporate strategies differ? *Business Strategy and the Environment* **19**(2), 77–89.

Wu, C.H., Kuo, T.C. and Lu, Y.Y., 2007. Environmental principles applicable to green supplier evaluation by using multi-objective decision analysis. *International Journal of Production Research* **45**(18–19), 4317–4331.

Zaim, S., Sevkii, M. and Tarim, M., 2003. Fuzzy analytic hierarchy base approach for supplier selection. *Logistics Information Management* **12**(3/4), 147–176.

Zhu, J., 2004. A buyer–seller game model for selection and negotiation of purchasing bids: Extensions and new models. *European Journal of Operational Research* **134**, 150–156.

Zhu, Q. and Geng, Y., 2001. Integrating environmental issues into supplier selection and management: A study of large and medium-sized state-owned enterprises in China. *Greener Management International* **35**, 27–39.

Chapter 39

ENVIRONMENTAL DECISION MAKING: ROAD MAPS AND MANAGEMENT SCIENCE TOOLS

Chu-hua Kuei[a] and Janice Winch[b]
Department of Management and Management Science,
Lubin School of Business, Pace University,
1 Pace Plaza, 10038 NY, USA
[a]*ckuei@pace.edu*
[b]*jwinch@pace.edu*

This chapter presents concepts and methods of environmental decision making (EDM). EDM can be divided into two broad categories: policy deployment at the strategic level and process management at the tactical level. Policy makers can use decision tools such as cognitive mapping and analytic hierarchy process to decide on the strategic direction in environmental sustainability. Once the strategic direction is laid out, the process managers can plan, execute, and control the environmental friendly practices with the help of tools such as simulation, linear programming, and six-sigma. Both the conceptual framework of the decision making and the examples of the decision tools are discussed.

1. Introduction

In making organizational decisions, the need to enhance efficiency and effectiveness typically takes the center stage. To an increasing number of organizations, however, meeting broader social goals such as sustaining the environment or developing the ability to maintain the qualities that are valued in our environment is equally important. Turner and Houston (2008, p. 99), for example, noted that

> "The US federal government has undertaken a multi-year renovation of the Pentagon that uses a great deal of recycled material, including more than 59,000 square feet of carpet tiles made from recycled material and 53,000 linear feet of recycled steel wall studs."

Using the Vietnamese catfish industry as an example, Cheong and Corbitt (2010, pp. 23, 24) also reported that "Catfish farming and transportation of live fish to processing factories are heavily dependent on the water environment which is very difficult to control and with high risks of epidemics. Environmental pollution may lead to the collapse of the industry."

The aim of today's organizations is thus to achieve economic development without negating the importance of sustainable development. This new social reality is also reflected in the following *remarks* in the Hewlett-Packard's annual corporate responsibility and sustainability reports:

> "Our global citizenship priorities — supply chain responsibility, climate and energy, and product reuse and recycling — are more critical than ever to our success." Mark Hurd, Chairman, CEO and President at Hewlett-Packard (http://www.hp.com/hpinfo/globalcitizenship/gcreport/).

To support such priorities is not simple. Hewlett-Packard's recycling, for example, is divided into open-loop and closed-loop processes. In the open-loop recycling process, the plastic from returned cartridges is shredded and sold for use in other industries, whereas in the closed-loop process, the shredded plastic is recycled back into new cartridges. Hewlett-Packard as a result needs to support sustainability through a three-step management cycle: analysis, planning and execution, and continuous improvement. Along the way, Hewlett-Packard relies on environmental decision making (EDM).

Anecdotal evidence from sources such as Time, Newsweek, Popular Science, and Scientific American (Earth 3.0) also suggests that decision makers in governments, businesses, and non-profit organizations are becoming increasingly aware of the strategic role of EDM in the overall success of environmental sustainability. Turner and Houston (2008, p. 103) further noted that there are two major groups of EDM.

(1) *Strategic*: Policy and procedures, innovation, staff development, investment in eco-technology, supplier collaboration, and adoption of targets and metrics.
(2) *Tactical*: Life-cycle analysis, trade-off evaluation, demand management, procurement, and contract negotiation

Good strategic/tactical decision making is fundamental for achieving sustainable development (Cheong and Corbitt, 2010; Epstein, 2010; Cruz, 2008; Sarkis, 2003). To this effect, it is most important to choose the best set of decision making tools for both policy deployment and strategy formulation. Thus, the goal of this chapter, EDM, is to present the ideas, concepts, and methods of decision analyses in the context of environmental management at a level that can be understood by anyone who does not have an extensive background in management/decision sciences. A systemic perspective of EDM is adopted in this chapter to help decision makers spot problems early, and develop plans to better serve the stakeholders. This chapter is divided into four parts. Section 2 starts with a discussion of the aspects of decision making needed to bring up issues involved in the context of environmental problem exploration (Fig. 1). Section 3 provides an overview of implementation (Fig. 2), highlighting the different worldviews between policy makers and process decision makers (Fig. 3a,b). Policy maker is usually involved in environmental policy deployment, whereas the process decision maker ensures that the policies are formulated into strategies that are achievable. In Section 4, we define key strategic dimensions from the perspectives of policy makers and describe two possible

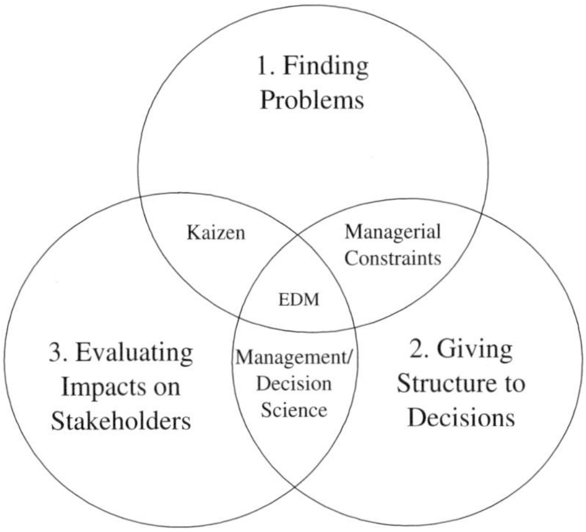

Fig. 1. Aspects of EDM.

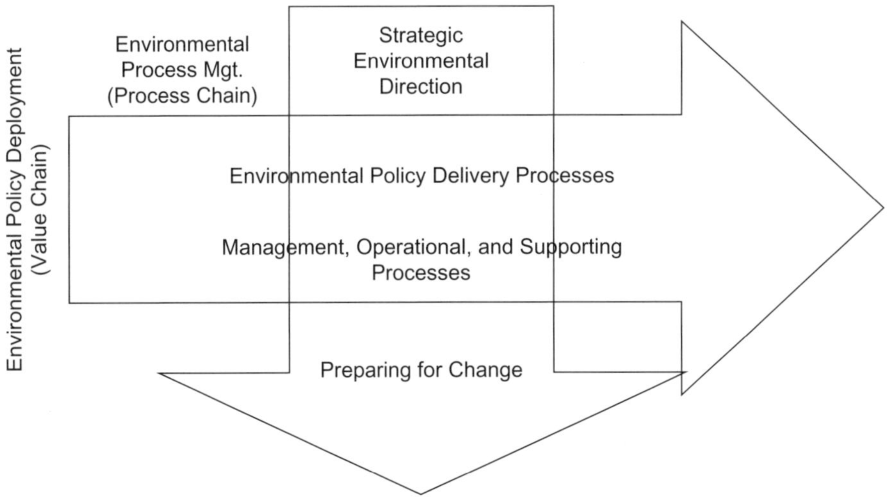

Fig. 2. Big picture for implementation.

environmental policies (Fig. 4). As examples in point, cognitive mapping (Fig. 5a, b) and the analytic hierarchy process (AHP, Fig. 5c) are introduced as the primary decision-making tools. Three other types of decision tools will be discussed fully in Section 5 to see how to plan, understand, schedule, monitor, and control process activities by process decision makers (Fig. 6). They are capability maturity model integration (CMMI Fig. 7), Monte Carlo simulation (Fig. 8a, b), and Six Sigma (Fig. 9).

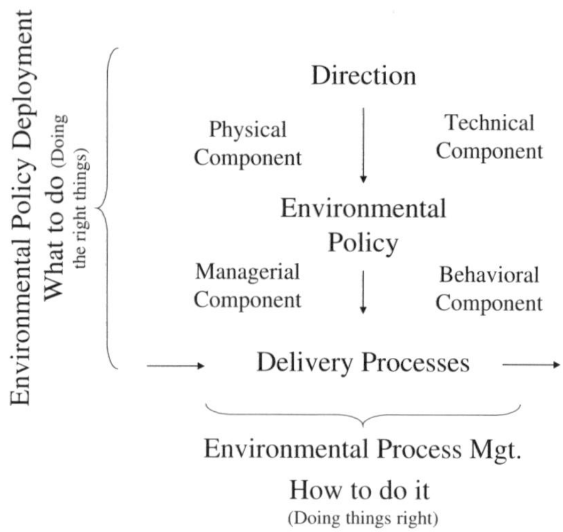

(a)

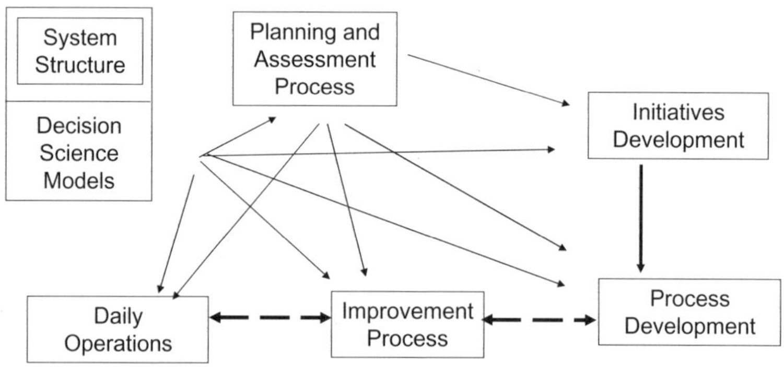

(b)

Fig. 3. Big picture for implementation: The worldview of (a) policy maker and (b) process decision maker.

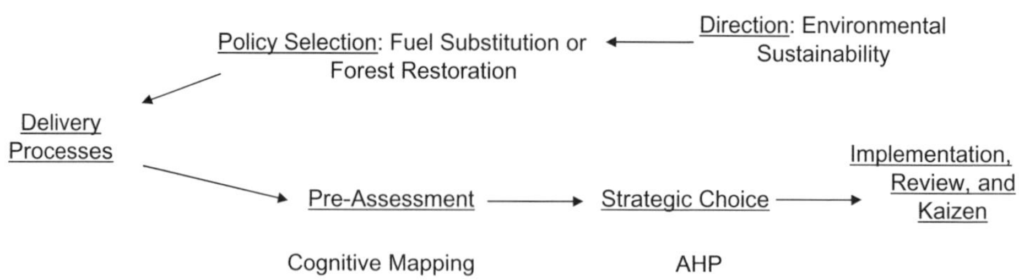

Fig. 4. Policy maker — Strategic route map and decision making tools.

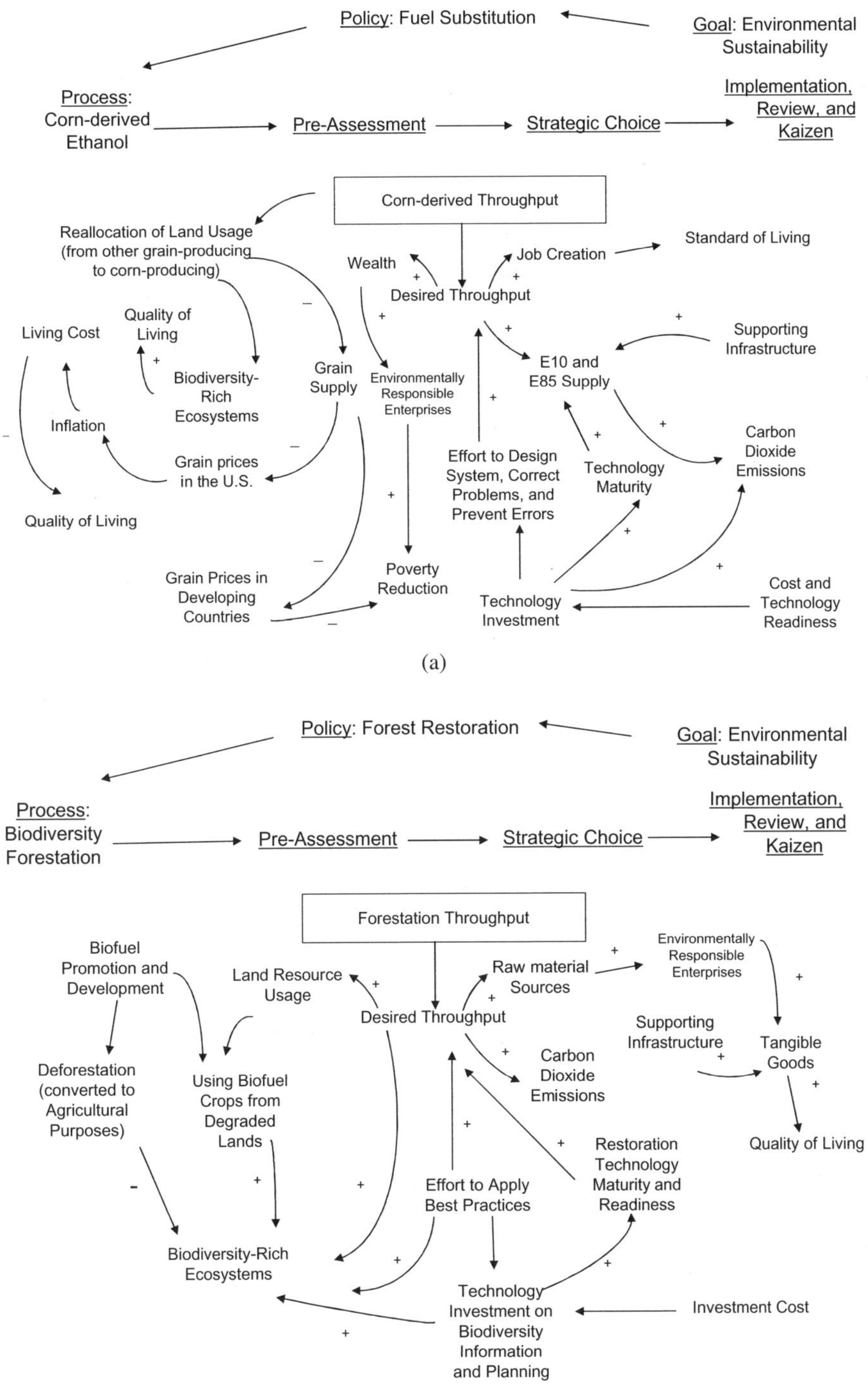

Fig. 5. Cognitive mapping (CP)-based pre-assessment: (a) fuel substitution and (b) forest restoration. (c) policy selection — analytic hierarchy process.

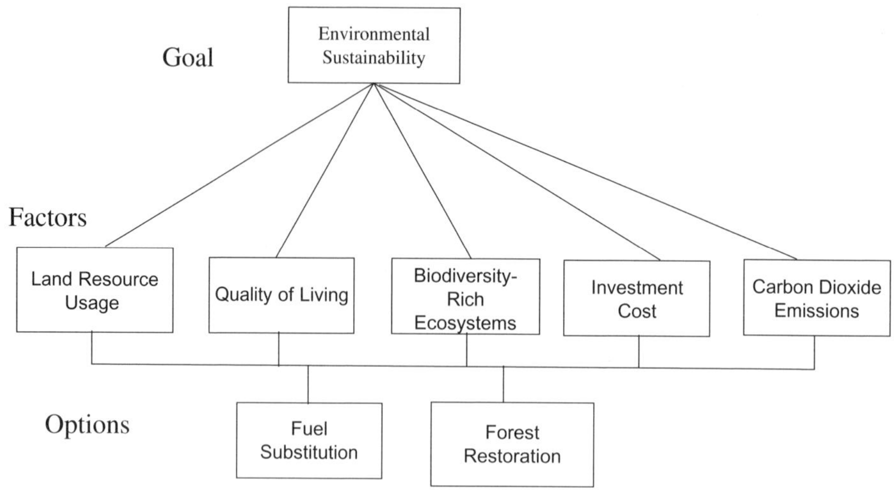

(c)

Fig. 5. (*Continued.*)

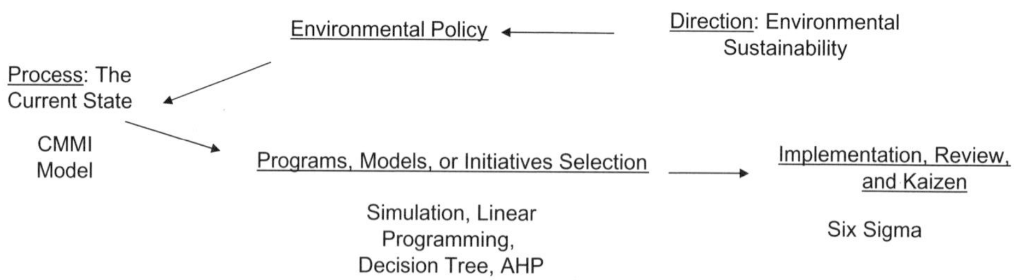

Fig. 6. Process decision maker — Tactical route map and decision making tools.

2. Aspects of EDM

The concepts of EDM in this particular chapter developed as a result of concerns about environmental sustainability. Figure 1 depicts the major activities and collection of interrelated tasks to solve a specific environmental problem. They can be classified into three groups: (1) finding problems and understanding managerial constraints, (2) giving structure to decisions and using decision science models, and (3) evaluating impacts on stakeholders and committing to Kaizen (continuous improvement). This framework assists decision makers in managing environmental issues in three important areas: problem finding, problem solving, and problem context. They are reviewed as follows.

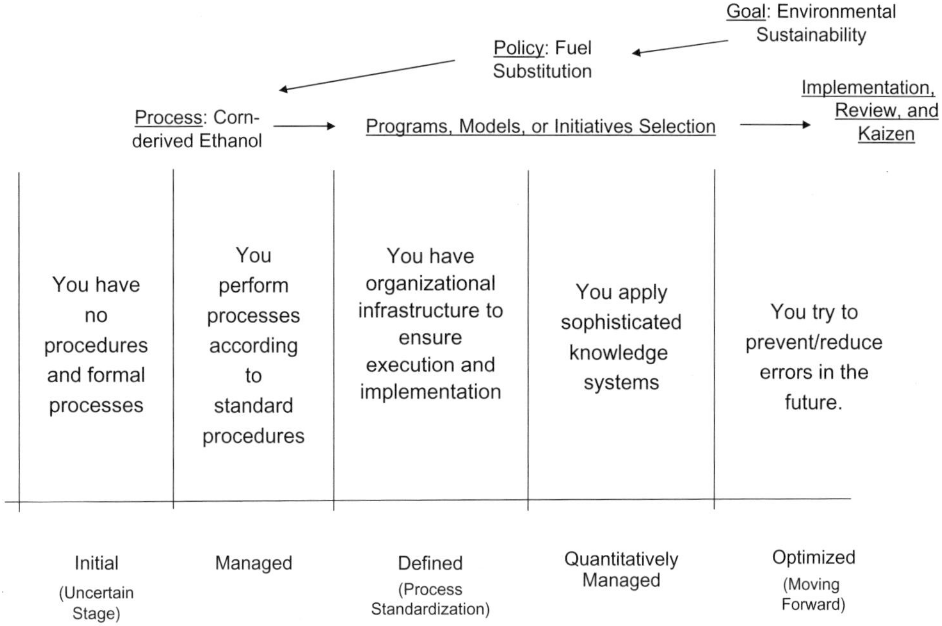

Fig. 7. CMMI-based process maturity analysis — Fuel substitution.

2.1. *Finding problems and understanding managerial constraints*

Identifying a decision opportunity is the primary concern and activity at this stage. It also involves a problem classification task from which the problem is viewed and categorized as an unstructured environmental problem (non-programmed decisions normally made in response to unique, unstructured environmental situations such as forest restoration), a semi-structured environmental problem, or a structured environmental problem (programmed decisions made in response to routine environmental situations such as recycling programs and operations). The constraint identified here defines the complexity of the circumstances decision makers face and the ways in which decision makers can approach decision making. For unstructured environmental problems, for example, unpredictability is the norm and Strategic Benchmarking is helpful in principle. For semi-structured environmental problems, cause-and-effect relationships can be sensed and analyzed by experts. Cognitive mapping, as an example in point, is thus useful in this complicated situation where the interactions between events and possible action consequences can be discovered but are not immediately apparent. As for more structured environmental problems, circumstances are quite straightforward and solutions usually exist. A broad range of tools such as linear programming, experimental designs, and computer simulation can be used to find the right solution. In this chapter, we will concentrate particularly on semi-structured and structured environmental problems. Illustrative examples will be presented in the following sections.

4	▪Establish interval of random numbers (R.N.) for each variable			
5	**Utility Expense Distribution**			
6	Prob.	Lower R.N.	Upper R.N.	Monthly U. Exp.
7	0.2	0	0.2	$400
8	0.2	0.2	0.4	$500
9	0.2	0.4	0.6	$600
10	0.2	0.6	0.8	$700
11	0.2	0.8	1	$800
12				
13	▪Construct the Excel Simulation Model with Excel functions (e.g. VLOOKUP, IF)			
14	**Simulation Trials**			
15	Month	Random Number	Simulated U. Exp.	
16	1	0.6953	$700	
17	2	0.0082	$400	
18	3	0.6799	$700	
19	4	0.8898	$800	
20	5	0.6515	$700	
21	6	0.3976	$500	
22	7	0.0642	$400	
23	8	0.0377	$400	
24	9	0.5739	$600	
25	10	0.5827	$600	
26	11	0.0508	$400	
27	12	0.4757	$600	
28		**Average:**	$567	
29		Total:	$6,800	

(a)

	A	B	C
4	**Utility Expense Distribution (Normal Distribution)**		
5	Mean	$600	
6	Standard Deviation	$100	
7			
8	**Simulation Trials**		
9	Month	Random Numbe	Simulated U. Exp.
10	1	0.06	$447.89
11	2	0.89	$725.20
12	3	0.12	$482.81
13	4	0.77	$673.88
14	5	0.36	$563.55
15	6	0.67	$644.02
16	7	0.18	$509.82
17	8	0.16	$500.43
18	9	0.60	$625.44
19	10	0.73	$662.66
20	11	0.24	$528.53
21	12	0.65	$638.42
22		**Average:**	$584
23		Total:	$7,003

(b)

Fig. 8. Simulation solutions for (a) and (b).

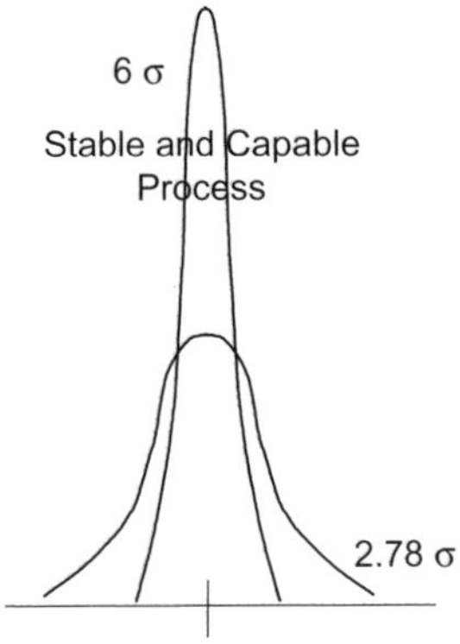

Fig. 9. Six Sigma-based process review.

2.2. *Giving structure to decisions and using management/decision science models*

Presenting all significant factors and possible consequences with respect to environmental problems is the primary challenge and activity at this stage. Simple diagrams or charts such as Pareto charts, flow charts, and cause-and-effect diagrams will normally do the trick for a straightforward environmental problem. More advanced diagrams such as Poka Yoke and cognitive mapping are also available for semi-structured environmental problems. These advanced tools allow decision makers to see things from different perspectives, assimilate complex concepts, understand causal relationships between concepts, identify environmental problems, and address potential flaws. The relevant task here is to make expert use of statistics and modeling to improve a wide variety of environmental functions. In general, there are two groups of environmental decision models.

Descriptive model: This type of model is used to describe an environmental system, understand action consequences, and better comprehend functional relationships in the system of concern and the outside world. Probably the best known tool for this category is computer simulation. The cognitive mapping method is often used by decision makers to get succinct cause-and-effect relationships among major factors. Process decision makers can also develop *Monte Carlo* simulations in *Excel for the concern of energy use, resource management, sustainable construction, waste management, and sustainable agriculture and forest management.*

Prescriptive model: If decision makers want to optimize a system and/or identity the best option, they need to rely on prescriptive models. An important example of prescriptive models is linear programming. Linear programming is a widely used optimization method that seeks to maximize or minimize some objective function while satisfying a set of constraints. Process decision makers can also develop *inventory models* in *Excel for the reverse logistics operations* (e.g., Hewlett-Packard's open-loop and closed-loop processes) *and sustainable resource use.*

Environmental managers often face problems of understanding the system behaviors and allocating scarce resources in the best possible way. Equipped with the knowledge of decision science, they can explore the implications of decision options and bounded rationality (i.e., constrained optimization) for decision making.

2.3. *Evaluating impacts on stakeholders and committing to Kaizen (continuous improvement)*

The process of EDM should not be considered complete without the full inclusion of concerns of stakeholders and tasks for continuous improvement (i.e., Kaizen). At this stage, critical-to-environmental-quality (CTEQ) or the concerns of stakeholders must be probed and connected to the other two aspects of EDM. This stage thus begins with the identification of qualities that are valued by stakeholders in our physical environment. To most societies, for example, CTEQ consists of a suitable climate, clean water and air, the reduction of non-renewable resource depletion, and the livability and beauty of the environment. Damage done to the CTEQ can create threats of serious decline in environmental sustainability. Given dimensions of CTEQ, it is therefore very important for environmental managers to figure out when and where threats will have their effects if certain decisions are made. Traditionally, environmental managers tend to focus on reducing the damage and working out corrective actions. Since the turn of this century, many environmental managers have had a new mindset of prevention and continuous improvement. They adopt strategies aimed to prevent damages from being done in the first place.

These three aspects of EDM in Fig. 1 need to be well connected. Without staying connected to each other, the entire EDM process will be seriously flawed in terms of conceptualization and operationalization.

3. Implementation Issues

3.1. *Big picture for implementation*

Once environmental managers embrace the concept that an integrated EDM with a special focus on CTEQ could make a difference in environmental sustainability, and that the entire environmental system of concern could be managed and/or designed better, they then have to plan implementation strategies to realize the full potential of EDM. Two major areas of interest with respect to implementation issues are policy deployment and process management. Policy deployment, in the context of environmental management, involves a strategic environmental direction and a policy development loop. The emphasis is on doing the right things. Process management, on the other hand, involves planning and execution processes. The focus is on doing things right and linking everyone in the environmental process chain.

Figure 2 depicts these two areas of interest. The vertical area, labeled environmental policy deployment, encompasses prescriptive and strategic direction that guides the operational aspects of the environmental system. Such strategic direction may be as simple as a phrase. For example, Wal-Mart Stores have rolled out a new advertising campaign with the slogan "Save Money. Live Better." To be truly effective, a system must also be ready for change. This can be done through the preparation of the following three elements: the management components, the environmental policy delivery processes, and the organizational structure. Environmental managers thus need to take steps to ensure high performances through changes. Wal-Mart, for example, assigns its suppliers a score based on the amount of recycled content used in the package. As noted by Turner and Houston (2008, p. 106), "about 97,000 products are being evaluated in this way. The company [Wal-Mart] aims to put 160,000 through the process." The horizontal area, labeled environmental process management, represents a collection of interrelated tasks designed to deliver a specific result. An environmental policy delivery process, such as a business process, is made up of several sub-processes. For the purpose of this chapter, we classify environmental policy delivery processes into three major categories: strategic environmental management processes, operational processes, and supporting processes. Strategic environmental management processes deal with strategic aspects of physical/technical management components as well as managerial/behavioral management components. Operational processes are the core of operations or the engine that moves the environmental system. Supporting processes, such as information systems and human resources, provide the enabling conditions needed for operational processes. These three sub-processes need to be working together to achieve the goal of the environmental system.

3.2. *The worldview of decision makers*

The worldview, according to Magee (1998), refers to the framework of ideas and beliefs through which an individual interprets the world and interacts with it. In a simple EDM circumstance such as buying a hybrid car, we might only have one decision maker who needs to assume the role of planning and execution. Most EDM situations today, however, are complex because the nature of the environmental context such as developing and promoting an eco-friendly fuel is dynamic and uncertain. Nevertheless, as we all know, when running a business, we have executives (people who establish strategies to guide the work of their organizations) and managers (people who run their organizations). In a complex environmental situation, along this line of thinking, we also have policy makers and process decision makers. The former is usually involved in strategy and environmental policy deployment, whereas the jobs of the latter entail the scope of the delivery process, plan for initiating the best practices, and secure adequate resources for the implementation.

Figure 3a shows the worldview of policy makers in an environmental context. This model is built around the following two common themes: What to do (or content) and How to do it (or process). As shown in Fig. 3a, the concepts along the vertical line (i.e., direction, environmental policy, and management components) represent the contents needed to attain a long-term environmental vision. The alignment of such entities enables policy makers to deploy the environmental vision through the development of strategy portfolios and the training of the hierarchy, which in turn, could help to institutionalize the organizational, operational, and behavioral changes needed to manage the entire environmental system. When organizations begin the vertical implementation task, it is also a good idea to acknowledge the importance of preparation for process change. The vertical guideline dictates everything a policy maker should do.

The worldview of process decision makers is different from that of policy makers. As shown in Fig. 3b, there are five core processes in their reference framework of ideas and beliefs: planning and assessment, initiatives development, process development, daily operations, and improvement. These critical process steps, from planning/assessment to daily operations, establish the context within which strategic, tactical, and operational choices are made. Process decision makers must also develop a better understanding on system structure and decision science models and take a lead to instill quality into the environmental policy delivery system. To meet the challenges mentioned, process decision makers must address the following: determine the scope of routine operations and process change, institute the change, determine how to overcome resistance to change, understand their role and the impact of their EDM style, institute control, and conduct periodical review. The aim of any environmental policy delivery system in the end is to develop capability and cultural forces to positively influence people and human systems. A well-designed system will have a meaningful impact and achieve results regarding environmental sustainability.

4. Policy Maker — Strategic Route Map and Decision Making Tools

As environmental managers at the strategic level adopt the concept of worldviews of policy makers, they understand right strategic choices are of great importance when establishing strategies to guide the work of their environmental systems. However, due to information overload, it usually becomes difficult to comprehend and compare all ideas, concepts, and solution options at one time. Further, environmental policy decision making (EPDM) normally involves un- or semi-structured environmental problems and subjective judgment. As a result, an inability to achieve rationality is expected. To overcome implementation barriers, we develop a strategic route map (Fig. 4) that shows how an environmental policy can be effectively deployed. As an example in point, we assume that two environmental policies, namely, fuel substitution and forest restoration, are under consideration by environmental managers at the policy level. The ultimate goal of this strategic choice is environmental sustainability. The essence of this strategic choice exercise is to get the opportunity to think things through. Along the way, two decision making tools, namely, cognitive mapping and the AHP, can be utilized effectively in this knowledge acquisition and strategic choice process. Each decision tool and how it will affect the EPDM are discussed in more depth.

4.1. *Cognitive mapping*

Policy development should be based on the best available knowledge and analyses. A cognitive mapping approach, a graphical presentation of linked concepts or issues (Fig. 5a,b), is instrumental in achieving this goal. The cognitive mapping method, as noted by Eden and Ackermann (2004), Madu (1996), and Madu *et al.* (1991), provides policy makers with a defined way to construct a thematic network with concepts and links. Policy makers can thus think more deeply about the issue under discussion. Figure 5a, entitled "Cognitive mapping (CP)-based pre-assessment — fuel substitution," depicts the development of such a network model. It starts with initial conditions, the throughput of corn-derived ethanol, and then is expanded to capture all relevant concepts and possible issues regarding fuel substitution. When constructed, it is helpful in understanding the linked concepts on a causal basis. Links, in this case, represent causal relationships between concepts/issues. For the purpose of this chapter, three kinds of link are adopted. The positive (+) signs next to the links indicate positive association, negative (−) signs indicate negative association, and links without signs indicate "may lead to."

Before making the strategic choice, policy makers can rely on the cognitive mapping approach to master a decent level of forethought. It is observed that, for example, from Fig. 5a, some resources such as lands may deteriorate if the value extracted is invested in producing more corn-derived ethanol. Dealing with extreme poverty around the globe and raising the standard of living throughout the world are also major emerging issues in sustainability. As evident in Fig. 5a, a high quality of living cannot be maintained without the producing capacity of supporting lands and the carrying capacity of biodiversity-rich ecosystems. Further, the carbon footprints could be shrunk and the environment could be greener with environmental initiatives such as the usage of corn-derived ethanol. Technological investment also seems to play a vital role in developing sustainability. It is thus clear that the exercise of cognitive map development can be used as a focus for discussion at the policy pre-assessment stage. It can also give policy makers an opportunity to question his/her assumptions about the associations between concepts/issues. Figure 5b shows how such maps may be created for the issue of forest restoration. From the map, information on cause–effect can be derived. For example, the map shows that the restoration of forest will lead to an increase in raw material resources. These positive impacts on the environment may also lead to more green-minded enterprises, which, in turn, may be more willing to approach environmentalism in concrete terms.

As shown in Fig. 5a,b, five common themes recur across our network models: land resource usage, quality of living, biodiversity-rich ecosystems, investment, and carbon dioxide emission. They should be considered as our CTEQ and linked with strategic choice approaches.

4.2. *Analytic hierarchy process*

One of the best known strategic choice approaches is the AHP. AHP is a multi-criteria decision model that develops priority weights for items based on decision makers' professional evaluation (Saaty, 1980). AHP is based on hierarchy building, as presented in Fig. 5c. Basically, there are three main levels in our recommended AHP structure. The highest level denotes the *goal*, the middle level denotes the *factors*, and the bottom level the *options*. A good understanding of the problem about environmental sustainability at the previous pre-assessment stage helps structure our AHP model. As shown in Fig. 5c, five factors derived from the cognitive mapping exercises are considered as *factors*. Two decision alternatives under review are fuel substitution and forest restoration. As the AHP has been widely applied (Madu, 2006, Madu and Kuei, 2004; Madu *et al.*, 2002), we shall limit our discussions on it. However, one must note that it is based on a pairwise comparison of two items at a time (see Appendix, Tables A1–A6). Further, we must note that "policy makers" or participants are asked to use the following 9-point scale system to assign their judgments:

When comparing items i and j, select a numerical value A_{ij} from the following set:

1: items i and j are of equal importance;
3: item i is weakly more important than item j;
5: item i is strongly more important than item j;
7: item i is very strongly more important than item j;
9: item i is absolutely more important than item j; and
2, 4, 6, 8: intermediate values between the two adjacent judgments.

Further, $A_{ii} = 1$. If $A_{ij} = y$, then $A_{ji} = 1/y$.

Once policy makers fill out the forms (see Appendix, Tables A1–A2), a decision support system (expert choice or super decisions) can be used to obtain priority indices for both factors and options. As the goal of this chapter is not to collect empirical data, but to show how to make strategic choices with the help of widely used decision making tools, we shall direct interested readers to Saaty and Vargas (2006), Madu (1996, 2006), and Saaty (1980) for the computing details of getting priority weights.

For a dynamic non-linear multi-attribute decision setting, as noted by Sarkis and Talluri (2002), the analytical network process (ANP) can be used as an effective and more realistic modeling approach. For details on how to establish the

ANP-based model for decision making within the green supply network, refer Sarkis (2003).

5. Process Decision Maker — Tactical Route Map and Decision Making Tools

The need for EPDM must be identified at the strategic level. This importance becomes pronounced through the strategic route map analyses (see Figs. 4 and 5). Once the need for a specific environmental policy is accepted, a "process" decision making team should be formed with active participation of policy makers to chart the course of environmental delivery processes. The team should follow the tactical route map shown in Fig. 6 and pay attention to dimensions of CTEQ identified at the policy pre-assessment stage. Decision makers at the process level should screen the current state of environmental delivery processes, break a set of complex overwhelming tasks into small manageable tasks, make the most out of decision science discipline, select mission-critical initiatives, accomplish those predetermined implementation goals, and commit to the process of continuous improvement.

Given the environmental policy, tactical action plans should be implemented through the following three stages: appraising current processes, programs/models/initiatives selection, and implementation/review/Kaizen. To show that process level goals can be feasibly realized at those three stages through the use of decision science methods, three models are introduced in this section: CMMI, simulation, and Six Sigma.

5.1. *Capability maturity model integration*

As noted by Software Engineering Institute at Carnegie Mellon University (http://www.sei.cmu.edu/cmmi/general), CMMI "is a process improvement approach that provides guidance for quality processes, and a point of reference for appraising current processes." The ultimate goal of this process change and development is process maturity. In the world of CMMI, there are five levels of maturity: initial, managed, defined, quantitatively managed, and optimized. Using our corn-derived ethanol process as an example, we present a typical CMMI analysis with a special focus on screening the current state of the process of concern (see Fig. 7).

The data gathered from the CMMI stage can be used to guide efforts and deal with getting to high process maturity. We also characterized CMMI in the context of process EDM using "4W + 1H." The definitions shown in Table 1 outline the process of CMMI and how it leads to process change and improvement. Process decision makers and practicing environmental managers who are interested in continuous improvement may want to adopt our proposed CMMI framework as shown in this section.

Table 1. Process EDM and CMMI.

	Level 1: Initial	Level 2: Managed	Level 3: Defined	Level 4: Quantitatively managed	Level 5: Optimized
Why	Ignorance	Awareness	Acting on the vision	Competing on analytics	Satisfaction of stakeholders
Who	Heroes and people with insufficient skills and resources	People with adequate skills and resources	Highly trained people and team	People and team with the very best analytic skills	Everyone
What	No environmental procedures and formal processes	Standard environmental procedures in place	A powerful organizational structure	Sophisticated information systems and performance metrics	Innovative and incremental improvements
How	Putting out fires	Monitoring and controlling environmental delivery processes	Empowering personnel to follow through	Making expert use of statistics and modeling	Carrying everyone along; preventing errors from happening
When	Getting a sense of urgency	People understanding process priorities and strategic directions	Institutionalizing tailoring environmental guidelines and processes	People working smarter and making better use of their expertise and EDM skills	World-class environmental delivery processes

5.2. *Simulation*

A simulation model evaluates process EDM situations by simulating the operation of the process for an extended period of time. The expected impact of policy changes and risk involved in process EDM can be evaluated through the simulation modeling. It is a descriptive model, not a prescriptive model. We shall focus on the Monte Carlo simulation, one of the best known simulation tools, in this section. Monte Carlo simulation deals with a system where the values of one or more input variables are unknown. By assuming a specific probability distribution for each unknown input variable (random variable) and generating numerous scenarios, one can build a distribution of the desired output. The Monte Carlo simulation approach consists of the following six steps:

1. Determine a probability distribution for each random variable.
2. Build a cumulative distribution for each variable.
3. Establish the interval of random numbers (RN) for each variable.
4. Generate RN between 0 and 1 using an RN generator/use them to generate values for each random variable.
5. Simulate a (sufficient) number of trials to create a distribution of results.
6. Study the system behavior given a set of tactical choices.

To illustrate and develop the Monte Carlo simulation, a hypothetical problem about the cost of energy usage is considered:

5.2.1. *Problem*

Henry, the owner of local cafe, is interested in using a simulation to estimate his utility expenses. The empirical distribution of concern is shown below:

Monthly utility expense $	Relative frequency %	Probability
400	20	0.2
500	20	0.2
600	20	0.2
700	20	0.2
800	20	0.2

(a) Simulate the entire year using the following RNs: 0.6953, 0.0082, 0.6799, 0.8898, 0.6515, 0.3976, 0.0642, 0.0377, 0.5739, 0.5827, 0.0508, and 0.4757. What are the total expenses over the 12-month period?
(b) Suppose that instead, the normal probability distribution with a mean of $600 and a standard deviation of $100 is a good description of the utility expense.

What are the total expenses over the 12-month period? Use the RAND() function to generate RNs.

5.2.2. *Solutions*

(a) An Excel simulation model is established according to the first five steps of Monte Carlo simulation. The utility expense is generated according to the interval that contains the RN. For instance, the RN for the first month, 0.6953, is between 0.6 and 0.8; hence, utility expense of \$700 is generated. It is observed that the total expense over the 12-month period is \$6800 (Fig. 8a). Note that only one test run is presented here. In practice, multiple simulation runs are required.

(b) As for the case of normal distribution with a mean of \$600 and a standard deviation of \$100, the monthly utility expenses can be generated using the formula: =NORMINV(RAND(), μ,σ). With =RAND() in the cells B10 to B21, the specific formulas used in cells C10 to C21 are as follows: =NORMINV(B10, \$B\$5, \$B\$6),..., =NORMINV(B21, \$B\$5, \$B\$6). An Excel simulation model is established according to those general steps presently before. It is observed that the total expense over the 12-month period is \$7003 (Fig. 8b). Note that simulation results will vary when utilizing Excel's RAND() functions. Also note that only one test run is presented here. In practice, multiple simulation runs are required. In many situations, Crystal Ball (http://www.crystalball.com) can also be used to automate the complex tasks required (e.g., multiple runs in Monte Carlo simulations).

From the preceding illustration, it is observed that, simulations can be adopted as a pragmatic way to explore the effectiveness and efficiency of process operations and/or change. For example, Henry can study the impact of any intended tactical choices with the help of simulation models. In the pursuit of sustainability with a focus on environmental and social goals such as cutting waste and using less energy, simulation can be used to describe problems at hand and develop solutions. Simulation is a powerful tool that helps process decision makers evaluate the impact of tactical decisions on system performance in an uncertain environment. For more advanced simulation reports and examples, refer Kuei *et al.* (2008) and Waters (2008).

5.3. *Six Sigma*

Six Sigma is meeting a specific goal provided by the 6σ metric. When a process is operating at Six Sigma level, the variability of the process (measured by standard deviation, σ) is so small that at least 6σ can fit in the distance between the mean and the nearest specification limit. In practice, Six Sigma implies about 3.4 defects per million process outputs, meaning that the capability of a process of concern is defect-free 99.99966% of the time. Defects, in the language of Six Sigma, can be defined as failures to deliver what the customer/stakeholder wants. The essence of Six Sigma needs to be fully integrated into core environmental operations, supporting processes, and strategic planning processes. The engine of Six Sigma is project management via the DMAIC (define, measure, analyze, improve, and control) cycle. The project should also be clearly connected to the strategic objectives of the green systems. To be implemented effectively, one of the key questions is as follows: What is the current state of our system in terms of Sigma level? Relatively low sigma level indicates a need for improvement. To illustrate the Six Sigma way, consider the following example:

Problem: A city college has set an environmental standard that daily paper consumption must be within 2000 lbs. If, out of a sample of 100 days, 10 fail to meet this requirement (i.e., daily paper consumption >2000 lbs), at what sigma level is this process operating?

Solution: The solution procedure is as follows:

Step 1: Determine the number of CTEQs). In this case, there is only one CTEQ: City College's daily paper consumption.
Step 2: Determine the defect rate (DR): $10/100 = 0.1$.
Step 3: Determine the DR per CTEQ characteristic (DRC). DR/no. of CTEQs = $0.1/1 = 0.1$.
Step 4: Compute the Defects Per Million Opportunities. DRC *1,000,000 = $0.1{*}10^6 = 100{,}000$.
Step 5: Determine the Sigma level (z value) by computing the number of sigmas that correspond to the DRC and adjusting it by adding 1.5: Using Excel function, =NORMSINV(1-DRC)+1.5 =NORMSINV(1–0.1)+1.5 = $1.28 + 1.5 = 2.78$

We can see in this example that the current performance is only $2.78\,\sigma$. Once the process is out of favor, process decision makers need to take charge of managing for process change and improvement. The DMAIC cycle is normally adopted here as a change vehicle. One of the primary goals at this stage is always trying to have a stable and capable process (see Fig. 9). Policy makers also need to take a leading role in influencing a system's social and technical components to accomplish what is important as outlined by CTEQs. The fulfillment of those responsibilities should be guided by and rooted in statistical thinking, environmental thinking, and strategic thinking. For details on Six Sigma, refer to chapter eight of Madu (2006). In that chapter, Six Sigma and QFD, Madu (2006) identified some key foci of Six Sigma.

6. Conclusions

This chapter first describes a conceptual framework for the understanding of effective EDM. The framework (Fig. 1) integrates three major aspects of EDM. It is important that decision makers understand the need for a holistic view, the constraints of the decision making, the role of decision science models, and the essence of continuous improvement. This chapter also describes a big picture for implementation (Fig. 2). Both policy- and process-driven change approaches are discussed to evaluate not only the implementation issues but also their potential countermeasures. To get a feel for the different implementation issues involved during policy deployment and process management, we also discuss both policy and process decision makers' worldviews (Fig. 3a,b). On that note, a strategic route map is developed (Fig. 4). The framework highlights the components that are composed of a vertical implementation strategy. With this roadmap, policy makers can quickly incorporate the key components to form a unique environmental strategy and develop a unique capability along with the implementation process. Two decision tools that go with the strategic route map are illustrated in the same section: cognitive mapping and the AHP (Fig. 5a–c). With respect to process management, our thesis is also simple and direct: environmental systems of all sizes need operationally validated frameworks. Such a sound, validated framework involves process practices and the usage of appropriate decision science tools. A detailed blueprint, called a tactical route map, is used in this chapter to complement the existing physical structure and delivery processes. Three decision models are elaborated at this stage: CMMI (Fig. 7), Monte Carlo simulation (Fig. 8a,b), and Six Sigma (Fig. 9). The goal in this chapter is to ensure that sustainable systems are achievable though the adoption of our EDM frameworks. Following our conceptual models and worked examples, environmental leaders can appreciate the major issues involved in building environmental systems, and the techniques used to address them in order to transform the entire delivery process and build the green capability.

References

Cheong, F. and Corbitt, B.J., 2010. From childhood poverty to catfish: A conceptual participatory modeling framework for strategic decision making. *International Journal of Strategic Decision Sciences* **1**(3), 15–33.

Cruz, J.M., 2008. Dynamics of supply chain networks with corporate social responsibility through integrated environmental decision-making. *European Journal of Operational Research* **184**(3), 1005–1031.

Eden, C. and Ackermann, F., 2004. *Making Strategy: The Journey of Strategic Management.* Sage, Thousand Oaks, CA.

Epstein, M.J., 2010. Thinking straight about sustainability. *Stanford Social Innovation Review* **8**(3), 51–55.

Kuei, C., Madu, C.N. and Winch, J.K., 2008. Supply chain quality management: A simulation study. *International Journal of Information and Management Sciences* **19**(1), 131–151.

Madu, C.N., Kuei, C. and Madu, A.N., 1991. Establishing priorities for the information technology industry in taiwan — A Delphi approach. *Long Range Planning* **24**(5), 105–118.

Madu, C.N., 1996. *Managing Green Technologies for Global Competitiveness.* Quorum Books, Westport, CT.

Magee, B., 1998. *The Story of Philosophy.* DK Publishing, NY.

Madu, C.N., 2006. *House of Quality in a Minute.* Chi Publishers, Fairfield, CT.

Madu, C.N., Kuei, C. and Madu, I.E., 2002. A hierarchic metric approach for integration of green issues in manufacturing: A paper recycling application. *Journal of Environmental Management* **64**, 261–272.

Madu, C.N. and Kuei, C., 2004. *ERP and Supply Chain Management.* Chi Publishers, Fairfield, Connecticut.

Saaty, T.L., 1980. *The Analytic Hierarchy Process.* McGraw Hill, New York.

Sarkis, J., and Talluri, S., 2002. A model for strategic supplier selection. *Journal of Supply Chain Management.* **38**(1), 18–28.

Sarkis, J., 2003. A strategic decision framework for green supply chain management. *Journal of Cleaner Production* **11**, 397–409.

Saaty, T.L and Vargas, L.G., 2006. *Decision Making with the Analytic Network Process.* Springer, NY.

Turner, M. and Houston, H., 2008. Start with sourcing. *Strategy + Business* **Autumn**, 96–107.

Waters, D., 2008. *Quantitative Methods for Business.* Prentice Hall, NY.

Appendix

With respect to *environmental sustainability*, make 10 judgments about the relative importance of the items *i* and *j* (Table A1).

Table A1. Pairwise comparison table — Goal.

Land resource usage	Quality of living	Biodiveristy-Rich ecosystems	Investment cost	Carbon dioxide emissions	
Land Resource Usage	1				
Quality of Living	X	1			
Biodiveristy Rich Ecosystems	X	X	1		
Investment Cost	X	X	X	1	
Carbon Dioxide Emissions	X	X	X	X	1

With respect to *land resource usage*, make your judgments about the relative importance of the items *i* and *j* (Table A2).

Table A2. Pairwise comparison table — Factor 1.

	Fuel substitution	Forest restoration
Fuel substitution	1	
Forest restoration	X	1

With respect to *quality of living*, make your judgments about the relative importance of the items *i* and *j* (Table A3).

Table A3. Pairwise comparison table — Factor 2.

	Fuel substitution	Forest restoration
Fuel substitution	1	
Forest restoration	X	1

With respect to *biodiversity-rich ecosystems*, make your judgments about the relative importance of the items i and j (Table A4).

Table A4. Pairwise comparison table — Factor 3.

	Fuel substitution	Forest restoration
Fuel substitution	1	
Forest restoration	X	1

With respect to *investment cost*, make your judgments about the relative importance of the items i and j (Table A5).

Table A5. Pairwise comparison table — Factor 4.

	Fuel substitution	Forest restoration
Fuel substitution	1	
Forest restoration	X	1

With respect to *carbon dioxide emissions*, make your judgments about the relative importance of the items i and j (Table A6).

Table A6. Pairwise comparison table — Factor 5.

	Fuel substitution	Forest restoration
Fuel substitution	1	
Forest restoration	X	1

Chapter 40

CORPORATE SUSTAINABLE DEVELOPMENT: A REVIEW AND DEVELOPMENT OF A RESEARCH MODEL

Wing S. Chow[a] and Yang Chen
School of Business, Hong Kong Baptist University, Hong Kong
[a]*vwschow@net1.hkbu.edu.hk*

This chapter presents a review of the literature on corporate sustainable development (CSD) and proposes a framework for future CSD research. We first review three key CSD constructs, namely environmental, social, and economic development. We then develop a research framework that includes four components: first, CSD modeling, followed by the antecedent, consequent, and contingency constructs, respectively, of CSD.

1. Introduction

A formal definition of sustainable development (SD) was first proposed by the Brundtland Commision, as "development that meets the needs of the present without compromising the ability of future generations to meet their own needs" (WCED, 1987). Since then, the study of SD has mainly taken the form of epistemological studies focusing on the concept (Townsend, 2004). It is thus not surprising to note that many research papers have offered different definitions. Most, if not all, these studies have regarded SD as being orchestrated by three constructs of sustainability, namely environmental, economic, and social systems (Gladwin *et al.*, 1995; Goodland and Daly, 1996; Hopwood *et al.*, 2005; Ko, 2005; Székely and Knirsch, 2005). These three SD constructs have fundamentally evolved from the concept of the triple bottom line (Elkington, 1998).

Corporate sustainable development (CSD) is a crucial branch of SD research and practice (Hart, 1997). The corporation, as a form of the most important constituent of business and society, faces the challenge of moving business strategy from conventional development to SD. Dyllick and Hockerts (2002) define CSD as "meeting the needs of a firm's direct and indirect stakeholders (such as shareholders, employees, clients, pressure groups, communities, etc.), without compromising its ability to meet the needs of future stakeholders as well." Townsend (2004) suggests that firms engaged in CSD are those whose do "not use natural resources more quickly than they can be renewed naturally, than ecosystems' regenerative rates are exceeded before new technologies or sustainable resources can replace, or than recycling rates."

Previous research on CSD has focused primarily on theoretical and empirical studies describing how organizations can survive alongside the natural environment. Recently, scholars have begun to advocate that social development should also be considered as an important factor in CSD strategies (Sharma and Ruud, 2003). This observation has been echoed by the UN Climate Summit held in Copenhagen on December 2009, which proposed that CSD strategy should strike a balance between the three key constructs. Sharma (2002) describes these constructs as environment development, achieved through corporate environmental management; social development, stemming from corporate social responsibility (CSR); and economic development, emerging from value creation.

To date, no research paper has attempted to develop an integrative model bringing these CSD studies together. Accordingly, the objectives of this paper are firstly to shed light on this area by identifying relevant measurement constructs, and then to propose a research framework that can be applied to the further study of CSD models.

In the following section, we first discuss the streams of extant CSD research focusing on measurement constructs, followed by a review of the areas that can be addressed for future research.

2. CSD Research

This section reviews CSD research surrounding the three constructs of environmental, social, and economic development.

2.1. *Environmental development*

Azapagic (2004) claims that environmental development refers to organizations' efforts to manage their operations in such a way that their final products do little harm to our environment, including land, air, and water (Azapagic, 2004). Gladwin (1992) recommends that the objective of environmental development should be to avoid polluting our environment, either directly or indirectly. Organizations could progress this through innovative operations which, for example, waste-proof production processes. There are two approaches, namely the strategic and operational levels.

McDaniel and Gitman (2008) suggest that the strategic level involves "creating long-range (one to five years) broad goals for the organizations and determining what resources will be needed to accomplish those goals." In contrast, the operational level means "creating specific standards, methods, policies, and procedures that are used in specific functional areas of the organizations." We discuss each of these concepts further below.

2.1.1. *The strategic level of environmental development*

The literature documents three different approaches to this: proactive to reactive (Sharma and Vredenburg, 1998), pollution prevention to pollution control (Hart, 1995; Hart and Ahuja, 1996), and compliance to beyond-compliance (Sharma, 2000). The first of these approaches is considered the more extensive, and we elaborate it further in this section.

A reactive environmental strategy means that the organization is complying with the regulations outlined by the local community. For example, organizations should take initiatives to control and reduce their pollution of air, land, and water. One example of this is that firms should manage their emissions and effluents, and ensure that these wastes are stored, treated, and disposed of using end-of-the-pipe pollution-control equipment that incorporates clean-up technologies and processes. Porter and van der Linde (1995) point out that firms practicing CSD need to commit financially when working at the strategic level.

González-Benito and González-Benito (2006) describe a proactive environmental strategy as one where organizations involved in CSD engage voluntarily in such efforts. Hart and Ahuja (1996) explain that a proactive strategy improves production processes, with the end result being to minimize damage to the environment. The benefits of a proactive environmental strategy have been described as including production efficiency (Christmann, 2000), cost reduction (Aragón-Correa, 1998), regulatory compliance (Berry and Rondinelli, 1998), and capital improvement (Anderson, 1999). These benefits can also be treated as competitive advantages to a firm.

2.1.2. *The operational level of environmental development*

Gilley *et al.* (2000) state that this is related to changes made at the production level that reduce harm to the environment. There are two approaches to achieving this, namely product- and process-related development.

Product-related environmental development means the design of environmentally-friendly products, such as packaging, that reduce environmental pollution. Bansal (2005) proposes that this type of development can minimize harm to the environment without compromising product quality. Hart (1995) suggests that product stewardship is a good approach to achieving product-related environmental development. Product stewardship refers to a process where product design and operational processes are considered together such that the final product has a positive environmental impact. Eco-labeling is another application of this concept; it can elicit consumers' awareness of the effect of the product on their health and the environment of the products they are using.

Shrivastava (1995) shows that product-related development increases organizations' awareness of their responsibilities to recycle, reuse, and remanufacture their product wastes. Firms engaged in CSD could evaluate the harm their products are causing to the environment by assessing impact, from the original source of the raw materials throughout the lifespan of their products.

González-Benito and González-Benito (2006) define process-related environmental development as the "development and implementation of more environmentally conscious manufacturing and operational methods and processes." Gilley *et al.* (2000) explain that operational activities in process-related development include recycling waste and redesigning products in ways that will reduce waste. Prahinski and Kocabasoglu (2006) claim that process-related environmental development can be based on the reverse logistic concept, whereas Zsidisin and Siferd (2001) suggest that it can be better achieved by exercising the practice of green purchasing.

2.2. *Social development*

Social development refers to managing human relationships within a company in ways that would reduce social inequality and divisions, improve quality of life, and strengthen relationships with its various stakeholders.

Social development is an important construct in CSD (Sharma and Ruud, 2003). Lafferty and Langhelle (1999) claim that it should be considered as an ethical code that facilitates human survival and progress, on a par with other high-minded ideas such as democracy, freedom, and human rights. Bansal (2005) states that social development is based on the concept of CSR. Social development has evolved from CSR, through which corporations can develop strong relationships with employees through operational transparency and distribution of the value created (Wood, 1991; Hillman and Keim, 2001). There are four CSR theories: the instrumental, political, integrative, and ethical (Garriga and Melé, 2004).

2.2.1. *Instrumental theories*

Garriga and Melé (2004) show that social development is intended to achieve economic performance in organizations. McWilliams and Siegel (2001) suggest that the distribution of wealth creation to stakeholders is one example of how a firm engaged in CSD could achieve social development. McGuire *et al.* (1988) and Cochran and Wood (1984) conclude that social development has a direct and positive effect on the productivity of a firm.

2.2.2. *Political theories*

Political theories suggest that firms engaged in CSD should abide by local laws and the legal system. Wood and Lodgson (2002) and Carroll (1999) also recommend that organizations should be responsible for their actions and participate in activities which engage the goodwill of society. In social development, Bansal and Roth (2000) suggest that organizations should conduct business transactions and operations lawfully. Davenport (2000) shares this view and adds that social development should strike a balance of the interests of stakeholders and the local community.

2.2.3. *Integrative theories*

Integrative theories refer to ways of conducting business based on social demands. Garriga and Melé (2004) define the latter as those activities which are "generally considered to be the way in which society interacts with business and gives it a certain legitimacy and prestige." Social development is thus related to social demands. Organizations practicing CSD should consider social demands as part of their operations. This should not only be limited to a consideration of the interests of their stakeholders, but also encompass the benefits and constraints experienced by suppliers, customers, and the local community. Hillman and Keim (2003) claim that a social development-oriented focus on social demands would enhance the reputation and financial performance of a firm.

2.2.4. *Ethical theories*

Garriga and Melé (2004) state that organizations should conduct their business ethically and develop a good relationship with the local community. McWilliams *et al.* (2006) claim that social development involves engagement with local ethical issues, such as human and labor rights, and respect for the environment. Alford and Naughton (2002) suggest that firms engaged in CSD can implement social development by promoting the harmony of society by donating cash, equipment, and products to those in need.

2.3. *Economic development*

Makadok (2003) claims that firms engaged in CSD should be able to survive as durable participants in the business arena, with sound financial performance. López-Gamero *et al.* (2009) claim that economic development can be promoted by engaging operations such as product innovation and differentiation, and cost reduction. Steurer *et al.* (2005) suggest that the objectives of economic development could be based on improvements in share earnings and competitiveness. Bansal (2005) suggests that economic development is activated by value creation, which is calculated by subtracting capital investment from market value (Hillman and Keim, 2001). Other ways to measure value creation include the reduction of operations costs (Farrell, 2005), revenue generation (Seth, 1990), stimulating sales growth (Chirstmann, 2000), improving the relationship between buyers and sellers (Ulaga, 2003), improving the production process (Porter and van der Linde, 1995), and the enhancement of government regulations (Makadok, 2001).

3. CSD Research Framework

In this section, we propose a framework to shape and model future research on CSD. This is depicted in Fig. 1. As can be seen, the framework consists of four main components: first, CSD modeling, followed by the antecedent, consequent, and contingency constructs, respectively, of CSD.

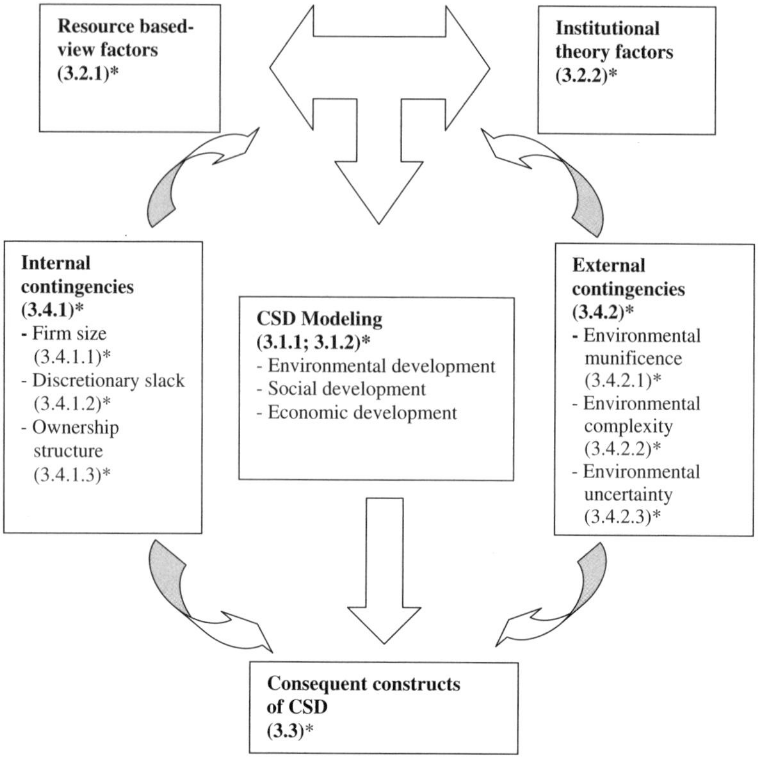

Fig. 1. A framework for CSD research.

3.1. *CSD modeling*

Our approach to this consists of two different concepts. The first is the development of measurement items for CSD, and the second is the determination of approaches to the use of structural equation modeling (SEM) for CSD.

3.1.1. *CSD measurement items*

CSD measurements can be based on either the most conventional method — that is, an extensive review of the relevant literature — or the study of operational systems. Table 1 outlines a group of measurement items gathered using the former approach. For ease of reference, we have grouped all items based on the three underpinning constructs of environmental, social, and economic development. One possible way to extend the work represented in Table 1 is to apply confirmatory factor analysis to quantify the critical success factors involved in each.

A methodology for the latter approach can be found in the work of Ravichandran and Rai (1999, 2000), who collected measurement items about one information technology system by analyzing operational performance based on the concept of total quality management.

Table 1. Measurement items for CSD constructs.

CSD constructs	Possible measurement items
Environmental development	Product and packaging recovery; emissions to air, water, and land; impact on ecosystem; air resources use; water consumption; safe disposal of wastes; toxic waste handled responsibility; energy use; renewable energy use; material recycling; logistics and transportation; supply chain management; environmentally sensitive locations; land use; disposal and treatment of hazardous/toxic wastes; environmental monitoring; environmental legislation; monitor and report; and public disclosure
Social development	Employee participation; stakeholder involvement; transparency; employee health and safety; community health and safety; supporting local community; equal treatment; treatment to discrimination; harassment; violence; social impact of operation; investment in social projects; customer comfort; customer awareness; and customer satisfactions
Economic development	Payment made to employees; payment per share; tax paid; environmental cost; innovation capability and R&D expenditure; production; revenue and payment; competitiveness; prices; supply efficiency; financial health; economic performance; financial benefit; trading opportunity; profit; growth; economic resources; capital distribution; and financial responsibility

3.1.2. *Structural Equation Modeling (SEM)*

SEM in this context can be considered as an extension of the research described in Section 3.1.1. It can be based on both the first and second order (see Fig. 2). The first order refers to a study of the relationships between CSD constructs. The second order denotes the study of the constructs as an integrative model; that is, they are not all treated as independent. Figure 2 depicts these two different types of SEM. Each model consists of the three underpinning CSD constructs, namely environmental, social, and economic development. The second-order approach is considered as better and also more economical than the first-order approach, because it facilitates better measurement of CSD performance. However, additional work has to be done to determine the existence of the second order because it may not be a statistically valid model. Readers who are interested in second-order modeling can refer to the work of Shi *et al.* (2005), Tojib *et al.* (2008), and Smith *et al.* (2009).

3.2. *Antecedent constructs of CSD*

The antecedent constructs of CSD refer to the factors which influence via a direct effect on the three underpinning constructs. They can be based on a resource-based view and institutional factors.

3.2.1. *Resource-based view factors*

Aragón-Correa and Sharma (2003) and Luo and Bhattacharya (2006) show that organizational resources have a direct effect on the level to which firms have implemented CSD practices. Chan (2005) claims that there are two types of factors within

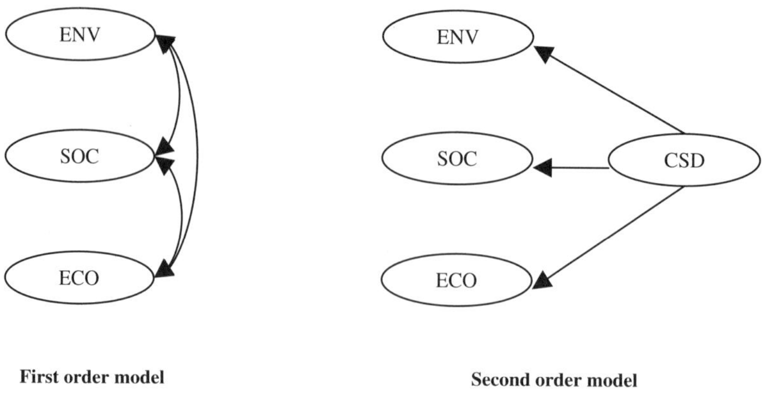

ECO: Economicdevelopment; SOC: Social development; ENV: Environmental development
CSD: Corporate sustainable development.

Fig. 2. Two structural models for CSD research.

the resource-based view, namely the availability of technology, and the availability of financial commitment. Sharma *et al.* (2007) suggest that organizations can treat continuous innovation of their products and services as part of their organizational resources. Bansal (2003) and Sharma (2000) propose that the level to which organizational resources have a significant effect on CSD practices is also subject to firms' responsibility to the natural environment and the competitiveness of their services.

3.2.2. *Institutional factors*

Institutional factors denote "the social context within which firms operate" (Bansal, 2005). This is related to a commitment to the environmental standards that regulate the local community. Examples of these are the ISO 14000 and ISO 9000 series. Other institutional factors may also be included, such as the behavior of customers (Porter, 1990), media attention (Bansal, 2005), and the requirements of the community (Henriques and Sadorsky, 1999).

3.3. *Consequent constructs of CSD*

These refer to measures of CSD performance. Chang and Kuo (2008) show that one such construct can be based on the firm's financial performance. Elkington (1997) and Székely and Knirsch (2005) extend this to include factors such as environmental and social performance. Enticott and Walker (2008) claim that public organizations place more emphasis on sustainable performance.

3.4. *Contingency factors in CSD*

Contingency factors are the moderator or control factors in CSD practices. There are two types, namely internal and external.

3.4.1. *Internal contingency factors*

These refer to the demographic data of CSD organizations, which include firm size, discretionary slack, and ownership structure.

3.4.2. *Firm size*

Many studies have shown that the decision variable of firm size has a direct and positive correlation with the CSD practices of a firm (Russo and Fouts, 1997; Aragón-Correa, 1998; Judge and Douglas, 1998; Bansal, 2005; González-Benito and González-Benito, 2005). Stanwick and Stanwick (1998) and Chan (2005) report that large firms encounter societal pressure to implement CSD practices: this is unlikely to be the case for smaller firms (Hunt and Auster, 1990).

3.4.3. *Discretionary slack*

Discretionary slack refers to the unused capacities of organizational resources (Etzioin, 2007). Extant studies have confirmed that firms with a large amount of slack have a tendency to adopt CSD practices (Aragon-Correa *et al.*, 2004; Bowen and Sharma, 2005; Bansal, 2005). Love and Nohria (2005) support this view and explain the reason as the fact that such firms can transfer their slack resources to promoting CSD activities such as production innovation and R&D.

3.4.4. *Ownership structure*

Chan (2005) shows that the ownership structure of organizations has played a significant role in CSD adoption in China. Darnall and Edwards (2006) demonstrate that firms like public companies are more likely to adopt CSD practices because they are better able to secure the necessary financial commitment. Other ownership structures that may contribute to CSD adoption include having a multinational scope (Bansal and Roth, 2000; Etzioin, 2007), international experience (Bansal, 2005), and top management commitment (Wade and Hulland, 2004).

3.4.5. *External contingency factors*

External contingency factors are those related to the business environments in which firms are engaged, such as environmental munificence, complexity, and uncertainty.

3.4.5.1 *Environmental munificence*: Environmental munificence is "the extent to which a business environment can support sustained growth" (Wade and Hulland, 2004). Aragón-Correa and Sharma (2003) state that firms operating in munificent environments will be likely to adopt CSD practices because this factor promotes experimentation with the concepts, and learning from them.

3.4.5.2 *Environmental complexity*: Wade and Hulland (2004) define this as "the heterogeneity and range of an industry and/or an organization's activities."

Aragón-Correa and Sharma (2003) suggest that the complexity of a firm's operational environment has a direct and negative effect on its engagement with CSD practices.

3.4.5.3 *Environmental uncertainty*: The unpredictable nature of the business environment will cause some uncertainty when it comes to a firm's options for the future. Aragón-Correa and Sharma (2003) suggest that environmental uncertainty will affect the relationship between environmental development and a firm's competitive advantage. Chan (2005) has shown statistically that environmental uncertainty has a contingency effect on the adoption of CSD practices.

4. Conclusions

CSD has recently become a focal point of research because of the widespread interest in sustainability management. Firms undertake research in CSD as a means of promoting a business venture while doing less harm to our natural environment.

This chapter serves two objectives. First, we review CSD practices based on three key constructs of environmental, social, and economic development. Second, we present a research framework that can be used to conduct future studies of CSD. It has four main components: CSD modeling, the antecedent, consequent, and contingency constructs of CSD. This chapter presents evidences from the literature in support of each of these components.

References

Alford, H. and Naugthon, M., 2002. Beyond the shareholder model of the firm: Working toward the common good of a business. In: Cortright, S.A. and Naugthon, M. (Eds.), *Rethinking the Purpose of Business Interdisciplinary Essays from the Catholic Social Tradition*. Notre Dame University Press, Notre Dame, pp. 27–47.

Anderson, D.R., 1999. Incorporating risk management into environmental management systems. *CPCU Journal* **52**, 115–124.

Aragón-Correa, J.A., 1998. Strategic proactivity and firm approach to the natural environment. *Academy of Management Journal* **41**(5), 556–567.

Aragón-Correa, J.A. and Sharma, S., 2003. A contingent resource-based view of proactive corporate environmental strategy. *Academy of Management Review* **28**, 71–88.

Aragón-Correa, J.A., Matias-Reche, F. and Senise-Barrio, M.E., 2004. Managerial discretion and corporate commitment to the natural environment. *Journal of Business Research* **57**(9), 964–975.

Azapagic, A., 2004. Developing a framework for sustainable development indicators for the mining and minerals industry. *Journal of Cleaner Production* **12**, 639–662.

Bansal, P., 2003. From issues to actions: The importance of individual concerns and organizational values in responding to natural environmental issues. *Organization Science* **14**(5), 510–527.

Bansal, P., 2005. Evolving sustainably: A longitudinal study of corporate sustainable development. *Strategic Management Journal* **26**, 197–218.

Bansal, P. and Roth, K., 2000. Why companies go green: A model of ecological responsiveness. *Academy of Management Journal* **43**(4), 717–736.

Berman, S.L., Wicks, A.C., Kotha, S. and Jones, T.M., 1999. Does stakeholder orientation matter? The relationship between stakeholder management models and firm financial performance. *Academy of Management Journal* **42**(5), 488–506.

Berry, M.A. and Rondinelli, D.A. 1998. Proactive corporate environmental management: A new industrial revolution. *Academy of Management Executive* **12**(2), 38–50.

Bowen, F.E. and Sharma, S., 2005. Resourcing corporate environmental strategy: Behavioral and resource-based perspectives. Paper Presented at the *Conference of the Academy of Management, Honolulu, HI*, 2005.

Carroll, A.B., 1999. Corporate social responsibility. Evolution of definitional construct. *Business and Society* **38**(3), 268–295.

Chan, R.Y.K., 2005. Does the natural-resource-based view of the firm apply in an emerging economy? A survey of foreign invested enterprise in China. *Journal of Management Studies* **42**(3), 625–672.

Chang, D. and Kuo, L.R., 2008. The effects of sustainable development in firms' financial performance — an empirical approach. *Sustainable Development* **16**, 365–380.

Christmann, P., 2000. Effects of 'best practices' of environmental management on cost advantage: The role of complementary assets. *Academy of Management Journal* **43**(3), 663–680.

Cochran, P.L. and Wood, R.A., 1984. Corporate social responsibility and financial performance. *Academy of Management Journal* **27**(1), 42–56.

Darnall, N. and Edwards, D. Jr., 2006. Predicting the cost of environmental management system adoption: The role of capabilities, resources and ownership structure. *Strategic Management Journal* **27**, 301–320.

Davenport, K., 2000. Corporate citizenship: A stakeholder approach for defining corporate social performance and identifying measures for assessing it. *Business and Society* **39**(2), 210–220.

Donaldson, T. and Preston, L., 1995. The stakeholder theory of the corporation: Concepts, evidence, and implications. *Academy of Management Review* **20**, 65–91.

Dyllick, T. and Hockerts, K., 2002. Beyond the business case for corporate sustainability. *Business Strategy and the Environment* **11**, 130–41.

Elkington, J., 1997. *Cannibals with forks. The Triple Bottom Line of 21st Century Business.* Capstone Publishing Ltd., Oxford, UK.

Elkington, J., 1998. *Cannibals with forks: The Triple Bottom Line of the 21st Century.* New Society Publishers, Stoney Creek, CT.

Enticott, G. and Walker, R.M., 2008. Sustainability, performance and organizational strategy: An empirical analysis of public organizations. *Business Strategy and the Environment* **17**(2), 79–92.

Etzioin, D., 2007. Research on organizations and the natural environment, 1992-Present: A review. *Journal of Management* **33**(4), 637–664.

Farrell D., 2005. Offshoring: Value creation through economic change. *Journal of Management Studies* **42**, 675–683.

Garriga, E. and Melé, D., 2004. Corporate social responsibility theories: Mapping the territory. *Journal of Business Ethics* **53**, 51–71.

Gilley, K.M., Worrell, D.L., Davidson, W.N. and El-Jelly, A., 2000. Corporate environmental initiatives and anticipated firm performance: The differential effects of process-driven versus product-driven greening initiatives. *Journal of Management* **26**(6), 1199–1216.

Gladwin, T.N., 1992. *Building the Sustainable Corporation: Creating Environmental Sustainability and Competitive Advantage.* National Wildlife Federation, Washington, DC.

Gladwin, T.N., Kennelly, J.J. and Krause, T., 1995. Shifting paradigms for sustainable development: Implications for management theory and research. *Academy of Management Review* **20**(4), 874–907.

González-Benito, J. and González-Benito, O., 2005. An analysis of the relationship between environmental motivations and ISO14001 certification. *British Journal of Management* **16**(2), 133–148.

González-Benito, J. and González-Benito, O., 2006. A review of determinant factors of environmental proactivity. *Business Strategy and the Environment* **15**, 87–102.

Goodland, R. and Daly, H., 1996. Environmental sustainability: Universal and non-negotiable. *Ecological Applications.*

Hart, S.L., 1995. A natural-resource-based view of the firm. *Academy of Management Review* **20**(4), 986–1014.

Hart, S.L., 1997. Beyond greening: Strategies for a sustainable world. *Harvard Business Review* **75**(1), 66–76.

Hart, S. and Ahuja, G., 1996. Does it pay to be green? An empirical examination of the relationship between emission reduction and firm performance. *Business Strategy and the Environment* **5**, 30–37.

Henriques, I. and Sadorsky, P., 1999. The relationship between environmental commitment and managerial perceptions of stakeholder importance. *Academy of Management Journal* **42**(1), 87–99.

Hillman, A.J. and Keim, G.D., 2001. Shareholder value, stakeholder management, and social issues: what's the bottom line? *Strategic Management Journal* **22**, 125–139.

Hopwood, B., Mellor, M. and O'Brien, G., 2005. Sustainable development: Mapping different approaches. *Sustainable Development* **13**, 38–52.

Hunt, C.B. and Auster, E.R., 1990. Proactive environmental management — Avoiding the toxic trap. *Sloan Management Review* **31**(2), 7–18.

Judge, W.Q., Jr. and Douglas, T.J., 1998. Performance implications of incorporating natural environmental issues into the strategic planning process: An empirical assessment. *Journal of Management Studies* **35**(2), 241–262.

Ko, T.G., 2005. Development of a tourism sustainability assessment procedure: A conceptual approach. *Tourism Management* **26**, 431–445.

Lafferty, W. and Langhelle, O., 1999. *Towards sustainable development: On the Goals of Development — and the Conditions of Sustainability.* Macmillan: London.

López-Gamero, M.D., Molina-Azorín, J.F. and Claver-Cortés, E., 2009. The whole relationship between environmental variables and firm performance: Competitive advantage and firm resources as mediator variables. *Journal of Environmental Management* **90**(10), 3110–3121.

Love, E.G. and Nohria, N., 2005. Reducing slack: The performance consequences of downsizing by large industrial firms, 1977–1993. *Strategic Management Journal* **26**, 1087–1108.

Luo, X. and Bhattacharya, C.B., 2006. Corporate social responsibility, customer satisfaction, and market value. *Journal of Marketing* **70**, 1–18.

Makadok, R., 2001. A pointed commentary on Priem and Butler. *Academy of Management Journal* **31**, 854–872.

Mansfield, E., Rapoport, J., Romeo, A., Wagner, S. and Beardsley, G., 1977. Social and private rates of return from industrial innovation. *Quarterly Journal of Economics* **91**, 221–240.

McDaniel, C. and Gitman, L.J., 2008. *The Future of Business: The Essentials.* Third ed. Thomson Higher Education, USA.

McGuire, J., Sundgren, A. and Schneeweis, T., 1988. Corporate social responsibility and firm financial performance. *Academy of Management Journal* **31**(4), 854–872.

McWilliams, A. and Siegel, D., 2001. Corporate social responsibility: A theory of the firm perspective. *Academy of Management Review* **26**(1), 117–127.

McWilliams, A., Siegel, D. and Wright, P.M., 2006. Corporate social responsibility: Strategic implications. *Journal of Management Studies* **43**(1), 1–18.

Porter, M.E., 1990. The Competitive Advantage of Nations. Free Press, New York.

Porter, M.E. and van der Linde, C., 1995. Green and competitive. *Harvard Business Review* **73**(5), 120–134.

Prahinski, C. and Kocabasoglu, C., 2006. Empirical research opportunities in reverse supply chain. *Omega* **34**(6), 519–532.

Ravichandran, T. and Rai, A., 1999/2000. Total quality management in information systems development: Key constructs and relationships. *Journal of Management Information Systems* **16**(3), 119–155.

Russo, M.V. and Fouts, P.A., 1997. A resource-based perspective on corporate environmental performance and profitability. *Academy of Management Journal* **40**(3), 534–559.

Seth, A., 1990. Value creation in acquisitions: A re-examination of performance issues. *Strategic Management Journal* **11**(2), 99–115.

Sharma, S., 2000. Managerial interpretations and organizational context as predictors of corporate choice of environmental strategy. *Academy of Management Journal* **43**(4), 681–697.

Sharma, S., 2002. What really matters: Research on corporate sustainability? In: Sharma, S. and Starik, M., (Eds.), *Research in Corporate Sustainability: The Evolving Theory and Practice of Organizations in the Natural Environment.* Elgar, Northampton, MA, 1–29.

Sharma, S. and Ruud, A., 2003. On the path to sustainability: Integrating social dimensions into the research and practice of environmental management. *Business Strategy and the Environment* **12**, 205–214.

Sharma, S., Aragón-Correa, J.A. and Rueda-Manzanares, A., 2007. The contingent influence of organizational capabilities on proactive environmental strategy in the service sector: An analysis of North American and European ski resorts. *Canadian Journal of Administrative Sciences* **24**, 268–283.

Sharma, S. and Vredenburg, H., 1998. Proactive corporate environmental strategy and the development of competitively valuable organizational capabilities. *Strategic Management Journal* **19**(8), 729–753.

Shi, Z., Kunnathur, A.S. and Ragu-Nathan, T.S., 2005. Is outsourcing management competence dimensions: Instrument development and relationship exploration. *Information and Management* **42**, 901–919.

Shrivastava, P., 1995. The Role of corporations in achieving ecological sustainability. *Academy of Management Review* **20**(4), 936–960.

Smith, J.S., Karwan, K.R. and Markland, R.E., 2009. An empirical examination of the structural dimensions of the service recovery system. *Decision Sciences* **40**, 165–186.

Stanwick, P.A. and Stanwick, S.D., 1998. The relationship between corporate social performance and organizational size, financial performance, and environmental performance: An empirical examination. *Journal of Business Ethics* **17**(2), 195–204.

Steurer, R., Langer, M.E., Konrad, A. and Martinuzzi, A., 2005. Corporations, stakeholders and sustainable development I: A theoretical exploration of business-society relations. *Journal of Business Ethics* **61**, 263–281.

Székely, F. and Knirsch, M., 2005. Responsible leadership and corporate social responsiblility: metrics for sustainable performance. *European Management Journal* **23**(6), 628–647.

Tojib, D.R., Sugianto, L. and Sendjaya, S., 2008. User satisfaction with business-to-employee portals: Conceptualization and scale development. *European Journal of Information Systems* **17**, 649–667.

Townsend, A.K., 2004. An assessment and critique of green business best practices. *Phd Thesis.*

Ulaga, W., 2003. Capturing value creation in business relationships: A customer perspective. *Industrial Marketing Management* **32**, 677–693.

Wade, M. and Hulland, J., 2004. Review: The resource-based view and information systems research: Review, extension, and suggestions for future research. *MIS Quarterly* **28**(1), 107–142.

Wood, D.J., 1991. Corporate social performance revisited. *Academy of Management Review* **16**(4), 691–718.

Wood, D.J. and Lodgson, J.M., 2002. Business citizenship: From individuals to organizations. *Business Ethics Quarterly* **3**, 59–94.

World Commission on Environment and Development, 1987. *Our common future: Brundtland report.* Oxford University Press, New York.

Zsidisin, G.A. and Siferd, S.P., 2001. Environmental purchasing: A framework for theory development. *European Journal of Purchasing and Supply Management* **7**(1), 61–73.

Chapter 41

DEVELOPING DESIGN SYSTEMS FOR SUSTAINABILITY

Chu-hua Kuei[a] and Chinho Lin[b]

[a]*Department of Management and Management Science,*
Lubin School of Business, Pace University,
1 Pace Plaza, New York, NY 10038, USA
ckuei@pace.edu

[b]*Department of Industrial and Information Management and*
Institute of Information Management,
College of Management Science,
National Cheng Kung University, Taiwan
linn@mail.ncku.edu.tw

The purpose of this chapter is threefold: first, to introduce the worldviews of modern designers; second, to discuss the issues of design science in the context of sustainability; and third, to propose a design system blueprint for responsible change. On the path to creating design systems for sustainability, designers need to better understand critical aspects of a sustainable design problem space. To ensure the quality of the design systems for sustainability, a modern design science framework is needed. The framework could be used as a guide in developing a complex operations network but most importantly, it emphasizes the need to address the context, interventions, mechanisms, outputs, and outcomes around sustainable development. To ensure the effectiveness of implementation, a unique design system blueprint is also established. Designers and project managers should focus their attention and resources on these critical areas to achieve and sustain organizational overall success in terms of economic development, environmental protection, and social equity.

1. Introduction

Sustainability is the best approach today to describe how to integrate economic, environmental, and social impacts into all organization decisions (Epstein, 2010; Nidumolu *et al.*, 2009; Esquer-Peralta *et al.*, 2008; Lo and Sheu, 2007). As noted by Nidumolu *et al.* (2009, p. 58), in the future, only organizations "that make sustainability a goal will achieve competitive advantage." Sustainability management focuses on economic development, environmental protection, and social equity. These three areas constitute what is referred to as the triple bottom line (TBL). To respond to the challenges of the developed/developing world in moving toward *the TBL*, design activities must be managed. What is design? This is a *question* which needs a satisfactory *answer*. Salustri and Eng (2007, p. 22) state that "one designs to cause an eventuality." This can be done through devising "courses of action aimed at changing existing situations into preferred ones (Votolato, 1998, p. 7)." Other definitions have also been proposed. Some of them are shown as follows:

- "Design makes, shapes, and transforms (McVeigh, 2007, p. 36)."
- "Design is the conception and realization of new things (Vaughan *et al.*, 2007, p. 33)."
- "Design is contextual, exploratory, aesthetic, and applied (Vaughan *et al.*, 2007, p. 33)."
- "Design is about desired change (McFall and Beacham, 2008, p. 68)."

McVeigh (2007, p. 36) further suggests that "design affects objects and our objects affect us," and "the process of design moves us from a given state to another realization of potential." This view is echoed by Votolato (1998). As noted by Votolato (1998, p. 5), design is "the process of conceiving, planning, and describing the form of an object to be made either by oneself or by someone else." Forms in this case are the explanation of things. Forms, in Aristotle's words, consist of aim (the final cause), function (the efficient cause), appearance (the formal cause), and component (the material cause; Magee, 1998).

The theoretical approach used to analyze different schools of design thinking in the 20th century is also developed by Votolato (1998). From a functional perspective, for example, design is composed of three activities: a rational analysis of need, construction techniques, and abstractions of forms (Votolato, 1998). Action can thus be seen as "the presence of function (Votolato 1998, p. 79)." An approach similar to that of Votolato is also presented by Brown (2008). In his contribution to the theoretical model of design thinking, Brown (2008) emphasizes three issues: inspiration, ideation (or idea generation), and implementation. Further, according to McVeigh (2007, p. 37), neither the designed (design-as-product) nor design (design-as-process) "can lay claim to any sense of superiority."

Design, as a result of these discussions, can be viewed as a set of product and process development activities aimed at achieving some meaningful results. How design can be best managed in order to meet organizations' overall goal thus is the issue at hand here. In the following section, we shall focus on the worldviews of modern designers. In Section 3, we present the birth of design science framework in the context of sustainability. In Section 4, we present the outline of modern design systems for sustainable development. Our models are helpful to organization around the globe to understand the key aspects of design systems for sustainability and enhance the current knowledge base in sustainability literature.

2. The Worldview of Designers — A Qualitative Comparative Approach

Researchers studying the designed (design-as-product) and design (design-as-process) have documented the importance of understanding the context of design problems (Esquivel, 2009; Brown, 2008; McFall and Beacham, 2008; McVeigh, 2007; Salustri and Eng, 2007). Institutional constraints and the social environment of the design process need to be taken into account in the *earliest stage* of design system development. In this section, we use a qualitative comparative approach to investigate 17 groups of constructs. In exploring those constructs, we seek to help designers better understand aspects of a design problem space in a complex setting. On the path to creating design systems for sustainability, designers and executive managers of design teams must be fully conversant with these 17 groups of constructs presented here.

2.1. *Analytical thinking and holistic thinking*

Sound rational *thinking* is the root of good designs. It is useful for modern designers to distinguish between two very different ways of thinking: analytical and holistic. The former style involves studying the parts and their relations and scientific approaches to produce (predetermined) system effects, whereas the latter tries to solve a problem by considering every aspect of one large-scale system (e.g., vision, thinking, structure, interventions, arrangements, functions, context, message, problems of uncertainty, the needs of others, mechanisms, decision making and support quality standards, and the awareness of design activities). The competitive environment of the future centering on economic vitality, resource/energy conservation, low carbon/pollution, and community development will require a good mix of these two thinking styles.

2.2. *Analysis and creativity*

Although designers are not always conscious about analytical analyses and skills, many routine parts of design rely on some particular method of reasoning or argumentation (e.g., forecasting, decision trees, quality control). One should not be surprised that designers, when facing sustainability challenges, also need to rely on both quantitative reasoning and qualitative analyses when trying to reach their decisions. Such challenges, as noted by Nidumolu *et al.* (2009), include (1) viewing compliance as opportunity, (2) making value chains sustainable, (3) designing sustainable products and services, (4) developing new business models, and (5) creating sustainability practice platforms. Creativity, on the other hand, always gets designers' attention. Traditionally, creativity is about coming up with new and useful initiatives and enhancing the ability to detect connections between ideas. In the modern, uncertain material world, creativity might need to broadly be defined as a purposeful act of shaping some aspect of the designed and design to make things better. Creativity in this context is therefore not an exercise of self-expression. *Creativity* is a catalyst for increased system sustainability and *innovation in learning*. On the path to creating design systems for sustainability, designers need to pay attention to innovation opportunities that go with those five challenges outlined by

Nidumolu *et al.* (2009). For making value chains sustainable, for example, Nidumolu *et al.* (2009, p. 60), propose three innovation opportunities: (1) innovative uses of returned goods, (2) innovative uses of clean energy sources such as wind, solar power, and nuclear fusion, and (3) finding sustainable sources of parts, components, and raw materials.

2.3. *Creativity engineering and design engineering*

The road to creativity might be divided into creativity engineering and design engineering. Creativity engineering is a matter of following through a proven path and coming up with creative solutions to design problems. Perhaps the best known example with regard to creativity engineering is Toyota Production System (TPS). With a focus on customer satisfaction and the needs of organizations, TPS outlines a flexible production system blueprint that encourages creative solutions and continuous improvement. Design engineering consists of the creation of design systems and transforming some aspect of the environment from a chaotic state into a more preferred (or ordered) one. Designers, in this context, will make attempts to model a design system with a non-linear structure approach using elements of our proposed design science framework (see Fig. 1). Designers acknowledge the fact that a built system is normally a non-linear, dynamic system that *exists* and *operates* within an environment with imperfect information, continual change, and stress and discomfort factors. The emphasis here is thus on consistent strategic directions, principles of sustainability, organized structures, system flexibility and behaviors, stimulus-response rules and forms, collaborative and self-organization capacities, and learning and growth capabilities. *Both approaches encourage* an expanded application of human creativity in meaningful, communicative contexts.

2.4. *Planning and designing*

Salustri and Eng (2007, p. 22) state that "while one usually plans for an eventuality, one designs to cause an eventuality." When modern designers look around, there are countless examples of planning activities. *Strategic planning, for example*, provides a logical framework for dealing with non-structured problems (e.g., changing markets, technological advances, new products and processes). In the long-term planning perspective, the chance of confronting such a problem is relatively high. Planning activities and processes can therefore be more specified, even standardized through using Dr. Deming's Plan–Do–Check–Act cycle to cope with highly *anticipated* upcoming *events*. In the case of designing, systems are created and developed to cause an eventuality. Although the overall processes used under both planning and designing perspectives will likely be very similar, as noted by Salustri and Eng (2007), designing is, however, generally *more preventive and flexible in nature*. The attempt of designing is to prevent repeated occurrences of problems and undesired behaviors, and offer workable principles and rules in advance. Nidumolu *et al.*

(2009) suggest five principles that may be necessary in responding to this challenge: (1) establishing the shape of things to come regarding sustainability, (2) focusing on experiments, pilot studies, competencies needed, and successful stories, (3) staying wedded to the goal of sustainability while constantly adjusting tactics, (4) forming alliances with other enterprises, governments, and non-governmental agencies, and (5) using a global presence to experiment and considering the international context of sustainable organizations.

2.5. *Technical proof and business case*

Sometimes the issue at hand is judged first by referring to technical proof and/or stakeholders' complaints and reactions. Six Sigma in the domain of quality management, for example, allows managers to readily describe process performances using a common metric: Defects Per Million Opportunities (DPMO). Generally, a Six Sigma system is defined as having 3.4 DPMO. One question that persists in this context is thus "at what sigma level is the process in question?" In practice, *finding* a *solution* to a Six Sigma *problem* is not difficult. Modern designers, when investigating the path to creating design systems for sustainability, should also ask: What, then, is the main function of technical proofs? How do essential facts fit in the big picture? Is it important to step back occasionally and ponder the business case at hand? Is it important to focus on technical proofs as well as business cases? What steps can organizations take today that will enable them to design sustainability strategies, sustainability structures, sustainability systems, sustainability programs, and sustainability projects. These are some of the questions that modern designers need to think through by themselves before applying sound designing thinking and practices for sustainability.

2.6. *Assembly-line production and craft production*

Chipotle Mexican Grill, a chain of restaurants specializing in burritos and tacos in the US, is known for its assembly-line production approach when processing customers' orders. Using assembly-line balancing techniques, large efficiency gains could thus be realized. On the other hand, in the education, housing, and musical instrument industries, *craft production* has still been a significant focus of many aspects of the designed and design. To Martin Luther King (MLK) Jr. Middle School, for example, craft production is central to delivering productive citizens and lifetime learners. To achieve this, other than instructional leadership, innovative curriculums, and infrastructure designs (e.g., edible schoolyard), MLK parents are also expected to be involved along the learning progress of their children. Designers today need to develop a well-balanced sustainable system by identifying aspects of managerial and policy challenges, evaluating the effectiveness of specific production strategies, and facilitating designing from the view point of Design for Sustainability. The aim is to meet sustainability goals.

2.7. *Project management and competence development*

Project management is concerned with scheduling, monitoring, and controlling the activities of a project involving interrelated activities. It is normally adopted to ensure accountability, prevent unwarranted deviation, and implement (continuous improvement and/or reengineering) strategies. Well-cited examples include business and organizational change projects, the Six Sigma quality project, the Enterprise Resource Planning project, and the Sustainable Operations Management project. Project management techniques in general allow the organization to combine the advantages of internal efficiency and effectiveness with external responsiveness. Designers need to pay early attention to such techniques. Competence is a portfolio of organizational, managerial, technical, and strategic capabilities and skills developed by organizations along the design systems over time. Competence development is ongoing and repetitive. It enables organizations to find *creative* ways of *accomplishing* sustainability goals with regard to the designed and design.

2.8. *Decision science and behavior science*

Many of our decisions concerning the designed and design actually display an underlying logic and principle based on decision science and behavior science. Decision science is the discipline of using scientific methods to solve individual, group, and organization problems. Analytical methods (e.g., mathematical programming) and simulation techniques are usually adopted to understand and improve decision making with respect to problems at hand. Behavior science is the field of studying the effect of human actions on relationships and decision making. It involves disciplines such as psychology, cognitive theory, anthropology, and organizational behavior. Both decision and behavior sciences require the following skills: communication, data analyses, critical thinking and problem formulating, and solving skills.

2.9. *Uni-objective function and multi-objective function*

When organizations prepare to make things, other than facing a set of constraints that limit the possible solutions, they have to be clear about their objective(s). Linear programming in this context can show modern designers how to formulate an aim (either maximizing or minimizing) and simultaneous equations to search for the best possible solutions to problems. It is, however, worth noting that identifying an objective function is often considered complicated, but not complex. Complex *problems* in the world of design, in many situations, require multiple objectives. In this domain, the *Analytic Hierarchy Process* (AHP) is perhaps the most intuitive approach to solving the *multi-objective* design problem. Goncu and Bayazit (2007), for example, use the AHP model for predicting the fashion trends of 2006–2007 suitable for young target groups of consumers. Multiple objectives are selected for determining the trends of styles and fabrics categories in the fashion business. As noted by Ayag and Ozdemir (2009, p. 369), however, “the AHP cannot accommodate the variety of interactions, dependencies, and feedback between higher- and lower-level elements." The *Analytic Network Process* (ANP), as a result, is a new model of choice when dealing with more complex decision-making problems (Ayag and Ozdemir, 2009; Liu and Lai, 2009; Sarkis, 2003). The ANP application to sustainability is thus needed today to address the multi-criteria and the multi-stakeholders that are involved in reaching TBL decisions. Examples can be found in Appendix A.

2.10. *Internal measure of success and external measure of success*

How do designers *measure success*? Once the sustainable initiatives are confirmed, *internal* and external performance metrics can be adopted to track progress and *measure* the *success* of each initiative. Since 2005, Hewlett-Packard (HP) has published its global citizenship report. In the report, six relevant performance variables are used to represent the concept of corporate responsibility. They are economic values, supply chain, climate and energy, product reuse and recycling, operations, and other (employees and social investment). Performance in the past thus can serve as the baseline for *performance* comparisons. Based on the global citizenship report, management at HP can further take kaizen (i.e., continuous improvement) initiatives to achieve corporate objectives. Emphasis here is on sustainability performance, stakeholder reactions, and long-term financial performance (Epstein, 2010). Examples of enterprise reporting on sustainable development can be found in Appendix B.

2.11. *Excel spreadsheets and graphic design software*

Some emerging questions to be highlighted in this construct include: Can technology be considered part of the world of design? What needs of designers must be addressed by technology? What key functions of technology are required in the modern design world? In this regard, Excel spreadsheets are often considered a *technology enabler* for production-oriented data and knowledge management. Other powerful enabling digital frontiers in the domain of design are also ready to be explored by the new generation of designers. Today's graphic design software, notably CorelDraw and SketchUp, offer a wide range of graphics functionality in the area of 3D modeling. The choice of best technology in the real world to initiate sustainable designs is a matter of personal/professional judgment and preference. Yet, the *hierarchy of design problems* and a production system with different degrees of complexity will surely affect the final decision. Examples of graphic design niches and opportunities in our current unsustainable environment due to irresponsible human/enterprise actions can be found in Esquivel (2009).

2.12. *Product life-cycle management (PLM) and strategic PLM*

Modern economy requires an extension of the current *notion* of PLM. While the concept of PLM was widely employed in the existing management and technology literature to track a product from its conception to its disposal or end of its useful life, it is taking a new meaning today in the effective running of purchasing and logistics. The key here is to recognize different types of product life cycles in *face of sourcing* scenarios and possible supply chain mix decisions for each. Along this line, modern designers also need to pay attention to a viewpoint called SPLM. This new paradigm consists of three types of PLM in today's material world: structural, functional, and innovative. *Structural product life cycle scenarios* result from changes in the basic requirements in the market place. These changes of market simultaneously open new opportunities and threats. Value chain partners need to share consistent information, plan ahead, and respond timely and profitably to new markets, new methods, and new technological breakthroughs. Structural change does not normally *happen* immediately but *over a considerable period of time. Functional product life cycle scenarios* are characterized *by product*-classes with constant demand and longer product life spans. In the new world of abundant choices and fragmented markets, products are designed on customer demands and at a price the customer wants. This *poses new challenges to modern enterprises with global partners. What they need collectively are critical market information, quality, and environmental guidelines, product* information at different *stages* of the *product life* cycle, and supply chain capabilities and capacities. *Innovative product life cycles are associated with products in an environment far from equilibrium. Supply chains as a result need to rely heavily on flexibility*, operational efficiencies, and *supply chain speed. This strategic focus reduces the risk of* supply imbalances that may expose supply chains to higher costs and inefficiencies. It is obvious from this discussion that there are different types of product life cycles. Yet, it is also necessary to note that all these product classes are made and realized for the same goal, which is to achieve cross-stage sustainability and enhance the competitive posture of the entire supply chain.

2.13. *Just-in-time (JIT) and minimal supply chain operations*

Toyota's production system popularized the use of JIT in the 1950s. The basic idea of JIT is to produce only what is needed by the next process. The central concern of such a system is as follows: how can a production system optimize the *efficiency and quality* of outputs of its continuous flow? It has *proved* itself extremely *useful* as a means of ensuring successes of internal production units. It is only in the recent past that modern enterprises with supply networks have come to recognize the vital impact that minimal manufacturing can have in the achievement of sustainability. Akira (2007) stresses the importance of minimal manufacturing as a foundation for developing next generation production systems and supply chain systems. The aim of minimal manufacturing is to develop systems with the *maximum performance* possible while using a minimum of energy and resources and producing a minimum of waste. *Minimization opportunities* or cost savings could be evaluated, according to Akira (2007), in detail using information from the following areas: energy, resource, size (e.g., flexible machine, micro-factory), time, cost, and complexity. The emphasis of minimal manufacturing is thus broadly *consistent with* the lean and Six Sigma manufacturing that the major charge of production as well as supply chain systems is zero waste (Madu, 2006; Carreira, 2005). In the new paradigm of sustainability management in supply chains, designers shall introduce "minimal supply chain operations" as an extension of minimal manufacturing.

2.14. *Supply chain management and green supply chain management (GSCM)*

Supply networks exist for the purpose of connection, transaction, and delivery (Madu and Kuei, 2004). Madu and Kuei (2004) use two equations to define supply chain management, where

$$\begin{aligned}\text{Supply Chain} &= \text{a production-distribution network,}\\ \text{Management} &= \text{enabling conditions for integrity, integration, process}\\ &\quad \text{optimization, operational efficiency, continuous improvement,}\\ &\quad \text{and sustainable competitive capabilities.}\end{aligned} \tag{1}$$

Recently, organizations along a supply chain have made attempts to go green. Feng and Ma (2009), for example, report the efforts made in a color TV industry in China in the area of materials purchasing, fleet operations, *environmentally-conscious* manufacturing, energy use, field use, and *product end*-of-*life management* programs. This new trend is called GSCM. Srivastava (2008, p. 536) defines GSCM as "integrating environmental thinking into supply chain management including product design, material sourcing and selection, manufacturing processes, delivery of the final product to the consumers, and end-of-life management of the product after its useful life." Zhu *et al.* (2008, p. 578) further note that GSCM is designed to "incorporate environmental considerations into decision making at each inbound logistics stage of materials management all the way through to the out-bound logistics stage of post-consumer disposal and the closing-the-loop concept of reverse logistics." The green supply chain process model described by Zhu *et al.* (2008) includes the following five elements:

- inputs (goods bought and inbound and outbound logistics activities);
- external drivers (regulations, suppliers, cost pressures, customer demands, industry levels of the relevant practices),
- enablers (Total Quality Management and environmental management systems),
- management support (support for GSCM ideas and practices and cooperation across organizational functions), and
- outputs and outcomes (economic and environmental performance).

2.15. *Technology capabilities and multidimensional competence*

Multidimensional competence in a supply chain setting is typically composed of three facilitators: contractual, knowledge, and good will. Contractual competence is a set of factors that facilitate and predict the conformance to contract specifications. Some of the most *notable factors are customer focus, quality of IT systems, supplier relationships, process integration, and global innovation leadership.* Knowledge competence is the sum total of the knowledge assets of supply chains. It is the collective learning of the modern enterprise and its value chain partners. Goodwill competence refers to characteristics of sustainable supply chains that view sustainability and continuous improvement as a norm. They listen to voices of society and show sensitivity to issues that are important to every member of the society. Along this line, Kuei *et al.* (2008) further note that supply chain competency studies generally can be classified into two groups: (1) examining the preconditions for competency development, and (2) identifying actual competencies. Both of them need to be managed effectively to realize the expected results. *Competitive advantage* over others, as suggested by Kuei *et al.* (2008), is a direct function of a supply chain's *ability* to sustain and successfully *integrate* all three facilitators (contractual, knowledge, and goodwill) across disciplines and product areas within the supply network. *When facing* an uncertain economic future, this line of paradigm change seems to pose another *big challenge* to the modern enterprise and design teams.

2.16. *Simplicity and complexity*

As noted by Gigch (1978, p. 373), "action is impossible unless we are in a position to simplify the world." Designers thus tend to model a design system with a linear structure approach so that they can offer practical, easy-to-understand explanations for events of concern. The actual, true design system is, however, far more intricate. The major attributes of a complex system thus need to be identified (Luisi and Houshmand, 2009; Gharajedaghi, 2006; Davenport, 2003; Leu, 2000; Cilliers, 1998; Holland, 1995; Gigch, 1978; Simon, 1962):

- Complex systems contain parts, structures, and arrangements.
- Structured complexity is possible. Instability is also possible. To support organized structures, a flow of energy is needed.
- Complex systems have (a hierarchy of) objectives. Domain *specialization must* be distinct. Initial conditions will make a difference.
- The relationships in Eq. (1) are not fixed.
- The dynamics of the patterns may affect the behavior of the aggregate system in question. Non-linear behaviors are possible. Control is dispersed.
- Experimental errors should be expected.

Knowledge acquisition is possible. A complex system can be viewed as a knowledge-producing system. Modern designers need to take this into account.

2.17. *Image building and creative philanthropy*

There is a difference between image building and creative philanthropy. When enterprises are interested in only building awareness and keeping a good corporate image, they take actions when it is newsworthy. This has been referred to as intrinsic good. When enterprises are committed to creative philanthropy, an act of instrumental good; however, they invest in good causes and show sensitivity to critical issues important to their employees and the community. National Instruments (NI), an Austin, Texas-based scientific equipment company, is a typical example of the success of philanthropy projects (Rubin, 2009). Over the past decade, NI has been working with the Lego group, the non-profit FIRST (For Inspiration and Recognition of Science and Technology), and the University of Texas at Austin's DTEACh (Design Technology and Engineering for America's Children) program on creative philanthropy initiatives. Through volunteering in regular classrooms and at the FIRST Lego League competition, NI employees feel fulfilled. Rubin (2009) further reports that as a result of this fulfillment, NI is recognized as one of Fortune's "100 Best Places to Work" for the ninth year in a row. To reduce strategic, organizational, and external risks, philanthropic success requires many moving parts along the supply chain to work together. New guidelines must be developed to meet the new challenges. These new guidelines should be community sensitive and capable of producing the expected results. Oracle, for example, aimed at "using its technology and resources to advance education in innovative ways, protect the environment, promote diversity, and enrich community life (http://www.oracle.com/corporate/community/corporate-citizenship-report-2008.pdf)." To move toward creative philanthropy, several activities are noted at Oracle. They are education technology programs, best practices (e.g., for data use among the medical community), community impact grants, volunteer programs, the Oracle academy, the Oracle education foundation, and career development programs. As in the case of NI, Oracle formally factors in a new paradigm in the development of both internal and external communities.

The qualitative comparative approach presented in this section allows designers and executive managers of design teams to identify which features are relevant to the modern design system development in question. It also helps designers point out which phenomena are not relevant to the job. Finding the *leverage* points for each construct presented in this section is critical to long-term success and responsible change.

3. The Birth of Design Science Framework in the Context of Sustainability

In the face of rising costs and environmental concerns, design (or design-as-process) can be seen as an engine of continued growth, a force for efficiency and cost savings, and the key to sustainability leadership. To move modern organizations (and their products/services) from a given state to another realization of potential, it is also

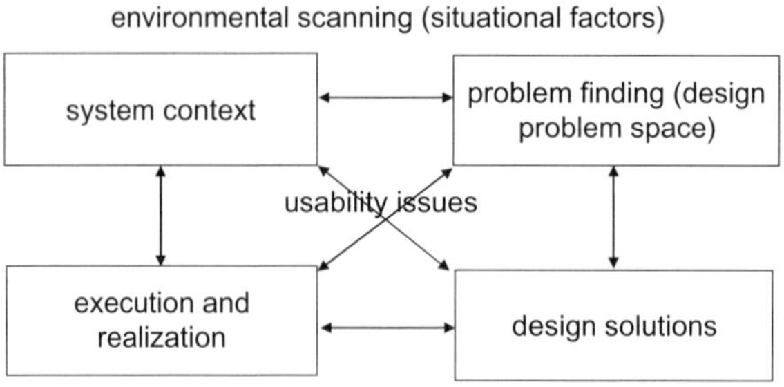

Fig. 1. Design science framework.

necessary to understand the framework of design science from the start. Figure 1 is a graphic depiction of a design science framework and its managerial components. We shall examine the conceptualization of each of the *dimensions* briefly with the city of Chicago as an example.

3.1. *Environmental scanning*

Besides referring to the worldviews of designers, design teams should also evaluate situational factors before proceeding to formulate their design strategies. Through environmental scanning, members of design teams can identify the objective reality, investigate stakeholders' expectations and regulatory requirements, evaluate human and financial resources, better define their missions and purpose, and proceed to provide needed outputs and outcomes. As noted by Boak (2008), for example, it is extremely difficult to try to transform Chicago into a greener, more sustainable city. In 2008, given aging infrastructures, hulking factories, old power grids, and mostly fossil fuel users in this old, windy city, Chicago authorities still unveiled an action plan to reduce carbon dioxide emissions by 25% by 2020 (http://www.chicagoclimateaction.org/). In it, the design team of Chicago describes a new paradigm for residents, businesses, utilities, and transportation, and states that it is time to move to the life values paradigm and to a sustainable world.

3.2. *Problem finding*

At the conclusion of environmental scanning, design is initiated through the phase of problem finding. As noted by English (2007, p. 21), members of the design teams must ponder about five constructs that frame the design problem space:

- The purpose of the system in question — why do we want it?
- The market place, cost/price — who is it for?
- The context of use — how does it fit in?
- Function — what does it do?
- Message — what does it say?

If a system has no mission or goal, there will be no clear direction to follow. This very system is in a position to satisfy a particular market niche. The actual conditions under which a given design product/process is used, from the physical domain to the domain of mind, are generally under continuous assessment. Market demands and functional requirements further form the foundation of system functions. Sending a consistent positive *message* to one's target market and value chain partners is the ultimate goal of the designed (design-as-product) and design (design-as-process). The ability to define and comprehend the space of design problem is crucial to high-level system performance. The Climate Action Plan issued by the city of Chicago consists of the following design problem space:

- the purpose of the system in question — making Chicago a greener, more sustainable city;
- the market place — the city of Chicago;
- the context of use — conditions for sustainability are created;
- function — energy efficient buildings and factories, waste and industrial pollution reduction, sustainable transportation infrastructures, clean and renewable sources; and
- message — carbon dioxide emissions reduction by 25% by 2020.

3.3. *System context*

After reviewing the design problem space, the evaluation of system context commences. Designers should pay attention to the following:

- membership planning — who are the members of interest?
- physical components — which physical parts of the institutional setting are of interest?
- technical platform — which technical parts of the institutional setting are of interest?
- behavioral and managerial components — which behavioral and managerial relationships are of interest? and
- design delivery processes — which aspects of design delivery processes are of interest?

The concept of system context also applies to the explanation of how members of the administration in the city of Chicago further develop their system components for design. System context is composed of five elements: membership planning, physical components, technical platform, behavioral and managerial components, and design delivery processes. System components are frequently arranged along a hierarchy. As noted by Gigch (1978, p. 375), "hierarchy implies a framework that permits complex systems to be built from simpler ones. In turn, the existence of a hierarchy allows complex systems to be broken into their component parts and subsystems." Hierarchy thus helps designers "to organize, to understand, to communicate, and to know about complexity (Gigch, 1978, p. 375)." In a modern socio-cultural environment, designers and executive managers of design teams need to understand hierarchical arrangements, restructure system dimensions if needed, and handle interdependent sets of variables when developing design system components. In the case of Chicago, city design teams need to work with their residents, businesses, utilities, and transportation to shape the city environment into a "greener," more sustainable one. Strategic partnerships also need to be established with banks and the foundations for sustainability. The former can provide all concerned parties with needed working capital to go green, whereas the latter can provide the expertise the city of Chicago *needs* to properly address implementation issues. For the physical domain in Chicago, as noted by Boak (2008), notable observations are as follows: buildings in Chicago account for 70% of greenhouse gas emissions in the Chicago area, motor vehicles in Chicago account for 21% of greenhouse gas emissions, 1/3 of the nation's rail traffic comes through Chicago, and Chicago is home to 13,000 manufacturers. Along with this goes the technical platform consisting of an energy-information system, green roofs, fuel-efficient vehicles, and wind farms. Further, city residents have to be motivated to go green. Civic leaders thus must be able to build a committed leadership team, and mobilize their residents around a focused agenda outlined in the Chicago Climate Action Plan. Civic leadership is the most important driver of design system performance. To drive delivery of sustainable development, carbon-saving recommendations and going-green initiatives are *posted* for review on the city's *web site* (http://www.chicagoclimateaction.org/).

3.4. *Design solutions*

Members of the design teams at the solution stage, with a focus on the "interventions" such as leadership styles, training and education, and planning/control systems, will review a whole range of possible enablers to direct design activities and movements toward a predefined order. Design enablers generally take the form of 5 M's: materials, machines, methods, mind, and money. In the case of Chicago, several initiatives have been introduced with the aim of improving the energy performance of all buildings: insulation materials (the material aspect), heating, hot water, and lighting systems (the machine aspect), cost and benefit analyses (the method aspect), loans being made available for building owners (the money aspect). The design team leaders in Chicago also believe that there is no turning back to the conditions of an earlier era (the mind aspect). Further, as noted by Gharajedaghi (2006) and Gigch (1978), the design enablers under consideration need to interact with holistic thinking (the design problem space and system components) and operational thinking (the hierarchy of problems and of problem-solving methods) to effectively get things made or organized.

3.5. *Execution and realization*

What are the appropriate means of delivering results? This is a broad question which is more likely to obtain a variety of viewpoints. In a context that may lead to the TBL outcomes, it is our belief that modern organizations need to find mechanisms to minimize wastes, maximize throughputs, and conserve resources. As members of the design teams consider their strategic and operational plans, they also need to prepare their systems for responsible delivery. This can be done through the preparation of the following four elements: operations planning, implementation, post-implementation, and job complexity (physical, mental, and information). Designers, in addition to setting up design stages, must take steps to ensure high performances through the delivery system, meet periodically to assess the effectiveness of decision making, and identify winning principles, critical success factors, major and minor conflict management issues, and best practices to deliver specific results. To meet the challenges of going green in Chicago, Boak (2008) highlighted specific tasks that have been carried out by the city of Chicago: introducing the Chicago Climate Action Plan as a basis for operations planning, following the *Pareto's* principle or the 80–20 rule to work on neighborhoods where emission cuts would have the greatest impact, always trying to lead people on issues that result in minimal environmental impact and better quality of city life, and recognizing the fact that meeting the 2020 targets will require at least 40% of residential and commercial owners to participate. Execution and realization strategies thus center on continuous engagement and on building long-term sustainability competence.

3.6. *Usability issues*

What are the relevant outputs and outcomes? The designed and design should always be viewed from the perspective of the end users. Maximizing usability through iterative design is what truly differentiates the exceptional. By referring to ISO 9241, Brouwer and Voort (2008, p. 149), for example, report that usability is "the extent to which a product can be used by specified users to achieve specified goals with efficiency, effectiveness, and satisfaction in a specified context of use." They described usability goals as follows:

- Relevance — Being relevant is critical. In the case of Chicago, for example, "the economic argument is critical (Boak 2008, p. 48)."
- Ease of use — Designers need to ensure that ease-of-use approaches in the field are adopted.
- Learnability — Emphasis is on the willingness of users to further *develop* his/her competencies.
- User's perception — Usability is in the eye of the beholder. Therefore, communication platforms need to be in place to capture user feedback and stakeholder reactions for usability improvements.

Usability improvement is a task that can not be delegated. This dimension will distinguish design systems in the context of sustainability.

To sum up, design thinking based on our proposed design science framework should be viewed as a way of understanding the fundamental issues of design system development in the context of sustainability. For the purpose of quick reference, case summaries relating to Chicago is also shown in Table 1. To show the robustness of our proposed framework, we also apply *our* model to the case of Ford Motor Company, another first mover in today's sustainable economics (see Appendix C). Aspiring to design excellence in an increasingly carbon-constrained economy, modern organizations need comparative data such as qualitative summaries of evidence presented here to gain an accurate understanding about sustainable design.

4. Design System as a Method of Responsible Change

Cilliers (1998, p. 9) states that "truly complex problems can only be approached with complex resources." It is our suggestion in this chapter that designers, during the development cycle of a new system for sustainability, have to monitor complex systems with a lot of components (i.e., X_i), the intricate relationships between those components, and possible non-linear system responses and behaviors. In being responsible to the needs of ourselves and others, designers and executive managers of design teams have to strive to define the design problem space, find solutions, and develop plans for systemic implementation in a complex environment. The aim of such an effort is enabling conditions for ongoing momentum and improvement. It is our belief that this can be done through a well-designed design system. Design systems are concerned with taking purposeful designing actions along a hierarchy of systems to make things better and/or get things made in the material world. We have suggested, throughout this entire chapter, that "design system as a method of responsible change" has an important role to play in modern society. We hope that designers of tomorrow will have a lot more influence on the development side of the systems for sustainability. To guide designers and executive managers of design teams on fundamental issues such as building powerful enabling infrastructures, and to tackle managerial concerns at a more operational level, we present a blueprint for responsible change in Table 2.

As modern designers adopt our proposed approaches, they must understand the new requirements in this complex material world, the enablers of effective design management for the designed (design-as-product) and/or design (design-as-process), and the deliverables of such a holistic, yet dynamic system. An organization aspiring to be successful must establish a vision based on modern design thinking that incorporates the variables presented in Table 2. If leadership support for developing design systems is present, the transformation will enable the system in question to move beyond the blame cycle to the life values paradigm. In such a new paradigm, instrumental good is the norm and should be expected.

Table 1. Design system for the greening of the environment — Chicago.

Design science	Variables	Chicago
Environmental scanning	Situational factors	Aging infrastructure, hulking factories, old power grids, fossil fuel users
The design problem space	Purpose	Making Chicago a greener, more sustainable city
	Market	City of Chicago
	The context of use	Conditions for sustainability are created
	Function	Energy efficient buildings and factories, waste and industrial pollution reduction, sustainable transportation infrastructures, clean and renewable sources
	Message	Carbon dioxide emissions reduction by 25% by 2020
System components	Membership planning	Residents, businesses, utilities, transportation, foundations, banks
	Physical components	(a) Buildings in Chicago account for 70% of greenhouse gas emissions, (b) motor vehicles in Chicago account for 21% of greenhouse gas emissions, (c) 1/3 of the nation's rail traffic comes through Chicago, (d) Chicago is the home to 13,000 manufacturers
	Technical platform	Energy-information *platform*, green roofs, fuel-efficient vehicles, and wind farms
	Behavioral/managerial	Resident motivation, civic leadership
	Design delivery process	Carbon-saving recommendations, the green office challenge, the waste to profit network, green Chicago restaurant co-op
The design solution space (e.g., buildings)	Materials	Insulation materials
	Machines	Heating, hot water, and lighting systems
	Methods	Cost and benefit analyses
	Mind	No turning back to the conditions of an earlier era
	Money	Loans being made available for building owners
Execution and realization (e.g., buildings)	Operations planning	Climate change agenda — Chicago climate action plan
	Implementation	Focusing on neighborhoods where emission cuts would have the greatest impact
	Post-implementation	Minimal environmental impact and better quality of city life

(*Continued*)

Table 1. (*Continued*)

Design science	Variables	Chicago
	Job complexity	Meeting the 2020 targets will require 40% of residential and commercial owners to participate
Usability	Relevance	Saving money by living in a more efficient structure
	Ease of use	Instructions on getting started (e.g., top 10 things businesses/residents can do)
	Learnability	Expecting everyone to act out of choice rather than through the force of law
	User's perception	Web site has been created to facilitate communication and collaboration between design teams and residents

Sources: Boak, 2008; http://www.chicagoclimateaction.org/.

Table 2. Design system as a method of responsible change.

Variables*	Main point(s)
Condition of the system	Design systems for sustainability; the worldviews of designers
Concerns	Identifying aspects of managerial and policy challenges (deciding to change, guiding change, supporting change, sustaining change); investigating competencies needed and innovation opportunities in the context of sustainability
Paradigms	TBL, GSCM, strategic PLM, minimal SC operations, creative philanthropy, multidimensional competence
Thinking process	Taking a systems view on the structure and self-organization; considering the context, interventions, mechanisms, outputs, and outcomes around sustainable development
Outputs	Sustainability performance and stakeholder reactions; functional outputs (e.g., functional specifications) and non-functional outputs (e.g., design metrics)
Method	Linking design problem space to design solution space; evaluating the effectiveness of specific interventions, mechanisms, and innovations; using a participatory modeling approach for strategic decision making
Emphasis	Design affects systems (i.e., the whole process) and our system affects us; design pays; consistent strategic directions, tactical flexibility
Outlook	The increase in complexity through development, sustainability leadership, regulatory requirements
Designer's role	Finding leverage points; leading with modern design thinking; influencing system dimensions and development

*variables adopted from van Gigch (1978, p. 11).

5. Conclusions

"The wise are instructed by reason, average minds by experience, the stupid by necessity and the brute by instinct." Marcus Tullius Cicero (106BC–43BC)

This chapter is written to provide a comprehensive introduction to the issues of design system for sustainability. Although such a system can be initiated through careful planning and execution, it must be refined and renewed through development practices and feedback given by the user about the relevance or irrelevance of the current initiatives for improvement. Within the context of complex systems, we have also suggested that, as designers, one cannot focus one's attention on just the constituents of a system. Modern designers must keep a wide focus all the time on the interactions of parts and non-linear system responses. Modern designers must also focus on the whole process and keep scanning the internal as well as the external environments. Modern designers need to be trained and educated in the art of modern design thinking. The conceptual models discussed in this chapter are aimed at providing new generation of designers with the knowledge and skill to design and develop responsible change plans along the transition to a more sustainable economy.

Acknowledgment

The authors would like *to thank* Dr. Janice Winch for her comments.

References

Akira, T., 2007. *Minimal Manufacturing, 2nd Europe–Japan Workshop on Collaborative R&D.* Available at ftp://ftp.cordis.europa.eu/pub/ims/docs/8-tezuka.pdf.

Ayag, Z. and Ozdemir, R.G., 2009. A hybrid approach to concept selection through fuzzy network process. *Computers and Industrial Engineering* **56**, 368–379.

Boak, J., 2008. Chicago goes green. *Scientific American — Earth3.0* **18**(5) 46–51.

Brouwer, M. and Voort, M.C.v.d., 2008. Designing for dynamic usability: Development of a design method that supports designing products for dynamic use situations. *International Journal of Design Principles and Practices* **2**(1), 149–156.

Brown, T., 2008. Design thinking. *Harvard Business Review* **June**, 85–92.

Carreira, B., 2005. *Lean Manufacturing that Works.* AMACOM, NY.

Cilliers, P., 1998. *Complexity and Postmodernism.* Routledge, NY.

Davenport, T.H., 2003. Whatever happened to complexity theory? ıTrends and Ideas **Fall**, 3–4.

English, S.G., 2007. Creating universal form: Using universals to describe design solution space. *International Journal of Design Principles and Practices* **1**(4), 21–30.

Epstein, M.J., 2010. Thinking straight about sustainability. *Stanford Social Innovation Review* **8**(3), 51–55.

Esquer-Peralta, J., Velazquez, L. and Nora Munguia, N., 2008. Perceptions of core elements for sustainability management systems (SMS). *Management Decision* **46**(7), 1027–1038.

Esquivel, L.C.V., 2009. Graphic design niches in unsustainable development. *International Journal of Design Principles and Practices* **3**(1), 303–307.

Feng, C. and Ma, X.Q., 2009. The energy consumption and environmental impacts of a color TV set in China. *Journal of Cleaner Production* **17**, 13–25.

Gharajedaghi, J., 2006. *Systems Thinking: Managing Chaos and Complexity.* Butterworth-Heinemann, Boston, MA.

Gigch, J.P.v., 1978. *Applied General Systems Theory.* Harper & Row, NY.

Goncu, G. and Bayazit, N., 2007. An application of AHP in the selection of fashion trends for specific target group. *International Journal of Design Principles and Practices* **1**(3), 53–63.

Holland, J.H., 1995. *Hidden Order — How Adaptation Builds Complexity.* Perseus Books, Cambridge, Massachusetts.

Levy, D.L., 2000. Applications and limitations of complexity theory in organization theory and strategy. In: Rabin, J., Miller, G.J. and Hildreth, W.B., *Handbook of Strategic Management.* Marcel Dekker, NY. 67–87.

Liu, K.F.R. and Lai, J.H., 2009. Decision-support for environmental impact assessment: A hybrid approach using fuzzy logic and fuzzy analytic network process. *Expert Systems with Applications* **36**(3), 5119–5136.

Lo, S. and Sheu, H., 2007, Is corporate sustainability a value-increasing strategy for business? *Journal Compilation* **15**(2), 345–358.

Luisi, P.L. and Houshmand, Z., 2009. *Mind and Life, Discussions with the Dalai Lama on the Nature of Reality.* Columbia University Press, NY.

Kuei, C., Madu, C.N. and Lin, C., 2008. Implementing supply chain quality management. *Total Quality Management* **19**(11–12), 1127–1141.

Madu, C.N., 2006. House of Quality in a Minute. Chi Publishers, Fairfield, Connecticut.

Madu, C.N. and Kuei, C., 2004. *ERP and Supply Chain Management.* Chi Publishers, Fairfield, Connecticut.

McFall, B. and Beacham, C., 2008. Human–centered design: Delimiting wicked problems with PRSM. *International Journal of Design Principles and Practices* **2**(3), 67–74.

Magee, B., 1998. *The Story of Philosophy.* DK Publishing, NY.

McVeigh, R., 2007. The conflict in design (or, the conflict of design). *International Journal of Design Principles and Practices* **1**(2), 35–41.

Nidumolu, R., Prahalad, C.K. and Rangaswami, M.R., 2009. Why sustainability is now the key driver of innovation. *Harvard Business Review* **87**(9), 57–64.

Rubin, A., 2009. Clicking for smart CSR. *Stanford Social Innovation Review* **7**(1), 65–66.

Salustri, F.A. and Eng, N.L., 2007. Design as...: Thinking about what design might be. *International Journal of Design Principles and Practices* **1**(1), 19–27.

Sarkis, J., 2003. A strategic decision framework for green supply chain management. *Journal of Cleaner Production* **11**, 397–409.

Simon, H.A., 1962. The architecture of complexity. *Proceedings of the American Philosophical Society* **106**(6), 467–482.

Srivastava, S.K., 2008. Network design for reverse logistics. *Omega — International Journal of Management Science* **36**, 535–548.

Vaughan, L., Viller, S., Simpson, M.R., Akama, Y., Yuille, J. and Cooper, R., 2007. Design plus people: An approach to interaction design projects. **1**(4), 31–37.

Votolato, G., 1998. *American Design in the Twentieth Century.* Manchester University Press, NY.

Zhu, Q., Sarkis, J., Cordeiro, J.J. and Lai, K., 2008. Firm-level correlates of emergent green supply chain management practices in the Chinese context. *Omega* **36**, 577–591.

Appendix A How Modern Designers Build Analytic Network Process (ANP) Models

The following figure shows how modern designers build analytic network process models.

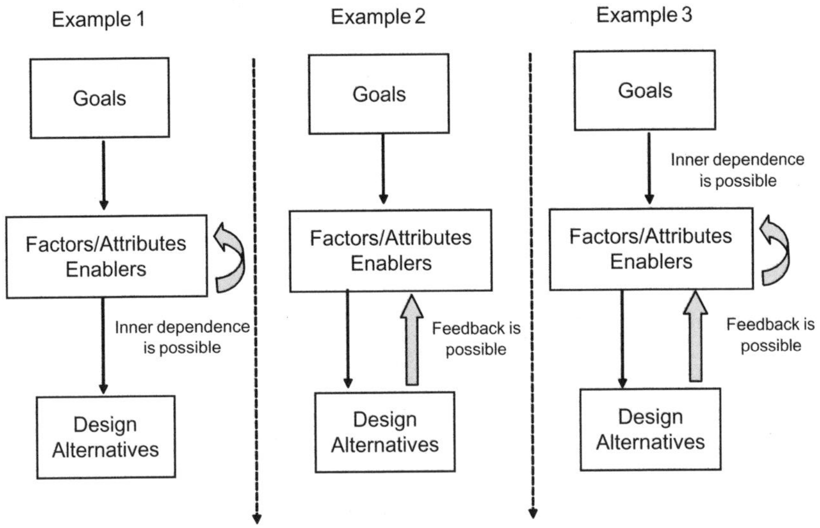

Appendix B Enterprise Reporting on Sustainability

The following table provides enterprise reporting on sustainability.

Enterprise (goal or core belief)	Environmental protection	Social development	Organizational structure/ conditions
Kellogg's (building trust in the Kellogg brand)	• Environmental impact reduction (energy use, greenhouse gas emissions, water use, waste) • Sustainable packaging • Sustainable agriculture and responsible sourcing	• Marketplace (health and nutrition, responsible marketing, consumer information and labeling, product quality & food safety, supplier diversity) • Community (strategic corporate philanthropy, community development, employee volunteerism) • Scholarship • Workplace	• Governance and ethics • Corporate responsibility across the value chain • Corporate responsibility strategy framework • Progress and Performance Reports • Corporate responsibility structure • Kellogg environmental management system • LEED-certified facility*
McDonald's (delivering responsible food for a sustainable future)	• Environmental responsibility • Sustainable supply chain	• Community • Nutrition and well-being • Employee experience	• Governance and ethics

(Continued)

(Continued)

Enterprise (goal or core belief)	Environmental protection	Social development	Organizational structure/ conditions
GE (investing and delivering in citizenship)	• Environment, health and safety • New products for ecomagination**	• Community and philanthropy • Employee data	• *A Leader's Guide to Integrity* (published in 2007) • Governance data
Deere & Company (pursuing sustainable growth and performance)	• Records of environmental stewardship • Brazil factory design for the environment	• Community and philanthropy • Scholarship (Tianjin U, China) • Diversity • Ergonomics and safe workplace	• Suppliers and the code of conduct • US EPA's Climate Leaders Program • Greenhouse-Gas Summit*** • EHS Programs and Mgt. System ****
ExxonMobil (taking on the world's toughest energy challenges)	• Environmental performance and protection	• Community development • International transparency, anti-corruption programs, and human rights • Work place (safety, health, and employment)	• Corporate governance (product stewardship, ethics, and political involvement) • Communication and engagement • Stakeholder feedback via Exxonmobil.com/citizenship

(Continued)

(*Continued*)

Enterprise (goal or core belief)	Environmental protection	Social development	Organizational structure/ conditions
			• Standards of business conduct • Integrity management system • Long-term eco solutions
Continental Airlines (promoting eco-skies)	• Eco-fleet • Eco-ground service equipment • Eco-airport facility • Eco-food service	• Community	• Carbon offset programs • Energy star***** • LEED • Sustainable operations and procedures
Nestle Waters North America (creating shared value)	• Resource use in manufacturing and distribution • Environmental metrics • Resource use in packages • Recycling • R&D-alternative packaging materials	• Health (product choices, water testing results) • Good neighbor policy (local economies, sustainable land use, water education and stewardship, grants, donations, volunteerism)	• Nestle environmental management system • LEED certification

(*Continued*)

(*Continued*)

Enterprise (goal or core belief)	Environmental protection	Social development	Organizational structure/ conditions
Pfizer (partnering for positive change)	• Environmental impact reduction (energy use, greenhouse gas emissions, water use, air and waste emissions) • Green chemistry • Global reporting initiative index (environment, product responsibility)	• Society's overall health • Patient safety • Employees • Supply chain (responsible contracting, supplier diversity, EHS performance of contract) • Global reporting initiative index (labor practices, human rights, society, product responsibility) • Research for diseases of the developing world • Ethical clinical trials • Animal welfare policy • Healthcare resources delivery (treat–teach–build–serve) • Workplace/process/fleet safety • Enterprise-wide diversity and inclusion	• Governance and compliancey • Standards and evaluation partners: Dow Jones Sustainable Asset Mgt; FTSE4Good∧; GAP∧∧; SEE Change and Climate Resolve∧ ∧ ∧; US EPA Climate Leaders • Cross-sector initiatives (Global Environmental Mgt. Initiative; UN Global Compact) • Scientific collaborations • Public policy • Sales and marketing (global marketing code of conduct) • Manufacturing/supply chain • Global reporting initiative index (economy, product responsibility)

(*Continued*)

(*Continued*)

Enterprise (goal or core belief)	Environmental protection	Social development	Organizational structure/ conditions
Hewlett-Packard (Product life cycle)	• Environmental impact reduction (energy use, greenhouse gas emissions, water use, air and waste emissions) • Operations and supply chain environmental responsibility • Product reuse and recycling • Supplier audit results • Material choices	• Education and entrepreneurship opportunities in underserved communities • Operations and supply chain social responsibility • Social investment • Privacy • Workplace • Supplier diversity	• Recovery service • Design for sustainability • Three priorities: supply chain, climate and energy, and product reuse and recycling
FedEx (connecting the world in responsible and resourceful way)	• Environmental impact reduction (aircraft emissions and efficiency, hybrid truck technology, energy use, greenhouse gas emissions, water use, air and waste emissions) • Facilities innovation (renewable energy) • Packaging innovation	• Community and disaster relief (philanthropy and volunteer hours) • People and workplace (full-time retention rate, women and minorities in mgt. roles, employee engagement) • Supplier diversity • Education and scholarship	• Corporate governance • Economics and access (physical and digital networks) • Hybrid truck timeline • ISO certification • Recycling • Alliances

(*Continued*)

(Continued)

Enterprise (goal or core belief)	Environmental protection	Social development	Organizational structure/ conditions
Abbott Laboratories (changing lives)	• EHS policies and programs • Progress reports • Green chemistry • Environmental impact reduction (energy use, greenhouse gas emissions, water use, waste, hazardous materials, air emissions, incident rate)	• Access to affordable medicines and quality products • Community engagement • Ethical clinical trials • Patient awareness • EHS policies and training programs • Employee safety • Diversity and inclusion • Leadership development • Culture and workplace • Supplier diversity	• Corporate governance • Ethics and compliance • Stakeholder engagement • Actions on priority issues • Dow Jones sustainability world index • Science and technology • Responsible advertising to consumers • e-Library on environmental health and safety • Alliance • Organizational change • Supplier code of conduct
IBM (A new model of global citizenship)	• Energy and climate • Pollution prevention • Chemical use and management • Water conservation	• Global contributions • Supplier diversity • Workforce diversity • Employee satisfaction • Training and learning	• Governance of corporate responsibility • Product stewardship • Product end-of-life management operations • Business conduct guidelines • Challenges–opportunities–strategy model

(Continued)

(*Continued*)

Enterprise (goal or core belief)	Environmental protection	Social development	Organizational structure/ conditions
			• Supply chain environmental management • Environmental partnerships • Global reporting initiative index (economy, environment, society)
Oracle (using Oracle's technology and resources to advance education in innovative ways, protect the environment, promote diversity, and enrich community life)	• Grants from Oracle (ConserveOnline (www.conserveonline.org)) • Environmental impact reduction (e.g., Energy-efficient buildings/data centers, event management) • Green business practices (suppliers' environmental policies and practices)	• Education technology programs • Best practices (for data use among the medical community) • Diversity • Community impact grants • Volunteers programs • Oracle academy • Oracle education foundation • Career development	• Corporate governance • Feedback (citizen-ship_ww@oracle.com) • Partnership with philanthropic organizations • A steering committee on environmental protection • A member of the climate savers computing initiative • ISO14001 guidelines • LEED • Asset retirement programs • Recycling • Code of ethics and business conduct

(*Continued*)

(*Continued*)

Enterprise (goal or core belief)	Environmental protection	Social development	Organizational structure/ conditions
Hilton Hotels Corporation+ (filling the earth with the light and warmth of hospitality; building sustainability into the core fabric of Hilton's businesses worldwide)	• Environmental impact reduction (e.g., sustainable buildings and operations, sustainable design and construction, chemical management, purchasing, renewable energy) • Centralized web content and training modules	• Scholarship programs • Hilton in the community foundation • Sponsor of selected organizations such as AIDS Walk Los Angeles and Save-A-Landmark • Environmental education through the "Teaching Kids to CARE' Program	• The Installation of a Commercial Fuel Cell Power System (atop the Hilton NY) • Mission Statement and Environmental Policy • Sustainability VP • Contribution Review Committee

*: Leadership in Energy and Environmental Design (LEED) — US Green Building Council.
**: Products with the highest degree of integrity.
***: John Deere's Greenhouse Gas Summit — Company-wide emissions reductions.
****: EHS: environmental, health and safety.
*****: Energy Star standards — US Environmental Protection Agency (EPA).
^ UK FTSE group — Corporate responsibility standards.
^^: Global Accountability Project (GAP) — One World Trust.
^^^: Social, Environmental, Economic (SEE) Change — Voluntary programs.
+: information derived from its web site.

References are shown based on the order of presentation. Reports were accessed in 2009:

Kellogg's Citizenship Report: http://www.kelloggcompany.com/corporate responsibility.aspx.

McDonald's Citizenship Report: http://www.crmcdonalds.com/publish/csr/home.html.

GE Citizenship Report: http://www.pewclimate.org/docUploads/GE_07_08_Citizenship_Report.pdf.

John Deere Citizenship Report: http://www.deere.com/en_US/compinfo/csr/report/index.html.

ExxonMobil Citizenship Report: http://www.exxonmobil.com/Corporate/files/Corporate/community_ccr_2007.pdf.

Continental Airlines Citizenship Report: http://www.continental.com/web/en-US/content/company/globalcitizenship/default.aspx.

Nestle Water Citizenship Report: http://www.nestle-watersna.com/pdf/Nestle_Corporate_Citizenship_Report_Final.pdf.

Pfizer Citizenship Report http://www.pfizer.com/responsibility/.

Hewlett-Packard (HP) Citizenship Report: http://www.hp.com/hpinfo/globalcitizenship/ gcreport/.

FedEx Citizenship Report: http://about.fedex.designcdt.com/corporate_res ponsibility.

Abbott Citizenship Report: http://www.socialfunds.com/csr/reports/Abbott_Laboratories_2005_Global_Citizenship_Report.pdf.

IBM Citizenship Report: http://www.ibm.com/ibm/responsibility/report-home.shtml.

Oracle Citizenship Report: http://www.oracle.com/corporate/community/corporate-citizenship-report-2008.pdf.

Hilton International: http://hiltonworldwide1.hilton.com/en_US/ww/fob/landing/Sustain/index.do.

Appendix C Design System for the Greening of the Environment — Ford Motor Company

The following table provides design system for the greening of the environment — Ford Motor Company.

Design science	Variables	Ford motor company
Environmental scanning	Situational factors	Climate change and energy security, energy-intensive global system, sustainable economics
The design problem space	Purpose	Achieving significant changes in auto industry's carbon footprint
	Market	200 markets across six continents
	The context of use	Field use (road transport), product life cycle, assembly operations, logistics
	Function	Hybrid lineup expansion, fuel efficient cars, energy efficient buildings and factories, clean and renewable sources, sustainable supply chain networks
	Message	Reducing road transport greenhouse gas (GHG) emissions, confronting new business challenges associated with climate change
System components	Membership planning	Consumers, fellow automakers, suppliers, energy firms, dealers, policy makers, non-governmental organizations
	Physical components	Motor vehicles, assembly plants, physical distribution networks

(Continued)

(*Continued*)

	Technical platform	Fuel-saving technologies, common platforms for product designs, wind turbines at Dagenham in the UK
	Behavioral/managerial	Market, policy, social, and technological enablers
	Design delivery process (e.g., product)	Ford's product GHG strategy (technology pilots, scaling up, mass marketing, and stabilization)
The design solution space (e.g., buildings)	Materials	Batteries, (ethanol) fuel, combinations of engines and transmissions
	Machines	Heating and air-conditioning systems, wind power projects, drying processes, and pollution control devices in paint shops
	Methods	Strategic road map centering around product, policy, plants, people, and partnerships
	Mind	Mulally (2009), Ford's CEO, states that we want to see the automobile industry be very vibrant going forward.*
	Money	Major capital investments needed due to long product development times
Execution and realization (e.g., buildings)	Operations planning	Ford's greenhouse gas emissions strategy
	Implementation	Principles (1: economically efficient decisions, 2: fuel+vehicle+driver=GHG emissions, 3: scenario planning, and 4: affordable steps), governance (1: office of the chairman and chief execution, 2: VP climate change task force, 3: sustainability

(*Continued*)

(*Continued*)

		mobility governance, and 4: climate change steering team)
	Post-implementation	Ford's sustainability report
	Job complexity	Meeting specific targets and milestones with the "fuel+vehicle+driver" formula
Usability	Relevance	Improved mpg, increased reliability and dependability, improved safety, alternative fuel vehicles, hybrid vehicles
	Ease of use	Field service, working with energy firms with regard to infrastructure
	Learnability	Customers' driving behavior, Ford's sustainability policy deployment
	User's perception	Fuel economy, climate change leadership, first mover

Sources: http://www.ford.com/doc/fordReptBusImpClimChg.pdf;*: Mulally, A., 2009. 10 *Questions*, Time, April 6, p. 6.

NOTES ON CONTRIBUTORS

Dr. John F. Affisco is Professor of Information Technology and Quantitative Methods at the Frank G. Zarb School of Business of Hofstra University. He received his Ph.D. in Business from the City University of New York Graduate School. Dr. Affisco's professional interests include quality management especially in health care organizations, experiential learning, environmental and energy management, and modeling of production systems. His research has appeared in a variety of academic and professional journals including, among others, *International Journal of Production Research, European Journal of Operational Research, ZOR-Methods and Models of Operations Research, Simulation and Gaming, Mathematical Methods of Operations Research, International Journal of Quality Science, Quality Engineering, Business Process Management, and Expert Systems with Applications.* Dr. Affisco research has been recognized by best paper awards in health care quality management, inventory modeling, and environmental issues in operations management. Dr. Affisco is a member and officer of the Northeast Decision Sciences Institute.

Javier Aguilera-Caracuel is a full member of the Innovation, Sustainability, and Development (ISDE) research group. His research interests include environmental management in multinational enterprises and corporate social responsibility. He has presented his works in multiple international colloquiums such as EGOS, EURAM, and Academy Management Conference.
Address: Department of Business and Management, Economics and Business School, University of Granada (Spain). Campus de Cartuja S/N, C. P. 18071 Granada, Spain. *Phone*: 0034 958249597. *E-mail*: javieraguilera@ugr.es

Elsie Anderberg is a Professor in education at Jönköping University. Her research primarily concerns individual learning in different settings, focusing the function of language in knowledge formation. She is the Scientific Director of the research program Language Use and Individual Learning (LUIL), which has obtained long-term funding from the Swedish National Research Council for a number of research projects. The program addresses learning issues in higher education, upper secondary school, compulsory school, and pre-school levels, as well as learning and teaching in global settings. She is a member of a research platform at Jönköping

University for capacity building in ESD among teachers, from pre-school to university level, particularly teacher education.

Robert Anderson is a Professor of Business Administration at the University of Regina, Canada. He is the Editor of the *Journal of Small Business and Entrepreneurship* and Co-editor, with Leo Paul Dana, of the *Journal of Enterprising Communities.*

Bob is President-Elect of the Administrative Sciences Association of Canada (ASAC) and Chair of the ASAC 2010 National Conference with the theme Enterprising Our World: Sustainable Business in Action (http://asac2010.uregina.ca/).

He has authored numerous peer-reviewed papers and made many conference presentations on economic development and entrepreneurship. More recent work has expanded to include corporate social responsibility; university innovation, entrepreneurship, and intellectual property mobilization; and sustainable development.

Dr. Juan Alberto Aragón-Correa received his Ph.D. in Business Administration and Economics, University of Seville. He is a full professor of Strategic Management at University of Granada, Spain, and visiting professor at U.C. in Berkeley, USA. His research interests include the connections between natural environment and business strategy. He has published in top tier journals such as *Academy of Management Journal*, *Academy of Management Review*, *British Journal of Management*, *Long Range Planning*, *Journal of Business Research*, *Ecological Economics*, *Sustainable Development*, among others. He has been Deputy Dean in the School of Economics and Business for seven years and Chair of Department of Management at University of Granada. He has lead different research projects supported by the European Union and the Spanish Ministry of Research analyzing how to integrate sustainability in the firms.
Address: Department of Business and Management, Economics and Business School, University of Granada, Spain. Campus de Cartuja s/n, C. P. 18071 Granada, Spain. *Phone*: 0034 958243705. *E-mail*: jaragon@ugr.es

Dr. Magnus Bengtsson is working for the Institute for Global Environmental Strategies (IGES), based in Japan, where he is currently the Director of the Sustainable Consumption and Production group. Before joining IGES in April 2007, he was a JST Post Doctoral Fellow at the University of Tokyo where his research focused on water demand scenario analysis and global water futures assessment. Magnus received his Ph.D. in Environmental Systems Analysis at Chalmers University of Technology in Göteborg, Sweden, where he was working on Life-Cycle Assessment (LCA), environmental valuation techniques, stakeholder consultation processes and sustainability controversies. While living in Sweden, Magnus was also actively engaged in two environmental NGOs where he was working on awareness raising through organizing roundtable dialogues and public seminars on sustainability. Magnus' current research covers a wide range of topics, mainly focusing on policy approaches toward Sustainable Materials Management and Sustainable Consumption and Production in Asia.

Yang Chen is a Ph.D. candidate in the school of Business, Hong Kong Baptist University. Mr. Chen obtained the Bachelor of Arts in Economics from Nanjing University in 2008. His research interests include sustainable management, sustainable development, and information system management.

Dr. In-Mo Cheong has been the Head of the Environment Strategy Planning Team of Hyundai Motor Company since its establishment in 2003. He received his doctorate in material engineering from Osaka Prefecture University, Japan. He has more than 15 years of experiences at Hyundai in corporate environmental strategy development, supply chain environmental management, life-cycle assessment, and sustainability reporting.

Dr. P.C. Chiang began engaging in teaching and research activities in National Taiwan University since he obtained Ph.D. degree from the Department of Civil Engineering, Purdue University, USA, in 1982. Currently, he is a distinguished professor of Graduate Institute of Environmental Engineering, National Taiwan University, and a Diplomat of the American Academy of Water Resources Engineers of the American Society of Civil Engineers (ASCE). He has been actively involved the international and national academic associations served as the Board of Director (1987–2007), Executive Committee (2001), Academic Committee (2008–present), WEF; Chairman, IAWQ Specialized Conference (2001); President, Chinese Institute of Environmental Engineering (2004–2006); and AIChE Local Chapter (2009–present).

Dr. Chiang is known for his work in physico-chemical treatment such as carbon adsorption, membrane, and ozonation processes. In addition, he was also devoted to the research projects in the area of carbon capture technology, integrated watershed management, and sustainability for energy and industrial development. He has received numerous awards for research achievements, including Outstanding Research Award, National Science Council (1988–1999), Distinguished Chinese Institute of Engineer Research Award (1993), Outstanding Chinese Institute of Environmental Engineering Research Award (1993, 1995), and Best Paper Award, Environmental and Water Resource Institute, ASCE (2005). Dr. Chiang has published more than 200 papers in the above area since 1990.

Dr. Wing S. Chow is an associate professor of MIS in the School of Business, Hong Kong Baptist University. Dr. Chow published more than 70 research papers in the journals and conference proceedings. He has contributed articles to Omega and International Journal of Production Economics. His latest edited book is entitled "Multimedia Information Systems in Practice" published by Springer, 1999. His current research interests include green supply chain management, supply chain quality management, and business/manufacturing information management.

Mr. Philip Cook is a Modernization and Retrofit Representative with Honeywell Inc. Mr. Cook has an over 25 year career in the Heating, Ventilating, and Air-Conditioning Industry and has held positions with Tyco International, Engineered Air and Cooper Crouse Hinds. He has been involved in sales, management, project management, and entrepreneurial endeavors throughout this time. A Certified Engineering Technician and most recently a graduate of the Athabasca University Master of Business Administration program. Mr. Cook has also coauthored a research paper on continuous commissioning. Mr. Cook's plans include the pursuit of further study in the areas of Sustainable Development, Systems Thinking, Organizations as Learning Systems, and Complexity Theory.

Dr. Crawford-Brown is Emeritus Professor in Environmental Sciences and Policy and Emeritus Director of the Institute for the Environment, at the University of North Carolina at Chapel Hill. He moved permanently to Cambridge, UK, in 2008 and is now Director of Sustainability for Pell Frischmann, an engineering consultancy in London; a Principal in EnviroTech, a business-for-business venture supporting growth of cleantech SMEs; and operates Crawford-Brown Energy and Environmental Consultancy through which he advises Mott MacDonald on climate change strategies. He received his degrees in physics (B.S., M.S.) and nuclear science (Ph.D.) from the Georgia Institute of Technology. His activities focus on sustainability in the public and private sectors, modeling human health risks from environmental contaminants, modeling alternative policies to tackle environmental problems, assessing the quality of scientific information, and developing tools of risk assessment. He is the author of 140+ scientific articles and 5 books on these subjects. He has served on the USEPA's National Drinking Water Advisory Committee, Clean Air Scientific Advisory Committee, Endocrine Disruptor Screening and Testing Advisory Committee, and National Pollution Prevention and Toxics Advisory Committee; on the American Water Works Association's Technical Advisory Workgroup; on the EC's Panel of Scientific Experts on Health and Risk Assessment; on the Legislative Global Climate Change Commission; and on an array of environmental workgroups in the UK, Austria, Mexico, France, Germany, and Thailand.

Giuseppe D'Alessandro is a mechanical engineer. He works as the General Manager of the Public Company for MSW management of the Municipality of Molfetta (Italy).

Dr. Leo Paul Dana is Professor at GSCM Montpellier, France, on leave from the University of Canterbury in New Zealand, where he has been tenured since 1999. He also holds the honorary title of Adjunct Professor at the University of Regina, Canada. He has served on the faculties of McGill University and INSEAD. He earned B.A. and M.B.A. degrees at McGill University, and a Ph.D. from the Ecole des Hautes Etudes Commerciales. He has published extensively in a variety of leading journals including the British Food Journal, Cornell Quarterly, Entrepreneurship: Theory & Practice, Journal of Small Business Management, Journal of World Business, and Small Business Economics. He is also a senior advisor to the World

Association for Small and Medium Enterprises and founding editor, Journal of International Entrepreneurship and Journal of Enterprising Communities.

Dr. Grace Ding is a senior academic in the School of Built Environment at the University of Technology Sydney, Australia. Her research areas include sustainable construction, sustainable building performance, life-cycle analysis, life-cycle costing, life-cycle energy analysis and have been published widely in journals, books, book chapters, and international conferences.

Dr. Vera Ferrón-Vílchez received his Ph.D. in Management, University of Granada. She is an assistant professor in the Business and Management Department at University of Granada, Spain. She is a full member of the research group ISDE and her work is oriented toward analyzing the relationship between advanced environmental strategies and the achievement of cost leadership. She has presented her works in multiple international conferences such as EGOS, EURAM, and Academy Management Conference. Vera works in joint projects with leading scholars at George Mason University (Virginia, USA) and she has been visiting there.
Address: Department of Business and Management, Economics and Business School, University of Granada, Spain. Campus de Cartuja S/N, C. P. 18071 Granada, Spain. *Tel.*: 0034 958249596. *E-mail*: vferron@ugr.es.

Dr. Johannes Freiesleben is vice president at Credit Suisse AG, Zurich. Previously, he was visiting scholar at the Department for Management Science and Engineering at Stanford University and head of strategic deployment at Ringier AG, Zurich, an international media company based in Switzerland. He earned his master degree in business administration and his Ph.D. in economics from the University of St. Gallen, Switzerland. As doctoral student of Soren Bisgaard, he specialized in the field of production economics, conducting research at the University of St. Gallen, the University of Wisconsin, and Stanford University. His main research interest is the economic rationale behind quality improvement. He was lecturer at the University of Zurich, Switzerland, Stanford University, and Universidad Catolica de Salta, Argentina.

Dr. Steven P. Frysinger is a Professor in the Integrated Science and Technology and Computer Science programs at James Madison University, located in Virginia's Shenandoah Valley. Serving in this capacity since 1995, he teaches courses in environmental science, environmental management, environmental information systems, and interaction design. He is the Director of two academic programs offering minors: one in Environmental Information Systems and the other in Environmental Management. Since 2003, he has also been a Guest Professor in the Faculty of Engineering Sciences at the Hochschule für Technik und Wirtschaft des Saarlandes, Saarbrücken, Germany.

Before coming to JMU, he was a Principal Technical Staff Member with AT&T's Environment, Health and Safety Organization, having first joined Bell Labs in 1982.

His most recent work at AT&T focused on efforts to integrate environmental information systems in order to enhance environmental management. Before that he was the lead investigator and systems engineer on several projects applying decision support system technology to various problems, such as Sandia National Laboratory's groundwater monitoring well network design. Dr. Frysinger received his Bachelor degree in Environmental Studies, concentrating in Physics, from William Paterson College. He completed Master degrees in Computer Science and Applied Psychology/Human Factors at Stevens Institute of Technology, and earned his Ph.D. in Environmental Sciences from Rutgers University. His research interests include environmental information and decision support systems, environmental systems engineering and management, and human/computer interface research, and he has authored numerous publications in these areas.

Maria Grazia Gnoni is Assistant Professor at the Engineering Faculty of the University of Salento ITALY. Her research interests are focused on sustainability and safety analysis of industrial processes and logistics and the RFID (Radio Frequency Identification) application for lean supply chain management.

Dr. Andreas Größler is Associate Professor at Nijmegen School of Management, Radboud University, the Netherlands. He teaches operations management and methodology courses at the undergraduate, master, and executive level. His research interests lie in the fields of strategic and international operations management, in the application of system dynamics modelling and simulation for business and not-for-profit organisations, and in the investigation of individual and organizational dynamic decision making. Dr. Größler is a managing editor of the *System Dynamics Review* and an area editor of *Operations Management Research.*

Dr. Rahul Hiremath has Ph.D. from Indian Institute of Science, Bangalore, and presently working as Dean, Academic & Research, Walchand Institute of Technology, Solapur 413006, India. His main area of research is decentralized energy planning.

George M. Hornberger is a Distinguished University Professor at Vanderbilt University, where he is the Director of the Vanderbilt Institute for Energy and the Environment. He has a shared appointment as the Craig E. Philip Professor of Engineering and as Professor of Earth and Environmental Sciences there. He was previously a professor at the University of Virginia for many years where he held the Ernest H. Ern Chair in Environmental Sciences. His research is aimed at understanding complex water–energy–climate interrelationships and at how hydrological processes affect the transport of dissolved and suspended constituents through catchments and aquifers. Hornberger is a fellow of the American Geophysical Union (AGU), a fellow of the Geological Society of America, and a fellow of the Association for Women in Science. Professor Hornberger won the Robert E. Horton Award (Hydrology Section) from the AGU in 1993. In 1995, he received the John Wesley Powell Award from the USGS. In 1999, he was presented with the Excellence in Geophysical Education Award by the AGU and in 2007 he was selected Virginia Outstanding Scientist. He is a member of the U.S. National Academy of Engineering.

Dr. Yasuhiko Hotta is a Deputy Director and Senior Policy Analyst for Sustainable Consumption and Production Group at the Institute for Global Environmental Strategies (IGES), Japan. He holds Ph.D. in International Relations from University of Sussex. His research focuses on the effects of globalization on reform in environmental policy. In IGES, he has been in charge for the projects to support Ministry of the Environment, Japan, for promotion of international cooperation for improving resource efficiency and waste management in developing countries such as G8's 3R Initiative and Regional 3R Forum in Asia. He is a member of Working Group on 3R Policies in East and Southeast Asia of Economic Research Institute for ASEAN and East Asia (ERIA). Among many contributions to the reports on resource efficiency and waste management in Asia, he was a main contributor to the ADB and IGES co-publication *Toward Resource Efficient Economies in Asia and the Pacific* in 2008.

Dr. Chia-Wei Hsu received his Ph.D. degree in engineering technology from the National Taipei University of Technology. He is currently an Assistant Professor in the Department of Environmental Management, Tungnan University, where he teaches courses in corporate environmental management, green design, and life-cycle assessment. He has published several journal articles in the area of green supply chain management, sustainable performance management, and supplier management. His research interests include green supply chain management, carbon management, corporate sustainability report, and corporate social responsibility.

Dr. Allen H. Hu is currently an associate professor and director of the Institute of Environmental Engineering and Management, at the National Taipei University of Technology (NTUT). Prior to joining the NTUT in 2003, he was a faculty member at the Graduate Institute of Environmental Management, Nanhua University from 1999 to 2003. Dr. Hu received his B.S. degree in civil engineering at National Central University, Taiwan, in 1983. After two years of military service, he joined the Department of Environmental Protection, Taipei City Government as an environmental inspector for three years. In 1998, he went to USA and received his both M.S. and Ph.D. in Environmental Engineering from the University of Connecticut. After that, Dr. Hu came back to Taiwan and joined the Industrial Technology Research Institute (ITRI)/National Center for Cleaner Production (NCCP) and served as a researcher in charge of promoting and disseminating CP in Taiwan. Since then, his study has been focusing mainly on corporate environmental management and sustainability issues.

Dr. Sc. (econ), Dr. rer.pol. Dmitry Ivanov is a research associate at the chair of production economics and industrial science at Chemnitz University of Technology, Germany. He graduated with major in production management and industrial organization in 2000 with distinction. In 2006, he gained the degree of Dr. rer. pol. at Chemnitz University of Technology. In 2008, he gained the degree of Doctor of Economics Science at Saint Petersburg University of Economics and Finance (FINEC). In 2005, he won a German Chancellor Scholarship and spent one year at Chemnitz University of Technology. Since 2006, he has worked at Chemnitz University of Technology. He is the (co-)author of more than 180 scientific works including papers in prestigious journals and six monographs in English, Russian, and German, incl. "Adaptive Supply Chain Management" published by Springer (co-author: Boris Sokolov). He is a presidium member of the Russian National Supply Chain Council. He leads the German–Russian Logistics Society DR-LOG. His research interests lie in the area of supply chain organization, planning and control, applied optimal control theory, operations research, and business information systems. He got an international reputation through the publication of numerous papers in various high-ranked academic journals, including European Journal of Operational Research, International Journal of Production Research, International Journal of Integrated Supply Management, International Journal of Manufacturing Technology and Management, International Journal of Agile Systems and Management, etc. *Homepages*: www.ivanov-scm.com; www.dr-log.org.

George Kassinis is an associate professor of Strategy in the Department of Public and Business Administration of the University of Cyprus. He was educated at Cornell, MIT and Princeton. His research interests revolve around issues of competitiveness, environmental management, social networks, and industrial ecology. He received grants from Fulbright and the MacArthur Foundation to support his research and worked at The World Bank and the Princeton Center for Energy and Environmental Studies. His research has been published in leading academic journals such as the Academy of Management Journal, Strategic Management Journal, and Production and Operations Management. He serves on the editorial boards of a variety of academic journals and collaborates widely with organizations internationally.

Dr. Tarja Ketola is Acting Professor of Sustainable Development in the Industrial Management Unit of the University of Vaasa, Finland, and Adjunct Professor of Environmental Management at the Turku School of Economics, Finland. She took her Ph.D. at Imperial College, University of London, and worked at Brunel University, UK, before returning to Finland. Her research interests include sustainable development, corporate sustainability, strategic environmental management, corporate responsibility, and responsible leadership psychology. She has written several books and published articles in a number of scientific journals. She can be contacted at tarja.ketola@uwasa.fi

Dr. Anshuman Khare is a Professor in Operations Management at Athabasca University, Canada. Dr. Khare is an Alexander von Humboldt Fellow. His research focuses on environmental regulation impacts on automobile industry and its supply chain. He is also a former Monbusho Scholar having completed a postdoc assignment at Ryukoku University in Kyoto, Japan. He has published five books and over 125 research papers and articles, and is listed as an Academic Expert in the areas of operations management, environmental strategy, sustainable development, environmental manufacturing, and responsible manufacturing on Environment Canada's Academic Expertise Database. He has published on a wide range of topics including JIT; Supply Chain Management; Sustainable Development related to public policies, regulations, and strategic developments due to climate change regulations, and initiatives; ecopreneurship; sustainable cities; and impact of environmental/climate change regulations on technology, innovation, and corporate strategy.

Dr. C. Kuei is professor of management and management science at the Lubin School of Business, Pace University, New York, NY. He received his Ph.D. in Business (Major: Management Planning Systems) from The City University of New York. He has contributed articles to Omega, *IIE Transaction, Decision Sciences, Long Range Planning, Computers and Operations Research, Design Principles and Practice, European Journal of Operational Research, International Journal of Production Economics, International Journal of Strategic Decision Sciences, Microelectronics and Reliability, International Journal of Production Research, Journal of Business and industrial Marketing, Journal of Operational Research Society, Total Quality Management and Business Excellence Journal, International Journal of Quality and Reliability Management, Journal of Environmental Management, Asia Pacific Management Review,* and others. Dr. Kuei is a *member* of the *Editorial Advisory Board* of International Journal of Quality & Reliability Management. He is also the co-author of Experimental Statistical Designs and Analysis in Simulation Modeling, Strategic Total Quality Management, and ERP and Supply Chain Management. His current research interests include sustainability management, Systems Thinking, Green Supply Networks, complexity theory, and the Analytic Network Process.

Dr. Bimlesh Kumar has Ph.D. from Indian Institute of Science, Bangalore, and is currently working as an assistant professor in Indian Institute of Technology, Guwahati, India. His main areas of research are environmental fluid mechanics, sediment transport, and energy planning for renewable sources.

Debra Lam is a senior policy consultant at Arup. She has several years of international experience in governance, sustainable and strategic development, best practices, government analysis and policy, and project management. She has crisscrossed sectors, from non-profit organizations to the federal and local governments.

Debra now works on bringing governance, policy and sustainability into city planning, stakeholder engagement, local capacity building, and project implementation. Her ongoing research and analysis includes low-carbon strategies for new built and retrofit, climate change adaptation and mitigation, and overall resilience of local governments.

Debra has worked and lived in the US, Hong Kong, UK, China, and Taiwan. She graduated in Foreign Service at Georgetown University and has a graduate degree in public policy at University of California, Berkeley.

Dr. Su-Yol Lee is an assistant professor at the College of Business Administration, the Chonnam National University, South Korea, where he teaches operations and supply chain management. He obtained a Ph.D. degree at the graduate school of management from Korea Advanced Institute of Science and Technology (KAIST), Korea. His current research interests include green and social issues in the supply chain, climate change and business, corporate carbon strategy competitiveness, and corporate social responsibility.

Dr. Chinho Lin. Editor-in-Chief of (1) International Journal of Commerce and Strategy and (2) International Journal of Management and Enterprise Development, is a distinguished professor in the Department of Industrial Management Science & Institute of Information Management at National Cheng Kung University, Taiwan, R.O.C. He received his Ph.D. in Business Administration from the City University of New York. His works were published in the Decision Sciences, Decision Support Systems, Information & Management, European Journal of Operational Research, Journal of the Operational Research Society, International Journal of Production Research, Journal of Computer Information Systems, TQM, Computers & OR, Industrial Management & Data Systems, Computers & IE, and others. His current research interests include knowledge management, supply chain management, quality and reliability management, technology management, and sustainability management.

Dr. Min H. Lu is an Associate Professor of Marketing at Leon Hess Business School, Monmouth University, West Long Branch, New Jersey, USA. Born in Shanghai, the People's Republic of China, Dr. Min H. Lu graduated with bachelor and master degrees from the University of International Business and Economics in Beijing. Dr. Lu received his doctorate in Marketing and International Business from the School of Business and Public Management, the George Washington University, Washington, DC, USA.

Dr. Lu is currently teaching and researching in the areas of Strategic Marketing Management, International Marketing, and Global Management. His current research interests focus on international entrepreneurship and global market sustainability. His works have been published in many journals and books. Dr. Lu is very active in research and presentation at various academic and professional conferences and seminars in the US and abroad. His articles have been published by Journal of Business and Industrial Marketing, International Journal of Quality and Reliability Management, International Journal of Quality Science, The Multinational Business Review, etc. His works have also been published in several books regarding international and global marketing strategies.

Dr. Christian N. Madu is Research Professor of Management Science and co-editor of the *International Journal of Quality and Reliability Management.* His area of research interest is in management science and he has published more than 120 research papers in journals such as *Decision Sciences, IIE Transactions, Socioeconomic Planning Sciences, Journal of Operational Research Society, European Journal of Operational Research, OMEGA, Journal of Environmental Management,* etc. He has published more than a dozen scholarly books including the Handbook of Total Quality Management, Handbook of Environmentally Conscious Manufacturing, Environmental Planning and Management, and House of Quality (QFD) in a Minute. Dr. Madu is currently writing a book on Process Change Management. He has received several research and teaching awards. He frequently conducts workshops for both the private and public sectors and he is presently running workshops on Process Change Initiatives.

Dr. Buddhi Marambe is currently a Professor of Crop Science and Director of the Agriculture Education Unit at The University of Peradeniya, Sri Lanka. He earned his D.Agr. from Hiroshima University, Japan. His recent publications include Biological diversity in lowland rice fields, Capacity building for conservation and utilization of farm animal genetic resources in Sri Lanka, Institutional coordination and policy framework for management of invasive alien species in Sri Lanka, and Contribution of plant genetic resources for food and agriculture (PGRFA) to food security and sustainable development. His research interests include sustainability in agriculture, biodiversity, and organic agriculture. He has published over 60 articles in the area of sustainable agricultural systems and contributed in preparing the national status reports on Biodiversity and Food Security in Sri Lanka.

Rodrigo Martín-Rojas is a Ph.D. candidate at University of Granada, Spain. He is a full member of the Innovation, Sustainability, and Development (ISDE) research group and he is working in the field of personal innovativeness and its influence on the performance of enterprises.
Address: Department of Business and Management, Economics and Business School, University of Granada, Spain. Campus de Cartuja S/N, C. P. 18071 Granada, Spain. *Tel.*: 0034 958249595. *E-mail*: rodrigomr@ugr.es

Evando Mirra is a member of the Royal Institute of Technology KTH. Evando can be contacted at Center for Strategic Studies and Management in Science, Technology and Innovation (CGEE), SCN Quadra 2, Ed. Corporate Financial Center, 70712-900 Brasília, Brazil. *E-mail*: emirra@cgee.org.br

Giovanni Mummolo is full-time professor of graduate and post-graduate industrial engineering courses at the Politecnico di Bari, Italy, where coordinates the first- and second-level degrees in Industrial and Management Engineering and is responsible for the Ph.D. course "Advanced Production Systems." His main fields of research are in production management and design of industrial plants. He contributes as referee and author for many international journals. He is a member of the European Academy of Industrial Management.

Birgitta Nordén has a master in educational science and a master in education with a professional orientation. Currently, she is a Ph.D. candidate at the Graduate School in Education and Sustainable Development (GRESD) coordinated by Uppsala University in Sweden. Her research focuses on the global dimension created in education and teaching toward sustainability, and in particular, on how the implementation process locally interacts with a global context. The variation of knowledge formation for sustainable development in a global setting is studied. "Global Learning for Sustainable Development" (GLSD) is introduced. In 1999–2007 she was the Director of Distance Education for the YMP at the International Institute for Industrial Environmental Economics at Lund University, which reached out worldwide to over 10000 school students in a program of networked learning on sustaining the environment in a developing world. She has been awarded several times, e.g., from the Royal Swedish Academy for Engineering Sciences for her commitment to IT-development in schools on pedagogical grounds, and the 12th Online Educa Berlin International Conference on Technology-supported learning and training for a best practices showcase.

Dr. Katia Regina Alves Nunes is Civil Engineer (Rio de Janeiro State University/1988), M.Sc. in Construction Management (Fluminense Federal University/1992), D.Sc. (Rio de Janeiro Federal University/2004) and Post-D.Sc. (Darmstadt University of Technology/2010). Her main research interests are resource and waste management, LCA (Life-Cycle Assessment), and corporate sustainability. Dr. Nunes have publications in many international journals, books, and proceedings. She has worked as academic researcher at Brazilian and German universities and also as engineer at the Rio de Janeiro Council Administration.
E-mail: k.nunes@iwar.tu-darmstadt.de

Dr. Costas P. Pappis is a Professor in Production Management at University of Piraeus, Greece. He holds a B.S. degree in Mechanical & Electrical Engineering from the National Technical University of Athens, a Diploma in Management Studies from the Polytechnic of Central London, and a Ph.D. in Engineering from the University of London. Prior to joining the University of Piraeus in 1993, he was an Associate Professor of Production/Operations Management at the University of Patras, Greece. He has also worked as an operations analyst, as an Engineer in the Technical Services of the National Bank of Greece, and as a Director of the Offsets Department of the Ministry of National Economy of Greece. He has been President of the Hellenic OR Society and has published papers in many international journals and books.

Dr. Jorge Verissimo Pereira is currently director of information technology (IT), Simulation and Supply Chain Management (SCM) group of Jove Logistics (jovelogistica.com.br). He holds a Ph.D. degree in Technology System from Warwick University, where his research focused on studying the use of IT in complex and dynamic systems through a System Dynamics-Discrete Event Simulation Approach. Dr. Pereira has supervised master students in Warwick University and has also reviewed and published papers in journals such as European Journal of Operational Research and International Journal of Information Management.

Dr. Robert J. Ries is a Rinker Professor of Construction, Assistant Professor, and Associate Director of the Powell Center for Construction and the Environment in the M. E. Rinker, Sr. School of Construction at the University of Florida. Dr. Ries's primary research work is focused on improving the environmental performance of buildings and the built environment. His work includes environmental life-cycle assessment (LCA) in the building domain, LCA studies of building systems, modeling construction processes, and building process modeling. His research also addresses developing LCA methodology such as incorporating optimization, managing uncertainty, and assessing impact at variable temporal and spatial scales. Dr. Ries has developed courses in green design and construction and sustainable development that are both required and elective courses in the undergraduate and graduate programs in the School.

Abeer Salem is a Ph.D. candidate in Sustainability Education at Prescott College for the Liberal Arts, the Environment, and Social Justice in Arizona, USA. She is an Egyptian, living in Cairo, and currently works in the field of International Development. She received her Bachelor's degree from Cairo University in Egypt and is a holder of a Master of Arts degree since 1989 in Communications Studies from the American University in Cairo. Her 20-year professional experience spans the business management, human development, inter-cultural, and development communication fields. This diverse scope of experience culminated in her current interest in living, livelihoods, and sustainability issues of the marginalized groups in her country as well as in other pockets of poverty around the globe.

Dr. Janya Sang-Arun is a policy researcher for Sustainable Consumption and Production Group at the Institute for Global Environmental Strategies (IGES), Japan. She holds a Ph.D. in Environmental Studies/International Studies from the University of Tokyo. Her current research focuses on sustainable solid waste management in developing Asian countries, particularly on achieving multiple benefits including climate change mitigation, food security, alternative energy, and poverty reduction. Some of her recent publications are accessible through the IGES homepage.

Rita Santos is a member of the Royal Institute of Technology KTH. Rita can be contacted at Dr Kristinas väg 30, Stockholm SE-100 44, Sweden. *E-mail*: rita.santos@abe.kth.se

Dr. Joseph Sarkis is currently a Professor of Management in The Graduate School of Management at Clark University. He earned his Ph.D. from the State University of New York at Buffalo. His research interests include supply chain management, management of technology, and business and the environment. He has published over 250 articles in a wide variety of outlets. He is currently editor of *Management Research Review* and Departmental Editor of the Models and Methodologies Department in *IEEE Transactions on Engineering Management*.

Dr. Liselotte Schebek is Director at Karlsruhe Institute of Technology, Germany, where she heads the Department of Technology-Induced Material Flow since 1999. She is Professor of Industrial Materials Cycles, Technical University of Darmstadt, Faculty of Civil Engineering and Geodesy, since 2000. Prof. Schebek received her Ph.D. at the Max-Planck Institute for Chemistry, Mainz, in 1990. Between 1990 and the entry in her current position she worked at ERM Lahmeyer International, Frankfurt, as Project Manager. *E-mail*: liselotte.schebek@kit.edu

Eunjung Shin is a Research Associate of the Science, Technology and Environment Research Lab and Ph.D. student in the Department of Public Administration at the University of Illinois at Chicago. She earned her master degree in environmental planning at Seoul National University in South Korea where her research focused on collaborative planning in local environmental management. After graduation, she worked as a researcher in the Korean Institute for Sustainable Society and the Korean Rural Economic Institute conducting research on environmental policy, rural development, and natural resource management. Her current research focuses on policy process and policy evaluation in the areas of environment, global genetic resource policy regimes for genetic resources, and information technology.

Dr. Pradeepa Silva received her Ph.D. in Genetics and Animal Breeding, University of Sydney, Australia. Pradeepa Silva is currently a senior lecturer in the University of Peradeniya, Sri Lanka. Her recent publications include Swine Industry of Sri Lanka, Indigenous Animal Genetic Resources in Sri Lanka, Assessment of invasiveness of alien species: a framework using statistical approach, and Agriculture Biodiversity in Sri Lanka (in Sinhala). Her research interests include sustainability in agriculture, agriculture biodiversity, and organic agriculture. She has contributed articles to Tropical Agricultural Research, BMC Evolutionary Biology, Animal Genetics, Journal of Animal Production, Indian Veterinary Journal, and others.

Folke Snickars is a member of the Royal Institute of Technology KTH. Folke can be contacted at Dr. Kristinas väg 30, Stockholm SE-100 44, Sweden. *E-mail*: folke.snickars@abe.kth.se

Dr. Eng. Boris Sokolov is a deputy director at the Russian Academy of Science, Saint Petersburg Institute of Informatics and Automation. He received his M.Sc., Ph.D. and Dr. Sc. Eng. in 1974, 1983, and 1993, respectively. From 1966 to 1999, he has developed his military career at the Mozaisky academy in Leningrad/St Petersburg, from a cadet to a colonel. Since 2000, he has been a professor in

St Petersburg State University of Aerospace Instrumentation. Professor Sokolov is the author of a new scientific lead: optimal control theory for structure dynamics of complex systems. Since 1985, he has been continuously developing this new scientific school. By now, his numerous followers got their Ph.D. and Dr. Science degrees under his supervision. The research interests of Prof. Sokolov are as follows: basic and applied research in mathematical modeling and mathematical methods in scientific research, optimal control theory, and mathematical models and methods of decision-making support in complex technical–organizational systems under uncertainties and multi-criteria. In 2008, he became an honored scientist of Russia. He is the author and co-author of three books on systems and control theory and of more than 220 scientific works published in various academic journals, including European Journal of Operational Research, International Journal of Manufacturing Technology and Management, Journal of Computer and Systems Sciences International, Differential Equations, and Automation and Remote Control. *Homepage: www.spiiras-grom.ru.*

Dr. Andreas Soteriou is an associate professor of Service Operations Management in the Department of Public and Business Administration of the University of Cyprus. He holds a Ph.D. degree from the University of Southern California. His research interests include service operations management, and quality and productivity issues in services. His work has appeared in leading academic journals such as Management Science, Manufacturing & Service Operations Management, Journal of Operations Management, Production and Operations Management, Interfaces and Decision Sciences. He is currently an Area Editor of Operations Management Research and serves on the editorial Board of a number of other academic journals.

Samir K. Srivastava is an Associate Professor and currently the Chairman of the Operations Management Area at the Indian Institute of Management Lucknow, India. He earned an undergraduate degree in Electrical Engineering from Institute of Technology, Banaras Hindu University, India, in 1989 and a postgraduate degree in management in 1999. He has over two decades of experience in research, teaching, and industry. He worked in various positions in industry (AV Birla Group) in technical as well as management roles. He served as an internal facilitator for TPM implementation at Indo Gulf Corporation Limited leading it to win the TPM Excellence Award from JIPM Japan in the year 2000. Prof. Samir received his doctorate (Fellow of Indian Institute of Management Lucknow) in 2004. He has designed and taught courses at both postgraduate and doctoral levels on topics such as Operations Management, Global Supply Chain Management, Operations Strategy, Logistics and Supply Chain Modeling, and System Dynamics Modeling. He has conducted on-campus and in-company Management Development Programmes for government and business executives. He has been member of TPM Club of India, Confederation of Indian Industry (CII), Technical Advisory Group (TAG), National Agricultural Innovation Project (Research on Production to Consumption Systems) and Sub-committee for Manufacturing Competitiveness, Confederation of

Indian Industry (CII). He is a life member of Society of Operations Management, India. Apart from teaching at Indian Institute of Management Lucknow, Prof. Samir has also taught at MDI Gurgaon and IMT Ghaziabad. His research has focused on Green Supply Chain Management, Reverse Logistics and other functional areas of management. His work on these and related issues has been published in refereed journals such as Asian Case Research Journal, Asian Journal of Management Cases, International Journal of Management Reviews, International Journal of Manufacturing Technology and Management, International Journal of Physical Distribution and Logistics Management, Omega and Total Quality Management and Business Excellence. He is the recipient of McGraw Hill Publishing Best Paper Award at the 3rd Global Conference on Business and Economics, Amsterdam, Netherlands, July 2004, and Best Student Paper Award at the 7th Annual Society of Operations Management International Conference, IIM Indore, December 2003. There are more than 200 citations (including 78 citations in Scopus) of his work. Prof. Samir is on the Editorial Board of Journal of Remanufacturing (a Springer Journal) and is a member of the Advisory Board of Indianjournals.com. He is an active referee for several other journals in the field. He has also co-authored two books — Business Logistics/Supply Chain Management with Ronald H. Ballou and Operations Management with Lee J. Krajewski, Larry P. Ritzman, and Manoj K. Malhotra.

Dr. Jürgen Strohhecker is Professor of Operations Management and Management Accounting at Frankfurt School of Finance & Management, Frankfurt/Main, Germany. He holds a diploma and Ph.D. from Mannheim University in Business Administration. His research explores how strategies and management control concepts affect an organization's performance. His research interests also include the analysis of behavioral dynamic decision making. He teaches operations management, management accounting and System Dynamics courses at the undergraduate, master, and executive level.

Dr. Noriko Sugiyama is an Associate Professor at the Graduate School of Environmental Studies, Nagoya University, Japan. She has a background in Environmental Policy and Meteorology, studying meteorology at Aichi University of Education and working at the Japan Weather Association for 10 years. She is a certificated weather forecaster and worked as a TV weathercaster from 1995 to 2002. She later went on to study Environmental Policy at Meijo University and Nagoya University, Japan, completing her doctorate (Doctor of Environment) on "Environmental Policy Study for Establishing 'Local Climate Policy' " in 2008. Her research interests are local and regional climate policy and environmental communication. She contributed to the "USA, Germany, and Japan Local Climate Partnership" project from 2007 to 2009.

Joo Hwa Tay Department of Environmental Science and Engineering, Fudan University, 220 Handan Road, Yangpu District, Shanghai, 200433, P.R. China.

Professor Tay Joo Hwa is a scholar, scientist, and engineer of international renown in the field of environmental science and engineering. Since 1971, he has held various research and teaching posts in University of Cincinnati, University of Toronto, University of Ottawa, George Washington University, and National University of Singapore of USA, Canada, and Singapore. After joining Nanyang Technological University of Singapore in 1983, Professor Tay has contributed tremendously to the development of the School of Civil and Environmental Engineering, and was made the Head of the Division of Water Resource and Environmental Engineering in 1991. Since then, he has led the division into world class research and development and has achieved leading position in areas such as aerobic granular sludge reactor.

Prof. Tay's current research interests focus on the hazardous and industrial waste management, membrane technology, cleaner production, waste minimization, recycling and reuse and biotechnological application on wastewater treatment. He has published more than 400 technical papers in the related journals and conferences, and filed several patents in these areas. He is the editor of the Journal of Hazardous Materials and International Journal on Performability Engineering, as well as an associate editor for the ASCE Journal of Environmental Engineering. Prof. Tay have also obtained various professional qualifications and served in various technical committees, e.g., the memberships of Water and Environmental Federation, Environmental Engineering Society of Singapore and International Water Association, and the Ministry of the Environment's Advisory Committee on Hazardous Substances & Toxic Wastes and the Resource Panel of the Government Parliamentary Committee.

In addition to being an outstanding researcher and educator, Prof. Tay also participates actively in industrial services. He has designed more than 30 treatment facilities for the water and wastewater management in the South-east Asia region, and provided consulting services to Asian Productivity Organization on more than 10 demonstration projects on green productivity and cleaner production in Asian countries. He has also acted as consultant to the World Bank, Asian Development Bank, some United Nations organizations (e.g., UNEP, UNDP, UNESCAP, UNCRD), World Health Organization (WHO), Asian Productivity Organization (APO), German Technical Assistant Programme (GTZ), and Carl Duisberg Gesellschaft (CDG) of Germany for various technical assistance programmes on environmental management in Asia/Pacific region for the past 20 years.

As recognition to his contribution, five national awards have been awarded to Professor Tay Joo Hwa, as well as one university level award. Prof. Tay has now joined Fudan University, Shanghai, China, to contribute his knowledge, experience, and insight to the development of the Department of Environmental Science and Engineering, as well as the environmental protection cause of China.

Dr. Lanka Thabrew is a Research Associate at the Vanderbilt Institute of Energy and Environment. She received her Bachelor of Science degree in civil engineering at the University of Moratuwa, Sri Lanka, in June 2001. She completed her Master of Science degree in civil and environmental engineering at the Carnegie Mellon University, USA, in May 2004. She completed her Doctor of Philosophy degree in

Design Construction and Planning at the University of Florida, USA, in August 2009. She has worked as an environmental engineer from 2001 to 2003, and also as a consultant to the German Technical Cooperation, Sri Lanka, from 2007 to 2008. Her focus of research has been life-cycle assessment, international sustainable development, transdisciplinary learning in development planning, and policy assessment. She is conducting research on triple-bottom-line life-cycle implications on marine transportation and nuclear energy, and demand-side water energy relationships and management strategies in a changing climate.

José M. de la Torre-Ruiz is a Ph.D. candidate and assistant professor in the School of Economics and Business at University of Granada, Spain. He is a full member of the research group Innovation, Sustainability, and Development (ISDE) and his primary research interests are human resource management and organizational behavior. Specifically, he is currently studying team design and newcomer adjustment and their consequences for the performance of teams. Additionally, he has also oriented his work toward environmental management focusing mainly on the impact of environmental management systems on firms' performance.
Address: Department of Business and Management, Economics and Business School, University of Granada, Spain. Campus de Cartuja S/N, C. P. 18071 Granada, Spain. *Tel.*: 0034 958249595. *E-mail*: jmtorre@ugr.es

Dr. Giannis T. Tsoulfas is Scientific Consultant at the Presidency of the Hellenic Republic and Tutor at the Department of Business Administration in the Hellenic Open University. He holds a B.S. degree in Mechanical Engineering from the University of Patras, Greece, and a Ph.D. in Industrial Management and Technology from the University of Piraeus, Greece. In the past, he has worked as a Consultant Engineer in various engineering companies. His published work has appeared in several international journals and books.

Marcus Wagner is a full-time professor at the University of Wuerzburg and holds the chair in entrepreneurship and corporate growth. His current research interests are strategic management (especially with regard to effects of corporate sustainability on economic performance), entrepreneurship, and innovation management. He is Associated Member of the Bureau d'Economie Théorique et Appliquée, Université de Strasbourg, and of the Centre for Sustainability Management, Leuphana University Lueneburg.

Dr. Eric W. Welch, is an Associate Professor and director of the Science, Technology, and Environment Research Lab in the Department of Public Administration at the University of Illinois at Chicago. He earned his Ph.D. in public administration at Syracuse University's Maxwell School of Citizenship and Public Affairs where his research focused on policy effects on the environmental behavior of firms. His research spans several areas including environmental policy and management, science and technology policy, technology processes in public organizations, and R&D

evaluation. Prof. Welch teaches courses in innovation and technology theory, decision analysis, organization theory, and environmental policy in both the masters of public administration and Ph.D. programs. His environmental research has been published in such journals as *Policy Sciences*, *Journal of Public Policy and Management*, *Environmental Behavior*, *Greener Management International: The Journal of Corporate Environmental Strategy and Practice*, *Journal of Public Ecological Economics*, *Journal of Environmental Planning and Management*, and *Business Strategy and the Environment*.

Dr. Janice Winch is an Associate Professor of Management Science in the Lubin School of Business, Pace University. She received her Ph.D. in operations research from Rutgers University in 1993. Her current research interests include decision modeling, simulation, and scheduling. Her articles have appeared in *Manufacturing and Service Operations Management*, *Probability in the Engineering and Informational Sciences*, *International Journal of Production Research*, and others.

Dr. Jianguo (Jingle) Wu is Dean's Distinguished Professor of Landscape Ecology and Sustainability Science, School of Life Sciences & Global Institute of Sustainability, Arizona State University, Tempe, Arizona, USA, the received his B.S. in biology from Inner Mongolia University in 1982, and M.S. and Ph.D. in ecology from Miami University in 1991. He is National Science Foundation (NSF) postdoctoral fellow at Cornell University (1991–1992) and Princeton University (1992–1993). His current research areas: landscape ecology, urban ecology, biodiversity and ecosystem functioning, and sustainability science. He authored/edited 10 books and more than 170 journal articles and book chapters. He is Editor-in-Chief of Landscape Ecology, and Editorial Board member for several other international journals. He is recipient of the 2006 American Association for the Advancement of Science (AAAS) Award for International Scientific Cooperation; Elected AAAS Fellow in 2007; Named a Leopold Leadership Fellow in 2009; Recipient of the Distinguished Landscape Ecologist Award in 2010, US Association for Landscape Ecology. He is Founding Director of the Sino-US Center of Conservation, Energy and Sustainability Science (SUCCESS) — a joint research center established by Inner Mongolia University and Arizona State University in 2007.

Tong Wu is currently a graduate researcher at the Ecological Restoration Institute and School of Forestry in Northern Arizona University. He previously attended the University of California-Berkeley, where he studied environmental economics and policy. Prior to his current academic placement, he worked as an environmental consultant and analyst in the private sector. Tong is the author or co-author of several journal articles and book chapters in the fields of environmental economics and sustainable development. His current research focuses on the economic and institutional factors underlying forest management.

Dr. Zhang Yi Department of Environmental Science and Engineering, Fudan University, 220 Handan Road, Yangpu District, Shanghai 200433, P.R. China.

Dr. Zhang Yi is an experienced environmental scientist and engineer. She obtained her Ph.D. degree in environmental engineering in the School of Civil and Environmental Engineering, Nanyang Technological University, Singapore. After serving the Institute of Environmental Science and Engineering, Singapore, for more than four years as a research scientist, she recently joined the Department of Environmental Science and Engineering, Fudan University. Her research expertise and focus have been wastewater treatment and environmental microbiology, but have extended into areas of clean energy, clean development mechanism, and cleaner production.

INDEX